Table F The *t* Distribution

d.f.	Confidence intervals	50%	80%	90%	95%	98%	99%
	One tail, α	0.25	0.10	0.05	0.025	0.01	0.005
	Two tails, α	0.50	0.20	0.10	0.05	0.02	0.01
1		1.000	3.078	6.314	12.706	31.821	63.657
2		.816	1.886	2.920	4.303	6.965	9.925
3		.765	1.638	2.353	3.182	4.541	5.841
4		.741	1.533	2.132	2.776	3.747	4.604
5		.727	1.476	2.015	2.571	3.365	4.032
6		.718	1.440	1.943	2.447	3.143	3.707
7		.711	1.415	1.895	2.365	2.998	3.499
8		.706	1.397	1.860	2.306	2.896	3.355
9		.703	1.383	1.833	2.262	2.821	3.250
10		.700	1.372	1.812	2.228	2.764	3.169
11		.697	1.363	1.796	2.201	2.718	3.106
12		.695	1.356	1.782	2.179	2.681	3.055
13		.694	1.350	1.771	2.160	2.650	3.012
14		.692	1.345	1.761	2.145	2.624	2.977
15		.691	1.341	1.753	2.131	2.602	2.947
16		.690	1.337	1.746	2.120	2.583	2.921
17		.689	1.333	1.740	2.110	2.567	2.898
18		.688	1.330	1.734	2.101	2.552	2.878
19		.688	1.328	1.729	2.093	2.539	2.861
20		.687	1.325	1.725	2.086	2.528	2.845
21		.686	1.323	1.721	2.080	2.518	2.831
22		.686	1.321	1.717	2.074	2.508	2.819
23		.685	1.319	1.714	2.069	2.500	2.807
24		.685	1.318	1.711	2.064	2.492	2.797
25		.684	1.316	1.708	2.060	2.485	2.787
26		.684	1.315	1.706	2.056	2.479	2.779
27		.684	1.314	1.703	2.052	2.473	2.771
28		.683	1.313	1.701	2.048	2.467	2.763
(z) ∞		.674	1.282[a]	1.645[b]	1.960	2.326[c]	2.576[d]

[a]This value has been rounded to 1.28 in the textbook.

[b]This value has been rounded to 1.65 in the textbook.

[c]This value has been rounded to 2.33 in the textbook.

[d]This value has been rounded to 2.58 in the textbook.

Source: Adapted from W. H. Beyer, *Handbook of Tables for Probability and Statistics,* 2nd ed., CRC Press, Boca Raton, Florida, 1986. Reprinted with permission.

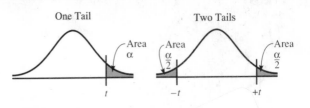

Elementary
Statistics
A Step by Step Approach

Elementary Statistics

A Step by Step Approach

Fourth Edition

Allan G. Bluman
Community College of Allegheny County

Boston   Burr Ridge, IL   Dubuque, IA   Madison, WI
New York   San Francisco   St. Louis
Bangkok   Bogotá   Caracas   Lisbon   London   Madrid   Mexico City
Milan   New Delhi   Seoul   Singapore   Sydney   Taipei   Toronto

McGraw-Hill Higher Education

A Division of The McGraw-Hill Companies

ELEMENTARY STATISTICS: A STEP BY STEP APPROACH
FOURTH EDITION

Published by McGraw-Hill, an imprint of The McGraw-Hill Companies, Inc., 1221 Avenue of the Americas, New York, NY 10020. Copyright © 2001, 1998, 1995, 1992 by The McGraw-Hill Companies, Inc. All rights reserved. No part of this publication may be reproduced or distributed in any form or by any means, or stored in a database or retrieval system, without the prior written consent of The McGraw-Hill Companies, Inc., including, but not limited to, in any network or other electronic storage or transmission, or broadcast for distance learning.

Some ancillaries, including electronic and print components, may not be available to customers outside the United States.

This book is printed on recycled, acid-free paper containing 10% postconsumer waste.

1 2 3 4 5 6 7 8 9 0 VNH/VNH 0 9 8 7 6 5 4 3 2 1 0

ISBN 0–07– 231694–2
ISBN 0–07–237593–0 (IE)
ISBN 0–07–118071–0 (ISE)

Vice president and editor-in-chief: *Kevin T. Kane*
Publisher: *JP Lenney*
Sponsoring editor: *Daryl Bruflodt*
Editorial assistant: *Jenni Lang*
Marketing manager: *Mary K. Kittell/Debra A. Besler*
Senior project manager: *Marilyn M. Sulzer*
Media technology project manager: *Steve Metz*
Production supervisor: *Sandy Ludovissy*
Designer: *K. Wayne Harms*
Cover design and illustration by: *Rokusek Design*
Senior photo research coordinator: *Lori Hancock*
Photo research: *Shirley Lanners*
Supplement coordinator: *Sandra M. Schnee*
Compositor: *GAC—Indianapolis*
Typeface: *10/12 Times Roman*
Printer: *Von Hoffmann Press, Inc.*

Portions of Minitab Statistical software input and/or output are reprinted with permission of Minitab Inc.

The credits section for this book begins on page 713 and is considered an extension of the copyright page.

Library of Congress Cataloging-in-Publication Data

Bluman, Allan G.
    Elementary statistics : a step by step approach / Allan G. Bluman. — 4th ed.
        p.   cm.
    Includes index.
    ISBN 0–07–231694–2 — ISBN 0–07–237593–0 (Instructor's edition)
    1. Statistics.   I. Title.
QA276.12 .B59   2001
519.5—dc21
                                                                    00–028272
                                                                    CIP

INTERNATIONAL EDITION ISBN 0–07–118071–0
Copyright © 2001. Exclusive rights by The McGraw-Hill Companies, Inc., for manufacture and export. This book cannot be re-exported from the country to which it is sold by McGraw-Hill. The International Edition is not available in North America.

www.mhhe.com

Brief Contents

Contents

chapter eight

chapter nine

chapter ten

chapter eleven

Preface

Approach

Elementary Statistics: A Step by Step Approach is a textbook for students in the beginning statistics course whose mathematical background is limited to basic algebra. The book uses a nontheoretical approach in which concepts are explained intuitively and supported by examples. There are no formal proofs in the book. The applications are general in nature, and the exercises include problems from business, economics, health, medicine, science, engineering, social science, education, and general interest.

About This Book

The learning system found in *Elementary Statistics* provides the student with a valuable framework in which to learn and apply concepts.

- Each chapter begins with an outline and a list of **learning objectives.** The objectives are repeated at the beginning of each section to help students focus on the concepts presented within that section.

2–3

Histograms, Frequency Polygons, and Ogives

Objective 2. Represent data in frequency distributions using histograms, frequency polygons, and ogives.

After the data have been organized into a frequency distribution, they can be presented in graphic forms. The purpose of graphs in statistics is to convey the data to the viewer in pictorial form. It is easier for most people to comprehend the meaning of data presented graphically than data presented numerically in tables or frequency distributions. This is especially true if they have little or no statistical knowledge.

Statistical graphs can be used to describe the data set or analyze it. Graphs are also useful in getting the audience's attention in a publication or a speaking presentation. They can be used to discuss an issue, reinforce a critical point, or summarize a data set. They can also be used to discover a trend or pattern in a situation over a period of time.

- The outline and learning objectives are followed by a feature titled **Statistics Today,** in which a real-life problem shows students the relevance of the material in the chapter. This problem is subsequently solved near the end of the chapter using the statistical techniques that were presented in the chapter.

Statistics Today

Why Are We Running Out of 800 Numbers?

Phone companies and other agencies that deal in numbers need to know how many phone numbers, ID tags, or license plates they can issue using certain combinations of various letters and numbers. The article shown below explains that the phone companies are running out of toll-free 800 numbers. The question is: How many phone numbers with the 800 prefix can be issued in the United States?

Toll-free call? Get ready to dial 888

By Becky Beyers
USA TODAY

Get ready to keep your finger on the 8 when you make a toll-free call.

Phone companies will run out of 800 numbers early next year and start issuing toll-free numbers beginning with 888.

Use of 800 numbers has grown so fast, "we're a victim of our own success," says Dennis Byrne of the U.S. Telephone Association.

Only about 1.7 million of the 7.6 million possible 800-prefix combinations are still available. Why so few are left:

▶ Demand has taken off since May 1993, when the government allowed users to keep 800 numbers if they changed long-distance carriers.

▶ 800 numbers aren't just for big companies anymore. Small businesses use them, as do residential customers so family members can call home more cheaply than collect.

Such customers may pay 25 cents a minute for each call plus a monthly fee of $5.

▶ Some toll-free numbers are hoarded for promotional value or occasional use.

The industry's numbering council — phone companies and associations that set phone-number policies — is asking that little-used numbers be returned so they can be reissued.

Setting up a new toll-free access code involves the entire phone industry, Byrne says.

All internal systems must upgrade switching equipment so they can handle 888 calls.

What happens when the 888s are used up? It's on to 877, 866, and all the way down to 822.

Source: *USA Today*, February 13, 1995. Copyright 1995 *USA TODAY.* Used with permission.

In this chapter, you will learn the rule for counting, the differences between permutations and combinations, and how to figure out how many different combinations for specific situations exist.

- Over 300 **examples** with detailed solutions are provided to help students learn to solve problems. Examples are solved by using a step by step explanation. Illustrations provide a clear display of results for students.

Example 3–23

Find the variance and standard deviation for the amount of European auto sales for a sample of six years shown. The data are in millions of dollars.

11.2, 11.9, 12.0, 12.8, 13.4, 14.3

Source: *USA Today,* March 22, 1999.

Solution

STEP 1   Find the sum of the values.

$$\Sigma X = 11.2 + 11.9 + 12.0 + 12.8 + 13.4 + 14.3 = 75.6$$

STEP 2   Square each value and find the sum.

$$\Sigma X^2 = 11.2^2 + 11.9^2 + 12.0^2 + 12.8^2 + 13.4^2 + 14.3^2 = 958.94$$

STEP 3   Substitute in the formulas and solve.

$$s^2 = \frac{\Sigma X^2 - [(\Sigma X)^2/n]}{n-1} = \frac{958.94 - [(75.6^2)/6]}{5}$$
$$s^2 = 1.28$$

The variance of the sample is 1.28

$$s = \sqrt{1.28} = 1.13$$

Hence, the sample standard deviation is 1.13.

- Numerous examples and exercises use **real data.**

36. The following data represent the attendance at seven Pittsburgh museums for 1997 and 1998. Draw two boxplots for the data and compare the distributions. The data are in thousands.

1997	1998
101	107
754	764
291	293
83	110
589	627
152	103
158	154

Source: *The Pittsburgh Tribune Review,* February 21, 1999.

- Numerous **Procedure Tables** summarize processes for the student. All use the step by step method.

Procedure Table

Finding the Sample Variance and Standard Deviation for Grouped Data

STEP 1 Make a table as shown, and find the midpoint of each class.

A Class	B Frequency	C Midpoint	D $f \cdot X_m$	E $f \cdot X_m^2$

STEP 2 Multiply the frequency by the midpoint for each class, and place the products in column D.

STEP 3 Multiply the frequency by the square of the midpoint, and place the products in column E.

STEP 4 Find the sums of columns B, D, and E. (The sum of column B is n. The sum of column D is $\Sigma f \cdot X_m$. The sum of column E is $\Sigma f \cdot X_m^2$.)

STEP 5 Substitute in the formula and solve to get the variance.

$$s^2 = \frac{\Sigma f \cdot X_m^2 - [(\Sigma f \cdot X_m)^2/n]}{n - 1}$$

STEP 6 Take the square root to get the standard deviation.

- The **Speaking of Statistics** sections invite students to think about poll results and other statistics-related news stories.

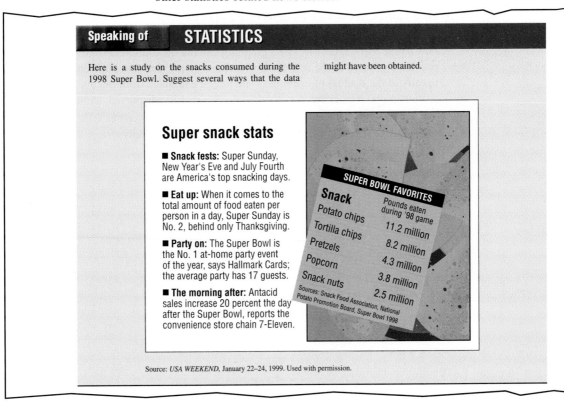

Speaking of STATISTICS

Here is a study on the snacks consumed during the 1998 Super Bowl. Suggest several ways that the data might have been obtained.

Super snack stats

■ **Snack fests:** Super Sunday, New Year's Eve and July Fourth are America's top snacking days.

■ **Eat up:** When it comes to the total amount of food eaten per person in a day, Super Sunday is No. 2, behind only Thanksgiving.

■ **Party on:** The Super Bowl is the No. 1 at-home party event of the year, says Hallmark Cards; the average party has 17 guests.

■ **The morning after:** Antacid sales increase 20 percent the day after the Super Bowl, reports the convenience store chain 7-Eleven.

SUPER BOWL FAVORITES

Snack	Pounds eaten during '98 game
Potato chips	11.2 million
Tortilla chips	8.2 million
Pretzels	4.3 million
Popcorn	3.8 million
Snack nuts	2.5 million

Sources: Snack Food Association, National Potato Promotion Board, Super Bowl 1998

Source: *USA WEEKEND,* January 22–24, 1999. Used with permission.

- **Historical Notes, Unusual Stats,** and **Interesting Facts,** located in the margins, make statistics come alive for the reader.

Unusual Stats

According to the *Statistical Abstract of the United States,* 52% of Americans live within 50 miles of a coastal shoreline.

- **Rules and definitions** are set off for easy referencing by the student.

Objective 2. Find the total number of outcomes in a sequence of events using the multiplication rule.

Multiplication Rule

In a sequence of n events in which the first one has k_1 possibilities and the second event has k_2 and the third has k_3, and so forth, the total number of possibilities of the sequence will be

$$k_1 \cdot k_2 \cdot k_3 \cdot \cdots \cdot k_n$$

Note: "And" in this case means to multiply.

- Over 1,600 **exercises** are located at the end of major sections within each chapter.
- At the end of appropriate sections, **Technology Step by Step** boxes show students how to use MINITAB, the TI-83 graphing calculator, and Excel to solve the types of problems covered in the section. Instructions are presented in numbered steps, usually in the context of examples—including examples from the main part of the section. Numerous computer or calculator screens are displayed, showing intermediate steps as well as the final answer. This feature, **new to the fourth edition,** will be valuable to students using any of these tools.

Technology Step by Step

MINITAB Step by Step

Finding the Mean and Standard Deviation

Example MT3–1

1. Type the data from Example 3–39 (in the following section) into C1 of MINITAB. Name the column **CARS-THEFT.**

   52 58 75 79 57 65 62 77 56 59 51 53 51 66 55
   68 63 78 50 53 67 65 69 66 69 57 73 72 75 55

2. Select **Stat>Basic Statistics>Display Descriptive Statistics.**

3. The cursor will be blinking in the Variables text box. Double-click C1.

4. Click [OK].

The results will be displayed in the Session Window as shown. The column label "CARS-THEFT" is truncated to 8 letters in the display. The standard deviation is the unbiased estimate, *s*. The trimmed mean or TrMean is the mean for the data after the lowest and highest 5% are discarded. If the trimmed mean is different from the mean, there may be outliers.

Session Window with Descriptive Statistics

Session

Descriptive Statistics: CARS-THEFT

Variable	N	Mean	Median	TrMean	StDev	SE Mean
CARS-THE	30	63.20	64.00	63.00	9.01	1.64

Variable	Minimum	Maximum	Q1	Q3
CARS-THE	50.00	79.00	55.00	69.75

• **Critical Thinking** sections at the end of each chapter challenge the students to apply what they have learned to new situations. The problems presented are designed to deepen conceptual understanding and/or to extend topical coverage.

Critical Thinking Challenges

1. A person decides to shake hands with six different people on a certain day. The next day, each of the six people will shake hands with six different people. The process continues until every person in the United States has shaken someone's hand. How many days will it take until everyone in the United States has shaken hands once? Assume that once a person shakes hands with six different people, he or she does not shake hands again. (*Hint:* The population of the United States is 248,709,873, according to the 1990 census.)

2. If it can be assumed that the maximum number of hairs on a human head is about 500,000, explain why at least two people living in Houston (population 1,629,902,

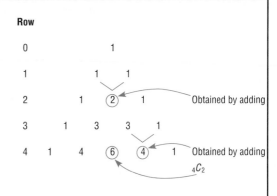

triangle is f... ...dding the two adja... ...mb...

- **Data projects** further challenge students' understanding and application of the material presented in the chapter. Many of these require the student to gather, analyze, and report on real data. These projects, which appear at the end of each chapter, may include a World Wide Web icon ⊕, indicating that websites are listed as possible sources of data.

⊕ Data Projects

Where appropriate, use MINITAB, the TI-83, Excel, or a computer program of your choice to complete the following exercises.

1. Select a categorical (nominal) variable, such as the colors of cars in the school's parking lot or the major fields of the students in statistics class, and collect data on this variable.
   a. State the purpose of the project.
   b. Define the population.
   c. State how the sample was selected.
   d. Show the raw data.
   e. Construct a frequency distribution for the variable.
   f. Draw some appropriate graphs (pie, Pareto, etc.) for the data.
   g. Analyze the results.

2. Using an almanac, select a variable that varies over a period of several years (e.g., silver production) and draw a time series graph for the data. Write a short paragraph interpreting the findings.

3. Select a variable (interval or ratio) and collect at least 30 values. For example, you may ask the students in your class how many hours they study per week or how old they are, etc..
   a. State the purpose of the project.
   b. Define the population.
   c. State how the sample was selected.
   d. Show the raw data.
   e. Construct a frequency distribution for the data.
   f. Draw a histogram, frequency polygon, and ogive for the data.
   g. Analyze the results.

You may use the following websites to obtain raw data:

   http://www.mhhe.com/math/stat/bluman/
   http://lib.stat.cmu.edu/DASL
   http://www.oecd.org/statlist.htm
   http://www.statcan.ca/english/

- **Hypothesis-Testing Summaries** are found at the end of Chapter 10 (z, t, χ^2, and F tests for testing means, proportions, and variances), Chapter 13 (correlation, chi-square, and ANOVA), and Chapter 14 (nonparametric tests) to show students the different types of hypotheses and the types of tests to use.

- A **Data Bank** listing various attributes (educational level, cholesterol level, gender, etc.) for 100 people is included and referenced in various exercises and projects throughout the book, including the projects presented in Data Projects sections.

- A **reference card** containing the formulas and the z, t, χ^2, and PPMC tables is included with this textbook.

- End-of-chapter **Summaries, Important Terms,** and **Important Formulas** give students a concise summary of the chapter topics and provide a good source for quiz or test preparation.

- **Review Exercises** are found at the end of each chapter.

- Special sections called **Data Analysis** require students to work with a data set to perform various statistical tests or procedures and then summarize the results. The data is included in the Data Bank following Appendix C and can be downloaded from the book's website at www.mhhe.com/math/stat/bluman.

- **Chapter quizzes,** found at the end of each chapter, include multiple choice, true–false, and completion questions along with exercises to test students' knowledge and comprehension of chapter content.

Content Changes for the Fourth Edition

To build on the success of the third edition, the content and format have been maintained in the fourth edition while changes based on suggestions of reviewers and the current thinking of those who are knowledgeable in the realm of statistical education have been integrated.

With this in mind, the major goals for this revision are to enable students to

1. Draw conclusions from data
2. Become more statistically literate
3. Have a better understanding of the "logic" of hypothesis testing
4. Know how to use the P-value method for hypothesis testing
5. Work with raw data as well as summary data

To achieve goal 1, questions based on interpreting the computational solutions for exercises have been included throughout the book. In addition, new problems that involve the comparisons of two data sets by frequency distributions, graphs, and summary statistics have been added. Examples and exercises comparing data sets using back-to-back stem and leaf plots and comparing data sets using two boxplots have been added. Also, students are asked to describe the nature of distributions with regard to their shape, spread, etc.

To achieve goal 2, topics in statistical literacy have been incorporated throughout the textbook where appropriate. These topics include

1. The nature of observational and experimental studies in Chapter 1
2. More material on misleading graphs in Chapter 2
3. Probability and risk-taking in Chapter 5
4. Expectation and gambling in Chapter 6
5. A new section on sampling techniques and questionnaire design in Chapter 15

To achieve goal 3, material has been added at the beginning of Chapter 9 on the logic of hypothesis testing. This information includes comparing two distributions, one when the null hypothesis is true and the other when the null hypothesis is false. Also, material has been added comparing the hypothesis testing situation to a jury trial.

To achieve goal 4, explanations, examples, and exercises on the P-value method of hypothesis testing have been included in Chapters 9 through 14. A five-step procedure for testing hypotheses using P-values has been added to Chapter 9. Finding intervals from tables for P-values have been shown for the t, χ^2, and F tests. To help students with this concept, the actual P-values obtained from a calculator have also been given with the interval values.

To achieve goal 5, examples and exercises using real data have been included in Chapters 2, 3, 8, 9, 10, 11, 12, 13, and 14.

Other changes have been made in the following chapters:

- Chapter 2: The purpose of relative frequency graphs has been explained. Note: Those wishing to teach scatter plots with the other graphs can teach Section 11–2 with this chapter. Coverage of ungrouped frequency distributions was streamlined.
- Chapter 3: Back-to-back stem and leaf plots have been added for the comparison of data sets. Quartiles instead of hinges are used in boxplots since most calculators and

computer programs and other textbooks use quartiles. An easy way to compute quartiles has been shown. Drawings for quartiles, deciles, and percentiles have been added. An explanation of how to compare boxplots has been added.

- Chapter 4: This chapter has been shortened considerably. There is only one multiplication rule, one permutation rule, and one combination rule. In addition, sections have been combined.
- Chapter 5: Bayes's theorem has been moved to an Appendix.
- Chapter 6: The title of this chapter has been changed from Probability Distributions to Discrete Probability Distributions.
- Chapter 7: The section on the finite population correction factor has been de-emphasized. The section on normal approximation to the binomial distribution has been rewritten and a procedure table has been added to clarify the concept.
- Chapter 8: The definition of *confidence level* now comes before the definition of *confidence interval.*
- Chapter 9: In addition to the changes mentioned previously, the formula

$$z = \frac{\hat{p} - p}{\sqrt{pq/n}} \quad \text{instead of} \quad z = \frac{X - \mu}{\sigma}$$

is used to test hypotheses for a proportion. Some of the section titles have been changed to more accurately describe the material contained in the sections. The material on the power of a test has been rewritten and simplified.
- Chapter 10: The formula for finding the confidence interval for the difference of two proportions has been changed.
- Chapter 11: As previously stated, Section 11–2 on drawing and analyzing scatter plots can be taught with the other graphs in Chapter 2. A subsection on prediction intervals using the *t* distribution has been added.
- Chapters 12 through 14: Examples and exercises for hypothesis testing using *P*-values and raw data values have been added.
- Chapter 15: A new section on surveys and questionnaire design has been added.
- Chapter 16, on quality control, has been removed.

Altogether, the pedagogical improvements discussed above have resulted in the following changes to the Fourth Edition:

5 new "Speaking of Statistics" boxes
Over 250 new or modified Exercises
17 new Critical Thinking Challenges
Many new Examples
24 additional Data Analysis Problems

Finally, the technology material (MINITAB, TI-83, and Excel) is covered in new "Technology Step by Step" subsections at the end of the appropriate chapter sections.

Supplements

The text is accompanied by an extensive set of supplements for use by you and your students, all of which are carefully coordinated with the text.

Website

The **website** for *Elementary Statistics: A Step by Step Approach,* Fourth Edition, provides the data sets used in examples and exercises in a variety of formats including

- MINITAB
- TI Graph Link files for TI-83
- Excel (for Windows and Macintosh)
- SPSS
- Comma-Delimited ASCII

This can save the student using a computer or calculator from having to enter data by hand, which takes up valuable time and increases the chances of error. The "Data Bank" statistics that are used in the Data Analysis assignments at the ends of chapters are provided for download as well.

The website also provides links to data sources referred to in many of the Data Projects and links to dozens of other statistics-related sites. A PowerPoint presentation also found on the website is available for use in lectures or as a student study aid.

The website address is www.mhhe.com/math/stat/bluman.

For the Instructor

- *Instructor's Solutions,* by Sally Robinson of South Plains College. This manual includes worked-out solutions to most of the exercises in the text.
- *Critical Thinking Workbook: Instructor's Edition,* contains solutions to the students' version of the *Critical Thinking Workbook* described below.
- The *Test Bank* contains a variety of questions, including true–false, multiple-choice, short answer, and short problems requiring analysis and written answers. The testing material is coded by type of question and level of difficulty.
- The computerized test bank enables you to efficiently select, add, and organize questions, such as by type of question or level of difficulty. It also allows for printing tests along with answer keys, as well as editing the original questions. The computerized test bank is available for Windows and Mac systems.
- Full-color lecture slides in PowerPoint format highlight chapter concepts, summarize main points, and illustrate examples. These files can be downloaded from the book's website at www.mhhe.com/math/stat/bluman. PowerPoint users can customize the slides to suit the specific needs of their course.
- *Against All Odds* and *Decisions through Data* are video series available to qualified adopters. Please contact your local sales representative for more information about these programs.

For the Student

- *Critical Thinking Workbook,* by James Condor of Manatee Community College, provides a number of additional challenging problems for students to solve that are drawn from real-world applications. Problems are keyed to each chapter and are designed to highlight and emphasize key concepts.
- *Student Study Guide* by Pat Foard of South Plains College will assist students in understanding and reviewing key concepts and preparing for exams. It emphasizes all important concepts contained in each chapter, includes explanations, and provides opportunities for students to test their understanding by completing related exercises and problems.
- *Student Solutions Manual,* by Sally Robinson of South Plains College, contains detailed solutions to all odd-numbered text problems.
- *MINITAB—Student Version.* This software and user manual provides the student with how-to information on data and file management, conducting various statistical analyses, and creating presentation-style graphics.

Acknowledgments

I would like to thank the following people and companies for granting permission to reprint their statistical tables and other material:

Addison-Wesley Publishing Company, Inc.
Benjamin/Cummings Publishing Company
CRC Press, Inc.
Institute of Mathematical Statistics
Prentice Hall, Inc.
Texas Instruments
Consumers Union (Copyright 1999 by Consumers Union of U.S., Yonkers, NY 10703-1057. Each data set used by permission from *Consumer Reports.* To subscribe, call 800-234-1645; www.ConsumerReports.org.)

I am also grateful to the many authors and publishers who granted me permission to use their articles and cartoons. I would like to acknowledge the cooperation of MINITAB, Inc., in the preparation of this textbook.

The most important and useful advice about *Elementary Statistics: A Step by Step Approach* comes from users of the book and other statistics instructors. I am grateful to the following reviewers for their recommendations.

William A. Ahroon, *Plattsburgh State University*
Anne G. Albert, *University of Findlay*
Abraham K. Biggs, *Broward Community College*
Cecil J. Coone, *State Technical Institute at Memphis*
Callie Harmon Daniels, *St. Charles County Community College*
Bill Dunn, *Las Positas College*
David J. French, *Tidewater Community College*
Kathleen Fritsch, *University of Tennessee at Martin*
James R. Fryxell, *College of Lake County*
Ashis K. Gangopadhyay, *Boston University*
Mark E. Glickman, *Boston University*
David R. Gurney, *Southeastern Louisiana University*
Barney Herron, *Muskegon Community College*
Robert L. Horvath, *El Camino College*
Rebecca M. Howard, *Roane State Community College*
Jane Keller, *Metropolitan Community College, Omaha*
Michael J. Keller, *St. Johns River Community College*
Michael Kelly, *State Technical Institute at Memphis*
Rhonda Magel, *North Dakota State University*
Donald K. Mason, *Elmhurst College*
Jeff Mock, *Diablo Valley College*
Carla Monticelli, *Camden County College*
Gerry Moultine, *Northwood University*
Sharon R. Neidert, *University of Tennessee*
Neal Rogness, *Grand Valley State University*
Susan C. Schott, *University of Central Florida*
Larry Snyder, *Ohio University*

David Stewart, *Community College of Baltimore County*

Richard H. Stockbridge, *University of Kentucky*

Donald B. White, *University of Toledo*

Laurie Sawyer Woodman, *University of New England*

I would also like to thank the more than 60 reviewers of the third edition, whose suggestions and insights have been a positive influence on every page of this book. They are:

Dan Abbey, *Broward Community College*

Randall Allbritton, *Daytona Beach Community College*

Michael S. Allen, *Glendale Community College*

Mostafa S. Aminzadeh, *Towson State University*

Raymond Badalian, *Los Angeles City College*

Carole Bernett, *William Rainey Harper College*

Rich Campbell, *Butte College*

Mark Carpenter, *Sam Houston State University*

Daniel Cherwien, *Cumberland County College*

James A. Condor, *Manatee Community College*

David T. Cooney, *Polk Community College*

James C. Curl, *Modesto Junior College*

Carol Curtis, *Fresno City College*

Steven Day, *Riverside Community College*

Nirmal Devi, *Embry-Riddle Aeronautical University*

Wayne Ehler, *Anne Arundel College*

Eugene Enneking, *Portland State University*

Michael Eurgubian, *Santa Rosa Junior College*

Ruby Evans, *Santa Fe Community College*

Jeff Gervasi, *Porterville College*

Dawit Getahew, *Chicago State University*

Gary Grimes, *Mt. Hood Community College*

Leslie Grunes, *Mercer County Community College*

Dianne Haber, *Westfield State College*

Ronald Hamill, *Community College of Rhode Island*

Mark Harbison, *Rio Hondo College*

Linda Harper, *Harrisburg Area Community College*

Susan Herring, *Sonoma State University*

Keith A. Hilmer, *Moorpark College*

Shu-ping Hodgson, *Central Michigan University*

Robert L. Horvath, *El Camino College*

K. G. Janardan, *Eastern Michigan University*

Steve Kahn, *Anne Arundel Community College*

David Kozlowski, *Triton College*

Don Krekel, *Southeastern Community College*

Marie Langston, *Palm Beach Community College*

Kaiyang Liang, *Miami-Dade Community College*

Rowan Lindley, *Westchester Community College*

Bill McClure, *Golden West College*

Caren McClure, *Ranch Santiago College*

Rhonda Magel, *North Dakota State University*

Rudy Maglio, *Oakton Community College*

Mary M. Marco, *Bucks County Community College*

Donald K. Mason, *Elmhurst College*

Ed Migliore, *Monterey Peninsula College*

Jeff Mock, *Diablo Valley College*

Charlene Moeckel, *Polk Community College*

Keith Oberlander, *Pasadena City College*

Orlan D. Ohlhausen, *Richland College*

Linda Padilla, *Joliet Junior College*

Marnie Pearson, *Foothill College*

Ronald E. Pierce, *Eastern Kentucky University*

Pervez Rahman, *Truman College*

Mohammed Rajah, *Mira Costa College*

Helen M. Roberts, *Montclair State University*

Martin Sade, *Pima Community College*

Arnold L. Schroeder, *Long Beach City College*

Bruce Sisko, *Belleville Area College*

Aileen Solomon, *Trident Technical College*

Charlotte Stewart, *Southeastern Louisiana University*

Joe Sukta, *Moraine Valley Community College*

James M. Sullivan, *Sierra College*

Mary M. Sullivan, *Curry College*

Arland Thompson, *Community College of Aurora*

Dave Wallach, *University of Findlay*

Sandra A. Weeks, *Johnson & Wales University*

Bob Wendling, *Ashland University*

I would also like to thank the authors of the supplements that have been developed to accompany the text. Sally Robinson updated the *Instructor's Solutions Manual* and the *Student Solutions Manual*. James Condor developed the *Critical Thinking Workbook*. Pat Foard revised the *Student Study Guide*.

Thanks to Gerry Moultine who provided the Minitab instructions, Michael Keller who provided TI-83 instructions, and Charles Seiter who provided Excel instructions. The people at Laurel Technical Services deserve thanks for their error checking and preparation of material for the website.

Finally, I would like to thank all the people at McGraw-Hill Higher Education for their efforts and support. Thanks go to Daryl Bruflodt, Sponsoring Editor; David Dietz, Senior Development Editor; and Marilyn Sulzer, Senior Project Manager.

Allan G. Bluman

chapter

1

The Nature of Probability and Statistics

Objectives

After completing this chapter, you should be able to

1. Demonstrate knowledge of statistical terms.

2. Differentiate between the two branches of statistics.

3. Identify types of data.

4. Identify the measurement level for each variable.

5. Identify the four basic sampling techniques.

6. Explain the difference between an observational and an experimental study.

7. Explain the importance of computers and calculators in statistics.

Are We Improving Our Diet?

Statistics Today

It has been determined that diets rich in fruits and vegetables are associated with a lower risk of chronic diseases such as cancer. Nutritionists recommend that by the year 2000 Americans should be consuming five or more servings of fruits and vegetables each day. Several researchers from the Division of Nutrition, the National Center for Chronic Disease Control and Prevention, the National Cancer Institute, and the National Institutes of Health decided to use statistical procedures to see how much progress is being made toward this goal.

The procedures they used and the results of the study will be explained in this chapter.

1–1

Introduction

Most people become familiar with probability and statistics through radio, television, newspapers, and magazines. For example, the following statements were found in newspapers.

> *A typical one-a-day vitamin and mineral pill boosted certain immune responses in older people by 64 percent, according to researchers at the University of Medicine and Dentistry of New Jersey. (USA Weekend, January 6–8, 1995.)*

Of 1,000 households polled nationwide, 40% said they owned at least one cordless phone; 9% had two or more. (Tribune-Review, Greensburg, PA, January 8, 1995, p. B3.)

In 1994 specialty ski shop retailers sold 760,000 parkas at an average price of $153, and 176,200 ski suits at an average price of $280. (Tribune-Review, Greensburg, PA, January 8, 1995, p. G1.)

The average price of real estate sold in Dormont [Pennsylvania] over the past 12 months was $64,304. (Tribune Review, Greensburg, PA, January 8, 1995, p. B3.)

On average, U.S. residents spend $193 a year on sports apparel. (USA Today, January 10, 1995.)

Statistics is used in almost all fields of human endeavor. In sports, for example, a statistician may keep records of the number of yards a running back gains during a football game, or the number of hits a baseball player gets in a season. In other areas, such as public health, an administrator would be concerned with the number of residents who contract a new strain of flu virus during a certain year. In education, the researcher might want to know if new methods of teaching are better than old ones. These are only a few examples of how statistics can be used in various occupations.

Furthermore, statistics is used to analyze the results of surveys and as a tool in scientific research to make decisions based on controlled experiments. Other uses of statistics include operations research, quality control, estimation, and prediction.

Statistics is the science of conducting studies to collect, organize, summarize, analyze, and draw conclusions from data.

Students study statistics for several reasons:

1. Students, like professional people, must be able to read and understand the various statistical studies performed in their field. To have this understanding, they must be knowledgeable about the vocabulary, symbols, concepts, and statistical procedures used in these studies.

2. Students and professional people may be called on to conduct research in their field, since statistical procedures are basic to research. To accomplish this, they must be able to design experiments; collect, organize, analyze, and summarize data; and possibly make reliable predictions or forecasts for future use. They must also be able to communicate the results of the study in their own words.

3. Students and professional people can also use the knowledge gained from studying statistics to become better consumers and citizens. For example, they can make intelligent decisions about what products to purchase based on consumer studies, government spending based on utilization studies, and so on.

These reasons can be considered the goals for students and professionals who study statistics.

It is the purpose of this chapter to introduce the student to the basic concepts of *probability* and statistics by answering questions such as the following:

What are the branches of statistics?

What are data?

How are samples selected?

This *USA Today* Snapshot shows the favorite toppings of hamburger lovers. Does it state that 93% of Americans eat hamburgers? Explain your answer.

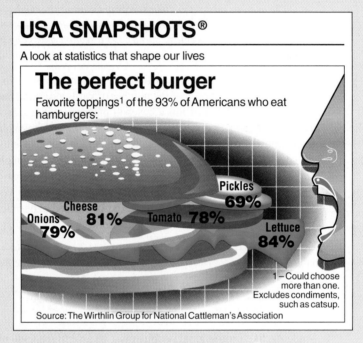

USA SNAPSHOTS®

A look at statistics that shape our lives

The perfect burger

Favorite toppings[1] of the 93% of Americans who eat hamburgers:

Pickles **69%**

Cheese **81%**

Onions **79%**

Tomato **78%**

Lettuce **84%**

1 – Could choose more than one. Excludes condiments, such as catsup.

Source: The Wirthlin Group for National Cattleman's Association

Source: Copyright 1995, *USA Today,* September, 1995. Used with permission.

1–2

Descriptive and Inferential Statistics

Objective 1. Demonstrate knowledge of statistical terms.

In order to gain knowledge about seemingly haphazard events, statisticians collect information for *variables,* which describe the event.

A **variable** is a characteristic or attribute that can assume different values.

Data are the values (measurements or observations) that the variables can assume. Variables whose values are determined by chance are called **random variables.**

Suppose that an insurance company studies its records over the past several years and determines that, on average, 3 out of every 100 automobiles the company insured were involved in an accident during a one-year period. Although there is no way to predict the specific automobiles that will be involved in an accident (random occurrence), the company can adjust its rates accordingly, since the company knows the general pattern over the long run. (That is, on average, 3% of the insured automobiles will be involved in an accident each year.)

A collection of data values forms a **data set.** Each value in the data set is called a **data value** or a **datum.**

Data can be used in different ways. The body of knowledge called statistics is sometimes divided into two main areas, depending on how data are used. The two areas are

STATISTICS

Here is a study on the snacks consumed during the 1998 Super Bowl. Suggest several ways that the data might have been obtained.

Super snack stats

■ **Snack fests:** Super Sunday, New Year's Eve and July Fourth are America's top snacking days.

■ **Eat up:** When it comes to the total amount of food eaten per person in a day, Super Sunday is No. 2, behind only Thanksgiving.

■ **Party on:** The Super Bowl is the No. 1 at-home party event of the year, says Hallmark Cards; the average party has 17 guests.

■ **The morning after:** Antacid sales increase 20 percent the day after the Super Bowl, reports the convenience store chain 7-Eleven.

SUPER BOWL FAVORITES

Snack	Pounds eaten during '98 game
Potato chips	11.2 million
Tortilla chips	8.2 million
Pretzels	4.3 million
Popcorn	3.8 million
Snack nuts	2.5 million

Sources: Snack Food Association, National Potato Promotion Board, Super Bowl 1998

Source: *USA WEEKEND,* January 22–24, 1999. Used with permission.

Objective 2. Differentiate between the two branches of statistics.

Historical Note

The origin of descriptive statistics can be traced to data collection methods used in censuses taken by the Babylonians and Egyptians between 4500 and 3000 B.C. In addition, the Roman Emperor Augustus (27 B.C.–A.D. 17) conducted surveys on births and deaths of the citizens of the empire, as well as the number of livestock each owned and the crops each citizen harvested yearly.

1. Descriptive statistics

2. Inferential statistics

In *descriptive statistics* the statistician tries to describe a situation. Consider the national census conducted by the U.S. government every 10 years. Results of this census give the average age, income, and other characteristics of the U.S. population. To obtain this information, the Census Bureau must have some means to collect relevant data. Once data are collected, the bureau must organize and summarize them. Finally, the bureau needs a means of presenting the data in some meaningful form, such as charts, graphs, or tables.

Descriptive statistics consists of the collection, organization, summarization, and presentation of data.

The second area of statistics is called *inferential statistics.* Here, the statistician tries to make inferences from *samples* to *populations.* Inferential statistics uses **probability**—i.e., the chance of an event occurring. Many people are familiar with the concepts of probability through various forms of gambling. People who play cards, dice, bingo, and lotteries win or lose according to the laws of probability. Probability theory is also used in the insurance industry and other areas.

It is important to distinguish between a sample and a population.

This *USA Today* Snapshot shows the common causes of accidental deaths in the home. Why might insurance companies be interested in this information?

USA SNAPSHOTS®

A look at statistics that shape our lives

Risks in the home

Most common causes of accidental death in the home:

Falls	Poisoning	Fire and burns	Suffocation	Others	Drowning/firearms
32%	25%	14%	12%	10%	7%

Source: Copyright 1995, *USA TODAY,* March 2, 1995. Used with permission.

A **population** consists of all subjects (human or otherwise) that are being studied.

Sometimes populations can be very large. In order to save time and money, statisticians may study only a part of the population. This part is called a *sample*.

A **sample** is a group of subjects selected from a population.

The area of inferential statistics called **hypothesis testing** is a decision-making process for evaluating claims about a population, based on information obtained from samples. For example, a researcher may wish to know if a new drug will reduce the number of heart attacks in men over 70 years of age. For this study, two groups of men over 70 would be selected. One group would be given the drug, and the other would be given a placebo (a substance with no medical benefits or harm). Later, the number of heart attacks occurring in each group of men would be counted, a statistical test would be run, and a decision would be made about the effectiveness of the drug.

Statisticians also use statistics to determine *relationships* among variables. For example, relationships were the focus of the most noted study in the past few decades,

Historical Note

Inferential statistics originated in the 1600s, when John Graunt published his book on population growth, *Natural and Political Observations Made upon the Bills of Mortality.* About the same time, another mathematician/ astronomer, Edmund Halley, published the first complete mortality tables. (Insurance companies use mortality tables to determine life insurance rates.)

"Smoking and Health," published by the surgeon general of the United States in 1964. He stated that after reviewing and evaluating the data, his group found a definite relationship between smoking and lung cancer. He did not say that cigarette smoking actually causes lung cancer but that there is a relationship between smoking and lung cancer. This conclusion was based on a study done in 1958 by Hammond and Horn. In this study, 187,783 men were observed over a period of 45 months. The death rate from lung cancer in this group of volunteers was 10 times as great for smokers as for nonsmokers.

Finally, by studying past and present data and conditions, statisticians try to make predictions based on this information. For example, a car dealer may look at past sales records for a specific month to decide what types of automobiles and how many of each type to order for that month next year.

Inferential statistics consists of generalizing from samples to populations, performing hypothesis tests, determining relationships among variables, and making predictions.

1–3

Variables and Types of Data

Objective 3. Identify types of data.

As stated in the previous section, statisticians gain information about a particular situation by collecting data for random variables. This section will explore in more detail the nature of variables and types of data.

Variables can be classified as qualitative or quantitative. **Qualitative variables** are variables that can be placed into distinct categories, according to some characteristic or attribute. For example, if subjects are classified according to gender (male or female), then the variable "gender" is qualitative. Other examples of qualitative variables are religious preference and geographic locations.

Quantitative variables are numerical in nature and can be ordered or ranked. For example, the variable "age" is numerical, and people can be ranked in order according to the value of their ages. Other examples of quantitative variables are heights, weights, and body temperatures.

Quantitative variables can be further classified into two groups, discrete or continuous. *Discrete variables* can be assigned values such as 0, 1, 2, 3, and are said to be countable. Examples of discrete variables are the number of children in a family, the number of students in a classroom, and the number of calls received by a switchboard operator each day for one month.

Discrete variables assume values that can be counted.

Continuous variables, by comparison, can assume all values between any two specific values. Temperature, for example, is a continuous variable, since the variable can assume all values between any two given temperatures.

Continuous variables can assume all values between any two specific values. They are obtained by measuring.

Since continuous data must be measured, answers must be rounded due to the limits of the measuring device. Usually, answers are rounded to the nearest given unit. For example, heights might be rounded to the nearest inch, weights to the nearest ounce, etc. Hence, a recorded height of 73 inches could mean any measure from 72.5 inches up to

Many times, researchers overlook factors that influence their statistical studies. In this article, researchers did not take into account the fact that as people age, they lose some height. When the original data were adjusted for height loss due to aging, the results were not significant.

Explain how the original study may have been conducted (i.e., what statistical comparisons or tests were used) and how the adjustments for height loss due to aging may have been calculated.

Height Fright Not Right

Turns out a recent study that claimed short men have a greater risk of heart attack may have been a tall tale. According to the scientific journal *Circulation*, the height loss that comes with age accounts for shorter men having an increased risk of coronaries. After the calculations are adjusted for this age-related shrinkage, the tall-guy advantage apparently disappears.

Source: "Height Fright Not Right," *Prime*, Winter 1995, p. 14. Used with permission.

---

but not including 73.5 inches. Thus, the boundary of this measure is given as 72.5–73.5 inches. *Boundaries are written for convenience as 72.5–73.5 but understood to mean all values up to but not including 73.5.* Actual data values of 73.5 would be rounded to 74 and would be included in a class with boundaries of 73.5 up to but not including 74.5, written as 73.5–74.5. As another example, if a recorded weight is 86 pounds, the exact boundaries are 85.5 up to but not including 86.5, written as 85.5–86.5 pounds. Table 1–1 helps to clarify this concept. The boundaries of a continuous variable are given in one additional decimal place and always end with the digit 5.

Table 1–1   Recorded Values and Boundaries

Variable	Recorded value	Boundaries
Length	15 centimeters (cm)	14.5–15.5 cm
Temperature	86 degrees Fahrenheit (°F)	85.5°–86.5°
Time	0.43 second (sec)	0.425–0.435 sec
Mass	1.6 grams (g)	1.55–1.65 g

In addition to being classified as qualitative or quantitative, variables can also be classified by how they are categorized, counted, or measured. For example, can the data be organized into specific categories, such as area of residence (rural, suburban, or urban)? Can

Unusual Stats

According to *The Book of Odds*, 63% of us say we would rather hear the bad news first, p. 10.

Objective 4. Identify the measurement level for each variable.

the data values be ranked, such as first place, second place, etc.? Or are the values obtained from measurement, such as heights, IQs, or temperature? This type of classification—i.e., how variables are categorized, counted, or measured—uses **measurement scales,** and four common types of scales are used: nominal, ordinal, interval, and ratio.

The **nominal level of measurement** classifies data into mutually exclusive (nonoverlapping), exhausting categories in which no order or ranking can be imposed on the data.

A sample of college instructors classified according to subject taught (e.g., English, history, psychology, or mathematics) is an example of nominal-level measurement. Classifying survey subjects as male or female is another example of nominal-level measurement. No ranking or order can be placed on the data. Classifying residents according to zip codes is also an example of the nominal level of measurement. Even though numbers are assigned as zip codes, there is no meaningful order or ranking. Other examples of nominal-level data are political party (Democratic, Republican, Independent), religion (Lutheran, Jewish, Catholic, Methodist, etc.), and marital status (married, divorced, widowed, separated).

The next level of measurement is called the *ordinal level*. Data measured at this level can be placed into categories, and these categories can be ordered, or ranked. For example, from student evaluations, guest speakers might be ranked as superior, average, or poor. Floats in a homecoming parade might be ranked as first place, second place, etc. *Note that precise measurement of differences in the ordinal level of measurement* does not *exist*. For instance, when people are classified according to their build (small, medium, or large), a large variation exists among the individuals in each class.

The **ordinal level of measurement** classifies data into categories that can be ranked; however, precise differences between the ranks do not exist.

Examples of ordinal-measured data are letter grades (A, B, C, D, F), rating scales, and rankings.

The third level of measurement is called the *interval level*. This level differs from the ordinal level in that precise differences do exist between units. For example, many standardized psychological tests yield values measured on an interval scale. IQ is an example of such a variable. There is a meaningful difference of one point between an IQ of 109 and an IQ of 110. Temperature is another example of interval measurement, since there is a meaningful difference of 1 degree between each unit, such as 72 degrees and 73 degrees. *One property is lacking in the interval scale: There is no true zero.* For example, IQ tests do not measure people who have no intelligence. For temperature, 0 degrees Fahrenheit does not mean no heat.

The **interval level of measurement** ranks data, and precise differences between units of measure do exist; however, there is no meaningful zero.

The highest level of measurement is called the *ratio level*. Examples of ratio scales are those used to measure height, weight, area, and number of phone calls received. Ratio scales have differences between units (1 inch, 1 pound, etc.) and a true zero. In addition, the ratio scale contains a true ratio between values. For example, if one person can lift 200 pounds and another can lift 100 pounds, then the ratio between them is 2 to 1. Put another way, the first person can lift twice as much as the second person.

The **ratio level of measurement** possesses all the characteristics of interval measurement, and there exists a true zero. In addition, true ratios exist when the same variable is measured on two different members of the population.

There is not complete agreement among statisticians about the classification of data into one of the four categories. For example, some researchers classify IQ data as ratio data rather than interval. Also, data can be altered so that they fit into a lower category. For instance, if the incomes of all professors of a college are classified into three categories—low, average, and high—then a ratio variable becomes an ordinal variable. Table 1–2 gives some examples of each type of data.

Table 1–2 Examples of Measurement Scales			
Nominal-level data	**Ordinal-level data**	**Interval-level data**	**Ratio-level data**
Zip code	Grade (A, B, C, D, F)	SAT score	Height
Gender (male, female)		IQ	Weight
Eye color (blue, brown, green, hazel)	Judging (first place, second place, etc.)	Temperature	Time
Political affiliation	Rating scale (poor, good, excellent)		Salary
Religious affiliation			Age
Major field (mathematics, computers, etc.)	Ranking of tennis players		
Nationality			

1–4

Data Collection and Sampling Techniques

Objective 5. Identify the four basic sampling techniques.

In research, statisticians use data in many different ways. As stated previously, data can be used to describe situations or events. For example, a manufacturer might want to know something about the consumers who will be purchasing his product so he can plan an effective marketing strategy. In another situation, the management of a company might survey its employees to assess their needs in order to negotiate a new contract with the employees' union. Data can be used to determine whether the educational goals of a school district are being met. Finally, trends in various areas, such as the stock market, can be analyzed, enabling prospective buyers to make more intelligent decisions concerning what stocks to purchase. These examples illustrate a few situations where collecting data will help people make better decisions on courses of action.

Data can be collected in a variety of ways. One of the most common methods is through the use of surveys. Surveys can be done by using a variety of methods. Three of the most common methods are the telephone survey, the mailed questionnaire, and the personal interview.

Telephone surveys have an advantage over personal interview surveys in that they are less costly. Also, people may be more candid in their opinions since there is no face-to-face contact. A major drawback to the telephone survey is that some people in the population will not have phones or will not be home when the calls are made; hence, not all people have a chance of being surveyed.

Mailed questionnaire surveys can be used to cover a wider geographic area than telephone surveys or personal interviews since they are less expensive to conduct. Also, respondents can remain anonymous if they desire. Disadvantages of mailed questionnaire surveys include a low number of responses and inappropriate answers to questions.

Another drawback is that some people may have difficulty reading or understanding the questions.

Personal interview surveys have the advantage of obtaining in-depth responses to questions from the person being interviewed. One disadvantage is that interviewers must be trained in asking questions and recording responses, which makes the personal interview survey more costly than the other two survey methods. Another disadvantage is that the interviewer may be biased in his or her selection of respondents.

Data can also be collected in other ways, such as *surveying records* or *direct observation* of situations.

As stated in Section 1–2, researchers use samples to collect data and information about a particular variable from a large population. Using samples saves time and money and, in some cases, enables the researcher to get more detailed information about a particular subject. Samples cannot be selected in haphazard ways because the information obtained might be biased. For example, interviewing people on a street corner would not include responses from people working in offices at that time or people attending school; hence, all subjects in a particular population would not have a chance of being selected.

In order to obtain samples that are unbiased—i.e., give each subject in the population an equally likely chance of being selected—statisticians use four basic methods of sampling: random, systematic, stratified, and cluster sampling.

Random Sampling

Random samples are selected using chance methods or random numbers. One such method is to number each subject in the population. Then place numbered cards in a bowl, mix them thoroughly, and select as many cards as needed. The subjects whose numbers are selected constitute the sample. Since it is difficult to mix the cards thoroughly, there is a chance of obtaining a biased sample. For this reason, statisticians use another method of obtaining numbers. They generate random numbers with a computer or calculator. Before the invention of computers, random numbers were obtained from a table of random numbers, similar to the one shown in Appendix C, Table D. State lottery numbers are selected at random. A more detailed explanation of random numbers is given in Chapter 15.

Systematic Sampling

Researchers obtain **systematic samples** by numbering each subject of the population and then selecting every kth number. For example, suppose there are 2000 subjects in the population and a sample of 50 subjects is needed. Since $2000 \div 50 = 40$, then

This *USA Today* Snapshot compares how men and women cope with bad days at work. Explain why the percentages do not add up to 100. Suggest some other ways people can survive a bad day at work. Do you think the survey should include your choices?

Source: Copyright 1995. *USA TODAY*, August 13, 1995. Used with permission.

$k = 40$, and every 40th subject would be selected; however, the first subject (numbered between 1 and 40) would be selected at random. Suppose subject 12 was the first subject selected; then the sample would consist of the subjects whose numbers were 12, 52, 92, etc., until 50 subjects were obtained. When using systematic sampling, one must be careful about how the subjects in the population are numbered. If subjects were arranged in a manner such as wife, husband, wife, husband, and every 40th subject was selected, the sample would consist of all husbands.

Stratified Sampling   Researchers obtain **stratified samples** by dividing the population into groups (called strata) according to some characteristic that is important to the study, then sampling from each group. For example, suppose the president of a two-year college wants to learn how students feel about a certain issue. Furthermore, the president wishes to see if the opinions of the freshmen differ from those of the sophomores. The president will select students from each group to use in the sample.

Cluster Sampling   Researchers select **cluster samples** by using intact groups called clusters. Suppose a researcher wishes to survey apartment dwellers in a large city. If there are 10 apartment buildings in the city, the researcher can select at random 2 buildings from the 10

Speaking of   STATISTICS

This *USA Today* Snapshot shows how often families eat dinner at home together. Can you see anything wrong with the responses? (*Hint:* How would you define frequently? Occasionally? Always?)

USA SNAPSHOTS®

A look at statistics that shape our lives

How often does your family eat dinner at home together?

Frequently
38%

Always
34%

Don't know
1%

Never
6%

Source:
The
Wirthlin
Group

Occasionally
21%

Source: Copyright 1995. *USA TODAY*, October 23, 1995. Used with permission.

Historical Note

In 1936, the *Literary Digest,* on the basis of a biased telephone sample of its subscribers, predicted that Alf Landon would defeat Franklin D. Roosevelt in the upcoming presidential election. Roosevelt won by a landslide. The magazine ceased publication the following year.

and interview all the residents of these buildings. Cluster sampling is used when the population is large or when it involves subjects residing in a large geographic area. For example, if one wanted to do a study involving the patients in the hospitals in New York City, it would be very costly and time-consuming to try to obtain a random sample of patients since they would be spread over a large area. Instead, a few hospitals could be selected at random and the patients in these hospitals would be interviewed in a cluster.

The four basic sampling methods are summarized in Table 1–3.

Table 1–3	Summary of Sampling Methods
Random	Subjects are selected by random numbers
Systematic	Subjects are selected by using every kth number after the first subject is randomly selected from 1 through k
Stratified	Subjects are selected by dividing up the population into groups (strata) and subjects within groups are randomly selected
Cluster	Subjects are selected by using an intact group that is representative of the population

Speaking of   STATISTICS

Statistical studies often have conflicting conclusions. When this occurs, the researchers attempt to cite reasons for the discrepancies. The studies discussed here suggest several reasons for the different conclusions. Which of the reasons stated do you think was responsible for the discrepancy? Explain your answer.

Shake and Wake

Two major earthquakes, two studies on how traumatic events affect the heart—and two completely different results? The answer may be a testament to what the mind can endure once it adapts to a stressful situation. Or it may hinge on the clock.

At Stanford University's School of Medicine, psychiatrist C. Barr Taylor, M.D., and colleagues looked at the prevalence of heart attacks after the 1989 quake that rocked San Francisco. Medical records at five Bay Area hospitals turned up 16 heart attacks in the week before the disaster—and only 12 in the seven days following.

Surprising, considering the cardiovascular trauma unleashed by the quake: Heart rates skyrocket, blood pressure soars, and breathing becomes labored. "It's as if we gave a stress test to everybody in the Santa Clara Valley," Taylor says of the quake. "Not

to see any demonstrated increase in heart attacks is remarkable."

But the findings were far more grim following the 1994 Northridge quake in Los Angeles. Checking 81 coronary care units around L.A., Good Samaritan Hospital cardiologists Robert A. Kloner, M.D., Ph.D., and Jonathan Leor, M.D., counted a 35 percent jump in heart attacks the week after the quake.

Hospitals within 15 miles of the epicenter were particularly likely to report more cardiac patients. While the physical strain of escaping the devastation was a factor in some cases, in others emotional stress was clearly the trigger, Kloner says.

Why the discrepancy between studies? Are San Franciscans simply more mentally prepared for the Big One—and thus less shocked when a quake hits? Support for this idea comes

courtesy of Israeli researchers who looked at Iraq's missile strikes on Israel during the 1991 Gulf War.

On the day of the first attack, deaths throughout Israel rose by 58 percent, mostly from cardiac-related causes. But in subsequent attacks there was no rise in mortality, suggesting most people adapted to higher levels of stress and fear once they knew more missile strikes were likely.

But Kloner has a different take on the L.A. quake's high coronary toll: It struck at 4 A.M. "Being awakened from a sound sleep," he contends, "is very different from already being up and about." Add to that the physiological reality that heart attacks are already most common around dawn and the killing potential of an early morning quake takes on power that doesn't show up on Mr. Richter's scale.

Source: Reprinted with permission from *Psychology Today* Magazine. Copyright © 1995 (Sussex Publishers, Inc.).

Other Sampling Methods

Interesting Facts

In addition to the four basic sampling methods, researchers use other methods to obtain samples. One such method is called a **convenience sample.** Here a researcher uses subjects that are convenient. For example, the researcher may interview subjects entering a local mall to determine the nature of their visit or perhaps what stores they will be patronizing. This sample is probably not representative of the general customers for several reasons. For one thing, it was probably taken at a specific time of day, so not all customers entering the mall have an equal chance of being selected since they were not there when the survey was being conducted. But convenience samples can be representative of the population. If the researcher investigates the characteristics of the population and determines that the sample is representative, then it can be used.

Other sampling techniques, such as *sequential sampling, double sampling,* and *multistage sampling,* are explained in Chapter 15, along with a more detailed explanation of the four basic sampling techniques.

Observational and Experimental Studies

Objective 6. Explain the difference between an observational and an experimental study.

There are several different ways to classify statistical studies. This section explains two types of studies: *observational studies* and *experimental studies.*

In an **observational study,** the researcher merely observes what is happening or what has happened in the past and tries to draw conclusions based on these observations.

For example, data from the Motorcycle Industry Council (*USA Today,* May 7, 1999) stated that "Motorcycle owners are getting older and richer." Data were collected on the ages and incomes of motorcycle owners for the years 1980 and 1998 and then compared. The findings showed considerable differences in the ages and incomes of motorcycle owners for the two years.

In this study, the researcher merely observed what had happened to the motorcycle owners over a period of time. There was no type of research intervention.

In an **experimental study,** the researcher manipulates one of the variables and tries to determine how the manipulation influences other variables.

For example, a study conducted at Virginia Polytechnic Institute and presented in *Psychology Today* divided female undergraduate students into two groups and had the students perform as many sit-ups as possible in 90 seconds. The first group was told only to "Do your best," while the second group was told to try to increase the actual number of situps they did each day by 10%. After four days, the subjects in the group that were given the vague instructions, "Do your best," averaged 43 sit-ups, while the group that was given the more specific instructions to increase the number of sit-ups by 10% averaged 56 sit-ups by the last day's session. The conclusion then was that athletes who were given specific goals perform better than those who were not given specific goals.

This study is an example of a statistical experiment since the researchers intervened in the study by manipulating one of the variables, namely, the type of instructions given to each group.

In a true experimental study, the subjects should be assigned to groups randomly. Also, the treatments should be assigned to the groups at random. In the sit-up study, the article did not mention whether the subjects were randomly assigned to the groups.

Sometimes when random assignment is not possible, researchers use intact groups. These types of studies are done quite often in education where already intact groups are available in the form of existing classrooms. When these groups are used, the study is said to be a **quasi-experimental study.** The treatments, though, should be assigned at random. Most articles do not state whether random assignment of subjects was used.

Statistical studies usually include one or more *independent variables* and one *dependent variable.*

The **independent variable** in an experimental study is the one that is being manipulated by the researcher. The independent variable is also called the **explanatory variable.** The resultant variable is called the **dependent variable** or the **outcome variable.**

The outcome variable is the variable that is studied to see if it has changed significantly due to the manipulation of the independent variable. For example, in the sit-up study, the researchers gave the groups two different types of instructions, general and specific. Hence, the independent variable is the type of instruction. The dependent variable, then, is the resultant variable, that is, the number of sit-ups each group was able to perform after four days of exercise. If the differences in the dependent or outcome variable are large and other factors are equal, these differences can be attributed to the manipulation of the independent variable. In this case, specific instructions were shown to increase athletic performance.

In the sit-up study, there were two groups. The group that received the special instruction is called the **treatment group** while the other is called the **control group.** The treatment group receives a specific treatment (in this case, instructions for improvement) while the control group does not.

Both types of statistical studies have advantages and disadvantages. Experimental studies have the advantage that the researcher can decide how to select subjects and how to assign them to specific groups. The researcher can also control or manipulate the independent variable. For example, in studies that require the subjects to consume a certain amount of medicine each day, the researcher can determine the precise dosages and, if necessary, vary the dosage for the groups.

There are several disadvantages to experimental studies. First, they may occur in unnatural settings, such as laboratories, special classrooms, etc. This can lead to several problems. One such problem is that the results might not apply to the natural setting. The age-old question then is, "This mouthwash may kill 10,000 germs in a test tube, but how many germs will it kill in my mouth?"

Another disadvantage with an experimental study is the so-called **"Hawthorne effect."** This effect was discovered in 1924 in a study of workers at the Hawthorne plant of the Western Electric Company. In this study, researchers found because the subjects knew they were participating in an experiment it actually changed their behavior in ways that affected the results of the study.

Another problem is called *confounding of variables.*

A **confounding variable** is one that influences the dependent or outcome variable but cannot be separated from the independent variable.

Researchers try to control most variables in a study, but this is not possible in some studies. For example, subjects who are put on an exercise program might also improve their diet unbeknownst to the researcher and perhaps improve their health in other ways not due to exercise alone.

Observational studies also have their advantages and disadvantages. One advantage of an observational study is that it usually occurs in a natural setting. For example, researchers can observe people's driving patterns on streets and highways in large cities. Another advantage of an observational study is that it can be done in situations where it would be unethical or outright dangerous to conduct an experiment. Using observational studies, researchers can study suicides, rapes, murders, etc. In addition, observational studies can be done using variables that cannot be manipulated by the researcher such as drug users versus nondrug users, right-handedness versus left-handedness, etc.

Observational studies have disadvantages, too. As mentioned previously, since the variables are not controlled by the researcher, a definite cause-and-effect situation cannot be shown since other factors may have had an effect on the results. Observational studies can be expensive and time-consuming. For example, if one wanted to study the habitat of lions in Africa, one would need a lot of time and money, and there would be a certain amount of danger involved. Finally, since the researcher may not be using his or her own measurements, the results could be subject to inaccuracies of those who collected the data. For example, if the researchers were doing a study of events that occurred in the 1800s, they would have to rely on information and records obtained by others from a previous era. There is no way to ensure the accuracy of these records.

When reading the results of statistical studies, decide if the study was observational or experimental in nature. Then see if the conclusion follows logically, based on the nature of these studies.

No matter what type of study is conducted, two studies on the same subject sometimes have conflicting conclusions. Why might this occur? An article entitled "Bottom Line: Is It Good for You?" (*USA Weekend,* February 26–28, 1999) states that in the 1960s studies suggested that margarine was better for the heart than butter since margarine contains less saturated fat and users had lower cholesterol levels. In a 1980 study, researchers found that butter was better than margarine since margarine contained trans-fatty acids, which are worse for the heart than butter's saturated fat. Then in a 1998 study, researchers found that margarine was better for a person's health. Now, what is to be believed? Should one use butter or margarine?

The answer here is to take a closer look at these studies. Actually, it is not a choice between butter or margarine that counts but the type of margarine used. In the 1980s, studies showed that solid margarine contains trans-fatty acids, and scientists believe that they are worse for the heart than butter's saturated fat. In the 1998 study, liquid margarine was used. It is very low in trans-fatty acids, and hence, it is more healthful than butter because these acids have been shown to lower cholesterol. Hence, the conclusion is to use liquid margarine instead of solid margarine or butter.

Before decisions based on research studies are made, it is important to get all the facts and examine them in light of the particular situation.

1–6

Computers and Calculators

Objective 7 Explain the importance of computers and calculators in statistics.

In the past, statistical calculations were done with pencil and paper. However, with the advent of calculators, numerical computations became much easier. Computers do all the numerical calculation. All one does is enter the data into the computer and use the appropriate command; the computer will print the answer or display it on-screen. Now the TI-83 Graphing Calculator accomplishes the same thing.

There are many statistical packages available; this book uses MINITAB and Excel. Instructions for using MINITAB, the TI-83 Graphing Calculator, and Excel have been placed at the end of each relevant section, in subsections entitled "Technology Step by Step."

Students should realize that the computer and calculator merely give numerical answers and save the time and effort of doing calculations by hand. The student is still responsible for understanding and interpreting each statistical concept. In addition, students should realize that the results come from the data and do not appear magically on the computer. Doing calculations using the procedure tables will help reinforce this idea.

The author has left it up to instructors to choose how much technology they will incorporate into the course.

Technology Step by Step

MINITAB
Step by Step

MINITAB statistical software provides a wide range of statistical analysis and graphing capabilities. When you start MINITAB, the screen will look like the one shown. You will see the Session Window, the Project Manager Window, and the Data Window (labeled "Worksheet 1" in screen shown). The Project Manager is new to MINITAB, Release 13. The other windows are common

The Minitab Program
Window

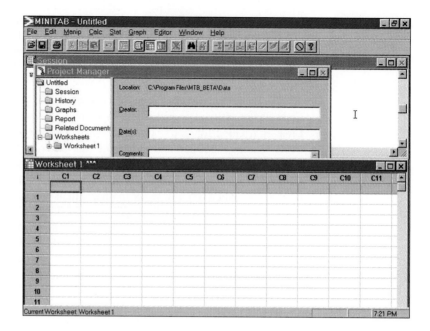

to Release 13 and Release 12; throughout this book, we will use these windows most of the time. The Info Window, the History Window, and the Graph Manager are part of the Project Manager.

Entering Data in MINITAB

In MINITAB, all the data for one variable are stored in a column. To enter data by typing:

1. Click `File` on the menu bar, then click `New` (from now on, we will write this instruction as **File>New**).
2. Select `Minitab Worksheet` and click `[OK]`.
3. Click the arrow in the upper left-hand corner of the Data Window so that it points down.
4. Type the data values in the first column and press **[Enter]** after each value.

To enter data into MINITAB by opening a file see the following example.

Example MT1–1

Data for 100 employees were collected. There is a row for each person's data and a column for each variable. The raw data is shown in Appendix D. We put the data into the worksheet by opening the Databank file available for download at the book's website.

1. Select **File>Open Worksheet.** The `Open Worksheet` dialog box will be displayed. You must check three items in this dialog box. Note that your files may not look exactly as shown.
2. The `Look in:` dialog box should show the directory where the file is located.
3. Make sure the `Files of Type:` shows the correct type. The default extension is MINITAB [*.mtw].
4. Double click the file name in the list box. The data will be copied into the worksheet of MINITAB.

Open Worksheet
Dialog Box

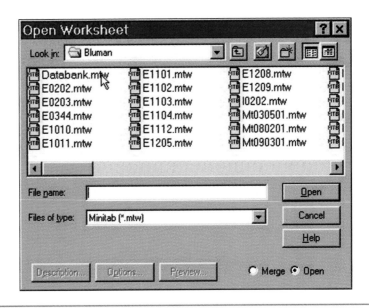

The TI-83 graphing calculator can be used for a variety of statistical graphs and tests.

TI-83
Step by Step

General Information

To turn calculator on:

Press **ON** key.

To turn calculator off:

Press **2nd [OFF]**.

To reset defaults only:

1. Press **2nd** then **[MEM]**.

2. Select **5** then **2** then **2**.

(Optional). To reset settings on calculator and clear memory: (*Note:* This will clear all settings and programs in the calculator's memory.)
Press **2nd** then **[MEM]**. Then press **5**, then **1**, then **2**.
(Note: Use 7 instead of 5 for the TI-83+.
Also, the contrast may need to be adjusted after this.)

To adjust contrast (if necessary):

Press **2nd.** Then press and hold ▲ to darken or ▼ to lighten contrast.

To clear screen:

Press **CLEAR.**

(*Note:* This will return you to the screen you were using.)

To display a menu:

Press appropriate menu key. Example: **STAT.**

To return to home screen:

Press **2nd,** then **[QUIT].**

To move around on the screens:

Use the arrow keys.

To select items on the menu:

Press the corresponding number or move the cursor to the item using the arrow keys. Then press **ENTER.**

(Note: In some cases, you do not have to press **ENTER,** and in other cases you may need to press **ENTER** twice.)

Entering Data

To enter single variable data (if necessary, clear the old list):

1. Press **STAT** to display the `Edit menu`.
2. Press **ENTER** to select `1:Edit`.
3. Enter the data in L_1 and press **ENTER** after each value.
4. After all data values are entered, press **STAT** to get back to the `Edit` menu or **2nd [QUIT]** to end.

Example TI1–1

Enter the following data values in L_1: **213, 208, 203, 215, 222.**

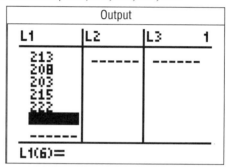

To enter multiple variable data:

The TI-83 will take up to six lists designated L_1, L_2, L_3, L_4, L_5, and L_6.

1. To enter more than one set of data values, complete the preceding steps. Then move the cursor to L_2 by pressing the ▶ key
2. Repeat the steps in the preceding part.

Editing Data

To correct a data value before pressing **ENTER,** use ◀ and retype the value and press **ENTER.**

To correct a data value in a list after pressing **ENTER,** move cursor to incorrect value in list and type in the correct value. Then press **ENTER.**

To delete a data value in a list:

Move cursor to value and press **DEL.**

To insert a data value in a list:

1. Move cursor to position where data value is to be inserted, then press **2nd [INS].**
2. Type data value; then press **ENTER.**

To clear a list:

1. Press **STAT** then **4.**
2. Enter list to be cleared. Example: to clear L_1, press **2nd [L₁].** Then press **ENTER.**

   (Note: To clear several lists, follow step 1, but enter each list to be cleared, separating them with a comma.)

Sorting Data

To sort the data in a list:

1. Enter the data in L_1.

2. Press **STAT 2** to get Sort A to sort list in ascending order.

3. Then press **2nd [L₁] ENTER.**

The calculator will display Done.

4. Press **STAT ENTER** to display sorted list.

(*Note:* The Sort D or **3** sorts the list in descending order.)

Example TI1–2

Sort in ascending order the data values entered in Example TI1–1.

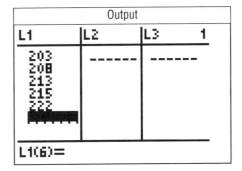

Excel
Step by Step

Microsoft Excel has two different ways to solve statistical problems. First, there are built-in functions, such as STDEV and CHITEST, available from the standard toolbar by clicking the f_x icon. For most of the problems in this textbook, however, it is easier to use the packaged tests in the Data Analysis Add-In.

To activate the Data Analysis Add-In:

1. Select Tools on the Worksheet menu bar.

2. Select Add-Ins from the Tools menu.

3. Check the box for Analysis ToolPak.

Excel's Analysis
ToolPak Add-In

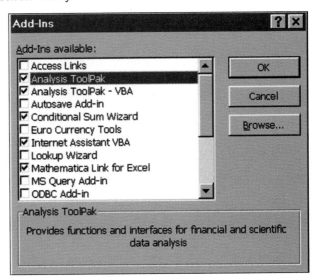

To enter data into Excel by typing:

1. Click the cell at the top of the column where you want to enter data. When working with values for a single variable, you usually will want to enter the values into a column.

2. Type each data value and press **[Enter]** or **[Tab]**.

You can also enter data into Excel by opening an Excel workbook file. Follow the steps in the example below to open the Databank file available for download at the book's website.

Example XL1–1

1. Click `File` on the menu bar, then click `Open . . .` (i.e., select **File>Open . . .**). The `Open` dialog box will be displayed.

2. In the `Look in` box, click the folder where the workbook file is located.

3. In the folder list, double-click folders until you open the folder containing the file you want.

4. Double-click the file name in the list box. The data will be copied into Excel.

1–7

Summary

The two major areas of statistics are descriptive and inferential. Descriptive statistics includes the collection, organization, summarization, and presentation of data. Inferential statistics includes making inferences from samples to populations, hypothesis testing, determining relationships, and making predictions. Inferential statistics is based on *probability theory.*

Since in most cases the populations under study are large, statisticians use subgroups called samples to get the necessary data for their studies. There are four basic methods used to obtain samples: random, systematic, stratified, and cluster.

Data can be classified as qualitative or quantitative. Quantitative data can be either discrete or continuous, depending on the values they can assume. Data can also be measured by various scales. The four basic levels of measurement are nominal, ordinal, interval, and ratio.

LAFF - A - DAY

"We've polled the entire populace, Your Majesty, and we've come up with exactly the results you ordered!"

Source: © 1993 King Features Syndicate, Inc. World Rights reserved. Reprinted with special permission of King Features Syndicate.

There are two basic types of statistical studies. They are observational studies and experimental studies. When conducting observational studies, researchers observe what is happening or what has happened and then draw conclusions based on these observations. They do not attempt to manipulate the variables in any way.

When conducting an experimental study, the researchers manipulate one or more of the independent or explanatory variables and see how this manipulation influences the dependent or outcome variable.

Finally, the applications of statistics are many and varied. People encounter them in everyday life, such as in reading newspapers or magazines, listening to the radio, or watching television. Since statistics is used in almost every field of endeavor, the educated individual should be knowledgeable about the vocabulary, concepts, and procedures of statistics.

Today computers and calculators are used extensively in statistics to facilitate the computations.

Important Terms

cluster sample   12

confounding variable   16

convenience sample   14

continuous variables   7

control group   16

data   4

data set   4

data value or datum   4

dependent variable   15

descriptive statistics   5

discrete variables   7

experimental study   15

explanatory variable   15

Hawthorne effect   16

hypothesis testing   6

independent variable   15

inferential statistics   7

interval level of measurement   9

measurement scales   9

nominal level of measurement   9

observational study   15

ordinal level of measurement   9

outcome variable   15

population   6

probability   5

qualitative variables   7

quantitative variables   7

quasi-experimental study   15

random sample   11

random variable   4

ratio level of measurement   10

sample   6

statistics   3

stratified sample   12

systematic sample   11

treatment group   16

variable   4

Review Exercises

Note: All odd-numbered problems and even-numbered problems marked with **(ans)** are included in the answer section at the end of this book.

Note: Problems with an asterisk (*) require more in-depth skills than other problems.

1–1. Name and define the two areas of statistics.

1–2. What is probability? Name two areas where probability is used.

1–3. Suggest some ways statistics can be used in everyday life.

1–4. Explain the differences between a sample and a population.

1–5. Why are samples used in statistics?

1–6. (ans) In each statement that follows, tell whether descriptive or inferential statistics have been used.

   a. In the year 2010, 148 million Americans will be enrolled in an HMO (Source: *USA Today,* December 30, 1997).

   b. Nine out of 10 on-the-job fatalities are men (Source: *USA Weekend,* January 2–4, 1998).

   c. Expenditures for the cable industry were 5.66 billion dollars in 1996 (Source: *USA Today,* December 30, 1997).

   d. The median household income for people ages 25–34 is $35,888 (Source: *USA Today,* December 8, 1997).

   e. "Allergy therapy makes bees go away." (*Prevention,* April 1995).

   f. Drinking decaffeinated coffee can raise cholesterol levels by 7%. (Source: American Heart Association).

   g. The national average annual medicine expenditure per person is $1,052 (Source: *The Tribune Review,* Greensburg, Pa., December 22, 1996).

   h. Experts say that mortgage rates may soon hit bottom (Source: *USA Today,* October 6, 1997).

1–7. Classify each as nominal-level, ordinal-level, interval-level, or ratio-level data.

a. Horsepower of motorcycle engines.

b. Ratings of newscasts in Houston (poor, fair, good, excellent).

c. Temperature of automatic popcorn poppers.

d. Time required by drivers to complete a course.

e. Salaries of cashiers of Day-Night grocery stores.

f. Marital status of respondents to a survey on savings accounts.

g. Ages of students enrolled in a martial arts course.

h. Weights of beef cattle fed a special diet.

i. Rankings of weight lifters.

j. Pages in the telephone book for the city of Los Angeles.

1–8. (ans) Classify each variable as qualitative or quantitative.

a. Colors of automobiles in the faculty parking lot.

b. Number of desks in classrooms.

c. Classification of children in a day care center (infant, toddler, preschool).

d. Weights of fish caught in Lake Emilie.

e. Number of pages in statistics textbooks.

f. Capacity (in gallons) of water in selected dams.

g. Number of off-road vehicles sold in the United States.

1–9. Classify each variable as discrete or continuous.

a. Number of loaves of bread baked each day at a local bakery.

b. Water temperature of the saunas at a given health spa.

c. Incomes of single parents who attend a community college.

d. Lifetimes of batteries in a tape recorder.

e. Weights of newborn infants at a certain hospital.

f. Capacity (in gallons) of water in swimming pools.

g. Number of pizzas sold last year in the United States.

1–10. Give the boundaries for each value.

a. 84.5 miles

b. 6 tons

c. 0.15 milliliters

d. 17 inches

e. 8.72 ounces

f. 97 feet

1–11. Name and define the four basic sampling methods.

1–12. (ans) Classify each sample as random, systematic, stratified, or cluster.

a. In a large school district, all teachers from two buildings are interviewed to determine whether they believe the students have less homework to do now than in previous years.

b. Every seventh customer entering a shopping mall is asked to select his or her favorite store.

c. Nursing supervisors are selected using random numbers in order to determine annual salaries.

d. Every hundredth hamburger manufactured is checked to determine its fat content.

e. Mail carriers of a large city are divided into four groups according to gender (male or female) and according to whether they walk or ride on their routes. Then 10 are selected from each group and interviewed to determine whether they have been bitten by a dog in the last year.

1–13. Give three examples each of nominal, ordinal, interval, and ratio data.

1–14. For each of the following statements, define a population and state how a sample might be obtained.

a. The average cost of an airline meal in 1993 was $4.55 (Source: *Everything Has Its Price,* Richard E. Donley, Simon and Schuster, 1995).

b. "More than 1 in 4 United States children have cholesterol levels of 180 milligrams or higher" (Source: The American Health Foundation).

c. "Every 10 minutes, 2 people die in car crashes and 170 are injured" (Source: National Safety Council estimates).

d. "When older people with mild to moderate hypertension were given the mineral salt for six months, the average blood pressure reading dropped by eight points systolic and three points diastolic" (Source: *Prevention,* March 1995, p. 19).

e. "The average amount spent per gift for Mom on Mother's Day is $25.95" (Source: The Gallup Organization).

1–15. Select a newspaper or magazine article that involves a statistical study and write a paper answering the following questions.

a. Is this study descriptive or inferential in nature? Explain your answer.

b. What are the variables used in the study? In your opinion, what level of measurement was used to obtain the data from the variables?

c. Does the article define the population? If so, how is it defined? If not, how could it be defined?

d. Does the article state the sample size and how the sample was obtained? If so, explain the size of the sample and how it was selected. If not, suggest a way it could have been obtained.

e. Explain *in your own words* what procedure (survey, comparison of groups, etc.) might have been used to determine the study's conclusions.

f. Do you agree or disagree with the conclusions? State your reasons.

1–16. Information from research studies is sometimes taken out of context. Explain why the claims of these studies might be suspect.

a. The average salary of the graduates of the class of 1980 is $32,500.

b. It is estimated that in Podunk there are 27,256 cats.

c. Only 3% of the men surveyed read *Cosmopolitan* magazine.

d. Based on a recent mail survey, 85% of the respondents favored gun control.

e. A recent study showed that high school dropouts drink more coffee than students who graduated; therefore, coffee dulls the brain.

f. Since most automobile accidents occur within 15 miles of a person's residence, it is safer to make long trips.

1–17. Identify each study as being either observational or experimental:

a. Subjects were randomly assigned to two groups and one group was given an herb and the other group a placebo. After six months, the number of respiratory tract infections each group had were compared.

b. A researcher stood at a busy intersection to see if the color of an automobile a person drives is related to running red lights.

c. A researcher finds that people who are more hostile have higher total cholesterol levels than those who are less hostile.

d. Subjects are randomly assigned to four groups. Each group is placed on one of four special diets—low-fat diet, a high-fish diet, a combination of low-fat diet and high-fish diet, and a regular diet. After six months, the blood pressures of the groups are compared to see if diet has any affect on blood pressure.

1–18. Identify the independent variable(s) and the dependent variable for each of the studies in the preceding exercise.

1–19. For each of the studies in Exercise 1–17, suggest possible confounding variables.

1–20. In the 1980s, a study linked coffee to a higher risk of heart disease and pancreatic cancer. In the early 1990s, studies showed that drinking coffee posed minimal health threats. However, in 1994, a study showed that pregnant women who drank three or more cups of tea daily may be at risk for spontaneous abortion. In 1998, a study claimed that women who drank more than a half-cup of caffeinated tea every day may actually increase their fertility. In 1998, a study showed that over a lifetime, a few extra cups of coffee a day can raise blood pressure, heart rate, and stress (Source: "Bottom Line: Is It Good for You? Or Bad?" by Monika Guttman, *USA Today Weekend,* February 26–28, 1999, pp. 8–9). Suggest some reasons why these studies appear to be conflicting.

*1–21. Find an article that describes a statistical study and identify the study as observational or experimental.

*1–22. For the article that you used in the previous exercise, identify the independent variable(s) and dependent variable for the study.

*1–23. For the article that you selected in Exercise 1–21, suggest some confounding variables that may have an effect on the results of the study.

Are We Improving Our Diet? Revisited

Researchers selected a *sample* of 23,699 adults in the United States, using phone numbers selected at *random,* and conducted a *telephone survey.* All respondents were asked six questions:

1. How often do you drink juices such as orange, grapefruit, or tomato?
2. Not counting juice, how often do you eat fruit?
3. How often do you eat green salad?
4. How often do you eat potatoes (not including french fries, fried potatoes, or potato chips)?
5. How often do you eat carrots?
6. Not counting carrots, potatoes, or salad, how many servings of vegetables do you usually eat?

Researchers found that men consumed fewer servings of fruits and vegetables per day (3.3) than women (3.7). Only 20% of the population consumed the recommended 5 or more daily servings. In addition, they found that youths and less-educated people consume an even lower amount than the average.

Based on this study, they recommend that greater educational efforts are needed to improve fruit and vegetable consumption by Americans and to provide environmental and institutional support to encourage increased consumption.

Source: Mary K. Sendula, M.D., et al., "Fruit and Vegetable Intake among Adults in 16 States: Results of a Brief Telephone Survey," *American Journal of Public Health* 85, no. 2 (February 1995).

Chapter Quiz

Determine whether each statement is true or false. If the statement is false, explain why.

1. Probability is used as a basis for inferential statistics.

2. The height of President Lincoln is an example of a variable.

3. The highest level of measurement is the interval level.

4. When the population of college professors is divided into groups according to their rank (instructor, assistant professor, etc.), and then several are selected from each group to make up a sample, the sample is called a cluster sample.

5. The variable of age is an example of a qualitative variable.

6. The weight of pumpkins is considered to be a continuous variable.

7. The boundary of a value such as 6 inches would be 5.9 inches to 6.1 inches.

Select the best answer.

8. The number of absences per year a worker has is an example of what type of data?
   a. Nominal
   b. Qualitative
   c. Discrete
   d. Continuous

9. What are the boundaries of 25.6 ounces?
   a. 25–26
   b. 25.55–25.65
   c. 25.5–25.7
   d. 20–39

10. A researcher divided subjects into two groups according to gender and then selected members from each group for his sample. What sampling method was the researcher using?
   a. Cluster
   b. Random
   c. Systematic
   d. Stratified

11. Data that can be classified according to color are measured on what scale?
   a. Nominal
   b. Ratio
   c. Ordinal
   d. Interval

12. A study that involves no researcher intervention is called
   a. an experimental study.
   b. a noninvolvement study.
   c. an observational study.
   d. a quasi-experimental study.

13. A variable that interferes with other variables in the study is called
   a. a confounding variable.
   b. an explanatory variable.
   c. an outcome variable.
   d. an interfering variable.

Use the best answer to complete the following statements.

14. Two major branches of statistics are ____ and ____ .

15. Two uses of probability are ____ and ____ .

16. The group of all subjects under a study is called a ____ .

17. A group of subjects selected from the group of all subjects under study is called a ____ .

18. Three reasons why samples are used in statistics are
   a. ____   b. ____   c. ____ .

19. The four basic sampling methods are
   a. ____   b. ____   c. ____   d. ____ .

20. A study that uses intact groups when it is not possible to randomly assign participants to the groups is called a(n) ____ study.

21. In a research study, participants should be assigned to groups using ____ methods if possible.

22. For each statement, decide whether descriptive or inferential statistics is used.
   a. A recent study showed that eating garlic can lower blood pressure.
   b. The average number of students in a class at White Oak University is 22.6.
   c. It is predicted that the average number of automobiles each household owns will increase next year.
   d. Last year's total attendance at Long Run High School's football games was 8235.
   e. The chance that a person will be robbed in a certain city is 15%.

23. Classify each as nominal, ordinal, interval, or ratio level.
   a. Number of exams given in a statistics course.

b. Ratings of word-processing programs as user-friendly.

c. Temperatures of a sample of automobile tires tested at 55 miles per hour for six minutes.

d. Weights of suitcases on a selected commercial airline flight.

e. Classification of students according to major field.

24. Classify each variable as discrete or continuous.

a. The time it takes to drive to work.

b. The number of credit cards a person has.

c. The number of employees working in a large department store.

d. The amount of a drug injected into a rat.

e. The amount of sodium contained in a bag of potato chips.

f. The number of cars stolen each week in a large city.

25. Give the boundaries of each.

a. 3.2 quarts

b. 18 pounds

c. 9 feet

d. 0.27 centimeter

e. 36 seconds

Critical Thinking Challenges

1. A study of the worlds' busiest airports was conducted by *Airports Council International.* Describe three variables that one could use to determine which airports are the busiest. What *units* would one use to measure these variables? Are these variables categorical, discrete, or continuous?

2. The results of a study published in *Archives of General Psychiatry* stated that male children born to women who smoke during pregnancy run a risk of violent and criminal behavior that lasts into adulthood. The results of this study were challenged by some people in the media. Give several reasons why the results of this study would be challenged.

3. The results of a study published in *Neurological Research* stated that second-graders who took piano lessons and played a computer math game more readily grasped math problems in fractions and proportions than a similar group who took an English class and played the same math game. What type of inferential study was this? Give several reasons why the piano lessons could improve a student's math ability.

4. A study of 2958 collegiate soccer players showed that in 46 anterior cruciate ligament (ACL) tears, 36 were in women. Calculate the percentages of tears for each gender. Can it be concluded that female athletes tear their knees more often than male athletes?

5. Read the article on the next page and answer the questions that follow.

The recent study reprinted below reached the following conclusion:

"A heart patient overly dependent on a spouse may have a harder time making necessary lifestyle changes in diet and exercise."

a. Do you agree or disagree with this statement? State your reasons.

b. Comment on how this study's conclusion might have been reached.

c. What are the variables used in the study and how might they have been defined?

d. How might the researcher measure the variables?

e. What would the population be for this study?

f. What factors other than dependence on a spouse might have influenced the results of the study?

g. Do you think the gender of the patients would make a difference in the results? Why or why not?

6. The effect of living near or working on power lines has been a very controversial issue. The article on page 28 states several conclusions.

a. Do you think that even if the risk of both cancers (leukemia and brain cancer) is greater, the workers should be more concerned with electrocution or other accidents? Explain your answer.

b. Do you think that "a big, relatively well-done study that doesn't yield persuasive evidence of harm" is a justifiable reason to build power lines near residential areas? Explain your answer.

c. Do you think that since the death rate of industrial workers is lower than in the general population, it is safer to work as an industrial worker than in some other occupation?

d. Do you consider the sample size (138,905 men) large enough to be used to state the conclusion? Explain.

e. What does the statement "But brain cancer risk was 50% higher for men who worked more than five years as a lineman or electrician" mean to you?

f. Do you think the fact that the study was funded by the Electric Power Research Institute could have any effect on its findings?

Heart to Heart

There is a flurry of new research findings about how heart heals the heart:

• A heart patient overly dependent on a spouse may have a harder time making necessary life-style changes in diet and exercise.

• For women heart attack victims, spousal support is critical—but hard to come by. "The family sometimes feels abandoned," explains Martin Sullivan, "and they don't want the woman to take time out of her duties as a wife and mother to make important lifestyle changes. Women are more willing to change for men."

• For men, a heart attack may shatter the sole definition of self (as family provider). The introduction of larger concepts of the self is therapeutic.

• Patients who feel a sense of self-efficacy and control over their disease do better than those who don't.

• Depression and anxiety affect pain perception and the capacity to function in the face of medical symptoms.

• In a study at Stanford University, behavioral counseling after heart attack, especially for hard-driving Type A individuals, lowered the rate of recurrent heart attacks by 45%—the same as the most powerful prescription drugs.

Such findings have led Martin Sullivan to introduce innovative techniques at the Duke Center. These include a program known as PAIRS (Practical Application of Intimacy Relationship Skills), which teaches couples healthy interactive skills, and a meditation program that teaches patients to freeze-frame a moment in time and look at the emotional content of what they are experiencing. Says Sullivan of the Duke Center's work, "We're trying to take the best of everything."

Source: Reprinted with permission from *PSYCHOLOGY TODAY MAGAZINE,* Copyright © 1994 (Sussex Publishers, Inc.).

Power Lines' Link to Brain Cancer but No Tie Found with Leukemia

Long-term exposure to power lines appears to slightly increase risk of brain cancer, but not leukemia, among utility workers, says a large study reported in the *American Journal of Epidemiology.*

Some earlier studies linked electro-magnetic fields to leukemia among utility workers, but no study has been conclusive on the risks.

But even if the risk is greater, both cancers remain rare enough that other dangers to those workers—such as electrocution or other work accidents—should be a greater concern, says the University of North Carolina's David Savitz, one of the study leaders.

Savitz is encouraged by "a big, relatively well-done study that doesn't yield persuasive evidence of harm." But the findings also may help explain why the rate of brain cancer in the general population,

though still low, has increased in recent decades, he says.

The study differs from earlier ones because of its size and because it used actual measurements of exposure levels. It looked at 138,905 men who worked for power companies between 1950 and 1986, including 20,733 who died.

Overall death risk among the men was lower than the general population, probably because industrial workers are most likely to be healthy, the researchers report.

But brain cancer risk was 50% higher for men who worked more than five years as a lineman or electrician. Those exposed to the highest levels of magnetic fields had more than double the risk.

There was no association found between amount of magnetic field exposure and leukemia. The research was funded by the Electric Power Research Institute.

Source: Doug Levy, "Power Lines' Link to Brain Cancer." *USA Today,* January 11, 1995. Copyright 1995, USA TODAY. Reprinted with permission.

Speaking of   STATISTICS

In these studies, behaviorists found that hotheadedness can lead to heart disease. After reading this article, decide if these studies are experimental or observational in nature. What is the independent variable and what is the dependent variable for each study?

Science sees heart trouble for hostile personalities

Two studies published today give scientific credence to what behaviorists have long suspected: Hotheadedness can lead to heart disease. Researchers who studied 1,081 older men found that those with high hostility scores (paranoid alienation, cynicism, agression, social avoidance) were more likely to be overweight and insulin-resistant–factors that lead to increased risk for heart disease. A smaller study, also in the January/February *Psychosomatic Medicine*, found that among 80 men and women, those with high hostility ratings had a bigger jump in blood pressure in mild, non-provoking social situations. Findings suggest that those people who have "high levels of mistrust and suspicion anticipate trouble in interpersonal situations, even before they have any overt indication that there is some cause for alarm," says researcher Mary Davis, author of the smaller study.

chapter

2

Frequency Distributions and Graphs

Outline

Objectives

After completing this chapter, you should be able to

1. Organize data using frequency distributions.

2. Represent data in frequency distributions graphically using histograms, frequency polygons, and ogives.

3. Represent data using Pareto charts, time series graphs, and pie graphs.

Are We Flying More?

Many executives need to use statistics to win support for their concerns. For example, David Henson, head of the Federal Aviation Administration, in an article entitled, "Airport Expansion: Doing More with Less," stated that increased air travel is causing congestion at most major airports in the United States. He suggested that airports will have to expand their capacities while facing cutbacks in government spending. In the article, he included the following table showing the increase in air traffic.

Because the data are stated in table form, they do not have as much impact on the reader as if they were presented using a statistical graph.

This chapter will show how to organize data and then construct appropriate graphs to represent the data in a concise, easy-to-understand form. An appropriate graph for this table appears near the end of the chapter.

Air Travel		
Here's how many millions of passengers used U.S. planes here and abroad; projections out Friday will show even greater increases in the next decade:	1980	287.9
	1981	274.7
	1982	286.1
	1983	308.2
	1984	334.0
	1985	370.1
	1986	404.7
	1987	441.2
	1988	441.2
	1989	443.6
	1990	456.6
	1991	445.7
	1992	463.0
	1993	468.1
	1994	509.0
	Source: FAA	

Source: *USA Today,* February 28, 1996. Copyright 1996, USA TODAY. Used with permission.

2–1

Introduction

When conducting a statistical study, the researcher must gather data for the particular variable under study. For example, if a researcher wishes to study the number of people who were bitten by poisonous snakes in a specific geographic area over the past several years, he or she would have to gather the data from various doctors, hospitals, or health departments.

In order to describe situations, draw conclusions, or make inferences about events, the researcher must organize the data in some meaningful way. The most convenient method of organizing data is to construct a *frequency distribution*.

After organizing the data, the researcher must present them so they can be understood by those who will benefit from reading the study. The most useful method of presenting the data is by constructing *statistical charts and graphs*. There are many different types of charts and graphs, and each one has a specific purpose.

This chapter explains how to organize data by constructing frequency distributions and how to present the data by constructing charts and graphs. The charts and graphs illustrated here are histograms, frequency polygons, ogives, pie graphs, Pareto charts, and time series graphs.

2–2

Organizing Data

Objective 1. Organize data using frequency distributions.

Suppose a researcher wished to do a study on the number of miles the employees of a large department store traveled to work each day. The researcher would first have to collect the data by asking each employee the approximate distance the store is from his or her home. When data are collected in original form, they are called **raw data.** In this case, the data are as follows:

1	2	6	7	12	13	2	6	9	5
18	7	3	15	15	4	17	1	14	5
4	16	4	5	8	6	5	18	5	2
9	11	12	1	9	2	10	11	4	10
9	18	8	8	4	14	7	3	2	6

Since little information can be obtained from looking at raw data, the researcher organizes the data by constructing a frequency distribution. The **frequency** is the number of values in a specific class of the distribution. For this data set, a frequency distribution is shown as follows:

Class limits (in miles)	Tally	Frequency
1–3	𝙸𝙸𝙸𝙸 𝙸𝙸𝙸𝙸	10
4–6	𝙸𝙸𝙸𝙸 𝙸𝙸𝙸𝙸 ////	14
7–9	𝙸𝙸𝙸𝙸 𝙸𝙸𝙸𝙸	10
10–12	𝙸𝙸𝙸𝙸 /	6
13–15	𝙸𝙸𝙸𝙸	5
16–18	𝙸𝙸𝙸𝙸	5
		Total 50

Now, some general observations can be obtained from looking at the data in the form of a frequency distribution. For example, the majority of employees live within nine miles of the store.

A **frequency distribution** is the organization of raw data in table form, using classes and frequencies.

The classes in this distribution are 1–3, 4–6, etc. These values are called class limits; the data values 1, 2, 3 can be tallied in the first class, 4, 5, 6 in the second class, and so on.

Two types of frequency distributions that are most often used are the *categorical frequency distribution* and the *grouped frequency distribution.* The procedures for constructing these distributions are shown next.

Categorical Frequency Distributions

The categorical frequency distribution is used for data that can be placed in specific categories, such as nominal- or ordinal-level data. For example, data such as political affiliation, religious affiliation, or major field of study would use categorical frequency distributions.

Example 2–1

Twenty-five army inductees were given a blood test to determine their blood type. The data set is as follows:

A	B	B	AB	O
O	O	B	AB	B
B	B	O	A	O
A	O	O	O	AB
AB	A	O	B	A

Construct a frequency distribution for the data.

Solution

Since the data are categorical, discrete classes can be used. There are four blood types: A, B, O, and AB. These types will be used as the classes for the distribution.

The procedure for constructing a frequency distribution for categorical data is given next.

STEP 1   Make a table as shown.

A Class	B Tally	C Frequency	D Percent
A			
B			
O			
AB			

STEP 2   Tally the data and place the results in column B.

STEP 3   Count the tallies and place the results in column C.

STEP 4   Find the percentage of values in each class by using the formula

$$\% = \frac{f}{n} \cdot 100\%$$

where

f = frequency of the class

n = total number of values

For example, in the class of type A blood, the percentage is

$$\% = \frac{5}{25} \cdot 100\% = 20\%$$

Percentages are not normally a part of a frequency distribution, but they can be added since they are used in certain types of graphic presentations, such as pie graphs.

STEP 5   Find the totals for columns C and D (see the completed table that follows).

Class	Tally	Frequency	Percent
A	𝖳𝖧𝖪	5	20
B	𝖳𝖧𝖪 //	7	28
O	𝖳𝖧𝖪 ////	9	36
AB	////	4	16
		Total 25	100

For the sample, more people have type O blood than any other type.

Grouped Frequency Distributions

When the range of the data is large, the data must be grouped into classes that are more than one unit in width. For example, a distribution of the number of hours boat batteries lasted is as follows:

Class limits	Class boundaries	Tally	Frequency	Cumulative frequency
24–30	23.5–30.5	///	3	3
31–37	30.5–37.5	/	1	4
38–44	37.5–44.5	𝖳𝖧𝖪	5	9
45–51	44.5–51.5	𝖳𝖧𝖪 ////	9	18
52–58	51.5–58.5	𝖳𝖧𝖪 /	6	24
59–65	58.5–65.5	/	1	25
			25	

The procedure for constructing the preceding frequency distribution is given in the next example; however, several things should be noted. In this distribution, the values 24 and 30 of the first class are called *class limits.* The **lower class limit** is 24; it represents the smallest data value that can be included in the class. The **upper class limit** is 30; it represents the largest data value that can be included in the class. The numbers in the second column are called **class boundaries.** These numbers are used to separate the classes so that there are no gaps in the frequency distribution. The gaps are due to the limits; for example, there is a gap between 30 and 31.

Students sometimes have difficulty finding class boundaries when given the class limits. The basic rule of thumb is that *the class limits should have the same decimal place value as the data, but the class boundaries have one additional place value and end in a 5.* For example, if the values in the data set are whole numbers, such as 24, 32, 18, the limits for a class might be 31–37, and the boundaries are 30.5–37.5. Find the boundaries by subtracting 0.5 from 31 (the lower class limit) and adding 0.5 to 37 (the upper class limit) (lower limit) $- 0.5 = 31 - 0.5 = 30.5 =$ (lower boundary)

(upper limit) $+ 0.5 = 37 + 0.5 = 37.5 =$ (upper boundary)

If the data are in tenths, such as 6.2, 7.8, 12.6, the limits for a class hypothetically might be 7.8–8.8, and the boundaries for that class would be 7.75–8.85. Find these values by subtracting 0.05 from 7.8 and adding 0.05 to 8.8.

Finally, the **class width** for a class in a frequency distribution is found by subtracting the lower (or upper) class limit of one class from the lower (or upper) class limit of the next class. For example, the class width in the preceding distribution on the duration of boat batteries is 7, found by subtracting $31 - 24 = 7$.

The class width can also be found by subtracting the lower boundary from the upper boundary for any given class. In this case, $30.5 - 23.5 = 7$.

Note: Do not subtract the limits of a single class. It will result in an incorrect answer.

The researcher must decide how many classes to use and the width of each class. To construct a frequency distribution, follow these rules.

1. *There should be between 5 and 20 classes.* A student would not be in error for having fewer than 5 classes or more than 20 classes; however, statisticians generally agree on these numbers.

2. *The class width should be an odd number.* This ensures that the midpoint of each class has the same place value as the data. The **class midpoint** X_m is obtained by adding the lower and upper boundaries and dividing by 2, or adding the lower and upper limits and dividing by 2:

$$X_m = \frac{\text{lower boundary} + \text{upper boundary}}{2}$$

or

$$X_m = \frac{\text{lower limit} + \text{upper limit}}{2}$$

For example, the midpoint of the first class in the example with boat batteries is

$$\frac{24 + 30}{2} = 27 \qquad \text{or} \qquad \frac{23.5 + 30.5}{2} = 27$$

The midpoint is the numerical location of the center of the class. Midpoints are necessary for graphing (see Section 2–3). If the class width is an even number, the midpoint is in tenths. For example, if the class width is 6 and the boundaries are 5.5–11.5, the midpoint is

$$\frac{5.5 + 11.5}{2} = \frac{17}{2} = 8.5$$

Rule 2 is only a suggestion, and it is not rigorously followed, especially when a computer is used to group data.

3. *The classes must be mutually exclusive.* Mutually exclusive classes have nonoverlapping class limits so that data cannot be placed into two classes. Many times, frequency distributions such as

Age
10–20
20–30
30–40
40–50

are found in the literature or in surveys. If a person is 40 years old, into which class should he or she be placed? A better way to construct a frequency distribution is to use classes such as

Age
10–20
21–31
32–42
43–53

4. *The classes must be continuous.* Even if there are no values in a class, the class must be included in the frequency distribution. There should be no gaps in a frequency distribution. The only exception occurs when the class with a zero frequency is the first or last class. A class with a zero frequency at either end can be omitted without affecting the distribution.

5. *The classes must be exhaustive.* There should be enough classes to accommodate all the data.

6. *The classes must be equal in width.* This avoids a distorted view of the data.

One exception occurs when a distribution is **open-ended**—i.e., it has no specific beginning value or no specific ending value. Following are the class limits for two open-ended distributions.

Age	Minutes
10–20	Below 110
21–31	110–114
32–42	115–119
43–53	120–124
54–64	125–129
65 and above	130–134

The frequency distribution for age is open-ended for the last class, which means that anybody who is 65 years or older will be tallied in the last class. The distribution for minutes is open-ended for the first class, meaning that any minute values below 110 will be tallied in that class.

Example 2–2 shows the procedure for constructing a grouped frequency distribution, i.e., when the classes contain more than one data value.

Example 2–2

The following data represent the record high temperatures for each of the 50 states. Construct a grouped frequency distribution for the data using 7 classes.

112	100	127	120	134	118	105	110	109	112
110	118	117	116	118	122	114	114	105	109
107	112	114	115	118	117	118	122	106	110
116	108	110	121	113	120	119	111	104	111
120	113	120	117	105	110	118	112	114	114

Source: Reprinted with permission from *The World Almanac and Book of Facts 1995,* Copyright © 1994 Primedia Reference Inc. All rights reserved.

Solution

The procedure for constructing a grouped frequency distribution for numerical data follows.

STEP 1   Determine the classes.

Find the highest value and lowest value: $H = 134$ and $L = 100$.

Find the range: $R =$ highest value $-$ lowest value $= H - L$

$$R = 134 - 100 = 34$$

Select the number of classes desired (usually between 5 and 20). In this case, 7 is arbitrarily chosen.

Find the class width by dividing the range by the number of classes.

$$\text{width} = \frac{R}{\text{number of classes}} = \frac{34}{7} = 4.9$$

Round the answer up to the nearest whole number if there is a remainder: $4.9 \approx 5$. (Rounding up is different from rounding off. A number is rounded up if there is any decimal remainder when dividing. For example, $85 \div 6 = 14.167$ and is rounded up to 15. Also, $53 \div 4 = 13.25$ and is rounded up to 14.)

Select a starting point for the lowest class limit. This can be the smallest data value or any convenient number less than the smallest data value. In this case, 100 is used. Add the width to the lowest score taken as the starting point to get the lower limit of the next class. Keep adding until there are 7 classes, as shown. 100, 105, 110, etc.

Subtract one unit from the lower limit of the second class to get the upper limit of the first class. Then add the width to each upper limit to get all the upper limits.

$$105 - 1 = 104$$

The first class is 100–104. The second class is 105–109, etc.

Find the class boundaries by subtracting 0.5 from each lower class limit and adding 0.5 to each upper class limit, as shown.

$$99.5–104.5, \quad 104.5–109.5, \text{ etc.}$$

STEP 2   Tally the data.

STEP 3   Find the numerical frequencies from the tallies.

STEP 4   Find the cumulative frequencies.

A cumulative frequency column can be added to the distribution by adding the frequency in each class to the total of the frequencies of the classes preceding that class, as shown. $0 + 2 = 2$, $2 + 8 = 10$, $10 + 18 = 28$, $28 + 13 = 41$, etc.

The completed frequency distribution follows:

Class limits	Class boundaries	Tally	Frequency	Cumulative frequency
100–104	99.5–104.5	//	2	2
105–109	104.5–109.5	7𝐻𝐿 ///	8	10
110–114	109.5–114.5	7𝐻𝐿 7𝐻𝐿 7𝐻𝐿 ///	18	28
115–119	114.5–119.5	7𝐻𝐿 7𝐻𝐿 ///	13	41
120–124	119.5–124.5	7𝐻𝐿 //	7	48
125–129	124.5–129.5	/	1	49
130–134	129.5–134.5	/	1	50

The frequency distribution shows that the class 109.5–114.5 contains the largest number of temperatures (18) followed by the class 114.5–119.5 with 13 temperatures. Hence, most of the temperatures (31) fall between 109.5° and 119.5°.

Cumulative frequencies are used to show how many data values are accumulated up to and including a specific class. In the preceding example, 28 of the total record high temperatures are less than or equal to 114°. Forty-eight of the total record high temperatures are less than or equal to 124°.

When the range of the data values is relatively small, a frequency distribution can be constructed using single data values for each class. This type of distribution is called an *ungrouped frequency distribution,* and is shown next.

Example 2–3

The data shown here represent the number of miles per gallon that 30 selected four-wheel drive sports utility vehicles obtained in city driving. Construct a frequency distribution.

12	17	12	14	16	18
16	18	12	16	17	15
15	16	12	15	16	16
12	14	15	12	15	15
19	13	16	18	16	14

Source: *Model Year 1999 Fuel Economy Guide.* United States Environmental Protection Agency, October 1998.

Solution

STEP 1 Determine the classes. Since the range of the data set is small ($19 - 12 = 7$), classes consisting of a single data value can be used. They are 12, 13, 14, 15, 16, 17, 18, 19.

Note: If the data are continuous, class boundaries can be used. Subtract 0.5 from each class value to get the lower class boundary and add 0.5 to each class value to get the upper class boundary.

STEP 2 Tally the data.

STEP 3 Find the numerical frequencies from the tallies.

STEP 4 Find the cumulative frequencies.

The completed ungrouped frequency distribution is shown next.

Class limits	Class boundaries	Tally	Frequency	Cumulative frequency
12	11.5–12.5	7H/ I	6	6
13	12.5–13.5	/	1	7
14	13.5–14.5	///	3	10
15	14.5–15.5	7H/ I	6	16
16	15.5–16.5	7H/ ///	8	24
17	16.5–17.5	//	2	26
18	17.5–18.5	///	3	29
19	18.5–19.5	/	1	30

In this case, almost half (14) of the vehicles get 15 or 16 miles per gallon.

The steps for constructing a grouped frequency distribution are summarized in the Procedure Table.

Procedure Table

Constructing a Grouped Frequency Distribution

STEP 1   Determine the classes.

Find the highest and lowest value.

Find the range.

Select the number of classes desired.

Find the width by dividing the range by the number of classes and rounding up.

Select a starting point (usually the lowest value or any convenient number less than the lowest value); add the width to get the lower limits.

Find the upper class limits.

Find the boundaries.

STEP 2   Tally the data.

STEP 3   Find the numerical frequencies from the tallies.

STEP 4   Find the cumulative frequencies.

When one is constructing a frequency distribution, the guidelines presented in this section should be followed. However, one can construct several different but correct frequency distributions for the same data by using a different class width, a different number of classes, or a different starting point.

Furthermore, the method shown here for constructing a frequency distribution is not unique, and there are other ways of constructing one. Slight variations exist, especially in computer packages. But regardless of what methods are used, classes should be mutually exclusive, continuous, exhaustive, and of equal width.

In summary, the different types of frequency distributions were shown in this section. The first type, shown in Example 2–1, is used when the data are categorical (nominal), such as blood type or political affiliation. This type is called a **categorical**

frequency distribution. The second type of distribution is used when the range is large and classes several units in width are needed. This type is called a **grouped frequency distribution** and is shown in Example 2–2. Another type of distribution is used for numerical data and when the range of data is small, as shown in Example 2–3. Since each class is only one unit, this distribution is called an **ungrouped frequency distribution.**

All the different types of distributions are used in statistics and are helpful when one is organizing and presenting data.

The reasons for constructing a frequency distribution follow:

1. To organize the data in a meaningful, intelligible way.

2. To enable the reader to determine the nature or shape of the distribution.

3. To facilitate computational procedures for measures of average and spread (shown in Sections 3–2 and 3–3).

4. To enable the researcher to draw charts and graphs for the presentation of data (shown in Section 2–3).

5. To enable the reader to make comparisons among different data sets.

The factors used to analyze a frequency distribution are essentially the same as those used to analyze histograms and frequency polygons, which are shown in the next section.

Exercises

2–1. List five reasons for organizing data into a frequency distribution.

2–2. Name the three types of frequency distributions and explain when each should be used.

2–3. Find class boundaries, midpoints, and widths for each class.
  a. 11–15
  b. 17–39
  c. 293–353
  d. 11.8–14.7
  e. 3.13–3.93

2–4. How many classes should frequency distributions have? Why should the class width be an odd number?

2–5. Shown here are four frequency distributions. Each is incorrectly constructed. State the reason why.

a.
Class	Frequency
27–32	1
33–38	0
39–44	6
45–49	4
50–55	2

b.
Class	Frequency
5–9	1
9–13	2
13–17	5
17–20	6
20–24	3

c.
Class	Frequency
123–127	3
128–132	7
138–142	2
143–147	19

d.
Class	Frequency
9–13	1
14–19	6
20–25	2
26–28	5
29–32	9

2–6. What are open-ended frequency distributions? Why are they necessary?

Nomina

2–7. The following zip codes were obtained from the respondents to a mail survey. Construct a frequency distribution for the data.

15132	15130	15132	15130
15130	15131	15134	15133
15131	15133	15133	15133
15130	15131	15132	15130
15133	15134	15133	15133

2–8. At a college financial aid office, students who applied for a scholarship were classified according to their class rank: Fr = freshman, So = sophomore, Jr = junior, Se = senior. Construct a frequency distribution for the data.

Fr	Fr	Fr	Fr	Fr
Jr	Fr	Fr	So	Fr
Fr	So	Jr	So	Fr
So	Fr	Fr	Fr	So
Se	Jr	Jr	So	Fr
Fr	Fr	Fr	Fr	So
Se	Se	Jr	Jr	Se
So	So	So	So	So

2–9. A survey taken in a restaurant shows the following number of cups of coffee consumed with each meal. Construct an ungrouped frequency distribution.

Unordered list

0	2	2	1	1	2
3	5	3	2	2	2
0	1	2	4	2	
0	1	0	1	4	4
2	2	0	1	1	5

2–10. In a survey of 20 patients who smoked, the following data were obtained. Each value represents the number of cigarettes the patient smoked per day. Construct a frequency distribution, using six classes.

10	8	6	14
22	13	17	19
11	9	18	14
13	12	15	15
5	11	16	11

2–11. The number of games won by the pitchers who were inducted into the Baseball Hall of Fame through 1992 are shown below. Construct a frequency distribution for the data using 12 classes. (The data for this exercise will be used for Exercise 2–21.)

373	254	237	243	308
210	266	253	201	266
239	114	224	373	286
329	236	284	247	273
198	361	416	207	243
326	251	160	360	311
215	189	344	268	363
21	270	165	240	48
150	300	207	314	197
209	210	260	327	

2–12. In a study of 32 student grade point averages (GPA), the following data were obtained. Construct a frequency distribution using seven classes. (*Note:* A = 4, B = 3, C = 2, D = 1, F = 0.) (The information in this exercise will be used again for Exercises 2–22 and 3–22.)

3.2	2.0	3.3	2.7	2.1	3.9
1.1	3.5	1.9	1.7	0.8	2.6
0.6	4.0	3.5	2.3	1.6	
2.8	2.6	1.6	1.6	2.4	
2.6	2.3	3.8	2.1	2.9	
3.0	1.7	4.0	1.2	3.1	

2–13. The ages of the signers of the Declaration of Independence are shown below. (Age is approximate since only the birth year appeared in the source, and one has been omitted since his birth year is unknown.) Construct a frequency distribution for the data using seven classes. (The data for this exercise will be used for Exercise 3–23.)

41	54	47	40	39	35	50	37	49	42	70	32
44	52	39	50	40	30	34	69	39	45	33	42
44	63	60	27	42	34	50	42	52	38	36	45
35	43	48	46	31	27	55	63	46	33	60	62
35	46	45	34	53	50	50					

2–14. The number of automobile fatalities in 27 states where the speed limits were raised in 1996 are shown below. Construct a frequency distribution using 8 classes. (The data for this exercise will be used for Exercises 2–24, 2–34, and 3–24.)

1100	460	85
970	480	1430
4040	405	70
620	690	180
125	1160	3630
2805	205	325
1555	300	875
260	350	705
1430	485	145

Source: *USA Today*, July 14, 1997

2–15. The data (in cents) is the cigarette tax per pack imposed by each state. Construct a frequency distribution. Use classes 0–19, 20–39, 40–59, etc. (*Note:* The class width is an even number because the data are in cents.) (The information in this exercise will be used for Exercises 2–27 and 3–25.)

16.5	12.0	76.0	56.0	41.0
100.0	80.0	75.0	5.0	51.1
58.0	28.0	48.0	44.0	44.0
32.5	58.0	18.0	24.0	2.5
37.0	15.5	17.0	23.0	82.5
20.0	36.0	18.0	68.0	17.0
50.0	24.0	34.0	31.0	59.0
24.0	3.0	35.0	71.0	12.0
20.0	37.0	7.0	33.0	36.0
33.9	74.0	80.0	21.0	13.0

Source: *USA Today*, March 17, 1998.

2–16. The acreage of the 39 U.S. National Parks under 900,000 acres (in thousands of acres) is shown here. Construct a frequency distribution for the data using eight classes. (The data in this exercise will be used in Exercise 2–29.)

41	66	233	775	169
36	338	233	236	64
183	61	13	308	77
520	77	27	217	5
650	462	106	52	52
505	94	75	265	402
196	70	132	28	220
760	143	46	539	

Source: THE UNIVERSAL ALMANAC © by John W. Wright. Reprinted with permission of Andrews McMeel Publishing. All rights reserved.

2–17. The heights in feet above sea level of the major active volcanoes in Alaska are given here. Construct a frequency distribution for the data using 10 classes. (The data in this exercise will be used in Exercises 3–9 and 3–59.)

4,265	3,545	4,025	7,050	11,413
3,490	5,370	4,885	5,030	6,830
4,450	5,775	3,945	7,545	8,450
3,995	10,140	6,050	10,265	6,965
150	8,185	7,295	2,015	5,055
5,315	2,945	6,720	3,465	1,980
2,560	4,450	2,759	9,430	
7,985	7,540	3,540	11,070	
5,710	885	8,960	7,015	

Source: THE UNIVERSAL ALMANAC © by John W. Wright. Reprinted with permission of Andrews McMeel Publishing. All rights reserved.

2–18. During the 1998 baseball season, Mark McGwire and Sammy Sosa both broke Roger Maris's home run record of 61. The distances in feet for each home run follow. Construct a frequency distribution for each player, using eight classes. (The information in this exercise will be used for Exercises 2–30, 3–10, and 3–56.)

McGwire				Sosa			
306	370	370	430	371	350	430	420
420	340	460	410	430	434	370	420
440	410	380	360	440	410	420	460
350	527	380	550	400	430	410	370
478	420	390	420	370	410	380	340
425	370	480	390	350	420	410	415
430	388	423	410	430	380	380	366
360	410	450	350	500	380	390	400
450	430	461	430	364	430	450	440
470	440	400	390	365	420	350	420
510	430	450	452	400	380	380	400
420	380	470	398	370	420	360	368
409	385	369	460	430	433	388	440
390	510	500	450	414	482	364	370
470	430	458	380	400	405	433	390
430	341	385	410	480	480	434	344
420	380	400	440	410	420		
377	370						

Source: *USA Today*, September 28, 1998.

Technology Step by Step

MINITAB
Step by Step

Make a Categorical Frequency Table
(Qualitative or Discrete Variable)

Example MT2–1

1. Type in all of the blood types from Example 2–1 down C1 of the worksheet.

    A B B AB O O O B AB B B B O A O A O O O AB AB A O B A

2. Click above row 1 and name the column **BloodType.**

3. From the menus, select **STAT>TABLES>TALLY.**

Tally Dialog Box

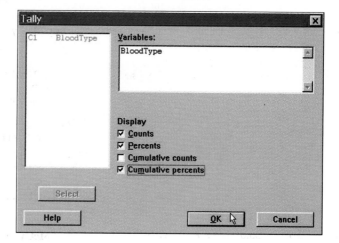

4. Double click C1 in the variables list.

5. Check the boxes for the statistics that you would like. Here Counts, Percents, and Cumulative percents are checked.

6. Click [OK].

The results will be displayed in the session window as shown.

Session Window with TALLY Results

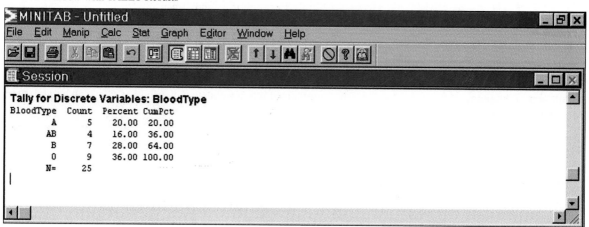

Make a Grouped Frequency Distribution (Quantitative Variable)

Example MT2–2

1. Type the data used in Example 2–2 into C1 or open the worksheet from the website.
2. Select **Graph>Character Graphs>Histogram**. The cursor should be blinking in the `Variables` dialog box. If not, click inside the box then double click C1 **TEMPERATURES** in the list. Only quantitative variables will be shown in this list since this menu option will only work with that data type.
3. Click in the box for `First midpoint:` then type in **102.**
4. Click in the box for `Interval width:` then type in **5.** You must calculate the class width (5) and the midpoint of the first class (102) using the instructions in text.
5. Click [OK].

Grouped Frequency Table

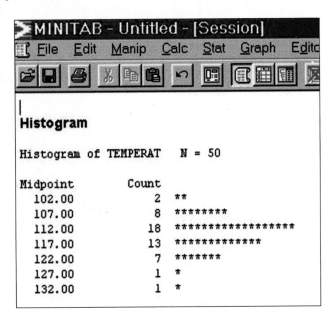

The table and a histogram will be displayed in the Session Window. The midpoint of each class is displayed, not the class limits or boundaries. "TEMPERATURES," the column name, is truncated to eight characters for display only.

2–3

Histograms, Frequency Polygons, and Ogives

Objective 2. Represent data in frequency distributions using histograms, frequency polygons, and ogives.

After the data have been organized into a frequency distribution, they can be presented in graphic forms. The purpose of graphs in statistics is to convey the data to the viewer in pictorial form. It is easier for most people to comprehend the meaning of data presented graphically than data presented numerically in tables or frequency distributions. This is especially true if they have little or no statistical knowledge.

Statistical graphs can be used to describe the data set or analyze it. Graphs are also useful in getting the audience's attention in a publication or a speaking presentation. They can be used to discuss an issue, reinforce a critical point, or summarize a data set. They can also be used to discover a trend or pattern in a situation over a period of time.

The three most commonly used graphs in research are

1. The histogram.

2. The frequency polygon.

3. The cumulative frequency graph, or ogive (pronounced o-jive).

An example of each type of graph is shown in Figure 2–1. The data for each graph are the distribution of the miles that 20 randomly selected runners ran during a given week.

Figure 2–1

Examples of Commonly Used Graphs

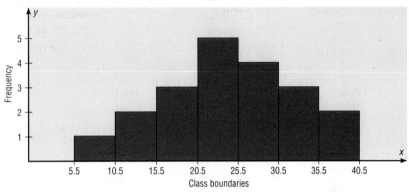

(a) Histogram

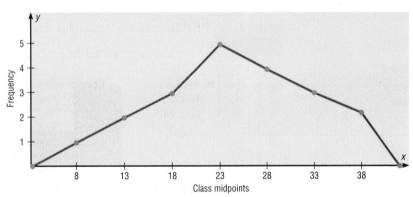

(b) Frequency polygon

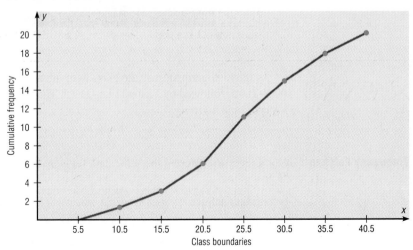

(c) Cumulative frequency graph

The Histogram

The **histogram** is a graph that displays the data by using vertical bars of various heights to represent the frequencies.

Construct a histogram to represent the data shown below for the record high temperatures for each of the 50 states (see Example 2–2).

Class boundaries	Frequency
99.5–104.5	2
104.5–109.5	8
109.5–114.5	18
114.5–119.5	13
119.5–124.5	7
124.5–129.5	1
129.5–134.5	1

Solution

STEP 1   Draw and label the x and y axes. The x axis is always the horizontal axis, and the y axis is always the vertical axis.

STEP 2   Represent the frequency on the y axis and the class boundaries on the x axis.

STEP 3   Using the frequencies as the heights, draw vertical bars for each class. See Figure 2–2.

Figure 2–2

Histogram for
Example 2–4

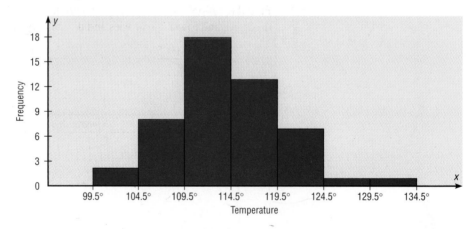

As the histogram shows, the class with the greatest number of data values (18) is 109.5–114.5, followed by 13 for 114.5–119.5. The graph also has one peak with the data clustering around it.

The Frequency Polygon

Another way to represent the same data set is by using a frequency polygon.

The **frequency polygon** is a graph that displays the data by using lines that connect points plotted for the frequencies at the midpoints of the classes. The frequencies are represented by the heights of the points.

The next example shows the procedure for constructing a frequency polygon.

Example 2–5

Using the frequency distribution given in Example 2–4, construct a frequency polygon.

Solution

STEP 1   Find the midpoints of each class. Recall that midpoints are found by adding the upper and lower boundaries and dividing by 2.

$$\frac{99.5 + 104.5}{2} = 102 \qquad \frac{104.5 + 109.5}{2} = 107$$

And so on. The midpoints are listed next.

Class boundaries	Midpoints	Frequency
99.5–104.5	102	2
104.5–109.5	107	8
109.5–114.5	112	18
114.5–119.5	117	13
119.5–124.5	122	7
124.5–129.5	127	1
129.5–134.5	132	1

STEP 2   Draw the x and y axes. Label the x axis with the midpoint of each class, and then use a suitable scale on the y axis for the frequencies.

STEP 3   Using the midpoints for the x values and the frequencies as the y values, plot the points.

STEP 4   Connect adjacent points with straight lines. Draw a line back to the x axis at the beginning and end of the graph, at the same distance that the previous and next midpoints would be located, as shown in Figure 2–3.

Figure 2–3

Frequency Polygon for
Example 2–5

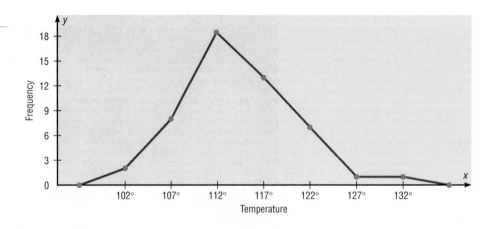

The frequency polygon and the histogram are two different ways to represent the same data set. The choice of which one to use is left to the discretion of the researcher.

The Ogive

The third type of graph that can be used represents the cumulative frequencies for the classes. This type of graph is called the cumulative frequency graph or ogive. The **cumulative frequency** is the sum of the frequencies accumulated up to the upper boundary of a class in the distribution.

The **ogive** is a graph that represents the cumulative frequencies for the classes in a frequency distribution.

Example 2–6 shows the procedure for constructing an ogive.

Example 2–6

Construct an ogive for the frequency distribution described in Example 2–4.

Solution

STEP 1   Find the cumulative frequency for each class.

Class boundaries	Cumulative frequency
99.5–104.5	2
104.5–109.5	10
109.5–114.5	28
114.5–119.5	41
119.5–124.5	48
124.5–129.5	49
129.5–134.5	50

STEP 2   Draw the x and y axes. Label the x axis with the class boundaries. Use an appropriate scale for the y axis to represent the cumulative frequencies. (Depending on the numbers in the cumulative frequency columns, scales such as 0, 1, 2, 3, . . . , or 5, 10, 15, 20, . . . , or 1000, 2000,

3000, ... can be used. Do *not* label the *y* axis with the numbers in the cumulative frequency column.) In this example, a scale of 0, 5, 10, 15, ... will be used.

STEP 3   Plot the cumulative frequency at each upper class boundary, as shown in Figure 2–4. Upper boundaries are used since the cumulative frequencies represent the number of data values accumulated up to the upper boundary of each class.

Figure 2–4

Plotting the Cumulative
Frequency for
Example 2–6

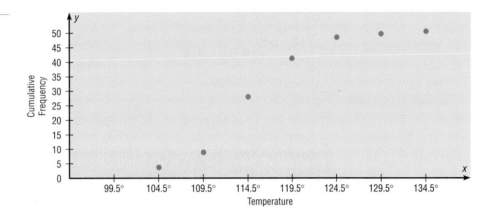

STEP 4   Starting with the first upper class boundary, 104.5, connect adjacent points with straight lines, as shown in Figure 2–5. Then extend the graph to the first lower class boundary, 99.5, on the *x* axis.

Figure 2–5

Ogive for Example 2–6

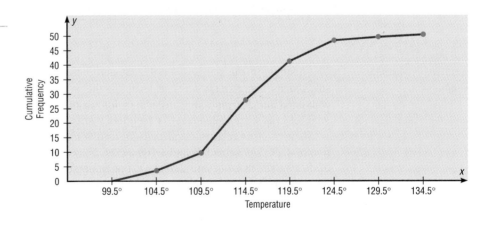

Cumulative frequency graphs are used to visually represent how many values are below a certain upper class boundary. For example, to find out how many record high temperatures are less than 114.5°, locate 114.5° on the *x* axis, draw a vertical line up until it intersects the graph, and then draw a horizontal line at that point to the *y* axis. The *y* axis value is 28, as shown in Figure 2–6.

Figure 2–6

Finding a Specific
Cumulative Frequency

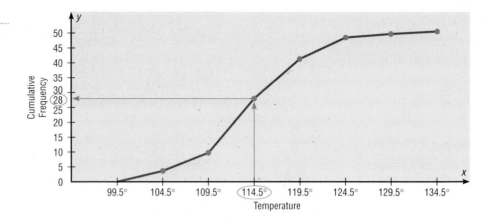

The steps for drawing the three types of graphs are shown in the Procedure Table.

Procedure Table		
Constructing Statistical Graphs		
STEP 1	Draw and label the x and y axes.	
STEP 2	Choose a suitable scale for the frequencies or cumulative frequencies and label it on the y axis.	
STEP 3	Represent the class boundaries for the histogram or ogive, or the midpoint for the frequency polygon, on the x axis.	
STEP 4	Plot the points and then draw the bars or lines.	

**Relative Frequency
Graphs**

The histogram, the frequency polygon, and the ogive shown previously were constructed by using frequencies in terms of the raw data. These distributions can be converted into distributions using *proportions* instead of raw data as frequencies. These types of graphs are called **relative frequency graphs.**

Graphs using relative frequencies instead of frequencies are used when the proportion of data values that fall into a given class is more important than the actual number of data values that fall into that class. For example, if one wanted to compare the age distribution of adults in the city of Philadelphia, Pennsylvania, with the age distribution of adults of Erie, Pennsylvania, one would use relative frequency distributions. The reason is that since the population of Philadelphia is 1,478,002 and the population of Erie is 105,270, the bars using the actual data values for Philadelphia would be much taller than those for the same classes for Erie.

To convert a frequency into a proportion or relative frequency, divide the frequency for each class by the total of the frequencies. The sum of the relative frequencies will always be 1. These graphs are similar to the ones that use raw data as frequencies, but the values on the y axis are in terms of proportions. The next example shows the three types of relative frequency graphs.

Example 2–7

Construct a histogram, frequency polygon, and ogive using relative frequencies for the distribution (shown here) of the miles 20 randomly selected runners ran during a given week.

Class boundaries	Frequency	Cumulative frequency
5.5–10.5	1	1
10.5–15.5	2	3
15.5–20.5	3	6
20.5–25.5	5	11
25.5–30.5	4	15
30.5–35.5	3	18
35.5–40.5	2	20
	20	

Solution

STEP 1   Convert each frequency to a proportion or relative frequency by dividing the frequency for each class by the total number of observations.

For class 5.5–10.5, the relative frequency is $\frac{1}{20} = 0.05$.

For class 10.5–15.5, the relative frequency is $\frac{2}{20} = 0.10$.

For class 15.5–20.5, the relative frequency is $\frac{3}{20} = 0.15$.

And so on.

STEP 2   Using the same procedure, find the relative frequencies for the cumulative frequency column. The relative frequencies are shown here.

Class boundaries	Midpoints	Relative frequency	Cumulative relative frequency
5.5–10.5	8	0.05	0.05
10.5–15.5	13	0.10	0.15
15.5–20.5	18	0.15	0.30
20.5–25.5	23	0.25	0.55
25.5–30.5	28	0.20	0.75
30.5–35.5	33	0.15	0.90
35.5–40.5	38	0.10	1.00
		1.00	

STEP 3   Draw each graph as shown in Figure 2–7. For the histogram and ogive, use the class boundaries along the x axis. For the frequency polygon, use the midpoints on the x axis. The scale on the y axis uses proportions.

When analyzing histograms and frequency polygons, look at the shape of the curve. For example, does it have one peak or two peaks, or is it relatively flat, or is it U-shaped? Are the data values spread out on the graph, or are they clustered around the

Figure 2–7

Graphs for Example 2–7

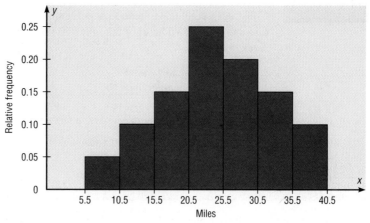

(a) Histogram

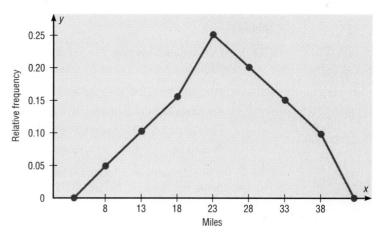

(b) Frequency polygon

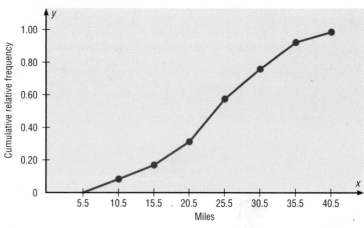

(c) Ogive

center? Are there data values in the extreme ends? These may be *outliers.* (See Section 3-4 for an explanation of outliers.) Are there any gaps in the histogram, or does the frequency polygon touch the *x* axis somewhere other than the ends? Finally, are the data clustered at one end or the other, indicating a *skewed distribution?* (See Section 3-2 for an explanation of skewness.)

For example, the histogram for the record high temperatures shown in Figure 2-2 (page 46) shows a single peaked distribution, with the class 109.5-114.5 containing the largest number of temperatures. The distribution has no gaps, and there are fewer temperatures in the highest class than in the lowest class.

Exercises

2-19. For 108 randomly selected college applicants, the following frequency distribution for entrance exam scores was obtained. Construct a histogram, frequency polygon, and ogive for the data. (The data for this exercise will be used for Exercise 2-31.)

Class limits	Frequency
90–98	6
99–107	22
108–116	43
117–125	28
126–134	9

Applicants who score above 107 need not enroll in a summer developmental program. In this group, how many students do not have to enroll in the developmental program?

2-20. For 75 employees of a large department store, the following distribution for years of service was obtained. Construct a histogram, frequency polygon, and ogive for the data. (The data for this exercise will be used for Exercise 2-32.)

Class limits	Frequency
1–5	21
6–10	25
11–15	15
16–20	0
21–25	8
26–30	6

A majority of the employees have worked for how many years or less?

2-21. Construct a histogram, frequency polygon, and ogive for the data in Exercise 2-11 and analyze the results.

2-22. Construct a histogram, frequency polygon, and ogive for the data in Exercise 2-12 and analyze the results.

2-23. Thirty automobiles were tested for fuel efficiency, in miles per gallon (mpg). The following frequency distribution was obtained. Construct a histogram, frequency polygon, and ogive for the data. (The data for this exercise will be used for Exercise 2-33.)

Class boundaries	Frequency
7.5–12.5	3
12.5–17.5	5
17.5–22.5	15
22.5–27.5	5
27.5–32.5	2

2-24. Construct a histogram, frequency polygon, and ogive for the data in Exercise 2-14 and analyze the results. (The data in this exercise will be used for Exercise 2-34.)

2-25. In a class of 35 students, the following grade distribution was found. Construct a histogram, frequency polygon, and ogive for the data. (A = 4, B = 3, C = 2, D = 1, F = 0.) (The data in this exercise will be used for Exercise 2-35.)

Grade	Frequency
0	3
1	6
2	9
3	12
4	5

A grade of C or better is required for the next level course. Were the majority of the students able to meet this requirement?

2-26. In a study of reaction times of dogs to a specific stimulus, an animal trainer obtained the following data, given in seconds. Construct a histogram, frequency polygon, and ogive for the data and analyze the results. (The histogram in this exercise will be used for Exercises 2-36, 3-16, and 3-68.)

Class limits	Frequency
2.3–2.9	10
3.0–3.6	12
3.7–4.3	6
4.4–5.0	8
5.1–5.7	4
5.8–6.4	2

2–27. Construct a histogram, frequency polygon, and ogive for the data in Exercise 2–15 and analyze the results.

2–28. To determine their lifetimes, 80 randomly selected batteries were tested. The following frequency distribution was obtained. The data values are in hours. Construct a histogram, frequency polygon, and ogive for the data and analyze the results.

Class boundaries	Frequency
63.5–74.5	10
74.5–85.5	15
85.5–96.5	22
96.5–107.5	17
107.5–118.5	11
118.5–129.5	5

2–29. Construct a histogram, frequency polygon, and ogive for the data in Exercise 2–16 and analyze the results.

2–30. For the data in Exercise 2–18, construct a histogram for the home run distances for each player and compare them. Are they basically the same, or are there any noticeable differences? Explain your answer.

2–31. For the data in Exercise 2–19, construct a histogram, frequency polygon, and ogive, using relative frequencies. What proportion of the applicants need to enroll in the summer developmental program?

2–32. For the data in Exercise 2–20, construct a histogram, frequency polygon, and ogive, using relative frequencies. What proportion of the employees have been with the store for more than 20 years?

2–33. For the data in Exercise 2–23, construct a histogram, frequency polygon, and ogive, using relative frequencies. What proportion of the automobiles had a fuel efficiency of 17.5 miles per gallon or higher?

2–34. For the data in Exercise 2–14, construct a histogram, frequency polygon, and ogive, using relative frequencies.

2–35. For the data in Exercise 2–25, construct a histogram, frequency polygon, and ogive, using relative frequencies. What proportion of the students cannot meet the requirement for enrollment in the next course?

2–36. The animal trainer in Exercise 2–26 selected another group of dogs that were much older than the first group and measured their reaction times to the same stimulus. Construct a histogram, frequency polygon, and ogive for the data.

Class limits	Frequency
2.3–2.9	1
3.0–3.6	3
3.7–4.3	4
4.4–5.0	16
5.1–5.7	14
5.8–6.4	4

Analyze the results and compare the histogram for this group with the one obtained in Exercise 2–26. Are there any differences in the histograms? (The data in this exercise will be used for Exercise 3–16 and 3–68.)

*2–37. Using the following histogram:
a. Construct a frequency distribution; include class limits, class frequencies, midpoints, and cumulative frequencies.
b. Construct a frequency polygon.
c. Construct an ogive.

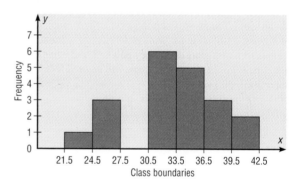

*2–38. Using the results from Exercise 2–37, answer the following questions.
a. How many values are in the class 27.5–30.5?
b. How many values fall between 24.5 and 36.5?
c. How many values are below 33.5?
d. How many values are above 30.5?

Technology Step by Step

Minitab
Step by Step

Constructing a Histogram

Example MT2–3

1. ˙Enter the data from Example 2–2, the high temperatures for the 50 states.
2. Select **Graph>Histogram**.
3. In the dialog box double click C1 **TEMPERATURES.**
4. Click [Options].
   a. In Type of Interval: check CutPoint: to use class boundaries instead of midpoints.
   b. In Definition of Intervals: click the button for Midpoint/cutpoint positions: then type the boundaries MINITAB will use. The format is Lowest:Highest/Increment.

Histogram Option

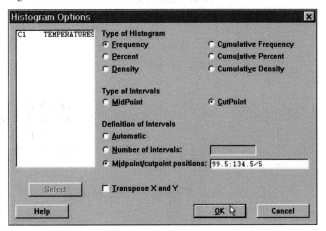

5. Click [OK]. The Histogram Dialog Box will be displayed.
6. Click [Edit Attributes]. Select Solid for Fill and Yellow for the Back color. This will make the bars of the histogram a solid yellow color.
7. Click [OK] twice.

The resulting graph window can be printed, copied to the clipboard or saved.

Histogram of
Temperatures

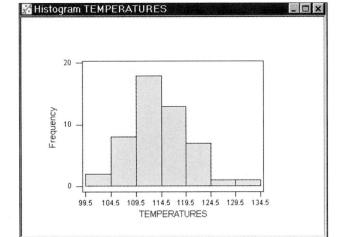

TI-83
Step by Step

Constructing a Histogram

In order to display the graphs on the screen, enter the appropriate values in the calculator using the Window menu. The default values are $X_{min} = -10$, $X_{max} = +10$, $Y_{min} = -10$, and $Y_{max} = +10$.

The X_{scl} changes the distance between the tick marks on the x axis and can be used to change the class width for the histogram.

To change the values in the `Window`:

1. Press **WINDOW**.
2. Move the cursor to the value that needs to be changed. Then type in the desired value and press **ENTER.**
3. Continue until all values are appropriate.
4. Press **[2nd] [QUIT]** to leave the `Window` menu.

To plot the histogram:

1. Enter the data in L_1.
2. Make sure Window values are appropriate for the histogram.
3. Press **[2nd] [STAT PLOT] ENTER.**
4. Press **ENTER** to turn the plot on, if necessary.
5. Move cursor to the `Histogram` symbol and press **ENTER,** if necessary.
6. Move cursor to L_1 on the `Xlist`. Press **ENTER,** if necessary.
7. Move cursor to 1 on `Freq`. Press **ENTER,** if necessary.
8. Press **GRAPH** to display the histogram.
9. To obtain various coordinates, press the **TRACE** key, followed by ◀ or ▶ keys.

Example TI2–1

Plot a histogram for the following data from Examples 2–2 and 2–4.

112	100	127	120	134	118	105	110	109	112
110	118	117	116	118	122	114	114	105	109
107	112	114	115	118	117	118	122	106	110
116	108	110	121	113	120	119	111	104	111
120	113	120	117	105	110	118	112	114	114

Set the Window values as follows:

$X_{min} = 100$
$X_{max} = 135$
$Y_{min} = -5$
$Y_{max} = 20$

Input

Input

Press **TRACE** and use the arrow keys to determine the number of values in each group.

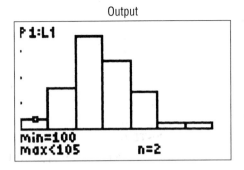

Excel
Step by Step

Constructing a Histogram

Example XL2–1

To make a histogram:

1. Press **[Ctrl]-N** for a new worksheet.

2. Enter the data below in column A, one number per cell.

68	80	69	81	72	100	101	73	102	93
91	92	93	88	82	83	75	75	89	96
103	83	84	85	89					

3. Select **Tools > Data Analysis**.

4. In **Data Analysis**, select **Histogram** and click the [OK] button.

5. In the **Histogram** dialog box, type **A1:A25** as the **Input Range**.

Histogram Dialog Box

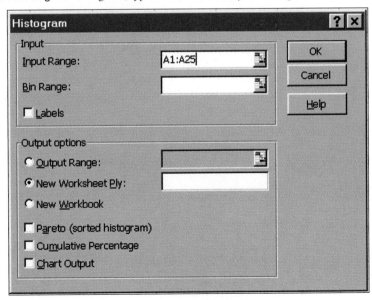

6. Select **New Worksheet Ply** and **Chart Output**. Click [OK].

Excel presents both a table and a chart on the new worksheet ply. It decides "bins" for the histogram itself (here it picked a bin size of seven units), but you can also define your own bin range on the data worksheet. Here is the histogram with Excel-selected bins.

Histogram for
Example XL2–1

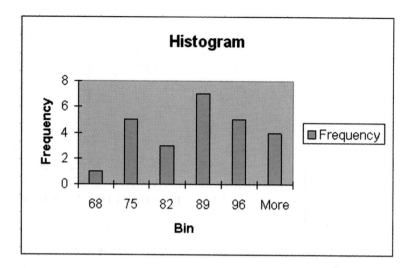

Other Types of Graphs

In addition to the histogram, the frequency polygon, and the ogive, several other types of graphs are often used in statistics. They are the Pareto chart, the time series graph, and the pie graph. Figure 2–8 shows an example of each type of graph.

Figure 2–8

Other Types of Graphs
Used in Statistics

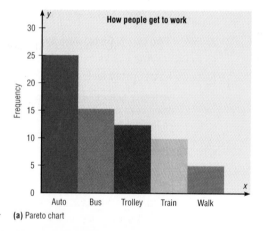

(a) Pareto chart

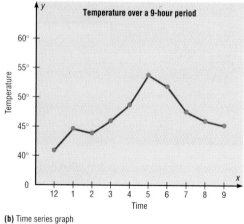

(b) Time series graph

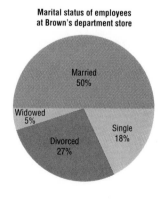

(c) Pie graph

Pareto Charts

Objective 3. Represent data using Pareto charts, time series graphs, and pie graphs.

In the previous section, graphs such as the histogram, frequency polygon, and ogive showed how data can be represented when the variable displayed on the horizontal axis is quantitative, such as heights and weights.

On the other hand, when the variable displayed on the horizontal axis is qualitative or categorical, a *Pareto chart* can be used.

A **Pareto chart** is used to represent a frequency distribution for a categorical variable, and the frequencies are displayed by the heights of vertical bars, which are arranged in order from highest to lowest.

Example 2–8

Historical Note

Vilfredo Pareto (1848–1923) was an Italian scholar who developed theories in economics, statistics, and social sciences. His contributions to statistics include the development of a mathematical function used in economics. This function has many statistical applications and is called the Pareto distribution. In addition, he researched income distribution, and his findings became known as Pareto's law.

The following table shows the number of crimes investigated by law enforcement officers in U.S. national parks during 1995. Construct a Pareto chart for the data.

Type	Number
Homicide	13
Rape	34
Robbery	29
Assault	164

Source: "Crime Rate in 1990–95," *USA Today*, June 5, 1995.

Solution

STEP 1   Arrange the data from the largest to smallest according to frequency.

Type	Number
Assault	164
Rape	34
Robbery	29
Homicide	13

STEP 2   Draw and label the *x* and *y* axes.

STEP 3   Draw the bars corresponding to the frequencies. See Figure 2–9. The Pareto chart shows that assaults had the highest frequency and homicides the lowest.

Figure 2–9

Pareto Chart for Example 2–8

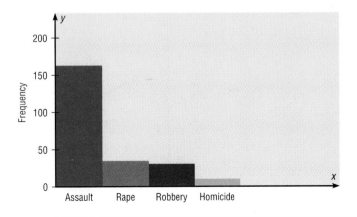

Historical Note

Time series graphs are over 1000 years old. The first ones were used to chart the movements of the planets and the sun.

Suggestions for Drawing Pareto Charts

1. Make the bars the same width.
2. Arrange the data from largest to smallest according to frequencies.
3. Make the units that are used for the frequency equal in size.

When analyzing a Pareto chart, make comparisons by looking at the heights of the bars.

The Time Series Graph

When data are collected over a period of time, they can be represented by a time series graph.

A **time series graph** represents data that occur over a specific period of time.

The next example shows the procedure for constructing a time series graph.

Example 2–9

A transit manager wishes to use the following data for a presentation showing how Port Authority Transit ridership has changed over the years. Draw a time series graph for the data and summarize the findings.

Year	Ridership (in millions)
1990	88.0
1991	85.0
1992	75.7
1993	76.6
1994	75.4

Source: Port Authority Transit, *Tribune-Review* (Greensburg, PA), January 28, 1995.

Solution

STEP 1   Draw and label the x and y axes.

STEP 2   Label the x axis for years and the y axis for the number of riders.

STEP 3   Plot each point according to the table.

STEP 4   Draw straight lines connecting adjacent points. Do not try to fit a smooth curve through the data points. See Figure 2–10.

Figure 2–10

Time Series Graph for Example 2–9

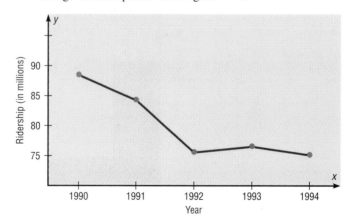

The graph shows a decline in ridership through 1992 and then a leveling off for the years 1993 and 1994.

When analyzing a time series graph, look for a trend or pattern that occurs over the time period. For example, is the line ascending (indicating an increase over time) or descending (indicating a decrease over time)? Another thing to look for is the slope, or steepness, of the line. A line that is steep over a specific time period indicates a rapid increase or decrease over that period.

Two data sets can be compared on the same graph (called a *compound time series graph*) if two lines are used, as shown in Figure 2–11. This graph shows the number of snow shovels sold at a store for two seasons.

Figure 2–11

Two Time Series Graphs
for Comparison

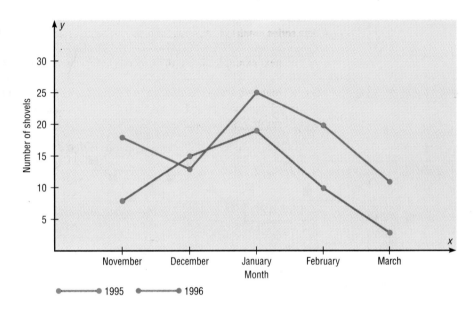

1995   1996

The Pie Graph

Pie graphs are used extensively in statistics. The purpose of the pie graph is to show the relationship of the parts to the whole by visually comparing the sizes of the sectors. Percentages or proportions can be used. The variable is nominal or categorical.

A **pie graph** is a circle that is divided into sections or wedges according to the percentage of frequencies in each category of the distribution.

The next example shows the procedure for constructing a pie graph.

Example 2–10

The following distribution shows the number of pounds of each snack food eaten during the 1998 Super Bowl. Construct a pie graph for the data.

Snack	Pounds (frequency)
Potato chips	11.2 million
Tortilla chips	8.2 million
Pretzels	4.3 million
Popcorn	3.8 million
Snack nuts	2.5 million
	Total $n = 30.0$ million

Source: *USA Weekend,* January 22–24, 1999, p. 10.

Speaking of   **STATISTICS**

This time series graph compares the coal production of Pennsylvania with that of the Untied States as a whole. Using the information presented in the graph, explain in words how the total coal production of the United States compares with the coal production of Pennsylvania over the years.

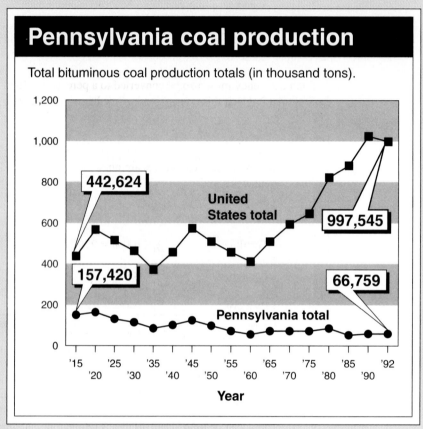

Pennsylvania coal production

Total bituminous coal production totals (in thousand tons).

Source: Pittsburgh *Tribune-Review,* June 5, 1994. Used with permission.

Solution

STEP 1   Since there are 360 degrees in a circle, the frequency for each class must be converted into a proportional part of the circle. This conversion is done by using the formula

$$\text{degrees} = \frac{f}{n} \cdot 360°$$

where

f = frequency for each class

n = sum of the frequencies

Hence, the following conversions are obtained. The degrees should sum to 360°.*

Potato chips	$\dfrac{11.2}{30} \cdot 360° = 134°$
Tortilla chips	$\dfrac{8.2}{30} \cdot 360° = 98°$
Pretzels	$\dfrac{4.3}{30} \cdot 360° = 52°$
Popcorn	$\dfrac{3.8}{30} \cdot 360° = 46°$
Snack nuts	$\dfrac{2.5}{30} \cdot 360° = \underline{30°}$
Total	$360°$

STEP 2   Each frequency must also be converted to a percentage. Recall from Example 2–1 that this conversion is done by using the formula

$$\% = \frac{f}{n} \cdot 100\%$$

Hence, the following percentages are obtained.
The percentages should sum to 100%.*

Potato chips	$\dfrac{11.2}{30} \cdot 100\% = 37.3\%$
Tortilla chips	$\dfrac{8.2}{30} \cdot 100\% = 27.3\%$
Pretzels	$\dfrac{4.3}{30} \cdot 100\% = 14.3\%$
Popcorn	$\dfrac{3.8}{30} \cdot 100\% = 12.7\%$
Snack nuts	$\dfrac{2.5}{30} \cdot 100\% = \underline{8.3\%}$
Total	99.9%

*Note: The degrees column does not always sum to 360° due to rounding, and the percent column does not always sum to 100% for the same reason.

STEP 3   Next, using a protractor and a compass, draw the graph and label each section with the name and percentages, as shown in Figure 2–12.

Figure 2–12

Pie Graph for
Example 2–10

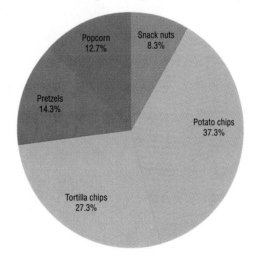

Construct a pie graph showing the blood types of the army inductees described in Example 2–1. The frequency distribution is repeated here.

Class	Frequency	Percent
A	5	20%
B	7	28%
O	9	36%
AB	4	16%
	25	100%

Solution

STEP 1   Find the number of degrees for each class, using the formula

$$\text{degrees} = \frac{f}{n} \cdot 360°$$

For each class, then, the following results are obtained.

A $\frac{5}{25} \cdot 360° = 72°$

B $\frac{7}{25} \cdot 360° = 100.8°$

O $\frac{9}{25} \cdot 360° = 129.6°$

AB $\frac{4}{25} \cdot 360° = 57.6°$

STEP 2   Find the percentages. (This was already done in Example 2–1.)

STEP 3   Using a protractor, graph each section and write its name and corresponding percentage, as shown in Figure 2–13.

Figure 2–13

Pie Graph for
Example 2–11

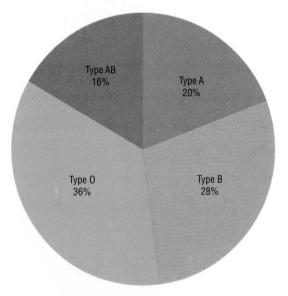

The graph shows that in this case the most common blood type is type O.

To analyze the nature of the data shown in the pie graph, compare the sectors. For example, are any sectors relatively large compared to the rest?

Figure 2–13 shows that among the inductees, type O blood is more prevalent than any other type. People who have type AB blood are in the minority. More than twice as many people have type O blood as type AB.

Misleading Graphs

Graphs give a visual representation that enables readers to analyze and interpret data more easily than they could simply by looking at numbers. However, inappropriately drawn graphs can misrepresent the data and lead the reader to false conclusions. For example, a car manufacturer's ad stated that 98% of the vehicles it had sold in the past 10 years were still on the road. The ad then showed a graph similar to the one in Figure 2–14. The graph shows the percentage of the manufacturer's automobiles still on the road and the percentage of its competitors' automobiles still on the road. Is there a large difference? Not necessarily.

Figure 2–14

Graph of Automaker's Claim, Using a Scale from 95% to 100%

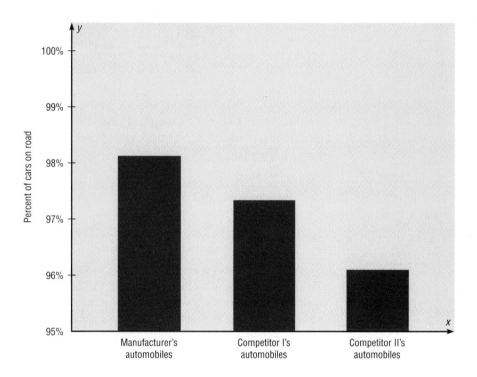

Notice the scale on the vertical axis in Figure 2–14. It has been cut off (or truncated), and it starts at 95%. When the graph is redrawn using a scale that goes from 0% to 100%, as in Figure 2–15, there is hardly a noticeable difference in the percentages. Thus, changing the units at the starting point on the y axis can convey a very different visual representation of the data.

It is not wrong to truncate an axis of the graph; many times it is necessary to do so (see Example 2–9). However, the reader should be aware of this fact and interpret the graph accordingly. Do not be misled if an inappropriate impression is given.

Figure 2–15

Graph in Figure 2–14
Redrawn, Using a Scale
from 0% to 100%

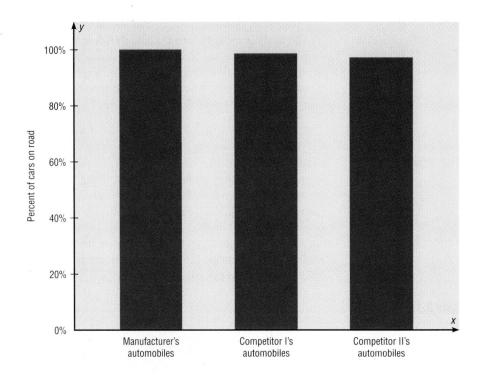

Let's consider another example. The percentage of the world's total motor vehicles produced by manufacturers in the United States declined from 25% in 1986 to 18% in 1991, as shown by the following data.

Year	1986	1987	1988	1989	1990	1991
Percent produced in United States	25	23.8	23.3	22.1	20.3	18.0

When one draws the graph, as shown in Figure 2–16a, a scale ranging from 0% to 100% shows a slight decrease. However, this decrease can be emphasized by using a scale that ranged from 15% to 25%, as shown in Figure 2–16b. Again, by changing the units or the starting point on the *y* axis, one can change the visual message.

Another misleading graphing technique sometimes used is exaggerating a one-dimensional increase by showing it in two dimensions. For example, the average cost of a 30-second Super Bowl commercial has increased from $40,000 in 1967 to $1 million in 1995. (Source: *USA Today* Snapshot by Cliff Vancura, January 30, 1995; based on information from Nielsen Media Research.)

The increase shown by the bar graph in Figure 2–17a represents the change by a comparison of the heights of the two bars in one dimension. The same data are shown two-dimensionally with circles in Figure 2–17b. Notice that the difference seems much larger because the eye is comparing the areas of the circles rather than the lengths of the diameters.

Note that it is not wrong to use the graphing techniques of truncating the scales or representing data by two-dimensional pictures. But when these techniques are used, the reader should be cautious of the conclusion drawn on the basis of the graphs.

Figure 2–16

Percent of World's Motor
Vehicles Produced by
Manufacturers in the
United States

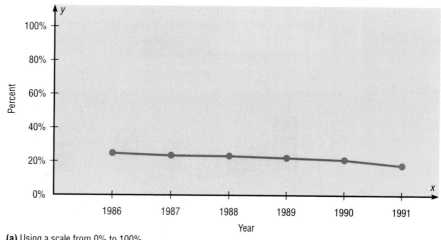

(a) Using a scale from 0% to 100%

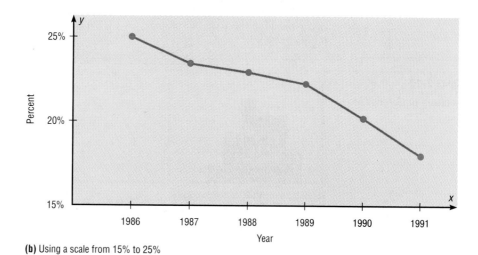

(b) Using a scale from 15% to 25%

Figure 2–17

Comparison of Costs for a
30-Second Super Bowl
Commercial

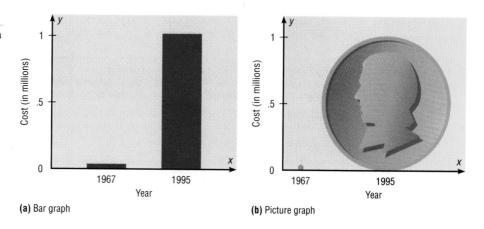

(a) Bar graph

(b) Picture graph

Another way to misrepresent data on a graph is by omitting labels or units on the
axes of the graph. The graph shown in Figure 2–18 compares the cost of living,

economic growth, population growth, etc. of four main geographical areas in the United States. However, since there are no numbers on the *y* axis, very little information can be gained from this graph, except a crude ranking of each factor. There is no way to decide the actual magnitude of the differences.

Figure 2–18

A Graph with No Units on the *y* Axis.

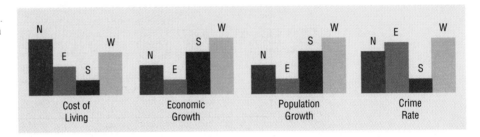

Finally, all graphs should contain a source for the information presented. The inclusion of a source for the data will enable you to check the reliability of the organization presenting the data. A summary of the types of graphs and their uses is shown in Figure 2–19.

Figure 2–19

Summary of Graphs and Uses of Each

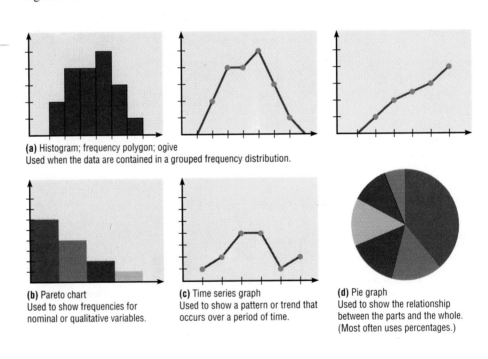

This *USA Today* Snapshot compares the costs of driving in several big cities. Can you see anything misleading about the graph? (Note the lengths of the bars for the first two cities in relationship to the numbers.)

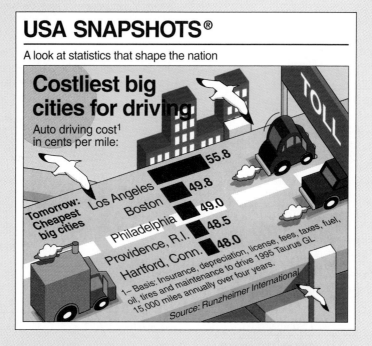

USA SNAPSHOTS®

A look at statistics that shape the nation

Costliest big cities for driving

Auto driving cost[1] in cents per mile:

Tomorrow: Cheapest big cities

Los Angeles   55.8
Boston   49.8
Philadelphia   49.0
Providence, R.I.   48.5
Hartford, Conn.   48.0

1– Basis: Insurance, depreciation, license, fees, taxes, fuel, oil, tires and maintenance to drive 1995 Taurus GL 15,000 miles annually over four years.

Source: Runzheimer International

Source: *USA Today,* May 3, 1995. Copyright 1995 *USA TODAY.* Used with permission.

Exercises

2–39. Construct a Pareto chart for the number of transplants of various types performed in 1994.

Type	Number
Kidney	11,390
Liver	3,653
Pancreas	844
Heart	2,340
Lung	737

Source: "Matchmaking at the Heart of Transplant Process," *USA Today.* October 19, 1995.

2–40. Construct a Pareto chart for the number of health conditions per 100 reported by the elderly in a survey.

Condition	Number
Arthritis	48
Hypertension	36
Heart disease	32
Hearing impairments	32
Orthopedic impairments	19
Cataracts	17
Sinusitis	16
Diabetes	11
Visual impairments	9

Source: "The Senior Profile," *USA Today,* July 7, 1995.

2–41. Construct a Pareto chart for the number of registered taxicabs in the selected cities.

City	Number
New York	11,787
Washington, D.C.	8,348
Chicago	5,300
Philadelphia	1,480
Baltimore	1,151

Source: "Yellow Cab Service at a Glance," Pittsburgh *Tribune-Review.* Used with permission.

2–42. Construct a Pareto chart for the number of unemployed people in the selected states for June of 1995.

State	Number
Texas	605,000
New York	494,000
Pennsylvania	364,000
Florida	363,000
Ohio	269,000

Source: "Employment in the USA," *USA Today,* July 10, 1995.

2–43. Construct a Pareto chart for the average number of hurricanes reported for the selected months from 1900 to 1995.

Month	Number
June	12
July	16
August	40
September	61
October	23
November	6

Source: *USA Today* Snapshots, September 6, 1995.

If you were planning a vacation during hurricane season in an area where hurricanes occur, what two months might you avoid?

2–44. Draw a time series graph to represent the data for the number of worldwide airline fatalities for the given years.

Year	1990	1991	1992	1993	1994	1995	1996
No. of fatalities	440	510	990	801	732	557	1,132

Reprinted with permission from *The World Almanac and Book of Facts,* 1999. Copyright © 1998. PRIMEDIA Reference, Inc. All rights reserved.

2–45. The data here represent the personal consumption expenditures for transportation for the United States (in billions of dollars). Draw a time series graph to represent the data.

Year	1991	1992	1993	1994	1995	1996	1997
Amount	$436.8	$471.5	$504.0	$542.2	$572.3	$611.6	$636.4

Reprinted with permission from *The World Almanac and Book of Facts,* 1999. Copyright © 1998. PRIMEDIA Reference, Inc. All rights reserved.

2–46. The number of operable nuclear power reactors in the United States for the given year is shown below. Draw a time series graph to represent the data and analyze the results.

Year	1986	1988	1990	1992	1994	1996
Number operable	100	108	111	109	109	110

Reprinted with permission from *The World Almanac and Book of Facts,* 1999. Copyright © 1998. PRIMEDIA Reference, Inc. All rights reserved.

2–47. A bacteriologist charted the growth of a certain bacterium over a period of 8 hours. The data are shown here. Construct a time series graph to represent the data and analyze the results.

Hour	1	2	3	4	5	6	7	8
No. of cells	2	5	8	13	17	22	30	38

2–48. The following data are based on a survey from the American Travel Survey on why people travel. Construct a pie graph for the data and analyze the results.

Purpose	Number
Personal Business	146
Visit friends or relatives	330
Work-related	225
Leisure	299

Source: *USA Today,* November 13, 1997.

2–49. A survey of 500 Philadelphia families asked the question "Where are you planning to vacation this summer?" It resulted in the following distribution. Construct a pie graph for the data and summarize the results.

Area	Number vacationing
Great Lakes region	37
New England	104
East Coast	206
South	96
West Coast	57

2–50. The frequency distribution here shows the number of freshmen, sophomores, juniors, and seniors who have part-time jobs after school. Construct a pie graph for the data.

Rank	Frequency
Freshmen	12
Sophomores	25
Juniors	36
Seniors	17

Why is a pie graph better than a Pareto chart for the data in this exercise?

2–51. In an insurance company study of the causes of 1000 deaths, the following data were obtained. Construct a pie graph to represent the data.

Cause of death	Number of deaths
1. Heart disease	432
2. Cancer	227
3. Stroke	93
4. Accidents	24
5. Other	224
	1000

Could a Pareto chart be used to represent the data for this exercise?

2–52. State which graph (Pareto chart, time series graph, or pie graph) would most appropriately represent the given situation.

a. The number of students enrolled at a local college for each year during the last five years.

b. The budget for the student activities department at a certain college for each year during the last five years.

c. The means of transportation the students use to get to school.

d. The percentage of votes each of the four candidates received in the last election.

e. The record temperatures of a city for the last 30 years.

f. The frequency of each type of crime committed in a city during the year.

*2–53. The number of successful space launches by the United States and Japan for the years 1993–1997 is shown

here. Construct a compound time series graph for the data. What comparison can be made regarding the launches?

Year	1993	1994	1995	1996	1997
U.S.	29	27	24	32	37
Japan	1	4	2	1	2

Reprinted with permission from *The World Almanac and Book of Facts,* 1999. Copyright © 1998. PRIMEDIA Reference, Inc. All rights reserved.

*2–54. Meat production for veal and lamb for the years 1950–90 is shown here. (Data are in millions of pounds.) Construct a compound time series graph for the data. What comparison can be made regarding meat production?

Year	1950	1960	1970	1980	1990
Veal	1230	1109	588	400	327
Lamb	576	769	551	318	358

Reprinted with permission from *The World Almanac and Book of Facts,* 1999. Copyright © 1998. PRIMEDIA Reference, Inc. All rights reserved.

No doubt you'll each draw your own conclusions from this chart...

Source: Cartoon by Bradford Veley, Marquette, Michigan. Used with permission.

Technology Step by Step

MINITAB
Step by Step

Constructing a Pie Chart

Example MT2–4

1. Enter the blood types from Example 2–11 in C1 of the worksheet. Label the column **BloodType.**

2. Select **Graph>Pie Chart**.

Pie Chart Dialog Box

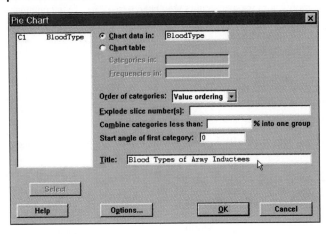

3. Click the button for Chart data in: then double-click C1 to select it.
4. In the Title text box, type in an appropriate title such as **Blood Types of Army Inductees.**
5. Click [OK]. The graph will be displayed in a graph window.

Speaking of **STATISTICS**

This *USA Today* Snapshot compares the budget deficit for fiscal 1995 with that for fiscal 1996. Can you see any-thing misleading about this graph?

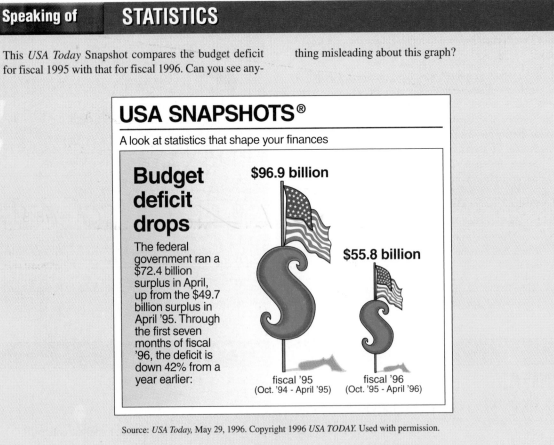

Source: *USA Today*, May 29, 1996. Copyright 1996 *USA TODAY*. Used with permission.

2–5

Summary

When data are collected, they are called raw data. Since very little knowledge can be obtained from raw data, they must be organized in some meaningful way. A frequency distribution using classes is the solution. Once a frequency distribution is constructed, the representation of the data by graphs is a simple task. The most commonly used graphs in research statistics are the histogram, frequency polygon, and ogive. Other graphs, such as the Pareto chart, time series graph, and pie graph, can also be used. Some of these graphs are seen frequently in newspapers, magazines, and various statistical reports.

Important Terms

categorical frequency
distribution   39

class boundaries   34

class midpoint   35

class width   35

cumulative
frequency   48

frequency   32

frequency
distribution   32

frequency polygon   46

grouped frequency
distribution   40

histogram   46

lower class limit   34

ogive   48

open-ended
distribution   36

Pareto chart   59

pie graph   61

range   37

raw data   32

relative frequency
graph   50

time series graph   60

ungrouped frequency
distribution   40

upper class limit   34

Important Formulas

Formula for the percentage of values in each class:

$$\% = \frac{f}{n} \cdot 100\%$$

where

f = **frequency of the class**

n = **total number of values**

Formula for the range:

$$R = \textbf{highest value} - \textbf{lowest value}$$

Formula for the class width:

$$\frac{\textbf{class}}{\textbf{width}} = \frac{\textbf{upper}}{\textbf{boundary}} - \frac{\textbf{lower}}{\textbf{boundary}}$$

Formula for the class midpoint:

$$X_m = \frac{\textbf{lower boundary + upper boundary}}{2}$$

or

$$X_m = \frac{\textbf{lower limit + upper limit}}{2}$$

Formula for the degrees for each section of a pie graph:

$$\textbf{degrees} = \frac{f}{n} \cdot 360°$$

Review   Exercises

2–55. A questionnaire about how people get news resulted in the following information from 25 respondents. Construct a frequency distribution for the data (N = newspaper, T = television, R = radio, M = magazine). (The data in this exercise will be used for Exercise 2–56.)

N	N	R	T	T
R	N	T	M	R
M	M	N	R	M
T	R	M	N	M
T	R	R	N	N

2–56. Construct a pie graph for the data in Exercise 2–55 and analyze the results.

2–57. A sporting-goods store kept a record of sales of five items for one randomly selected hour during a recent sale. Construct a frequency distribution for the data (B = baseballs, G = golf balls, T = tennis balls, S = soccer balls, F = footballs). (The data for this exercise will be used for Exercise 2–58.)

F	B	B	B	G	T	F
G	G	F	S	G	T	
F	T	T	T	S	T	
F	S	S	G	S	B	

2–58. Draw a pie graph for the data in Exercise 2–57 showing the sales of each item and analyze the results.

2–59. The blood urea nitrogen count (BUN) of 20 randomly selected patients is given here in mg/dl. Construct an ungrouped frequency distribution for the data. (The data for this exercise will be used for Exercise 2–60.)

17	18	13	14
12	17	11	20
13	18	19	17
14	16	17	12
16	15	19	22

2–60. Construct a histogram, frequency polygon, and ogive for the data in Exercise 2–59 and analyze the results.

2–61. The American Film Institute compiled a list of "100 Best American Movies of All Time" based on the ballots of screenwriters, directors, critics, etc. Shown is the year in which the film was released. Construct a frequency distribution for the data using classes 1910–1919, 1920–1929, etc. (*Note:* The class width is an even number since decades are being used.) (The data for this exercise will be used for Exercises 2–62 and 2–65.)

1941	1942	1972	1939	1962	1939	1967
1954	1993	1952	1946	1950	1957	1959
1977	1950	1951	1960	1974	1975	1940
1968	1941	1980	1982	1942	1964	1967
1979	1939	1948	1977	1974	1952	1962
1934	1969	1946	1944	1965	1959	1961
1954	1933	1915	1951	1971	1976	1975
1937	1969	1940	1953	1984	1930	1965
1970	1949	1940	1955	1981	1958	1982
1939	1977	1991	1976	1962	1951	1953
1971	1994	1959	1939	1925	1990	1931
1973	1976	1978	1969	1936	1956	1986
1996	1933	1935	1931	1969	1970	1927
1964	1951	1960	1990	1994	1956	1938
1992	1967					

2–62. Construct a histogram, frequency polygon, and ogive for the data in Exercise 2–61 and analyze the results.

2–63. The data (in millions of dollars) are the values of the 30 National Football League franchises. Construct a frequency distribution for the data using eight classes. (The data for this exercise will be used for Exercises 2–64 and 2–66.)

170	191	171	235	173	187	181	191
200	218	243	200	182	320	184	239
186	199	186	210	209	240	204	193
211	186	197	204	188	242		

Source: *Pittsburgh Post-Gazette,* October 26, 1997

2–64. Construct a histogram, frequency polygon, and ogive for the data in Exercise 2–63 and analyze the results.

2–65. Construct a histogram, frequency polygon, and ogive by using relative frequencies for the data in Exercise 2–61.

2–66. Construct a histogram, frequency polygon, and ogive by using relative frequencies for the data in Exercise 2–63.

2–67. Construct a Pareto chart for the number of homicides reported in 1995 for the following cities.

City	Number
New Orleans	363
Washington, D.C.	352
Chicago	824
Baltimore	323
Atlanta	184

Source: "Homicide, Rape, Robbery: The Numbers Are Fewer," *USA Today,* May 6, 1996.

2–68. Construct a Pareto chart for the number of trial-ready civil action and equity cases decided in less than six months for the selected counties in southwestern Pennsylvania.

County	Number
Westmoreland	427
Washington	298
Green	151
Fayette	106
Somerset	87

Source: "Quick Cases," Pittsburgh *Tribune-Review,* May 14, 1995, p. B1. Used with permission.

2–69. The given data represent the federal minimum hourly wage in the years shown. Draw a time series graph to represent the data and analyze the results.

Year	Wage
1960	$1.00
1965	$1.25
1970	$1.60
1975	$2.10
1980	$3.10
1985	$3.35
1990	$3.80
1995	$4.25

Source: Reprinted with permission from *The World Almanac and Book of Facts,* 1999. Copyright © 1998. PRIMEDIA Reference, Inc. All rights reserved.

2–70. The number of bank failures in the United States during the years 1986–1997 is shown below. Draw a time series graph to represent the data and analyze the results.

Year	Number of failures
1986	145
1987	203
1988	221
1989	207
1990	169
1991	127
1992	122
1993	41
1994	13
1995	6
1996	5
1997	1

Source: Reprinted with permission from *The World Almanac and Book of Facts,* 1999. Copyright © 1998. PRIMEDIA Reference, Inc. All rights reserved.

2–71. The following data are the numbers (in millions) of new sports cards sold in the United States. Draw a time series graph for the data. Analyze the results.

Year	Number
1988	250
1989	450
1990	800
1991	1200
1992	1100
1993	850
1994	750
1995	700
1996	650
1997	600

Source: *USA Today,* October 3, 1997.

2–72. In a study of 100 women, the numbers shown here indicate the major reason why each woman surveyed worked outside the home. Construct a pie graph for the data and analyze the results.

Reason	Number
To support self/family	62
For extra money	18
For something different to do	12
Other	8

2–73. A survey of the students in the school of education of a large university obtained the following data for students enrolled in specific fields. Construct a pie graph for the data and analyze the results.

Major field	Number
Preschool	893
Elementary	605
Middle	245
Secondary	1096

Statistics Today

Are We Flying More? Revisited

The type of graph that can be used to show the increase in the number of passengers on U.S. planes is the time series graph. An example is shown here. The ascending nature of the line over the years shows a major increase in the number of passengers using air travel in the United States. Comparing the height of the beginning point

(1980) with the height of the ending point (1994) shows that the number of passengers almost doubled.

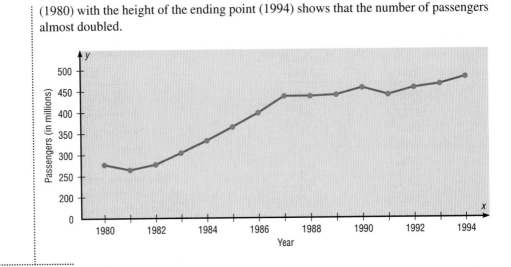

Data Analysis

A Data Bank is found in Appendix D, or on the world wide web by following links from www.mhhe.com/math/stat/bluman

1. From the Data Bank located in Appendix D, choose one of the following variables: age, weight, cholesterol level, systolic pressure, IQ, or sodium level. Select at least 30 values. For these values, construct a grouped frequency distribution. Draw a histogram, frequency polygon, and ogive for the distribution. Describe briefly the shape of the distribution.

2. From the Data Bank, choose one of the following variables: educational level, smoking status, or exercise. Select at least 20 values. Construct an ungrouped frequency distribution for the data. For the distribution, draw a Pareto chart and describe briefly the nature of the chart.

3. From the Data Bank, select at least 30 subjects and construct a categorical distribution for their marital status. Draw a pie graph and describe briefly the findings.

4. Using the data from Data Set IV in Appendix D, construct a grouped frequency distribution and draw a histogram. Describe briefly the shape of the distribution of the heights of the 130 tallest buildings in New York City.

5. Using the data from Data Set XII in Appendix D, draw a compound time series graph for the deaths related to extreme cold and snow or blizzards for the years 1988 to 1993. Compare the two results.

6. Using the data from Data Set IX in Appendix D, divide the United States into four regions, as follows:

Northeast CT ME MA NH NJ NY PA RI VT
Midwest   IL IN IA KS MI MN MS NE ND OH SD WI
South   AL AR DE DC FL GA KY LA MD NC OK
           SC TN TX VA WV
West   AK AZ CA CO HI ID MT NV NM OR UT
           WA WY

Find the total population for each region and draw a Pareto chart and a pie graph for the data. Analyze the results. Explain which chart might be a better representation for the data.

Chapter Quiz

Determine whether each statement is true or false. If the statement is false, explain why.

1. In the construction of a frequency distribution, it is a good idea to have overlapping class limits, such as 10–20, 20–30, 30–40.

2. Histograms can be drawn by using vertical or horizontal bars.

3. It is not important to keep the width of each class the same in a frequency distribution.

4. Frequency distributions can aid the researcher in drawing charts and graphs.

5. The type of graph used to represent data is determined by the type of data collected and by the researcher's purpose.

6. In construction of a frequency polygon, the class limits are used for the x axis.

7. Data collected over a period of time can be graphed by using a pie graph.

Select the best answer.

8. What is another name for the ogive?
   a. histogram
   b. frequency polygon
   c. cumulative frequency graph
   d. Pareto chart

9. What are the boundaries for 8.6–8.8?
   a. 8–9
   b. 8.5–8.9
   c. 8.55–8.85
   d. 8.65–9.75

10. What graph should be used to show the relationship between the parts and the whole?
   a. histogram
   b. pie graph
   c. Pareto chart
   d. ogive

11. Except for rounding errors, relative frequencies should add up to what sum?
   a. 0
   b. 1
   c. 50
   d. 100

Complete the following statements with the best answers.

12. The three types of frequency distributions are _____, _____, and _____.

13. In a frequency distribution, the number of classes should be between _____ and _____.

14. Data such as blood types (A, B, AB, O) can be organized into a _____ frequency distribution.

15. Data collected over a period of time can be graphed using a _____ graph.

16. When data are first collected, they are called _____ data.

17. On a Pareto chart, the frequencies should be represented on the _____ axis.

18. A questionnaire on housing arrangements showed the following information obtained from 25 respondents. Construct a frequency distribution for the data (H = house, A = apartment, M = mobile home, C = condominium).

H	C	H	M	H	A	C	A	M
C	M	C	A	M	A	C	C	M
C	C	H	A	H	H	M		

19. Construct a pie graph for the data in the previous problem.

20. When 30 randomly selected customers left a convenience store, each was asked the number of items he or she purchased. Construct an ungrouped frequency distribution for the data.

2	9	4	3	6
6	2	8	6	5
7	5	3	8	6
6	2	3	2	4
6	9	9	8	9
4	2	1	7	4

21. Construct a histogram, a frequency polygon, and an ogive for the data in the previous problem.

22. During June, a local theater recorded the following number of patrons per day. Construct a frequency distribution for the data. Use six classes.

102	116	113	132	128	117
156	182	183	171	168	179
170	160	163	187	185	158
163	167	168	186	117	108
171	173	161	163	168	182

23. Construct a histogram, frequency polygon, and ogive for the data in the previous problem. Analyze the histogram.

24. Construct a Pareto chart for the number of tons (in millions) of trash recycled per year by Americans based on an Environmental Protection Agency Study.

Type	Amount
Paper	320
Iron/Steel	292
Aluminum	276
Yard waste	242.4
Glass	196
Plastics	41.6

Source: *USA Today* Snapshot, November 14, 1997.

25. These data show the expenses of Chemistry Lab, Inc., for research and development for the years indicated. Each number represents thousands of dollars. Draw a time series graph of the data.

Year	Amount
1995	$ 8,937
1996	9,388
1997	11,271
1998	13,877
1999	19,203

What can you conclude about the amount of money spent for research and development?

Critical Thinking Challenges

1. Using the information given in the table below, draw an appropriate graph or graphs comparing the phone rates. Then write an analysis of the data.

2. The figure at the bottom of the page shows a comparison of the lawsuits filed by inmates in U.S. District Court, Western District of Pennsylvania, and the total number of lawsuits for the years 1982 to 1993.
   a. For *each* year, find the proportion of lawsuits filed by the inmates.

   b. In which year was the proportion highest?
   c. In which year was the proportion lowest?
   d. For the years 1982 through 1993, find the proportion of suits filed by the inmates. (Use the *total* number of lawsuits filed by the inmates divided by the *total* lawsuits filed.)
   e. How many years were above the proportion found in step *d?* How many were below the proportion?
   f. Using the proportions instead of the numbers, draw a time series graph for the data.

NEW YORK CALLING L.A.

How typical prepaid phone cards compare with a calling card and a coin phone for three calls.

Length of call	AT&T prepaid card	MCI, Sprint prepaid cards	7-Eleven $20 Phone Card	Regular AT&T Calling Card		Coin Phone (AT&T)	
				Business hrs.	Night/weekend	Business hrs.	Night/weekend
1 min.	$0.45	$0.60	$0.33	$1.08	$0.98	$2.85	$2.45
3 min.	1.35	1.80	1.00	1.64	1.34	2.85	2.45
10 min.	4.50	6.00	3.30	3.60	2.60	5.00	3.85

Source: "New Telephone Calling Cards Let You Pay Before You Dial." Copyright 1995 by Consumers Union of U.S., Inc., Yonkers, NY 10703-1057. Reprinted by permission from CONSUMER REPORTS, January 1995.

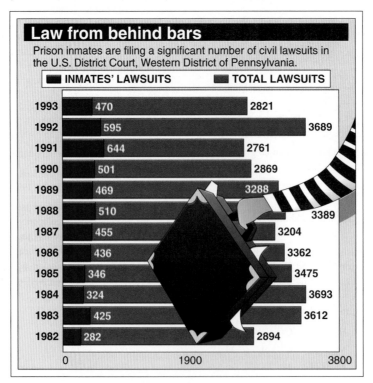

Law from behind bars

Prison inmates are filing a significant number of civil lawsuits in the U.S. District Court, Western District of Pennsylvania.

■ INMATES' LAWSUITS   ■ TOTAL LAWSUITS

Year	Inmates' Lawsuits	Total Lawsuits
1993	470	2821
1992	595	3689
1991	644	2761
1990	501	2869
1989	469	3288
1988	510	3389
1987	455	3204
1986	436	3362
1985	346	3475
1984	324	3693
1983	425	3612
1982	282	2894

0   1900   3800

Source: *Tribune-Review* (Greensburg, PA), December 11, 1994. Used with permission.

g. Which graph do you think better represents the data? Why?

3. Shown below are various statistics about the Great Lakes. Using appropriate graphs (your choice) and summary statements, write a report analyzing the data.

	Superior	**Michigan**	**Huron**	**Erie**	**Ontario**
Length (miles)	350	307	206	241	193
Breadth (miles)	160	118	183	57	53
Depth (feet)	1,330	923	750	210	802
Volume (cubic miles)	2,900	1,180	850	116	393
Area (square miles)	31,700	22,300	23,000	9,910	7,550
Shoreline (U.S., miles)	863	1,400	580	431	300

Source: Reprinted with permission from *The World Almanac and Book of Facts,* 1999. Copyright © 1998. PRIMEDIA Reference, Inc. All rights reserved.

4. In 1995, strokes were responsible for 96,428 deaths for women and 61,563 deaths for men (Source: Center for Disease Control and Prevention). Draw two vertical bar graphs for the data, one starting at zero and one starting at 50,000. Decide which one is more representative of the data and explain why you feel that way.

5. The data shown are for the number of alcohol-related highway fatalities for two counties in Pennsylvania.

Draw a time series graph for each county and explain how each might be used to draw different conclusions.

County	1991	1992	1993	1994	1995	1996
Allegheny	40	50	38	37	29	34
Westmoreland	24	24	24	21	13	26

Source: *Pittsburgh Tribune Review,* February 15, 1998.

Data Projects

Where appropriate, use MINITAB, the TI-83, Excel, or a computer program of your choice to complete the following exercises.

1. Select a categorical (nominal) variable, such as the colors of cars in the school's parking lot or the major fields of the students in statistics class, and collect data on this variable.
  a. State the purpose of the project.
  b. Define the population.
  c. State how the sample was selected.
  d. Show the raw data.
  e. Construct a frequency distribution for the variable.
  f. Draw some appropriate graphs (pie, Pareto, etc.) for the data.
  g. Analyze the results.

2. Using an almanac, select a variable that varies over a period of several years (e.g., silver production) and draw a time series graph for the data. Write a short paragraph interpreting the findings.

3. Select a variable (interval or ratio) and collect at least 30 values. For example, you may ask the students in your class how many hours they study per week or how old they are, etc..
  a. State the purpose of the project.
  b. Define the population.
  c. State how the sample was selected.
  d. Show the raw data.
  e. Construct a frequency distribution for the data.
  f. Draw a histogram, frequency polygon, and ogive for the data.
  g. Analyze the results.

You may use the following websites to obtain raw data:

http://www.mhhe.com/math/stat/bluman/
http://lib.stat.cmu.edu/DASL
http://www.oecd.org/statlist.htm
http://www.statcan.ca/english/

chapter

3 Data Description

Objectives

After completing this chapter, you should be able to

1. Summarize data using the measures of central tendency, such as the mean, median, mode, and midrange.

2. Describe data using the measures of variation, such as the range, variance, and standard deviation.

3. Identify the position of a data value in a data set using various measures of position, such as percentiles, deciles, and quartiles.

4. Use the techniques of exploratory data analysis, including stem and leaf plots, boxplots, and five-number summaries to discover various aspects of data.

Who Is a Typical First-Time Home Buyer?

Real estate agents, bankers, and insurance company executives can use statistics to obtain a profile of the typical first-time home buyer. With this knowledge, they can tailor their advertising to target certain groups and provide their best services to their customers. In order to devise this profile, descriptive statistics such as the means, medians, modes, ranges, variances, and standard deviations are used.

This chapter will show you how to obtain these descriptive statistics and explain how each can be used to create a profile of a typical first-time home buyer.

3–1

Introduction

The previous chapter showed how one can gain useful information from raw data by organizing it into a frequency distribution, then presenting the data by using various graphs. This chapter shows the statistical methods that can be used to summarize data. The most familiar of these methods is finding averages.

For example, one may read that the average speed of a car crossing midtown Manhattan during the day is 5.3 miles per hour or that the average number of minutes an American father of a four-year-old spends alone with his child each day is 42.[1]

[1]"Harper's Index," *Harper's* magazine, June 1995, p. 11. All rights reserved.

In the book *American Averages* by Mike Feinsilber and William B. Meed, the authors state:

"Average" when you stop to think of it is a funny concept. Although it describes all of us it describes none of us. . . . While none of us wants to be the average American, we all want to know about him or her.

The authors go on to give examples of averages:

The average American man is five feet, nine inches tall; the average woman is five feet, 3.6 inches.
The average American is sick in bed seven days a year missing five days of work.
On the average day, 24 million people receive animal bites.
By his or her 70th birthday, the average American will have eaten 14 steers, 1050 chickens, 3.5 lambs, and 25.2 hogs.[2]

In these examples, the word *average* is ambiguous, since several different methods can be used to obtain an average. Loosely stated, the average means the center of the distribution or the most typical case. Measures of average are also called *measures of central tendency* and include the *mean, median, mode,* and *midrange.*

Knowing the average of a data set is not enough to describe the data set entirely. Even though a shoe-store owner knows that the average size of a man's shoe is size 10, she would not be in business very long if she ordered only size 10 shoes.

As this example shows, in addition to knowing the average, one must know how the data values are dispersed. That is, do the data values cluster around the mean, or are they spread more evenly throughout the distribution? The measures that determine the spread of the data values are called *measures of variation* or *measures of dispersion.* These measures include the *range, variance,* and *standard deviation.*

Finally, another set of measures is necessary to describe data. These measures are called *measures of position.* They tell where a specific data value falls within the data set or its relative position in comparison with other data values. The most common position measures are *percentiles, deciles,* and *quartiles.* These measures are used extensively in psychology and education. Sometimes they are referred to as *norms.*

The measures of central tendency, variation, and position explained in this chapter are part of what is called *traditional statistics.*

The last section of this chapter shows the techniques of what is called *exploratory data analysis.* These techniques include the *stem and leaf plot,* the *boxplot,* and the *five-number summary.* They can be used to explore data to see what they show (as opposed to the traditional techniques, which are used to confirm conjectures about the data).

3–2

Measures of Central Tendency

Objective 1. Summarize data using measures of central tendency, such as the mean, median, mode, and midrange.

Chapter 1 stated that statisticians use samples taken from populations; however, when populations are small, it is not necessary to use samples since the entire population can be used to gain information. For example, suppose an insurance manager wanted to know the average weekly sales of all the company's representatives. If the company employed a large number of salespeople, say nationwide, he would have to use a sample and make an inference to the entire sales force. But if the company had only a few salespeople, say only 87 agents, he would be able to use all representatives' sales for a randomly chosen week and thus use the entire population.

---

[2]Mike Feinsilber and William B. Meed, *American Averages* (New York: Bantam Doubleday Dell, 1980). Used with permission.

Historical Note

In 1796, Adolphe Quetelet investigated the characteristics (heights, weights, etc.) of French conscripts to determine the "average man." Florence Nightingale was so influenced by Quetelet's work that she began collecting and analyzing medical records in the military hospitals during the Crimean War. Based on her work, hospitals began keeping accurate records on their patients.

Measures taken by using all the data values in the populations are called *parameters*. Measures obtained by using the data values of samples are called *statistics*. Hence, the average of the sales from a sample of representatives is called a *statistic*, and the average of sales obtained from the entire population is called a *parameter*.

A **statistic** is a characteristic or measure obtained by using the data values from a sample.

A **parameter** is a characteristic or measure obtained by using all the data values for a specific population.

These concepts as well as the symbols used to represent them will be explained in detail in this chapter.

General Rounding Rule   In statistics the basic rounding rule is that when computations are done in the calculation, rounding should not be done until the final answer is calculated. When rounding is done in the intermediate steps, it tends to increase the difference between that answer and the exact one. But in the textbook and solutions manual, it is not practical to show long decimals in the intermediate calculations; hence, the values in the examples are carried out enough places (usually three or four) to obtain the same answer a calculator would give after rounding on the last step.

The Mean

The *mean,* also known as the arithmetic average, is found by adding the values of the data and dividing by the total number of values. For example, the mean of 3, 2, 6, 5, and 4 is found by adding $3 + 2 + 6 + 5 + 4 = 20$ and dividing by 5; hence, the mean of the data is $20 \div 5 = 4$. The values of the data are represented by X's. In this data set, $X_1 = 3$, $X_2 = 2$, $X_3 = 6$, $X_4 = 5$, and $X_5 = 4$. To show a sum of the total X values, the symbol Σ (the capital Greek letter sigma) is used, and ΣX means to find the sum of the X values in the data set. The summation notation is explained in Appendix A–2.

The **mean** is the sum of the values divided by the total number of values. The symbol $\bar{X}$ represents the sample mean.

$$\bar{X} = \frac{X_1 + X_2 + X_3 + \cdots + X_n}{n} = \frac{\Sigma X}{n}$$

where n represents the total number of values in the sample.
   For a population, the Greek letter μ (mu) is used for the mean.

$$\mu = \frac{X_1 + X_2 + X_3 + \cdots + X_N}{N} = \frac{\Sigma X}{N}$$

where N represents the total number of values in the population.

In statistics, Greek letters are used to denote parameters and Roman letters are used to denote statistics. Assume that the data are obtained from samples unless otherwise specified.

Example 3–1

The data represent the number of different plans 10 HMO systems offer their enrollees. Find the mean: 84, 12, 27, 15, 40, 18, 33, 33, 14, 4.

Source: *USA Today,* December 10, 1997.

	Calories:	2,000	2,500
Total Fat	Less than	65g	80g
Sat Fat	Less than	20g	25g
Cholesterol	Less than	300mg	300mg
Sodium	Less than	2,400mg	2,400mg
Total Carbohydrate		300g	375g
Dietary Fiber		25g	30g

Niacin 6%

** Contains less than 2 percent of the daily value of these nutrients.

* Percent Daily Values are based on a 2,000 calorie diet. Your daily values may be higher or lower depending on your calorie needs.

INGREDIENTS: ENRICHED FLOUR (FLOUR, NIACIN, REDUCED IRON, THIAMIN MONONITRATE [VITAMIN B₁], RIBOFLAVIN [VITAMIN B₂]), VEGETABLE SHORTENING (CONTAINS ONE OR MORE OF THE FOLLOWING PARTIALLY HYDROGENATED OILS: SOYBEAN, CANOLA), CHEDDAR AND PARMESAN CHEESES (MILK, CHEESE CULTURES, SALT, ENZYMES), SALT, WHEY, MALTODEXTRIN, DRY CREAM, WHEY PROTEIN CONCENTRATE, MONOSODIUM GLUTAMATE, GU-

Solution

$$\bar{X} = \frac{\Sigma X}{n} = \frac{84 + 12 + 27 + 15 + 40 + 18 + 33 + 33 + 14 + 4}{10}$$

$$\bar{X} = \frac{280}{10} = 28$$

The mean is 28 plans.

....

Example 3–2

The fat contents in grams for one serving of 11 brands of packaged foods, as determined by the U.S. Department of Agriculture, are given as follows. Find the mean.

6.5, 6.5, 9.5, 8.0, 14.0, 8.5, 3.0, 7.5, 16.5, 7.0, 8.0

Source: *Consumer Reports,* June 1995.

Solution

$$\bar{X} = \frac{\Sigma X}{n} = \frac{6.5 + 6.5 + 9.5 + 8.0 + 14.0 + 8.5 + 3.0 + 7.5 + 16.5 + 7.0 + 8.0}{11}$$

$$= \frac{95}{11} = 8.64 \text{ grams}$$

Hence, the mean fat content is 8.64 grams.

....

The mean, in most cases, is not an actual data value.

Rounding Rule for the Mean The mean should be rounded to one more decimal place than occurs in the raw data. For example, if the raw data are given in whole numbers, the mean should be rounded to the nearest tenth. If the data are given in tenths, the mean should be rounded to the nearest hundredth, and so on.

The procedure for finding the mean for grouped data uses the midpoints of the classes. This procedure is shown next.

Example 3–3

Using the frequency distribution for Example 2–7 in Chapter 2, find the mean. The data represent the number of miles run during one week for a sample of 20 runners.

Solution

The procedure for finding the mean for grouped data is given here.

STEP 1   Make a table as shown.

A Class	B Frequency (f)	C Midpoint (X_m)	D $f \cdot X_m$
5.5–10.5	1		
10.5–15.5	2		
15.5–20.5	3		
20.5–25.5	5		
25.5–30.5	4		
30.5–35.5	3		
35.5–40.5	2		
	$n = 20$		

STEP 2   Find the midpoints of each class and enter them in column C.

$$X_m = \frac{5.5 + 10.5}{2} = 8, \qquad \frac{10.5 + 15.5}{2} = 13, \qquad \text{etc.}$$

STEP 3   For each class, multiply the frequency by the midpoint, as shown below, and place the product in column D.

$$1 \cdot 8 = 8, \qquad 2 \cdot 13 = 26, \qquad \text{etc.}$$

The completed table is shown here.

A Class	B Frequency (f)	C Midpoint (X_m)	D $f \cdot X_m$
5.5–10.5	1	8	8
10.5–15.5	2	13	26
15.5–20.5	3	18	54
20.5–25.5	5	23	115
25.5–30.5	4	28	112
30.5–35.5	3	33	99
35.5–40.5	2	38	76
	$n = 20$		$\Sigma f \cdot X_m = 490$

STEP 4   Find the sum of column D, as shown above.

STEP 5   Divide the sum by n to get the mean.

$$\overline{X} = \frac{\Sigma f \cdot X_m}{n} = \frac{490}{20} = 24.5 \text{ miles}$$

The procedure for finding the mean for grouped data assumes that the mean of all of the raw data values in each class is equal to the midpoint of the class. In reality, this is not true, since the average of the raw data values in each class usually will not be exactly equal to the midpoint. However, using this procedure will give an acceptable approximation of the mean, since some values fall above the midpoint and some values fall below the midpoint for each class, and the midpoint represents an estimate of all values in the class.

The steps for finding the mean for grouped data are summarized in the next Procedure Table.

Procedure Table

Finding the Mean for Grouped Data

STEP 1   Make a table as shown.

A	B	C	D
Class	Frequency (f)	Midpoint (X_m)	$f \cdot X_m$

STEP 2   Find the midpoints of each class and place them in column C.

STEP 3   Multiply the frequency by the midpoint for each class and place the product in column D.

STEP 4   Find the sum of column D.

STEP 5   Divide the sum obtained in column D by the sum of the frequencies obtained in column B.

The formula for the mean is

$$\overline{X} = \frac{\Sigma f \cdot X_m}{n}$$

The Median

An article recently reported that the median income for college professors was $43,250. This measure of average means that half of all the professors surveyed earned more than $43,250, and half earned less than $43,250.

The *median* is the halfway point in a data set. Before one can find this point, the data must be arranged in order. When the data set is ordered, it is called a **data array.** The median either will be a specific value in the data set or will fall between two values, as shown in the following examples.

The **median** is the midpoint of the data array. The symbol for the median is MD.

Steps in computing the median of a data array

STEP 1   Arrange the data in order.

STEP 2   Select the middle point.

Example 3–4

The weights (in pounds) of seven army recruits are 180, 201, 220, 191, 219, 209, and 186. Find the median.

Solution

STEP 1   Arrange the data in order.

180, 186, 191, 201, 209, 219, 220

This snapshot shows average holiday spending for selected age groups. Survey your class and using the age groups in the snapshot, compare your results with the ones given here.

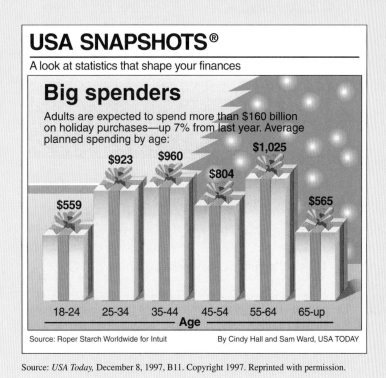

USA SNAPSHOTS®

A look at statistics that shape your finances

Big spenders

Adults are expected to spend more than $160 billion on holiday purchases—up 7% from last year. Average planned spending by age:

$923 $960 $1,025 $804 $559 $565

18-24   25-34   35-44   45-54   55-64   65-up
Age

Source: Roper Starch Worldwide for Intuit   By Cindy Hall and Sam Ward, USA TODAY

Source: *USA Today,* December 8, 1997, B11. Copyright 1997. Reprinted with permission.

STEP 2   Select the middle value.

$$180, 186, 191, \widehat{201,} 209, 219, 220$$

$$\uparrow$$

Median

Hence, the median weight is 201 pounds.

Example 3–5

Find the median for the ages of seven preschool children. The ages are 1, 3, 4, 2, 3, 5, and 1.

Solution

$$1, 1, 2, \widehat{3,} 3, 4, 5$$

$$\uparrow$$

Median

Hence, the median age is 3 years.

Each of these examples had an odd number of values in the data set; hence, the median was an actual data value. When there is an even number of values in the data set, the median will fall between two given values, as illustrated in the following examples.

Example 3–6

The number of tornadoes that have occurred in the United States over an eight-year period follows. Find the median.

684, 764, 656, 702, 856, 1133, 1132, 1303

Source: *THE UNIVERSAL ALMANAC* © by John W. Wright. Reprinted with permission of Andrews McMeel Publishing. All rights reserved.

Solution

656, 684, 702, 764, 856, 1132, 1133, 1303

Median

Since the middle point falls halfway between 764 and 856, find the median by adding the two values and dividing by 2.

$$MD = \frac{764 + 856}{2} = \frac{1620}{2} = 810$$

The median number of tornadoes is 810.

Example 3–7

The ages of 10 college students are given below. Find the median.

18, 24, 20, 35, 19, 23, 26, 23, 19, 20

Solution

18, 19, 19, 20, 20, 23, 23, 24, 26, 35

↑

Median

$$MD = \frac{20 + 23}{2} = 21.5$$

Hence, the median age is 21.5 years.

Example 3–8

Six customers purchased the following number of magazines: 1, 7, 3, 2, 3, 4. Find the median.

Solution

1, 2, 3, 3, 4, 7, $MD = \dfrac{3 + 3}{2} = 3$

Median

Hence, the median number of magazines purchased is 3.

The Mode

The third measure of average is called the *mode*. The mode is the value that occurs most often in the data set. It is sometimes said to be the most typical case.

The value that occurs most often in a data set is called the **mode.**

A data set can have more than one mode or no mode at all. These situations will be shown in some of the examples that follow.

Example 3–9

The following data represent the duration (in days) of U.S. space shuttle voyages for the years 1992–94. Find the mode.

$$8, 9, 9, 14, 8, 8, 10, 7, 6, 9, 7, 8, 10, 14, 11, 8, 14, 11$$

Source: *THE UNIVERSAL ALMANAC* © by John W. Wright. Reprinted with permission of Andrews McMeel Publishing. All rights reserved.

Solution

It is helpful to arrange the data in order, although it is not necessary.

$$6, 7, 7, 8, 8, 8, 8, 8, 9, 9, 9, 10, 10, 11, 11, 14, 14, 14$$

Since 8-day voyages occurred five times—a frequency larger than any other number—the mode for the data set is 8.

Example 3–10

Find the mode for the number of coal employees per county for 10 selected counties in Southwestern Pennsylvania.

$$110, 731, 1031, 84, 20, 118, 1162, 1977, 103, 752$$

Source: *Pittsburgh Tribune Review,* February 15, 1998

Solution

Since each value occurs only once, there is no mode.

Note: Do not say that the mode is zero. That would be incorrect, because in some data, such as temperature, zero can be an actual value.

Example 3–11

Eleven different automobiles were tested at a speed of 15 miles per hour for stopping distances. The data, in feet, are shown below. Find the mode.

$$15, 18, 18, 18, 20, 22, 24, 24, 24, 26, 26$$

Solution

Since 18 and 24 both occur three times, the modes are 18 and 24 feet. This data set is said to be *bimodal.*

The mode for grouped data is the modal class. The **modal class** is the class with the largest frequency.

Example 3–12

Find the modal class for the frequency distribution of miles 20 runners ran in one week, used in Example 2–7 in Chapter 2.

Class	Frequency
5.5–10.5	1
10.5–15.5	2
15.5–20.5	3
20.5–25.5	5 ← Modal class
25.5–30.5	4
30.5–35.5	3
35.5–40.5	2

Solution

The modal class is 20.5–25.5, since it has the largest frequency. Sometimes the midpoint of the class is used rather than the boundaries; hence, the mode could also be given as 23 miles per week.

The mode is the only measure of central tendency that can be used in finding the most typical case when the data are nominal or categorical.

Example 3–13

A survey showed the following distribution for the number of students enrolled in each field. Find the mode.

Business	1425
Liberal arts	878
Computer science	632
Education	471
General studies	95

Solution

Since the category with the highest frequency is business, the most typical case is a business major.

For a data set, the mean, median, and mode can be quite different. Consider the following example.

Example 3–14

A small company consists of the owner, the manager, the salesperson, and two technicians, all of whose annual salaries are listed here. (Assume that this is the entire population.)

Staff	Salary
Owner	$50,000
Manager	20,000
Salesperson	12,000
Technician	9,000
Technician	9,000

Many statistical studies involve typical individuals. Here is a profile of a "typical" bus tour traveler. What descrip-tive statistics are used here to estimate the parameters of the population and hence describe those travelers?

USA SNAPSHOTS®

A look at statistics that shape our lives

Leave the driving to . . .

Profile of typical bus tour traveler:

Female
Married
Age 35-54
College educated
Income $25,000-plus

Source: United Bus Owners of America

By Cindy Hall and Cliff Vancura, USA TODAY

Source: *USA Today,* April 28, 1995, p. 1D. Copyright 1995, USA TODAY. Reprinted with permission.

Find the mean, median, and mode.

Solution

$$\mu = \frac{\Sigma X}{N} = \frac{50{,}000 + 20{,}000 + 12{,}000 + 9{,}000 + 9{,}000}{5} = \$20{,}000$$

Hence, the mean is $20,000, the median is $12,000, and the mode is $9,000.

....

In this example, the mean is much higher than the median or the mode. This is because the extremely high salary of the owner tends to raise the value of the mean. In this and similar situations, the median should be used as the measure of central tendency.

The Midrange

The *midrange* is a rough estimate of the middle. It is found by adding the lowest and highest values in the data set and dividing by 2. It is a very rough estimate of the aver-age and can be affected by one extremely high or low value.

The **midrange** is defined as the sum of the lowest and highest values in the data set divided by 2. The symbol MR is used for the midrange.

$$MR = \frac{\text{lowest value} + \text{highest value}}{2}$$

Example 3–15

In the last two winter seasons, the city of Brownsville, Minnesota, reported the following number of water-line breaks per month. Find the midrange.

2, 3, 6, 8, 4, 1

Solution

$$MR = \frac{1 + 8}{2} = \frac{9}{2} = 4.5$$

Hence, the midrange is 4.5.

If the data set contains one extremely large value or one extremely small value, a higher or lower midrange value will result and may not be a typical description of the middle.

Example 3–16

Suppose the number of water-line breaks was as follows: 2, 3, 6, 16, 4, and 1. Find the midrange.

Solution

$$MR = \frac{1 + 16}{2} = \frac{17}{2} = 8.5$$

Hence, the midrange is 8.5. The value 8.5 is not typical of the average monthly number of breaks, since an excessively high number of breaks, 16, occurred in one month.

In statistics, several measures can be used for an average. The most common measures are the mean, the median, the mode, and the midrange. Each has its own specific purpose and use. Exercises 3–39 through 3–41 show examples of other averages—such as the harmonic mean, the geometric mean, and the quadratic mean. Their applications are limited to specific areas, as shown in the exercises.

The Weighted Mean

Sometimes, one must find the mean of a data set in which not all values are equally represented. Consider the case of finding the average cost of a gallon of gasoline for three taxis. Suppose the drivers buy gasoline at three different service stations at a cost of $1.19, $1.27, and $1.32 per gallon. One might try to find the average by using the formula

$$\bar{X} = \frac{\Sigma X}{n}$$

$$= \frac{1.19 + 1.27 + 1.32}{3} = \$1.26$$

But not all drivers purchased the same number of gallons. Hence, to find the true average cost per gallon, one must take into consideration the number of gallons each driver purchased.

The type of mean that considers an additional factor is called the *weighted mean*, and it is used when the values are not all equally represented.

Find the **weighted mean** of a variable X by multiplying each value by its corresponding weight and dividing the sum of the products by the sum of the weights.

$$\bar{X} = \frac{w_1 X_1 + w_2 X_2 + \cdots + w_n X_n}{w_1 + w_2 + \cdots + w_n} = \frac{\Sigma wX}{\Sigma w}$$

where $w_1, w_2, \ldots, w_n$ are the weights and $X_1, X_2, \ldots, X_n$ are the values.

The next example shows how the weighted mean is used to compute a grade point average. Since courses vary in their credit value, the number of credits must be used as weights.

Example 3–17

A student received an A in English Composition I (3 credits), a C in Introduction to Psychology (3 credits), a B in Biology I (4 credits), and a D in Physical Education (2 credits). Assuming A = 4 grade points, B = 3 grade points, C = 2 grade points, D = 1 grade point, and F = 0 grade points, find the student's grade point average.

Solution

Course	Credits (w)	Grade (X)
Eng Comp I	3	A (4 points)
Intro to Psych	3	C (2 points)
Biology I	4	B (3 points)
Phys Ed	2	D (1 point)

$$\bar{X} = \frac{\Sigma wX}{\Sigma w} = \frac{3 \cdot 4 + 3 \cdot 2 + 4 \cdot 3 + 2 \cdot 1}{3 + 3 + 4 + 2} = \frac{32}{12} = 2.7$$

The grade point average is 2.7.

Table 3–1 summarizes the measures of central tendency.

Table 3–1   Summary of Measures of Central Tendency

Measure	Definition	Symbol(s)
Mean	Sum of values divided by total number of values	μ, $\bar{X}$
Median	Middle point in data set that has been ordered	MD
Mode	Most frequent data value	none
Midrange	(Lowest value plus highest value) divided by 2	MR

Researchers and statisticians must know which measure of central tendency is being used and when to use each measure of central tendency. The properties and uses of the three measures of central tendency are summarized here.

Properties and Uses of Central Tendency

The Mean

1. One computes the mean by using all the values of the data.
2. The mean varies less than the median or mode when samples are taken from the same population and all three measures are computed for these samples.
3. The mean is used in computing other statistics, such as the variance.
4. The mean for the data set is unique, and not necessarily one of the data values.
5. The mean cannot be computed for an open-ended frequency distribution.
6. The mean is affected by extremely high or low values and may not be the appropriate average to use in these situations.

The Median

1. The median is used when one must find the center or middle value of a data set.
2. The median is used when one must determine whether the data values fall into the upper half or lower half of the distribution.
3. The median is used to find the average of an open-ended distribution.
4. The median is affected less than the mean by extremely high or extremely low values.

The Mode

1. The mode is used when the most typical case is desired.
2. The mode is the easiest average to compute.
3. The mode can be used when the data are nominal, such as religious preference, gender, or political affiliation.
4. The mode is not always unique. A data set can have more than one mode, or the mode may not exist for a data set.

The Midrange

1. The midrange is easy to compute.
2. The midrange gives the midpoint.
3. The midrange is affected by extremely high or low values in a data set.

Distribution Shapes

Frequency distributions can assume many shapes. The three most important shapes are positively skewed, symmetrical, and negatively skewed. Figure 3–1 shows histograms of each.

In a **positively skewed or right skewed distribution,** the majority of the data values fall to the left of the mean and cluster at the lower end of the distribution; the "tail" is to the right. Also, the mean is to the right of the median, and the mode is to the left of the median.

For example, if an instructor gave an examination and most of the students did poorly, their scores would tend to cluster on the left side of the distribution. A few high scores would constitute the tail of the distribution, which would be on the right side. Another example of a positively skewed distribution is the incomes of the population of the United States. Most of the incomes cluster about the low end of the distribution; those with high incomes are in the minority and are in the tail at the right of the distribution.

In a **symmetrical distribution,** the data values are evenly distributed on both sides of the mean. In addition, when the distribution is unimodal, the mean, median, and mode are the same and are at the center of the distribution. Examples of symmetrical distributions are IQ scores and heights of adult males.

Figure 3–1

Types of Distributions

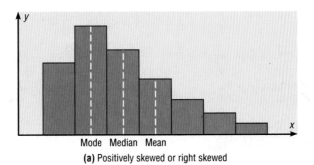

(a) Positively skewed or right skewed

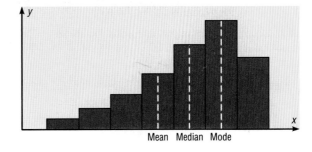

(b) Symmetrical   **(c)** Negatively skewed or left skewed

When the majority of the data values fall to the right of the mean and cluster at the upper end of the distribution, with the tail to the left, the distribution is said to be **negatively skewed or left skewed.** Also, the mean is to the left of the median, and the mode is to the right of the median. As an example, a negatively skewed distribution results if the majority of students score very high on an instructor's examination. These scores will tend to cluster to the right of the distribution.

When a distribution is extremely skewed, the value of the mean will be pulled toward the tail, but the majority of the data values will be greater than the mean or less than the mean (depending on which way the data are skewed); hence, the median rather than the mean is a more appropriate measure of central tendency. An extremely skewed distribution can also affect other statistics.

A measure of skewness for a distribution is discussed in Exercise 3–88 in Section 3–3.

Exercises

For Exercises 3–1 through 3–8, find (a) the mean, (b) the median, (c) the mode, and (d) the midrange.

3–1. Twelve secretaries were given a typing test, and the times (in minutes) to complete it were as follows:

8, 12, 15, 9, 6, 8, 10, 9, 8, 6, 7, 8

3–2. Ten novels were randomly selected, and the numbers of pages were recorded as follows:

415, 398, 402, 399, 400, 405, 395, 401, 412, 407

3–3. The following data are the number of burglaries reported in 1996 for nine western Pennsylvania universities. Which measure of average might be the best in this case? Explain your answer.

61, 11, 1, 3, 2, 30, 18, 3, 7,

Source: *Pittsburgh Post Gazette,* May 3, 1998, A12.

3–4. The number of hospitals for the five largest hospital systems is shown below.

340, 75, 123, 259, 151

Source: *USA Today,* November 18, 1997.

3–5. The calories per serving of 11 fruit juices are shown below.

150, 110, 100, 25, 60, 130,
40, 140, 120, 160, 110

Source: *Consumer Reports,* February 1995, p. 79.

3–6. The exam scores of 18 English composition students were recorded as follows:

78, 62, 98, 90, 88, 73, 79, 86, 81,
84, 93, 97, 63, 59, 78, 82, 87, 93

3–7. During 1993, the major earthquakes had Richter magnitudes as shown below.

7.0, 6.2, 7.7, 8.0, 6.4, 6.2,
7.2, 5.4, 6.4, 6.5, 7.2, 5.4

Source: *THE UNIVERSAL ALMANAC* © by John W. Wright. Reprinted with permission of Andrews McMeel Publishing. All rights reserved.

3–8. The Land Trust Alliance reported the following number of acres in trusts for each state.

22,077	178	156,747	18,751
737	4,644	27,497	125,070
1,386	44,230	1,692	35,203
0	1,946	1,722	4,180
484,271	27,273	5,299	12,357
45,419	16	176,573	0
44,220	9,163	15,665	12,569
33,062	14,528	98,896	326,616
97,197	119,052	1,067,227	9,784
7,188	66,159	65,789	47,483
947	18,928	7,116	489,381
89,266	105,318	26,909	155
15,080	44,314		

Source: *USA Today,* September 17, 1997

3–9. Find the (a) mean, (b) median, (c) mode, and (d) midrange for the data in Exercise 2–17. Is the distribution symmetrical or skewed? Use the individual data values.

3–10. Find the (a) mean, (b) median, (c) mode, and (d) midrange for the distances of the homeruns for McGwire and Sosa using the data in Exercise 2–18.

Compare the means. Decide if the means are approximately equal or is one of the players hitting longer homeruns? Use the individual data values.

3–11. The following data represent the number of traffic fatalities for the years 1995 and 1996 for 27 selected states.

Find the (a) mean, (b) median, (c) mode, and (d) midrange for each data set. Are the four measures of average for fatalities for 1996 the same as those for 1995? (The data in this exercise will be used in Exercise 3–57.)

1995			1996		
1113	1488	868	1100	260	205
1031	262	1109	970	1430	300
4192	1586	215	4040	460	350
645	527	254	620	480	485
121	442	313	125	405	85
2805	444	485	2805	690	1430
900	653	170	1555	1160	70
74	1480	69	180	3360	325
158	3181	326	875	705	145

Source: *USA Today,* July 14, 1997.

For Exercises 3–12 through 3–21, find the (a) mean, and the (b) modal class.

3–12. For 108 randomly selected college students, the following exam score frequency distribution was obtained. (The data in this exercise will be used in Exercise 3–60.)

Class limits	Frequency
90–98	6
99–107	22
108–116	43
117–125	28
126–134	9

3–13. For 50 antique car owners, the following distribution of the cars' ages was obtained.

Class limits	Frequency
16–18	20
19–21	18
22–24	8
25–27	4

3–14. Thirty automobiles were tested for fuel efficiency (in miles per gallon). The following frequency distribution was obtained. (The data in this exercise will be used in Exercise 3–62.)

Class boundaries	Frequency
7.5–12.5	3
12.5–17.5	5
17.5–22.5	15
22.5–27.5	5
27.5–32.5	2

3–15. The following numbers of books were read by each of the 28 students in a literature class.

Number of books	Frequency
0–2	2
3–5	6
6–8	12
9–11	5
12–14	3

3–16. Find the mean and modal class for the two frequency distributions in Exercises 2–26 and 2–36. Are the "average" reactions the same? Explain your answer.

3–17. Eighty randomly selected lightbulbs were tested to determine their lifetimes (in hours). The following frequency distribution was obtained. (The data in this exercise will be used in Exercise 3–65.)

Class boundaries	Frequency
52.5–63.5	6
63.5–74.5	12
74.5–85.5	25
85.5–96.5	18
96.5–107.5	14
107.5–118.5	5

3–18. The following data represent the net worth (in millions of dollars) of 45 national corporations.

Class limits	Frequency
10–20	2
21–31	8
32–42	15
43–53	7
54–64	10
65–75	3

3–19. The cost per load (in cents) of 35 laundry detergents tested by a consumer organization is shown below. (The data in this exercise will be used for Exercise 3–61.)

Class limits	Frequency
13–19	2
20–26	7
27–33	12
34–40	5
41–47	6
48–54	1
55–61	0
62–68	2

Source: *Consumer Reports,*
February 1995.

3–20. The following frequency distribution represents the commission earned (in dollars) by 100 salespeople employed at several branches of a large chain store.

Class limits	Frequency
150–158	5
159–167	16
168–176	20
177–185	21
186–194	20
195–203	15
204–212	3

3–21. This frequency distribution represents the data obtained from a sample of 75 copying machine service technicians. The values represent the days between service calls for various copying machines.

Class boundaries	Frequency
15.5–18.5	14
18.5–21.5	12
21.5–24.5	18
24.5–27.5	10
27.5–30.5	15
30.5–33.5	6

3–22. Find the mean and modal class for the data in Exercise 2–12, Chapter 2.

3–23. Find the mean and modal class for the data in Exercise 2–13, Chapter 2.

3–24. Find the mean and modal class for the data in Exercise 2–14, Chapter 2.

3–25. Find the mean and modal class for the data in Exercise 2–15, Chapter 2.

3–26. Find the weighted mean price of three models of automobiles sold. The number and price of each model sold are shown in the following list.

Model	Number	Price
A	8	$10,000
B	10	$12,000
C	12	$ 8,000

3–27. Using the weighted mean, find the average number of grams of fat in meat or fish a person would consume over a five-day period if he ate the following:

Meat or Fish	Fat (grams/oz.)
3 oz. fried shrimp	3.33
3 oz. veal cutlet (broiled)	3.00
2 oz. roast beef (lean)	2.50
2.5 oz. fried chicken drumstick	4.40
4 oz. tuna (canned in oil)	1.75

Source: Reprinted with permission from *The World Almanac and Book of Facts 1995.* Copyright © 1994 PRIMEDIA Reference. All rights reserved.

3–28. A recent survey of a new diet cola reported the following percentages of people who liked the taste. Find the weighted mean of the percentages.

Area	% Favored	Number surveyed
1	40	1000
2	30	3000
3	50	800

3–29. The costs of three models of helicopters are shown below. Find the weighted mean of the costs of the models.

Model	Number sold	Cost
Sunscraper	9	$427,000
Skycoaster	6	365,000
Highflyer	12	725,000

3–30. An instructor grades as follows: exams, 20%; term paper, 30%; final exam, 50%. A student had grades of 83, 72, and 90, respectively, for exams, term paper, and final exam. Find the student's final average. Use the weighted mean.

3–31. Another instructor gives four 1-hour exams and one final exam, which counts as two 1-hour exams. Find a student's grade if she received 62, 83, 97, and 90 on the hour exams and 82 on the final exam.

3–32. For the following situations, state which measure of central tendency—mean, median, or mode—should be used.
a. The most typical case is desired.
b. The distribution is open-ended.
c. There is an extreme value in the data set.
d. The data are categorical.
e. Further statistical computations will be needed.
f. The values are to be divided into two approximately equal groups, one group containing the larger values and one containing the smaller values.

3–33. Describe which measure of central tendency—mean, median, or mode—was probably used in each situation.
a. Half of the factory workers make more than $5.37 per hour and half make less than $5.37 per hour.
b. The average number of children per family in the Plaza Heights Complex is 1.8.
c. Most people prefer red convertibles over any other color.
d. The average person cuts the lawn once a week.
e. The most common fear today is fear of speaking in public.
f. The average age of college professors is 42.3 years.

3–34. What types of symbols are used to represent sample statistics? Give an example.

3–35. What types of symbols are used to represent population parameters? Give an example.

*3–36. If the mean of five values is 64, find the sum of the values.

*3–37. If the mean of five values is 8.2, and four of the values are 6, 10, 7, and 12, find the fifth value.

*3–38. Find the mean of 10, 20, 30, 40, and 50.
a. Add 10 to each value, and find the mean.
b. Subtract 10 from each value, and find the mean.
c. Multiply each value by 10, and find the mean.
d. Divide each value by 10, and find the mean.
e. Make a general statement about each situation.

*3–39. The **harmonic mean** (HM) is defined as the number of values divided by the sum of the reciprocals of each value. The formula is

$$HM = \frac{n}{\Sigma 1/X}$$

For example, the harmonic mean of 1, 4, 5, and 2 is

$$HM = \frac{4}{\frac{1}{1} + \frac{1}{4} + \frac{1}{5} + \frac{1}{2}} = 2.05$$

This mean is useful for finding the average speed. Suppose a person drove 100 miles at 40 miles per hour and returned driving 50 miles per hour. The average miles per hour is *not* 45 miles per hour, which is found by adding 40 and 50 and dividing by 2. The average is found as follows.

Since

$$\text{time} = \text{distance} \div \text{rate}$$

then

$$\text{time 1} = \frac{100}{40} = 2.5 \text{ hours to make the trip}$$

$$\text{time 2} = \frac{100}{50} = 2 \text{ hours to return}$$

Hence, the total time is 4.5 hours and the total miles driven are 200. Now, the average speed is

$$\text{rate} = \frac{\text{distance}}{\text{time}} = \frac{200}{4.5} = 44.44 \text{ miles per hour}$$

This value can also be found by using the harmonic mean formula:

$$HM = \frac{2}{\frac{1}{40} + \frac{1}{50}} = 44.44$$

Using the harmonic mean, find each of the following.
a. A salesperson drives 300 miles round trip at 30 miles per hour going to Chicago and 45 miles per hour returning home. Find the average miles per hour.
b. A bus driver drives the 50 miles to West Chester at 40 miles per hour and returns driving 25 miles per hour. Find the average miles per hour.

c. A carpenter buys $500 worth of nails at $50 per pound and $500 worth of nails at $10 per pound. Find the average cost of a pound of nails.

*3–40. The *geometric mean* (GM) is defined as the *n*th root of the product of *n* values. The formula is

$$GM = \sqrt[n]{(X_1)(X_2)(X_3) \cdots (X_n)}$$

The geometric mean of 4 and 16 is

$$GM = \sqrt{(4)(16)} = \sqrt{64} = 8$$

The geometric mean of 1, 3, and 9 is

$$GM = \sqrt[3]{(1)(3)(9)} = \sqrt[3]{27} = 3$$

The geometric mean is useful in finding the average of percentages, ratios, indexes, or growth rates. For example, if a person receives a 20% raise after one year of service and a 10% raise after the second year of service, the average percentage raise per year is not 15% but 14.89%, as shown.

$$GM = \sqrt{(1.2)(1.1)} = 1.1489$$

or

$$GM = \sqrt{(120)(110)} = 114.89\%$$

His salary is 120% at the end of the first year and 110% at the end of the second year. This is equivalent to an average of 14.89%, since $114.89\% - 100\% = 14.89\%$.

This answer can also be shown by assuming that the person makes $10,000 to start and receives two raises of 20% and 10%.

$$\text{raise } 1 = 10,000 \cdot 20\% = \$2000$$
$$\text{raise } 2 = 12,000 \cdot 10\% = \$1200$$

His total salary raise is $3200. This total is equivalent to

$$\$10,000 \cdot 14.89\% = \$1489.00$$
$$\$11,489 \cdot 14.89\% = \underline{\$1710.71}$$
$$\$3199.71 \approx \$3200$$

Find the geometric mean of each of the following.
a. The growth rates of the Living Life Insurance Corporation for the past three years were 35%, 24%, and 18%.
b. A person received the following percentage raises in salary over a four-year period: 8%, 6%, 4%, and 5%.
c. A stock increased each year for five years at the following percentages: 10%, 8%, 12%, 9%, and 3%.
d. The price increases, in percentages, for the cost of food in a specific geographic region for the past three years were 1%, 3%, and 5.5%.

*3–41. A useful mean in the physical sciences is the *quadratic mean* (QM), which is found by taking the square root of the average of the squares of each value. The formula is

$$QM = \sqrt{\frac{\sum X^2}{n}}$$

The quadratic mean of 3, 5, 6, and 10 is

$$QM = \sqrt{\frac{3^2 + 5^2 + 6^2 + 10^2}{4}}$$
$$= \sqrt{42.5} = 6.52$$

Find the quadratic mean of 8, 6, 3, 5, and 4.

3–3

Measures of Variation

In statistics, in order to describe the data set accurately, statisticians must know more than the measures of central tendency. Consider the following example.

Example 3–18

Objective 2. Describe data using measures of variation, such as the range, variance, and standard deviation.

A testing lab wishes to test two experimental brands of outdoor paint to see how long each would last before fading. The testing lab makes six gallons of each paint to test. Since different chemical agents are added to each group and only six cans are involved, these two groups constitute two small populations. The results (in months) follow. Find the mean of each group.

Brand A	Brand B
10	35
60	45
50	30
30	35
40	40
20	25

Solution

The mean for brand A is

$$\mu = \frac{\Sigma X}{N} = \frac{210}{6} = 35 \text{ months}$$

The mean for brand B is

$$\mu = \frac{\Sigma X}{N} = \frac{210}{6} = 35 \text{ months}$$

Since the means are equal in Example 3–18, one might conclude that both brands of paint last equally well. However, when the data sets are examined graphically, a somewhat different conclusion might be drawn. See Figure 3–2.

Figure 3–2

Examining Data Sets Graphically

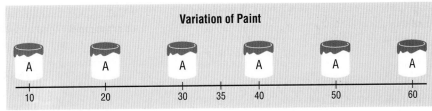

(a) Brand A

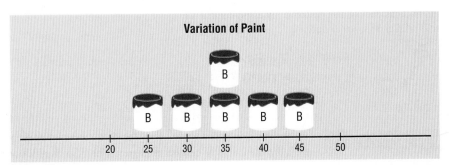

(b) Brand B

As Figure 3–2 shows, even though the means are the same for both brands, the spread, or variation, is quite different. The figure shows that brand B performs more consistently; it is less variable. For the spread or variability of a data set, three measures are commonly used: *range, variance,* and *standard deviation.* Each measure will be discussed in this section.

Range

The range is the simplest of the three measures and is defined next.

The **range** is the highest value minus the lowest value. The symbol R is used for the range.

$$R = \text{highest value} - \text{lowest value}$$

Example 3–19

Find the ranges for the paints in Example 3–18.

Solution

For brand A, the range is

$$R = 60 - 10 = 50 \text{ months}$$

For brand B, the range is

$$R = 45 - 25 = 20 \text{ months}$$

Make sure the range is given as a single number.

The range for brand A shows that 50 months separate the largest data value from the smallest data value. For brand B, 20 months separate the largest data value from the smallest data value, which is less than half of brand A's range.

....

One extremely high or one extremely low data value can affect the range markedly, as shown in the next example.

....

Example 3–20

The salaries for the staff of the XYZ Manufacturing Co. follow. Find the range.

Staff	Salary
Owner	$100,000
Manager	40,000
Sales representative	30,000
Workers	25,000
	15,000
	18,000

Solution

The range is $R = \$100,000 - \$15,000 = \$85,000$.

....

Since the owner's salary is included in the data for Example 3–20, the range is a large number. In order to have a more meaningful statistic to measure the variability, statisticians use measures called the *variance* and *standard deviation.*

Variance and Standard Deviation

Before the variance and standard deviation are defined formally, the computational procedure will be shown, since the definition is derived from the procedure.

Rounding Rule for the Standard Deviation   The rounding rule for the standard deviation is the same as for the mean. The final answer should be rounded to one more decimal place than the original data.

....

Example 3–21

Find the variance and standard deviation for the data set for brand A paint in Example 3–18.

$$10, 60, 50, 30, 40, 20$$

Solution

STEP 1   Find the mean for the data.

$$\mu = \frac{\Sigma X}{N} = \frac{10 + 60 + 50 + 30 + 40 + 20}{6} = \frac{210}{6} = 35$$

STEP 2   Subtract the mean from each data value.

$$10 - 35 = -25 \qquad 50 - 35 = +15 \qquad 40 - 35 = +5$$
$$60 - 35 = +25 \qquad 30 - 35 = -5 \qquad 20 - 35 = -15$$

STEP 3   Square each result.

$$(-25)^2 = 625 \qquad (+15)^2 = 225 \qquad (+5)^2 = 25$$
$$(+25)^2 = 625 \qquad (-5)^2 = 25 \qquad (-15)^2 = 225$$

STEP 4   Find the sum of the squares.

$$625 + 625 + 225 + 25 + 25 + 225 = 1750$$

STEP 5   Divide the sum by N to get the variance.

$$\text{variance} = 1750 \div 6 = 291.7$$

STEP 6   Take the square root of the variance to get the standard deviation.

Hence, the standard deviation equals $\sqrt{291.7}$, or 17.1. It is helpful to make a table, as follows:

A Values (X)	B $X - \mu$	C $(X - \mu)^2$
10	-25	625
60	$+25$	625
50	$+15$	225
30	-5	25
40	$+5$	25
20	-15	225
		1750

Column A contains the raw data, X. Column B contains the differences obtained in Step 2, $X - \mu$. Column C contains the squares of the differences obtained in Step 3.

....

The preceding computational procedure reveals several things. First, the square root of the variance gives the standard deviation; vice versa, squaring the standard deviation gives the variance. Second, the variance is actually the average of the square of the distance that each value is from the mean. Therefore, if the values are near the mean, the variance will be small. In contrast, if the values are farther away from the mean, the variance will be large.

One might wonder why the squared distances are used instead of the actual distances. One reason is that the sum of the distances will always be zero. To verify this result for a specific case, add the values in column B of the table in Example 3–21. When each value is squared, the negative signs are eliminated.

Finally, why is it necessary to take the square root? The reason is that since the distances were squared, the units of the resultant numbers are the squares of the units of the original raw data. Taking the square root of the variance puts the standard deviation in the same units as the raw data.

When taking the square root, always use its positive or principal value, since the variance and standard deviation of a data set can never be negative.

The **variance** is the average of the squares of the distance each value is from the mean. The symbol for the population variance is σ^2 (σ is the Greek lowercase letter sigma).

The formula for the population variance is

$$\sigma^2 = \frac{\Sigma (X - \sigma)^2}{N}$$

where
X = individual value
μ = population mean
N = population size

The **standard deviation** is the square root of the variance. The symbol for the population standard deviation is σ.

The corresponding formula for the population standard deviation is

$$\sigma = \sqrt{\sigma^2} = \sqrt{\frac{\Sigma (X - \mu)^2}{N}}$$

Example 3–22

Find the variance and standard deviation for brand B paint data in Example 3–18. The months were

$$35, 45, 30, 35, 40, 25$$

Solution

STEP 1   Find the mean.

$$\mu = \frac{\Sigma X}{N} = \frac{35 + 45 + 30 + 35 + 40 + 25}{6} = \frac{210}{6} = 35$$

STEP 2   Subtract the mean from each value and place the result in column B of the table.

STEP 3   Square each result and place the squares in column C of the table.

A	B	C
X	$X - \mu$	$(X - \mu)^2$
35	0	0
45	10	100
30	−5	25
35	0	0
40	5	25
25	−10	100

STEP 4   Find the sum of the squares in column C.

$$\Sigma (X - \mu)^2 = 0 + 100 + 25 + 0 + 25 + 100 = 250$$

STEP 5   Divide the sum by N to get the variance.

$$\sigma^2 = \frac{\Sigma (X - \mu)^2}{N} = \frac{250}{6} = 41.7$$

STEP 6 Take the square root to get the standard deviation.

$$\sigma = \sqrt{\frac{\Sigma (X - \mu)^2}{N}} = \sqrt{41.7} = 6.5$$

Hence, the standard deviation is 6.5.

Since the standard deviation of brand A is 17.1 (see Example 3–21), and the standard deviation of brand B is 6.5, the data are more variable for brand A. *In summary, when the means are equal, the larger the variance or standard deviation is, the more variable the data are.*

The Unbiased Estimator

When computing the variance for a sample, one might expect the following formula would be used:

$$\frac{\Sigma (X - \overline{X})^2}{n}$$

where $\overline{X}$ is the sample mean and n is the sample size. *This formula is not usually used, however, since in most cases the purpose of calculating the statistic is to estimate the corresponding parameter.* For example, the sample mean $\overline{X}$ is used to estimate the population mean μ. The formula

$$\frac{\Sigma (X - \overline{X})^2}{n}$$

does not give the best estimate of the population variance because when the population is large and the sample is small (usually less than 30), the variance computed by this formula usually underestimates the population variance. Therefore, instead of dividing by n, find the variance of the sample by dividing by $n - 1$, giving a slightly larger value and an *unbiased* estimate of the population variance.

The formula for the sample variance, denoted by s^2, is

$$s^2 = \frac{\Sigma (X - \overline{X})^2}{n - 1}$$

where

$\overline{X}$ = sample mean
n = sample size

To find the standard deviation of a sample, one must take the square root of the sample variance, which was found by using the preceding formula.

Formula for the Sample Standard Deviation

The standard deviation of a sample (denoted by s) is

$$s = \sqrt{s^2} = \sqrt{\frac{\Sigma (X - \overline{X})^2}{n - 1}}$$

where

X = individual value
$\overline{X}$ = sample mean
n = sample size

Shortcut formulas for computing the variance and standard deviation are presented next and will be used in the remainder of the chapter and in the exercises. These formulas are mathematically equivalent to the preceding formulas and do not involve using the mean. They save time when repeated subtracting and squaring occur in the original formulas. They are also more accurate when the mean has been rounded.

Shortcut or Computational Formulas for s^2 and s

The shortcut formulas for computing the variance and standard deviation for data obtained from samples are as follows.

Variance

$$s^2 = \frac{\Sigma X^2 - [(\Sigma X)^2/n]}{n - 1}$$

Standard deviation

$$s = \sqrt{\frac{\Sigma X^2 - [(\Sigma X)^2/n]}{n - 1}}$$

The next two examples explain how to use the shortcut formulas.

Example 3–23

Find the variance and standard deviation for the amount of European auto sales for a sample of six years shown. The data are in millions of dollars.

$$11.2, \ 11.9, \ 12.0, \ 12.8, \ 13.4, \ 14.3$$

Source: *USA Today,* March 22, 1999.

Solution

STEP 1   Find the sum of the values.

$$\Sigma X = 11.2 + 11.9 + 12.0 + 12.8 + 13.4 + 14.3 = 75.6$$

STEP 2   Square each value and find the sum.

$$\Sigma X^2 = 11.2^2 + 11.9^2 + 12.0^2 + 12.8^2 + 13.4^2 + 14.3^2 = 958.94$$

STEP 3   Substitute in the formulas and solve.

$$s^2 = \frac{\Sigma X^2 - [(\Sigma X)^2/n]}{n - 1} = \frac{958.94 - [(75.6^2)/6]}{5}$$

$$s^2 = 1.28$$

The variance of the sample is 1.28

$$s = \sqrt{1.28} = 1.13$$

Hence, the sample standard deviation is 1.13.

Note that ΣX^2 is not the same as $(\Sigma X)^2$. The notation ΣX^2 means to square the values first, then sum; $(\Sigma X)^2$ means to sum the values first, then square.

Variance and Standard Deviation for Grouped Data

The procedure for finding the variance and standard deviation for grouped data is similar to that for finding the mean for grouped data, and it uses the midpoints of each class.

Example 3–24

Find the variance and the standard deviation for the frequency distribution of the data in Example 2–7 in Chapter 2. The data represent the number of miles 20 runners ran during one week.

Class	Frequency	Midpoint
5.5–10.5	1	8
10.5–15.5	2	13
15.5–20.5	3	18
20.5–25.5	5	23
25.5–30.5	4	28
30.5–35.5	3	33
35.5–40.5	2	38

Solution

STEP 1 Make a table as shown, and find the midpoint of each class.

A Class	B Frequency (f)	C Midpoint (X_m)	D $f \cdot X_m$	E $f \cdot X_m^2$
5.5–10.5	1	8		
10.5–15.5	2	13		
15.5–20.5	3	18		
20.5–25.5	5	23		
25.5–30.5	4	28		
30.5–35.5	3	33		
35.5–40.5	2	38		

STEP 2 Multiply the frequency by the midpoint for each class and place the products in column D.

$$1 \cdot 8 = 8 \qquad 2 \cdot 13 = 26 \qquad \ldots \qquad 2 \cdot 38 = 76$$

STEP 3 Multiply the frequency by the square of the midpoint and place the products in column E.

$$1 \cdot 8^2 = 64 \qquad 2 \cdot 13^2 = 338 \qquad \ldots \qquad 2 \cdot 38^2 = 2888$$

STEP 4 Find the sum of columns B, D, and E. The sum of column B is n, the sum of column D is $\Sigma f \cdot X_m$, and the sum of column E is $\Sigma f \cdot X_m^2$. The completed table follows.

A Class	B Frequency	C Midpoint	D $f \cdot X_m$	E $f \cdot X_m^2$
5.5–10.5	1	8	8	64
10.5–15.5	2	13	26	338
15.5–20.5	3	18	54	972
20.5–25.5	5	23	115	2,645
25.5–30.5	4	28	112	3,136
30.5–35.5	3	33	99	3,267
35.5–40.5	2	38	76	2,888
	$n = 20$		$\Sigma f \cdot X_m = 490$	$\Sigma f \cdot X_m^2 = 13{,}310$

STEP 5   Substitute in the formula and solve for s^2 to get the variance.

$$s^2 = \frac{\Sigma f \cdot X_m^2 - [(\Sigma f \cdot X_m)^2/n]}{n - 1}$$

$$= \frac{13{,}310 - [(490)^2/20]}{20 - 1} = 68.7$$

STEP 6   Take the square root to get the standard deviation.

$$s = \sqrt{68.7} = 8.3$$

Be sure to use the number found in the sum of column B (i.e., the sum of the frequencies) for n. Do not use the number of classes.

The steps for finding the variance and standard deviation for grouped data are summarized in the Procedure Table.

Procedure Table

Finding the Sample Variance and Standard Deviation for Grouped Data

STEP 1   Make a table as shown, and find the midpoint of each class.

A	B	C	D	E
Class	Frequency	Midpoint	$f \cdot X_m$	$f \cdot X_m^2$

STEP 2   Multiply the frequency by the midpoint for each class, and place the products in column D.

STEP 3   Multiply the frequency by the square of the midpoint, and place the products in column E.

STEP 4   Find the sums of columns B, D, and E. (The sum of column B is n. The sum of column D is $\Sigma f \cdot X_m$. The sum of column E is $\Sigma f \cdot X_m^2$.)

STEP 5   Substitute in the formula and solve to get the variance.

$$s^2 = \frac{\Sigma f \cdot X_m^2 - [(\Sigma f \cdot X_m)^2/n]}{n - 1}$$

STEP 6   Take the square root to get the standard deviation.

Table 3–2   Summary of Measures of Variation

Measure	Definition	Symbol(s)
Range	Distance between highest value and lowest value	R
Variance	Average of the squares of the distance each value is from the mean	σ^2, s^2
Standard deviation	Square root of the variance	σ, s

Uses of the Variance and Standard Deviation

1. As previously stated, variances and standard deviations can be used to determine the spread of the data. If the variance or standard deviation is large, the data are more dispersed. This information is useful in comparing two or more data sets to determine which is more (most) variable.

2. The measures of variance and standard deviation are used to determine the consistency of a variable. For example, in the manufacture of fittings, like nuts and bolts, the variation in the diameters must be small or the parts will not fit together.

3. The variance and standard deviation are used to determine the number of data values that fall within a specified interval in a distribution. For example, Chebyshev's theorem (explained later) shows that for any distribution, at least 75% of the data values will fall within two standard deviations of the mean.

4. Finally, the variance and standard deviation are used quite often in inferential statistics. These uses will be shown in later chapters of this textbook.

Coefficient of Variation

Whenever two samples have the same units of measure, the variance and standard deviation for each can be compared directly. For example, suppose an automobile dealer wanted to compare the standard deviation of miles driven for the cars he received as trade-ins on new cars. He found that for a specific year, the standard deviation for Buicks was 422 miles and the standard deviation for Cadillacs was 350 miles. He could say that the variation in mileage was greater in the Buicks. But what if a manager wanted to compare the standard deviations of two different variables, like the number of sales per salesperson over a three-month period and the commissions made by these salespeople?

A statistic that allows one to compare standard deviations when the units are different, as in this example, is called the *coefficient of variation*.

The **coefficient of variation** is the standard deviation divided by the mean. The result is expressed as a percentage.

For samples,

$$\text{CVar} = \frac{s}{\overline{X}} \cdot 100\%$$

For populations,

$$\text{CVar} = \frac{\sigma}{\mu} \cdot 100\%$$

Example 3–25

The mean of the number of sales of cars over a three-month period is 87 and the standard deviation is 5. The mean of the commissions is $5225 and the standard deviation is $773. Compare the variations of the two.

Solution

The coefficients of variation are

$$\text{CVar} = \frac{s}{\overline{X}} = \frac{5}{87} \cdot 100\% = 5.7\% \qquad \text{sales}$$

$$\text{CVar} = \frac{773}{5225} \cdot 100\% = 14.8\% \qquad \text{commissions}$$

Since the coefficient of variation is larger for commissions, the commissions are more variable than the sales.

Example 3–26

The mean for the number of pages of a sample of women's fitness magazines is 132, with a variance of 23; the mean for the number of advertisements of a sample of women's fitness magazines is 182, with a variance of 62. Compare the variations.

Solution

The coefficients of variation are

$$\text{CVar} = \frac{\sqrt{23}}{132} \cdot 100\% = 3.6\% \qquad \text{pages}$$

$$\text{CVar} = \frac{\sqrt{62}}{182} \cdot 100\% = 4.3\% \qquad \text{advertisements}$$

The number of advertisements is more variable than the number of pages since the co-efficient of variation is larger for advertisements.

Chebyshev's Theorem

As stated previously, the variance and standard deviation of a variable can be used to de-termine the spread or dispersion of a variable. That is, the larger the variance or standard deviation, the more the data values are dispersed. For example, if two variables meas-ured in the same units have the same mean, say 70, and variable one has a standard de-viation of 1.5 while variable two has a standard deviation of 10, then the data for variable two will be more spread out than the data for variable one. *Chebyshev's theo-rem,* developed by the Russian mathematician Chebyshev, specifies the proportions of the spread in terms of the standard deviation.

Chebyshev's theorem   The proportion of values from a data set that will fall within k standard deviations of the mean will be at least $1 - 1/k^2$, where k is a number greater than 1 (k is not necessarily an integer).

This theorem states that at least $\frac{3}{4}$, or 75%, of the data values will fall within two standard deviations of the mean of the data set. This result is found by substituting $k = 2$ in the formula.

$$1 - \frac{1}{k^2} \qquad \text{or} \qquad 1 - \frac{1}{2^2} = 1 - \frac{1}{4} = \frac{3}{4} = 75\%$$

Referring to the example in which variable one has a mean of 70 and a standard de-viation of 1.5, at least $\frac{3}{4}$, or 75%, of the data values fall between 67 and 73. These values are found by adding two standard deviations to the mean and subtracting two standard deviations from the mean, as shown.

$$70 + 2(1.5) = 70 + 3 = 73$$

and

$$70 - 2(1.5) = 70 - 3 = 67$$

For variable two, at least $\frac{3}{4}$, or 75%, of the data values fall between 50 and 90. Again, these values are found by adding and subtracting, respectively, two standard deviations to and from the mean.

$$70 + 2(10) = 70 + 20 = 90$$

and

$$70 - 2(10) = 70 - 20 = 50$$

Furthermore, the theorem states that at least $\frac{8}{9}$, or 88.89%, of the data values will fall within three standard deviations of the mean. This result is found by letting $k = 3$ and substituting in the formula.

$$1 - \frac{1}{k^2} \quad \text{or} \quad 1 - \frac{1}{3^2} = 1 - \frac{1}{9} = \frac{8}{9} = 88.89\%$$

For variable one, at least $\frac{8}{9}$, or 88.89%, of the data values fall between 65.5 and 74.5, since

$$70 + 3(1.5) = 70 + 4.5 = 74.5$$

and

$$70 - 3(1.5) = 70 - 4.5 = 65.5$$

For variable two, at least $\frac{8}{9}$, or 88.89%, of the data values fall between 40 and 100.

This theorem can be applied to any distribution regardless of its shape (see Figure 3–3).

Figure 3–3

Chebyshev's Theorem

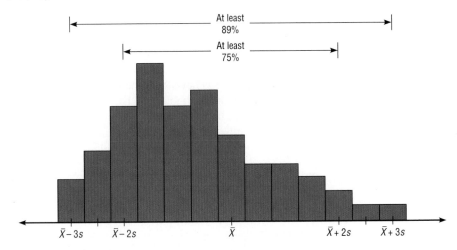

The next two examples illustrate the application of Chebyshev's theorem.

Example 3–27

The mean price of houses in a certain neighborhood is $50,000, and the standard deviation is $10,000. Find the price range for which at least 75% of the houses will sell.

Solution

Chebyshev's theorem states that $\frac{3}{4}$, or 75%, of the data values will fall within two standard deviations of the mean. Thus,

$$\$50,000 + 2(\$10,000) = \$50,000 + \$20,000 = \$70,000$$

and

$$\$50,000 - 2(\$10,000) = \$50,000 - \$20,000 = \$30,000$$

Hence, at least 75% of all homes sold in the area will have a price range from $30,000 to $70,000.

Chebyshev's theorem can be used to find the minimum percentage of data values that will fall between any two given values. The procedure is shown in the next example.

Example 3–28

A survey of local companies found that the mean amount of travel allowance for executives was $0.25 per mile. The standard deviation was $0.02. Using Chebyshev's theorem, find the minimum percentage of the data values that will fall between $0.20 and $0.30.

Solution

STEP 1   Subtract the mean from the larger value.

$$\$0.30 - \$0.25 = \$0.05$$

STEP 2   Divide the difference by the standard deviation to get k.

$$k = \frac{0.05}{0.02} = 2.5$$

STEP 3   Use Chebyshev's theorem to find the percentage.

$$1 - \frac{1}{k^2} = 1 - \frac{1}{2.5} = 1 - \frac{1}{6.25} = 1 - 0.16 = 0.84 \quad \text{or} \quad 84\%$$

Hence, at least 84% of the data values will fall between $0.20 and $0.30

The Empirical (Normal) Rule

Chebyshev's theorem applies to any distribution regardless of its shape. However, when a distribution is *bell-shaped* (or what is called *normal*), the following statements, which make up the **empirical rule,** are true.

Approximately 68% of the data values will fall within one standard deviation of the mean.

Approximately 95% of the data values will fall within two standard deviations of the mean.

Approximately 99.7% of the data values will fall within three standard deviations of the mean.

For example, suppose that the scores on a national achievement exam have a mean of 480 and a standard deviation of 90. If these scores are normally distributed, then approximately 68% will fall between 390 and 570 (480 + 90 = 570 and 480 − 90 = 390). Approximately 95% of the scores will fall between 300 and 660 (480 + 2 · 90 = 660 and 480 − 2 · 90 = 300). Approximately 99.7% will fall between 210 and 750 (480 + 3 · 90 = 750 and 480 − 3 · 90 = 210). See Figure 3–4.

Figure 3–4

The Empirical Rule

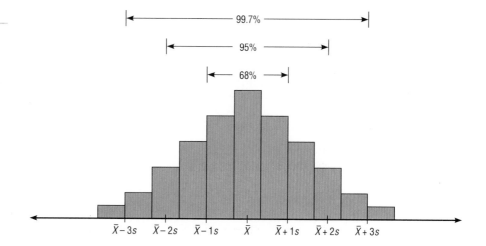

Exercises

3–42. Name three measures of variation.

3–43. What is the relationship between the variance and the standard deviation?

3–44. Why might the range *not* be the best estimate of variability?

3–45. What are the symbols used to represent the population variance and standard deviation?

3–46. What are the symbols used to represent the sample variance and standard deviation?

3–47. Why is the unbiased estimator of variance used?

3–48. The three data sets below have the same mean and range, but is the variation the same? Prove your answer by computing the standard deviation. Assume the data were obtained from samples.
  a. 5, 7, 9, 11, 13, 15, 17
  b. 5, 6, 7, 11, 15, 16, 17
  c. 5, 5, 5, 11, 17, 17, 17

For Exercises 3–49 through 3–55, find the range, variance, and standard deviation. Assume the data represent samples, and use the shortcut formula for the unbiased estimator to compute the variance and standard deviation.

3–49. Twelve students were given an arithmetic test, and the times (in minutes) to complete it were as follows:

  10, 9, 12, 11, 8, 15, 9, 7, 8, 6, 12, 10

3–50. Ten used trail bikes are randomly selected, and the odometer reading of each is recorded as follows.

  1902, 103, 653, 1901, 788,
  361, 216, 363, 223, 656

3–51. Fifteen students were selected and asked how many hours each studied for the final exam in statistics. Their answers are recorded here.

  8, 6, 3, 0, 0, 5, 9, 2, 1, 3, 7, 10, 0, 3, 6

3–52. The weights of nine football players are recorded as follows.

  206, 215, 305, 297, 265, 282, 301, 255, 261

3–53. Shown below are the numbers of stories in the 11 tallest buildings in St. Paul, Minnesota.

  32, 36, 46, 20, 32, 18, 16, 34, 26, 27, 26

Source: Reprinted with permission from *The World Almanac and Book of Facts 1995,* Copyright © 1994. PRIMEDIA Reference Inc. All rights reserved.

3–54. The following data are the prices of a gallon of premium gasoline in U.S. dollars in seven foreign countries.

  $3.80, $3.80, $3.20, $3.57, $3.62, $3.74, $3.69

Source: *Pittsburgh Post Gazette,* November 11, 1995

3–55. The manufacturer's ratings (in amps) of 22 upright vacuum cleaners are shown here.

  7.8, 7, 12, 10, 12, 7, 7.2, 7.3, 11, 7, 12,
  11, 9.5, 9, 9, 10, 7.2, 9.5, 11, 10, 7.5, 10

Source: *Consumer Reports,* April 1995.

3–56. Find the range, variance, and standard deviation for the distances of the homeruns for McGwire and Sosa using the data in Exercise 2–18. Compare the ranges and standard deviations. Decide which is more variable or if the variability is about the same (Use individual data.).

3–57. Find the range, variance, and standard deviation for each data set in Exercise 3–11. Based on the results, which data set is more variable?

3–58. The Federal Highway Administration reported the number of deficient bridges in each state. Find the range, variance, and standard deviation.

15,458	1,055	5,008	3,598	8,984
1,337	4,132	10,618	17,361	6,081
6,482	25,090	12,681	16,286	18,832
12,470	17,842	16,601	4,587	47,196
23,205	25,213	23,017	27,768	2,686
7,768	25,825	4,962	22,704	2,694
4,131	13,144	15,582	7,279	12,613
810	13,350	1,208	22,242	7,477
10,902	2,343	2,333	2,979	6,578
14,318	4,773	6,252	734	13,220

Source: *USA Today,* October 3, 1997, A18.

3–59. Find the range, variance, and standard deviation for the data in Exercise 2–17.

For Exercises 3–60 through 3–69, find the variance and standard deviation.

3–60. For 108 randomly selected college students, the following exam score frequency distribution was obtained.

Class limits	Frequency
90–98	6
99–107	22
108–116	43
117–125	28
126–134	9

3–61. The costs per load (in cents) of 35 laundry detergents tested by a consumer organization are shown below.

Class limits	Frequency
13–19	2
20–26	7
27–33	12
34–40	5
41–47	6
48–54	1
55–61	0
62–68	2

Source: *Consumer Reports*, February 1995.

3–62. Thirty automobiles were tested for fuel efficiency (in miles per gallon). The following frequency distribution was obtained.

Class boundaries	Frequency
7.5–12.5	3
12.5–17.5	5
17.5–22.5	15
22.5–27.5	5
27.5–32.5	2

3–63. In a class of 29 students, the following distribution of quiz scores was recorded:

Grade	Frequency
0–2	1
3–5	3
6–8	5
9–11	14
12–14	6

3–64. In a study of reaction times to a specific stimulus, a psychologist recorded the following data (in seconds).

Class limits	Frequency
2.1–2.7	12
2.8–3.4	13
3.5–4.1	7
4.2–4.8	5
4.9–5.5	2
5.6–6.2	1

3–65. Eighty randomly selected lightbulbs were tested to determine their lifetimes (in hours). The following frequency distribution was obtained.

Class boundaries	Frequency
52.5–63.5	6
63.5–74.5	12
74.5–85.5	25
85.5–96.5	18
96.5–107.5	14
107.5–118.5	5

3–66. The following data represent the net worth (in millions of dollars) of 50 businesses in a large city.

Class limits	Frequency
10–20	5
21–31	10
32–42	3
43–53	7
54–64	18
65–75	7

3–67. The following data represent the scores (in words per minute) of 25 typists on a speed test.

Class limits	Frequency
54–58	2
59–63	5
64–68	8
69–73	0
74–78	4
79–83	5
84–88	1

3–68. Find the variance and standard deviation for the two distributions in Exercises 2–26 and 2–36. Compare the variation of the data sets. Decide if one data set is more variable than the other.

3–69. This frequency distribution represents the data obtained from a sample of word-processor repairers. The values are the days between service calls on 80 machines.

Class boundaries	Frequency
25.5–28.5	5
28.5–31.5	9
31.5–34.5	32
34.5–37.5	20
37.5–40.5	12
40.5–43.5	2

3–70. The average score of the students in one calculus class is 110, with a standard deviation of 5; the average score of students in a statistics class is 106, with a standard deviation of 4. Which class is more variable in terms of scores?

3–71. The average price of the Panther convertible is $40,000, with a standard deviation of $4000. The average price of the Suburban station wagon is $20,000, with a standard deviation of $2000. Compare the variability of the two prices.

3–72. The average score on an English final examination was 85, with a standard deviation of 5; the average score on a history final exam was 110, with a standard deviation of 8. Which class was more variable?

3–73. The average age of the accountants at Three Rivers Corp. is 26, with a standard deviation of 6; the average salary of the accountants is $31,000, with a standard deviation of $4000. Compare the variations of age and income.

3–74. Using Chebyshev's theorem, solve the following problems for a distribution with a mean of 80 and a standard deviation of 10.
  a. At least what percentage of values will fall between 60 and 100?
  b. At least what percentage of values will fall between 65 and 95?

3–75. The mean of a distribution is 20 and the standard deviation is 2. Answer each. Use Chebyshev's theorem.
  a. At least what percentage of the values will fall between 10 and 30?
  b. At least what percentage of the values will fall between 12 and 28?

3–76. In a distribution of 200 values, the mean is 50 and the standard deviation is 5. Answer each. Use Chebyshev's theorem.
  a. At least how many values will fall between 30 and 70?
  b. At most how many values will be less than 40 or more than 60?

3–77. A sample of the hourly wages of employees who work in restaurants in a large city has a mean of $5.02 and a standard deviation of $0.09. Using Chebyshev's theorem, find the range in which at least 75% of the data values will fall.

3–78. A sample of the labor costs per hour to assemble a certain product has a mean of $2.60 and a standard deviation of $0.15. Using Chebyshev's theorem, find the values in which at least 88.89% of the data will lie.

3–79. A survey of a number of the leading brands of cereal shows that the mean content of potassium per serving is 95 milligrams, and the standard deviation is 2 milligrams. Find the values in which at least 88.89% of the data will fall. Use Chebyshev's theorem.

3–80. The average score on a special test of knowledge of wood refinishing has a mean of 53 and a standard deviation of 6. Using Chebyshev's theorem, find the range of values in which at least 75% of the scores will lie.

3–81. The average of the number of trials it took a sample of mice to learn to traverse a maze was 12. The standard deviation was 3. Using Chebyshev's theorem, find the minimum percentage of data values that will fall in the range of 4 to 20 trials.

3–82. The average cost of a certain type of grass seed is $4.00 per box. The standard deviation is $0.10. Using Chebyshev's theorem, find the minimum percentage of data values that will fall in the range of $3.82 to $4.18.

*3–83. For the following data set, find the mean and standard deviation of the variable. The data represent the serum cholesterol level of 30 individuals. Count the number of data values that fall within two standard deviations of the mean. Compare this with the number obtained from Chebyshev's theorem. Comment on the answer.

211	240	255	219	204
200	212	193	187	205
256	203	210	221	249
231	212	236	204	187
201	247	206	187	200
237	227	221	192	196

*3–84. For the following data set, find the mean and standard deviation of the variable. The data represent the ages of 30 customers who ordered a product advertised on television. Count the number of data values that fall within two standard deviations of the mean. Compare this with the number obtained from Chebyshev's theorem. Comment on the answer.

42	44	62	35	20
30	56	20	23	41
55	22	31	27	66
21	18	24	42	25
32	50	31	26	36
39	40	18	36	22

*3–85. Using Chebyshev's theorem, complete the table to find the minimum percentage of the data values that fall within k standard deviations of the mean.

k	1.5	2	2.5	3	3.5
Percent					

*3–86. Use the following data set: 10, 20, 30, 40, 50.
  a. Find the standard deviation.
  b. Add 5 to each value, and then find the standard deviation.

c. Subtract 5 from each value, and find the standard deviation.

d. Multiply each value by 5, and find the standard deviation.

e. Divide each value by 5, and find the standard deviation.

f. Generalize the results of parts b through e.

g. Compare these results with those in Exercise 3–38.

*3–87. The **mean deviation** is found by using the following formula:

$$\text{mean deviation} = \frac{\Sigma|X - \overline{X}|}{n}$$

where

X = value

$\overline{X}$ = mean

n = number of values

$|\ |$ = absolute value

Find the mean deviation for the following data.

5, 9, 10, 11, 11, 12, 15, 18, 20, 22

*3–88. A measure to determine the skewness of a distribution is called the ***Pearson coefficient of skewness.*** The formula is

$$\text{skewness} = \frac{3(\overline{X} - MD)}{s}$$

The values of the coefficient usually range from -3 to $+3$. When the distribution is symmetrical, the coefficient is zero; when the distribution is positively skewed, it is positive; and when the distribution is negatively skewed, it is negative.

Using the formula, find the coefficient of skewness for each distribution, and describe the shape of the distribution.

a. Mean = 10, median = 8, standard deviation = 3.

b. Mean = 42, median = 45, standard deviation = 4.

c. Mean = 18.6, median = 18.6, standard deviation = 1.5.

d. Mean = 98, median = 97.6, standard deviation = 4.

3–4

Measures of Position

Objective 3. Identify the position of a data value in a data set using various measures of position such as percentiles, deciles, and quartiles.

Standard Scores

In addition to measures of central tendency and measures of variation, there are also measures of position or location. These measures include standard scores, percentiles, deciles, and quartiles. They are used to locate the relative position of a data value in the data set. For example, if a value is located at the 80th percentile, it means that 80% of the values fall below it in the distribution and 20% of the values fall above it. The *median* is the value that corresponds to the 50th percentile, since half of the values fall below it and half of the values fall above it. This section discusses these measures of position.

There is an old saying that states, "You can't compare apples and oranges." But with the use of statistics, it can be done to some extent. Suppose that a student scored 90 on a music test and 45 on an English exam. Direct comparison of raw scores is impossible, since the exams might not be equivalent in terms of number of questions, value of each question, and so on. However, a comparison of a relative standard similar to both can be made. This comparison uses the mean and standard deviation and is called a standard score or z score. (We also use z scores in later chapters.)

A **standard score** or **z score** for a value is obtained by subtracting the mean from the value and dividing the result by the standard deviation. The symbol for a standard score is z. The formula is

$$z = \frac{\text{value} - \text{mean}}{\text{standard deviation}}$$

For samples, the formula is

$$z = \frac{X - \overline{X}}{s}$$

For populations, the formula is

$$z = \frac{X - \mu}{\sigma}$$

The z score represents the number of standard deviations a data value falls above or below the mean.

Example 3–29

A student scored 65 on a calculus test that had a mean of 50 and a standard deviation of 10; she scored 30 on a history test with a mean of 25 and a standard deviation of 5. Compare her relative positions on the two tests.

Solution

First, find the z scores. For calculus the z score is

$$z = \frac{X - \overline{X}}{s} = \frac{65 - 50}{10} = 1.5$$

For history the z score is

$$z = \frac{30 - 25}{5} = 1.0$$

Since the z score for calculus is larger, her relative position in the calculus class is higher than her relative position in the history class.

Note that if the z score is positive, the score is above the mean. If the z score is 0, the score is the same as the mean. And if the z score is negative, the score is below the mean.

Example 3–30

Find the z score for each test and state which is higher.

Test A	$X = 38$	$\overline{X} = 40$	$s = 5$
Test B	$X = 94$	$\overline{X} = 100$	$s = 10$

Solution

For test A,

$$z = \frac{X - \overline{X}}{s} = \frac{38 - 40}{5} = -0.4$$

For test B,

$$z = \frac{94 - 100}{10} = -0.6$$

The score for test A is relatively higher than the score for test B.

When all data for a variable are transformed into z scores, the resulting distribution will have a mean of 0 and a standard deviation of 1. A z score, then, is actually the number of standard deviations each variable is from the mean for a specific distribution. In Example 3–29, the calculus score of 65 was actually 1.5 standard deviations above the mean of 50. This will be explained in more detail in Chapter 7.

Percentiles

Percentiles are position measures used in educational and health-related fields to indicate the position of an individual in a group.

A **percentile** P is an integer ($1 \leq P \leq 99$) such that Pth percentile is a value where $P\%$ of the data values are less than or equal to the value and $100 - P\%$ of the data values are greater than or equal to the value.

In many situations, the graphs and tables showing the percentiles for various measures such as test scores, heights, or weights have already been completed. Table 3–3 shows the percentile ranks for scaled scores on the Test of English as a Foreign Language. If a student had a scaled score of 58 for Section 1 (listening and comprehension), that student would have a percentile rank of 81. Hence, that student did better than 81% of the students who took Section 1 of the exam.

Table 3–3	Percentile Ranks and Scaled Scores on the Test of English as a Foreign Language*				
Scaled score	Section 1: Listening comprehension	Section 2: Structure and written expression	Section 3: Vocabulary and reading comprehension	Total scaled score	Percentile rank
68	99	98			
66	98	96	98	660	99
64	96	94	96	640	97
62	92	90	93	620	94
60	87	84	88	600	89
→58	81	76	81	580	82
56	73	68	72	560	73
54	64	58	61	540	62
52	54	48	50	520	50
50	42	38	40	500	39
48	32	29	30	480	29
46	22	21	23	460	20
44	14	15	16	440	13
42	9	10	11	420	9
40	5	7	8	400	5
38	3	4	5	380	3
36	2	3	3	360	1
34	1	2	2	340	1
32		1	1	320	
30		1	1	300	
Mean	51.5	52.2	51.4	Mean	517
S.D.	7.1	7.9	7.5	S.D.	68

*Based on the total group of 1,178,193 examinees tested from July 1989 through June 1991.

Source: Reprinted by permission of Educational Testing Service, the copyright owner.

Figure 3–5 shows percentiles in graphic form of weights of girls from ages 2 to 18. To find the percentile rank of an 11-year-old who weighs 82 pounds, start at the 82-pound weight on the left axis and move horizontally to the right. Find the 11 on the horizontal axis and move up vertically. The two lines meet at the 50th percentile curved line; hence, an 11-year-old girl who weighs 82 pounds is in the 50th percentile for her age group. If the lines do not meet exactly on one of the curved percentile lines, then the percentile rank must be approximated.

Figure 3–5

Weights of Girls by Age
and Percentile Rankings

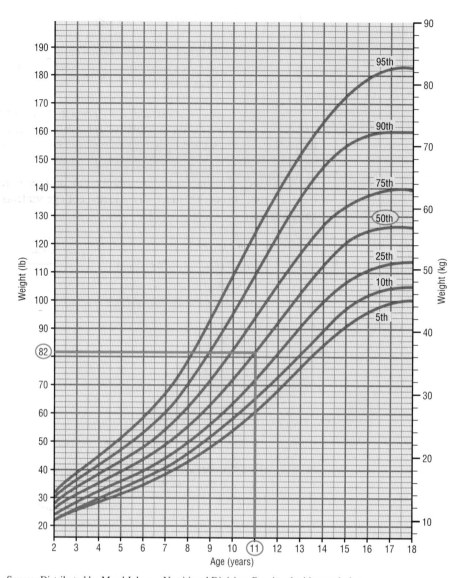

Source: Distributed by Mead Johnson Nutritional Division. Reprinted with permission.

Percentiles are also used to compare an individual's test score with the national norm. For example, tests such as the National Educational Development Test (NEDT) are taken by students in ninth or tenth grade. A student's scores are compared with those of other students locally and nationally by using percentile ranks. A similar test for elementary school students is called the California Achievement Test.

Percentiles are not the same as percentages. That is, if a student gets 72 correct answers out of a possible 100, she obtains a percentage score of 72. There is no indication of her position with respect to the rest of the class. She could have scored the highest, the lowest, or somewhere in between. On the other hand, if a raw score of 72 corresponds to the 64th percentile, then she did better than 64% of the students in her class.

Percentiles are symbolized by

$$P_1, P_2, P_3, \ldots, P_{99}$$

and divide the distribution into 100 groups.

Percentile graphs can be constructed as shown in the next example. Percentile graphs use the same values as the cumulative relative frequency graphs described in Section 2–3, except that the proportions have been converted to percents.

Example 3–31

The frequency distribution for the systolic blood pressure readings (in millimeters of mercury, mm Hg) of 200 randomly selected college students follows. Construct a percentile graph.

A Class boundaries	B Frequency	C Cumulative frequency	D Cumulative percent
89.5–104.5	24		
104.5–119.5	62		
119.5–134.5	72		
134.5–149.5	26		
149.5–164.5	12		
164.5–179.5	4		
	200		

Solution

STEP 1   Find the cumulative frequencies and place them in column C.

STEP 2   Find the cumulative percentages and place them in column D. To do this step, use the formula

$$\text{cumulative \%} = \frac{\text{cumulative frequency}}{n} \cdot 100\%$$

For the first class,

$$\text{cumulative \%} = \frac{24}{200} \cdot 100\% = 12\%$$

The completed table is shown next.

A Class boundaries	B Frequency	C Cumulative frequency	D Cumulative percent
89.5–104.5	24	24	12
104.5–119.5	62	86	43
119.5–134.5	72	158	79
134.5–149.5	26	184	92
149.5–164.5	12	196	98
164.5–179.5	4	200	100
	200		

STEP 3 Graph the data, using class boundaries for the x axis and the percentages for the y axis, as shown in Figure 3–6.

Figure 3–6

Percentile Graph for Example 3–31.

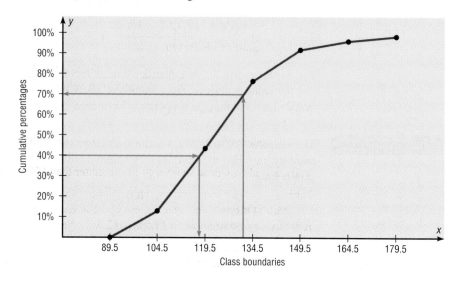

Once a percentile graph has been constructed, one can find the approximate corresponding percentile ranks for given blood pressure values and find approximate blood pressure values for given percentile ranks.

For example, to find the percentile rank of a blood pressure reading of 130, find 130 on the x axis of Figure 3–6, and draw a vertical line to the graph. Then move horizontally to the value on the y axis. Note that a blood pressure of 130 corresponds to approximately the 70th percentile.

If the value that corresponds to the 40th percentile is desired, start on the y axis at 40 and draw a horizontal line to the graph. Then draw a vertical line to the x axis, and read the value. In Figure 3–6, the 40th percentile corresponds to a value of approximately 118. Thus, if a person has a blood pressure of 118, he or she is at the 40th percentile.

Finding values and the corresponding percentile ranks by using a graph yields only approximate answers. Several mathematical methods exist for computing percentiles for data. They can be used to find the approximate percentile rank of a data value or to find a data value corresponding to a given percentile. When the data set is large (100 or more), these methods yield better results. The next several examples show these methods.

Percentile Formula

The percentile corresponding to a given value (X) is computed by using the following formula:

$$\text{percentile} = \frac{(\text{number of values below } X) + 0.5}{\text{total number of values}} \cdot 100\%$$

Example 3–32

A teacher gives a 20-point test to 10 students. The scores are shown below. Find the percentile rank of a score of 12.

18, 15, 12, 6, 8, 2, 3, 5, 20, 10

Solution

Arrange the data in order from lowest to highest.

2, 3, 5, 6, 8, 10, 12, 15, 18, 20

Then substitute in the formula.

$$\text{percentile} = \frac{(\text{number of values below } X) + 0.5}{\text{total number of values}} \cdot 100\%$$

Since there are six values below a score of 12, the solution is

$$\text{percentile} = \frac{6 + 0.5}{10} \cdot 100\% = 65\text{th percentile}$$

Thus, a student whose score was 12 did better than 65% of the class.

Note: One assumes that a score of 12 in Example 3–32, for instance, means theoretically any value between 11.5 and 12.5.

Example 3–33

Using the data in Example 3–32, find the percentile rank for a score of 6.

Solution

There are three values below 6. Thus

$$\text{percentile} = \frac{3 + 0.5}{10} \cdot 100\% = 35\text{th percentile}$$

A student who scored 6 did better than 35% of the class.

The next two examples show a procedure for finding a value corresponding to a given percentile.

Example 3–34

Using the scores in Example 3–32, find the value corresponding to the 25th percentile.

Solution

STEP 1   Arrange the data in order from lowest to highest.

2, 3, 5, 6, 8, 10, 12, 15, 18, 20

STEP 2   Compute

$$c = \frac{n \cdot p}{100}$$

where

n = total number of values

p = percentile

Thus,

$$c = \frac{10 \cdot 25}{100} = 2.5$$

STEP 3   If c is not a whole number, round it up to the next whole number; in this case, $c = 3$. (If c is a whole number, see the next example.) Start at the lowest value and count over to the third value, which is 5. Hence, the value 5 corresponds to the 25th percentile.

Example 3–35

Using the data set in Example 3–32, find the value that corresponds to the 60th percentile.

Solution

STEP 1   Arrange the data in order from smallest to largest.

$$2, 3, 5, 6, 8, 10, 12, 15, 18, 20$$

STEP 2   Substitute in the formula.

$$c = \frac{n \cdot p}{100} = \frac{10 \cdot 60}{100} = 6$$

STEP 3   If c is a whole number, use the value halfway between the c and $c + 1$ values when counting up from the lowest value—in this case, the 6th and 7th values.

$$2, 3, 5, 6, 8, 10, 12, 15, 18, 20$$

        6th value   7th value

The value halfway between 10 and 12 is 11. Find it by adding the two values and dividing by 2.

$$\frac{10 + 12}{2} = 11$$

Hence, 11 corresponds to the 60th percentile. Anyone scoring 11 would have done better than 60% of the class.

The steps for finding a value corresponding to a given percentile are summarized in the Procedure Table.

Procedure Table
Finding a Data Value Corresponding to a Given Percentile

STEP 1   Arrange the data in order from lowest to highest.

STEP 2   Substitute in the formula

$$c = \frac{n \cdot p}{100}$$

where

        n = total number of values
        p = percentile

STEP 3A   If c is not a whole number, round up to the next whole number. Starting at the lowest value, count over to the number that corresponds to the rounded-up value.

STEP 3B   If c is a whole number, use the value halfway between c and $c + 1$ when counting up from the lowest value.

Quartiles and Deciles

Quartiles divide the distribution into four groups, denoted by Q_1, Q_2, Q_3.

Note that Q_1 is the same as the 25th percentile; Q_2 is the same as the 50th percentile or the median; Q_3 corresponds to the 75th percentile, as shown.

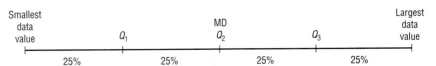

Quartiles can be computed using the formulas given for percentiles; however, it is much easier to arrange the data in order from smallest to largest and find the median. This is Q_2. To find Q_1, find the median of the data values less than the median. To find Q_3, find the median of the data values that are larger than the median.

Example 3–36

Find Q_1, Q_2, and Q_3 for the data set 15, 13, 6, 5, 12, 50, 22, 18.

Solution

STEP 1   Arrange the data in order:

$$5, 6, 12, 13, 15, 18, 22, 50$$

STEP 2   Find the median (Q_2).

$$5, 6, 12, 13, 15, 18, 22, 50$$
$$\uparrow$$
$$\text{MD}$$

$$\text{MD} = \frac{13 + 15}{2} = 14$$

STEP 3   Find the median of the data values less than 14.

$$5, 6, 12, 13$$
$$\uparrow$$
$$Q_1$$

$$Q_1 = \frac{6 + 12}{2} = 9$$

Q_1 is 9.

STEP 4   Find the median of the data values greater than 14.

$$15, 18, 22, 50$$
$$\uparrow$$
$$Q_3$$

$$Q_3 = \frac{18 + 22}{2} = 20$$

Here Q_3 is 20. Hence, $Q_1 = 9$, $Q_2 = 14$, $Q_3 = 20$.

Deciles divide the distribution into 10 groups as shown. They are denoted by D_1, D_2, etc.

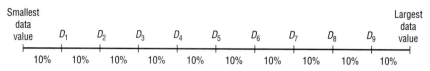

Note that D_1 corresponds to P_{10}; D_2 corresponds to P_{20}, etc. Deciles can be found using the formulas given for percentiles. Taken altogether then, these are the relationships among percentiles, deciles, and quartiles.

Deciles are denoted by $D_1, D_2, D_3, \ldots, D_9$
and they correspond to $P_{10}, P_{20}, P_{30}, \ldots, P_{90}$
Quartiles are denoted by Q_1, Q_2, Q_3
and they correspond to P_{25}, P_{50}, P_{75}
The median is the same as P_{50} or Q_2 or D_5

The position measures are summarized in Table 3–4.

Table 3–4 Summary of Position Measures

Measure	Definition	Symbol(s)
Standard score or z score	Number of standard deviations a data value is above or below the mean	z
Percentile	Position in hundredths a data value is in the distribution	P_n
Decile	Position in tenths a data value is in the distribution	D_n
Quartile	Position in fourths a data value is in the distribution	Q_n

Outliers

A data set should be checked for extremely high or extremely low values. These values are called *outliers*.

An **outlier** is an extremely high or an extremely low data value when compared with the rest of the data values.

There are several ways to check for outliers. One method is shown in the next example.

Example 3–37

Check the following data set for outliers.

5, 6, 12, 13, 15, 18, 22, 50

Solution

The data value 50 is extremely suspect. The steps in checking for an outlier follow.

STEP 1 Find Q_1 and Q_3. This was done in the previous example; Q_1 is 9 and Q_3 is 20.

STEP 2 Find the **interquartile range (IQR),** which is $Q_3 - Q_1$.

$$\text{IQR} = Q_3 - Q_1 = 20 - 9 = 11$$

STEP 3 Multiply this value by 1.5.

$$1.5(11) = 16.5$$

STEP 4 Subtract the value obtained in Step 3 from Q_1 and add the value obtained in Step 3 to Q_3.

$$9 - 16.5 = -7.5 \qquad \text{and} \qquad 20 + 16.5 = 36.5$$

STEP 5   Check the data set for any data values that fall outside the interval from −7.5 to 36.5. The value 50 is outside this interval; hence, it can be considered an outlier.

There are several reasons to check a data set for outliers. First, the data value may have resulted from a measurement or observational error. Perhaps the researcher measured the variable incorrectly. Second, the data value may have resulted from a recording error. That is, it may have been written or typed incorrectly. Third, the data value may have been obtained from a subject that is not in the defined population. For example, suppose test scores were obtained from a seventh-grade class, but a student in that class was actually in the sixth grade and had special permission to attend the class. This student might have scored extremely low on that particular exam on that day. Fourth, the data value might be a legitimate value that occurred by chance (although the probability is extremely small).

There are no hard-and-fast rules on what to do with outliers, nor is there complete agreement among statisticians on ways to identify them. Obviously, if they occurred as a result of an error, an attempt should be made to correct the error or else the data value should be omitted entirely. When they occur naturally by chance, the statistician must make a decision about whether to include them in the data set.

When a distribution is normal or bell-shaped, data values that are beyond three standard deviations of the mean can be considered suspected outliers.

Exercises

3–89. What is a z score?

3–90. Define *percentile rank*.

3–91. What is the difference between a percentage and a percentile?

3–92. Define *quartile*.

3–93. What is the relationship between quartiles and percentiles?

3–94. What is a decile?

3–95. How are deciles related to percentiles?

3–96. To which percentile, quartile, and decile does the median correspond?

3–97. If a history test has a mean of 100 and a standard deviation of 10, find the corresponding z score for each test score.

a. 115   d. 100
b. 124   e. 85
c. 93

3–98. The reaction time to a stimulus for a certain test has a mean of 2.5 seconds and a standard deviation of 0.3 second. Find the corresponding z score for each reaction time.

a. 2.7   d. 3.1
b. 3.9   e. 2.2
c. 2.8

3–99. A final examination for a psychology course has a mean of 84 and a standard deviation of 4. Find the corresponding z score for each raw score.

a. 87   d. 76
b. 79   e. 82
c. 93

3–100. An aptitude test has a mean of 220 and a standard deviation of 10. Find the corresponding z score for each exam score.

a. 200   d. 212
b. 232   e. 225
c. 218

3–101. Which of the following exam grades has a better relative position?

a. A grade of 43 on a test with $\bar{X} = 40$ and $s = 3$.
b. A grade of 75 on a test with $\bar{X} = 72$ and $s = 5$.

3–102. A student scores 60 on a mathematics test that has a mean of 54 and a standard deviation of 3, and she scores 80 on a history test with a mean of 75 and a standard deviation of 2. On which test did she do better than the rest of the class?

3–103. Which score indicates the highest relative position?

a. A score of 3.2 on a test with $\bar{X} = 4.6$ and $s = 1.5$.
b. A score of 630 on a test with $\bar{X} = 800$ and $s = 200$.
c. A score of 43 on a test with $\bar{X} = 50$ and $s = 5$.

3–104. The following distribution represents the data for weights of fifth-grade boys. Find the approximate weights corresponding to each percentile given by constructing a percentile graph.

Weight (pounds)	Frequency
52.5–55.5	9
55.5–58.5	12
58.5–61.5	17
61.5–64.5	22
64.5–67.5	15

a. 25th   c. 80th
b. 60th   d. 95th

3–105. For the data in Exercise 3–104, find the approximate percentile ranks of the following weights.
a. 57 pounds   c. 64 pounds
b. 62 pounds   d. 59 pounds

3–106. (ans) The data below represent the scores on a national achievement test for a group of tenth-grade students. Find the approximate percentile ranks of the following scores by constructing a percentile graph.
a. 220   d. 280
b. 245   e. 300
c. 276

Score	Frequency
196.5–217.5	5
217.5–238.5	17
238.5–259.5	22
259.5–280.5	48
280.5–301.5	22
301.5–322.5	6

3–107. For the data in Exercise 3–106, find the approximate scores that correspond to the following percentiles.
a. 15th   d. 65th
b. 29th   e. 80th
c. 43rd

3–108. (ans) The airborne speeds in miles per hour of 21 planes are shown next. Find the approximate values that correspond to the given percentiles by constructing a percentile graph.

Class	Frequency
366–386	4
387–407	2
408–428	3
429–449	2
450–470	1
471–491	2
492–512	3
513–533	4
	21

a. 9th   d. 60th
b. 20th   e. 75th
c. 45th

3–109. Using the data in Exercise 3–108, find the approximate percentile ranks of the following miles per hour.
a. 380 mph   d. 505 mph
b. 425 mph   e. 525 mph
c. 455 mph

3–110. Find the percentile ranks of each weight in the data set. The weights are in pounds.

78, 82, 86, 88, 92, 97

3–111. In Exercise 3–110, what value corresponds to the 30th percentile?

3–112. Find the percentile rank for each test score in the data set.

12, 28, 35, 42, 47, 49, 50

3–113. In Exercise 3–112, what value corresponds to the 60th percentile?

3–114. Find the percentile rank for each test score in the data set.

5, 12, 15, 16, 20, 21

3–115. What test score in Exercise 3–114 corresponds to the 33rd percentile?

3–116. Using the procedure shown in Example 3–37, check each data set for outliers.
a. 16, 18, 22, 19, 3, 21, 17, 20
b. 24, 32, 54, 31, 16, 18, 19, 14, 17, 20
c. 321, 343, 350, 327, 200
d. 88, 72, 97, 84, 86, 85, 100
e. 145, 119, 122, 118, 125, 116
f. 14, 16, 27, 18, 13, 19, 36, 15, 20

*3–117. Another measure of average is called the *midquartile;* it is the numerical value halfway between Q_1 and Q_3, and the formula is

$$\text{midquartile} = \frac{Q_1 + Q_3}{2}$$

Using this formula and other formulas, find Q_1, Q_2, Q_3, the midquartile, and the interquartile range for each data set.
a. 5, 12, 16, 25, 32, 38
b. 53, 62, 78, 94, 96, 99, 103

Technology Step by Step

MINITAB
Step by Step

Finding the Mean and Standard Deviation

Example MT3–1

1. Type the data from Example 3–39 (in the following section) into C1 of MINITAB. Name the column **CARS-THEFT.**

> 52 58 75 79 57 65 62 77 56 59 51 53 51 66 55
> 68 63 78 50 53 67 65 69 66 69 57 73 72 75 55

2. Select **Stat>Basic Statistics>Display Descriptive Statistics.**
3. The cursor will be blinking in the Variables text box. Double-click C1.
4. Click [OK].

The results will be displayed in the Session Window as shown. The column label "CARS-THEFT" is truncated to 8 letters in the display. The standard deviation is the unbiased estimate, s. The trimmed mean or TrMean is the mean for the data after the lowest and highest 5% are discarded. If the trimmed mean is different from the mean, there may be outliers.

Session Window with
Descriptive Statistics

Session

Descriptive Statistics: CARS-THEFT

Variable	N	Mean	Median	TrMean	StDev	SE Mean
CARS-THE	30	63.20	64.00	63.00	9.01	1.64

Variable	Minimum	Maximum	Q1	Q3
CARS-THE	50.00	79.00	55.00	69.75

TI-83
Step by Step

To calculate various descriptive statistics:
1. Enter data into L_1.
2. Press **STAT** to get the menu.
3. Press ▶ to move cursor to CALC; then press **1** for 1 Var Stats
4. Press **2nd** [L_1] then **ENTER.**

The calculator will display

$\bar{x}$	sample mean
Σx	sum of the data values
Σx^2	sum of the squares of the data values
S_x	sample standard deviation
σ_x	population standard deviation
n	number of data values
min X	smallest data value
Q_1	lower quartile
Med	median
Q_3	upper quartile
max X	largest data value

Example TI3–1

Find the various descriptive statistics for the auto sales data from Example 3–23:

> 11.2   11.9   12.0   12.8   13.4   14.3

Output

```
1-Var Stats
 x̄=12.6
 Σx=75.6
 Σx²=958.94
 Sx=1.1296017
 σx=1.031180553
↓n=6
```

Output

```
1-Var Stats
↑n=6
 minX=11.2
 Q₁=11.9
 Med=12.4
 Q₃=13.4
 maxX=14.3
```

Following the steps above, we obtain the following results, as shown on the screen:

The mean is 12.6.

The sum is 75.6.

The sum of x^2 is 958.94.

The unbiased estimator of the standard deviation S_x is 1.1296017.

The population standard deviation σ_x is 1.031180553.

The sample size n is 6.

The smallest data value is 11.2.

Q_1 is 11.9.

The median is 12.4.

Q_3 is 13.4.

The largest data value is 14.3.

Excel
Step by Step

Finding the Central Tendency

Example XL3–1

To find the mean, mode, and median of a data set:

1. Enter the numbers in a range of cells (here shown as the numbers in cells A2 to A12). We use the data from Example 3–11 on stopping distances:

    15 18 18 18 20 22 24 24 24 26 26

2. For the mean, enter =AVERAGE(A2:A12) in a blank cell.

3. For the mode, enter =MODE(A2:A12) in a blank cell.

4. For the median, enter =MEDIAN(A2:A12) in a blank cell.

These three functions are available from the standard toolbar by clicking the f_x icon and scrolling down the list of statistical functions. *Note:* for distributions that are *bimodal,* like this one, the Excel MODE function reports the first mode only. A better practice is to use the Histogram routine from the Data Analysis Add-In, which reports actual counts in a table.

	A	B	C	D
1	Stopping distance			
2	15		21.36364	mean
3	18		18	mode
4	18		22	median
5	18			
6	20			
7	22			
8	24			
9	24			
10	24			
11	26			
12	26			

Finding Measures of Variation

Example XL3–2

To find values that estimate the spread of a distribution of numbers:

1. Enter the numbers in a range (here **A1:A6**). We use the data from Example 3–23 on European automobile sales.
2. For the sample variance, enter =VAR(A1:A6) in a blank cell.
3. For the sample standard deviation, enter =STDEV(A1:A6) in a blank cell.
4. For the range, you can compute the value =MAX(A1:A6) − MIN(A1:A6).

	A	B	C	D	E
1	11.2		*European auto sales in millions*		
2	11.9				
3	12		1.129602	standard deviation	
4	12.8		1.276	variance	
5	13.4				
6	14.3				
7					

There are also functions STDEVP for population standard deviation and VARP for population variances

---

Descriptive Statistics in Excel

Example XL3–3

Excel's Data Analysis options include an item called Descriptive Statistics that reports all the standard measures of a data set.

1. Enter the data set shown (9 numbers) in column A of a new worksheet.

    12 17 15 16 16 14 18 13 10

2. Select **Tools>Data Analysis**.
3. Use this data (A1:A9) as the Input Range in the Descriptive Statistics dialog box.
4. Check the Summary statistics option, and click [OK].

Descriptive Statistics
Dialog Box

Here's the summary output for this data set. Note that this one operation reports most of the statistics used in this chapter.

Column1	
Mean	14.55555556
Standard Error	0.85165054
Median	15
Mode	16
Standard Deviation	2.554951619
Sample Variance	6.527777778
Kurtosis	-0.3943866
Skewness	-0.51631073
Range	8
Minimum	10
Maximum	18
Sum	131
Count	9
Confidence Level(95.0%)	1.963910937

3–5

Exploratory Data Analysis

Objective 4. Use the techniques of exploratory data analysis, including stem and leaf plots, boxplots, and five-number summaries to discover various aspects of data.

In traditional statistics, data are organized by using a frequency distribution. From this distribution various graphs such as the histogram, frequency polygon, and ogive can be constructed to determine the shape or nature of the distribution. In addition, various statistics such as the mean and standard deviation can be computed to summarize the data.

The purpose of traditional analysis is to confirm various conjectures about the nature of the data. For example, from a carefully designed study, a researcher might want to know if the proportion of Americans who are exercising today has increased from 10 years ago. This study would contain various assumptions about the population, various definitions such as exercise, and so on.

In **exploratory data analysis (EDA),** data are organized using a *stem and leaf plot.* The summary statistics used are the *median* and *interquartile range.* Finally, a *boxplot* can be constructed to determine visually the nature of the distribution. The purpose of exploratory data analysis is to examine data in order to find out what information can be discovered. For example, are there any gaps in the data? Can any patterns be discerned? Here the researcher starts out with few or no assumptions.

Exploratory data analysis was developed by John Tukey and presented in his book *Exploratory Data Analysis* (Addison-Wesley, 1977).

Stem and Leaf Plots

The stem and leaf plot is a method of organizing data and is a combination of sorting and graphing. It has the advantage over grouped frequency distribution of retaining the actual data while showing them in graphic form.

A **stem and leaf plot** is a data plot that uses part of a data value as the stem and part of the data value as the leaf to form groups or classes.

Example 3–38 shows the procedure for constructing a stem and leaf plot.

Example 3–38

At an outpatient testing center, a sample of 20 days showed the following number of cardiograms done each day. Construct a stem and leaf plot for the data.

25	31	20	32	13
14	43	02	57	23
36	32	33	32	44
32	52	44	51	45

Solution

STEP 1   Arrange the data in order:

02, 13, 14, 20, 23, 25, 31, 32, 32, 32,

32, 33, 36, 43, 44, 44, 45, 51, 52, 57

Note: Arranging the data in order is not essential, but it is helpful in construction of the plot.

STEP 2   Separate the data according to the first digit, as shown.

02   13, 14   20, 23, 25   31, 32, 32, 32, 32, 33, 36

43, 44, 44, 45   51, 52, 57

STEP 3   A display can be made by using the leading digit as a *stem* and the trailing digit as the *leaf.* For example, for the value 32, the leading digit, 3, is the stem and the trailing digit, 2, is the leaf. For the value 14, the 1 is the stem and the 4 is the leaf. Now a plot can be constructed as shown in Figure 3–7.

Figure 3–7

Stem and Leaf Plot for Example 3–38

Leading digit (stem)	Trailing digit (leaf)
0	2
1	3 4
2	0 3 5
3	1 2 2 2 2 3 6
4	3 4 4 5
5	1 2 7

0	2
1	3 4
2	0 3 5
3	1 2 2 2 2 3 6
4	3 4 4 5
5	1 2 7

Figure 3–7 shows that the distribution peaks in the center and that there are no gaps in the data. For 7 of the 20 days, the number of patients receiving cardiograms was between 31 and 36. The plot also shows that the testing center treated from a minimum of 2 patients to a maximum of 57 patients in any one day.

Example 3–39

An insurance company researcher conducted a survey on the number of car thefts in a large city for a period of 30 days last summer. The raw data are shown below. Construct a stem and leaf plot by using classes 50–54, 55–59, 60–64, 65–69, 70–74, and 75–79.

52	62	51	50	69
58	77	66	53	57
75	56	55	67	73
79	59	68	65	72
57	51	63	69	75
65	53	78	66	55

Solution

STEP 1 Arrange the data in order.

50, 51, 51, 52, 53, 53, 55, 55, 56, 57, 57, 58, 59, 62, 63,

65, 65, 66, 66, 67, 68, 69, 69, 72, 73, 75, 75, 77, 78, 79

STEP 2 Separate the data according to the classes.

50, 51, 51, 52, 53, 53   55, 55, 56, 57, 57, 58, 59

62, 63   65, 65, 66, 66, 67, 68, 69, 69   72, 73

75, 75, 77, 78, 79

Figure 3–8

Stem and Leaf Plot for
Example 3–39

STEP 3 Plot the data as shown here.

Leading digit (Stem)	Trailing digit (Leaf)
5	0 1 1 2 3 3
5	5 5 6 7 7 8 9
6	2 3
6	5 5 6 6 7 8 9 9
7	2 3
7	5 5 7 8 9

5	0 1 1 2 3 3
5	5 5 6 7 7 8 9
6	2 3
6	5 5 6 6 7 8 9 9
7	2 3
7	5 5 7 8 9

The graph for this plot is shown in Figure 3–8.

When the values are in the hundreds, such as 325, the stem is 32 and the leaf is 5. When data are grouped into classes, such as 20–24, 25–29, and so on, a stem and leaf plot might look like this:

Class	Leading digit (Stem)	Trailing digit (Leaf)	Value
20–24	2	0 1 3 4	20, 21, 23, 24
25–29	2	5 6	25, 26

Of course, the class and value columns would be left out of the final plot.

When analyzing a stem and leaf plot, look for peaks and gaps in the distribution. See if the distribution is symmetrical or skewed. Check the variability of the data by looking at the spread.

Related distributions can be compared using a back-to-back stem and leaf plot. The back-to-back stem and leaf plot uses the same digits for the stems of both distributions, but the digits that are used for the leaves are arranged in order out from the stems on both sides. The next example shows a back-to-back stem and leaf plot.

Example 3–40

The number of stories in two selected samples of tall buildings in Atlanta and Philadelphia are shown. Construct a back-to-back stem and leaf plot and compare the distribution.

	Atlanta					Philadelphia			
55	70	44	36	40	61	40	38	32	30
63	40	44	34	38	58	40	40	25	30
60	47	52	32	32	54	40	36	30	30
50	53	32	28	31	53	39	36	34	33
52	32	34	32	50	50	38	36	39	32
26	29								

Source: *The World Almanac and Book of Facts, 1999.* Copyright 1998. PRIMEDIA INC.

Solution

STEP 1 Arrange the data for both data sets in order.

STEP 2 Construct a stem and leaf plot using the same digits as stems. Place the digits for the leaves for Atlanta on the left side of the stem and the digits for the leaves for Philadelphia on the right side as shown.

Atlanta		Philadelphia
9 8 6	2	5
8 6 4 4 2 2 2 2 2 1	3	0 0 0 0 2 2 3 4 6 6 6 8 8 9 9
7 4 4 0 0	4	0 0 0 0
5 3 2 2 0 0	5	0 3 4 8
3 0	6	1
0	7	

STEP 3 Compare the distributions. The buildings in Atlanta have a larger variation in the number of stories per building. Although both distributions are peaked in the 30- to 39-story class, Philadelphia has more buildings in this class. Atlanta has more buildings that have 40 or more stories than Philadelphia.

Boxplots

When the data set contains a small number of values, a **boxplot** is used to graphically represent the data set. These plots involve five specific values:

1. The lowest value of the data set (i.e., minimum)
2. Q_1
3. The median
4. Q_3
5. The highest value of the data set (i.e., maximum)

These values are called a **five-number summary** of the data set.

Example 3–41

A stockbroker recorded the number of clients she saw each day over an 11-day period. The data are shown below. Construct a boxplot for the data.

33, 38, 43, 30, 29, 40, 51, 27, 42, 23, 31

Solution

STEP 1 Arrange the data in order.

23, 27, 29, 30, 31, 33, 38, 40, 42, 43, 51

STEP 2   Find the median.

$$23, 27, 29, 30, 31, 33, 38, 40, 42, 43, 51$$
$$\uparrow$$
Median

STEP 3   Find Q_1.

$$23, 27, 29, 30, 31$$
$$\uparrow$$
$$29$$

STEP 4   Find Q_3.

$$38, 40, 42, 43, 51$$
$$\uparrow$$
$$42$$

STEP 5   Draw a scale for the data on the x axis.

STEP 6   Locate the lowest value, Q_1, the median, Q_3, and the highest value on the scale.

STEP 7   Draw a box around Q_1 and Q_3, draw a vertical line through the median, and connect the upper and lower values, as shown in Figure 3–9.

Figure 3–9

Boxplot for
Example 3–41

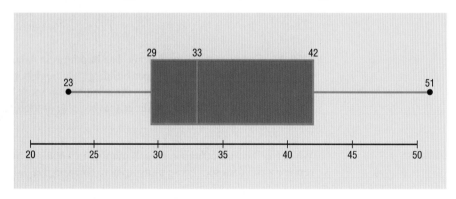

The box in Figure 3–9 represents the middle 50% of the data, and the lines represent the lower and upper ends of the data.

Information Obtained from a Boxplot

1. *a.* If the median is near the center of the box, the distribution is approximately symmetric.
    b. If the median falls to the left of the center of the box, the distribution is positively skewed.
    c. If the median falls to the right of the center, the distribution is negatively skewed.
2. *a.* If the lines are about the same length, the distribution is approximately symmetric.
    b. If the right line is larger than the left line, the distribution is positively skewed.
    c. If the left line is larger than the right line, the distribution is negatively skewed.

The boxplot in Figure 3–9 indicates that the distribution is slightly positively skewed.

Boxplots can be used to compare distributions as shown in the next example.

Example 3–42

Using boxplots, compare the two distributions of the sodium contents of a sample of real cheese and a cheese substitute.

Real cheese				Cheese substitute			
310	420	45	40	270	180	250	290
220	240	180	90	130	260	340	310

Source: *The Complete Book of Food Counts,* Corrine T. Netzer, Dell Publishing, New York, 1997.

Solution

STEP 1   Find Q_1, MD, and Q_3 for the real cheese data.

$$40 \quad 45 \quad 90 \quad 180 \quad 220 \quad 240 \quad 310 \quad 420$$
$$ \uparrow \uparrow \uparrow$$
$$ Q_1 MD Q_3$$

$$Q_1 = \frac{45 + 90}{2} = 67.5 \qquad MD = \frac{180 + 220}{2} = 200$$

$$Q_3 = \frac{240 + 310}{2} = 275$$

STEP 2   Find Q_1, MD, and Q_3 for the cheese substitute data.

$$130 \quad 180 \quad 250 \quad 260 \quad 270 \quad 290 \quad 310 \quad 340$$
$$ \uparrow \uparrow \uparrow$$
$$ Q_1 MD Q_3$$

$$Q_1 = \frac{180 + 250}{2} = 215 \qquad MD = \frac{260 + 270}{2} = 265$$

$$Q_3 = \frac{290 + 310}{2} = 300$$

STEP 3   Draw the boxplots for each distribution on the same graph. See Figure 3–10.

Figure 3–10

Boxplots for Example 3–42

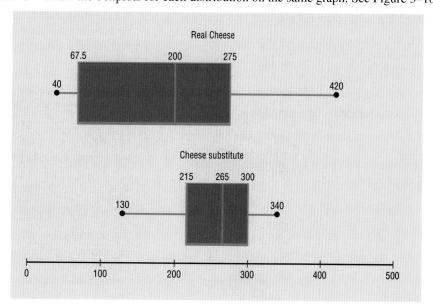

STEP 4 Compare the plots. It is quite apparent that the distribution for the cheese substitute data has a higher median than the median for the distribution for the real cheese data. The variation or spread for the distribution of the real cheese data is larger than the variation for the distribution of the cheese substitute data.

In exploratory data analysis, *hinges* are used instead of quartiles to construct box-plots. When the data set consists of an even number of values, hinges are the same as quartiles. Hinges for a data set with an odd number of values differ somewhat from quartiles. However, since most calculators and computer programs use quartiles, they will be used in this textbook.

Another important point to remember is that the summary statistics (median and interquartile range) used in exploratory data analysis are said to be resistant statistics. A **resistant statistic** is relatively less affected by outliers than a *nonresistant statistic*. The mean and standard deviation are nonresistant statistics. Sometimes when a distribution is skewed or contains outliers, the median and interquartile range may more accurately summarize the data than the mean and standard deviation, since the mean and standard deviation are more affected in this case. Table 3–5 compares the traditional versus the exploratory data analysis approach.

Table 3–5 Traditional versus EDA Techniques

Traditional	Exploratory data analysis
Frequency distribution	Stem and leaf plot
Histogram	Boxplot
Mean	Median
Standard deviation	Interquartile range

Exercises

3–118. What is the purpose of exploratory data analysis?

3–119. What four statistical techniques are used in exploratory data analysis?

3–120. What five statistics are used in the five-number summaries?

3–121. What is meant by resistant statistics?

3–122. Twenty-nine executives reported the following number of telephone calls made during a randomly selected week. Construct a stem and leaf plot for the data and analyze the results.

22	14	12	9	54	12
16	12	14	49	10	14
8	21	37	28	36	22
9	33	58	31	41	19
3	18	25	28	52	

3–123. The National Insurance Crime Bureau reported that the following data represent the number of registered vehicles per car stolen for 35 selected cities in the United States. For example, in Miami, one automobile is stolen for every 38 registered vehicles in the City. Construct a stem and leaf plot for the data and analyze the distribution. (The data have been rounded to the nearest whole number.)

38	53	53	56	69	89	94
41	58	68	66	69	89	52
50	70	83	81	80	90	74
50	70	83	59	75	78	73
92	84	87	84	85	84	89

Source: *USA Today,* March 22, 1999.

3–124. The growth (in centimeters) of two varieties of plant after 20 days is shown below. Construct a back-to-back stem and leaf plot for the data and compare the distributions.

Variety 1				Variety 2			
20	12	39	38	18	45	62	59
41	43	51	52	53	25	13	57
59	55	53	59	42	55	56	38
50	58	35	38	41	36	50	62
23	32	43	53	45	55		

3–125. The data shown represent the percentage of unemployed males and females in 1995 for a sample of countries of the world. Using the whole numbers as stems and the decimals as leaves, construct a back-to-back stem and leaf plot and compare the distributions of the two groups.

Females					Males				
8.0	3.7	8.6	5.0	7.0	8.8	1.9	5.6	4.6	1.5
3.3	8.6	3.2	8.8	6.8	2.2	5.6	3.1	5.9	6.6
9.2	5.9	7.2	4.6	5.6	9.8	8.7	6.0	5.2	5.6
5.3	7.7	8.0	8.7	0.5	4.4	9.6	6.6	6.0	0.3
6.5	3.4	3.0	9.4		4.6	3.1	4.1	7.7	

Source: *The TIME Almanac, 1999 edition,* 162.

3–126. Shown next are the number of farms in six counties in western Pennsylvania. Construct a boxplot for the data and comment on the shape of the distribution.

338, 767, 633, 747, 1369, 1139

Source: *Pittsburgh Tribune-Review,* August 28, 1994.

3–127. Shown next are the sizes of the police forces in the 10 largest cities in the United States in 1993 (the numbers represent hundreds). Construct a boxplot for the data and comment on the shape of the distribution.

29.3, 7.6, 12.1, 4.7, 6.2, 1.9, 3.9, 2.8, 2.0, 1.7

Source: *USA Today,* February 17, 1995.

3–128. Construct a boxplot for the number of calculators sold during a randomly selected week. Comment on the shape of the distribution.

8, 12, 23, 5, 9, 15, 3

3–129. The following data represent the divorce rates per 1000 married women for a sample of countries of the world for the years 1970 and 1993. Construct a boxplot for the data for each year and compare the distributions.

1970	6	8	5	4	3	7	5	15
1993	11	13	7	6	9	13	13	21

Source: *The TIME Almanac, 1999 edition,* 158.

3–130. The following data represent the volumes in cubic yards of the largest dams in the United States and in South America. Construct a boxplot of the data for each region and compare the distributions.

USA	South America
125,628	311,539
92,000	274,026
78,008	105,944
77,700	102,014
66,500	56,242
62,850	46,563
52,435	
50,000	

Source: *The TIME Almanac, 1999 edition,* 480.

Technology Step by Step

Minitab
Step by Step

Constructing a Stem and Leaf Plot

Example MT3–2

1. Enter the data for Example 3–39. Label the column **CARS-THEFT.**
2. Select **STAT>EDA>Stem-and-Leaf.**
3. Double-click CARS-THEFT
4. Click in the Increment text box and enter the class width of **5.**
5. Click [OK].

This character graph will be displayed in the Session Window.

Stem and Leaf Dialog Box and Plot

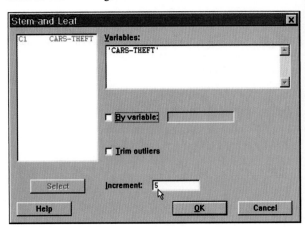

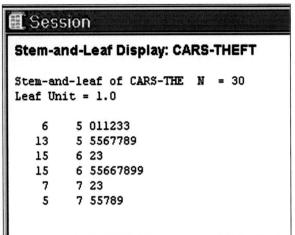

Constructing a Boxplot

Example MT3–3

1. Enter the data for Example 3–39. Label the column **CARS-THEFT.**
2. Select **Stat>EDA>Box Plot.**
3. Double click CARS-THEFT to select it for the Y variable.
4. Click on the drop down arrow for Annotation.
5. Click on Title then enter an appropriate title such as **"Car Thefts for a Large City, U.S.A."**

Boxplot Dialog Box and Boxplot

6. Click [OK] twice.

The graph will be displayed in a graph window.

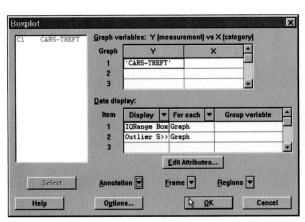

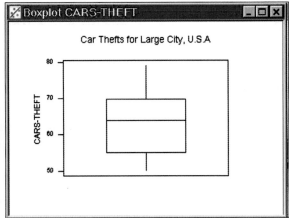

TI-83
Step by Step

Constructing a Boxplot

To draw a boxplot:

1. Enter data into L_1.
2. Change values in WINDOW menu, if necessary. (*Note:* Make X_{min} somewhat smaller than the smallest data value and X_{max} somewhat larger than the largest data value.) Change Y_{min} to 0 and Y_{max} to 10.

3. Press **[2nd] [STAT PLOT]** then **1** for `Plot 1`.
4. Press **ENTER** to turn `Plot 1` on.
5. Move cursor to `Boxplot` symbol on the `Type:` line then press **ENTER.**
6. Move cursor to `Xlist: L₁` line, and press **ENTER.**
7. Move cursor to `1 Freq:` line and press **ENTER.**
8. Press **GRAPH** to display the boxplot.
9. Press **TRACE** followed by ◄ or ▶ to obtain the values or the boxplot.

To display two boxplots on the same display, follow the above steps and use the `2: Plot 2` and L_2 symbols.

Example TI3–2

Construct a boxplot for the data values in Example 3–41:

33 38 43 30 29 40 51 27 42 23 31

Change the settings as follows:

$$X_{min} = 20$$
$$X_{max} = 55$$
$$X_{scl} = 5$$
$$Y_{min} = 0$$
$$Y_{max} = 1$$
$$Y_{scl} = 1$$

Input

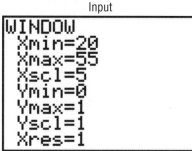

Input

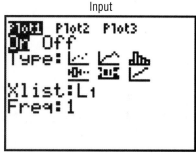

Using the **TRACE** key along with the ◄ and ▶ keys we obtain the five number summary. The minimum value is 23.

Q_1 is 29.

The median is 33.

Q_3 is 42.

The maximum value is 51.

Output

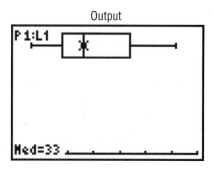

Excel
Step by Step

"Do-It-Yourself" Boxplots in Excel

Excel's built-in charts do not offer boxplots directly, but one type of Excel chart gives you a plot that is similar to a boxplot.

1. Enter this 10-number data set in column A:

    12 17 15 16 16 14 18 13 10 16

2. In a blank cell enter =QUARTILE(A1:A10,1).
3. In the next cell under this one, enter =MAX(A1:A10).
4. In the next cell under this one, enter =MIN(A1:A10).
5. In a blank cell enter =QUARTILE(A1:A10,3).
6. Drag-select these four cells.
7. Select Insert, Chart, and pick the Stock chart type (the second option).
8. Click Finish.

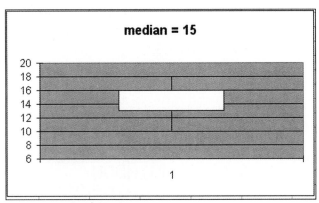

Because this chart type only plots four quantities, it does not include a value for median. In the plot displayed here, the median is included in the chart's title.

3-6

Summary

This chapter explains the basic ways to summarize data. These include measures of central tendency, measures of variation or dispersion, and measures of position. The three most commonly used measures of central tendency are the mean, median, and mode. The midrange is also used occasionally to represent an average. The three most commonly used measurements of variation are the range, variance, and standard deviation.

The most common measures of position are percentiles, quartiles, and deciles. This chapter explains how data values are distributed according to Chebyshev's theorem and the empirical rule. The coefficient of variation is used to describe the standard deviation in relationship to the mean. These methods are commonly called traditional statistical methods and are primarily used to confirm various conjectures about the nature of the data.

Other methods, such as the stem and leaf plot, the box plot, and five-number summaries, are part of exploratory data analysis; they are used to examine data to see what they reveal.

After learning the techniques presented in Chapter 2 and this chapter, students will have a substantial knowledge of descriptive statistics. That is, they will be able to collect, organize, summarize, and present data.

Important Terms

boxplot   133

Chebyshev's theorem   109

coefficient of variation   108

data array   86

decile   123

empirical rule   111

exploratory data analysis   130

five-number summary   133

interquartile range   124

mean   83

median   86

midrange   92

modal class   89

mode   89

negatively skewed distribution   95

outlier   124

parameter   83

percentile   116

positively skewed distribution   94

quartile   123

range   100

resistant statistic   136

standard deviation   103

statistic   83

stem and leaf plot   130

symmetrical distribution   94

unbiased estimator   104

variance   103

weighted mean   93

z score or standard score   115

Important Formulas

Formula for the mean for individual data:

$$\bar{X} = \frac{\Sigma X}{n} \qquad \mu = \frac{\Sigma X}{N}$$

Formula for the mean for grouped data:

$$\bar{X} = \frac{\Sigma f \cdot X_m}{n}$$

Formula for the weighted mean:

$$\bar{X} = \frac{\Sigma wX}{\Sigma w}$$

Formula for the midrange:

$$\text{MR} = \frac{\text{lowest value} + \text{highest value}}{2}$$

Formula for the range:

$$R = \text{highest value} - \text{lowest value}$$

Formula for the variance for population data:

$$\sigma^2 = \frac{\Sigma (X - \mu)^2}{N}$$

Formula for the variance for sample data (shortcut formula for the unbiased estimator):

$$s^2 = \frac{\Sigma X^2 - [(\Sigma X)^2/n]}{n - 1}$$

Formula for the variance for grouped data:

$$s^2 = \frac{\Sigma f \cdot X_m^2 - [(\Sigma f \cdot X_m)^2/n]}{n - 1}$$

Formula for the standard deviation for population data:

$$\sigma = \sqrt{\frac{\Sigma (X - \mu)^2}{N}}$$

Formula for the standard deviation for sample data (shortcut formula):

$$s = \sqrt{\frac{\Sigma X^2 - [(\Sigma X)^2/n]}{n - 1}}$$

Formula for the standard deviation for grouped data:

$$s = \sqrt{\frac{\Sigma f \cdot X_m^2 - [(\Sigma f \cdot X_m)^2/n]}{n - 1}}$$

Formula for the coefficient of variation:

$$\text{CVar} = \frac{s}{\bar{X}} \cdot 100\% \quad \text{or} \quad \text{CVar} = \frac{\sigma}{\mu} \cdot 100\%$$

Formula for Chebyshev's theorem: The proportion of values from a data set that will fall within k standard deviations of the mean will be at least

$$1 - \frac{1}{k^2}$$

where k is a number greater than 1

Formula for the z score (standard score):

$$z = \frac{X - \mu}{\sigma} \quad \text{or} \quad z = \frac{X - \bar{X}}{s}$$

Formula for the cumulative percentage:

$$\text{cumulative } \% = \frac{\text{cumulative frequency}}{n} \cdot 100\%$$

Formula for the percentile rank of a value X:

$$\text{percentile} = \frac{\left(\begin{array}{c}\text{number of values} \\ \text{below } X + 0.5\end{array}\right)}{\begin{array}{c}\text{total number} \\ \text{of values}\end{array}} \cdot 100\%$$

Formula for finding a value corresponding to a given percentile:

$$c = \frac{n \cdot p}{100}$$

Formula for interquartile range:

$$\text{IQR} = Q_3 - Q_1$$

Review   Exercises

3–131. The following data represent the number of deer killed by motor vehicles during 1994 for eight counties in southwestern Pennsylvania.

2343, 1240, 1088, 600, 497, 1925, 1480, 458

Source: *Pittsburgh Post-Gazette,* June 6, 1995, p. 1.

Find each of the following.
a. Mean   e. Range
b. Median   f. Variance
c. Mode   g. Standard deviation
d. Midrange

3–132. The following data represent the area in square miles of major islands in the Caribbean Sea and the Mediterranean Sea.

Caribbean Sea			Mediterranean Sea	
108	926	436	1,927	1,411
75	100	3,339	229	95
5,382	171	116	3,189	540
2,300	290	1,864	3,572	9,301
166	687	59	86	9,926
42,804	4,244	134		
29,389				

Source: *The World Almanac and Book of Facts, 1999 edition,* 454.

Find each of the following.
a. Mean   e. Range
b. Median   f. Variance
c. Mode   g. Standard deviation
d. Midrange

Are the averages and variations of the areas approximately equal?

3–133. Twelve batteries were tested to see how many hours they would last. The frequency distribution is shown below.

Hours	Frequency
1–3	1
4–6	4
7–9	5
10–12	1
13–15	1

Find each of the following.
a. Mean   c. Variance
b. Modal class   d. Standard deviation

3–134. A survey of 40 clothing stores reported the following number of sales held during randomly selected year.

Number of sales	Frequency
11–15	3
16–20	5
21–25	12
26–30	9
31–35	8
36–40	3

Find each of the following.
a. Mean   c. Variance
b. Modal class   d. Standard deviation

3–135. Shown below is a frequency distribution for the rise in tides at 30 selected locations in the United States.

Rise in tides (inches)	Frequency
12.5–27.5	6
27.5–42.5	3
42.5–57.5	5
57.5–72.5	8
72.5–87.5	6
87.5–102.5	2

Find each of the following.
a. Mean   c. Variance
b. Modal class   d. Standard deviation

3–136. The fuel capacity in gallons of 50 randomly selected 1995 cars is shown below.

Class	Frequency
10–12	6
13–15	4
16–18	14
19–21	15
22–24	8
25–27	2
28–30	1
	50

Source: *Consumer Reports,* April 1995.

Find each of the following.
a. Mean   c. Variance
b. Modal class   d. Standard deviation

3–137. In a dental survey of third-grade students, the following distribution was obtained for the number of cavities found. Find the average number of cavities for the class. Use the weighted mean.

Number of students	Number of cavities
12	0
8	1
5	2
5	3

3–138. An investor calculated the following percentages of each of three stock investments with payoffs as shown. Find the average payoff. Use the weighted mean.

Stock	Percent	Payoff
A	30%	$10,000
B	50%	$ 3,000
C	20%	$ 1,000

3–139. In an advertisement, a transmission service center stated that the average years of service of its employees was 13. The distribution is shown below. Using the weighted mean, calculate the correct average.

Number of employees	Years of service
8	3
1	6
1	30

3–140. The average number of textbooks in professors' offices is 16, and the standard deviation is 5. The average age of the professors is 43, with a standard deviation of 8. Which data set is more variable?

3–141. A survey of bookstores showed that the average number of magazines carried is 56, with a standard deviation of 12. The same survey showed that the average length of time each store had been in business was 6 years, with a standard deviation of 2.5 years. Which is more variable, the number of magazines or the number of years?

3–142. The number of previous jobs held by each of six applicants is shown here.

2, 4, 5, 6, 8, 9

a. Find the percentile for each value.
b. What value corresponds to the 30th percentile?
c. Construct a boxplot and comment on the nature of the distribution.

3–143. The weights in pounds of 30 one-room air conditioners are shown below.

Class	Frequency
46–54	3
55–63	12
64–72	10
73–81	3
82–90	1
91–99	0
100–108	1
	30

Source: *Consumer Reports,* June 1995.

a. Construct a percentile graph.
b. Find the values that correspond to the 35th, 65th, and 85th percentiles.
c. Find the percentile of values 48, 54, and 62.

3–144. Check each data set for outliers.
a. 506, 511, 517, 514, 400, 521
b. 3, 7, 9, 6, 8, 10, 14, 16, 20, 12
c. 14, 18, 27, 26, 19, 13, 5, 25
d. 112, 157, 192, 116, 153, 129, 131

3–145. A survey of car rental agencies shows that the average cost of a car rental is $0.32 per mile. The standard deviation is $0.03. Using Chebyshev's theorem, find the range in which at least 75% of the data values will fall.

3–146. The average cost of a certain type of seed per acre is $42. The standard deviation is $3. Using Chebyshev's theorem, find the range in which at least 88.89% of the data values will fall.

3–147. The average labor charge for automobile mechanics is $54 per hour. The standard deviation is $4. Find the minimum percentage of data values that will fall within the range of $48 to $60. Use Chebyshev's theorem.

3–148. For a certain type of job, it costs a company an average of $231 to train an employee to perform the task. The standard deviation is $5. Find the minimum percentage of data values that will fall in the range of $219 to $243. Use Chebyshev's theorem.

3–149. The average delivery charge for a refrigerator is $32. The standard deviation is $4. Find the minimum percentage of data values that will fall in the range of $20 to $44. Use Chebyshev's theorem.

3–150. Which of the following exam grades has a better relative position?
a. A grade of 37 on a test with $\overline{X} = 42$ and $s = 5$
b. A grade of 72 on a test with $\overline{X} = 80$ and $s = 6$

3–151. The number of visitors to the Railroad Museum during 24 randomly selected hours is shown here. Construct a stem and leaf plot for the data.

67	62	38	73	34	43	72	35
53	55	58	63	47	42	51	62
32	29	47	62	29	38	36	41

3–152. The data set shown below represents the number of hours 25 part-time employees worked at the Sea Side Amusement Park during a randomly selected week in June. Construct a stem and leaf plot for the data and summarize the results.

16	25	18	39	25	17	29	14	37
22	18	12	23	32	35	24	26	
20	19	25	26	38	38	33	29	

3–153. A special aptitude test is given to job applicants. The data shown below represent the scores of 30

applicants. Construct a stem and leaf plot for the data and summarize the results.

204	210	227	218	254
256	238	242	253	227
251	243	233	251	241
237	247	211	222	231
218	212	217	227	209
260	230	228	242	200

3–154. The data shown here represent the number of hours 12 part-time employees at a toy store worked during the week before and after Christmas. Construct two boxplots and compare the distributions.

Before	38	16	18	24	12	30	35	32	31	30	24	35
After	26	15	12	18	24	32	14	18	16	18	22	12

Statistics Today

Who Is a Typical First-Time Home Buyer?

The *USA Today* Snapshot shown below uses several measures of average and two percentages (63% and 82%). Although it is not possible to know which measures were used, one could make an educated guess. For example, the mean was probably used for family size (2.7 people) since it is in decimal form. The age of the buyer (32) is probably a median. The time (5 months) it took to find a home is probably a mode.

The information presented in this Snapshot is very vague. After mastering the concepts in this chapter, you should be able to present a more accurate profile if given the raw data.

Source: *USA Today,* February 22, 1995. Copyright 1995, USA TODAY. Reprinted with permission.

Data Projects

The Data Bank is Appendix D in this book, or on the World Wide Web by following links from www.mhhe.com/math/stat/bluman/.

1. From the Data Bank, choose one of the following variables: age, weight, cholesterol level, systolic pressure, IQ, or sodium level. Select at least 30 values and find the mean, median, mode, and midrange. State which measurement of central tendency best describes the average and why.

2. Find the range, variance, and standard deviation for the data selected in Exercise 1.

3. From the Data Bank, choose 10 values from any variable, construct a boxplot, and interpret the results.

4. Using the data in Data Set V in Appendix D, construct a stem and leaf plot for the life expectancies in the African and Asian countries. Compare the distributions.

5. Using the data from Data Set V in Appendix D, construct a boxplot and a five-number summary for the life expectancies in the former Soviet Union countries. Comment on the nature of the data.

6. Using the data from Data Set V in Appendix D, find the mean, median, mode, range, variance, and standard deviation of the life expectancies of the people in the European countries and the Latin American Countries. Compare the averages and variability of the data.

Chapter Quiz

Determine whether each statement is true or false. If the statement is false, explain why.

1. When the mean is computed for individual data, all values in the data set are used.

2. The mean cannot be found for grouped data when there is an open class.

3. A single extremely large value can affect the median more than the mean.

4. Half of all the data values will fall above the mode and half will fall below the mode.

5. In a data set, the mode will always be unique.

6. The range and midrange are both measures of variation.

7. One disadvantage of the median is that it is not unique.

8. The mode and midrange are both measures of variation.

9. If a person's score on an exam corresponds to the 75th percentile, then that person obtained 75 correct answers out of 100 questions.

Select the best answer.

10. What is the value of the mode when all values in the data set are different?
    a. Zero
    b. One
    c. There is no mode.
    d. It cannot be determined unless the data values are given.

11. When data are categorized as, for example, places of residence (rural, suburban, urban), the most appropriate measure of central tendency is
    a. Mean   c. Mode
    b. Median   d. Midrange

12. P_{50} corresponds to
    a. Q_2 c. IQR
    b. D_5 d. Midrange

13. Which is not part of the five-number summary?
    a. Q_1 and Q_3
    b. The mean
    c. The median
    d. The smallest and the largest data values

14. A statistic that tells the number of standard deviations a data value is above or below the mean is called
    a. A quartile
    b. A percentile
    c. A coefficient of variation
    d. A z score

15. When a distribution is bell-shaped, approximately what percentage of data values will fall within one standard deviation of the mean?
    a. 50%   c. 95%
    b. 68%   d. 99.7%

Complete the following statements with the best answer.

16. A measure obtained from sample data is called a _____.

17. Generally, Greek letters are used to represent _____, and Roman letters are used to represent _____.

18. The positive square root of the variance is called the _____.

19. The symbol for the population standard deviation is _____.

20. When the sum of the lowest data value plus the highest data value is divided by 2, the measure is called _____.

21. If the mode is to the left of the median and the mean is to the right of the median, then the distribution is _____ skewed.

22. An extremely high or extremely low data value is called an _____ .

23. The following temperatures were recorded in Pasadena for a week in April.

$$87, 85, 80, 78, 83, 86, 90$$

Find each of the following.

a. Mean e. Range
b. Median f. Variance
c. Mode g. Standard deviation
d. Midrange

24. The distribution of the number of errors 10 students made on a typing test is shown.

Errors	Frequency
0–2	1
3–5	3
6–8	4
9–11	1
12–14	1

Find each of the following.

a. Mean c. Variance
b. Modal class d. Standard deviation

25. Shown here is a frequency distribution for the number of inches of rain received in one year in 25 selected cities in the United States.

Number of inches	Frequency
5.5–20.5	2
20.5–35.5	3
35.5–50.5	8
50.5–65.5	6
65.5–80.5	3
80.5–95.5	3

Find each of the following.

a. Mean c. Variance
b. Modal class d. Standard deviation

26. A survey of 36 selected recording companies showed the following number of days it took to receive a shipment from the day it was ordered.

Days	Frequency
1–3	6
4–6	8
7–9	10
10–12	7
13–15	0
16–18	5

Find each of the following.

a. Mean c. Variance
b. Modal class d. Standard deviation

27. In a survey of third-grade students, the following distribution was obtained for the number of "best friends" each had.

Number of students	Number of best friends
8	1
6	2
5	3
3	0

Find the average number of best friends for the class. Use the weighted mean.

28. In an advertisement, a retail store stated that its employees averaged nine years of service. The distribution is shown here.

Number of employees	Years of service
8	2
2	6
3	10

Using the weighted mean, calculate the correct average.

29. The average number of newspapers for sale in an airport newsstand is 12, and the standard deviation is 4. The average age of the pilots is 37, with a standard deviation of 6. Which data set is more variable?

30. A survey of grocery stores showed that the average number of brands of toothpaste carried was 16, with a standard deviation of 5. The same survey showed the average length of time each store was in business was 7 years, with a standard deviation of 1.6 years. Which is more variable, the number of brands or the number of years?

31. A student scored 76 on a general science test where the class mean and standard deviation were 82 and 8, respectively; he also scored 53 on a psychology test where the class mean and standard deviation were 58 and 3, respectively. In which class was his relative position higher?

32. Which score has the highest relative position?

a. $X = 12$ $\bar{X} = 10$ $s = 4$
b. $X = 170$ $\bar{X} = 120$ $s = 32$
c. $X = 180$ $\bar{X} = 60$ $s = 8$

33. The number of credits in business courses eight job applicants had is shown here.

$$9, 12, 15, 27, 33, 45, 63, 72$$

a. Find the percentile for each value.
b. What value corresponds to the 40th percentile?
c. Construct a boxplot and comment on the nature of the distribution.

34. On a philosophy comprehensive exam, the following distribution was obtained from 25 students.

Score	Frequency
40.5–45.5	3
45.5–50.5	8
50.5–55.5	10
55.5–60.5	3
60.5–65.5	1

a. Construct a percentile graph.
b. Find the values that correspond to the 22nd, 78th, and 99th percentiles.
c. Find the percentile of the values 52, 43, and 64.

35. The number of visitors to the Historic Museum for 25 randomly selected hours is shown below. Construct a stem and leaf plot for the data.

15	53	48	19	38
86	63	98	79	38
62	89	67	39	26
28	35	54	88	76
31	47	53	41	68

36. The following data represent the attendance at seven Pittsburgh museums for 1997 and 1998. Draw two boxplots for the data and compare the distributions. The data are in thousands.

1997	1998
101	107
754	764
291	293
83	110
589	627
152	103
158	154

Source: *The Pittsburgh Tribune Review,* February 21, 1999.

Critical Thinking Challenges

1. Consider the following problem: Mary and Bill play basketball. In the first game, Mary had a foul shot average of 0.60 (She made 12 of 20 shots), and Bill had a foul shot average of 0.50 (He made 5 of 10 shots). In the second game, Mary had a foul shot average of 0.80 (She made 8 out of 10 shots), and Bill had a foul shot average of 0.75 (He made 15 out of 20 shots). Now, who do you think has the best overall average? The answer may surprise you. Hint: Compute the averages for both games based on 30 shots for each player.

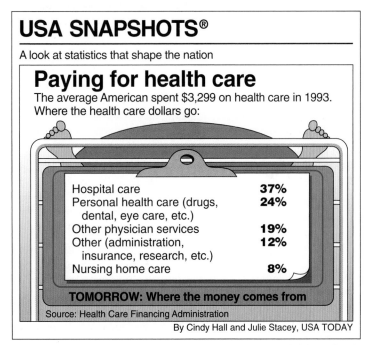

Source: *USA Today,* January 12, 1995. Copyright © 1995, USA TODAY. Reprinted with permission.

2. A *USA Today* Snapshot, on the previous page, states that, "The average American spent $3,299 on health care in 1993."
 a. Explain what is meant by an "average American."
 b. Why would it be better to state "Americans spent an average of $3,299 on health care in 1993"?
 c. How might this average have been derived?
 d. Is it representative of what you spent on health care? Explain.

3. The following graph indicates the number of hours students spent studying during four years of high school. Use it to answer the following questions.
 a. Which average do you think was used (mean, median, or mode)? Why do you think this average was used?
 b. If a mean was used, how would it be calculated?
 c. Can you estimate the average number of hours per day the students studied? (*Note:* You would not use 1 year = 365 days, since school is not in session 365 days per year.)
 d. Do you think the length of the school year in each country is the same? How could you find out?

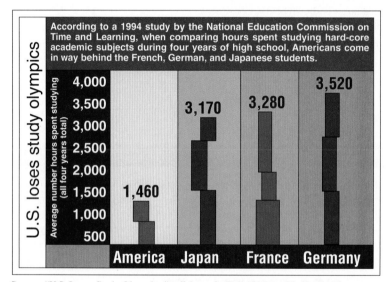

Source: "U.S. Loses Study Olympics," *tell* 4, no. 3, (Fall 1994), p. 23. Used with permission.

 e. Do you think the length of the school year would affect the results of the study? Explain your answer.

 f. In your own words, write a brief summary of the study results.

Data Projects

Where appropriate, use MINITAB, the TI-83, or a computer program of your choice to complete the following exercises.

1. Select a variable and collect about 10 values for two groups. (For example, you may want to ask 10 men how many cups of coffee they drink per day and 10 women the same question.)
 a. Define the variable.
 b. Define the populations.
 c. Describe how the samples were selected.
 d. Write a paragraph describing the similarities and differences between the two groups, using appropriate descriptive statistics such as means, standard deviations, and so on.

2. Collect data consisting of at least 30 values.
 a. State the purpose of the project.
 b. Define the population.
 c. State how the sample was selected.
 d. Using appropriate descriptive statistics, write a paragraph summarizing the data.

You may use the following websites to obtain raw data:

 http://www.mhhe.com/math/stat/bluman
 http://lib.stat.cmu.edu/DASL
 http://www.oecd.org/statlist.htm
 http://www.statcan.ca/english/

chapter

4

Counting Techniques

Outline

Objectives

After completing this chapter, you should be able to

1. Determine the number of outcomes of a sequence of events using a tree diagram.

2. Find the total number of outcomes in a sequence of events using the multiplication rule.

3. Find the number of ways r objects can be selected from n objects using the permutation rule.

4. Find the number of ways r objects can be selected from n objects without regard to order using the combination rule.

Why Are We Running Out of 800 Numbers?

Phone companies and other agencies that deal in numbers need to know how many phone numbers, ID tags, or license plates they can issue using certain combinations of various letters and numbers. The article shown below explains that the phone companies are running out of toll-free 800 numbers. The question is: How many phone numbers with the 800 prefix can be issued in the United States?

Toll-free call? Get ready to dial 888

By Becky Beyers
USA TODAY

Get ready to keep your finger on the 8 when you make a toll-free call.

Phone companies will run out of 800 numbers early next year and start issuing toll-free numbers beginning with 888.

Use of 800 numbers has grown so fast, "we're a victim of our own success," says Dennis Byrne of the U.S. Telephone Association.

Only about 1.7 million of the 7.6 million possible 800-prefix combinations are still available. Why so few are left:

▶ Demand has taken off since May 1993, when the government allowed users to keep 800 numbers if they changed long-distance carriers.

▶ 800 numbers aren't just for big companies anymore. Small businesses use them, as do residential customers so family members can call home more cheaply than collect.

Such customers may pay 25 cents a minute for each call plus a monthly fee of $5.

▶ Some toll-free numbers are hoarded for promotional value or occasional use.

The industry's numbering council — phone companies and associations that set phone-number policies — is asking that little-used numbers be returned so they can be reissued.

Setting up a new toll-free access code involves the entire phone industry, Byrne says.

All internal systems must upgrade switching equipment so they can handle 888 calls.

What happens when the 888s are used up? It's on to 877, 866, and all the way down to 822.

Source: *USA Today,* February 13, 1995. Copyright 1995 *USA TODAY.* Used with permission.

In this chapter, you will learn the rule for counting, the differences between permutations and combinations, and how to figure out how many different combinations for specific situations exist.

4–1

Introduction

Many problems in probability and statistics require a careful analysis of the outcomes of a sequence of events. A sequence of events occurs when one or more events follow one another. For example, a sales representative may wish to select the most efficient way to visit several different stores in four cities.

Sometimes, in order to determine costs or rates, a manager must know all possible outcomes of a classification scheme. Or an insurance company may wish to classify its drivers according to the following classes:

1. Gender (male, female).
2. Age (under 25, between 25 and 60, over 60).
3. Area of residence (rural, suburban, urban).
4. Distance driven to work (under 4 miles, between 4 and 10 miles, over 10 miles).
5. Value of the vehicle (under $5000, between $5000 and $10,000, over $10,000).

We can think of falling into one of these classes or categories as an "event." Then the particular set of categories (female, under 25, suburban, etc.) that a person is in is like a sequence of events.

In a psychological study, a researcher might attempt to train a rat to run a maze. To determine the rat's success, the researcher must know the number of possible choices the rat can make to traverse the maze. Then the researcher can differentiate between actual learning and chance successes.

On a game show, a contestant might be required to arrange five digits correctly to guess the exact price of a new car. To determine the probability of his guessing the correct answer, one must know the number of possible ways the five digits can be arranged.

The vice president of a company might wish to know the number of different possible ways four employees can be selected from a group of 10 in order to be transferred to a new location.

Sometimes the total number of possible outcomes is enough; other times a list of all outcomes is needed. One can use several methods of counting here: the multiplication rule, the permutation rule, and the combination rule.

4–2

Tree Diagrams and the Multiplication Rule for Counting

Tree Diagrams

Objective 1. Determine the number of outcomes of a sequence of events using a tree diagram.

Many times one wishes to list each possibility of a sequence of events. For example, it would be difficult to list all possible outcomes of the options available on a new automobile by guessing alone. Rather than do this listing in a haphazard way, one can use a tree diagram.

A **tree diagram** is a device used to list all possibilities of a sequence of events in a systematic way.

Tree diagrams are also useful in determining the probabilities of events, as will be shown in the next chapter.

Example 4–1

Suppose a sales rep can travel from New York to Pittsburgh by plane, train, or bus, and from Pittsburgh to Cincinnati by bus, boat, or automobile. List all possible ways he can travel from New York to Cincinnati.

Solution

A tree diagram can be drawn to show the possible ways. First, the salesman can travel from New York to Pittsburgh by three methods. The tree diagram for this situation is shown in Figure 4–1.

Figure 4–1

Tree Diagram for New York–Pittsburgh Trips in Example 4–1

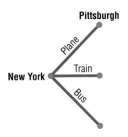

Then the salesman can travel from Pittsburgh to Cincinnati by bus, boat, or automobile. This tree diagram is shown in Figure 4–2.

Figure 4–2

Tree Diagram for Pittsburgh–Cincinnati Trips in Example 4–1

Next, the second branch is paired up with the first branch in three ways, as shown in Figure 4–3.

Figure 4–3

Complete Tree Diagram for Example 4–1

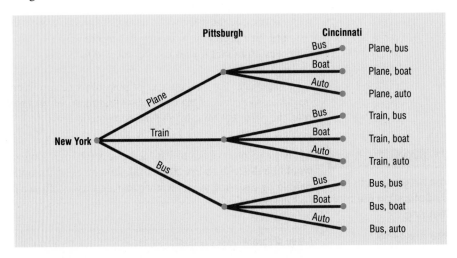

Finally, all outcomes can be listed by starting at New York and following the branches to Cincinnati, as shown at the right end of the tree in Figure 4–3. There are nine different ways.

Example 4–2

A coin is tossed and a die is rolled. Find all possible outcomes of this sequence of events.

Solution

Since the coin can land either heads up or tails up, and since the die can land with any one of six numbers shown face up, the outcomes can be represented as shown in Figure 4–4.

Figure 4–4

Complete Tree Diagram
for Example 4–2

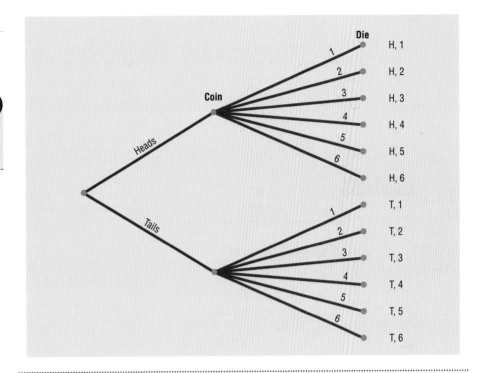

**The Multiplication
Rule for Counting**

In order to determine the total number of outcomes in a sequence of events, the *multiplication* rule can be used.

Objective 2. Find the total
number of outcomes in a
sequence of events using the
multiplication rule.

Multiplication Rule

In a sequence of *n* events in which the first one has k_1 possibilities and the second event has k_2 and the third has k_3, and so forth, the total number of possibilities of the sequence will be

$$k_1 \cdot k_2 \cdot k_3 \cdot \cdots \cdot k_n$$

Note: "And" in this case means to multiply.

The next examples illustrate the multiplication rule.

Example 4–3

A paint manufacturer wishes to manufacture several different paints. The categories include

Color	Red, blue, white, black, green, brown, yellow
Type	Latex, oil
Texture	Flat, semigloss, high gloss
Use	Outdoor, indoor

How many different kinds of paint can be made if a person can select one color, one type, one texture, and one use?

Solution

A person can choose one color and one type and one texture and one use. Since there are seven color choices, two type choices, three texture choices, and two use choices, the total number of possible different paints is

Color		**Type**		**Texture**		**Use**	
7	·	2	·	3	·	2	= 84

Example 4–4

There are four blood types, A, B, AB, and O. Blood can also be Rh+ and Rh−. Finally, a blood donor can be classified as either male or female. How many different ways can a donor have his or her blood labeled?

Solution

Since there are four possibilities for blood type, two possibilities for the Rh factor, and two possibilities for the gender of the donor, there are $4 \cdot 2 \cdot 2$, or 16, different classification categories as shown.

Blood type		**Rh**		**Gender**	
4	·	2	·	2	= 16

When determining the number of different possibilities of a sequence of events, one must know whether repetitions are permissible.

Example 4–5

The digits 0, 1, 2, 3, and 4 are to be used in a four-digit ID card. How many different cards are possible if repetitions are permitted?

Solution

Since there are four spaces to fill and five choices for each space, the solution is

$$5 \cdot 5 \cdot 5 \cdot 5 = 5^4 = 625$$

Now, what if repetitions are not permitted? For Example 4–5, the first digit can be chosen in five ways. But the second digit can be chosen in only four ways, since there are only four digits left; etc. Thus, the solution is

$$5 \cdot 4 \cdot 3 \cdot 2 = 120$$

The same situation occurs when one is drawing balls from an urn or cards from a deck. If the ball or card is replaced before the next one is selected, then repetitions are permitted, since the same one can be selected again. But if the selected ball or card is not replaced, then repetitions are not permitted, since the same ball or card cannot be selected the second time.

These examples illustrate the multiplication rule. In summary: if repetitions are permitted, then the numbers stay the same going from left to right. If repetitions are "not" permitted, then the numbers decrease by one for each place left to right.

Exercises

4–1. By means of a tree diagram, find all possible outcomes for the genders of the children in a family that has three children.

4–2. Bill's Burger Palace sells hot dogs, hamburgers, cheeseburgers, root beer, cola, lemon soda, french fries, and baked potatoes. If a customer selects one sandwich, one drink, and one potato, how many possible selections can the customer make? Draw a tree diagram to show the possibilities.

4–3. A quiz consists of four true–false questions. How many possible answer keys are there? Use a tree diagram.

4–4. Students are classified according to eye color (blue, brown, green), gender (male, female), and major (chemistry, mathematics, physics, business). How many possible different classifications are there? Use a tree diagram.

4–5. A box contains a $1 bill, a $5 bill, and a $10 bill. Two bills are selected in succession, without the first bill being replaced. Draw a tree diagram and represent all possible amounts of money that can be selected.

4–6. The Eagles and the Hawks play three games of hockey. Draw a tree diagram to represent the outcomes of the victories.

4–7. An inspector selects three batteries from a lot, then tests each to see whether each is overcharged, normal, or undercharged. Draw a tree diagram to represent all possible outcomes.

4–8. Draw a tree diagram to represent the outcomes when two players flip coins to see whether or not they match.

4–9. A coin is tossed. If it comes up heads, it is tossed again. If it lands tails, a die is rolled. Find all possible outcomes of this sequence of events.

4–10. A person has a chance of obtaining a degree from each category listed below. Draw a tree diagram showing all possible ways a person could obtain these degrees.

Bachelor's	Master's	Doctor's
B.S.	M.S.	Ph.D.
B.A.	M.Ed.	D.Ed.
	M.A.	

4–11. If blood types can be A, B, AB, and O, and Rh+ and Rh−, draw a tree diagram for the possibilities.

4–12. A woman has three skirts, five blouses, and four scarves. How many different outfits can she wear, assuming that they are color-coordinated?

4–13. How many five-digit zip codes are possible if digits can be repeated? If there cannot be repetitions?

4–14. How many ways can a baseball manager arrange a batting order of nine players?

4–15. How many different ways can seven floral arrangements be arranged in a row on a single display shelf?

4–16. How many different ways can six radio commercials be played during a one-hour radio program?

4–17. A store manager wishes to display eight different brands of shampoo in a row. How many ways can this be done?

4–18. There are eight different statistics books, six different geometry books, and three different trigonometry books. A student must select one book of each type. How many different ways can this be done?

4–19. At a local cheerleaders' camp, five routines must be practiced. A routine may not be repeated. In how many different orders can these five routines be presented?

4–20. The call letters of a radio station must have four letters. The first letter must be a *K* or a *W*. How many different station call letters can be made if repetitions are not allowed? If repetitions are allowed?

4–21. How many different three-digit identification tags can be made if the digits can be used more than once? If the first digit must be a 5 and repetitions are not permitted?

4–22. How many different ways can nine trophies be arranged on a shelf?

4–23. If a baseball manager has five pitchers and two catchers, how many different possible pitcher–catcher combinations can he field?

4–24. There are two major roads from city *X* to city *Y*, and four major roads from city *Y* to city *Z*. How many different trips can be made from city *X* to city *Z* passing through city *Y*?

*4–25. Pine Pizza Palace sells pizza plain or with one or more of the following toppings: pepperoni, sausage, mushrooms, olives, onions, or anchovies. How many different pizzas can be made? (*Hint:* A person can select or not select each item.)

*4–26. Generalize Exercise 4–25 for *n* different toppings. (*Hint:* For example, there are two ways to select pepperoni: either take it or not take it. For two toppings, a person can select none, both, or either one. Continue this reasoning for three toppings, etc.)

*4–27. How many different ways can a person select one or more coins if he has two nickels, one dime, and one half-dollar?

*4–28. A photographer has five photographs that she can mount on a page in her portfolio. How many different ways can she mount her photographs?

*4–29. In a barnyard there is an assortment of chickens and cows. Counting heads, one gets 15; counting legs, one gets 46. How many of each are there?

*4–30. How many committees of two or more people can be formed from four people? (*Hint:* Make a list using the letters A, B, C, and D to represent the people.)

4–3

Permutations and Combinations

Factorial Notation

Two other rules that can be used to determine the total number of possibilities of a sequence of events are the permutation rule and the combination rule.

These rules use *factorial notation.* The factorial notation uses the exclamation point.

$$5! \text{ means } 5 \cdot 4 \cdot 3 \cdot 2 \cdot 1$$
$$9! = 9 \cdot 8 \cdot 7 \cdot 6 \cdot 5 \cdot 4 \cdot 3 \cdot 2 \cdot 1$$

In order to use the formulas in the permutation and combination rules, a special definition of 0! is needed. $0! = 1$.

> **Historical Note**
>
> In 1808 Christian Kramp first used the factorial notation.

Factorial Formulas

For any counting n

$$n! = n(n - 1)(n - 2) \ldots 1$$
$$0! = 1$$

Permutations

A **permutation** is an arrangement of n objects in a specific order.

The next two examples illustrate permutations.

Example 4–6

Suppose a business owner has a choice of five locations in which to establish her business. She decides to rank each location according to certain criteria, such as price of the store and parking facilities. How many different ways can she rank the five locations?

Solution

There are

$$5! = 5 \cdot 4 \cdot 3 \cdot 2 \cdot 1 = 120$$

different possible rankings. The reason is that she has five choices for the first location, four choices for the second location, three choices for the third location, etc.

In the previous example, all objects were used up. But what happens when all objects are not used up? The answer to this question is given in Example 4–7.

Example 4–7

Suppose the business owner in Example 4–6 wishes to rank only the top three of the five locations. How many different ways can she rank them?

Solution

Using the multiplication rule, she can select any one of the five for first choice, then any one of the remaining four locations for her second choice, and finally, any one of the remaining three locations for her third choice, as shown.

First choice		Second choice		Third choice	
5	·	4	·	3	= 60

The solutions in Examples 4–6 and 4–7 are permutations.

Permutation Rule

Objective 3. Find the number of ways r objects can be selected from n objects using the permutation rule.

The arrangement of n objects in a specific order using r objects at a time is called a *permutation of n objects taking r objects at a time*. It is written as $_nP_r$, and the formula is

$$_nP_r = \frac{n!}{(n-r)!}$$

The notation $_nP_r$ is used for permutations.

$$_6P_4 \text{ means } \frac{6!}{(6-4)!} \text{ or } \frac{6!}{2!} = \frac{6 \cdot 5 \cdot 4 \cdot 3 \cdot \cancel{2} \cdot \cancel{1}}{\cancel{2} \cdot \cancel{1}} = 360$$

Although Examples 4–6 and 4–7 were solved by the multiplication rule, they can now be solved by the permutation rule.

In Example 4–6, five locations were taken and then arranged in order; hence,

$$_5P_5 = \frac{5!}{(5-5)!} = \frac{5!}{0!} = \frac{5 \cdot 4 \cdot 3 \cdot 2 \cdot 1}{1} = 120$$

(Recall 0! = 1)

In Example 4–7, three locations were selected from five locations, so $n = 5$ and $r = 3$; hence

$$_5P_3 = \frac{5!}{(5-3)!} = \frac{5!}{2!} = \frac{5 \cdot 4 \cdot 3 \cdot \cancel{2} \cdot \cancel{1}}{\cancel{2} \cdot \cancel{1}} = 60$$

The next two examples illustrate the permutation rule.

....

Example 4–8

A television news director wishes to use three news stories on an evening show. One story will be the "lead story," one will be the second story, and the last will be a "closing story." If the director has a total of eight stories to choose from, how many possible ways can the program be set up?

Solution

Since order is important, the solution is

$$_8P_3 = \frac{8!}{(8-3)!} = \frac{8!}{5!} = 336$$

Hence, there would be 336 ways to set up the program.

Example 4–9

How many different ways can a chairperson and an assistant chairperson be selected for a research project if there are seven scientists available?

Solution

$$_7P_2 = \frac{7!}{(7 - 2)!} = \frac{7!}{5!} = 42$$

Combinations

Objective 4. Find the number of ways *r* objects can be selected from *n* objects without regard to order using the combination rule.

Suppose a dress designer wishes to select two colors of material to design a new dress, and she has on hand four colors. How many different possibilities can there be in this situation?

This type of problem differs from previous ones in that the order of selection is not important. That is, if the designer selects yellow and red, this selection is the same as the selection red and yellow. This type of selection is called a *combination*. The difference between a permutation and a combination is that in a combination, the order or arrangement of the objects is not important; by contrast, order *is* important in a permutation. The next example illustrates this difference.

A selection of distinct objects without regard to order is called a **combination.**

Example 4–10

Given the letters A, B, C, and D, list the permutations and combinations for selecting two letters.

Solution

The listings follow:

	Permutations			Combinations	
AB	BA	CA	DA	AB	BC
AC	BC	CB	DB	AC	BD
AD	BD	CD	DC	AD	CD

Note that in permutations, AB is different from BA. But in combinations, AB is the same as BA, so only AB is listed. (Alternatively BA could be listed instead of AB.)

The elements of a combination are usually listed alphabetically.

Combinations are used when the order or arrangement is not important, as in the selecting process. Suppose a committee of 5 students is to be selected from 25 students. The five selected students represent a combination, since it does not matter who is selected first, second, etc.

Combination Rule

The number of combinations of r objects selected from n objects is denoted by $_nC_r$ and is given by the formula

$$_nC_r = \frac{n!}{(n - r)!r!}$$

Example 4–11

How many combinations of four objects are there taken two at a time?

Solution

Since this is a combination problem, the answer is

$$_4C_2 = \frac{4!}{(4 - 2)!2!} = \frac{4!}{2!2!} = \frac{\overset{2}{\cancel{4}} \cdot 3 \cdot \cancel{2!}}{2 \cdot 1 \cdot \cancel{2!}} = 6$$

This is the same result shown in Example 4–10.

Notice that the formula for $_nC_r$ is

$$\frac{n!}{(n - r)!r!}$$

which is the formula for permutations,

$$\frac{n!}{(n - r)!}$$

with an $r!$ in the denominator. This $r!$ divides out the duplicates from the number of permutations, as shown in Example 4–10. For each two letters, there are two permutations but only one combination. Hence, dividing the number of permutations by $r!$ eliminates the duplicates. This result can be verified for other values of n and r. *Note:* $_nC_n = 1$.

Example 4–12

A bicycle shop owner has 12 mountain bicycles in the showroom. The owner wishes to select 5 of them to display at a bicycle show. How many different ways can a group of 5 be selected?

Solution

$$_{12}C_5 = \frac{12!}{(12 - 5)!5!} = \frac{12!}{7!5!} = \frac{12 \cdot 11 \cdot \overset{3}{\cancel{10}} \cdot \overset{}{\cancel{9}} \cdot \overset{2}{\cancel{8}} \cdot \cancel{7!}}{\cancel{7!} \cdot \cancel{5} \cdot \cancel{4} \cdot \cancel{3} \cdot 2 \cdot 1} = 792$$

Example 4–13

In a club there are 7 women and 5 men. A committee of 3 women and 2 men is to be chosen. How many different possibilities are there?

Solution

Here, one must select 3 women from 7 women, which can be done in $_7C_3$, or 35, ways. Next, 2 men must be selected from 5 men, which can be done in $_5C_2$, or 10, ways. Finally, by the multiplication rule, the total number of different ways is $35 \cdot 10 = 350$, since one is choosing both men and women. Using the formula

$$_7C_3 \cdot {}_5C_2 = \frac{7!}{(7-3)!3!} \cdot \frac{5!}{(5-2)!2!} = 350$$

Table 4–1 summarizes the counting rules.

Table 4–1 Summary of Counting Rules		
Rule	**Definition**	**Formula**
Multiplication rule	The number of ways a sequence of n events can occur if the first event can occur in k_1 ways, the second event can occur in k_2 ways, etc.	$k_1 \cdot k_2 \cdot k_3 \cdots \cdot k_n$
Permutation rule	The number of permutations of n objects taking r objects at a time (order is important)	$_nP_r = \dfrac{n!}{(n-r)!}$
Combination rule	The number of combinations of r objects taken from n objects (order is not important)	$_nC_r = \dfrac{n!}{(n-r)!r!}$

Exercises

4–31. Evaluate each:

a. 8!

b. 10!

c. 4!

d. 1!

e. 0!

4–32. (ans) Evaluate each expression.

a. $_8P_2$

b. $_7P_5$

c. $_{12}P_4$

d. $_5P_3$

e. $_6P_0$

f. $_6P_6$

g. $_8P_0$

h. $_8P_8$

i. $_{11}P_3$

j. $_6P_2$

4–33. How many different four-letter permutations can be formed from the letters in the word *decagon?*

4–34. In a board of directors composed of eight people, how many ways can a chief executive officer, a director, and a treasurer be selected?

4–35. How many different ID cards can be made if there are six digits on a card and no digit can be used more than once?

4–36. How many ways can seven different types of soaps be displayed on a shelf in a grocery store?

4–37. How many different four-color code stripes can be made on a sports car if each code consists of the colors green, red, blue, and white? All colors are used only once.

4–38. An inspector must select three tests to perform in a certain order on a manufactured part. He has a choice of seven tests. How many ways can he perform three different tests?

4–39. The Anderson Research Co. decides to test-market a product in six areas. How many different ways can three areas be selected in a certain order for the first test?

4–40. How many different ways can a city building inspector visit six buildings in the city if she visits all of them in one day?

4–41. How many different ways can five radio commercials be run during an hour of time?

4–42. How many different ways can four tickets be selected from 50 tickets if each ticket wins a different prize?

4–43. How many different ways can a researcher select five rats from 20 rats and assign each to a different test?

4–44. How many different signals can be made by using at least three distinct flags if there are five different flags from which to select?

4–45. An investigative agency has seven cases and five agents. How many different ways can the cases be assigned if only one case is assigned to each agent?

4–46. **(ans)** Evaluate each expression.

a. $_5C_2$ d. $_6C_2$ g. $_3C_3$ j. $_4C_3$
b. $_8C_3$ e. $_6C_4$ h. $_9C_7$
c. $_7C_4$ f. $_3C_0$ i. $_{12}C_2$

4–47. How many ways can 3 cards be selected from a standard deck of 52 cards disregarding the order of selection?

4–48. How many ways are there to select 3 bracelets from a box of 10 bracelets disregarding the order of selection?

4–49. How many ways can 4 baseball players and 3 basketball players be selected from 12 baseball players and 9 basketball players?

4–50. How many ways can a committee of 4 people be selected from a group of 10 people?

4–51. If a person can select 3 presents from 10 presents under a Christmas tree, how many different combinations are there?

4–52. How many different tests can be made from a test bank of 20 questions if the test consists of 5 questions?

4–53. The general manager of a fast-food restaurant chain must select 6 restaurants from 11 for a promotional program. How many different possible ways can this selection be done?

4–54. How many ways can 3 cars and 4 trucks be selected from 8 cars and 11 trucks to be tested for a safety inspection?

4–55. In a train yard there are 4 tank cars, 12 boxcars, and 7 flatcars. How many ways can a train be made up consisting of 2 tank cars, 5 boxcars, and 3 flatcars? (In this case order is not im portant.)

4–56. There are seven women and five men in a department. How many ways can a committee of four people be selected? How many ways can this committee be

selected if there must be two men and two women on the committee? How many ways can this committee be selected if there must be at least two women on the committee?

4–57. Wake Up cereal comes in two types, crispy and crunchy. If a researcher has 10 boxes of each, how many ways can she select 3 boxes of each for a quality control test?

4–58. How many ways can a dinner patron select three appetizers and two vegetables if there are six appetizers and five vegetables on the menu?

4–59. How many ways can a jury of 6 men and 6 women be selected from 12 men and 10 women?

4–60. How many ways can a foursome of 2 men and 2 women be selected from 10 men and 12 women in a golf club?

4–61. The state narcotics bureau must form a 5 member investigative team. If it has 25 agents from which to choose, how many different possible teams can be formed?

4–62. How many different ways can an instructor select 2 textbooks from a possible 17?

4–63. The Environmental Protection Agency must investigate nine mills for complaints of air pollution. How many different ways can a representative select five of these to investigate this week?

4–64. How many ways can a person select 7 television commercials from 11 television commercials?

4–65. How many ways can a person select 8 videotapes from 10 tapes?

4–66. A buyer decides to stock 8 different posters. How many ways can she select these 8 if there are 20 from which to choose?

4–67. An advertising manager decides to have an ad campaign in which 8 special calculators will be hidden at various locations in a shopping mall. If he has 17 locations from which to pick, how many different possible combinations can he choose?

*4–68. How many different ways can five people—A, B, C, D, and E—sit in a row at a movie theater if (a) A and B must sit together; (b) C must sit to the right of, but not necessarily next to, B; (c) D and E will not sit next to each other?

*4–69. Using combinations, calculate the number of each poker hand in a deck of cards. (A poker hand consists of five cards dealt in any order.)

a. Royal flush   c. Four of a kind
b. Straight flush   d. Full house

Technology Step by Step

TI-83 Step by Step

Permutations, Combinations, and Factorials

A. To find the value of a permutation: Example $_5P_3$
1. Enter **5**.
2. Press **MATH** and mover the cursor to PRB.
3. Press **2**, then **3**, then **ENTER**.

The calculator will display the answer, 60.

B. To find the value of a combination: Example $_8C_5$
1. Enter **8**.
2. Press **MATH** move the cursor to PRB, then press 3.
3. Then press **5** and **ENTER**.

The calculator will display the answer, 56.

C. To find a factorial of a number: Example 5!
1. Enter **5**.
2. Press **MATH** and move the cursor to PRB.
3. Press 4, then **ENTER**.

The calculator will display the answer, 120.

4–4

Summary

This chapter illustrates how one can count or list all possible outcomes of a sequence of events. A tree diagram can be used when a list of all possible outcomes is necessary. When only the total number of outcomes is needed, the multiplication rule, the permutation rule, and the combination rule can be used.

Using these rules, statisticians can find the solutions to a variety of problems in which they must know the number of possibilities that can occur. These rules will be used in the next chapter to determine the probabilities of events.

Important Terms

combination   158   permutation   156   tree diagram   151

Important Formulas

Multiplication rule: In a sequence of n events in which the first one has k_1 possibilities, the second event has k_2 possibilities, the third has k_3 possibilities, etc., the total number possibilities of the sequence will be

$$k_1 \cdot k_2 \cdot k_3 \cdot \cdots \cdot k_n$$

Permutation rule: The number of permutations of n objects taking r objects at a time when order is important is

$$_nP_r = \frac{n!}{(n-r)!}$$

Combination rule: The number of combinations of r objects selected from n objects when order is not important

$$_nC_r = \frac{n!}{(n-r)!r!}$$

Review Exercises

4–70. An automobile license plate consists of three letters followed by four digits. How many different plates can be made if repetitions are allowed? If repetitions are not allowed? If repetitions are allowed in the letters but not in the digits?

4–71. How many different arrangements of the letters in the word *bread* can be made?

4–72. How many different three-digit combinations can be made by using the numbers 1, 3, 5, 7, and 9 without repetitions if the "right" combination can be used to open a safe? (*Hint:* Does a combination lock really use combinations?)

4–73. How many two-card pairs are there in a poker deck?

4–74. A quiz consists of six multiple-choice questions. Each question has three possible answer choices. How many different answer keys can be made?

4–75. How many ways can 5 different television programs be selected from 12 programs?

4–76. How many different ways can a buyer select four television models from a possible choice of six models?

4–77. Draw a tree diagram to show all possible outcomes when a coin is flipped four times.

4–78. How many ways can three outfielders and four infielders be chosen from five outfielders and seven infielders?

4–79. How many different ways can eight computer operators be seated in a row?

4–80. How many ways can a student select 2 electives from a possible choice of 10 electives?

4–81. There are six Republican, five Democrat, and four Independent candidates. How many different ways can a committee of three Republicans, two Democrats, and one Independent be selected?

4–82. Using Exercise 4–81, how many ways can a committee of four people be selected if they are all from the same party?

4–83. Employees can be classified according to gender (male, female), income (low, medium, high), and rank (assistant, instructor, dean). Draw a tree diagram and show all possible outcomes.

4–84. A disc jockey can select 4 records from 10 to play in one segment. How many ways can this selection be done? (*Note:* Order is important.)

4–85. A judge is to rank six brands of cookies according to their flavor. How many different ways can this ranking be done?

4–86. A vending machine servicer must restock and collect money from 20 machines, each one at a different location.

How many different ways can he select 5 machines to service in one day?

4–87. How many different ways can four floral centerpieces be arranged in a display case? (Assume four spaces are available.)

4–88. How many different ways can 3 fraternity members be selected from 10 members if one must be president, one must be vice president, and one must be secretary/treasurer?

4–89. How many different ways can two balls be drawn from a bag containing five balls? Each ball is a different color, and the first ball is replaced before the second one is selected. How many different ways are there if the first ball is not replaced before the second one is selected?

4–90. A restaurant offers three choices of meat, two choices of potatoes, four choices of vegetables, and five choices of dessert. How many different possible meals can be made if a customer must select one item from each category?

4–91. If someone wears a blouse or a sweater and a pair of slacks or a skirt, how many different outfits can she wear?

4–92. How many different computer passwords are possible if each consists of four symbols and if the first one must be a letter and the other three must be digits?

4–93. If a student has a choice of five computers, three printers, and two monitors, how many ways can she select a computer system?

4–94. A combination lock consists of the numbers 0 to 39. If no number can be used twice, how many different combinations are possible using three numbers? Remember, a combination lock is really a permutation lock.

4–95. There are 12 students who wish to enroll in a particular course. There are only four seats left in the classroom. How many different ways can 4 students be selected to attend the class?

4–96. A candy store allows customers to select three different candies to be packaged and mailed. If there are 13 varieties available, how many possible selections can be made?

4–97. If a student can select 5 novels from a reading list of 20 for a course in literature, how many different possible ways can this selection be done?

4–98. If a student can select one of three language courses, one of five mathematics courses, and one of four history courses, how many different schedules can be made?

Statistics Today

Why Are We Running Out of 800 Numbers? Revisited

Since each 800 number is followed by a seven-digit number, multiplication rule 1 can be used to determine the total number of 800 phone numbers that are available.

Since there are 10 digits (0 through 9) that can be used for each digit of the seven-digit number, the answer is 10^7, or 10,000,000, starting with 800-0000000 and ending with 800-9999999. (Since the telephone industry forbids usage of certain numbers, the article that opens this chapter refers to a smaller number of "possible" 800 numbers.)

Chapter Quiz

Determine whether each statement is true or false. If the statement is false, explain why.

1. If there are 5 contestants in a race, the number of different ways the first- and second-place winners can be selected is 25.

2. If a true–false exam contains 10 questions, there are 20 different ways to answer all the questions.

3. Some permutation problems can be solved by the multiplication rule.

4. The arrangement ABC is the same as BAC for combinations.

5. When objects are arranged in a specific order, the arrangement is called a combination.

Select the best answer.

6. What is $_nP_0$?
   a. 0
   b. 1
   c. n
   d. It cannot be determined.

7. What is the number of permutations of six different objects taken all together?
   a. 0   c. 36
   b. 1   d. 720

8. How many permutations of the letters in the word *tide* are there?
   a. 1   c. 12
   b. 6   d. 24

9. What is 0!?
   a. 0   c. undefined
   b. 1   d. 10

10. What is $_nC_n$?
    a. 0   c. n
    b. 1   d. It cannot be determined.

Complete the following statements with the best answer.

11. A device that is helpful in listing the outcomes of a sequence of events is called a _____.

12. When a coin is tossed and a die is rolled, there are _____ outcomes.

13. If in a sequence of k events each event can occur the same number of ways (n), then the total number of outcomes is _____.

14. The number of permutations of n objects taken all together is _____.

15. Telephone numbers are examples of _____.

16. One company's ID cards consist of five letters followed by two digits. How many cards can be made if repetitions are allowed? If repetitions are not allowed?

17. How many different arrangements of the letters in the word *number* can be made?

18. A physics test consists of 25 true–false questions. How many different possible answer keys can be made?

19. How many different ways can four radios be selected from a total of seven radios?

20. The National Bridge Association can select one of four cities for its playoff tournament next year. The cities are Pasadena, Wilmington, Chicago, and Charleston. The following year, it can hold the tournament in Hyattsville or Green Springs. How many different possibilities are there for the next two years? Draw a tree diagram and show all possibilities.

21. How many ways can five sopranos and four altos be selected from seven sopranos and nine altos?

22. How many different ways can eight speakers be seated on a stage?

23. Employees can be classified according to gender (male, female), income (low, medium, high), and rank (staff nurse, charge nurse, head nurse). Draw a tree diagram and show all possible outcomes.

24. A soda machine servicer must restock and collect money from 15 machines, each one at a different location. How many ways can she select four machines to service in one day?

25. How many different ways can three cubes be drawn from a bag containing four cubes? Each cube is a different color, and the first cube is replaced before the second one is selected. How many different ways are there if the first cube is not replaced before the second one is selected?

26. If a man can wear a shirt or a sweater and a pair of dress slacks or a pair of jeans, how many different outfits can he wear?

27. A beachfront candy store allows customers to select three different candies to be packaged and mailed. If there are 10 varieties available, how many possible selections can be made?

28. How many different ways can an avid reader select 6 novels from a shelf containing 15?

29. If a woman can select one of three nail-polish colors, one of five lipstick shades, and one of four blush shades, how many different combinations can she make?

30. Find the number of different snacks, consisting of one or more of these crackers—whole wheat, onion, buttercrisp, or sesame—that can be served to guests.

Critical Thinking Challenges

1. A person decides to shake hands with six different people on a certain day. The next day, each of the six people will shake hands with six different people. The process continues until every person in the United States has shaken someone's hand. How many days will it take until everyone in the United States has shaken hands once? Assume that once a person shakes hands with six different people, he or she does not shake hands again. (*Hint:* The population of the United States is 248,709,873, according to the 1990 census.)

2. If it can be assumed that the maximum number of hairs on a human head is about 500,000, explain why at least two people living in Houston (population 1,629,902, according to the 1990 census) have the same number of hairs on their heads.

3. A mathematician named Pascal wrote a treatise showing how combinations can be derived from a triangular array of numbers. This triangle became known as Pascal's triangle, although there is evidence that the triangle existed in China as early as the 1300s. The

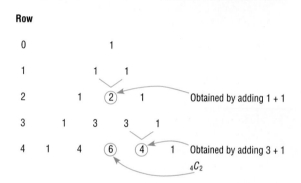

triangle is formed by adding the two adjacent numbers and writing the sum below in a triangular fashion.

Each number in the triangle represents the number of combinations of n objects taken r at a time. For example, $_4C_2 = 6$, which is the third value found in row 4.

Complete Pascal's triangle for nine rows and verify the answers using combinations.

chapter

5 Probability

Objectives

After completing this chapter, you should be able to

1. Determine sample spaces and find the probability of an event using classical probability or empirical probability.

2. Find the probability of compound events using the addition rules.

3. Find the probability of compound events using the multiplication rules.

4. Find the conditional probability of an event.

5. Find the probability of an event using the counting rules.

Statistics Today

Would You Bet Your Life?

Humans not only bet money when they gamble, but they also bet their lives by engaging in unhealthy activities such as smoking, drinking, using drugs, and exceeding the speed limit when driving. Many people don't care about the risks involved in these activities since they do not understand the concepts of probability. On the other hand, people may fear activities that involve little risk to health or life because these activities have been sensationalized by the press and media.

In his book *Probabilities in Everyday Life* (Ivy Books, 1986, p. 191), John D. McGervey states,

> *When people have been asked to estimate the frequency of death from various causes, the most overestimated categories are those involving pregnancy, tornadoes, floods, fire, and homicide. The most underestimated categories include deaths from diseases such as diabetes, strokes, tuberculosis, asthma, and stomach cancer (although cancer in general is overestimated).*

The question then is: Would you feel safer if you flew across the United States on a commercial airline or if you drove? How much greater is the risk of one way to travel over the other?

In this chapter, you will learn about probability—its meaning, how it is computed, and how to evaluate it in terms of the likelihood of an event actually happening.

5–1

Introduction

A cynical person once said, "The only two sure things are death and taxes." This philosophy no doubt arose because so much in people's lives is affected by chance. From the time a person awakes until he or she goes to bed, that person makes decisions regarding the possible events that are governed at least in part by chance. For example, should I carry an umbrella to work today? Will my car battery last until spring? Should I accept that new job?

Probability as a general concept can be defined as the chance of an event occurring. Many people are familiar with probability from observing or playing games of chance, such as card games, slot machines, or lotteries. In addition to being used in games of chance, probability theory is used in the fields of insurance, investments, and weather forecasting, and in various other areas. Finally, as stated in Chapter 1, probability is the basis of inferential statistics. For example, predictions are based on probability, and hypotheses are tested by using probability.

The basic concepts of probability are explained in this chapter. These concepts include *probability experiments, sample spaces,* the *addition and multiplication rules,* and the *probabilities of complementary events.* Section 5–5 explains how the counting rules of Chapter 4 and the probability rules can be used together to solve a wide variety of problems.

5–2

Sample Spaces and Probability

The theory of probability grew out of the study of various games of chance using coins, dice, and cards. Since these devices lend themselves well to the application of concepts of probability, they will be used in this chapter as examples. This section begins by explaining some basic concepts of probability. Then the types of probability and probability rules are discussed.

Basic Concepts

Objective 1. Determine sample spaces and find the probability of an event using classical probability or empirical probability.

Processes such as flipping a coin, rolling a die, or drawing a card from a deck are called *probability experiments.*

A **probability experiment** is a chance process that leads to well-defined results called outcomes.

An **outcome** is the result of a single trial of a probability experiment.

A trial means flipping a coin once, rolling one die once, or the like. When a coin is tossed, there are two possible outcomes: head or tail. (*Note:* We exclude the possibility of a coin landing on its edge.) In the roll of a single die, there are six possible outcomes: 1, 2, 3, 4, 5, or 6. In any experiment, the set of all possible outcomes is called the *sample space.*

A **sample space** is the set of all possible outcomes of a probability experiment.

Some sample spaces for various probability experiments are shown here.

Experiment	Sample space
Toss one coin	Head, tail
Roll a die	1, 2, 3, 4, 5, 6
Answer a true–false question	True, false
Toss two coins	Head-head, tail-tail, head-tail, tail-head

It is important to realize that when two coins are tossed, there are *four* possible outcomes, as shown in the fourth experiment above. Both coins could fall heads up. Both coins could fall tails up. Coin 1 could fall heads up and coin 2 tails up. Or coin 1 could fall tails up and coin 2 heads up. Heads and tails will be abbreviated as H and T throughout this chapter.

Example 5–1

Find the sample space for rolling two dice.

Solution

Since each die can land in six different ways, and two dice are rolled, the sample space can be presented by a rectangular array, as shown in Figure 5–1. The sample space is the list of pairs of numbers in the chart.

Figure 5–1

Sample Space for Rolling Two Dice (Example 5–1)

Die 1	Die 2					
	1	2	3	4	5	6
1	(1, 1)	(1, 2)	(1, 3)	(1, 4)	(1, 5)	(1, 6)
2	(2, 1)	(2, 2)	(2, 3)	(2, 4)	(2, 5)	(2, 6)
3	(3, 1)	(3, 2)	(3, 3)	(3, 4)	(3, 5)	(3, 6)
4	(4, 1)	(4, 2)	(4, 3)	(4, 4)	(4, 5)	(4, 6)
5	(5, 1)	(5, 2)	(5, 3)	(5, 4)	(5, 5)	(5, 6)
6	(6, 1)	(6, 2)	(6, 3)	(6, 4)	(6, 5)	(6, 6)

Example 5–2

Find the sample space for drawing one card from an ordinary deck of cards.

Solution

Since there are four suits (hearts, clubs, diamonds, and spades) and 13 cards for each suit (ace through king), there are 52 outcomes in the sample space. See Figure 5–2.

Figure 5–2

Sample Space for Drawing a Card (Example 5–2)

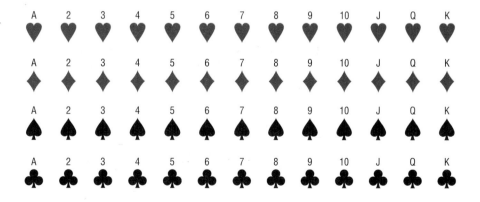

Example 5–3

Find the sample space for the gender of the children if a family has three children. Use B for boy and G for girl.

Solution

There are two genders, male and female, and each child could be either gender. Hence, there are eight possibilities, as shown here.

BBB   BBG   BGB   GBB   GGG   GGB   GBG   BGG

In the previous examples, the sample spaces were found by observation and reasoning; however, a tree diagram can also be used. In Chapter 4, the tree diagram was used to show all possible outcomes in a sequence of events. *The tree diagram can also be used as a systematic way to find all possible outcomes of a probability experiment.*

Example 5–4

Use a tree diagram to find the sample space for the gender of three children in a family, as in Example 5–3.

Solution

There are two possibilities for the first child, two for the second, and two for the third. Hence, the tree diagram can be drawn as shown in Figure 5–3.

Figure 5–3

Tree Diagram for Example 5–4

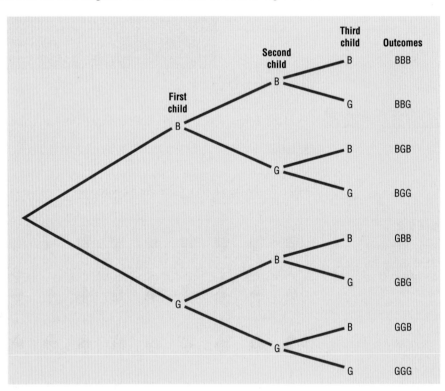

An outcome was defined previously as the result of a single trial of a probability experiment. In many problems, one must find the probability of two or more outcomes. For this reason, it is necessary to distinguish between an outcome and an event.

Historical Note

A mathematician named Jerome Cardan (1501–1576) used his talents in mathematics and probability theory to make his living as a gambler. He is thought to be the first person to formulate the definition of classical probability.

An **event** consists of the outcomes of a probability experiment.

An event can be one outcome or more than one outcome. For example, if a die is rolled and a 6 shows, this result is called an *outcome,* since it is a result of a single trial. An event with one outcome is called a **simple event.** The event of getting an odd number when a die is rolled is called a **compound event,** since it consists of three outcomes or three simple events. In general, a compound event consists of two or more outcomes or simple events.

There are three basic types of probability:

1. Classical probability
2. Empirical or relative frequency probability
3. Subjective probability

Classical Probability

Historical Note

During the mid-1600s, a professional gambler named Chevalier de Méré made a considerable amount of money on a gambling game. He would bet unsuspecting patrons that in four rolls of a die, he could get at least one 6. He was so successful at the game that some people refused to play. He decided that a new game was necessary to continue his winnings. By reasoning, he figured he could roll at least one double 6 in 24 rolls of two dice, but his reasoning was incorrect and he lost systematically. Unable to figure out why, he contacted a mathematician named Blaise Pascal (1623–1662) to find out why.
  Pascal became interested and began studying probability theory. He corresponded with a French government official, Pierre de Fermat (1601–1665), whose hobby was mathematics. Together the two formulated the beginnings of probability theory.

Classical probability uses sample spaces to determine the numerical probability that an event will happen. One does not actually have to perform the experiment to determine that probability. Classical probability is so named because it was the first type of probability studied formally by mathematicians in the 17th and 18th centuries.

Classical probability assumes that all outcomes in the sample space are equally likely to occur. For example, when a single die is rolled, each outcome has the same probability of occurring. Since there are six outcomes, each outcome has a probability of $\frac{1}{6}$. When a card is selected from an ordinary deck of 52 cards, one assumes that the deck has been shuffled, and each card has the same probability of being selected. In this case, it is $\frac{1}{52}$.

Equally likely events are events that have the same probability of occurring.

Formula for Classical Probability

The probability of any event E is

$$\frac{\text{number of outcomes in } E}{\text{total number of outcomes in the sample space}}$$

This probability is denoted by

$$P(E) = \frac{n(E)}{n(S)}$$

This probability is called *classical probability,* and it uses the sample space S.

Probabilities can be expressed as fractions, decimals, or—where appropriate—percentages. If one asks, "What is the probability of getting a head when a coin is tossed?" typical responses can be any of the following three.

"One-half."
"Point five."
"Fifty percent."

These answers are all equivalent. In most cases, the answers to examples and exercises given in this chapter are expressed as fractions or decimals, but percentages are used where appropriate.

Rounding Rule for Probabilities   Probabilities should be expressed as reduced fractions or rounded to two or three decimal places. When the probability of an event is an extremely small decimal, it is permissible to round the decimal to the first nonzero digit after the point. For example, 0.0000587 would be 0.00006. When obtaining probabilities from one of the tables in Appendix C, use the number of decimal places given in the table. If decimals are converted to percentages to express probabilities, move the point two places to the right and add a percent sign.

Example 5–5

For a card drawn from an ordinary deck, find the probability of getting a king.

Solution

Since there are 52 cards in a deck and there are 4 kings, $P(\text{king}) = \frac{4}{52} = \frac{1}{13}$.

Example 5–6

If a family has three children, find the probability that all the children are girls.

Solution

The sample space for the gender of children for a family that has three children is BBB, BBG, BGB, GBB, GGG, GGB, GBG, and BGG (see Examples 5–3 and 5–4). Since there is one way in eight possibilities for all three children to be girls,

$$P(\text{GGG}) = \frac{1}{8}$$

Example 5–7

A card is drawn from an ordinary deck. Find these probabilities.

a. Of getting a jack.

b. Of getting the 6 of clubs.

c. Of getting a 3 or a diamond.

Solution

a. Refer to the sample space in Figure 5–2. There are 4 jacks and 52 possible outcomes. Hence,

$$P(\text{jack}) = \frac{4}{52} = \frac{1}{13}$$

b. Since there is only one 6 of clubs, the probability of getting a 6 of clubs is

$$P(\text{6 of clubs}) = \frac{1}{52}$$

c. There are four 3s and 13 diamonds, but the 3 of diamonds is counted twice in this listing. Hence, there are 16 possibilities of drawing a 3 or a diamond, so

$$P(\text{3 or diamond}) = \frac{16}{52} = \frac{4}{13}$$

There are four basic probability rules. These rules are helpful in solving probability problems, in understanding the nature of probability, and in deciding if your answers to the problems are correct.

Probability Rule 1

The probability of any event E is a number (either a fraction or decimal) between and including 0 and 1. This is denoted by $0 \leq P(E) \leq 1$.

Rule 1 states that probabilities cannot be negative or greater than one.

Probability Rule 2

If an event E cannot occur (i.e., the event contains no members in the sample space), the probability is zero.

Example 5–8

When a single die is rolled, find the probability of getting a 9.

Solution

Since the sample space is 1, 2, 3, 4, 5, and 6, it is impossible to get a 9. Hence, the probability is $P(9) = \frac{0}{6} = 0$.

Probability Rule 3

If an event E is certain, then the probability of $E = 1$.

In other words, if $P(E) = 1$, then the event E is certain to occur. This rule is illustrated in the next example.

Example 5–9

When a single die is rolled, what is the probability of getting a number less than 7?

Solution

Since all outcomes, 1, 2, 3, 4, 5, and 6, are less than 7, the probability is

$$P(\text{number less than 7}) = \frac{6}{6} = 1$$

The event of getting a number less than 7 is certain.

Probability Rule 4

The sum of the probabilities of the outcomes in the sample space is 1.

For example, in the roll of a fair die, each outcome in the sample space has a probability of $\frac{1}{6}$. Hence, the sum of the probabilities of the outcomes is as shown.

Outcome	1	2	3	4	5	6
Probability	$\frac{1}{6}$	$\frac{1}{6}$	$\frac{1}{6}$	$\frac{1}{6}$	$\frac{1}{6}$	$\frac{1}{6}$
Sum	$\frac{1}{6}$ +	$\frac{1}{6}$ +	$\frac{1}{6}$ +	$\frac{1}{6}$ +	$\frac{1}{6}$ +	$\frac{1}{6} = \frac{6}{6} = 1$

Complementary Events

Another important concept in probability theory is that of *complementary events*. When a die is rolled, for instance, the sample space consists of the outcomes 1, 2, 3, 4, 5, and 6. The event E of getting odd numbers consists of the outcomes 1, 3, and 5. The event of not getting an odd number is called the *complement* of event E, and it consists of the outcomes 2, 4, and 6.

The **complement of an event** E is the set of outcomes in the sample space that are not included in the outcomes of event E. The complement of E is denoted by $\bar{E}$ (read "E bar").

The next example further illustrates the concept of complementary events.

Example 5–10

Find the complement of each event.

a. Rolling a die and getting a 4.

b. Selecting a letter of the alphabet and getting a vowel.

c. Selecting a month and getting a month that begins with a *J*.

d. Selecting a day of the week and getting a weekday.

Solution

a. Getting a 1, 2, 3, 5, or 6.

b. Getting a consonant (assume *y* is a consonant).

c. Getting February, March, April, May, August, September, October, November, or December.

d. Getting Saturday or Sunday.

The outcomes of an event and the outcomes of the complement make up the entire sample space. For example, if two coins are tossed, the sample space is HH, HT, TH, and TT. The complement of "getting all heads" is not "getting all tails," since the event "all heads" is HH, and the complement of HH is HT, TH, and TT. Hence, the complement of the event "all heads" is the event "getting at least one tail."

Since the event and its complement make up the entire sample space, it follows that the sum of the probability of the event and the probability of its complement will equal 1. That is, $P(E) + P(\bar{E}) = 1$. In the previous example, let E = all heads, or HH, and let $\bar{E}$ = at least one tail, or HT, TH, TT. Then $P(E) = \frac{1}{4}$ and $P(\bar{E}) = \frac{3}{4}$; hence, $P(E) + P(\bar{E}) = \frac{1}{4} + \frac{3}{4} = 1$.

The rule for complementary events can be stated algebraically in three ways.

Rule for Complementary Events

$$P(\bar{E}) = 1 - P(E) \text{ or } P(E) = 1 - P(\bar{E}) \text{ or } P(E) + P(\bar{E}) = 1$$

Stated in words, the rule is: *If the probability of an event or the probability of its complement is known, then the other can be found by subtracting the probability from 1.* This rule is important in probability theory because at times the best solution to a problem is to find the probability of the complement of an event and then subtract from 1 to get the probability of the event itself.

Example 5–11

If the probability that a person lives in an industrialized country of the world is $\frac{1}{5}$, find the probability that a person does not live in an industrialized country.

Source: *Harper's Index* 289, no. 1737 (February 1995), p. 11.

Solution

P (not living in an industrialized country) = $1 - P$ (living in an industrialized country) = $1 - \frac{1}{5} = \frac{4}{5}$.

Probabilities can be represented pictorially by **Venn diagrams.** Figure 5–4(a) shows the probability of a simple event E. The area inside the circle represents the probability of event E—that is, $P(E)$. The area inside the rectangle represents the probability of all the events in the sample space, $P(S)$.

The Venn diagram that represents the probability of the complement of an event $P(\bar{E})$ is shown in Figure 5–4(b). In this case, $P(\bar{E}) = 1 - P(E)$, which is the area inside the rectangle but outside the circle representing $P(E)$. Recall that $P(S) = 1$ and $P(E) = 1 - P(\bar{E})$. The reasoning is that $P(E)$ is represented by the area of the circle and $P(\bar{E})$ is the probability of the events that are outside the circle.

Figure 5–4

Venn Diagram for the
Probability and
Complement

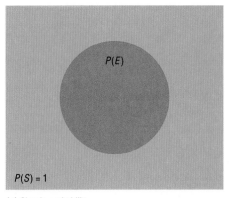

(a) Simple probability   **(b)** $P(\bar{E}) = 1 - P(E)$

Empirical Probability

The difference between classical and **empirical probability** is that classical probability assumes that certain outcomes are equally likely (such as the outcomes when a die is rolled) while empirical probability relies on actual experience to determine the likelihood of outcomes. In empirical probability, one might actually roll a given die 6000 times and observe the various frequencies and use these frequencies to determine the probability of an outcome. Suppose, for example, that a researcher asked 25 people if they liked the taste of a new soft drink. The responses were classified as "yes," "no," or "undecided." The results were categorized in a frequency distribution, as shown.

Response	Frequency
Yes	15
No	8
Undecided	2
Total	25

Probabilities now can be compared for various categories. For example, the probability of selecting a person who liked the taste is $\frac{15}{25}$, or $\frac{3}{5}$, since 15 out of 25 people in the survey answered "yes."

Formula for Empirical Probability

Given a frequency distribution, the probability of an event being in a given class is

$$P(E) = \frac{\text{frequency for the class}}{\text{total frequencies in the distribution}} = \frac{f}{n}$$

This probability is called *empirical probability* and is based on observation.

Example 5–12

In the soft-drink survey just described, find the probability that a person responded "no."

Solution

$$P(E) = \frac{f}{n} = \frac{8}{25}$$

Note: This is the same relative frequency explained in Chapter 2.

Example 5–13

In a sample of 50 people, 21 had type O blood, 22 had type A blood, 5 had type B blood, and 2 had type AB blood. Set up a frequency distribution and find the following probabilities:

a. A person has type O blood.

b. A person has type A or type B blood.

c. A person has neither type A nor type O blood.

d. A person does not have type AB blood.

Source: Based on American Red Cross figures presented in *The Book of Odds* by Michael D. Shook and Robert L. Shook (New York: Penguin Putnam, Inc., 1991), p. 33.

Solution

Type	Frequency
A	22
B	5
AB	2
O	21
Total	50

a. $P(O) = \dfrac{f}{n} = \dfrac{21}{50}$

b. $P(A \text{ or } B) = \dfrac{22}{50} + \dfrac{5}{50} = \dfrac{27}{50}$

(Add the frequencies of the two classes.)

c. $P(\text{neither A nor O}) = \dfrac{5}{50} + \dfrac{2}{50} = \dfrac{7}{50}$

(Neither A nor O means that a person has either type B or type AB blood.)

d. $P(\text{not AB}) = 1 - P(\text{AB}) = 1 - \dfrac{2}{50} = \dfrac{48}{50} = \dfrac{24}{25}$

(Find the probability of not AB by subtracting the probability of type AB from 1.)

Example 5–14

Hospital records indicated that maternity patients stayed in the hospital for the number of days shown in the distribution.

Number of days stayed	Frequency
3	15
4	32
5	56
6	19
7	5
	127

Find these probabilities.

a. A patient stayed exactly 5 days.

b. A patient stayed less than 6 days.

c. A patient stayed at most 4 days.

d. A patient stayed at least 5 days.

Solution

a. $P(5) = \dfrac{56}{127}$

b. $P(\text{less than 6 days}) = \dfrac{15}{127} + \dfrac{32}{127} + \dfrac{56}{127} = \dfrac{103}{127}$

(Less than 6 days means either 3, or 4, or 5 days.)

c. $P(\text{at most 4 days}) = \dfrac{15}{127} + \dfrac{32}{127} = \dfrac{47}{127}$

(At most 4 days means 3 or 4 days.)

d. $P(\text{at least 5 days}) = \dfrac{56}{127} + \dfrac{19}{127} + \dfrac{5}{127} = \dfrac{80}{127}$

(At least 5 days means either 5, or 6, or 7 days.)

Empirical probabilities can also be found using a relative frequency distribution, as shown in Section 2–3.

For example, the relative frequency distribution of the soft-drink data shown before is

Response	Frequency	Relative frequency
Yes	15	0.60
No	8	0.32
Undecided	2	0.08
	25	1.00

Hence, the probability that a person responded "no" is 0.32, which is equal to $\frac{8}{25}$.

Law of Large Numbers

When a coin is tossed one time, it is common knowledge that the probability of getting a head is $\frac{1}{2}$. But what happens when the coin is tossed 50 times? Will it come up heads

25 times? Not all of the time. One should expect about 25 heads if the coin is fair. But due to the chance variation, 25 heads will not occur most of the time.

If the empirical probability of getting a head is computed using a small number of trials, it is usually not exactly $\frac{1}{2}$. However, as the number of trials increases, the empirical probability of getting a head will approach the theoretical probability of $\frac{1}{2}$, if in fact the coin is fair (i.e., balanced). This phenomenon is an example of the **law of large numbers.** In other words, if one tosses a coin enough times, the number of heads and tails will tend to "even out." This law holds for any type of gambling game—tossing dice, playing roulette, and so on.

It should be pointed out that the probabilities that the proportions steadily approach may or may not agree with those theorized in the classical model. If not, it can have important implications, such as "the die is not fair." Pit bosses in Las Vegas watch for empirical trends that do not agree with classical theories, and they will sometimes take a set of dice out of play if observed frequencies are too far out of line with classical expected frequencies.

Subjective Probability

The third type of probability is called *subjective probability.* **Subjective probability** uses a probability value based on an educated guess or estimate, employing opinions and inexact information.

In subjective probability, a person or group makes an educated guess at the chance that an event will occur. This guess is based on the person's experience and evaluation of a solution. For example, a sportswriter may say that there is a 70% probability that the Pirates will win the pennant next year. A physician might say that on the basis of her diagnosis, there is a 30% chance the patient will need an operation. A seismologist might say there is an 80% probability that an earthquake will occur in a certain area. These are only a few examples of how subjective probability is used in everyday life.

All three types of probability (classical, empirical, and subjective) are used to solve a variety of problems in business, engineering, and other fields.

Probability and Risk Taking

An area where people fail to understand probability is in risk taking. Actually, people fear situations or events that have a relatively small probability of happening rather than those events that have a greater likelihood of occurring. For example, a recent *USA Weekend* magazine poll (August 22–24, 1997) showed that 9 out of 10 Americans answered "No" when asked the question, "Is the world a safer place than when you were growing up?" However, in his book entitled *How Risk Affects Your Everyday Life* (Merritt Publishing, Santa Monica, California, 1996), author James Walsh states: "Despite widespread concern about the number of crimes committed in the United States, FBI and Justice Department statistics show that the national crime rate has remained fairly level for 20 years. It even dropped slightly in the early 1990s."

He further states, "Today most media coverage of risk to health and well-being focuses on shock and outrage." Shock and outrage make good stories and can scare us about the wrong dangers. For example, the author states that if a person is 20% overweight, the loss of life expectancy is 900 days (about 3 years), but loss of life expectancy from exposure to radiation emitted by nuclear power plants is 0.02 days. As you can see, being overweight is much more of a threat than being exposed to radioactive emission.

Many people gamble daily with their lives—for example, using tobacco, drinking and driving, riding motorcycles, etc. When people are asked to estimate the probabilities or frequencies of death from various causes, they tend to overestimate causes such as accidents, fires, and floods and underestimate the probabilities of death from diseases (other than cancer), strokes, etc. For example, most people think that their chances of dying of a heart attack are 1 in 20, when in fact it is almost 1 in 3; the chances of dying by pesticide poisoning are 1 in 200,000 (*True Odds* by James Walsh). The reason people think this way is that the news media sensationalize deaths resulting from catastrophic events and rarely mention deaths from disease.

When dealing with life-threatening catastrophes such as hurricanes, floods, automobile accidents, or smoking, it is important to get the facts. That is, get the actual numbers from accredited statistical agencies or reliable statistical studies, and then compute the probabilities and make decisions based on your knowledge of probability and statistics.

In summary, then, when you make a decision or plan a course of action based on probability, make sure that you understand the true probability of the event occurring. Also, find out how the information was obtained (i.e., from a reliable source). Weigh the cost of the action and decide if it is worth it. Finally, look for other alternatives or courses of action with less risk involved.

Exercises

5–1. What is a probability experiment?

5–2. Define *sample space*.

5–3. What is the difference between an outcome and an event?

5–4. What are equally likely events?

5–5. What is the range of the values of the probability of an event?

5–6. When an event is certain to occur, what is its probability?

5–7. If an event cannot happen, what value is assigned to its probability?

5–8. What is the sum of the probabilities of all of the outcomes in a sample space?

5–9. If the probability that it will snow tomorrow is 0.85, what is the probability that it will not snow tomorrow?

5–10. A probability experiment is conducted. Which of the following cannot be considered a probability of an outcome?

a. $\frac{1}{3}$ d. −0.59   g. 1
b. $-\frac{1}{5}$ e. 0   h. 33%
c. 0.80   f. 1.45   i. 112%

5–11. Classify each statement as an example of classical probability, empirical probability, or subjective probability.

a. The probability that a person will watch the 6:00 evening news is 0.15.
b. The probability of winning at a chuck-a-luck game is $\frac{5}{36}$.
c. The probability that a bus will be in an accident on a specific run is about 6%.
d. The probability of getting a royal flush when five cards are selected at random is $\frac{1}{649,740}$.

e. The probability that a student will get a C or better in a statistics course is about 70%.
f. The probability that a new fast-food restaurant will be a success in Chicago is 35%.
g. The probability that interest rates will rise in the next six months is 0.50.

5–12. (ans) If a die is rolled one time, find these probabilities.
a. Of getting a 4
b. Of getting an even number
c. Of getting a number greater than 4
d. Of getting a number less than 7
e. Of getting a number greater than 0
f. Of getting a number greater than 3 or an odd number
g. Of getting a number greater than 3 and an odd number

5–13. If two dice are rolled one time, find the probability of getting these results.
a. A sum of 6
b. Doubles
c. A sum of 7 or 11
d. A sum greater than 9
e. A sum less than or equal to 4

5–14. (ans) If one card is drawn from a deck, find the probability of getting these results.
a. An ace
b. A diamond
c. An ace of diamonds
d. A 4 or a 6
e. A 4 or a club
f. A 6 or a spade
g. A heart or a club
h. A red queen
i. A red card or a 7
j. A black card and a 10

5–15. A box contains five red, two white, and three green marbles. If a marble is selected at random, find these probabilities.

a. That it is red.

b. That it is green.

c. That it is red or white.

d. That it is not green.

e. That it is not red. ·

5–16. In an office there are five women and four men. If one person is selected, find the probability that the person is a woman.

5–17. If there are 50 tickets sold for a raffle and one person buys 7 tickets, what is the probability of that person winning the prize?

5–18. There are 7 cans of cola and 9 cans of ginger ale in a cooler. If a person selects one can of soda at random, find the probability it is a can of cola.

5–19. A survey found that 53% of Americans think U.S. military forces should be used to "protect the interest of U.S. corporations" in other countries. If an American is selected at random, find the probability that he or she will disagree or have no opinion on the issue.

5–20. A certain brand of grass seed has an 86% probability of germination. If a lawn care specialist plants 9000 seeds, find the number that should germinate.

5–21. A couple has three children. Find each probability.

a. Of all boys

b. Of all girls or all boys

c. Of exactly two boys or two girls

d. Of at least one child of each gender

5–22. In the game craps using two dice, a person wins on the first roll if a 7 or an 11 is rolled. Find the probability of winning on the first roll.

5–23. In a game of craps, a player loses on the roll if a 2, 3, or 12 is tossed on the first roll. Find the probability of losing on the first roll.

5–24. The U.S. Bureau of Justice Statistics reported that in 1995, 12,249 men and women escaped from state prisons. Of these, 12,166 were captured. Find the probability that an escapee was captured in 1995.

5–25. A roulette wheel has 38 spaces numbered 1 through 36, 0, and 00. Find the probability of getting these results.

a. An odd number

b. A number greater than 25

c. A number less than 15 not counting 0 and 00

5–26. Thirty-nine of 50 states are currently under court order to alleviate overcrowding and poor conditions in one or more of their prisons. If a state is selected at random, find the probability that it is currently under such a court order.

Source: *Harper's Index* 289, no. 1736 (January 1995), p. 11.

5–27. A baseball player's batting average is 0.331. If she is at bat 53 times during the season, find the approximate number of times she gets to first base safely. Walks do not count.

5–28. In a survey, 16 percent of American children said they use flattery to get their parents to buy them things. If a child is selected at random, find the probability that the child said he or she does not use parental flattery.

Source: *Harper's Index* 289, no. 1735 (December 1994), p. 13.

5–29. If three dice are rolled, find the probability of getting triples—e.g., 1, 1, 1; 2, 2, 2.

5–30. Among 100 students at a small school, 50 are mathematics majors, 30 are English majors, and 20 are history majors. If a student is selected at random, find the probability that she is neither a math major nor an English major.

5–31. The distribution of ages of CEOs is as follows:

Age	Frequency
21–30	1
31–40	8
41–50	27
51–60	29
61–70	24
71–up	11

Source: Information based on *USA Today Snapshot,* November 13, 1997.

If a CEO is selected at random, find the probability that his or her age is

a. Between 31 and 40

b. Under 31

c. Over 30 and under 51

d. Under 31 or over 60

*5–32. A person flipped a coin 100 times and obtained 73 heads. Can the person conclude that the coin was unbalanced?

*5–33. A medical doctor stated that with a certain treatment, a patient has a 50% chance of recovering without surgery. That is, "Either he will get well or he won't get well." Comment on his statement.

*5–34. The wheel spinner shown on the next page is spun twice. Find the sample space, and then determine the probability of the following events.

a. An odd number on the first spin and an even number on the second spin. (*Note:* 0 is considered even.)

b. A sum greater than 4

c. Even numbers on both spins

d. A sum that is odd

e. The same number on both spins

*5–35. Roll a die 180 times and record the number of 1s, 2s, 3s, 4s, 5s, and 6s. Compute the probabilities of each, and compare these probabilities with the theoretical results.

*5–36. Toss two coins 100 times and record the number of heads (0, 1, 2). Compute the probabilities of each outcome, and compare these probabilities with the theoretical results.

*5–37. Odds are used in gambling games to make them fair. For example, if a person rolled a die and won every time he or she rolled a 6, then the person would win on the average of once every 6 times. So that the game is fair, the odds of 5 to 1 are given. This means that if the person bet $1 and won, he or she could win $5. On the average, the

player would win $5 once in 6 rolls and lose $1 on the other 5 rolls—hence the term *fair game.*

In most gambling games, the odds given are not fair. For example, if the odds of winning are really 20 to 1, the house might offer 15 to 1 in order to make a profit.

Odds can be expressed as a fraction or as a ratio, such as $\frac{5}{1}$, 5:1, or 5 to 1. Odds are computed in favor of the event or against the event. The formulas for odds are

$$\text{odds in favor} = \frac{P(E)}{1 - P(E)}$$

$$\text{odds against} = \frac{P(\bar{E})}{1 - P(\bar{E})}$$

In the die example,

$$\text{odds in favor of a 6} = \frac{\frac{1}{6}}{\frac{5}{6}} = \frac{1}{5} \text{ or } 1{:}5$$

$$\text{odds against a 6} = \frac{\frac{5}{6}}{\frac{1}{6}} = \frac{5}{1} \text{ or } 5{:}1.$$

Find the odds in favor of and against each event.

a. Rolling a die and getting a 2

b. Rolling a die and getting an even number

c. Drawing a card from a deck and getting a spade

d. Drawing a card and getting a red card

e. Drawing a card and getting a queen

f. Tossing two coins and getting two tails

g. Tossing two coins and getting one tail

5–3

The Addition Rules for Probability

Objective 2. Find the probability of compound events using the addition rules.

Many problems involve finding the probability of two or more events. For example, at a large political gathering, one might wish to know, for a person selected at random, the probability that the person is a female or is a Republican. In this case, there are three possibilities to consider:

1. The person is a female.

2. The person is a Republican.

3. The person is both a female and a Republican.

Consider another example. At the same gathering there are Republicans, Democrats, and Independents. If a person is selected at random, what is the probability that the person is a Democrat or an Independent? In this case, there are only two possibilities:

1. The person is a Democrat.

2. The person is an Independent.

The difference between the two examples is that in the first case, the person selected can be a female and a Republican at the same time. In the second case, the person

This *Snapshot* gives the distribution of ages for CEOs. Do you see anything wrong with the categories? Exercise 5–31 uses the distribution shown here.

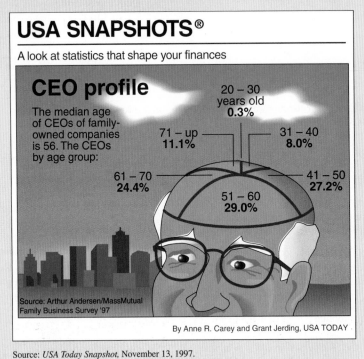

USA SNAPSHOTS®

A look at statistics that shape your finances

CEO profile

The median age of CEOs of family-owned companies is 56. The CEOs by age group:

20 – 30 years old
0.3%

71 – up
11.1%

31 – 40
8.0%

61 – 70
24.4%

41 – 50
27.2%

51 – 60
29.0%

Source: Arthur Andersen/MassMutual Family Business Survey '97

By Anne R. Carey and Grant Jerding, USA TODAY

Source: *USA Today Snapshot*, November 13, 1997.

selected cannot be both a Democrat and an Independent at the same time. In the second case, the two events are said to be mutually exclusive; in the first case, they are not mutually exclusive.

Two events are **mutually exclusive** if they cannot occur at the same time (i.e., they have no outcomes in common).

In another situation, the events of getting a 4 and getting a 6 when a single card is drawn from a deck are mutually exclusive events, since a single card cannot be both a 4 and a 6. On the other hand, the events of getting a 4 and getting a heart on a single draw are not mutually exclusive, since one can select the 4 of hearts when drawing a single card from an ordinary deck.

Example 5–15

Determine which events are mutually exclusive and which are not when a single die is rolled.

a. Getting an odd number and getting an even number.

b. Getting a 3 and getting an odd number.

c. Getting an odd number and getting a number less than 4.

d. Getting a number greater than 4 and getting a number less than 4.

Solution

a. The events are mutually exclusive, since the first event can be 1, 3, or 5, and the second event can be 2, 4, or 6.

b. The events are not mutually exclusive, since the first event is a 3 and the second can be 1, 3, or 5. Hence, 3 is contained in both events.

c. The events are not mutually exclusive, since the first event can be 1, 3, or 5, and the second can be 1, 2, or 3. Hence, 1 and 3 are contained in both events.

d. The events are mutually exclusive, since the first event can be 5 or 6, and the second event can be 1, 2, or 3.

Example 5–16

Determine which events are mutually exclusive and which are not when a single card is drawn from a deck.

a. Getting a 7 and getting a jack.
b. Getting a club and getting a king.
c. Getting a face card and getting an ace.
d. Getting a face card and getting a spade.

Solution

Only the events in parts *a* and *c* are mutually exclusive.

The probability of two or more events can be determined by the *addition rules*. The first addition rule is used when the events are mutually exclusive.

Addition Rule 1

When two events A and B are mutually exclusive, the probability that A or B will occur is

$$P(A \text{ or } B) = P(A) + P(B)$$

Example 5–17

A restaurant has 3 pieces of apple pie, 5 pieces of cherry pie, and 4 pieces of pumpkin pie in its dessert case. If a customer selects a piece of pie for dessert, find the probability that it will be either cherry or pumpkin.

Solution

Since there is a total of 12 pieces of pie,

$$P(\text{cherry or pumpkin}) = P(\text{cherry}) + P(\text{pumpkin}) = \frac{5}{12} + \frac{4}{12} = \frac{9}{12} = \frac{3}{4}$$

The events are mutually exclusive.

Example 5–18

At a political rally, there are 20 Republicans, 13 Democrats, and 6 Independents. If a person is selected at random, find the probability that he or she is either a Democrat or an Independent.

Solution

$$P(\text{Democrat or Independent}) = P(\text{Democrat}) + P(\text{Independent})$$
$$= \tfrac{13}{39} + \tfrac{6}{39} = \tfrac{19}{39}$$

Example 5–19

A day of the week is selected at random. Find the probability that it is a weekend day.

Solution

$$P(\text{Saturday or Sunday}) = P(\text{Saturday}) + P(\text{Sunday}) = \tfrac{1}{7} + \tfrac{1}{7} = \tfrac{2}{7}$$

When two events are not mutually exclusive, we must subtract one of the two probabilities of the outcomes that are common to both events, since they have been counted twice. This technique is illustrated in the next example.

Example 5–20

A single card is drawn from a deck. Find the probability that it is a king or a club.

Solution

Since the king of clubs is counted twice, one of the two probabilities must be subtracted, as shown.

$$P(\text{king or club}) = P(\text{king}) + P(\text{club}) - P(\text{king of clubs})$$
$$= \tfrac{4}{52} + \tfrac{13}{52} - \tfrac{1}{52} = \tfrac{16}{52} = \tfrac{4}{13}$$

When events are not mutually exclusive, addition rule 2 can be used to find the probability of the events.

Addition Rule 2

If A and B are *not* mutually exclusive, then

$$P(A \text{ or } B) = P(A) + P(B) - P(A \text{ and } B)$$

Note: This rule can also be used when the events are mutually exclusive, since $P(A$ and $B)$ will always equal 0. However, it is important to make a distinction between the two situations.

Example 5–21

In a hospital unit there are eight nurses and five physicians. Seven nurses and three physicians are females. If a staff person is selected, find the probability that the subject is a nurse or a male.

Solution

The sample space is shown below.

Staff	Females	Males	Total
Nurses	7	1	8
Physicians	3	2	5
Total	10	3	13

The probability is

$$P(\text{nurse or male}) = P(\text{nurse}) + P(\text{male}) - P(\text{male nurse})$$
$$= \tfrac{8}{13} + \tfrac{3}{13} - \tfrac{1}{13} = \tfrac{10}{13}$$

Example 5–22

On New Year's Eve, the probability of a person driving while intoxicated is 0.32, the probability of a person having a driving accident is 0.09, and the probability of a person having a driving accident while intoxicated is 0.06. What is the probability of a person driving while intoxicated or having a driving accident?

Solution

$$P(\text{intoxicated or accident}) = P(\text{intoxicated}) + P(\text{accident})$$
$$- P(\text{intoxicated and accident})$$
$$= 0.32 + 0.09 - 0.06 = 0.35$$

The probability rules can be extended to three or more events. For three mutually exclusive events, A, B, and C,

$$P(A \text{ or } B \text{ or } C) = P(A) + P(B) + P(C)$$

For three events that are *not* mutually exclusive,

$$P(A \text{ or } B \text{ or } C) = P(A) + P(B) + P(C) - P(A \text{ and } B) - P(A \text{ and } C)$$
$$- P(B \text{ and } C) + P(A \text{ and } B \text{ and } C)$$

See Exercises 5–61, 5–62, and 5–64.

Figure 5–5(a) shows a Venn diagram that represents two mutually exclusive events, A and B. In this case, $P(A \text{ or } B) = P(A) + P(B)$, since these events are mutually exclusive and do not overlap. In other words, the probability of event A or event B occurring is the sum of the areas of the two circles.

Figure 5–5

Venn Diagrams for the Addition Rules

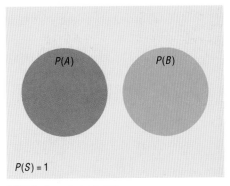
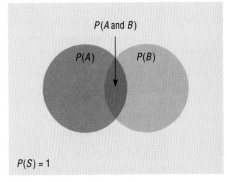

(a) Mutually exclusive events
$P(A \text{ or } B) = P(A) + P(B)$

(b) Non-mutually exclusive events
$P(A \text{ or } B) = P(A) + P(B) - P(A \text{ and } B)$

Figure 5–5(b) represents the probability of two events that are *not* mutually exclusive. In this case, $P(A \text{ or } B) = P(A) + P(B) - P(A \text{ and } B)$. The area in the intersection or overlapping part of both circles corresponds to $P(A \text{ and } B)$, and when the area of circle A is added to the area of circle B, the overlapping part is counted twice. It must therefore be subtracted once to get the correct area or probability.

Note: Venn diagrams were developed by mathematician John Venn (1834–1923) and are used in set theory and symbolic logic. They have been adapted to probability theory also. In set theory, the symbol $\cup$ represents the union of two sets and $A \cup B$ corresponds to A or B. The symbol $\cap$ represents the intersection of two sets, and $A \cap B$ corresponds to A and B. Venn diagrams show only a general picture of the probability rules and do not portray all situations, such as $P(A) = 0$, accurately.

Exercises

5–38. Define mutually exclusive events.

5–39. Give an example of two events that are mutually exclusive and two events that are not mutually exclusive.

5–40. Determine whether the following events are mutually exclusive.

a. Roll a die: Get an even number, and get a number less than 3.

b. Roll a die: Get a prime number (2, 3, 5), and get an odd number.

c. Roll a die: Get a number greater than 3, and get a number less than 3.

d. Select a student in your class: The student has blond hair, and the student has blue eyes.

e. Select a student in your college: The student is a sophomore, and the student is a business major.

f. Select any course: It is a calculus course, and it is an English course.

g. Select a registered voter: The voter is a Republican, and the voter is a Democrat.

5–41. A furniture store decides to select a month for its annual sale. Find the probability that it will be April or May. Assume that all months have an equal probability of being selected.

5–42. In a large department store, there are two managers, four department heads, 16 clerks, and four stock persons. If a person is selected at random, find the probability that the person is either a clerk or a manager.

5–43. At a convention there are seven mathematics instructors, five computer science instructors, three statistics instructors, and four science instructors. If an instructor is selected, find the probability of getting a science instructor or a math instructor.

5–44. An automobile dealer has 10 Fords, 7 Buicks, and 5 Plymouths on his used-car lot. If a person purchases a used car, find the probability that it is a Ford or Buick.

5–45. On a small college campus, there are five English professors, four mathematics professors, two science professors, three psychology professors, and three history professors. If a professor is selected at random, find the probability that the professor is the following:

a. An English or psychology professor

b. A mathematics or science professor

c. A history, science, or mathematics professor

d. An English, mathematics, or history professor

5–46. The probability of a California teenager owning a surfboard is 0.43, of owning a skateboard is 0.38, and of owning both is 0.28. If a California teenager is selected at random, find the probability that he or she owns a surfboard or a skateboard.

5–47. On any given day, the probability of a tourist visiting Indian Caverns is 0.80 and of visiting the Safari Zoo is 0.55. The probability of visiting both places on the same day is 0.42. Find the probability that a tourist visits Indian Caverns or visits Safari Zoo on any given day.

5–48. A single card is drawn from a deck. Find the probability of selecting the following:

a. A 4 or a diamond

b. A club or a diamond

c. A jack or a black card

5–49. In a statistics class there are 18 juniors and 10 seniors; 6 of the seniors are females, and 12 of the juniors are males. If a student is selected at random, find the probability of selecting the following:

a. A junior or a female

b. A senior or a female

c. A junior or a senior

5–50. A large department store purchases 400 boomboxes for a sale. Two hundred fifty of them have CD players and 150 of them have cassette players; 140 of the boomboxes with CD players are black, and the rest are white; 80 of the boomboxes with cassette players are black, and the rest are white. If a boombox is selected at random, find the probability that it will

a. Be black or have a CD player

b. Be white or have a cassette player

5–51. A woman's clothing store owner buys from three companies: A, B, and C. The most recent purchases are shown here.

Product	Company A	Company B	Company C
Dresses	24	18	12
Blouses	13	36	15

If one item is selected at random, find the following probabilities.

a. It was purchased from company A or is a dress.

b. It was purchased from company B or company C.

c. It is a blouse or was purchased from company A.

5–52. In a recent study, the following data were obtained in response to the question, "Do you favor the proposal of the school's combining the elementary and middle school students in one building?"

	Yes	No	No opinion
Males	72	81	5
Females	103	68	7

If a person is selected at random, find these probabilities.

a. The person has no opinion.
b. The person is a male or is against the issue.
c. The person is a female or favors the issue.

5–53. A grocery store employs cashiers, stock clerks, and deli personnel. The distribution of employees according to marital status is shown next.

Marital status	Cashiers	Stock clerks	Deli personnel
Married	8	12	3
Not married	5	15	2

If an employee is selected at random, find these probabilities:
a. The employee is a stock clerk or married.
b. The employee is not married.
c. The employee is a cashier or is not married.

5–54. In a certain geographic region, newspapers are classified as being published daily morning, daily evening, and weekly. Some have a comics section and some do not. The distribution is shown next.

Have comics section	Morning	Evening	Weekly
Yes	2	3	1
No	3	4	2

If a newspaper is selected at random, find these probabilities.
a. The newspaper is a weekly publication.
b. The newspaper is a daily morning publication or has comics.
c. The newspaper is published weekly or does not have comics.

5–55. Three cable channels (6, 8, and 10) have quiz shows, comedies, and dramas. The number of each is shown here.

Type of show	Channel 6	Channel 8	Channel 10
Quiz show	5	2	1
Comedy	3	2	8
Drama	4	4	2

If a show is selected at random, find these probabilities.
a. The show is a quiz show or it is shown on channel 8.
b. The show is a drama or a comedy.
c. The show is shown on channel 10 or it is a drama.

5–56. A local postal carrier distributes first-class letters, advertisements, or magazines. For a certain day, she distributed the following number of each type of item.

Delivered to	First-class letters	Ads	Magazines
Home	325	406	203
Business	732	1021	97

If an item of mail is selected at random, find these probabilities.
a. The item went to a home.
b. The item was an ad or it went to a business.
c. The item was a first-class letter or it went to a home.

5–57. The frequency distribution shown here illustrates the number of medical tests conducted on 30 randomly selected emergency patients.

Number of tests performed	Number of patients
0	12
1	8
2	2
3	3
4 or more	5

If a patient is selected at random, find these probabilities.
a. The patient has had exactly two tests done.
b. The patient has had at least two tests done.
c. The patient has had at most three tests done.
d. The patient has had three or fewer tests done.
e. The patient has had one or two tests done.

5–58. The following distribution represents the length of time a patient spends in a hospital.

Days	Frequency
0–3	2
4–7	15
8–11	8
12–15	6
16+	9

If a patient is selected, find these probabilities.
a. The patient spends 3 days or less in the hospital.
b. The patient spends less than 8 days in the hospital.
c. The patient spends 16 or more days in the hospital.
d. The patient spends a maximum of 11 days in the hospital.

5–59. A sales representative who visits customers at home finds she sells 0, 1, 2, 3, or 4 items according to the following frequency distribution.

Items sold	Frequency
0	8
1	10
2	3
3	2
4	1

Find the probability that she sells the following.
a. Exactly one item
b. More than two items

c. At least one item

d. At most three items

5–60. A recent study of 300 patients found that of 100 alcoholic patients, 87 had elevated cholesterol levels, and of 200 nonalcoholic patients, 43 had elevated cholesterol levels. If a patient is selected at random, find the probability that the patient is the following.

a. An alcoholic with elevated cholesterol level.

b. A nonalcoholic.

c. A nonalcoholic with nonelevated cholesterol level.

5–61. If one card is drawn from an ordinary deck of cards, find the probability of getting the following.

a. A king or a queen or a jack

b. A club or a heart or a spade

c. A king or a queen or a diamond

d. An ace or a diamond or a heart

e. A 9 or a 10 or a spade or a club

5–62. Two dice are rolled. Find the probability of getting the following.

a. A sum of 6 or 7 or 8

b. Doubles or a sum of 4 or 6

c. A sum greater than 9 or less than 4 or a 7

5–63. An urn contains six red balls, two green balls, one blue ball, and one white ball. If a ball is drawn, find the probability of getting a red or a white ball.

5–64. Three dice are rolled. Find the probability of getting the following:

a. Triples   *b.* A sum of 5

*5–65. The probability that a customer selects a pizza with mushrooms or pepperoni is 0.55, and the probability that

the customer selects mushrooms only is 0.32. If the probability that he or she selects pepperoni only is 0.17, find the probability of the customer selecting both items.

*5–66. In building new homes, a contractor finds that the probability of a home buyer selecting a two-car garage is 0.70 and of selecting a one-car garage is 0.20. Find the probability that the buyer will select no garage. The builder does not build houses with three or more car garages.

*5–67. In Exercise 5–66, find the probability that the buyer will not want a two-car garage.

"I know you haven't had an accident in thirteen years. We're raising your rates because you're about due one."

Source: © King Features Syndicate, Inc. Reprinted with special permission of King Features Syndicate, Inc. World rights reserved.

5–4

The Multiplication Rules and Conditional Probability

The Multiplication Rules

Objective 3. Find the probability of compound events using the multiplication rules.

The previous section showed that the addition rules are used to compute probabilities for mutually exclusive and not mutually exclusive events. This section introduces two more rules, the multiplication rules.

The multiplication rules can be used to find the probability of two or more events that occur in sequence. For example, if a coin is tossed and then a die is rolled, one can find the probability of getting a head on the coin *and* a 4 on the die. These two events are said to be *independent* since the outcome of the first event (tossing a coin) does not affect the probability outcome of the second event (rolling a die).

Two events *A* and *B* are **independent** if the fact that *A* occurs does not affect the probability of *B* occurring.

Here are other examples of independent events:

Rolling a die and getting a 6, and then rolling a second die and getting a 3.

Drawing a card from a deck and getting a queen, replacing it, and drawing a second card and getting a queen.

In order to find the probability of two independent events that occur in sequence, one must find the probability of each event occurring separately and then multiply the answers. For example, if a coin is tossed twice, the probability of getting two heads is $\frac{1}{2} \cdot \frac{1}{2} = \frac{1}{4}$. This result can be verified by looking at the sample space, HH, HT, TH, TT. Then $P(\text{HH}) = \frac{1}{4}$.

Multiplication Rule 1

When two events are independent, the probability of both occurring is

$$P(A \text{ and } B) = P(A) \cdot P(B)$$

Example 5–23

A coin is flipped and a die is rolled. Find the probability of getting a head on the coin and a 4 on the die.

Solution

$$P(\text{head and } 4) = P(\text{head}) \cdot P(4) = \frac{1}{2} \cdot \frac{1}{6} = \frac{1}{12}$$

Note that the sample space for the coin is H, T; and for the die it is 1, 2, 3, 4, 5, 6.

The problem in Example 5–23 can also be solved by using the sample space:

$$\text{H1} \quad \text{H2} \quad \text{H3} \quad \text{H4} \quad \text{H5} \quad \text{H6} \quad \text{T1} \quad \text{T2} \quad \text{T3} \quad \text{T4} \quad \text{T5} \quad \text{T6}$$

The solution is $\frac{1}{12}$, since there is only one way to get the head-4 outcome.

Example 5–24

A card is drawn from a deck and replaced; then a second card is drawn. Find the probability of getting a queen and then an ace.

Solution

The probability of getting a queen is $\frac{4}{52}$, and since the card is replaced, the probability of getting an ace is $\frac{4}{52}$. Hence, the probability of getting a queen and an ace is

$$P(\text{queen and ace}) = P(\text{queen}) \cdot P(\text{ace}) = \frac{4}{52} \cdot \frac{4}{52} = \frac{16}{2704} = \frac{1}{169}$$

Example 5–25

An urn contains three red balls, two blue balls, and five white balls. A ball is selected and its color noted. Then it is replaced. A second ball is selected and its color noted. Find the probability of each of the following.

a. Selecting two blue balls.

b. Selecting a blue ball and then a white ball.

c. Selecting a red ball and then a blue ball.

Solution

a. $P(\text{blue and blue}) = P(\text{blue}) \cdot P(\text{blue}) = \frac{2}{10} \cdot \frac{2}{10} = \frac{4}{100} = \frac{1}{25}$

b. $P(\text{blue and white}) = P(\text{blue}) \cdot P(\text{white}) = \frac{2}{10} \cdot \frac{5}{10} = \frac{10}{100} = \frac{1}{10}$

c. $P(\text{red and blue}) = P(\text{red}) \cdot P(\text{blue}) = \frac{3}{10} \cdot \frac{2}{10} = \frac{6}{100} = \frac{3}{50}$

Multiplication rule 1 can be extended to three or more independent events by using the formula

$$P(A \text{ and } B \text{ and } C \text{ and } \dots \text{ and } K) = P(A) \cdot P(B) \cdot P(C) \cdot \dots \cdot P(K)$$

When a small sample is selected from a large population and the subjects are not replaced, the probability of the event occurring changes so slightly that for the most part, it is considered to remain the same. The next two examples illustrate this concept.

Example 5–26

A Harris poll found that 46% of Americans say they suffer great stress at least once a week. If three people are selected at random, find the probability that all three will say that they suffer great stress at least once a week.

Source: *100% American* by Daniel Evan Weiss (Poseidon Press, 1988).

Solution

Let S denote stress. Then

$$
\begin{aligned}
P(S \text{ and } S \text{ and } S) &= P(S) \cdot P(S) \cdot P(S) \\
&= (0.46)(0.46)(0.46) \approx 0.097
\end{aligned}
$$

Example 5–27

Approximately 9% of men have a type of color blindness that prevents them from distinguishing between red and green. If 3 men are selected at random, find the probability that all of them will have this type of red-green color blindness.

Source: *USA Today Snapshot,* April 28, 1997.

Solution

Let C denote red-green color blindness. Then,

$$
\begin{aligned}
P(C \text{ and } C \text{ and } C) &= P(C) \cdot P(C) \cdot P(C) \\
&= (0.09)(0.09)(0.09) \\
&= 0.000729
\end{aligned}
$$

Hence, the rounded probability is 0.0007.

In the previous examples, the events were independent of each other, since the occurrence of the first event in no way affected the outcome of the second event. On the other hand, when the occurrence of the first event changes the probability of the occurrence of the second event, the two events are said to be *dependent*. For example, suppose a card is drawn from a deck and *not* replaced, and then a second card is drawn. What is the probability of selecting an ace on the first card and a king on the second card?

Before an answer to the question can be given, one must realize that the events are dependent. The probability of selecting an ace on the first draw is $\frac{4}{52}$. If that card is *not* replaced, the probability of selecting a king on the second card is $\frac{4}{51}$, since there are 4 kings and 51 cards remaining. The outcome of the first draw has affected the outcome of the second draw.

Dependent events are formally defined next.

When the outcome or occurrence of the first event affects the outcome or occurrence of the second event in such a way that the probability is changed, the events are said to be **dependent.**

Here are some examples of dependent events:

Drawing a card from a deck, not replacing it, and then drawing a second card.

Selecting a ball from an urn, not replacing it, and then selecting a second ball.

Being a lifeguard and getting a suntan.

Having high grades and getting a scholarship.

Parking in a no-parking zone and getting a parking ticket.

In order to find probabilities when events are dependent, use the multiplication rule with a modification in notation. For the problem just discussed, the probability of getting an ace on the first draw is $\frac{4}{52}$, and the probability of getting a king on the second draw is $\frac{4}{51}$. By the multiplication rule, the probability of both events occurring is

$$\frac{4}{52} \cdot \frac{4}{51} = \frac{16}{2652} = \frac{4}{663}$$

The event of getting a king on the second draw *given* that an ace was drawn the first time is called a *conditional probability*.

The **conditional probability** of an event B in relationship to an event A is the probability that event B occurs after event A has already occurred. The notation for conditional probability is $P(B|A)$. This notation does not mean that B is divided by A; rather, it means the probability that event B occurs given that event A has already occurred. In the card example, $P(B|A)$ is the probability that the second card is a king given that the first card is an ace, and it is equal to $\frac{4}{51}$ since the first card was *not* replaced.

Multiplication Rule 2

When two events are dependent, the probability of both occurring is

$$P(A \text{ and } B) = P(A) \cdot P(B|A)$$

Example 5–28

In a shipment of 25 microwave ovens, 2 are defective. If two ovens are randomly selected and tested, find the probability that both are defective if the first one is not replaced after it has been tested.

Solution

Since the events are dependent,

$$P(D_1 \text{ and } D_2) = P(D_1) \cdot P(D_2|D_1) = \frac{2}{25} \cdot \frac{1}{24} = \frac{2}{600} = \frac{1}{300}$$

Example 5–29

The World Wide Insurance Company found that 53% of the residents of a city had homeowner's insurance with the company. Of these clients, 27% also had automobile insurance with the company. If a resident is selected at random, find the probability that the resident has both homeowner's and automobile insurance with the World Wide Insurance Company.

Solution

$$P(H \text{ and } A) = P(H) \cdot P(A|H) = (0.53)(0.27) = 0.1431$$

This multiplication rule can be extended to three or more events, as shown in the next example.

Example 5–30

Three cards are drawn from an ordinary deck and not replaced. Find the probability of the following.

a. Getting three jacks.
b. Getting an ace, a king, and a queen in order.
c. Getting a club, a spade, and a heart in order.
d. Getting three clubs.

Solution

a. $P(3 \text{ jacks}) = \dfrac{4}{52} \cdot \dfrac{3}{51} \cdot \dfrac{2}{50} = \dfrac{24}{132,600} = \dfrac{1}{5525}$

b. $P(\text{ace and king and queen}) = \dfrac{4}{52} \cdot \dfrac{4}{51} \cdot \dfrac{4}{50} = \dfrac{64}{132,600} = \dfrac{8}{16,575}$

c. $P(\text{club and spade and heart}) = \dfrac{13}{52} \cdot \dfrac{13}{51} \cdot \dfrac{13}{50} = \dfrac{2197}{132,600} = \dfrac{169}{10,200}$

d. $P(3 \text{ clubs}) = \dfrac{13}{52} \cdot \dfrac{12}{51} \cdot \dfrac{11}{50} = \dfrac{1716}{132,600} = \dfrac{11}{850}$

Tree diagrams can be used as an aid to finding the solution to probability problems when the events are sequential. The next example illustrates the use of tree diagrams.

Example 5–31

Box 1 contains two red balls and one blue ball. Box 2 contains three blue balls and one red ball. A coin is tossed. If it falls heads up, box 1 is selected and a ball is drawn. If it falls tails up, box 2 is selected and a ball is drawn. Find the probability of selecting a red ball.

Solution

With the use of a tree diagram, the sample space can be determined as shown in Figure 5–6. First, assign probabilities to each branch. Next, using the multiplication rule, multiply the probabilities for each branch.

Figure 5–6

Tree Diagram for
Example 5–31

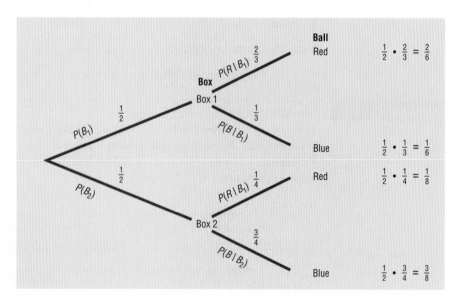

Finally, use the addition rule, since a red ball can be obtained from box 1 or box 2.

$$P(\text{red}) = \frac{2}{6} + \frac{1}{8} = \frac{8}{24} + \frac{3}{24} = \frac{11}{24}$$

(*Note:* The sum of all final probabilities will always be equal to 1.)

Tree diagrams can be used when the events are independent or dependent, and they can also be used for sequences of three or more events.

Conditional Probability

Objective 4. Find the conditional probability of an event.

The conditional probability of an event B in relationship to an event A was defined as the probability that event B occurs after event A has already occurred.

The conditional probability of an event can be found by dividing both sides of the equation for multiplication rule 2 by $P(A)$, as shown:

$$P(A \text{ and } B) = P(A) \cdot P(B|A)$$

$$\frac{P(A \text{ and } B)}{P(A)} = \frac{\cancel{P(A)} \cdot P(B|A)}{\cancel{P(A)}}$$

$$\frac{P(A \text{ and } B)}{P(A)} = P(B|A)$$

Formula for Conditional Probability

The probability that the second event B occurs given that the first event A has occurred can be found by dividing the probability that both events occurred by the probability that the first event has occurred. The formula is

$$P(B|A) = \frac{P(A \text{ and } B)}{P(A)}$$

The next three examples illustrate the use of this rule.

Example 5–32

A box contains black chips and white chips. A person selects two chips without replacement. If the probability of selecting a black chip *and* a white chip is $\frac{15}{56}$, and the probability of selecting a black chip on the first draw is $\frac{3}{8}$, find the probability of selecting the white chip on the second draw, *given* that the first chip selected was a black chip.

Solution

Let

$$B = \text{selecting a black chip} \qquad W = \text{selecting a white chip}$$

Then

$$P(W|B) = \frac{P(B \text{ and } W)}{P(B)} = \frac{\frac{15}{56}}{\frac{3}{8}}$$

$$= \frac{15}{56} \div \frac{3}{8} = \frac{15}{56} \cdot \frac{8}{3} = \frac{\overset{5}{\cancel{15}}}{\underset{7}{\cancel{56}}} \cdot \frac{\overset{1}{\cancel{8}}}{\underset{1}{\cancel{3}}} = \frac{5}{7}$$

Hence, the probability of selecting a white chip on the second draw given that the first chip selected was black is $\frac{5}{7}$.

Example 5–33

The probability that Sam parks in a no-parking zone *and* gets a parking ticket is 0.06, and the probability that Sam cannot find a legal parking space and has to park in the no-parking zone is 0.20. On Tuesday, Sam arrives at school and has to park in a no-parking zone. Find the probability that he will get a parking ticket.

Solution

Let

$$N = \text{parking in a no-parking zone} \qquad T = \text{getting a ticket}$$

Then

$$P(T|N) = \frac{P(N \text{ and } T)}{P(N)} = \frac{0.06}{0.20} = 0.30$$

Hence, Sam has a 0.30 probability of getting a parking ticket, given that he parked in a no-parking zone.

The conditional probability of events occurring can also be computed when the data is given in table form, as shown in the next example.

Example 5–34

A recent survey asked 100 people if they thought women in the armed forces should be permitted to participate in combat. The results of the survey are shown in the table.

Gender	Yes	No	Total
Male	32	18	50
Female	8	42	50
Total	40	60	100

Find these probabilities.

a. The respondent answered "yes," given that the respondent was a female.

b. The respondent was a male, given that the respondent answered "no."

Solution

Let

$$M = \text{respondent was a male} \qquad Y = \text{respondent answered "yes"}$$
$$F = \text{respondent was a female} \qquad N = \text{respondent answered "no"}$$

a. The problem is to find $P(Y|F)$. The rule states

$$P(Y|F) = \frac{P(F \text{ and } Y)}{P(F)}$$

The probability $P(F \text{ and } Y)$ is the number of females who responded "yes" divided by the total number of respondents:

$$P(F \text{ and } Y) = \frac{8}{100}$$

The probability $P(F)$ is the probability of selecting a female:

$$P(F) = \frac{50}{100}$$

Then

$$P(Y|F) = \frac{P(F \text{ and } Y)}{P(F)} = \frac{8/100}{50/100}$$

$$= \frac{8}{100} \div \frac{50}{100} = \frac{\overset{4}{\cancel{8}}}{\cancel{100}} \cdot \frac{\overset{1}{\cancel{100}}}{\underset{25}{\cancel{50}}} = \frac{4}{25}$$

b. The problem is to find $P(M|N)$.

$$P(M|N) = \frac{P(N \text{ and } M)}{P(N)} = \frac{18/100}{60/100}$$

$$= \frac{18}{100} \div \frac{60}{100} = \frac{\overset{3}{\cancel{18}}}{\cancel{100}} \cdot \frac{\overset{1}{\cancel{100}}}{\underset{10}{\cancel{60}}} = \frac{3}{10}$$

The Venn diagram for conditional probability is shown in Figure 5–7. In this case,

$$P(B|A) = \frac{P(A \text{ and } B)}{P(A)}$$

which is represented by the area in the intersection or overlapping part of the circles A and B divided by the area of circle A. The reasoning here is that if one assumes A has occurred, then A becomes the sample space for the next calculation and is the denominator of the probability fraction $\dfrac{P(A \text{ and } B)}{P(A)}$. The numerator $P(A \text{ and } B)$ represents the probability of the part of B that is contained in A. Hence, $P(A \text{ and } B)$ becomes the numerator of the probability fraction $\dfrac{P(A \text{ and } B)}{P(A)}$. Imposing a condition reduces the sample space.

Figure 5–7

Venn Diagram for
Conditional Probability

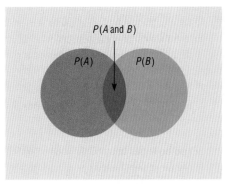

$$P(B|A) = \frac{P(A \text{ and } B)}{P(A)}$$

Probabilities for "At Least"

The multiplication rules can be used with the complementary event rule (Section 5–2) to simplify solving probability problems involving "at least." The next three examples illustrate how this is done.

Example 5–35

A game is played by drawing four cards from an ordinary deck and replacing each card after it is drawn. Find the probability of winning if at least one ace is drawn.

Solution

It is much easier to find the probability that no aces are drawn (i.e., losing) and then subtract from 1 than to find the solution directly, because that would involve finding the probability of getting one ace, two aces, three aces, and four aces and then adding the results.

Let E = at least one ace is drawn and $\bar{E}$ = no aces drawn. Then

$$P(\bar{E}) = \frac{48}{52} \cdot \frac{48}{52} \cdot \frac{48}{52} \cdot \frac{48}{52}$$

$$= \frac{12}{13} \cdot \frac{12}{13} \cdot \frac{12}{13} \cdot \frac{12}{13} = \frac{20{,}736}{28{,}561}$$

Hence,

$$P(E) = 1 - P(\bar{E})$$

$$P(\text{winning}) = 1 - P(\text{losing}) = 1 - \frac{20{,}736}{28{,}561} = \frac{7{,}825}{28{,}561} \approx 0.27$$

or a hand with at least one ace will win about 27% of the time.

Example 5–36

A coin is tossed five times. Find the probability of getting at least one tail.

Solution

It is easier to find the probability of the complement of the event, which is "all heads," and then subtract the probability from 1 to get the probability of at least one tail.

$$P(E) = 1 - P(\bar{E})$$

$$P(\text{at least 1 tail}) = 1 - P(\text{all heads})$$

$$P(\text{all heads}) = (\tfrac{1}{2})^5 = \tfrac{1}{32}$$

Hence,

$$P(\text{at least 1 tail}) = 1 - \tfrac{1}{32} = \tfrac{31}{32}$$

Example 5–37

The Neckware Association of America reported that 3% of ties sold in the United States are bow ties. If 4 customers who purchased a tie are randomly selected, find the probability that at least one purchased a bow tie.

Solution

Let E = at least one bow tie is purchased and $\bar{E}$ = no bow ties are purchased. Then

$$P(E) = 0.03 \text{ and } P(\bar{E}) = 1 - 0.03 = 0.97$$

$P(\text{no bow ties are purchased}) = (0.97)(0.97)(0.97)(0.97) \approx 0.885$; hence, P(at least one bow tie is purchased) = $1 - 0.885 = 0.115$.

Similar methods can be used for problems involving "at most."

Exercises

5–68. What is the difference between independent and dependent events? Give an example of each.

5–69. State which events are independent and which are dependent.

  a. Tossing a coin and drawing a card from a deck.
  b. Drawing a ball from an urn, not replacing it, and then drawing a second ball.
  c. Getting a raise in salary and purchasing a new car.

d. Driving on ice and having an accident.

e. Having a large shoe size and having a high IQ.

f. A father being left-handed and a daughter being left-handed.

g. Smoking excessively and having lung cancer.

h. Eating an excessive amount of ice cream and smoking an excessive amount of cigarettes.

5–70. If 18% of all Americans are underweight, find the probability that if three Americans are selected at random, all will be underweight.

Source: *100% American* by Daniel Evan Weiss (New York: Poseidon Press, 1988).

5–71. A survey found that 68% of book buyers are 40 or older. If two book buyers are selected at random, find the probability that both are 40 or older.

Source: *USA Today Snapshot,* March 24, 1999.

5–72. The Gallup Poll reported that 52% of Americans used a seat belt the last time they got into a car. If four people are selected at random, find the probability that they all used a seat belt the last time they got into a car.

Source: *100% American* by Daniel Evan Weiss (New York: Poseidon Press, 1988).

5–73. An automobile saleswoman finds that the probability of making a sale is 0.23. If she talks to four customers today, find the probability that she will sell four cars.

5–74. If 25% of U.S. federal prison inmates are not U.S. citizens, find the probability that two randomly selected federal prison inmates will not be U.S. citizens.

Source: *Harper's Index* 290, no. 1740 (May 1995), p. 11.

5–75. Find the probability of selecting two people at random who were born in the same month.

5–76. If two people are selected at random, find the probability that they have the same birthday (both month and day).

5–77. If three people are selected, find the probability that all three were born in March.

5–78. If half of Americans believe that the federal government should take "primary responsibility" for eliminating poverty, find the probability that three randomly selected Americans will agree that it is the federal government's responsibility to eliminate poverty.

Source: *Harper's Index* 289, no. 1735 (December 1994), p. 13.

5–79. What is the probability that a husband, wife, and daughter have the same birthday?

5–80. A flashlight has six batteries, two of which are defective. If two are selected at random without replacement, find the probability that both are defective.

5–81. In Exercise 5–80, find the probability that the first battery tests good and the second one is defective.

5–82. The U.S. Department of Justice reported that 6% of all American murders are committed without a weapon. If three murder cases are selected at random, find the probability that a weapon was not used in any one of them.

Source: *100% American* by Daniel Evan Weiss (New York: Poseidon Press, 1988).

5–83. In a department store there are 120 customers, 90 of whom will buy at least one item. If five customers are selected at random, one by one, find the probability that all will buy at least one item.

5–84. Three cards are drawn from a deck *without* replacement. Find these probabilities.

a. All are jacks.

b. All are clubs.

c. All are red cards.

5–85. In a scientific study there are eight guinea pigs, five of which are pregnant. If three are selected at random without replacement, find the probability that all are pregnant.

5–86. In Exercise 5–85, find the probability that none are pregnant.

5–87. In a class containing 12 men and 18 women, two students are selected at random to give an impromptu speech. Find the probability that both are women.

5–88. In Exercise 5–87, find the probability that both speeches are given by men.

5–89. A manufacturer makes two models of an item: model I, which accounts for 80% of unit sales, and model II, which accounts for 20% of unit sales. Because of defects, the manufacturer has to replace (or exchange) 10% of its model I and 18% of its model II. If a model is selected at random, find the probability that it will be defective.

5–90. An automobile manufacturer has three factories, A, B, and C. They produce 50%, 30%, and 20%, respectively, of a specific model of car. Thirty percent of the cars produced in factory A are white, 40% of those produced in factory B are white, and 25% produced in factory C are white. If an automobile produced by the company is selected at random, find the probability that it is white.

5–91. An insurance company classifies drivers as low-risk, medium-risk, and high-risk. Of those insured, 60% are low-risk, 30% are medium-risk, and 10% are high-risk. After a study, the company finds that during a one-year period, 1% of the low-risk drivers had an accident, 5% of the medium-risk drivers had an accident, and 9% of the high-risk drivers had an accident. If a driver is selected at random, find the probability that the driver will have had an accident during the year.

5–92. In a certain geographic location, 25% of the wage earners have a college degree and 75% do not. Of those who have a college degree, 5% earn more than $100,000 a year. Of those who do not have a college degree, 2% earn more than $100,000 a year. If a wage earner is selected at random, find the probability that he or she earns more than $100,000 a year.

5–93. Urn 1 contains five red balls and three black balls. Urn 2 contains three red balls and one black ball. Urn 3 contains four red balls and two black balls. If an urn is selected at random and a ball is drawn, find the probability it will be red.

5–94. At a small college, the probability that a student takes physics and sociology is 0.092. The probability that a student takes sociology is 0.73. Find the probability that the student is taking physics, given that he or she is taking sociology.

5–95. In a certain city, the probability that an automobile will be stolen and found within one week is 0.0009. The probability that an automobile will be stolen is 0.0015. Find the probability that a stolen automobile will be found within one week.

5–96. A circuit to run a model railroad has eight switches. Two are defective. If a person selects two switches at random and tests them, find the probability that the second one is defective, given that the first one is defective.

5–97. At the Avonlea Country Club, 73% of the members play bridge and swim, and 82% play bridge. If a member is selected at random, find the probability that the member swims, given that the member plays bridge.

5–98. At a large university, the probability that a student takes calculus *and* is on the dean's list is 0.042. The probability that a student is on the dean's list is 0.21. Find the probability that the student is taking calculus, given that he or she is on the dean's list.

5–99. In Rolling Acres Housing Plan, 42% of the houses have a deck and a garage; 60% have a deck. Find the probability that a home has a garage, given that it has a deck.

5–100. In a pizza restaurant, 95% of the customers order pizza. If 65% of the customers order pizza and a salad, find the probability that a customer who orders pizza will also order a salad.

5–101. At an exclusive country club, 68% of the members play bridge and drink champagne, and 83% play bridge. If a member is selected at random, find the probability that the member drinks champagne, given that he or she plays bridge.

5–102. Eighty students in a school cafeteria were asked if they favored a ban on smoking in the cafeteria. The results of the survey are shown in the table.

Class	Favor	Oppose	No opinion
Freshman	15	27	8
Sophomore	23	5	2

If a student is selected at random, find these probabilities.
a. Given that the student is a freshman, he or she opposes the ban.
b. Given that the student favors the ban, the student is a sophomore.

5–103. In a large shopping mall, a marketing agency conducted a survey on credit cards. The results are shown in the table.

Employment status	Owns a credit card	Does not own a credit card
Employed	18	29
Unemployed	28	34

If a person is selected at random, find these probabilities.
a. The person owns a credit card, given that the person is employed.
b. The person is unemployed, given that the person owns a credit card.

5–104. A study of graduates' average grades and degrees showed the following results.

Degree	Grade		
	C	**B**	**A**
B.S.	5	8	15
B.A.	7	12	8

If a graduate is selected at random, find these probabilities.
a. The graduate has a B.S. degree, given that he or she has an A average.
b. Given that the graduate has a B.A. degree, the graduate has a C average.

5–105. In a lab there are eight technicians. Three are male and five are female. If three technicians are selected, find the probability that at least one is female.

5–106. There are five chemistry instructors and six physics instructors at a college. If a committee of four instructors is selected, find the probability of at least one of them being a physics instructor.

5–107. On a surprise quiz consisting of five true–false questions, an unprepared student guesses each answer. Find the probability that he gets at least one correct.

5–108. A lot of portable radios contains 15 good radios and three defective ones. If two are selected and tested, find the probability that at least one will be defective.

5–109. If a family has 5 children, find the probability that at least one child is a boy.

5–110. A carpool contains three kindergartners and five first-graders. If two children are ill, find the probability that at least one of them is a kindergartner.

5–111. If four cards are drawn from a deck and not replaced, find the probability of getting at least one club.

5–112. At a local clinic there are eight men, five women, and three children in the waiting room. If three patients are randomly selected, find the probability that there is at least one child among them.

5–113. It has been found that 6% of all automobiles on the road have defective brakes. If five automobiles are stopped and checked by the state police, find the probability that at least one will have defective brakes.

5–114. A medication is 75% effective against a bacterial infection. Find the probability that if 12 people take the medication, at least one person's infection will not improve.

5–115. A coin is tossed six times. Find the probability of getting at least one tail.

5–116. If three digits are randomly selected, find the probability of getting at least one 7. Digits can be used more than once.

5–117. If a die is rolled five times, find the probability of getting at least one 6.

5–118. At a teachers' conference, there were four English teachers, three mathematics teachers, and five science teachers. If four teachers are selected for a committee, find the probability that at least one is a science teacher.

5–119. If a die is rolled three times, find the probability of getting at least one even number.

5–120. At a faculty meeting, there were seven full professors, five associate professors, six assistant professors, and 12 instructors. If four people are selected at random to attend a conference, find the probability that at least one is a full professor.

5–5

Probability and Counting Techniques (Optional)

Objective 5. Find the probability of an event using the counting rules.

The counting rules in Chapter 4 can be combined with the probability rules in this chapter to solve many types of probability problems. By using the multiplication rules, the permutations rules, and the combination rule, one can compute the probability of outcomes of many experiments, such as getting a full house when five cards are dealt or selecting a committee of three women and two men from a club consisting of 10 women and 10 men.

Example 5–38

Find the probability of getting four aces when five cards are drawn from an ordinary deck of cards.

Solution

There are $_{52}C_5$ ways to draw five cards from a deck. There is only one way to get four aces (i.e., $_4C_4$), but there are 48 possibilities to get the fifth card. Therefore, there are 48 ways to get four aces and one other card. Hence,

$$P(4 \text{ acres}) = \frac{_4C_4 \cdot 48}{_{52}C_5} = \frac{1 \cdot 48}{2,598,960} = \frac{48}{2,598,960} = \frac{1}{54,145}$$

Example 5–39

A box contains 24 transistors, four of which are defective. If four are sold at random, find the following probabilities.

a. Exactly two are defective.
b. None is defective.
c. All are defective.
d. At least one is defective.

Solution

There are $_{24}C_4$ ways to sell four transistors, so the denominator in each case will be 10,626.

a. Two defective transistors can be selected as $_4C_2$ and two nondefective ones as $_{20}C_2$. Hence,

$$P(\text{exactly 2 defectives}) = \frac{_4C_2 \cdot _{20}C_2}{_{24}C_4} = \frac{1140}{10,626} = \frac{190}{1771}$$

b. The number of ways to choose no defectives is $_{20}C_4$. Hence,

$$P(\text{no defectives}) = \frac{_{20}C_4}{_{24}C_4} = \frac{4845}{10,626} = \frac{1615}{3542}$$

c. The number of ways to choose four defectives from four is $_4C_4$, or 1. Hence,

$$P(\text{all defective}) = \frac{1}{_{24}C_4} = \frac{1}{10,626}$$

d. To find the probability of at least one defective transistor, find the probability that there are no defective transistors, and then subtract that probability from 1.

$$P(\text{at least 1 defective}) = 1 - P(\text{no defectives})$$

$$= 1 - \frac{_{20}C_4}{_{24}C_4} = 1 - \frac{1615}{3542} = \frac{1927}{3542}$$

Example 5–40

A store has 6 *TV Graphic* magazines and 8 *Newstime* magazines on the counter. If two customers purchased a magazine, find the probability that one of each magazine was purchased.

Solution

$$P(1 \text{ } TV \text{ } Graphic \text{ and } 1 \text{ } Newstime) = \frac{_6C_1 \cdot _8C_1}{_{14}C_2} = \frac{6 \cdot 8}{91} = \frac{48}{91}$$

Example 5–41

A combination lock consists of the 26 letters of the alphabet. If a three-letter combination is needed, find the probability that the combination will consist of the letters ABC in that order. The same letter can be used more than once. (*Note:* A combination lock is really a permutation lock.)

Solution

Since repetitions are permitted, there are $26 \cdot 26 \cdot 26 = 17{,}576$ different possible combinations. And since there is only one ABC combination, the probability is $P(\text{ABC}) = 1/26^3 = 1/17{,}576$.

Example 5–42

There are eight married couples in a tennis club. If one man and one woman are selected at random to plan the summer tournament, find the probability that they are married to each other.

Solution

Since there are eight ways to select the man and eight ways to select the woman, there are $8 \cdot 8$ or 64, ways to select one man and one woman. Since there are eight married couples, the solution is $\frac{8}{64} = \frac{1}{8}$.

As indicated at the beginning of this section, the rules in this chapter and the previous chapter can be used to solve a large variety of probability problems found in business, gambling, economics, biology, and other fields.

Exercises

5–121. Find the probability of getting two face cards (king, queen, or jack) when two cards are drawn from a deck without replacement.

5–122. A parent–teacher committee consisting of four people is to be formed from 20 parents and five teachers. Find the probability that the committee will consist of the following. (Assume that the selection will be random.)
  a. All teachers
  b. Two teachers and two parents
  c. All parents
  d. One teacher and three parents

5–123. In a company there are seven executives: four women and three men. Three are selected to attend a management seminar. Find these probabilities.
  a. All three selected will be women.
  b. All three selected will be men.
  c. Two men and one woman will be selected.
  d. One man and two women will be selected.

5–124. A city council consists of 10 members. Four are Republicans, three are Democrats, and three are Independents. If a committee of three is to be selected, find the probability of selecting the following.
  a. All Republicans

  b. All Democrats
  c. One of each party
  d. Two Democrats and one Independent
  e. One Independent and two Republicans

5–125. In a class of 18 students, there are 11 men and seven women. Four students are selected to present a demonstration on the use of the calculator. Find the probability that the group consists of the following.
  a. All men
  b. All women
  c. Three men and one woman
  d. One man and three women
  e. Two men and two women

5–126. A package contains 12 resistors, 3 of which are defective. If four are selected, find the probability of getting the following.
  a. No defective resistors
  b. One defective resistor
  c. Three defective resistors

5–127. If 50 tickets are sold and two prizes are to be awarded, find the probability that one person will win two prizes if that person buys 2 tickets.

5–128. Find the probability of getting a full house (three cards of one denomination and two of another) when five cards are dealt from an ordinary deck.

5–129. A committee of four people is to be formed from six doctors and eight dentists. Find the probability that the committee will consist of the following:
  a. All dentists
  b. Two dentists and two doctors
  c. All doctors
  d. Three doctors and one dentist
  e. One doctor and three dentists

5–130. An insurance sales representative selects three policies to review. The group of policies he can select from contains eight life policies, five automobile policies, and two homeowner's policies. Find the probability of selecting the following:
  a. All life policies
  b. Both homeowner's policies
  c. All automobile policies
  d. One of each policy
  e. Two life and one automobile policies

5–131. Find the probability of getting any triple-digit number, where all the digits are the same, on a lotto that consists of selecting a three-digit number.

5–132. Find the probability of selecting three science books and four math books from eight science books and nine math books. The books are selected at random.

5–133. When three dice are rolled, find the probability of getting a sum of 7.

5–134. Find the probability of randomly selecting two mathematics books and three physics books from four mathematics books and eight physics books from a box.

5–135. Find the probability that if five different-sized washers are arranged in a row, they will be arranged in order of size.

5–136. Using the information in Exercise 4–69, Chapter 4, find the probability of each poker hand.
  a. Royal flush
  b. Straight flush
  c. Four of a kind

5–6 Summary

In this chapter, the basic concepts and rules of probability are explained. The three types of probability are classical, empirical, and subjective. Classical probability uses sample spaces. Empirical probability uses frequency distributions and is based on observation. In subjective probability, the researcher makes an educated guess about the chance of an event occurring.

A probability event consists of one or more outcomes of a probability experiment. Two events are said to be mutually exclusive if they cannot occur at the same time. Events can also be classified as independent or dependent. If events are independent, whether or not the first event occurs does not affect the probability of the next event occurring. If the probability of the second event occurring is changed by the occurrence of the first event, then the events are dependent. The complement of an event is the set of outcomes in the sample space that are not included in the outcomes of the event itself. Complementary events are mutually exclusive.

Probability problems can be solved by using the addition rules, the multiplication rules, and the complementary event rules. Finally, when the number of outcomes of the sample space is large, probability problems can be solved by using the counting rules in Chapter 4.

Important Terms

classical probability   171

complement of an event   174

compound event   171

conditional probability   191

dependent events   190

empirical probability   175

equally likely events   171

event   171

independent events   188

law of large numbers   178

mutually exclusive events   182

outcome   168

probability   168

probability experiment   168

sample space   168

simple event   171

subjective probability   178

Venn diagrams   175

Important Formulas

Formula for classical probability:

$$P(E) = \frac{\text{number of outcomes in } E}{\text{total number of outcomes in the sample space}} = \frac{n(E)}{n(S)}$$

Formula for empirical probability:

$$P(E) = \frac{\text{frequency for the class}}{\text{total frequencies in the distribution}} = \frac{f}{n}$$

Addition rule 1, for two mutually exclusive events:

$$P(A \text{ or } B) = P(A) + P(B)$$

Addition rule 2, for events that are not mutually exclusive:

$$P(A \text{ or } B) = P(A) + P(B) - P(A \text{ and } B)$$

Multiplication rule 1, for independent events:

$$P(A \text{ and } B) = P(A) \cdot P(B)$$

Multiplication rule 2, for dependent events:

$$P(A \text{ and } B) = P(A) \cdot P(B|A)$$

Formula for conditional probability:

$$P(B|A) = \frac{P(A \text{ and } B)}{P(A)}$$

Formula for complementary events:

$$P(\bar{E}) = 1 - P(E) \quad \text{or} \quad P(E) = 1 - P(\bar{E})$$
$$\text{or} \quad P(E) + P(\bar{E}) = 1$$

Review Exercises

5–137. When a die is rolled, find the probability of getting
a. A 5
b. A 6
c. A number less than 5

5–138. When a card is selected from a deck, find the probability of getting
a. A club
b. A face card or a heart
c. A 6 and a spade
d. A king
e. A red card

5–139. In a survey conducted at a local restaurant during breakfast hours, 20 people preferred orange juice, 16 preferred grapefruit juice, and 9 preferred apple juice with breakfast. If a person is selected at random, find the probability that he or she prefers grapefruit juice.

5–140. If a die is rolled one time, find these probabilities:
a. Getting a 5
b. Getting an odd number
c. Getting a number less than 3

5–141. A recent survey indicated that in a town of 1500 households, 850 had cordless telephones. If a household is randomly selected, find the probability that it has a cordless telephone.

5–142. During a sale at a men's store, 16 white sweaters, 3 red sweaters, 9 blue sweaters, and 7 yellow sweaters were purchased. If a customer is selected at random, find the probability that he bought the following:
a. A blue sweater

b. A yellow or a white sweater
c. A red, a blue, or a yellow sweater
d. A sweater that was not white

5–143. At a swimwear store, the managers found that 16 women bought white bathing suits, 4 bought red suits, 3 bought blue suits, and 7 bought yellow suits. If a customer is selected at random, find the probability that she bought the following.
a. A blue suit
b. A yellow or a red suit
c. A white or a yellow or a blue suit
d. A suit that was not red

5–144. When two dice are rolled, find the probability of getting
a. A sum of 5 or 6
b. A sum greater than 9
c. A sum less than 4 or greater than 9
d. A sum that is divisible by 4
e. A sum of 14
f. A sum less than 13

5–145. The probability that a person owns a car is 0.80, that a person owns a boat is 0.30, and that a person owns both a car and a boat is 0.12. Find the probability that a person owns either a boat or a car, but not both.

5–146. There is a 0.39 probability that John will purchase a new car, a 0.73 probability that Mary will purchase a new car, and a 0.36 probability that both will purchase a new car. Find the probability that neither will purchase a new car.

5–147. A Gallup Poll found that 78% of Americans worry about the quality and healthfulness of their diet. If five people are selected at random, find the probability that all five worry about the quality and healthfulness of their diet.

Source: *The Book of Odds* by Michael D. Shook and Robert C. Shook (New York: Penguin Putnam, Inc., 1991), p. 33.

5–148. Twenty-five percent of the engineering graduates of a university received a starting salary of $25,000 or more. If three of the graduates are selected at random, find the probability that all had a starting salary of $25,000 or more.

5–149. Three cards are drawn from an ordinary deck *without* replacement. Find the probability of getting
a. All black cards
b. All spades
c. All queens

5–150. A coin is tossed and a card is drawn from a deck. Find the probability of getting
a. A head and a 6
b. A tail and a red card
c. A head and a club

5–151. A box of candy contains six chocolate-covered cherries, three peppermint patties, two caramels, and two strawberry creams. If a piece of candy is selected, find the probability of getting a caramel or a peppermint patty.

5–152. A manufacturing company has three factories: X, Y, and Z. The daily output of each is shown below.

Product	Factory X	Factory Y	Factory Z
TVs	18	32	15
Stereos	6	20	13

If one item is selected at random, find these probabilities.
a. It was manufactured at factory X or is a stereo.
b. It was manufactured at factory Y or factory Z.
c. It is a TV or was manufactured at factory Z.

5–153. A vaccine has a 90% probability of being effective in preventing a certain disease. The probability of getting the disease if a person is not vaccinated is 50%. In a certain geographic region, 25% of the people get vaccinated. If a person is selected at random, find the probability that he or she will contract the disease.

5–154. A manufacturer makes three models of a television set, models A, B, and C. A store sells 40% of model A sets, 40% of model B sets, and 20% of model C sets. Of model A sets, 3% have stereo sound; of model B sets, 7% have stereo sound; and of model C sets, 9% have stereo sound. If a set is sold at random, find the probability that it has stereo sound.

5–155. The probability that Sue will live on campus and buy a new car is 0.37. If the probability that she will live on campus is 0.73, find the probability that she will buy a new car, given that she lives on campus.

5–156. The probability that a customer will buy a television set and buy an extended warranty is 0.03. If the probability that a customer will purchase a television set is 0.11, find the probability that the customer will also purchase the extended warranty.

5–157. Of the members of the Blue River Health Club, 43% have a lifetime membership and exercise regularly (three or more times a week). If 75% of the club members exercise regularly, find the probability that a randomly selected member is a life member, given that he or she exercises regularly.

5–158. The probability that it snows and the bus arrives late is 0.023. John hears the weather forecast, and there is a 40% chance of snow tomorrow. Find the probability that the bus will be late, given that it snows.

5–159. A number of students were grouped according to their reading ability and education. The table shows the results.

Education	Low	Average	High
	Reading ability		
Graduated high school	6	18	43
Did not graduate	27	16	7

If a student is selected at random, find these probabilities.
a. The student has a low reading ability, given that the student is a high school graduate.
b. The student has a high reading ability, given that the student did not graduate.

5–160. At a large factory, the employees were surveyed and classified according to their level of education and whether or not they smoked. The data are shown in the table.

Smoking habit	Not high school graduate	High school graduate	College graduate
	Educational level		
Smoke	6	14	19
Do not smoke	18	7	25

If an employee is selected at random, find these probabilities.
a. The employee smokes, given that he or she graduated from college.
b. Given that the employee did not graduate from high school, he or she is a smoker.

5–161. A survey done for *Prevention* magazine found that 77% of bike riders sometimes ride without a helmet. If four bike riders are randomly selected, find the probability that at least one of the riders does not wear a helmet all the time.

Source: Snapshot, *USA Today,* May 26, 1995.

5–162. A coin is tossed five times. Find the probability of getting at least one tail.

5–163. The U.S. Department of Health and Human Services reports that 15% of Americans have chronic sinusitis. If five people are selected at random, find the probability that at least one has chronic sinusitis.

Source: *100% American* by Daniel Evans Weiss (New York: Poseidon Press, 1988).

5–164. A person has six bonds, three stocks, and two mutual funds. If three investments are selected, find the probability that one of each type is selected.

5–165. A newspaper advertises five different movies, three plays, and two baseball games for the weekend. If a couple selects three activities, find the probability that they attend two plays and one movie.

5–166. In an office there are three secretaries, four accountants, and two receptionists. If a committee of three is to be formed, find the probability that one of each will be selected.

Statistics Today

Would You Bet Your Life? Revisited

In his book *Probabilities in Everyday Life,* John D. McGervey states that the chance of being killed on any given commercial airline flight is almost 1 in 1 million and that the chance of being killed during a transcontinental auto trip is about 1 in 8000. The corresponding probabilities are 1/1,000,0000 = 0.000001 as compared to 1/8000 = 0.000125. Since the second number is 125 times greater than the first number, you have a much higher risk driving than flying across the United States.

Chapter Quiz

Determine whether each statement is true or false. If the statement is false, explain why.

1. Subjective probability has little use in the real world.

2. Classical probability uses a frequency distribution to compute probabilities.

3. In classical probability, all outcomes in the sample space are equally likely.

4. When two events are not mutually exclusive, $P(A \text{ or } B) = P(A) + P(B)$.

5. If two events are dependent, they must have the same probability of occurring.

6. An event and its complement can occur at the same time.

Select the best answer.

7. The probability that an event happens is 0.42. What is the probability that the event won't happen?
   a. −0.42
   b. 0.58
   c. 0
   d. 1

8. When a meteorologist says that there is a 30% chance of showers, what type of probability is the person using?
   a. Classical
   b. Empirical
   c. Relative
   d. Subjective

9. The sample space for tossing three coins consists of how many outcomes?
   a. 2
   b. 4
   c. 6
   d. 8

10. The complement of guessing five correct answers on a five-question true–false exam is
   a. Guessing five incorrect answers
   b. Guessing at least one incorrect answer
   c. Guessing at least one correct answer
   d. Guessing no incorrect answers

11. When two dice are rolled, the sample space consists of how many events?
   a. 6

b. 12

c. 36

d. 54

Complete the following statements with the best answer.

12. The set of all possible outcomes of a probability experiment is called the _____ .

13. The probability of an event can be any number between and including _____ and _____ .

14. If an event cannot occur, its probability is _____ .

15. The sum of the probabilities of the events in the sample space is _____ .

16. When two events cannot occur at the same time, they are said to be _____ .

17. When a card is drawn, what is the probability of getting
    a. A jack
    b. A 4
    c. A card less than 6 (an ace is considered above 6)

18. When a card is drawn from a deck, find the probability of getting
    a. A diamond
    b. A 5 or a heart
    c. A 5 and a heart
    d. A king
    e. A red card

19. At a men's clothing store, 12 men purchased blue golf sweaters, 8 purchased green sweaters, 4 purchased gray sweaters, and 7 bought black sweaters. If a customer is selected at random, find the probability that he purchased
    a. A blue sweater
    b. A green or gray sweater
    c. A green or black or blue sweater
    d. A sweater that was not black

20. When two dice are rolled, find the probability of getting
    a. A sum of 6 or 7
    b. A sum greater than 8
    c. A sum less than 3 or greater than 8
    d. A sum that is divisible by 3
    e. A sum of 16
    f. A sum less than 11

21. The probability that a person owns a microwave oven is 0.75, that a person owns a compact disk player is 0.25, and that a person owns both a microwave and a CD player, 0.16. Find the probability that a person owns either a microwave or a CD player, but not both.

22. Of the physics graduates of a university, 30% received a starting salary of $30,000 or more. If five of the graduates are selected at random, find the probability that all had a starting salary of $30,000 or more.

23. Five cards are drawn from an ordinary deck *without* replacement. Find the probability of getting
    a. All red cards
    b. All diamonds
    c. All aces

24. The probability that Sam will be accepted by the college of his choice *and* obtain a scholarship is 0.35. If the probability that he is accepted by the college is 0.65, find the probability that he will obtain a scholarship given that he is accepted by the college.

25. The probability that a customer will buy a car and an extended warranty is 0.16. If the probability that a customer will purchase a car is 0.30, find the probability that the customer will also purchase the extended warranty.

26. Of the members of the Spring Lake Bowling Lanes, 57% have a lifetime membership and bowl regularly (three or more times a week). If 70% of the club members bowl regularly, find the probability that a randomly selected member is a lifetime member given that he or she bowls regularly.

27. The probability that John has to work overtime and it rains is 0.028. John hears the weather forecast, and there is a 50% chance of rain. Find the probability that he will have to work overtime given that it rains.

28. At a large factory, the employees were surveyed and classified according to their level of education and whether or not they attend a sports event at least once a month. The data are shown in the table.

| | Educational level | | |
Sports event	High school graduate	Two-year college degree	Four-year college degree
Attend	16	20	24
Do not attend	12	19	25

If an employee is selected at random, find the probability that
    a. The employee attends sports events regularly given that he or she graduated from college (two- or four-year degree).
    b. Given that the employee is a high school graduate, he or she does not attend sports events regularly.

29. In a certain high-risk group, the chances of a person having suffered a heart attack are 55%. If six people are chosen, find the probability that at least one will have had a heart attack.

30. A single die is rolled four times. Find the probability of getting at least one 5.

31. If 85% of all people have brown eyes and six people are selected at random, find the probability that at least one of them has brown eyes.

32. On a lunch counter, there are three oranges, five apples, and two bananas. If three pieces of fruit are selected, find the probability that one orange, one apple, and one banana are selected.

33. A cruise director schedules four different movies, two bridge games, and three tennis games for a two-day period. If a couple selects three activities, find the probability that they attend two movies and one tennis game.

34. At a sorority meeting, there are six seniors, four juniors, and two sophomores. If a committee of three is to be formed, find the probability that one of each will be selected.

Critical Thinking Challenges

1. Consider the following problem: A con man has three coins. One coin has been specially made and has a head on each side. A second coin has been specially made and on each side it has a tail. Finally, a third coin has a head and a tail on it. All coins are of the same denomination. The con man places the three coins in his pocket, selects one, and shows you one side. It is heads. He is willing to bet you even money that it is the two-headed coin. His reasoning is that it can't be the two-tailed coin since a head is showing; therefore, there is a 50-50 chance of it being the two-headed coin. Would you take the bet? (*Hint:* See Exercise 1 in Data Projects.)

2. Chevalier de Méré won money when he bet unsuspecting patrons that in four rolls of a die, he could get at least one 6, but he lost money when he bet that in 24 rolls of two dice, he could get at least a double 6. Using the probability rules, find the probability of each event and explain why he won the majority of the time on the first game but lost the majority of the time when playing the second game. (*Hint:* Find the probabilities of losing each game and subtract from one.)

3. How many people do you think need to be in a room so that two people will have the same birthday (month and day)? You might think it is 366. This would, of course, guarantee it (excluding leap year), but how many people would need to be in a room so that there would be a 90% probability that two people would be born on the same day? What about a 50% probability?

   Actually, the number is much smaller than you may think. For example, if you have 50 people in a room, the probability that two people will have the same birthday is 97%. If you have 23 people in a room, there is a 50% probability that two people were born on the same day!

   The problem can be solved by using the probability rules. It must be assumed that all birthdays are equally likely, but this assumption will have little effect on the answers. The way to find the answer is by using the complementary event rule as P (two people having the same birthday) $= 1 - P$ (all have different birthdays).

For example, suppose there were three people in the room. The probability that each had a different birthday would be

$$\left(\frac{365}{365}\right) \cdot \left(\frac{364}{365}\right) \cdot \left(\frac{363}{365}\right) = \frac{{}_{365}P_3}{365^3} = 0.992$$

Hence, the probability that at least two of the three people will have the same birthday will be

$$1 - 0.992 = 0.008$$

Hence, for k people, the formula is

P(at least 2 people have the same birthday) $=$

$$1 - \frac{{}_{365}P_k}{365^k}$$

Using your calculator, complete the following table and verify that for at least a 50% chance of two people having the same birthday, 23 or more people will be needed.

Number of people	Probability that at least two have the same birthday
1	0.000
2	0.003
5	0.027
10	
15	
20	
21	
22	
23	

4. We know that if the probability of an event happening is 100%, then the event is a certainty. Can it be concluded that if there is a 50% chance of contracting a communicable disease through contact with an infected person, there would be a 100% chance of contracting the disease if two contacts were made with the infected person? Explain your answer.

Data Projects

1. Make a set of three cards—one with a red star on both sides, one with a black star on both sides, and one with a black star on one side and a red star on the other side. With a partner, play the game described in the first critical thinking challenge on page 207 one hundred times and record the results of how many times you win and how many times your partner wins. (*Note:* Do not change options during the 100 trials.)
   a. Do you think the game is fair (i.e., does one person win approximately 50% of the time)?
   b. If you think the game is unfair, explain what the probabilities might be and why.

2. Take a coin and tape a small weight (e.g., part of a paper clip) to one side. Flip the coin 100 times and record the results. Do you think you have changed the probabilities of the results of flipping the coin? Explain.

3. This game is called "Diet Fractions." Roll two dice and use the numbers to make a fraction less than or equal to one. Player A wins if the fraction cannot be reduced; otherwise, player B wins.
   a. Play the game 100 times and record the results.
   b. Decide if the game is fair or not. Explain why or why not.
   c. Using the sample space for two dice, compute the probabilities of player A winning and player B winning. Do these agree with the results obtained in part a?

Source: George W. Bright, John G. Harvey, and Margariete Montague Wheeler, "Fair Games, Unfair Games." Chapter 8, *Teaching Statistics and Probability.* NCTM 1981 Yearbook. Reston, Virginia: The National Council of Teachers of Mathematics, Inc., 1981, p. 49. Used with permission.

4. Often when playing gambling games or collecting items in cereal boxes, one wonders how long will it be before one achieves a success. For example, suppose there are six different types of toys with one toy packaged at random in a cereal box. If a person wanted a certain toy, about how many boxes would that person have to buy on average before obtaining that particular toy? Of course, there is a possibility that the particular toy would be in the first box opened or that the person might never obtain the particular toy. These are the extremes.
   a. To find out, simulate the experiment using dice. Start rolling dice until a particular number, say 3, is obtained and keep track of how many rolls are necessary. Repeat 100 times. Then find the average.
   b. You may decide to use another number, such as 10 different items. In this case, use 10 playing cards (ace through 10 of diamonds), select a particular card (say an ace), shuffle the deck each time, deal the cards, and count how many cards are turned over before the ace is obtained. Repeat 100 times, then find the average.
   c. Summarize the findings for both experiments.

c h a p t e r

6

Discrete Probability Distributions

Objectives

After completing this chapter, you should be able to

1. Construct a probability distribution for a random variable.

2. Find the mean, variance, and expected value for a discrete random variable.

3. Find the exact probability for X successes in n trials of a binomial experiment.

4. Find the mean, variance, and standard deviation for the variable of a binomial distribution.

5. Find probabilities for outcomes of variables using the Poisson, hypergeometric, and multinomial distributions.

Statistics Today

Are the Rockets Hitting the Targets?

During the latter days of World War II, the Germans developed flying rocket bombs. These bombs were used to attack London. Allied military intelligence didn't know whether these bombs were fired at random or had a sophisticated aiming device. To determine the answer, they used a probability distribution called the Poisson distribution.

This chapter explains the Poisson distribution as well as the concepts of a probability distribution and gives examples of several other common distributions used in probability and statistics.

Source: *Elementary Probability and Statistical Reasoning* by Howard E. Reinhart and Don O. Loftsgaarden (Lexington, MA: D. C. Heath & Co., 1977).

6–1

Introduction

Many decisions in business, insurance, and other real-life situations are made by assigning probabilities to all possible outcomes pertaining to the situation and then evaluating the results. For example, a saleswoman can compute the probability that she will make 0, 1, 2, or 3 or more sales in a single day. An insurance company might be able to assign probabilities to the number of vehicles a family owns. A self-employed speaker might be able to compute the probabilities for giving 0, 1, 2, 3, or 4 or more speeches each week. Once these probabilities are assigned, statistics such as the mean, variance, and standard deviation can be computed for these events. With these statistics, various decisions can be made. The saleswoman will be able to compute the average number of sales she makes per week, and if she is working on commission, she will be able to approximate her weekly income over a period of time, say monthly. The public speaker will be able to plan ahead and approximate his average income and expenses. The insurance company can use its information to design special computer forms and programs to accommodate its customers' future needs.

This chapter explains the concepts and applications of what is called a *probability distribution*. In addition, special probability distributions, such as the *binomial, multinomial, Poisson,* and *hypergeometric* distributions, are explained.

6–2

Probability Distributions

Objective 1. Construct a probability distribution for a random variable.

Before probability distribution is defined formally, the definition of variable is reviewed. In Chapter 1, a *variable* was defined as a characteristic or attribute that can assume different values. Various letters of the alphabet, such as X, Y, or Z, are used to represent variables. Since the variables in this chapter are associated with probability, they are called *random variables*.

For example, if a die is rolled, a letter such as X can be used to represent the outcomes. Then the value X can assume is 1, 2, 3, 4, 5, or 6, corresponding to the outcomes of rolling a single die. If two coins are tossed, a letter, say Y, can be used to represent the number of heads, in this case 0, 1, or 2. As another example, if the temperature at 8:00 A.M. is 43° and at noon it is 53°, then the values (T) the temperature assumes are said to be random, since they are due to various atmospheric conditions at the time the temperature was taken.

A **random variable** is a variable whose values are determined by chance.

Also recall from Chapter 1 that one can classify variables as discrete or continuous by observing the values the variable can assume. If a variable can assume only a specific number of values, such as the outcomes for the roll of a die or the outcomes for the toss of a coin, then the variable is called a *discrete variable*. *Discrete variables have values that can be counted.* For example, the number of joggers in Riverview Park each day and the number of phone calls received after a TV commercial airs are examples of discrete variables since they can be counted.

Variables that can assume all values in the interval between any two given values are called *continuous variables*. For example, if the temperature goes from 62° to 78° in a 24-hour period, it has passed through every possible number from 62 to 78. *Continuous random variables are obtained from data that can be measured rather than counted.* Examples of continuous variables are heights, weights, temperatures, and time. In this chapter only discrete random variables are used; Chapter 7 explains continuous random variables.

The procedure shown here for constructing a probability distribution for a discrete random variable uses the probability experiment of tossing three coins. Recall that when three coins are tossed, the sample space is represented as TTT, TTH, THT, HTT, HHT, HTH, THH, HHH, and if X is the random variable for the number of heads, then X assumes the value 0, 1, 2, or 3.

Probabilities for the values of X can be determined as follows:

No heads	One head			Two heads			Three heads
TTT	TTH	THT	HTT	HHT	HTH	THH	HHH
$\frac{1}{8}$	$\frac{1}{8}$	$\frac{1}{8}$	$\frac{1}{8}$	$\frac{1}{8}$	$\frac{1}{8}$	$\frac{1}{8}$	$\frac{1}{8}$
$\frac{1}{8}$		$\frac{3}{8}$			$\frac{3}{8}$		$\frac{1}{8}$

Hence, the probability of getting no heads is $\frac{1}{8}$, one head is $\frac{3}{8}$, two heads is $\frac{3}{8}$, and three heads is $\frac{1}{8}$. From these values, a probability distribution can be constructed by listing the outcomes and assigning the probability of each outcome, as shown.

Number of heads, X	0	1	2	3
Probability, $P(X)$	$\frac{1}{8}$	$\frac{3}{8}$	$\frac{3}{8}$	$\frac{1}{8}$

A **discrete probability distribution** consists of the values a random variable can assume and the corresponding probabilities of the values. The probabilities are determined theoretically or by observation.

Example 6–1

Construct a probability distribution for rolling a single die.

Solution

Since the sample space is 1, 2, 3, 4, 5, 6, and each outcome has a probability of $\frac{1}{6}$, the distribution is as follows:

Outcome, X	1	2	3	4	5	6
Probability, $P(X)$	$\frac{1}{6}$	$\frac{1}{6}$	$\frac{1}{6}$	$\frac{1}{6}$	$\frac{1}{6}$	$\frac{1}{6}$

Probability distributions can be shown graphically by representing the values of X on the x axis and the probabilities $P(X)$ on the y axis.

Example 6–2

Represent graphically the probability distribution for the sample space for tossing three coins.

Number of heads, X	0	1	2	3
Probability, $P(X)$	$\frac{1}{8}$	$\frac{3}{8}$	$\frac{3}{8}$	$\frac{1}{8}$

Solution

The values X assumes are located on the x axis, and the values for $P(X)$ are located on the y axis. The graph is shown in Figure 6–1.

Note that for visual appearances, it is not necessary to start with 0 at the origin.

The preceding examples are illustrations of *theoretical* probability distributions. One did not need to actually perform the experiments to compute the probabilities. In contrast, to construct actual probability distributions, one must observe the variable over a period of time. They are empirical, as shown in the next example.

Figure 6–1

Probability Distribution
for Example 6–2

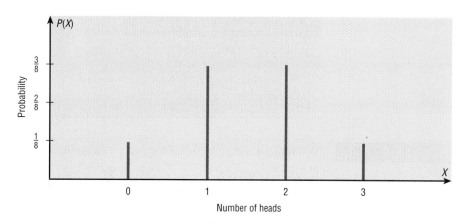

Example 6–3

During the summer months, a rental agency keeps track of the number of chain saws it rents each day during a period of 90 days. The number of saws rented per day is represented by the variable X. The results are shown here. Compute the probability $P(X)$ for each X, and construct a probability distribution and graph for the data.

X	Number of days
0	45
1	30
2	15
	Total 90

Solution

The probability $P(X)$ can be computed for each X by dividing the number of days X saws were rented by total days.

for 0 saws: $\frac{45}{90} = 0.50$ for 1 saw: $\frac{30}{90} = 0.33$ for 2 saws: $\frac{15}{90} = 0.17$

The distribution is as follows:

Number of saws rented, X	0	1	2
Probability, $P(X)$	0.50	0.33	0.17

The graph is shown in Figure 6–2.

Figure 6–2

Probability Distribution
for Example 6–3

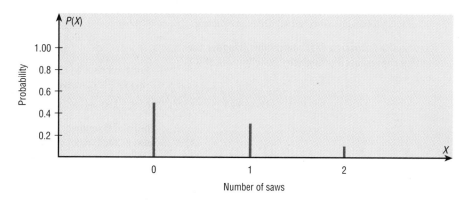

Two Requirements for a Probability Distribution

1. The sum of the probabilities of all the events in the sample space must equal 1; that is, $\Sigma P(X) = 1$.

2. The probability of each event in the sample space must be between or equal to 0 and 1. That is, $0 \leq P(X) \leq 1$.

Example 6–4

Determine whether each distribution is a probability distribution.

a.
X	0	5	10	15	20
P(X)	$\frac{1}{5}$	$\frac{1}{5}$	$\frac{1}{5}$	$\frac{1}{5}$	$\frac{1}{5}$

c.
X	1	2	3	4
P(X)	$\frac{1}{4}$	$\frac{1}{8}$	$\frac{1}{16}$	$\frac{9}{16}$

b.
X	0	2	4	6
P(X)	-1.0	1.5	0.3	0.2

d.
X	2	3	7
P(X)	0.5	0.3	0.4

Solution

a. Yes, it is a probability distribution.

b. No, it is not a probability distribution, since $P(X)$ cannot be 1.5 or -1.0.

c. Yes, it is a probability distribution.

d. No, it is not, since $\Sigma P(X) = 1.2$.

Many variables in business, education, engineering, and other areas can be analyzed by using probability distributions. The next section shows methods for finding the mean and standard deviation for a probability distribution.

Exercises

6–1. Define and give three examples of a random variable.

6–2. Explain the difference between a discrete and a continuous random variable.

6–3. Give three examples of a discrete random variable.

6–4. Give three examples of a continuous random variable.

6–5. What is a probability distribution? Give an example.

For Exercises 6–6 through 6–11, determine whether the distribution represents a probability distribution. If it is not, state why.

6–6.
X	1	3	5	7	9	11
P(X)	$\frac{1}{6}$	$\frac{1}{6}$	$\frac{1}{6}$	$\frac{1}{6}$	$\frac{1}{6}$	$\frac{1}{6}$

6–7.
X	3	6	9	12	15
P(X)	$\frac{4}{9}$	$\frac{2}{9}$	$\frac{1}{9}$	$\frac{1}{9}$	$\frac{1}{9}$

6–8.
X	3	6	8
P(X)	-0.3	0.6	0.7

6–9.
X	1	2	3	4	5
P(X)	$\frac{3}{10}$	$\frac{1}{10}$	$\frac{1}{10}$	$\frac{2}{10}$	$\frac{3}{10}$

6–10.
X	20	30	40	50
P(X)	1.1	0.2	0.9	0.3

6–11.
X	5	10	15
P(X)	1.2	0.3	0.5

For Exercises 6–12 through 6–18, state whether the variable is discrete or continuous.

6–12. The speed of a race car.

6–13. The number of cups of coffee a fast-food restaurant serves each day.

6–14. The number of people who play the state lottery each day.

6–15. The weight of a rhinoceros.

6–16. The time it takes to complete an exercise session.

6–17. The number of mathematics majors in your school.

6–18. The blood pressures of all patients admitted to a hospital on a specific day.

For Exercises 6–19 through 6–26, construct a probability distribution for the data and draw a graph for the distribution.

6–19. The probabilities that a patient will have 0, 1, 2, or 3 medical tests performed on entering a hospital are $\frac{6}{15}, \frac{5}{15}, \frac{3}{15}$, and $\frac{1}{15}$, respectively.

6–20. The probabilities of a return on an investment of $1000, $2000, and $3000 are $\frac{1}{2}, \frac{1}{4}$, and $\frac{1}{4}$, respectively.

6–21. The probabilities of a machine manufacturing 0, 1, 2, 3, 4, or 5 defective parts in one day are 0.75, 0.17, 0.04, 0.025, 0.01, and 0.005, respectively.

6–22. The probabilities that a customer will purchase 0, 1, 2, or 3 books are 0.45, 0.30, 0.15, and 0.10, respectively.

6–23. A die is loaded in such a way that the probabilities of getting 1, 2, 3, 4, 5, and 6 are $\frac{1}{2}, \frac{1}{6}, \frac{1}{12}, \frac{1}{12}, \frac{1}{12}$, and $\frac{1}{12}$, respectively.

6–24. The probabilities that a customer selects 1, 2, 3, 4, or 5 items at a convenience store are 0.32, 0.12, 0.23, 0.18, and 0.15, respectively.

6–25. The probabilities that a tutor sees 1, 2, 3, 4, or 5 students in any one day are 0.10, 0.25, 0.25, 0.20, and 0.20, respectively.

6–26. Three patients are given a headache relief tablet. The probabilities for 0, 1, 2, or 3 successes are 0.18, 0.52, 0.21, and 0.09, respectively.

6–27. A box contains three $1 bills, two $5 bills, one $10 bill, and one $20 bill. Construct a probability distribution for the data.

6–28. Construct a probability distribution for a family of three children. Let X represent the number of boys.

6–29. Construct a probability distribution for drawing a card from a deck of 40 cards consisting of 10 cards numbered 1, 10 cards numbered 2, 15 cards numbered 3, and 5 cards numbered 4.

6–30. Using the sample space for tossing two dice, construct a probability distribution for the sums 2 through 12.

A probability distribution can be written in formula notation as $P(X) = 1/X$, where $X = 2, 3, 6$. The distribution is shown as follows:

X	2	3	6
$P(X)$	$\frac{1}{2}$	$\frac{1}{3}$	$\frac{1}{6}$

For Exercises 6–31 through 6–36, write the distribution for the formula and determine whether it is a probability distribution.

*6–31. $P(X) = X/6$ for $X = 1, 2, 3$

*6–32. $P(X) = X$ for $X = 0.2, 0.3, 0.5$

*6–33. $P(X) = X/6$ for $X = 3, 4, 7$

*6–34. $P(X) = X + 0.1$ for $X = 0.1, 0.02, 0.04$

*6–35. $P(X) = X/7$ for $X = 1, 2, 4$

*6–36. $P(X) = X/(X + 2)$ for $X = 0, 1, 2$

6–3

Mean, Variance, and Expectation

The mean, variance, and standard deviation for a probability distribution are computed differently from the mean, variance, and standard deviation for samples. This section explains how these measures—as well as a new measure called the *expectation*—are calculated for probability distributions.

Mean

In Chapter 3, the means for a sample or population were computed by adding the values and dividing by the total number of values, as shown in the formulas:

$$\bar{X} = \frac{\Sigma X}{n} \qquad \mu = \frac{\Sigma X}{N}$$

Objective 2. Find the mean, variance, and expected value for a discrete random variable.

But how would one compute the mean of the number of spots that show on top when a die is rolled? One could try rolling the die, say, 10 times, recording the number of spots, and finding the mean; however, this answer would only approximate the true mean. What about 50 rolls or 100 rolls? Actually, the more times the die is rolled, the better the approximation. One might ask then, "How many times must the die be rolled to get the exact answer?" *It must be rolled an infinite number of times.* Since this task is impossible, the above formulas cannot be used because the denominators would be infinity.

Historical Note

A professor, Augustin Louis Cauchy (1789–1857), wrote a book on probability. While he was teaching at the Military School of Paris, one of his students was Napoleon Bonaparte.

Hence, a new method of computing the mean is necessary. This method gives the exact theoretical value of the mean as if it were possible to roll the die an infinite number of times.

Before the formula is stated, an example will be used to explain the concept. Suppose two coins are tossed repeatedly, and the number of heads that occurred is recorded. What will the mean of the number of heads be? The sample space is

HH, HT, TH, TT

and each outcome has a probability of $\frac{1}{4}$. Now, in the long run, one would *expect* two heads (HH) to occur approximately $\frac{1}{4}$ of the time, one head to occur approximately $\frac{1}{2}$ of the time (HT or TH), and no heads (TT) to occur approximately $\frac{1}{4}$ of the time. Hence, on average, one would expect the number of heads to be

$$\tfrac{1}{4} \cdot 2 + \tfrac{1}{2} \cdot 1 + \tfrac{1}{4} \cdot 0 = 1$$

That is, if it were possible to toss the coins many times or an infinite number of times, the *average* of the number of heads would be 1.

Hence, in order to find the mean for a probability distribution, one must multiply each possible outcome by its corresponding probability and find the sum of the products.

Formula for the Mean of a Probability Distribution

The mean of the random variable with a discrete probability distribution is

$$\mu = X_1 \cdot P(X_1) + X_2 \cdot P(X_2) + X_3 \cdot P(X_3) + \cdots + X_n \cdot P(X_n)$$
$$= \Sigma X \cdot P(X)$$

where $X_1, X_2, X_3, \ldots, X_n$ are the outcomes and $P(X_1), P(X_2), P(X_3), \ldots, P(X_n)$ are the corresponding probabilities.

Note: $\Sigma X \cdot P(X)$ means to sum the products.

Rounding Rule for the Mean, Variance, and Standard Deviation for a Probability Distribution
The rounding rule for the mean, variance, and standard deviation for variables of a probability distribution is this: The mean, variance, and standard deviation should be rounded to one more decimal place than the outcome, X. When fractions are used, they should be reduced to the lowest terms.

The next four examples illustrate the use of the formula.

Example 6–5

Find the mean of the number of spots that appear when a die is tossed.

Solution

In the toss of a die, the mean can be computed as follows.

Outcome, X	1	2	3	4	5	6
Probability, $P(X)$	$\frac{1}{6}$	$\frac{1}{6}$	$\frac{1}{6}$	$\frac{1}{6}$	$\frac{1}{6}$	$\frac{1}{6}$

$$\mu = \Sigma X \cdot P(X) = 1 \cdot \tfrac{1}{6} + 2 \cdot \tfrac{1}{6} + 3 \cdot \tfrac{1}{6} + 4 \cdot \tfrac{1}{6} + 5 \cdot \tfrac{1}{6} + 6 \cdot \tfrac{1}{6}$$
$$= \tfrac{21}{6} = 3\tfrac{1}{2} \text{ or } 3.5$$

That is, when a die is tossed many times, the theoretical mean will be 3.5. Note that even though the die cannot show a 3.5, the theoretical average is 3.5.

The reason why this formula gives the theoretical mean is that in the long run, each outcome would occur approximately $\frac{1}{6}$ of the time. Hence, multiplying the outcome by its corresponding probability and finding the sum would yield the theoretical mean. In other words, the outcome 1 would occur approximately $\frac{1}{6}$ of the time, the outcome 2 would occur approximately $\frac{1}{6}$ of the time, etc.

Also, since a mean is not a probability, it can be a number greater than one.

Example 6–6

In a family with two children, find the mean of the number of children who will be girls.

Solution

The probability distribution is as follows:

Number of girls, X	0	1	2
Probability, $P(X)$	$\frac{1}{4}$	$\frac{1}{2}$	$\frac{1}{4}$

Hence, the mean is

$$\mu = \Sigma X \cdot P(X) = 0 \cdot \tfrac{1}{4} + 1 \cdot \tfrac{1}{2} + 2 \cdot \tfrac{1}{4} = 1$$

Example 6–7

If three coins are tossed, find the mean of the number of heads that occur. (See the table preceding Example 6–1.)

Solution

The probability distribution is as follows:

Number of heads, X	0	1	2	3
Probability, $P(X)$	$\frac{1}{8}$	$\frac{3}{8}$	$\frac{3}{8}$	$\frac{1}{8}$

The mean is

$$\mu = \Sigma X \cdot P(X) = 0 \cdot \tfrac{1}{8} + 1 \cdot \tfrac{3}{8} + 2 \cdot \tfrac{3}{8} + 3 \cdot \tfrac{1}{8} = \tfrac{12}{8} = 1\tfrac{1}{2} \text{ or } 1.5$$

The value 1.5 cannot occur as an outcome. Nevertheless, it is the long-run or theoretical average.

Example 6–8

In a restaurant, the following probability distribution was obtained for the number of items a person ordered for a large pizza.

Number of items, X	0	1	2	3	4
Probability, $P(X)$	0.3	0.4	0.2	0.06	0.04

Find the mean for the distribution.
   Can one conclude that people like pizza with a lot of toppings?

Solution

$$\mu = \Sigma X \cdot P(X)$$
$$= (0)(0.3) + (1)(0.4) + (2)(0.2) + (3)(0.06) + (4)(0.04) = 1.14 \approx 1.1$$

No. The average is only one topping per pizza.

Variance

Historical Note

The Fey Manufacturing Co., located in San Francisco, invented the first three-reel, automatic payout slot machine in 1895.

For a probability distribution, the mean of the random variable describes the measure of the so-called long-run or theoretical average, but it does not tell anything about the spread of the distribution. Recall from Chapter 3 that in order to measure this spread or variability, statisticians use the variance and standard deviation. The following formulas were used:

$$\sigma^2 = \frac{\Sigma(X - \mu)^2}{N} \qquad \text{or} \qquad \sigma = \sqrt{\frac{\Sigma(X - \mu)^2}{N}}$$

These formulas cannot be used for a random variable of a probability distribution since N is infinite, so the variance and standard deviation must be computed differently.

To find the variance for the random variable of a probability distribution, subtract the theoretical mean of the random variable from each outcome and square the difference. Then multiply each difference by its corresponding probability and add the products. The formula is

$$\sigma^2 = \Sigma [(X - \mu)^2 \cdot P(X)]$$

Finding the variance by using this formula is somewhat tedious. So for simplified computations, a shortcut formula can be used. This formula is algebraically equivalent to the longer one and will be used in the examples that follow.

Formula for the Variance of a Probability Distribution

Find the variance of a probability distribution by multiplying the square of each outcome by its corresponding probability, summing those products, and subtracting the square of the mean. The formula for the variance of a probability distribution is

$$\sigma^2 = \Sigma [X^2 \cdot P(X)] - \mu^2$$

The standard deviation of a probability distribution is

$$\sigma = \sqrt{\sigma^2}$$

Remember that the variance and standard deviation cannot be negative.

Example 6–9

Compute the variance and standard deviation for the probability distribution in Example 6–5.

Solution

Recall that the mean is $\mu = 3.5$, as computed in Example 6–5. Square each outcome and multiply by the corresponding probability, sum those products, and then subtract the square of the mean.

$$\sigma^2 = [1^2 \cdot \tfrac{1}{6} + 2^2 \cdot \tfrac{1}{6} + 3^2 \cdot \tfrac{1}{6} + 4^2 \cdot \tfrac{1}{6} + 5^2 \cdot \tfrac{1}{6} + 6^2 \cdot \tfrac{1}{6}] - (3.5)^2 = 2.9$$

To get the standard deviation, find the square root of the variance.

$$\sigma = \sqrt{2.9} = 1.7$$

Example 6–10

Five balls numbered 0, 2, 4, 6, and 8 are placed in a bag. After the balls are mixed, one is selected, its number is noted, and then it is replaced. If this experiment is repeated many times, find the variance and standard deviation of the numbers on the balls.

Historical Note

In 1657 a Dutch mathematician, Huggens, wrote a treatise on the Pascal–Fermat correspondence and introduced the idea of "mathematical expectation."

Solution

Let X be the number on each ball. The probability distribution is as follows:

Number on ball, X	0	2	4	6	8
Probability, $P(X)$	$\frac{1}{5}$	$\frac{1}{5}$	$\frac{1}{5}$	$\frac{1}{5}$	$\frac{1}{5}$

The mean is

$$\mu = \Sigma\, X \cdot P(X) = 0 \cdot \tfrac{1}{5} + 2 \cdot \tfrac{1}{5} + 4 \cdot \tfrac{1}{5} + 6 \cdot \tfrac{1}{5} + 8 \cdot \tfrac{1}{5} = 4.0$$

The variance is

$$\sigma^2 = \Sigma\, [X^2 \cdot P(X)] - \mu^2$$
$$= [0^2 \cdot (\tfrac{1}{5}) + 2^2 \cdot (\tfrac{1}{5}) + 4^2 \cdot (\tfrac{1}{5}) + 6^2 \cdot (\tfrac{1}{5}) + 8^2 \cdot (\tfrac{1}{5})] - 4^2$$
$$= [0 + \tfrac{4}{5} + \tfrac{16}{5} + \tfrac{36}{5} + \tfrac{64}{5}] - 16$$
$$= \tfrac{120}{5} - 16$$
$$= 24 - 16 = 8$$

The standard deviation is $\sigma = \sqrt{8} = 2.8$.

The mean, variance, and standard deviation can also be found by using vertical columns, as shown. (0.2 is used for $P(X)$ since $\tfrac{1}{5} = 0.2$.)

X	$P(X)$	$X \cdot P(X)$	$X^2 \cdot P(X)$
0	0.2	0	0
2	0.2	0.4	0.8
4	0.2	0.8	3.2
6	0.2	1.2	7.2
8	0.2	1.6	12.8
		$\Sigma\, X \cdot P(X) = 4.0$	$\Sigma\, X^2 \cdot P(X) = 24$

Find the mean by summing the $X \cdot P(X)$ column and the variance by summing the $X^2 \cdot P(X)$ column and subtracting the square of the mean:

$$\sigma^2 = 24 - 4^2 = 8$$

Example 6–11

The probability that 0, 1, 2, 3, or 4 people will be placed on hold when they call a radio talk show is shown in the distribution. Find the variance and standard deviation for the data. The radio station has four phone lines. When all lines are full, a busy signal is heard.

X	0	1	2	3	4
$P(X)$	0.18	0.34	0.23	0.21	0.04

Should the station have considered getting more phone lines installed?

Solution

The mean is

$$\mu = \Sigma\, X \cdot P(X)$$
$$= 0 \cdot (0.18) + 1 \cdot (0.34) + 2 \cdot (0.23) + 3 \cdot (0.21) + 4 \cdot (0.04)$$
$$= 1.6$$

The variance is

$$\sigma^2 = [\Sigma X^2 \cdot P(X)] - \mu^2$$
$$= [0^2 \cdot (0.18) + 1^2 \cdot (0.34) + 2^2 \cdot (0.23)$$
$$\quad + 3^2 \cdot (0.21) + 4^2 \cdot (0.04)] - 1.6^2$$
$$= [0 + 0.34 + 0.92 + 1.89 + 0.64] - 2.56$$
$$= 3.79 - 2.56$$
$$= 1.2$$

The standard deviation is $\sigma = \sqrt{\sigma^2}$, or $\sigma = \sqrt{1.2} = 1.1$.

No. The mean number of people on hold is 1.6. Since the standard deviation is 1.1, most callers would be accommodated by having four phone lines because $\mu + 2\sigma$ would be $1.6 + 2(1.1) = 1.6 + 2.2 = 3.8$. Very few callers would get a busy signal since at least 75% of the callers would either get through or be put on hold. (See Chebyshev's theorem in Section 3–3.)

Expectation

Another concept related to the mean for a probability distribution is the concept of expected value or expectation. Expected value is used in various types of games of chance, in insurance, and in other areas, such as decision theory.

The **expected value** of a discrete random variable of a probability distribution is the theoretical average of the variable. The formula is

$$\mu = E(X) = \Sigma X \cdot P(X)$$

The symbol $E(X)$ is used for the expected value.

The formula for the expected value is the same as the formula for the theoretical mean. The expected value, then, is the theoretical mean of the probability distribution. That is, $E(X) = \mu$.

When expected value problems involve money, it is customary to round the answer to the nearest cent.

Example 6–12

One thousand tickets are sold at $1 each for a color television valued at $350. What is the expected value of the gain if a person purchases one ticket?

Solution

The problem can be set up as follows:

	Win	Lose
Gain, X	$349	−$1
Probability, $P(X)$	$\dfrac{1}{1000}$	$\dfrac{999}{1000}$

Two things should be noted. First, for a win, the net gain is $349, since the person does not get the cost of the ticket ($1) back. Second, for a loss, the gain is represented by a negative number, in this case −$1. The solution, then, is

$$E(X) = \$349 \cdot \frac{1}{1000} + (-\$1) \cdot \frac{999}{1000} = -\$0.65$$

Expected value problems of this type can also be solved by finding the overall gain (i.e., the value of the prize won or the amount of money won not considering the cost of the ticket for the prize or the cost to play the game) and subtracting the cost of the tickets or the cost to play the game, as shown:

$$E(X) = \$350 \cdot \frac{1}{1000} - \$1 = -\$0.65$$

Here, the overall gain ($350) must be used.

Note that the expectation is $-\$0.65$. This does not mean that a person loses $0.65, since the person can only win a television set valued at $350 or lose $1 on the ticket. What this expectation means is that the average of the losses is $0.65 for each of the 1000 ticket holders. Here is another way of looking at this situation: If a person purchased one ticket each week over a long period of time, the average loss would be $0.65 per ticket, since theoretically, on average, that person would win the set once for each 1000 tickets purchased.

Example 6–13

A ski resort loses $70,000 per season when it does not snow very much and makes $250,000 profit when it does snow a lot. The probability of it snowing at least 75 inches (i.e., a good season) is 40%. Find the expectation for the profit.

Solution

Profit, X	$250,000	$-$70,000
Probability, $P(X)$	0.40	0.60

$$E(X) = (\$250{,}000)(0.40) + (-\$70{,}000)(0.60) = \$58{,}000$$

Example 6–14

One thousand tickets are sold at $1 each for four prizes of $100, $50, $25, and $10. What is the expected value if a person purchases two tickets?

Gain, X	$98	$48	$23	$8	$-$2
Probability, $P(X)$	$\dfrac{2}{1000}$	$\dfrac{2}{1000}$	$\dfrac{2}{1000}$	$\dfrac{2}{1000}$	$\dfrac{992}{1000}$

Solution

$$E(X) = \$98 \cdot \frac{2}{1000} + \$48 \cdot \frac{2}{1000} + \$23 \cdot \frac{2}{1000} + \$8 \cdot \frac{2}{1000} + (-\$2) \cdot \frac{992}{1000}$$

$$= -\$1.63$$

An alternate solution is

$$E(X) = \$100 \cdot \frac{2}{1000} + \$50 \cdot \frac{2}{1000} + \$25 \cdot \frac{2}{1000} + \$10 \cdot \frac{2}{1000} - \$2$$

$$= -\$1.63$$

In gambling games, if the expected value of the game is zero, the game is said to be fair. If the expected value of a game is positive, then the game is in favor of the player.

That is, the player has a better-than-even chance of winning. If the expected value of the game is negative, then the game is said to be in favor of the house. That is, in the long run, the players will lose money.

In his book *Probabilities in Everyday Life* (Ivy Books, 1986), author John D. McGervy gives the expectations for various casino games. For keno, the "house" wins $0.27 on every $1.00 bet. For chuck-a-luck, the "house" wins about $0.52 on every dollar bet. For roulette, the "house" wins about $0.90 on every dollar bet. For craps, the "house" wins about $0.88 on every dollar bet. The bottom line here is that if you gamble long enough, sooner or later, you will end up losing money.

Exercises

6–37. From past experience, a company has found that in cartons of transistors, 92% contain no defective transistors, 3% contain one defective transistor, 3% contain two defective transistors, and 2% contain three defective transistors. Find the mean, variance, and standard deviation for the defective transistors.

About how many extra transistors per day would the company need to replace the defective ones if it uses 10 cartons per day?

6–38. The number of suits sold per day at a retail store is shown in the table, with the corresponding probabilities. Find the mean, variance, and standard deviation of the distribution.

Number of suits sold, X	19	20	21	22	23
Probability, $P(X)$	0.2	0.2	0.3	0.2	0.1

If the manager of the retail store wanted to be sure that he has enough suits for the next five days, how many should the manager purchase?

6–39. A bank vice-president feels that each savings account customer has, on average, three credit cards. The following distribution represents the number of credit cards people own. Find the mean, variance, and standard deviation. Is the vice president correct?

Number of cards, (X)	0	1	2	3	4
Probability, $P(X)$	0.18	0.44	0.27	0.08	0.03

6–40. The probability distribution for the number of customers per day at the Sunrise Coffee Shop is shown below. Find the mean, variance, and standard deviation of the distribution.

Number of customers, X	50	51	52	53	54
Probability, $P(X)$	0.10	0.20	0.37	0.21	0.12

6–41. A public speaker computes the probabilities for the number of speeches she gives each week. Compute the mean, variance, and standard deviation of the distribution shown.

Number of speeches, X	0	1	2	3	4	5
Probability, $P(X)$	0.06	0.42	0.22	0.12	0.15	0.03

If she receives $100 per speech, about how much would she earn per week?

6–42. A recent survey by an insurance company showed the following probabilities for the number of automobiles each policyholder owned. Find the mean, variance, and standard deviation for the distribution.

Number of automobiles, X	1	2	3	4
Probability, $P(X)$	0.4	0.3	0.2	0.1

6–43. A concerned parents group determined the number of commercials shown in each of five children's programs over a period of time. Find the mean, variance, and standard deviation for the distribution shown.

Number of commercials, X	5	6	7	8	9
Probability, $P(X)$	0.2	0.25	0.38	0.10	0.07

6–44. A study conducted by a TV station showed the number of televisions per household and the corresponding probabilities for each. Find the mean, variance, and standard deviation.

Number of televisions, X	1	2	3	4
Probability, $P(X)$	0.32	0.51	0.12	0.05

If you were taking a survey on the programs that were watched on television, how many program diaries would you send to each household in the survey?

6–45. The following distribution shows the number of students enrolled in CPR classes offered by the local fire department. Find the mean, variance, and standard deviation for the distribution.

Number of students, X	12	13	14	15	16
Probability, $P(X)$	0.15	0.20	0.38	0.18	0.09

6–46. A florist determines the probabilities for the number of flower arrangements she delivers each day. Find the mean, variance, and standard deviation for the distribution shown.

Number of arrangements, X	6	7	8	9	10
Probability, $P(X)$	0.2	0.2	0.3	0.2	0.1

6–47. The Lincoln Fire Department wishes to raise $5000 to purchase some new equipment. They decide to conduct a raffle. A cash prize of $5000 is to be awarded. If 2500 tickets are sold at $5.00 each, find the expected value of the gain. Are they selling enough tickets to make their goal?

6–48. A box contains ten $1 bills, five $2 bills, three $5 bills, one $10 bill, and one $100 bill. A person is charged $20 to select one bill. Find the expectation. Is the game fair?

6–49. If a person rolls doubles when he tosses two dice, he wins $5. For the game to be fair, how much should the person pay to play the game?

6–50. If a player rolls two dice and gets a sum of 2 or 12, she wins $20. If the person gets a 7, she wins $5. The cost to play the game is $3. Find the expectation of the game.

6–51. A lottery offers one $1000 prize, one $500 prize, and five $100 prizes. One thousand tickets are sold at $3 each. Find the expectation if a person buys one ticket.

6–52. In Exercise 6–51, find the expectation if a person buys two tickets. Assume that the player's ticket is replaced after each draw and the same ticket can win more than one prize.

6–53. For a daily lottery, a person selects a three-digit number. If the person plays for $1, she can win $500. Find the expectation. In the same daily lottery, if a person boxes a number, she will win $80. Find the expectation if the number 123 is played for $1 and boxed. (When a number is "boxed," it can win when the digits occur in any order.)

6–54. If a 60-year-old buys a $1000 life insurance policy at a cost of $60 and has a probability of 0.972 of living to age 61, find the expectation of the policy.

6–55. A roulette wheel has 38 numbers: 1 through 36, 0, and 00. Half of the numbers from 1 through 36 are red and half are black. A ball is rolled, and it falls into one of the 38 slots, giving a number and a color. Green is the color for 0 and 00. When a player wins, the player gets his dollar back in addition to the amount of the payoff. The payoffs for a $1 bet are as follows:

Red or black	$1	0	$35
Odd or even	$1	00	$35
1–18	$1	Any single number	$35
19–36	$1	0 or 00	$17

If a person bets $1 on any one of these before the ball is rolled, find the expected value for each.

a. Red
b. Even (exclude 0 and 00)
c. 0
d. Any single value
e. 0 or 00

*6–56. Construct a probability distribution for the sum shown on the faces when two dice are rolled. Find the mean, variance, and standard deviation of the distribution.

*6–57. When one die is rolled, the expected value of the number of spots is 3.5. In Exercise 6–56, the mean number of spots was found for rolling two dice. What is the mean number of spots if three dice are rolled?

*6–58. The formula for finding the variance for a probability distribution is

$$\sigma^2 = \Sigma\,[(X - \mu)^2 \cdot P(X)]$$

Verify algebraically that this formula gives the same result as the shortcut formula shown in this section.

*6–59. Roll a die 100 times. Compute the mean and standard deviation. How does the result compare with the theoretical results of Example 6–5?

*6–60. Roll two dice 100 times and find the mean, variance, and standard deviation of the sum of the spots. Compare the result with the theoretical results obtained in Exercise 6–56.

*6–61. Conduct a survey of the number of television sets in each of your classmates' homes. Construct a probability distribution and find the mean, variance, and standard deviation.

*6–62. In a recent promotional campaign, a company offered the following prizes and the corresponding probabilities. Find the expected value of winning. The tickets are free.

Number of Prizes	Amount	Probability
1	$100,000	$\dfrac{1}{1,000,000}$
2	$10,000	$\dfrac{1}{50,000}$
5	$1,000	$\dfrac{1}{10,000}$
10	$100	$\dfrac{1}{1,000}$

If the winner has to mail in the winning ticket to claim the prize, what will be the expectation if the cost of the stamp is considered? Use the current cost of a stamp for a first class letter.

6–4

The Binomial Distribution

Objective 3. Find the exact probability for *X* successes in *n* trials of a binomial experiment.

Many types of probability problems have only two outcomes, or they can be reduced to two outcomes. For example, when a coin is tossed, it can land heads or tails. When a baby is born, it will be either male or female. In a basketball game, a team either wins or loses. A true–false item can be answered in only two ways, true or false. Other situations can be reduced to two outcomes. For example, a medical treatment can be classified as effective or ineffective, depending on the results. A person can be classified as having normal or abnormal blood pressure, depending on the measure of the blood pressure gauge. A multiple-choice question, even though there are four or five answer choices, can be classified as correct or incorrect. Situations like these are called *binomial experiments*.

Historical Note

In 1653, Blaise Pascal created a triangle of numbers called Pascal's triangle that can be used in the binomial distribution.

A **binomial experiment** is a probability experiment that satisfies the following four requirements:

1. Each trial can have only two outcomes or outcomes that can be reduced to two outcomes. These outcomes can be considered as either success or failure.
2. There must be a fixed number of trials.
3. The outcomes of each trial must be independent of each other.
4. The probability of a success must remain the same for each trial.

A binomial experiment and its results give rise to a special probability distribution called the *binomial distribution.*

The outcomes of a binomial experiment and the corresponding probabilities of these outcomes are called a **binomial distribution.**

In binomial experiments, the outcomes are usually classified as successes or failures. For example, the correct answer to a multiple-choice item can be classified as a success, but any of the other choices would be incorrect and hence classified as a failure. The notation that is commonly used for binomial experiments and the binomial distribution is defined next.

Notation for the Binomial Distribution

$P(S)$	The symbol for the probability of success
$P(F)$	The symbol for the probability of failure
p	The numerical probability of a success
q	The numerical probability of a failure

$$P(S) = p \quad \text{and} \quad P(F) = 1 - p = q$$

n	The number of trials
X	The number of successes

Note that $0 \le X \le n$.

The probability of a success in a binomial experiment can be computed with the following formula.

Binomial Probability Formula

In a binomial experiment, the probability of exactly X successes in n trials is

$$P(X) = \frac{n!}{(n - X)!X!} \cdot p^X \cdot q^{n-X}$$

An explanation of why the formula works will be given following Example 6–15.

Example 6–15

A coin is tossed three times. Find the probability of getting exactly two heads.

Solution

This problem can be solved by looking at the sample space. There are three ways to get two heads.

HHH, <u>HHT, HTH, THH,</u> TTH, THT, HTT, TTT

The answer is $\frac{3}{8}$, or 0.375.

....

Looking at the problem in Example 6–15 from the standpoint of a binomial experiment, one can show that it meets the four requirements.

1. There are only two outcomes for each trial, heads or tails.
2. There is a fixed number of trials (three).
3. The outcomes are independent of each other (the outcome of one toss in no way affects the outcome of another toss).
4. The probability of a success (heads) is $\frac{1}{2}$ in each case.

In this case, $n = 3, X = 2, p = \frac{1}{2}$, and $q = \frac{1}{2}$. Hence, substituting in the formula gives

$$P(2 \text{ heads}) = \frac{3!}{(3-2)!2!} \cdot \left(\frac{1}{2}\right)^2\left(\frac{1}{2}\right)^1 = \frac{3}{8} = 0.375$$

which is the same answer obtained by using the sample space.

The same example can be used to explain the formula. First, note that there are three ways to get exactly two heads and one tail from a possible eight ways. They are HHT, HTH, and THH. In this case, then, the number of ways of obtaining two heads from three coin tosses is $_3C_2$, or 3, as shown in Chapter 4. In general, the number of ways to get X successes from n trials without regard to order is

$$_nC_X = \frac{n!}{(n-X)!X!}$$

This is the first part of the binomial formula. (Some calculators can be used for this.)

Next, each success has a probability of $\frac{1}{2}$, and can occur twice. Likewise, each failure has a probability of $\frac{1}{2}$ and can occur once, giving the $(\frac{1}{2})^2(\frac{1}{2})^1$ part of the formula. To generalize, then, each success has a probability of p and can occur X times, and each failure has a probability of q and can occur $(n - X)$ times. Putting it all together yields the binomial probability formula.

....

Example 6–16

If a student randomly guesses at five multiple-choice questions, find the probability that the student gets exactly three correct. Each question has five possible choices.

Solution

In this case $n = 5, X = 3$, and $p = \frac{1}{5}$, since there is one chance in five of guessing a correct answer. Then,

$$P(3) = \frac{5!}{(5-3)!3!} \cdot \left(\frac{1}{5}\right)^3\left(\frac{4}{5}\right)^2 = 0.05$$

....

Example 6–17

A survey from Teenage Research Unlimited (Northbrook, Ill.) found that 30% of teenage consumers receive their spending money from part-time jobs. If five teenagers are selected at random, find the probability that at least three of them will have part-time jobs.

Solution

To find the probability that at least three have a part-time job, it is necessary to find the individual probabilities for either 3, or 4, or 5, and then add them to get the total probability.

$$P(3) = \frac{5!}{(5-3)!(3!)} \cdot (0.3)^3(0.7)^2 = 0.132$$

$$P(4) = \frac{5!}{(5-4)!(4)!} \cdot (0.3)^4(0.7)^1 = 0.028$$

$$P(5) = \frac{5!}{(5-5)!(5)!} \cdot (0.3)^5(0.7)^0 = 0.002$$

Hence,

P(at least three teenagers have part-time jobs)

$= 0.132 + 0.028 + 0.002 = 0.162$

Computing probabilities using the binomial probability formula can be quite tedious at times, so tables have been developed for selected values of n and p. Table B in Appendix C gives the probabilities for individual events. The next example shows how to use Table B to compute probabilities for binomial experiments.

Example 6–18

Solve the problem in Example 6–15 by using Table B.

Solution

Since $n = 3$, $X = 2$, and $p = 0.5$, the value 0.375 is found as shown in Figure 6–3.

Figure 6–3

Using Table B for Example 6–18

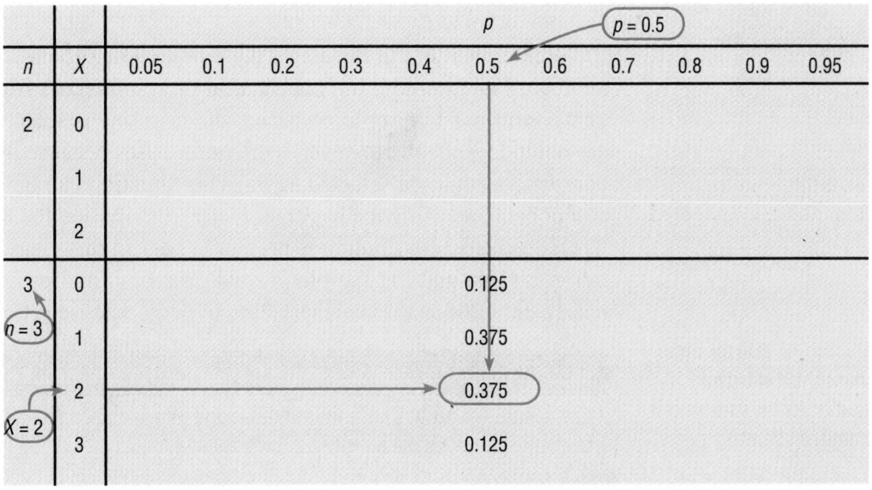

n	X	0.05	0.1	0.2	0.3	0.4	0.5	0.6	0.7	0.8	0.9	0.95
2	0											
	1											
	2											
3	0						0.125					
	1						0.375					
	2						0.375					
	3						0.125					

Example 6–19

Public Opinion reported that 5% of Americans are afraid of being alone in a house at night. If a random sample of 20 Americans is selected, find these probabilities using the binomial table:

a. There are exactly five people in the sample who are afraid of being alone at night.

b. There are at most three people in the sample who are afraid of being alone at night.

c. There are at least three people in the sample who are afraid of being alone at night.

Source: *100% American* by Daniel Evan Weiss (New York: Poseidon Press, 1988).

Solution

a. $n = 20$, $p = 0.05$, and $X = 5$. From the table, one gets 0.002.

b. $n = 20$ and $p = 0.05$. "At most three people" means 0, or 1, or 2, or 3.

Hence, the solution is

$$P(0) + P(1) + P(2) + P(3) = 0.358 + 0.377 + 0.189 + 0.060$$
$$= 0.984$$

c. $n = 20$ and $p = 0.05$. "At least three people" means 3, 4, 5, . . . , 20. This problem can best be solved by finding $P(0) + P(1) + P(2)$ and subtracting from 1.

$$P(0) + P(1) + P(2) = 0.358 + 0.377 + 0.189 = 0.924$$
$$1 - 0.924 = 0.076$$

Example 6–20

A report from the Secretary of Health and Human Services stated that 70% of single-vehicle traffic fatalities that occur at night on weekends involve an intoxicated driver. If a sample of 15 single-vehicle traffic fatalities that occur at night on a weekend is selected, find the probability that exactly 12 involve a driver who is intoxicated.

Source: *100% American* by Daniel Evan Weiss (New York: Poseidon Press, 1988).

Solution

$n = 15$, $p = 0.70$, and $X = 12$. From Table B, $P(12) = 0.170$. Hence, the probability is 0.17.

Remember that in the use of the binomial distribution, the outcomes must be independent. For example, in the selection of components from a batch to be tested, each component must be replaced before the next one is selected. Otherwise, the outcomes are not independent. However, a dilemma arises because there is a chance that the same component could be selected again. This situation can be avoided by not replacing the component and using the hypergeometric distribution to calculate the probabilities. The hypergeometric distribution is presented later in this chapter. Note that when the population is large and the sample is small, the binomial probabilities can be shown to be nearly the same as the corresponding hypergeometric probabilities.

Objective 4. Find the mean, variance, and standard deviation for the variable of a binomial distribution.

Mean, Variance, and Standard Deviation for the Binomial Distribution

The mean, variance, and standard deviation of a variable that has the *binomial distribution* can be found by using the following formulas.

mean $\mu = n \cdot p$

variance $\sigma^2 = n \cdot p \cdot q$

standard deviation $\sigma = \sqrt{n \cdot p \cdot q}$

These formulas are algebraically equivalent to the formulas for the mean, variance, and standard deviation of the variables for probability distributions, but because they are for variables of the binomial distribution, they have been simplified using algebra. The algebraic derivation is omitted here, but their equivalence is shown in the next example.

Example 6–21

A coin is tossed four times. Find the mean, variance, and standard deviation of the number of heads that will be obtained.

Solution

With the formulas for the binomial distribution and $n = 4$, $p = \frac{1}{2}$, and $q = \frac{1}{2}$, the results are

$$\mu = n \cdot p = 4 \cdot \tfrac{1}{2} = 2$$
$$\sigma^2 = n \cdot p \cdot q = 4 \cdot \tfrac{1}{2} \cdot \tfrac{1}{2} = 1$$
$$\sigma = \sqrt{1} = 1$$

From Example 6–21, when four coins are tossed many, many times, the average of the number of heads that appear is two, and the standard deviation of the number of heads is one. Note that these are theoretical values.

As stated previously, this problem can be solved by using the expected value formulas. The distribution is shown as follows:

No. of heads, X	0	1	2	3	4
Probability, $P(X)$	$\frac{1}{16}$	$\frac{4}{16}$	$\frac{6}{16}$	$\frac{4}{16}$	$\frac{1}{16}$

$$\mu = E(X) = \Sigma X \cdot P(X) = 0 \cdot \tfrac{1}{16} + 1 \cdot \tfrac{4}{16} + 2 \cdot \tfrac{6}{16} + 3 \cdot \tfrac{4}{16} + 4 \cdot \tfrac{1}{16} = \tfrac{32}{16} = 2$$
$$\sigma^2 = \Sigma X^2 \cdot P(X) - \mu^2$$
$$= 0^2 \cdot \tfrac{1}{16} + 1^2 \cdot \tfrac{4}{16} + 2^2 \cdot \tfrac{6}{16} + 3^2 \cdot \tfrac{4}{16} + 4^2 \cdot \tfrac{1}{16} - 2^2 = \tfrac{80}{16} - 4 = 1$$
$$\sigma = \sqrt{1} = 1$$

Hence, the simplified binomial formulas give the same results.

Example 6–22

A die is rolled 480 times. Find the mean, variance, and standard deviation of the number of 2 s that will be rolled.

Solution

$n = 480$, $p = \frac{1}{6}$. This is a binomial situation, where getting a 2 is a success and not getting a 2 is a failure; hence,

$$\mu = n \cdot p = 480 \cdot \left(\tfrac{1}{6}\right) = 80$$
$$\sigma^2 = n \cdot p \cdot q = 480 \cdot \left(\tfrac{1}{6}\right)\left(\tfrac{5}{6}\right) = 66.7$$
$$\sigma = \sqrt{n \cdot p \cdot q} = \sqrt{66.7} = 8.2$$

On average, there will be 80 twos. The standard deviation is 8.2.

Example 6–23

The *Statistical Bulletin* published by Metropolitan Life Insurance Co. reported that 2% of all American births result in twins. If a random sample of 8000 births is taken, find the mean, variance, and standard deviation of the number of births that would result in twins.

Source: *100% American* by Daniel Evan Weiss (New York: Poseidon Press, 1988).

Solution

This is a binomial situation, since a birth can result in either twins or not twins (i.e., two outcomes).

$$\mu = n \cdot p = (8000)(0.02) = 160$$
$$\sigma^2 = n \cdot p \cdot q = (8000)(0.02)(0.98) = 156.8$$
$$\sigma = \sqrt{n \cdot p \cdot q} = \sqrt{156.8} = 12.5$$

For the sample, the average number of births that would result in twins is 160, the variance is 156.8, or 157, and the standard deviation is 12.5, or 13 if rounded.

Exercises

6–63. Which of the following are binomial experiments or can be reduced to binomial experiments?

a. Surveying 100 people to determine if they like Sudsy Soap.

b. Tossing a coin 100 times to see how many heads occur.

c. Drawing a card from a deck and getting a heart.

d. Asking 1000 people which brand of cigarettes they smoke.

e. Testing four different brands of aspirin to see which brands are effective.

f. Testing one brand of aspirin using 10 people to determine whether it is effective.

g. Asking 100 people if they smoke.

h. Checking 1000 applicants to see whether they were admitted to White Oak College.

i. Surveying 300 prisoners to see how many different crimes they were convicted of.

j. Surveying 300 prisoners to see whether this is their first offense.

6–64. (ans) Compute the probability of X successes, using Table B in Appendix C.

a. $n = 2, p = 0.30, X = 1$

b. $n = 4, p = 0.60, X = 3$

c. $n = 5, p = 0.10, X = 0$

d. $n = 10, p = 0.40, X = 4$

e. $n = 12, p = 0.90, X = 2$

f. $n = 15, p = 0.80, X = 12$

g. $n = 17, p = 0.05, X = 0$

h. $n = 20, p = 0.50, X = 10$

i. $n = 16, p = 0.20, X = 3$

6–65. Compute the probability of X successes, using the binomial formula.

a. $n = 6, X = 3, p = 0.03$

b. $n = 4, X = 2, p = 0.18$

c. $n = 5, X = 3, p = 0.63$

d. $n = 9, X = 0, p = 0.42$

e. $n = 10, X = 5, p = 0.37$

For Exercises 6–66 through 6–75, assume all variables are binomial. (*Note:* If values are not found in Table B (Appendix C), use the binomial formula.)

6–66. A burglar alarm system has six fail-safe components. The probability of each failing is 0.05. Find these probabilities.

a. Exactly three will fail.

b. Fewer than two will fail.

c. None will fail.

d. Compare the answers for parts a, b, and c, and explain why the results are reasonable.

6–67. A student takes a 10-question, true–false exam and guesses on each question. Find the probability of passing if the lowest passing grade is 6 correct out of 10. Based on your answer, would it be a good idea not to study and depend on guessing?

6–68. If the quiz in Exercise 6–67 was a multiple-choice quiz with five choices for each question, find the probability of guessing at least 6 correct out of 10.

6–69. In a survey, 30% of the people interviewed said that they bought most of their books during the last three months of the year (October, November, December). If nine people are selected at random, find the probability that exactly three of these people bought most of their books during October, November, and December.

Source: USA Snapshot, *USA Today*, May 24, 1995.

6–70. In a Gallup Survey, 90% of the people interviewed were unaware that maintaining a healthy weight could reduce the risk of stroke. If 15 people are selected at random, find the probability that at least 9 are unaware that maintaining a proper weight could reduce the risk of stroke.

Source: USA Snapshot, *USA Today*, June 24, 1995.

6–71. In a survey, three of four students said the courts show "too much concern" for criminals. Find the probability that at most three out of seven randomly selected students will agree with this statement.

Source: *Harper's Index* 290, no. 1739 (April 1995), p. 13.

6–72. It was found that 60% of American victims of health care fraud are senior citizens. If 10 victims are randomly selected, find the probability that exactly 3 are senior citizens.

Source: *100% American* by Daniel Evan Weiss (New York: Poseidon Press, 1988).

6–73. R. H. Bruskin Associates Market Research found that 40% of Americans do not think having a college education is important to succeed in the business world. If a random sample of five Americans is selected, find these probabilities.
a. Exactly two people will agree with that statement.
b. At most three people will agree with that statement.
c. At least two people will agree with that statement.
d. Fewer than three people will agree with that statement.

Source: *100% American* by Daniel Evans Weiss (New York: Poseidon Press, 1988).

6–74. If 30% of the people in a community use the library in one year, find these probabilities for a sample of 15 people:
a. At most 7 used the library.
b. Exactly 7 used the library.
c. At least 5 used the library.

6–75. If 20% of the people in a community use the emergency room at a hospital in one year, find these probabilities for a sample of 10 people.
a. At most three used the emergency room.
b. Exactly three used the emergency room.
c. At least five used the emergency room.

6–76. (ans) Find the mean, variance, and standard deviation for each of the values of n and p when the conditions for the binomial distribution are met.
a. $n = 100, p = 0.75$
b. $n = 300, p = 0.3$
c. $n = 20, p = 0.5$
d. $n = 10, p = 0.8$
e. $n = 1000, p = 0.1$
f. $n = 500, p = 0.25$
g. $n = 50, p = \frac{2}{5}$
h. $n = 36, p = \frac{1}{6}$

6–77. A study found that 1% of Social Security recipients are too young to vote. If 800 Social Security recipients are randomly selected, find the mean, variance, and standard deviation of the number of recipients who are too young to vote.

Source: *Harper's Index* 289, no. 1737 (February 1995), p. 11.

6–78. Find the mean, variance, and standard deviation for the number of heads when 15 coins are tossed.

6–79. If 2% of automobile carburetors are defective, find the mean, variance, and standard deviation of a lot of 500 carburetors.

6–80. It has been reported that 83% of federal government employees use e-mail. If a sample of 200 federal government employees is selected, find the mean, variance, and standard deviation of the number who use e-mail.

Source: *USA Today,* June 14, 1995.

6–81. A survey found that 21% of Americans watch fireworks on television on July 4. Find the mean, variance, and standard deviation of the number of individuals who watch fireworks on television on July 4 if a random sample of 1000 Americans is selected.

Source: USA Snapshot, *USA Today,* July 3–4, 1995.

6–82. In a restaurant, a study found that 42% of all patrons smoked. If the seating capacity of the restaurant is 80 people, find the mean, variance, and standard deviation of the number of smokers.
About how many seats should be available for smoking customers?

6–83. A survey found that 25% of pet owners had their pets bathed professionally rather than doing it themselves. If 18 pet owners are randomly selected, find the probability that exactly five people have their pets bathed professionally.

Source: USA Snapshot, *USA Today,* June 15, 1995.

6–84. In a survey, 63% of Americans said they own an answering machine. If 14 Americans are selected at random, find the probability that exactly 9 own an answering machine.

Source: USA Snapshot, *USA Today,* May 22, 1995.

6–85. One out of every three Americans believes that the U.S. government should take "primary responsibility" for eliminating poverty in the United States. If 10 Americans are selected, find the probability that at most 3 will believe that the U.S. government should take primary responsibility for eliminating poverty.

Source: *Harper's Index* 289, no. 1735 (December 1994), p. 13.

6–86. In a survey, 58% of American adults said they had never heard of the Internet. If 20 American adults are selected at random, find the probability that exactly 12 will say they have never heard of the Internet.

Source: *Harper's Index* 289, no. 1737 (February 1995), p. 11.

6–87. In the past year, 13% of businesses have eliminated jobs. If five businesses are selected at random, find the probability that at least three have eliminated jobs during the last year.

Source: *USA Today,* June 15, 1995.

6–88. Of graduating high school seniors, 14% said that their generation will be remembered for their social concerns. If seven graduating seniors are selected at random, find the probability that either two or three will agree with that statement.

Source: *USA Today,* June 8, 1995.

6–89. A survey found that 86% of Americans have never been a victim of violent crime. If a sample of 12 Americans is selected at random, find the probability that 10 or more have never been victims of violent crime. Does it seem reasonable that 10 or more have never been victims of violent crime?

Source: *Harper's Index* 290, no 1741 (June 1995), p. 11.

*6–90. The graph shown here represents the probability distribution for the number of girls in a family of three children. From this graph, construct a probability distribution.

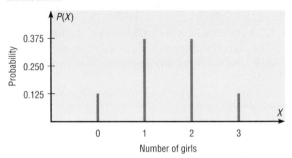

*6–91. Construct a binomial distribution graph for the number of defective computer chips in a lot of five if $p = 0.2$.

Technology Step by Step

MINITAB
Step by Step

Making a Binomial Distribution Table and Graph

These instructions will show how MINITAB computes binomial probabilities, using $n = 20$ and $p = 0.05$.

1. To enter the integers from 0 to 20 in C1 select **Calc>Make Patterned Data>Simple Set of Numbers.**

2. You must enter 3 items:
   a. Enter **X** in the box for `Store patterned data in:`. Minitab will use the first empty column of the active worksheet and name it **X.** Press [TAB].
   b. Enter the value of **0** for the first value. Press [TAB].
   c. Enter 20 for the last value.

3. Click [OK].

4. Select **Calc>Probability Distributions>Binomial.**

Binomial Distribution
Dialog Box

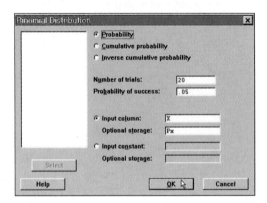

5. In the dialog box you must enter 5 items.
   a. Click the button for `Probability`.
   b. In the box for `Number of trials` enter **20,** the value of *n*.
   c. Enter **.05** the numerical `Probability of success`.

d. Check the button for `Input columns`. Click in the box and type the column name, **X.**

e. Click in the box for `Optional storage` type **Px.** The first available column will be named Px and the calculated probabilities will be stored in it.

6. Click [`OK`].

The results in the worksheet are shown.

Worksheet

C1	C2
X	Px
0	0.358486
1	0.377354
2	0.188677
3	0.059582
4	0.013328
5	0.002245
6	0.000295
7	0.000031
8	0.000003
9	0.000000
10	0.000000
11	0.000000
12	0.000000
13	0.000000
14	0.000000
15	0.000000
16	0.000000
17	0.000000
18	0.000000
19	0.000000
20	0.000000

TI-83
Step by Step

The Binomial Distribution

To find the probability for a binomial variable:

Example: $n = 20$ $X = 5$ $p = 0.05$

1. Press **2nd [DISTR]** then **0** to get `binompdf(`.

2. Enter **20, .05, 5)** then press **ENTER.**

The calculator will display .002244646.

Compare that result with that for Example 6–19 part *a.*

To find the probability for several values of a binomial variable:

Example: $n = 20$ $x = 0, 1, 2, 3$ $p = 0.05$

1. Press **2nd [DISTR] 0** to get `binompdf(`.

2. Enter **20, .05, {0, 1, 2, 3})** then press **ENTER.**

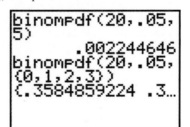

The calculator will display {.3584859224, .3773536025, .1886768013, .0595821478}

Use the arrow keys to view the entire display. Compare with Example 6–19 part *b*.

To find the cumulative probability for a binomial variable:

Example: $n = 20$ $X = 0, 1, 2, 3$ $p = 0.05$

1. Press **2nd [DISTR] A** to get `binomcdf(`.
2. Enter **20, .05, 3)** then press **ENTER**.

The calculator will display .984098474.

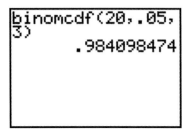

```
binomcdf(20,.05,
3)
          .984098474
```

Compare with Example 6–19 part *b*.

6–5

Other Types of Distributions (Optional)

In addition to the binomial distribution, other types of distributions are used in statistics. Three of the most commonly used distributions are the multinomial distribution, the Poisson distribution, and the hypergeometric distribution. They are described in the following subsections.

The Multinomial Distribution

Objective 5. Find probabilities for outcomes of variables using the Poisson, hypergeometric, and multinomial distributions.

Recall that in order for an experiment to be binomial, two outcomes are required for each trial. But if each trial in an experiment has more than two outcomes, a distribution called the **multinomial distribution** must be used. For example, a survey might require the responses of "approve," "disapprove," or "no opinion." In another situation, a person may have a choice of one of five activities for Friday night, such as a movie, dinner, baseball game, play, or party. Since these situations have more than two possible outcomes for each trial, the binomial distribution cannot be used to compute probabilities.

The multinomial distribution can be used for such situations if the probabilities for each trial remain constant and the outcomes are independent for a fixed number of trials. The events must also be mutually exclusive.

Formula for the Multinomial Distribution

If X consists of events E_1, E_2, E_3, . . . , E_k, which have corresponding probabilities p_1, p_2, p_3, . . . , p_k of occurring, and X_1 is the number of times E_1 will occur, X_2 is the number of times E_2 will occur, X_3 is the number of times E_3 will occur, etc., then the probability that X will occur is

$$P(X) = \frac{n!}{X_1! \cdot X_2! \cdot X_3! \cdot \ldots \cdot X_k!} \cdot p_1^{X_1} \cdot p_2^{X_2} \cdot \ldots \cdot p_k^{X_k}$$

where $X_1 + X_2 + X_3 + \cdots + X_k = n$, and $p_1 + p_2 + p_3 + \cdots + p_k = 1$

Example 6–24

In a large city, 50% of the people choose a movie, 30% choose dinner and a play, and 20% choose shopping as a leisure activity. If a sample of five people is randomly selected, find the probability that three are planning to go to a movie, one to a play, and one to a shopping mall.

Solution

$n = 5$, $X_1 = 3$, $X_2 = 1$, $X_3 = 1$, $p_1 = 0.50$, $p_2 = 0.30$, and $p_3 = 0.20$. Substituting in the formula gives

$$P(X) = \frac{5!}{3! \cdot 1! \cdot 1!} \cdot (0.50)^3(0.30)^1(0.20)^1 = 0.15$$

Again, note that the multinomial distribution can be used even though replacement is not done, provided that the sample is small in comparison with the population.

Example 6–25

In a music store, a manager found that the probabilities that a person buys zero, one, or two or more CDs are 0.3, 0.6, and 0.1, respectively. If six customers enter the store, find the probability that one won't buy any CDs, three will buy one CD, and two will buy two or more CDs.

Solution

$n = 6$, $X_1 = 1$, $X_2 = 3$, $X_3 = 2$, $p_1 = 0.3$, $p_2 = 0.6$, and $p_3 = 0.1$. Then,

$$P(X) = \frac{6!}{1! \, 3! \, 2!} \cdot (0.3)^1(0.6)^3(0.1)^2$$
$$= 60 \cdot (0.3)(0.216)(0.01) = 0.03888$$

Example 6–26

A box contains four white balls, three red balls, and three blue balls. A ball is selected at random, and its color is written down. It is replaced each time. Find the probability that if five balls are selected, two are white, two are red, and one is blue.

Solution

$n = 5$, $X_1 = 2$, $X_2 = 2$, $X_3 = 1$; $p_1 = \frac{4}{10}$, $p_2 = \frac{3}{10}$, and $p_3 = \frac{3}{10}$; hence,

$$P(X) = \frac{5!}{2! \, 2! \, 1!} \cdot \left(\frac{4}{10}\right)^2\left(\frac{3}{10}\right)^2\left(\frac{3}{10}\right)^1 = \frac{81}{625}$$

Thus, the multinomial distribution is similar to the binomial distribution but has the advantage of allowing one to compute probabilities when there are more than two outcomes for each trial in the experiment. That is, the multinomial distribution is a general distribution, and the binomial distribution is a special case of the multinomial distribution.

The Poisson Distribution

A discrete probability distribution that is useful when n is large and p is small and when the independent variables occur over a period of time is called the **Poisson distribution.** In addition to being used for the stated conditions (i.e., n is large, p is small, and the variables occur over a period of time), the Poisson distribution can be used when a density of items is distributed over a given area or volume, such as the number of plants growing per acre or the number of defects in a given length of videotape.

Formula for the Poisson Distribution

The probability of X occurrences in an interval of time, volume, area, etc., for a variable where λ (Greek letter lambda) is the mean number of occurrences per unit (area, time, volume, etc.) is

$$P(X; \lambda) = \frac{e^{-\lambda}\lambda^X}{X!} \qquad \text{where} \qquad X = 0, 1, 2, \ldots$$

The letter e is a constant approximately equal to 2.7183.

Round the answers to four decimal places.

Example 6–27

If there are 200 typographical errors randomly distributed in a 500-page manuscript, find the probability that a given page contains exactly three errors.

Solution

First, find the mean number (λ) of errors. Since there are 200 errors distributed over 500 pages, each page has an average of

$$\lambda = \frac{200}{500} = \frac{2}{5} = 0.4$$

or 0.4 error per page. Since $X = 3$, substituting into the formula yields

$$P(X; \lambda) = \frac{e^{-\lambda}\lambda^X}{X!} = \frac{(2.7183)^{-0.4}(0.4)^3}{3!} = 0.0072$$

Thus, there is less than a 1% probability that any given page will contain exactly three errors.

Since the mathematics involved in computing Poisson probabilities is somewhat complicated, tables have been compiled for these probabilities. Table C in Appendix C gives P for various values for λ and X.

In Example 6–27, where X is 3 and λ is 0.4, the table gives the value 0.0072 for the probability. See Figure 6–4.

Figure 6–4

Using Table C

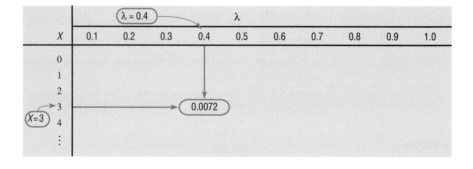

Example 6–28

A sales firm receives, on the average, three calls per hour on its toll-free number. For any given hour, find the probability that it will receive the following.

a. At most three calls   b. At least three calls   c. Five or more calls

Solution

a. "At most three calls" means 0, 1, 2, or 3 calls. Hence,

$$P(0; 3) + P(1; 3) + P(2; 3) + P(3; 3)$$
$$= 0.0498 + 0.1494 + 0.2240 + 0.2240$$
$$= 0.6472$$

b. "At least three calls" means 3 or more calls. It is easier to find the probability of 0, 1, and 2 calls and then subtract this answer from 1 to get the probability of at least 3 calls.

$$P(0; 3) + P(1; 3) + P(2; 3) = 0.0498 + 0.1494 + 0.2240 = 0.4232$$

and

$$1 - 0.4232 = 0.5768$$

c. For the probability of five or more calls, it is easier to find the probability of getting 0, 1, 2, 3, or 4 calls and subtract this answer from 1. Hence,

$$P(0; 3) + P(1; 3) + P(2; 3) + P(3; 3) + P(4; 3)$$
$$= 0.0498 + 0.1494 + 0.2240 + 0.2240 + 0.1680$$
$$= 0.8152$$

and

$$1 - 0.8152 = 0.1848$$

Thus, for the events described, the part *a* event is most likely to occur and the part *c* event is least likely to occur.

The Poisson distribution can also be used to approximate the binomial distribution when the expected value, $\lambda = n \cdot p$, is less than 5, as shown in the next example. (The same is true when $n \cdot q < 5$.)

Example 6–29

If approximately 2% of the people in a room of 200 people are left-handed, find the probability that exactly five people there are left-handed.

Solution

Since $\lambda = n \cdot p$, then $\lambda = (200)(0.02) = 4$. Hence,

$$P(X; \lambda) = \frac{(2.7183)^{-4}(4)^5}{5!} = 0.1563$$

which is verified by the formula $_{200}C_5(0.02)^5(0.98)^{195} \approx 0.1579$. The difference between the two answers is based on the fact that the Poisson distribution is an approximation and rounding has been used.

The Hypergeometric Distribution

When sampling is done *without* replacement, the binomial distribution does not give exact probabilities, since the trials are not independent. The smaller the size of the population, the less accurate the binomial probabilities will be.

For example, suppose a committee of four people is to be selected from seven women and five men. What is the probability that the committee will consist of three women and one man?

To solve this problem, one must find the number of ways a committee of three women and one man can be selected from seven women and five men. This answer can be found by using combinations; it is

$$_7C_3 \cdot {}_5C_1 = 35 \cdot 5 = 175$$

Next, find the total number of ways a committee of four people can be selected from 12 people. Again, by the use of combinations, the answer is

$$_{12}C_4 = 495$$

Finally, the probability of getting a committee of three women and one man from seven women and five men is

$$P(X) = \frac{175}{495} = \frac{35}{99}$$

The results of the problem can be generalized by using a special probability distribution called the hypergeometric distribution. The **hypergeometric distribution** is a distribution of a variable that has two outcomes when sampling is done without replacement.

The probabilities for the hypergeometric distribution can be calculated by using the formula given next.

Formula for the Hypergeometric Distribution

Given a population with only two types of objects (females and males, defective and nondefective, successes or failures, etc.), such that there are a items of one kind and b items of another kind and $a + b$ equals the total population, the probability $P(X)$ of selecting without replacement a sample of size n with X items of type a and $n - X$ items of type b is

$$P(X) = \frac{{}_aC_X \cdot {}_bC_{n-X}}{{}_{(a+b)}C_n}$$

The basis of the formula is that there are $_aC_X$ ways of selecting the first type of items, $_bC_{n-X}$ ways of selecting the second type of items, and $_{(a+b)}C_n$ ways of selecting n items from the entire population.

....

Example 6–30

Ten people apply for a job as assistant manager of a restaurant. Five have completed college and five have not. If the manager selects three applicants at random, find the probability that all three are college graduates.

Solution

Assigning the values to the variables gives

$$a = 5 \text{ college graduates} \qquad n = 3$$
$$b = 5 \text{ nongraduates} \qquad X = 3$$

and $n - X = 0$. Substituting in the formula gives

$$P(X) = \frac{{}_5C_3 \cdot {}_5C_0}{{}_{10}C_3} = \frac{10}{120} = \frac{1}{12}$$

Example 6–31

A recent study found that four out of nine houses were underinsured. If five houses are selected from the nine houses, find the probability that exactly two are underinsured.

Solution

In this problem

$$a = 4 \qquad b = 5 \qquad n = 5 \qquad X = 2 \qquad n - X = 3$$

Then,

$$P(X) = \frac{_4C_2 \cdot {_5C_3}}{_9C_5} = \frac{60}{126} = \frac{10}{21}$$

In many situations where objects are manufactured and shipped to a company, the company selects a few items and tests them to see whether they are satisfactory or defective. If a certain percentage is defective, the company then can refuse the whole shipment. This procedure saves the time and cost of testing every single item. In order to make the judgment about whether to accept or reject the whole shipment based on a small sample of tests, the company must know the probability of getting a specific number of defective items. To calculate the probability, the company uses the hypergeometric distribution.

Example 6–32

A lot of 12 compressor tanks is checked to see whether there are any defective tanks. Three tanks are checked for leaks. If one or more of the three is defective, the lot is rejected. Find the probability that the lot will be rejected if there are actually three defective tanks in the lot.

Solution

Since the lot is rejected if at least one tank is found to be defective, it is necessary to find the probability that none are defective and subtract this probability from 1.
 Here, $a = 3, b = 9, n = 3, X = 0$; so

$$P(X) = \frac{_3C_0 \cdot {_9C_3}}{_{12}C_3} = \frac{1 \cdot 84}{220} = 0.38$$

Hence,

$$P(\text{at least one defective}) = 1 - P(\text{no defectives}) = 1 - 0.38 = 0.62$$

There is a 0.62, or 62%, probability that the lot will be rejected when three of the 12 tanks are defective.

A summary of the discrete distributions used in this chapter is shown in Table 6–1.

Table 6–1   Summary of Discrete Distributions

1. Binomial distribution

$$P(X) = \frac{n!}{(n - X)!X!} \cdot p^X \cdot q^{n-X}$$

$$\mu = n \cdot p \qquad \sigma = \sqrt{n \cdot p \cdot q}$$

(continued)

Table 6–1 *(concluded)*

Used when there are only two independent outcomes for a fixed number of independent trials and the probability for each success remains the same for each trial.

2. Multinomial distribution

$$P(X) = \frac{n!}{X_1! \cdot X_2! \cdot X_3! \cdot \ldots \cdot X_k!} \cdot p_1^{X_1} \cdot p_2^{X_2} \cdot \ldots \cdot p_k^{X_k}$$

where

$$X_1 + X_2 + X_3 + \cdots + X_k = n \text{ and } p_1 + p_2 + p_3 + \cdots + p_k = 1$$

Used when the distribution has more than two outcomes, the probabilities for each trial remain constant, outcomes are independent, and there is a fixed number of trials.

3. Poisson distribution

$$P(X; \lambda) = \frac{e^{-\lambda}\lambda^X}{X!} \qquad \text{where} \qquad X = 0, 1, 2, \ldots$$

Used when n is large and p is small, the independent variable occurs over a period of time, or a density of items is distributed over a given area or volume.

4. Hypergeometric distribution

$$P(X) = \frac{{}_aC_X \cdot {}_bC_{n-X}}{{}_{(a+b)}C_n}$$

Used when there are two outcomes and sampling is done without replacement.

Exercises

6–92. What is the relationship between the multinomial distribution and the binomial distribution?

6–93. Use the multinomial formula and find the probabilities for each.
  a. $n = 6, X_1 = 3, X_2 = 2, X_3 = 1, p_1 = 0.5, p_2 = 0.3, p_3 = 0.2$
  b. $n = 5, X_1 = 1, X_2 = 2, X_3 = 2, p_1 = 0.3, p_2 = 0.6, p_3 = 0.1$
  c. $n = 4, X_1 = 1, X_2 = 1, X_3 = 2, p_1 = 0.8, p_2 = 0.1, p_3 = 0.1$
  d. $n = 3, X_1 = 1, X_2 = 1, X_3 = 1, p_1 = 0.5, p_2 = 0.3, p_3 = 0.2$
  e. $n = 5, X_1 = 1, X_2 = 3, X_3 = 1, p_1 = 0.7, p_2 = 0.2, p_3 = 0.1$

6–94. The probabilities that a textbook page will have 0, 1, 2, or 3 typographical errors are 0.79, 0.12, 0.07, and 0.02, respectively. If eight pages are randomly selected, find the probability that four will contain no errors, two will contain 1 error, one will contain 2 errors, and one will contain 3 errors.

6–95. The probabilities are 0.25, 0.40, and 0.35 that an 18-wheel truck will have no violations, 1 violation, or 2 or more violations when it is given a safety inspection.

If eight trucks are inspected, find the probability that three will have no violations, two will have 1 violation, and three will have 2 or more violations.

6–96. When a customer enters a pharmacy, the probability that he or she will have 0, 1, 2, or 3 prescriptions filled are 0.60, 0.25, 0.10, and 0.05, respectively. For a sample of six people who enter the pharmacy, find the probability that two will have no prescriptions, two will have 1 prescription, one will have 2 prescriptions, and one will have 3 prescriptions.

6–97. A die is rolled four times. Find the probability of two 1s, one 2, and one 3.

6–98. According to Mendel's theory, if tall and colorful plants are crossed with short and colorless plants, the corresponding probabilities are $\frac{9}{16}, \frac{3}{16}, \frac{3}{16}$, and $\frac{1}{16}$ for tall and colorful, tall and colorless, short and colorful, and short and colorless, respectively. If eight plants are selected, find the probability that one will be tall and colorful, three will be tall and colorless, three will be short and colorful, and one will be short and colorless.

6–99. Find each probability, $P(X; \lambda)$, using Table C in Appendix C.
  a. $P(5; 4)$ b. $P(2; 4)$ c. $P(6; 3)$
  d. $P(10; 7)$ e. $P(9; 8)$

6–100. If 2% of the batteries manufactured by a company are defective, find the probability that in a case of 144 batteries, there are 3 defective ones.

6–101. A recent study of robberies for a certain geographic region showed an average of one robbery per 20,000 people. In a city of 80,000 people, find the probability of the following.
a. No robberies
b. One robbery
c. Two robberies
d. Three or more robberies

6–102. In a 400-page manuscript, there are 200 randomly distributed misprints. If a page is selected, find the probability that it has one misprint.

6–103. A telephone-soliciting company obtains an average of five orders per 1000 solicitations. If the company reaches 250 potential customers, find the probability of obtaining at least two orders.

6–104. A mail-order company receives an average of five orders per 500 solicitations. If it sends out 100 advertisements, find the probability of receiving at least two orders.

6–105. A videotape has an average of one defect every 1000 feet. Find the probability of at least one defect in 3000 feet.

6–106. If 3% of all cars fail the emissions inspection, find the probability that in a sample of 90 cars, three will fail. Use the Poisson approximation.

6–107. The average number of phone inquiries per day at the poison control center is four. Find the probability it will receive five calls on a given day. Use the Poisson approximation.

6–108. In a batch of 2000 calculators, there are, on average, eight defective ones. If a random sample of 150 is selected, find the probability of five defective ones.

6–109. In a camping club of 18 members, nine prefer hoods and nine prefer hats and earmuffs. On a recent winter outing attended by six members, find the probability that exactly three members wore earmuffs and hats.

6–110. A bookstore owner examines 5 books from each lot of 25 to check for missing pages. If he finds at least 2 books with missing pages, the entire lot is returned. If, indeed, there are 5 books with missing pages, find the probability that the lot will be returned.

6–111. Shirts are packed at random in two sizes, regular and extra large. Four shirts are selected from a box of 24 and checked for size. If there are 15 regular shirts in the box, find the probability that all 4 will be regular size.

6–112. A shipment of 24 computer keyboards is rejected if 4 are checked for defects and at least 1 is found to be defective. Find the probability that the shipment will be returned if there are actually 6 defective keyboards.

6–113. A shipment of 24 electric typewriters is rejected if 3 are checked for defects and at least 1 is found to be defective. Find the probability that the shipment will be returned if there are actually 6 typewriters that are defective.

6–6

Summary

Many variables have special probability distributions. This chapter presented several of the most common probability distributions, including the binomial distribution, the multinomial distribution, the Poisson distribution, and the hypergeometric distribution.

The binomial distribution is used when there are only two outcomes for an experiment, there is a fixed number of trials, the probability is the same for each trial, and the outcomes are independent of each other. The multinomial distribution is an extension of the binomial distribution and is used when there are three or more outcomes for an experiment. The hypergeometric distribution is used when sampling is done without replacement. Finally, the Poisson distribution is used in special cases when independent events occur over a period of time, area, or volume.

A probability distribution can be graphed, and the mean, variance, and standard deviation can be found. The mathematical expectation can also be calculated for a probability distribution. Expectation is used in insurance and games of chance.

Important Terms

binomial distribution   225

binomial experiment   225

expected value   220

hypergeometric distribution   238

multinomial distribution   234

Poisson distribution   235

probability distribution   212

random variable   211

Important Formulas

Formula for the mean of a probability distribution:

$$\mu = \Sigma X \cdot P(X)$$

Formula for the variance of a probability distribution:

$$\sigma^2 = \Sigma [X^2 \cdot P(X)] - \mu^2$$

Formula for expected value:

$$E(X) = \Sigma X \cdot P(X)$$

Binomial probability formula:

$$P(X) = \frac{n!}{(n-X)!X!} \cdot p^X \cdot q^{n-X}$$

Formula for the mean of the binomial distribution:

$$\mu = n \cdot p$$

Formulas for the variance and standard deviation of the binomial distribution:

$$\sigma^2 = n \cdot p \cdot q \qquad \sigma = \sqrt{n \cdot p \cdot q}$$

Formula for the multinomial distribution:

$$P(X) = \frac{n!}{X_1! \cdot X_2! \cdot X_3! \cdots X_k!} \cdot$$
$$p_1^{X_1} \cdot p_2^{X_2} \cdots p_k^{X_k}$$

Formula for the Poisson distribution:

$$P(X; \lambda) = \frac{e^{-\lambda}\lambda^X}{X!} \quad \text{where} \quad X = 0, 1, 2, \ldots$$

Formula for the hypergeometric distribution:

$$P(X) = \frac{{}_aC_X \cdot {}_bC_{n-X}}{{}_{(a+b)}C_n}$$

Review Exercises

For Exercises 6–114 through 6–117, determine whether the distribution represents a probability distribution. If it does not, state why.

6–114.

X	1	2	3	4	5
P(X)	$\frac{3}{10}$	$\frac{1}{10}$	$\frac{2}{10}$	$\frac{3}{10}$	$\frac{1}{10}$

6–115.

X	5	10	15	20
P(X)	0.5	0.3	0.1	0.4

6–116.

X	100	200	300
P(X)	0.3	0.3	0.1

6–117.

X	8	12	16	20
P(X)	$\frac{5}{6}$	$\frac{1}{12}$	$\frac{1}{12}$	$\frac{1}{12}$

6–118. The number of rescue calls a helicopter ambulance service receives per 24-hour period is distributed as shown here. Construct a graph for the data.

Number of calls, X	6	7	8	9	10
Probability, P(X)	0.12	0.31	0.40	0.15	0.02

6–119. A study was conducted to determine the number of radios each household has. The data are shown here. Construct a probability distribution and draw a graph for the data.

Number of radios	Probability
0	0.05
1	0.30
2	0.45
3	0.12
4	0.08

6–120. A box contains five pennies, three dimes, one quarter, and one half-dollar. Construct a probability distribution and draw a graph for the data.

6–121. At Tyler's Tie Shop, Tyler found the probabilities that a customer will buy 0, 1, 2, 3, or 4 ties, as shown. Construct a graph for the distribution.

Number of ties, X	0	1	2	3	4
Probability, P(X)	0.30	0.50	0.10	0.08	0.02

6–122. A bank has a drive-through service. The number of customers arriving during a 15-minute period is distributed as shown. Find the mean, variance, and standard deviation for the distribution.

Number of customers, X	0	1	2	3	4
Probability, P(X)	0.12	0.20	0.31	0.25	0.12

6–123. At a small community library, the number of visitors per hour during the day has the distribution shown. Find the mean, variance, and standard deviation for the data.

Number of visitors, X	8	9	10	11	12
Probability, $P(X)$	0.15	0.25	0.29	0.19	0.12

6–124. During a recent paint sale at Corner Hardware, the number of cans of paint purchased was distributed as shown. Find the mean, variance, and standard deviation of the distribution.

Number of cans, X	1	2	3	4	5
Probability, $P(X)$	0.42	0.27	0.15	0.10	0.06

6–125. The number of inquiries received per day for a college catalog is distributed as shown. Find the mean, variance, and standard deviation for the data.

Number of inquiries, X	22	23	24	25	26	27
Probability, $P(X)$	0.08	0.19	0.36	0.25	0.07	0.05

6–126. There are five envelopes in a box. One envelope contains a penny, one a nickel, one a dime, one a quarter, and one a half-dollar. A person selects an envelope. Find the expected value of the draw.

6–127. A person selects a card from a deck. If it is a red card, he wins $1. If it is a black card between or including 2 and 10, he wins $5. If it is a black face card, he wins $10, and if it is a black ace, he wins $100. Find the expectation of the game. How much should a person bet if the game is to be fair?

6–128. If 30% of all commuters ride the train to work, find the probability that if 10 workers are selected, 5 will ride the train.

6–129. If 90% of all people between the ages of 30 and 50 drive a car, find these probabilities for a sample of 20 people in that age group.
  a. Exactly 20 drive a car.
  b. At least 15 drive a car.
  c. At most 15 drive a car.

6–130. If 10% of the people who are given a certain drug experience dizziness, find these probabilities for a sample of 15 people who take the drug.
  a. At least two people will become dizzy.
  b. Exactly three people will become dizzy.
  c. At most four people will become dizzy.

6–131. If 75% of nursing students are able to pass a drug calculation test, find the mean, variance, and standard deviation of the number of students who pass the test in a sample of 180 nursing students.

6–132. A club has 50 members. If there is a 10% absentee rate per meeting, find the mean, variance, and standard deviation of the number of people who will be absent from each meeting.

6–133. The chance that a U.S. police chief believes the death penalty "significantly reduces the number of homicides" is one in four. If a random sample of eight police chiefs is selected, find the probability that at most three believe that the death penalty significantly reduces the number of homicides.
Source: *Harper's Index* 290, no. 1741 (June 1995), p. 11.

6–134. *American Energy Review* reported that 27% of American households burn wood. If a random sample of 500 American households is selected, find the mean, variance, and standard deviation of the number of households that burn wood.
Source: *100% American* by Daniel Evan Weiss (New York: Poseidon Press, 1988).

6–135. Three out of four American adults under 35 have eaten pizza for breakfast. If a random sample of 20 adults under 35 is selected, find the probability that exactly 16 have eaten pizza for breakfast.
Source: *Harper's Index* 290, no. 1738 (March 1995), p. 9.

6–136. One out of four Americans over 55 has eaten pizza for breakfast. If a sample of 10 Americans over 55 is selected at random, find the probability that at most three have eaten pizza for breakfast.
Source: *Harper's Index* 290, no. 1738 (March 1995), p. 9.

6–137. (Opt.) The probabilities that a person will make 0, 1, 2, or 3 errors on an insurance claim are 0.70, 0.20, 0.08, and 0.02, respectively. If 20 claims are selected, find the probability that 12 will contain no errors, 4 will contain 1 error, 3 will contain 2 errors, and 1 will contain 3 errors.

6–138. (Opt.) Before a VCR leaves the factory, it is given a quality control check. The probabilities that a VCR contains 0, 1, or 2 defects are 0.90, 0.06, and 0.04, respectively. In a sample of 12 recorders, find the probability that 8 have no defects, 3 have 1 defect, and 1 has 2 defects.

6–139. (Opt.) In a Christmas display, the probability that all lights are the same color is 0.50; that 2 colors are used is 0.40; and that 3 or more colors are used is 0.10. If a sample of 10 displays is selected, find the probability that 5 have only 1 color of light, 3 have 2 colors, and 2 have 3 or more colors.

6–140. (Opt.) If 4% of the population carries a certain genetic trait, find the probability that in a sample of 100 people, there are exactly 8 people who have the trait. Assume the distribution is approximately Poisson.

6–141. (Opt.) Computer Help Hot Line receives, on the average, six calls per hour asking for assistance. The distribution is Poisson. For any randomly selected hour, find the probability that the company will receive the following.
  a. At least six calls
  b. Four or more calls
  c. At most five calls

6–142. (Opt.) The number of boating accidents on Lake Emilie follows a Poisson distribution. The probability of an accident is 0.003. If there are 1000 boats on the lake during a summer month, find the probability that there will be six accidents.

6–143. (Opt.) If five cards are drawn from a deck, find the probability that two will be hearts.

6–144. (Opt.) Of the 50 automobiles in a used-car lot, 10 are white. If five automobiles are selected to be sold at an auction, find the probability that exactly two will be white.

6–145. (Opt.) A board of directors consists of seven men and five women. If a slate of three officers is selected, find these probabilities.
  a. Exactly two are men.
  b. All three are women.
  c. Exactly two are women.

Statistics Today

Are the Rockets Hitting the Targets? Revisited

In order to assess the accuracy of these bombs, London was divided up into 576 square regions. Each region was $\frac{1}{4}$ square kilometer in area. The distribution of hits was recorded as follows:

Hits	0	1	2	3	4	5 or more
Regions	229	211	93	35	7	1

With the Poisson distribution, the expected number of hits is

Hits	0	1	2	3	4	5 or more
Regions	227.3	211.3	98.3	30.5	7.1	1.6

It was determined that the bombs were falling at random, since the observed frequencies were very close to those expected using the Poisson distribution. This conclusion was later verified by military intelligence.

Chapter Quiz

Determine whether each statement is true or false. If the statement is false, explain why.

1. The expected value of a random variable can be thought of as a long-run average.

2. The number of courses a student is taking this semester is an example of a continuous random variable.

3. When the multinomial distribution is used, the outcomes must be dependent.

4. A binomial experiment has a fixed number of trials.

Complete the following statements with the best answer.

5. Random variable values are determined by _____.

6. The mean for a binomial variable can be found by using the formula _____.

7. One requirement for a probability distribution is that the sum of all the events in the sample space must equal _____.

Select the best answer.

8. What is the sum of the probabilities of all outcomes in a probability distribution?
  a. 0
  b. 1/2
  c. 1
  d. It cannot be determined.

9. How many outcomes are there in a binomial experiment?
  a. 0
  b. 1
  c. 2
  d. It varies.

10. The number of plants growing in a specific area can be approximated by what distribution?
  a. Binomial
  b. Multinomial
  c. Hypergeometric
  d. Poisson

For questions 11 through 14, determine if the distribution represents a probability distribution. If not, state why.

11.

X	1	2	3	4	5
P(X)	$\frac{1}{7}$	$\frac{2}{7}$	$\frac{2}{7}$	$\frac{3}{7}$	$\frac{2}{7}$

12.

X	3	6	9	12	15
P(X)	0.3	0.5	0.1	0.08	0.02

13.

X	50	75	100
P(X)	0.5	0.2	0.3

14.

X	4	8	12	16
P(X)	$\frac{1}{6}$	$\frac{3}{12}$	$\frac{1}{2}$	$\frac{1}{12}$

15. The number of fire calls the Conestoga Valley Fire Company receives per day is distributed as follows:

Number X	5	6	7	8	9
Probability P(X)	0.28	0.32	0.09	0.21	0.10

Construct a graph for the data.

16. A study was conducted to determine the number of telephones each household has. The data are shown here.

Number of telephones	Frequency
0	2
1	30
2	48
3	13
4	7

Construct a probability distribution and draw a graph for the data.

17. During a recent cassette sale at Matt's Music Store, the number of tapes customers purchased was distributed as follows:

Number X	0	1	2	3	4
Probability P(X)	0.10	0.23	0.31	0.27	0.09

Find the mean, variance, and standard deviation of the distribution.

18. The number of calls received per day at a crisis hot line is distributed as follows:

Number X	30	31	32	33	34
Probability P(X)	0.05	0.21	0.38	0.25	0.11

Find the mean, variance, and standard deviation of the distribution.

19. There are six playing cards placed face down in a box. They are the 4 of diamonds, the 5 of hearts, the 2 of clubs, the 10 of spades, the 3 of diamonds, and the 7 of hearts. A person selects a card. Find the expected value of the draw.

20. A person selects a card from an ordinary deck of cards. If it is a black card, she wins $2. If it is a red card between or including 3 and 7, she wins $10. If it is a red face card, she wins $25, and if it is a black jack, she wins $100. Find the expectation of the game.

21. If 40% of all commuters ride to work in carpools, find the probability that if eight workers are selected, five will ride in carpools.

22. If 60% of all women are employed outside the home, find the probability that in a sample of 20 women,
   a. Exactly 15 are employed.
   b. At least 10 are employed.
   c. At most five are not employed outside the home.

23. If 80% of the applicants are able to pass a driver's proficiency road test, find the mean, variance, and standard deviation of the number of people who pass the test in a sample of 300 applicants.

24. A history class has 75 members. If there is a 12% absentee rate per class meeting, find the mean, variance, and standard deviation of the number of students who will be absent from each class.

25. The probability that a person will make zero, one, two, or three errors on his or her income tax return is 0.50, 0.30, 0.15, and 0.05, respectively. If 30 claims are selected, find the probability that 15 will contain no errors, 8 will contain one error, 5 will contain two errors, and 2 will contain three errors.

26. Before a television set leaves the factory, it is given a quality control check. The probability that a television contains zero, one, or two defects is 0.88, 0.08, and 0.04, respectively. In a sample of 16 televisions, find the probability that 9 will have no defects, 4 will have one defect, and 3 will have two defects.

27. Among the teams in a bowling league, the probability that the uniforms are all one color is 0.45, that two colors are used is 0.35, and that three or more colors are used is 0.20. If a sample of 12 uniforms is selected, find the probability that 5 contain only one color, 4 contain two colors, and 3 contain three or more colors.

28. If 8% of the population of trees are elm trees, find the probability that in a sample of 100 trees, there are exactly six elm trees. Assume the distribution is approximately Poisson.

29. Sports Scores Hot Line receives, on the average, eight calls per hour requesting the latest sports scores. The

distribution is Poisson in nature. For any randomly selected hour, find the probability that the company will receive

a. At least eight calls

b. Three or more calls

c. At most seven calls

30. There are 48 raincoats for sale at a local men's clothing store. Twelve are black. If six raincoats are selected to be marked down, find the probability that exactly three will be black.

31. A youth group has eight boys and six girls. If a slate of four officers is selected, find the probability that exactly

a. Three are girls

b. Two are girls

c. Four are boys

Critical Thinking Challenges

1. Pennsylvania has a lottery entitled "Big 4." In order to win, a player must correctly match four digits from a daily lottery in which four digits are selected. Find the probability of winning.

2. In the Big 4 lottery, for a bet of $100, the payoff is $5000. What is the expected value of winning? Is it worth it?

3. If you played the same 4 digit number every day (or any four digit number for that matter) in the Big 4, how often (in years) would you win, assuming you have average luck?

4. In the game "Chuck-a-Luck," three dice are rolled. A player bets a certain amount (say $1.00) on a number from one to six. If the number appears on one die, the person wins $1.00. If it appears on two dice, the person wins $2.00, and if it appears on all three dice, the person wins $3.00. What are the chances of winning $1.00? $2.00? $3.00?

5. What is the expected value of the game of "Chuck-a-Luck" if a player bets $1.00 on one number?

Data Projects

Probability Distributions   Roll three dice 100 times, recording the sum of the spots on the faces as you roll. Then find the average of the spots. How close is this to the theoretical average? Refer to Exercise 6–57 on page 223 in the textbook.

chapter

7

The Normal Distribution

Objectives

After completing this chapter, you should be able to

1. Identify distributions as symmetrical or skewed.

2. Identify the properties of the normal distribution.

3. Find the area under the standard normal distribution, given various z values.

4. Find probabilities for a normally distributed variable by transforming it into a standard normal variable.

5. Find specific data values for given percentages using the standard normal distribution.

6. Use the central limit theorem to solve problems involving sample means for large and small samples.

7. Use the normal approximation to compute probabilities for a binomial variable.

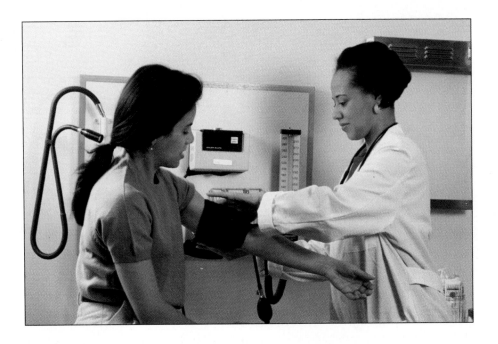

What Is Normal?

Medical researchers have determined so-called normal intervals for a person's blood pressure, cholesterol, triglycerides, and the like. For example, the normal range of systolic blood pressure is 110 to 140. The normal interval for a person's triglycerides is from 30 to 200 milligrams per deciliter (mg/dl). By measuring these variables, a physician can determine if a patient's vital statistics are within the normal interval, or if some type of treatment is needed to correct a condition and avoid future illnesses. The question then is, "How does one determine the so-called normal intervals?"

In this chapter, you will learn how researchers determine normal intervals for specific medical tests using the normal distribution. You will see how the same methods are used to determine the lifetimes of batteries, the strength of ropes, and many other traits.

7–1

Introduction

Random variables can be either discrete or continuous. Discrete variables and their distributions were explained in Chapter 6. Recall that a discrete variable cannot assume all values between any two given values of the variables. On the other hand, a continuous variable can assume all values between any two given values of the variables. Examples of continuous variables are the heights of adult men, body temperatures of rats, and cholesterol levels of adults. Many continuous variables, such as the examples just mentioned, have distributions that are bell-shaped and are called *approximately normally distributed variables.* For example, if a researcher selects a random sample of 100 adult women, measures their heights, and constructs a histogram, the researcher gets a graph similar to the one shown in Figure 7–1(a). Now, if the researcher increases the sample size and decreases the width of the classes, the histograms will look like the ones shown in Figure 7–1(b) and 7–1(c). Finally, if it were possible to measure exactly the heights of all adult females in the United States and plot them, the histogram would approach what is called the *normal distribution,* shown in Figure 7–1(d). This distribution is also

Figure 7–1

Histograms for the
Distribution of Heights of
Adult Women

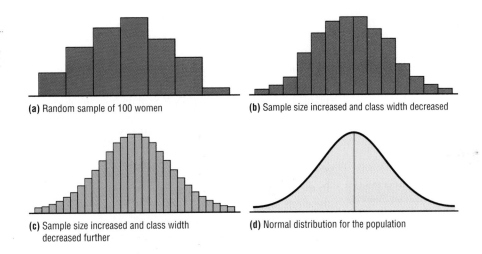

(a) Random sample of 100 women

(b) Sample size increased and class width decreased

(c) Sample size increased and class width
decreased further

(d) Normal distribution for the population

known as the *bell curve* or the *Gaussian distribution,* named for the German mathematician Carl Friedrich Gauss (1777–1855), who derived its equation.

No variable fits the normal distribution perfectly, since the normal distribution is a theoretical distribution. However, the normal distribution can be used to describe many variables, because the deviations from the normal distribution are very small. This concept will be explained further in the next section.

Objective 1. Identify
distributions as symmetrical
or skewed.

When the data values are evenly distributed about the mean, the distribution is said to be **symmetrical.** Figure 7–2(a) shows a symmetrical distribution. When the majority of the data values fall to the left or right of the mean, the distribution is said to be *skewed.* When the majority of the data values fall to the right of the mean, the distribution is said to be **negatively or left skewed.** The mean is to the left of the median, and the mean and the median are to the left of the mode. See Figure 7–2(b). When the majority of the data values fall to the left of the mean, the distribution is said to be **positively or right skewed.** The mean falls to the right of the median and both the mean and the median fall to the right of the mode. See Figure 7–2(c).

Figure 7–2

Normal and Skewed Distributions

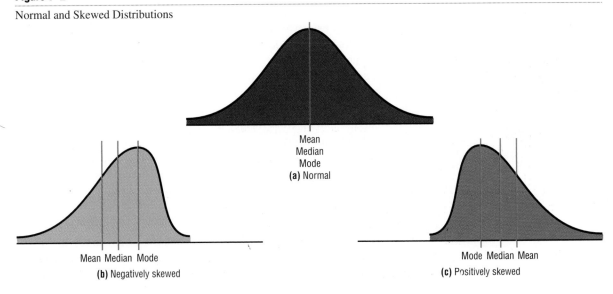

Mean
Median
Mode
(a) Normal

Mean Median Mode
(b) Negatively skewed

Mode Median Mean
(c) Positively skewed

The "tail" of the curve indicates the direction of skewness (right is positive, left negative). This distribution can be compared with the one in Figure 3–1. Both types follow the same principles.

This chapter will present the properties of the normal distribution and discuss its applications. Then a very important theorem called the *central limit theorem* will be explained. Finally, the chapter will explain how the normal curve distribution can be used as an approximation to other distributions, such as the binomial distribution. Since the binomial distribution is a discrete distribution, a correction for continuity may be employed when the normal distribution is used for its approximation.

7–2

Properties of the Normal Distribution

Objective 2. Identify the properties of the normal distribution.

In mathematics, curves can be represented by equations. For example, the equation of the circle shown in Figure 7–3 is $x^2 + y^2 = r^2$, where r is the radius. The circle can be used to represent many physical objects, such as a wheel or a gear. Even though it is not possible to manufacture a wheel that is perfectly round, the equation and the properties of the circle can be used to study the many aspects of the wheel, such as area, velocity, and acceleration. In a similar manner, the theoretical curve, called the *normal distribution curve,* can be used to study many variables that are not perfectly normally distributed but are nevertheless approximately normal.

The mathematical equation for the normal distribution is

$$y = \frac{e^{-(X-\mu)^2/(2\sigma^2)}}{\sigma\sqrt{2\pi}}$$

where

$e \approx 2.718$ ($\approx$ means "is approximately equal to")

$\pi \approx 3.14$

$\mu =$ population mean

$\sigma =$ population standard deviation

This equation may look formidable, but in applied statistics, tables are used for specific problems instead of the equation.

Another important aspect in applied statistics is that *the area under the normal distribution curve is more important than the frequencies.* Therefore, when the normal distribution is pictured, the y axis, which indicates the frequencies, is sometimes omitted.

Circles can be different sizes, depending on their diameters (or radii) and can be used to represent wheels of different sizes. Likewise, normal curves have different shapes and can be used to represent different variables.

The shape and position of the normal distribution curve depend on two parameters, the *mean* and the *standard deviation.* Each normally distributed variable has its own normal distribution curve, which depends on the values of the variable's mean and standard deviation. Figure 7–4(a) shows two normal distributions with the same mean values but different standard deviations. The larger the standard deviation, the more dispersed, or spread out, the distribution is. Figure 7–4(b) shows two normal distributions with the same standard deviation but with different means. These curves have the same shapes but are located at different positions on the x axis. Figure 7–4(c) shows two normal distributions with different means and different standard deviations.

Figure 7–3

Graph of a Circle and an Application

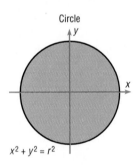

Circle

$x^2 + y^2 = r^2$

Wheel

The **normal distribution** is a continuous, symmetric, bell-shaped distribution of a variable.

Figure 7–4

Shapes of Normal Distributions

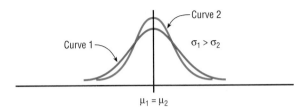

(a) Same means but different standard deviations

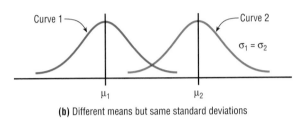

(b) Different means but same standard deviations

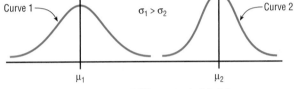

(c) Different means and different standard deviations

Historical Note

The discovery of the equation for the normal distribution can be traced to three mathematicians. In 1733, the French mathematician Abraham DeMoivre derived an equation for the normal distribution based on the random variation of the number of heads appearing when a large number of coins were tossed. Not realizing any connection with the naturally occurring variables, he showed this formula to only a few friends. About 100 years later, two mathematicians, Pierre Laplace in France and Carl Gauss in Germany, derived the equation of the normal curve independently and without any knowledge of DeMoivre's work. In 1924, Karl Pearson found that DeMoivre had discovered the formula before Laplace or Gauss.

The properties of the normal distribution, including those mentioned in the definition, are explained next.

Summary of the Properties of the Theoretical Normal Distribution

1. The normal distribution curve is bell-shaped.
2. The mean, median, and mode are equal and located at the center of the distribution.
3. The normal distribution curve is unimodal (i.e., it has only one mode).
4. The curve is symmetrical about the mean, which is equivalent to saying that its shape is the same on both sides of a vertical line passing through the center.
5. The curve is continuous—i.e., there are no gaps or holes. For each value of X, there is a corresponding value of Y.
6. The curve never touches the x axis. Theoretically, no matter how far in either direction the curve extends, it never meets the x axis—but it gets increasingly closer.
7. The total area under the normal distribution curve is equal to 1.00, or 100%. This fact may seem unusual, since the curve never touches the x axis, but one can prove it mathematically by using calculus. (The proof is beyond the scope of this textbook.)
8. The area under the normal curve that lies within one standard deviation of the mean is approximately 0.68, or 68%; within two standard deviations, about 0.95, or 95%; and within three standard deviations, about 0.997, or 99.7%. See Figure 7–5, which also shows the area in each region.

These values follow the *empirical rule* for data given in Section 3–3.

One must know these properties in order to solve problems using applications involving distributions that are approximately normal.

7–3

The Standard Normal Distribution

Since each normally distributed variable has its own mean and standard deviation, as stated earlier, the shape and location of these curves will vary. In practical applications, then, one would have to have a table of areas under the curve for each variable. To simplify this situation, statisticians use what is called the *standard normal distribution*.

Figure 7–5

Areas under the Normal
Distribution Curve

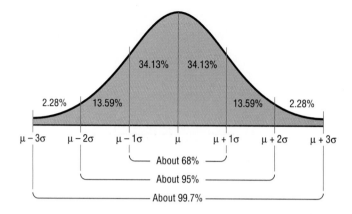

Objective 3. Find the area
under the standard normal
distribution, given various z
values.

The **standard normal distribution** is a normal distribution with a mean of 0 and a standard deviation of 1.

The standard normal distribution is shown in Figure 7–6.

Figure 7–6

Standard Normal
Distribution

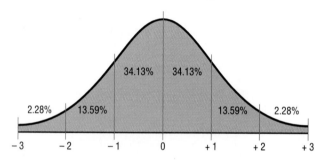

The values under the curve indicate the proportion of area in each section. For example, the area between the mean and one standard deviation above or below the mean is about 0.3413, or 34.13%.

The formula for the standard normal distribution is

$$y = \frac{e^{-X^2/2}}{\sqrt{2\pi}}$$

All normally distributed variables can be transformed into the standard normally distributed variable by using the formula for the standard score:

$$z = \frac{\text{value} - \text{mean}}{\text{standard deviation}} \quad \text{or} \quad z = \frac{X - \mu}{\sigma}$$

This is the same formula used in Section 3–4. The use of this formula will be explained in the next section.

As stated earlier, the area under the normal distribution curve is used to solve practical application problems, such as finding the percentage of adult women whose height is between 5 feet 4 inches and 5 feet 7 inches, or finding the probability that a new battery will last longer than four years. Hence, the major emphasis of this section will be to show the procedure for finding the area under the standard normal distribution curve for any z value. The applications will be shown in the next section. Once the X values are transformed using the preceding formula, they are called z values. The **z value** is actually the number of standard deviations that a particular X value is away from the mean. Table E in Appendix C gives the area (to four decimal places) under the standard normal curve for any *z* value from 0 to 3.09.

Finding Areas under the Standard Normal Distribution Curve

For the solution of problems using the normal distribution, a four-step procedure is recommended with the use of the Procedure Table shown below.

STEP 1   Draw a picture.

STEP 2   Shade the area desired.

STEP 3   Find the correct figure in the following Procedure Table (the figure that is similar to the one you've drawn).

STEP 4   Follow the directions given in the appropriate block of the Procedure Table to get the desired area.

There are seven basic types of problems and all seven are summarized in the Procedure Table. Note that this table is presented as an aid in understanding how to use the normal distribution table and in visualizing the problems. After learning the procedures, one should *not* find it necessary to refer to the procedure table for every problem.

Procedure Table

Finding the Area under the Normal Distribution Curve

1. Between 0 and any *z* value:
   Look up the *z* value in the table to get the area.

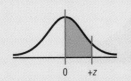

2. In any tail:
   a. Look up the *z* value to get the area.
   b. Subtract the area from 0.5000.

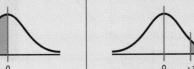

3. Between two *z* values on the same side of the mean:
   a. Look up both *z* values to get the areas.
   b. Subtract the smaller area from the larger area.

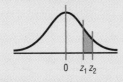

4. Between two *z* values on opposite sides of the mean:
   a. Look up both *z* values to get the areas.
   b. Add the areas.

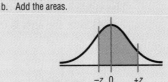

5. To the left of any *z* value, where *z* is greater than the mean:

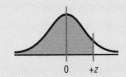

   a. Look up the *z* value to get the area.
   b. Add 0.5000 to the area

6. To the right of any *z* value, where *z* is less than the mean:

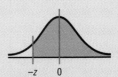

   a. Look up the *z* value in the table to get the area.
   b. Add 0.5000 to the area.

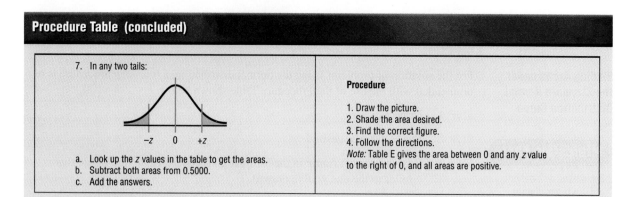

Procedure Table (concluded)

7. In any two tails:

-z   0   +z

a. Look up the z values in the table to get the areas.
b. Subtract both areas from 0.5000.
c. Add the answers.

Procedure

1. Draw the picture.
2. Shade the area desired.
3. Find the correct figure.
4. Follow the directions.
Note: Table E gives the area between 0 and any z value to the right of 0, and all areas are positive.

Situation 1   Find the area under the normal curve between 0 and any z value.

Example 7–1

Find the area under the normal distribution curve between $z = 0$ and $z = 2.34$.

Solution

Draw the figure and represent the area as shown in Figure 7–7.

Figure 7–7

Area under the Standard
Normal Curve for
Example 7–1

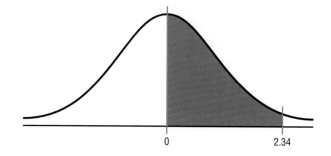

Since Table E gives the area between 0 and any z value to the right of 0, one need only look up the z value in the table. Find 2.3 in the left column and 0.04 in the top row. The value where the column and row meet in the table is the answer, 0.4904. See Figure 7–8. Hence, the area is 0.4904, or 49.04%.

Figure 7–8

Using Table E in
the Appendix for
Example 7–1

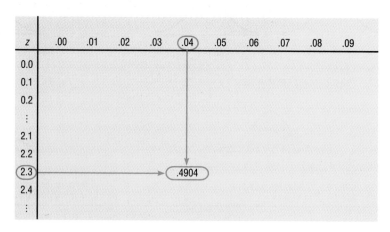

Example 7–2

Find the area between $z = 0$ and $z = 1.8$.

Solution

Draw the figure and represent the area as shown in Figure 7–9.

Figure 7–9

Area under the Standard
Normal Curve for
Example 7–2

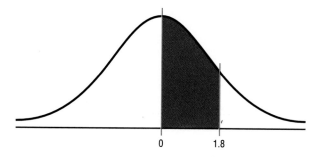

Find the area in Table E by finding 1.8 in the left column and 0.00 in the top row. The area is 0.4641 or 46.41%.

Next, one must be able to find the areas for values that are not in Table E. This is done by using the properties of the normal distribution described in Section 7–2.

Example 7–3

Find the area between $z = 0$ and $z = -1.75$.

Solution

Represent the area as shown in Figure 7–10.

Figure 7–10

Area under the Standard
Normal Curve for
Example 7–3

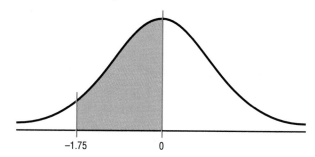

Table E does not give the areas for negative values of z. But since the normal distribution is symmetric about the mean, the area to the left of the mean (in this case, the mean is 0) is the same as the area to the right of the mean. Hence one need only look up the area for $z = +1.75$, which is 0.4599, or 45.99%. This solution is summarized in block 1 in the Procedure Table.

Remember that area is always a positive number, even if the z value is negative.

Situation 2   Find the area under the curve in either tail.

Example 7–4

Find the area to the right of $z = 1.11$.

Solution

Draw the figure and represent the area as shown in Figure 7–11.

Figure 7–11

Area under the Standard
Normal Curve for
Example 7–4

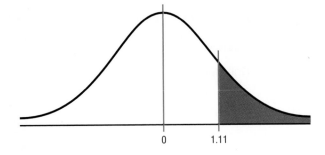

The required area is in the tail of the curve. Since Table E gives the area between $z = 0$ and $z = 1.11$, first find that area. Then subtract this value from 0.5000, since half of the area under the curve is to the right of $z = 0$. See Figure 7–12.

Figure 7–12

Finding the Area in the
Tail of the Curve
(Example 7–4)

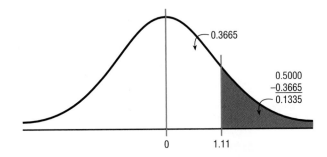

The area between $z = 0$ and $z = 1.11$ is 0.3665, and the area to the right of $z = 1.11$ is 0.1335, or 13.35%, obtained by subtracting 0.3665 from 0.5000.

Example 7–5

Find the area to the left of $z = -1.93$.

Solution

The desired area is shown in Figure 7–13.

Figure 7–13

Area under the Standard
Normal Curve for
Example 7–5

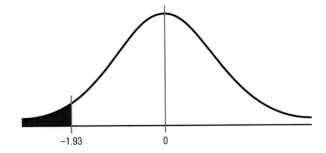

Again, Table E gives the area for positive z values. But from the symmetric property of the normal distribution, the area to the left of -1.93 is the same as the area to the right of $z = +1.93$, as shown in Figure 7–14.

Now find the area between 0 and $+1.93$ and subtract it from 0.5000, as shown:

$$
\begin{array}{r}
0.5000 \\
-0.4732 \\
\hline
0.0268, \text{ or } 2.68\%
\end{array}
$$

Figure 7–14

Comparison of Areas to
the Right of $+1.93$ and to
the Left of -1.93
(Example 7–5)

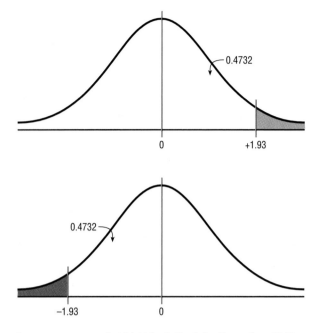

This procedure was summarized in block 2 of the Procedure Table.

Situation 3   Find the area under the curve between any two z values on the same side of the mean.

Example 7–6

Find the area between $z = 2.00$ and $z = 2.47$.

Solution

The desired area is shown in Figure 7–15.

Figure 7–15

Area under the Curve for
Example 7–6

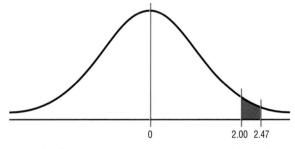

For this situation, look up the area from $z = 0$ to $z = 2.47$ and the area from $z = 0$ to $z = 2.00$. Then subtract the two areas, as shown in Figure 7–16.

Figure 7–16

Finding the Area
under the Curve for
Example 7–6

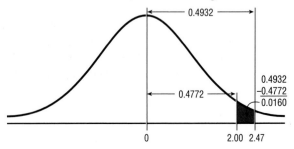

The area between $z = 0$ and $z = 2.47$ is 0.4932. The area between $z = 0$ and $z = 2.00$ is 0.4772. Hence, the desired area is $0.4932 - 0.4772 = 0.0160$, or 1.60%. This procedure is summarized in block 3 of the Procedure Table.

Two things should be noted here. First, the *areas,* not the z values, are subtracted. Subtracting the z values will yield an incorrect answer. Second, the procedure in Example 7–6 is used when both z values are on the same side of the mean.

Example 7–7

Find the area between $z = -2.48$ and $z = -0.83$.

Solution

The desired area is shown in Figure 7–17.

Figure 7–17

Area under the Curve for Example 7–7

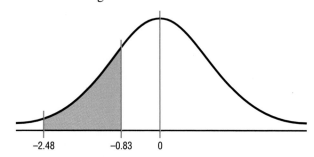

The area between $z = 0$ and $z = -2.48$ is 0.4934. The area between $z = 0$ and $z = -0.83$ is 0.2967. Subtracting yields $0.4934 - 0.2967 = 0.1967$, or 19.67%. This solution is summarized in block 3 of the Procedure Table.

Situation 4 Find the area under the curve between any two z values on opposite sides of the mean.

Example 7–8

Find the area between $z = +1.68$ and $z = -1.37$.

Solution

The desired area is shown in Figure 7–18.

Figure 7–18

Area under the Curve for Example 7–8

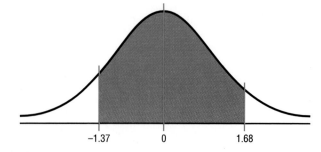

Now, since the two areas are on opposite sides of $z = 0$, one must find both areas and add them. The area between $z = 0$ and $z = 1.68$ is 0.4535. The area between $z = 0$ and $z = -1.37$ is 0.4147. Hence, the total area between $z = -1.37$ and $z = +1.68$ is $0.4535 + 0.4147 = 0.8682$, or 86.82%.

This type of problem is summarized in block 4 of the Procedure Table.

Situation 5   Find the area under the curve to the left of any z value, where z is greater than the mean.

Example 7–9

Find the area to the left of $z = 1.99$.

Solution

The desired area is shown in Figure 7–19.

Figure 7–19

Area under the Curve for Example 7–9

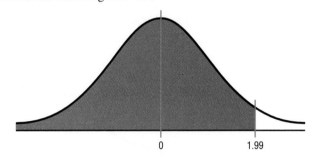

Since Table E gives only the area between $z = 0$ and $z = 1.99$, one must add 0.5000 to the table area, since 0.5000 (half) of the total area lies to the left of $z = 0$. The area between $z = 0$ and $z = 1.99$ is 0.4767, and the total area is $0.4767 + 0.5000 = 0.9767$, or 97.67%.

This solution is summarized in block 5 of the Procedure Table.

The same procedure is used when the z value is to the left of the mean, as shown in the next example.

Situation 6   Find the area under the curve to the right of any z value, where z is less than the mean.

Example 7–10

Find the area to the right of $z = -1.16$.

Solution

The desired area is shown in Figure 7–20.

Figure 7–20

Area under the Curve for Example 7–10

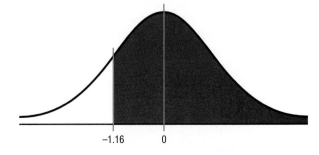

The area between $z = 0$ and $z = -1.16$ is 0.3770. Hence, the total area is $0.3770 + 0.5000 = 0.8770$, or 87.70%.

This type of problem is summarized in block 6 of the Procedure Table.

The final type of problem is that of finding the area in two tails. To solve it, find the area in each tail and add them, as shown in the next example.

Situation 7 Find the total area under the curve in any two tails.

Example 7–11

Find the area to the right of $z = +2.43$ and to the left of $z = -3.01$.

Solution

The desired area is shown in Figure 7–21.

Figure 7–21

Area under the Curve for Example 7–11

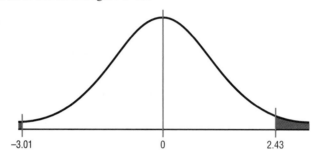

The area to the right of 2.43 is $0.5000 - 0.4925 = 0.0075$. The area to the left of $z = -3.01$ is $0.5000 - 0.4987 = 0.0013$. The total area, then, is $0.0075 + 0.0013 = 0.0088$, or 0.88%.

This solution is summarized in block 7 of the Procedure Table.

The Normal Distribution Curve as a Probability Distribution Curve

The normal distribution curve can be used as a probability distribution curve for normally distributed variables. Recall that the normal distribution is a *continuous distribution,* as opposed to a discrete probability distribution, as explained in Chapter 6. The fact that it is continuous means that there are no gaps in the curve. In other words, for every z value on the x axis, there is a corresponding height, or frequency value.

However, as stated earlier, the area under the curve is more important than the frequencies. *This area corresponds to a probability.* That is, if it were possible to select any z value at random, the probability of choosing one, say, between 0 and 2.00 would be the same as the area under the curve between 0 and 2.00. In this case, the area is 0.4772. Therefore, the probability of selecting any z value between 0 and 2.00 is 0.4772. The problems involving probability are solved in the same manner as the previous examples involving areas in this section. For example, if the problem is to find the probability of selecting a z value between 2.25 and 2.94, solve it by using the method shown in block 3 of the Procedure Table.

For probabilities, a special notation is used. For example, if the problem is to find the probability of any z value between 0 and 2.32, this probability is written as $P(0 < z < 2.32)$.

Example 7–12

Find the probability for each.

a. $P(0 < z < 2.32)$

b. $P(z < 1.65)$

c. $P(z > 1.91)$

Solution

a. $P(0 < z < 2.32)$ means to find the area under the normal distribution curve between 0 and 2.32. Look up the area in Table E corresponding to $z = 2.32$. It is 0.4898, or 48.98%. The area is shown in Figure 7–22.

Figure 7–22

Area under the Curve for Part *a* of Example 7–12

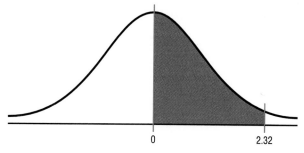

b. $P(z < 1.65)$ is represented in Figure 7–23.

Figure 7–23

Area under the Curve for Part *b* of Example 7–12

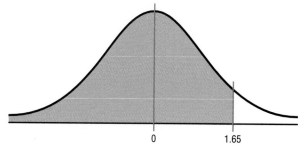

First, find the area between 0 and 1.65 in Table E. Then add it to 0.5000 to get $0.4505 + 0.5000 = 0.9505$, or 95.05%.

c. $P(z > 1.91)$ is shown in Figure 7–24.

Figure 7–24

Area under the Curve for Part *c* of Example 7–12

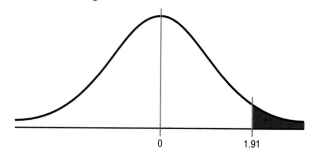

Since this area is a tail area, find the area between 0 and 1.91 and subtract it from 0.5000. Hence, $0.5000 - 0.4719 = 0.0281$, or 2.81%.

Sometimes, one must find a specific *z* value for a given area under the normal distribution. The procedure is to work backward, using Table E.

Example 7–13

Find the *z* value such that the area under the normal distribution curve between 0 and the *z* value is 0.2123.

Solution

Draw the figure. The area is shown in Figure 7–25.

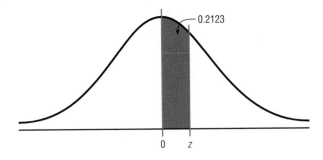

Next, find the area in Table E, as shown in Figure 7–26. Then read the correct z value in the left column as 0.5 and in the top row as 0.06 and add these two values to get 0.56.

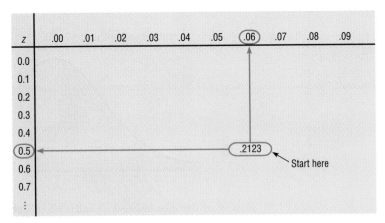

If the exact area cannot be found, use the closest value. For example, if one wanted to find the z value for an area 0.4241, the closest area is 0.4236, which gives a z value of 1.43. See Table E in Appendix C.

Finding the area under the standard normal distribution curve is the first step in solving a wide variety of practical applications in which the variables are normally distributed. Some of these applications will be presented in Section 7–4.

Exercises

7–1. What are the characteristics of the normal distribution?

7–2. Why is the normal distribution important in statistical analysis?

7–3. What is the total area under the normal distribution curve?

7–4. What percentage of the area falls below the mean? Above the mean?

7–5. What percentage of the area under the normal distribution curve falls within one standard deviation above and below the mean? Two standard deviations? Three standard deviations?

For Exercises 7–6 through 7–25, find the area under the normal distribution curve.

7–6. Between $z = 0$ and $z = 1.97$

7–7. Between $z = 0$ and $z = 0.56$

7–8. Between $z = 0$ and $z = -0.48$

7–9. Between $z = 0$ and $z = -2.07$

7–10. To the right of $z = 1.09$

7–11. To the right of $z = 0.23$

7–12. To the left of $z = -0.64$

7–13. To the left of $z = -1.43$

7–14. Between $z = 1.23$ and $z = 1.90$

7–15. Between $z = 0.79$ and $z = 1.28$

7–16. Between $z = -0.87$ and $z = -0.21$

7–17. Between $z = -1.56$ and $z = -1.83$

7–18. Between $z = 0.24$ and $z = -1.12$

7–19. Between $z = 2.47$ and $z = -1.03$

7–20. To the left of $z = 1.31$

7–21. To the left of $z = 2.11$

7–22. To the right of $z = -1.92$

7–23. To the right of $z = -0.18$

7–24. To the left of $z = -2.15$ and to the right of $z = 1.62$

7–25. To the right of $z = 1.92$ and to the left of $z = -0.44$

In Exercises 7–26 through 7–39, find probabilities for each, using the standard normal distribution.

7–26. $P(0 < z < 1.69)$

7–27. $P(0 < z < 0.67)$

7–28. $P(-1.23 < z < 0)$

7–29. $P(-1.57 < z < 0)$

7–30. $P(z > 2.59)$

7–31. $P(z > 2.83)$

7–32. $P(z < -1.77)$

7–33. $P(z < -1.51)$

7–34. $P(-0.05 < z < 1.10)$

7–35. $P(-2.46 < z < 1.74)$

7–36. $P(1.32 < z < 1.51)$

7–37. $P(1.46 < z < 2.97)$

7–38. $P(z > -1.39)$

7–39. $P(z < 1.42)$

For Exercises 7–40 through 7–45, find the z value that corresponds to the given area.

7–40.

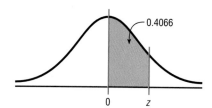

7–41.

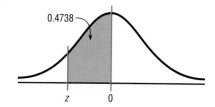

7–42.

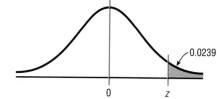

7–43.

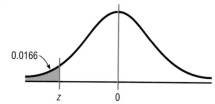

7–44.

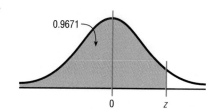

7–45.

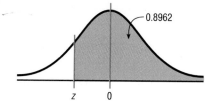

7–46. Find the z value to the right of the mean so that
a. 53.98% of the area under the distribution curve lies to the left of it.
b. 71.90% of the area under the distribution curve lies to the left of it.
c. 96.78% of the area under the distribution curve lies to the left of it.

7–47. Find the z value to the left of the mean so that
a. 98.87% of the area under the distribution curve lies to the right of it.
b. 82.12% of the area under the distribution curve lies to the right of it.
c. 60.64% of the area under the distribution curve lies to the right of it.

*7–48. Find two z values so that 40% of the middle area is bounded by them.

*7–49. Find two z values, one positive and one negative, so that the areas in the two tails total the following values.

a. 5%
b. 10%
c. 1%

*7–50. Find the z values that correspond to the 90th percentile, 80th percentile, 50th percentile, and 5th percentile.

*7–51. Draw a normal distribution with a mean of 100 and a standard deviation of 15.

*7–52. Find the equation for the standard normal distribution by substituting 0 for μ and 1 for σ in the equation

$$y = \frac{e^{-(X-\mu)^2/(2\sigma^2)}}{\sigma\sqrt{2\pi}}$$

*7–53. Graph the standard normal distribution by using the formula derived in Exercise 7–52. Let $\pi \approx 3.14$ and $e \approx 2.718$. Use X values of $-2, -1.5, -1, -0.5, 0, 0.5, 1, 1.5,$ and 2.

7–4

Applications of the Normal Distribution

Objective 4. Find probabilities for a normally distributed variable by transforming it into a standard normal variable.

The standard normal distribution curve can be used to solve a wide variety of practical problems. The only requirement is that the variable be normally or approximately normally distributed. There are several mathematical tests to determine whether a variable is normally distributed. See the Critical Thinking Challenge on page 294. For all the problems presented in this chapter, one can assume that the variable is normally or approximately normally distributed.

To solve problems by using the standard normal distribution, transform the original variable into a standard normal distribution variable by using the formula

$$z = \frac{\text{value} - \text{mean}}{\text{standard deviation}} \quad \text{or} \quad z = \frac{X - \mu}{\sigma}$$

This is the same formula presented in Section 3–4. This formula transforms the values of the variable into standard units or z values. Once the variable is transformed, then the Procedure Table and Table E in Appendix C can be used to solve problems.

For example, suppose that the scores for a standardized test are normally distributed, have a mean of 100, and have a standard deviation of 15. When the scores are transformed into z values, the two distributions coincide, as shown in Figure 7–27. (Recall that the z distribution has a mean of 0 and a standard deviation of 1.)

Figure 7–27

Test Scores and Their Corresponding z Values

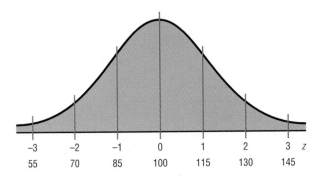

z	-3	-2	-1	0	1	2	3
	55	70	85	100	115	130	145

To solve the application problems in this section, transform the values of the variable into z values and then use the Procedure Table and Table E, as shown in the next examples.

Example 7–14

If the scores for the test have a mean of 100 and a standard deviation of 15, find the percentage of scores that will fall below 112.

Solution

STEP 1   Draw the figure and represent the area, as shown in Figure 7–28.

Figure 7–28

Area under the Curve for Example 7–14

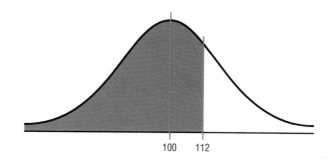

STEP 2   Find the z value corresponding to a score of 112.

$$z = \frac{X - \mu}{\sigma} = \frac{112 - 100}{15} = \frac{12}{15} = 0.8$$

Hence, 112 is 0.8 standard deviation above the mean of 100, as shown for the z distribution in Figure 7–29.

Figure 7–29

Area and z Values for Example 7–14

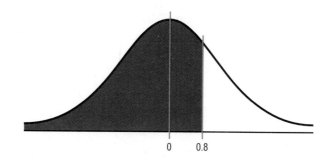

STEP 3   Find the area using Table E. The area between $z = 0$ and $z = 0.8$ is 0.2881. Since the area under the curve to the left of $z = 0.8$ is desired, add 0.5000 to 0.2881 (0.5000 + 0.2881 = 0.7881). Therefore, 78.81% of the scores fall below 112.

Example 7–15

Each month, an American household generates an average of 28 pounds of newspaper for garbage or recycling. Assume the standard deviation is two pounds. If a household is selected at random, find the probability of its generating

Historical Note

Astronomers in the late 1700s and 1800s used the principles underlying the normal distribution to correct measurement errors that occurred in charting the positions of the planets.

a. Between 27 and 31 pounds per month.

b. More than 30.2 pounds per month.

Assume the variable is approximately normally distributed.

Source: Michael D. Shook and Robert L. Shook, *The Book of Odds* (New York: Penguin Putnam Inc., 1991).

Solution *a*

STEP 1   Draw the figure and represent the area. See Figure 7–30.

Figure 7–30

Area under the Curve for Part *a* of Example 7–15

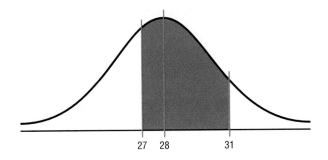

STEP 2   Find the two z values.

$$z_1 = \frac{X - \mu}{\sigma} = \frac{27 - 28}{2} = -\frac{1}{2} = -0.5$$

$$z_2 = \frac{X - \mu}{\sigma} = \frac{31 - 28}{2} = \frac{3}{2} = 1.5$$

STEP 3   Find the appropriate area, using Table E. The area between $z = 0$ and $z = -0.5$ is 0.1915. The area between $z = 0$ and $z = 1.5$ is 0.4332. Add 0.1915 and 0.4332 (0.1915 + 0.4332 = 0.6247). Thus, the total area is 62.47%. See Figure 7–31.

Figure 7–31

Area and z Values for Part *a* of Example 7–15

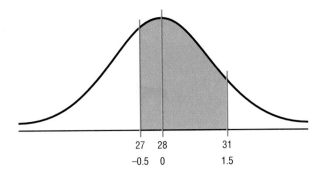

Hence, the probability that a randomly selected household generates between 27 and 31 pounds of newspapers per month is 62.47%.

Solution *b*

STEP 1   Draw the figure and represent the area, as shown in Figure 7–32.

Figure 7–32

Area under the Curve for
Part *b* of Example 7–15

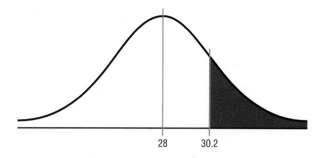

28   30.2

STEP 2   Find the *z* value for 30.2.

$$z = \frac{X - \mu}{\sigma} = \frac{30.2 - 28}{2} = \frac{2.2}{2} = 1.1$$

STEP 3   Find the appropriate area. The area between $z = 0$ and $z = 1.1$ obtained
from Table E is 0.3643. Since the desired area is in the right tail, subtract
0.3643 from 0.5000.

$$0.5000 - 0.3643 = 0.1357$$

Hence, the probability that a randomly selected household will accumulate
more than 30.2 pounds of newspapers is 0.1357, or 13.57%.

The normal distribution can also be used to answer questions of "How many?" This
application is shown in the next example.

Example 7–16

The American Automobile Association reports that the average time it takes to respond
to an emergency call is 25 minutes. Assume the variable is approximately normally distributed and the standard deviation is 4.5 minutes. If 80 calls are randomly selected, approximately how many will be responded to in less than 15 minutes?

Source: Michael D. Shook and Robert L. Shook, *The Book of Odds* (New York: Penguin Putnam Inc.,
1991), 70.

Solution

To solve the problem, find the area under the normal distribution curve to the left of 15.

STEP 1   Draw a figure and represent the area as shown in Figure 7–33.

Figure 7–33

Area under the Curve for
Example 7–16

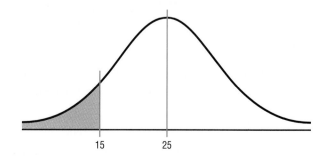

15   25

STEP 2   Find the z value for 15.

$$z = \frac{X - \mu}{\sigma} = \frac{15 - 25}{4.5} = -2.22$$

STEP 3   Find the appropriate area. The area obtained from Table E is 0.4868, which corresponds to the area between $z = 0$ and $z = -2.22$ (Use $+2.22$.)

STEP 4   Subtract 0.4868 from 0.5000 to get 0.0132.

STEP 5   To find how many calls will be made in less than 15 minutes, multiply the sample size (80) by the area (0.0132) to get 1.056. Hence, 1.056, or approximately one, call will be responded to in under 15 minutes.

Note: For problems using percentages, be sure to change the percentage to a decimal before multiplying. Also, round the answer to the nearest whole number, since it is not possible to have 1.056 calls.

Finding Data Values Given Specific Probabilities

The normal distribution can also be used to find specific data values for given percentages. This application is shown in the next example.

Example 7–17

Objective 5. Find specific data values for given percentages using the standard normal distribution.

In order to qualify for a police academy, candidates must score in the top 10% on a general abilities test. The test has a mean of 200 and a standard deviation of 20. Find the lowest possible score to qualify. Assume the test scores are normally distributed.

Solution

Since the test scores are normally distributed, the test value (X) that cuts off the upper 10% of the area under the normal distribution curve is desired. This area is shown in Figure 7–34.

Figure 7–34

Area under the Curve for Example 7–17

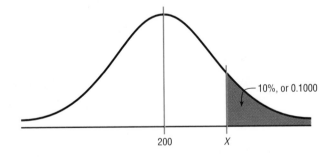

10%, or 0.1000

200   X

Work backward to solve this problem

STEP 1   Subtract 0.1000 from 0.5000 to get the area under the normal distribution between 200 and X: $0.5000 - 0.1000 = 0.4000$.

STEP 2   Find the z value that corresponds to an area of 0.4000 by looking up 0.4000 in the area portion of Table E. If the specific value cannot be found, use the closest value—in this case, 0.3997, as shown in Figure 7–35. The corresponding z value is 1.28. (If the area falls exactly halfway between two z values, use the larger of the two z values. For example, the area 0.4500 falls halfway between 0.4495 and 0.4505. In this case use 1.65 rather than 1.64 for the z value.)

Figure 7–35

Finding the z Value from Table E (Example 7–17)

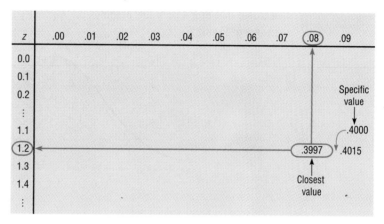

STEP 3   Substitute in the formula $z = (X - \mu)/\sigma$ and solve for X.

$$1.28 = \frac{X - 200}{20}$$

$$(1.28)(20) + 200 = X$$

$$25.60 + 200 = X$$

$$225.60 = X$$

$$226 = X$$

A score of 226 should be used as a cutoff. Anybody scoring 226 or higher qualifies.

Instead of using the formula shown in Step 3 one can use the formula $X = z \cdot \sigma + \mu$. This is obtained by solving

$$z = \frac{(X - \mu)}{\sigma} \quad \text{for } X$$

as shown.

$$z \cdot \sigma = X - \mu \qquad \text{Multiply both sides by } \sigma.$$

$$z \cdot \sigma + \mu = X \qquad \text{Add } \mu \text{ to both sides.}$$

$$X = z \cdot \sigma + \mu \qquad \text{Exchange both sides of the equation.}$$

When one must find the value of X, the following formula can be used:

$$X = z \cdot \sigma + \mu$$

Example 7–18

For a medical study, a researcher wishes to select people in the middle 60% of the population based on blood pressure. If the mean systolic blood pressure is 120 and the standard deviation is 8, find the upper and lower readings that would qualify people to participate in the study.

Solution

Assume that blood pressure readings are normally distributed; then cutoff points are as shown in Figure 7–36.

Figure 7–36

Area under the Curve for Example 7–18

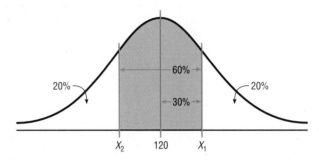

Note that two values are needed, one above the mean and one below the mean. Find the value to the right of the mean first. The closest z value for an area of 0.3000 is 0.84. Substituting in the formula $X = z\sigma + \mu$, one gets

$$X_1 = z\sigma + \mu = (0.84)(8) + 120 = 126.72$$

On the other side, $z = -0.84$; hence,

$$X_2 = (-0.84)(8) + 120 = 113.28$$

Therefore, the middle 60% will have blood pressure readings of $113.28 < X < 126.72$.

As shown in this section, the normal distribution is a useful tool in answering many questions about variables that are normally or approximately normally distributed.

Exercises

7–54. Explain why the standard normal distribution can be used to solve many real-life problems.

7–55. The average hourly wage of production workers in manufacturing is $11.76. Assume the variable is normally distributed. If the standard deviation of earnings is $2.72, find these probabilities for a randomly selected production worker.
  a. The production worker earns more than $12.55.
  b. The production worker earns less than $8.00.

Source: *Statistical Abstract of the United States 1994* (Washington, DC: U.S. Bureau of the Census).

7–56. The Speedmaster IV automobile gets an average 22.0 miles per gallon in the city. The standard deviation is 3 miles per gallon. Assume the variable is normally distributed. Find the probability that on any given day, the car will get more than 26 miles per gallon when driven in the city.

7–57. If the mean salary of telephone operators in the United States is $31,256, and the standard deviation is $3,000, find the following probabilities for a randomly selected telephone operator. Assume the variable is normally distributed.
  a. The operator earns more than $35,000.
  b. The operator earns less than $25,000.

7–58. For a specific year, Americans spent an average of $71.12 for books. Assume the variable is normally distributed. If the standard deviation of the amount spent on books is $8.42, find these probabilities for a randomly selected American.
  a. He or she spent more than $60 per year on books.
  b. He or she spent less than $80 per year on books.

Source: *Statistical Abstract of the United States 1994* (Washington, DC: U.S. Bureau of the Census).

7–59. A survey found that people keep their television sets an average of 4.8 years. The standard deviation is 0.89 year. If a person decides to buy a new TV set, find the probability that he or she has owned the old set for the following amount of time. Assume the variable is normally distributed.
 a. Less than 2.5 years
 b. Between 3 and 4 years
 c. More than 4.2 years

7–60. The average age of CEOs is 56 years. Assume the variable is normally distributed. If the standard deviation is four years, find the probability that the age of a randomly selected CEO will be in the following range.
 a. Between 53 and 59 years old
 b. Between 58 and 63 years old
 c. Between 50 and 55 years old

Source: Michael D. Shook and Robert L. Shook, *The Book of Odds* (New York: Penguin Putnam Inc., 1991), 49.

7–61. The average life of a brand of automobile tires is 30,000 miles, with a standard deviation of 2,000 miles. If a tire is selected and tested, find the probability that it will have the following lifetime. Assume the variable is normally distributed.
 a. Between 25,000 and 28,000 miles
 b. Between 27,000 and 32,000 miles
 c. Between 31,500 and 33,500 miles

7–62. The average time a visitor spends at the Renzie Park Art Exhibit is 62 minutes. The standard deviation is 12 minutes. If a visitor is selected at random, find the probability that he or she will spend the following amount of time at the exhibit. Assume the variable is normally distributed.
 a. At least 82 minutes
 b. At most 50 minutes

7–63. The average time for a courier to travel from Pittsburgh to Harrisburg is 200 minutes, and the standard deviation is 10 minutes. If one of these trips is selected at random, find the probability that the courier will have the following travel time. Assume the variable is normally distributed.
 a. At least 180 minutes
 b. At most 205 minutes

7–64. The average amount of rain per year in South Summerville is 49 inches. The standard deviation is 5.6 inches. Find the probability that next year South Summerville will receive the following amount of rain. Assume the variable is normally distributed.
 a. At most 51 inches of rain
 b. At least 58 inches of rain

7–65. The average waiting time for a drive-in window at a local bank is 9.2 minutes, with a standard deviation of 2.6 minutes. When a customer arrives at the bank, find the probability that the customer will have to wait the following amount of time. Assume the variable is normally distributed.
 a. Between 5 and 10 minutes
 b. Less than 6 minutes or more than 9 minutes

7–66. The average time it takes college freshmen to complete the Mason Basic Reasoning Test is 24.6 minutes. The standard deviation is 5.8 minutes. Find these probabilities. Assume the variable is normally distributed.
 a. It will take a student between 15 and 30 minutes to complete the test.
 b. It will take a student less than 18 minutes or more than 28 minutes to complete the test.

7–67. A brisk walk at 4 miles per hour burns an average of 300 calories per hour. If the standard deviation of the distribution is 8 calories, find the probability that a person who walks one hour at the rate of 4 miles per hour will burn the following calories. Assume the variable is normally distributed.
 a. More than 280 calories
 b. Less than 293 calories
 c. Between 285 and 320 calories

7–68. During September, the average temperature of Laurel Lake is 64.2° and the standard deviation is 3.2°. Assume the variable is normally distributed. For a randomly selected day, find the probability that the temperature will be as follows:
 a. Above 62°
 b. Below 67°
 c. Between 65° and 68°

7–69. If the systolic blood pressure for a certain group of obese people has a mean of 132 and a standard deviation of 8, find the probability that a randomly selected obese person will have the following blood pressure. Assume the variable is normally distributed.
 a. Above 130
 b. Below 140
 c. Between 131 and 136

7–70. In order to qualify for letter sorting, applicants are given a speed-reading test. The scores are normally distributed, with a mean of 80 and a standard deviation of 8. If only the top 15% of the applicants are selected, find the cutoff score.

7–71. The scores on a test have a mean of 100 and a standard deviation of 15. If a personnel manager wishes to select from the top 75% of applicants who take the test, find the cutoff score. Assume the variable is normally distributed.

7–72. For an educational study, a volunteer must place in the middle 50% on a test. If the mean for the population is 100 and the standard deviation is 15, find the two limits (upper and lower) for the scores that would enable a volunteer to participate in the study. Assume the variable is normally distributed.

7–73. A contractor decided to build homes that will include the middle 80% of the market. If the average size (in square feet) of homes built is 1,810, find the maximum and minimum sizes of the homes the contractor should build. Assume that the standard deviation is 92 square feet and the variable is normally distributed.

Source: Michael D. Shook and Robert L. Shook, *The Book of Odds* (New York: Penguin Putnam Inc., 1991), 15.

7–74. If the average price of a new home is $145,500, find the maximum and minimum prices of the houses a contractor will build to include the middle 80% of the market. Assume that the standard deviation of prices is $1,500 and the variable is normally distributed.

Source: Michael D. Shook and Robert L. Shook, *The Book of Odds* (New York: Penguin Putnam Inc., 1991), 15.

7–75. An athletic association wants to sponsor a footrace. The average time it takes to run the course is 58.6 minutes, with a standard deviation of 4.3 minutes. If the association decides to award certificates to the fastest 20% of the racers, what should the cutoff time be? Assume the variable is normally distributed.

7–76. In order to help students improve their reading, a school district decides to implement a reading program. It is to be administered to the bottom 5% of the students in the district, based on the scores on a reading achievement exam. If the average score for the students in the district is 122.6, find the cutoff score that will make a student eligible for the program. The standard deviation is 18. Assume the variable is normally distributed.

7–77. An automobile dealer finds that the average price of a previously owned vehicle is $8,256. He decides to sell cars that will appeal to the middle 60% of the market in terms of price. Find the maximum and minimum prices of the cars the dealer will sell. The standard deviation is $1,150 and the variable is normally distributed.

7–78. A small publisher wishes to publish self-improvement books. After a survey of the market, the publisher finds that the average cost of the type of book that she wishes to publish is $12.80. If she wants to price her books to sell in the middle 70% range, what should the maximum and minimum prices of the books be? The standard deviation is $0.83 and the variable is normally distributed.

7–79. A special enrichment program in mathematics is to be offered to the top 12% of students in a school district. A standardized mathematics achievement test given to all students has a mean of 57.3 and a standard deviation of 16. Find the cutoff score. Assume the variable is normally distributed.

7–80. A book store owner decides to sell children's talking books that will appeal to the middle 60% of his customers.

The owner reads in a study that the mean price of children's talking books is $10.52, with a standard deviation of $1.08. Find the maximum and minimum prices of talking books the owner should sell. Assume the variable is normally distributed.

7–81. An advertising company plans to market a product to low-income families. A study states that for a particular area, the average income per family is $24,596 and the standard deviation is $6,256. If the company plans to target the bottom 18% of the families based on income, find the cutoff income. Assume the variable is normally distributed.

7–82. If a one-person household spends an average of $40 per week on groceries, find the maximum and minimum dollar amount spent per week for the middle 50% of one-person households. Assume that the standard deviation is $5 and the variable is normally distributed.

Source: Michael D. Shook and Robert L. Shook, *The Book of Odds* (New York: Penguin Putnam Inc., 1991), 192.

7–83. The mean lifetime of a wristwatch is 25 months, with a standard deviation of 5 months. If the distribution is normal, for how many months should a guarantee be if the manufacturer does not want to exchange more than 10% of the watches? Assume the variable is normally distributed.

7–84. In order to quality for security officers' training, recruits are tested for stress tolerance. The scores are normally distributed, with a mean of 62 and a standard deviation of 8. If only the top 15% of recruits are selected, find the cutoff score.

7–85. In the distributions shown, state the mean and standard deviation for each. *Hint:* See Figures 7–5 and 7–6. *Hint:* The vertical lines are one standard deviation apart.

a.

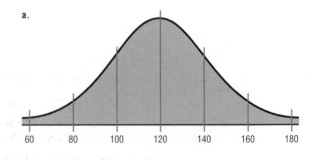

b.

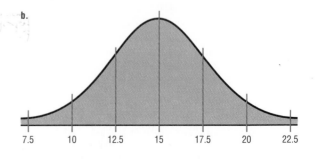

c.

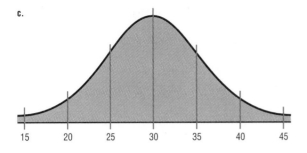

7–86. Suppose that the mathematics SAT scores for high school seniors for a specific year have a mean of 456 and a standard deviation of 100 and are approximately normally distributed. If a subgroup of these high school seniors, those who are in the National Honor Society, is selected, would you expect the distribution of scores to have the same mean and standard deviation? Explain your answer.

7–87. Given a data set, how could you decide if the distribution of the data was approximately normal?

7–88. If a distribution of raw scores were plotted and then the scores were transformed into z scores, would the shape of the distribution change? Explain your answer.

7–89. In a normal distribution, find σ when $\mu = 100$ and 2.68% of the area lies to the right of 105.

7–90. In a normal distribution, find μ when σ is 6 and 3.75% of the area lies to the left of 85.

7–91. In a certain normal distribution, 1.25% of the area lies to the left of 42 and 1.25% of the area lies to the right of 48. Find μ and σ.

7–92. An instructor gives a 100-point examination in which the grades are normally distributed. The mean is 60 and the standard deviation is 10. If there are 5% A's and 5% F's, 15% B's and 15% D's, and 60% C's, find the scores that divide the distribution into those categories.

Technology Step by Step

TI-83
Step by Step

The Normal Distribution

To find the area under the standard normal distribution curve between any two z values:

Example TI7–1

Find the area between $z = 2.00$ and $z = 2.47$, as in Example 7–6.

1. Press **2nd [DISTR]** then **2** to get `normalcdf(`.
2. Enter **2, 2.47)** then press **ENTER.**

The calculator will display .0159944012.
    To find the area under the normal distribution curve less than (or greater than) any z value, use −1E99 and 1E99 to specify infinity.

Example TI7–2

Find the area to the left of $z = -1.93$, as in Example 7–5.

1. Press **2nd [DISTR]** then **2** to get `normalcdf(`.
2. Enter **−1E99, −1.93)** then press **ENTER.** *Note:* Use **2nd [EE]** to paste E into the cursor location (this indicates the following number—here, 99—is an exponent).

The calculator will display .0268033499.

```
normalcdf(2,2.47
)
          .0159944012
normalcdf(-1E99,
-1.93)
          .0268033499
```

To find the area under a normal distribution curve between any two *z* values given a specific mean and standard deviation:

Example TI7–3

Find the area between 27 and 31 when $\mu = 28$ and $\sigma = 2$, as in Example 7–15a.

1. Press **2nd [DISTR]** then **2** to get `normalcdf(`.
2. Enter **27, 31, 28, 2**) then press **ENTER.**

The calculator will display .6246552391.

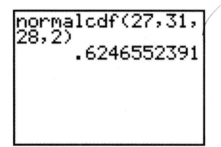

Excel
Step by Step

The Normal Distribution

Excel has a built-in table for the standard normal distribution, in the form of the function NORMSDIST. To find the area between two *z* values under the normal distribution curve:

1. Select a blank cell.
2. Click the f_x icon to call up the function list.
3. Select NORMSDIST from the `Statistical` function category.
4. Enter the smaller *z* value and click `[OK]`.
5. Repeat steps 1–4 for the larger *z* value.
6. Subtract the two computed values.

The Excel Function
NORMSDIST

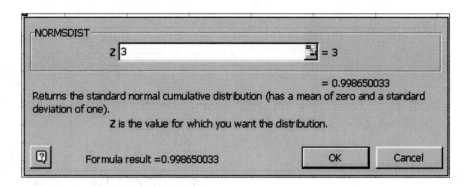

7–5

The Central Limit Theorem

In addition to knowing how individual data values vary about the mean for a population, statisticians are also interested in knowing about the distribution of the means of samples taken from a population. This topic is discussed in the subsections that follow.

Distribution of Sample Means

Objective 6. Use the central limit theorem to solve problems involving sample means for large and small samples.

Suppose a researcher selects 100 samples of a specific size from a large population and computes the mean of the same variable for each of the 100 samples. These sample means, $\bar{X}_1, \bar{X}_2, \bar{X}_3, \ldots, \bar{X}_{100}$, constitute a sampling distribution of sample means.

A **sampling distribution of sample means** is a distribution obtained by using the means computed from random samples of a specific size taken from a population.

If the samples are randomly selected with replacement, the sample means, for the most part, will be somewhat different from the population mean μ. These differences are caused by sampling error.

Sampling error is the difference between the sample measure and the corresponding population measure due to the fact that the sample is not a perfect representation of the population.

When all possible samples of a specific size are selected with replacement from a population, the distribution of the sample means for a variable has two important properties, which are explained next.

Properties of the Distribution of Sample Means

1. The mean of the sample means will be the same as the population mean.
2. The standard deviation of the sample means will be smaller than the standard deviation of the population, and it will be equal to the population standard deviation divided by the square root of the sample size.

The following example illustrates these two properties. Suppose a professor gave an eight-point quiz to a small class of four students. The results of the quiz were 2, 6, 4, and 8. For the sake of discussion, assume that the four students constitute the population. The mean of the population is

$$\mu = \frac{2 + 6 + 4 + 8}{4} = 5$$

The standard deviation of the population is

$$\sigma = \sqrt{\frac{(2-5)^2 + (6-5)^2 + (4-5)^2 + (8-5)^2}{4}} = 2.236$$

The graph of the original distribution is shown in Figure 7–37. This is called a *uniform distribution.*

Figure 7–37

Distribution of Quiz Scores

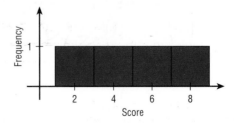

Now, if all samples of size 2 are taken with replacement, and the mean of each sample is found, the distribution is as shown next.

Sample	Mean	Sample	Mean
2, 2	2	6, 2	4
2, 4	3	6, 4	5
2, 6	4	6, 6	6
2, 8	5	6, 8	7
4, 2	3	8, 2	5
4, 4	4	8, 4	6
4, 6	5	8, 6	7
4, 8	6	8, 8	8

A frequency distribution of sample means is as follows.

$\overline{X}$	f
2	1
3	2
4	3
5	4
6	3
7	2
8	1

For the data from the example just discussed, Figure 7–38 shows the graph of the sample means. The graph appears to be somewhat normal, even though it is a histogram.

Figure 7–38

Distribution of
Sample Means

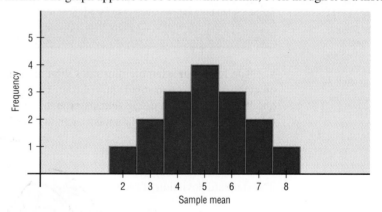

The mean of the sample means, denoted by $\mu_{\overline{X}}$, is

$$\mu_{\overline{X}} = \frac{2 + 3 + \cdots + 8}{16} = \frac{80}{16} = 5$$

which is the same as the population mean. Hence,

$$\mu_{\overline{X}} = \mu$$

The standard deviation of sample means, denoted by $\sigma_{\overline{X}}$, is

$$\sigma_{\overline{X}} = \sqrt{\frac{(2 - 5)^2 + (3 - 5)^2 + \cdots + (8 - 5)^2}{16}} = 1.581$$

which is the same as the population standard deviation divided by $\sqrt{2}$:

$$\sigma_{\overline{X}} = \frac{2.236}{\sqrt{2}} = 1.581$$

(*Note:* Rounding rules were not used here in order to show that the answers coincide.)

In summary, if all possible samples of size *n* are taken with replacement from the same population, the mean of the sample means, denoted by $\mu_{\bar{X}}$, equals the population mean μ; and the standard deviation of the sample means, denoted by $\sigma_{\bar{X}}$, equals $\sigma/\sqrt{n}$. The standard deviation of the sample means is called the **standard error of the mean.** Hence,

$$\sigma_{\bar{X}} = \frac{\sigma}{\sqrt{n}}$$

A third property of the sampling distribution of sample means pertains to the shape of the distribution and is explained by the **central limit theorem.**

> ### The Central Limit Theorem
>
> As the sample size *n* increases, the shape of the distribution of the sample means taken with replacement from a population with mean μ and standard deviation σ will approach a normal distribution. As previously shown, this distribution will have a mean μ and a standard deviation $\sigma/\sqrt{n}$.

The central limit theorem can be used to answer questions about sample means in the same manner that the normal distribution can be used to answer questions about individual values. The only difference is that a new formula must be used for the *z* values. It is

$$z = \frac{\bar{X} - \mu}{\sigma/\sqrt{n}}$$

Notice that $\bar{X}$ is the sample mean, and the denominator is the standard error of the mean.

If a large number of samples of a given size were selected from a large population, and the sample means computed, the distribution of sample means would look like the one shown in Figure 7–39. The percentages indicate the areas of the regions.

Figure 7–39

Distribution of Sample Means for Large Number of Samples

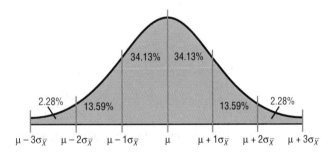

It's important to remember two things when using the central limit theorem:

1. When the original variable is normally distributed, the distribution of the sample means will be normally distributed, for any sample size *n*.
2. When the distribution of the original variable departs from normality, a sample size of 30 or more is needed to use the normal distribution to approximate the distribution of the sample means. The larger the sample, the better the approximation will be.

The next several examples show how the standard normal distribution can be used to answer questions about sample means.

Example 7–19

A. C. Neilsen reported that children between the ages of 2 and 5 watch an average of 25 hours of television per week. Assume the variable is normally distributed and the standard deviation is 3 hours. If 20 children between the ages of 2 and 5 are randomly selected, find the probability that the mean of the number of hours they watch television will be greater than 26.3 hours.

Source: Michael D. Shook and Robert L. Shook, *The Book of Odds* (New York: Penguin Putnam Inc., 1991), 161.

Solution

Since the variable is approximately normally distributed, the distribution of sample means will be approximately normal, with a mean of 25. The standard deviation of the sample means is

$$\sigma_{\bar{X}} = \frac{\sigma}{\sqrt{n}} = \frac{3}{\sqrt{20}} = 0.671$$

The distribution of the means is shown in Figure 7–40, with the appropriate area shaded.

Figure 7–40

Distribution of the Means for Example 7–19

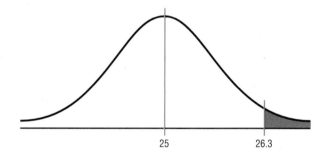

The *z* value is

$$z = \frac{\bar{X} - \mu}{\sigma/\sqrt{n}} = \frac{26.3 - 25}{3/\sqrt{20}} = \frac{1.3}{0.671} = 1.94$$

The area between 0 and 1.94 is 0.4738. Since the desired area is in the tail, subtract 0.4738 from 0.5000. Hence, $0.5000 - 0.4738 = 0.0262$, or 2.62%.

One can conclude that the probability of obtaining a sample mean larger than 26.3 hours is 2.62% (i.e., $P(\bar{X} > 26.3) = 2.62\%$).

Example 7–20

The average age of a vehicle registered in the United States is 8 years, or 96 months. Assume the standard deviation is 16 months. If a random sample of 36 vehicles is selected, find the probability that the mean of their age is between 90 and 100 months.

Source: *Harper's Index* 290, no. 1740 (May 1995), p. 11.

Solution

The desired area is shown in Figure 7–41.

Figure 7–41

Area under the Curve for Example 7–20

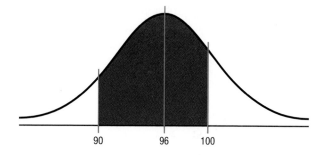

The two z values are

$$z_1 = \frac{90 - 96}{16/\sqrt{36}} = -2.25$$

$$z_2 = \frac{100 - 96}{16/\sqrt{36}} = 1.50$$

The two areas corresponding to the z values of -2.25 and 1.50, respectively, are 0.4878 and 0.4332. Since the z values are on opposite sides of the mean, find the probability by adding the areas: $0.4878 + 0.4332 = 0.921$, or 92.1%.

Hence, the probability of obtaining a sample mean between 90 and 100 months is 92.1% i.e., $P(90 < \overline{X} < 100) = 92.1\%$.

Since the sample size is 30 or larger, the normality assumption is not necessary, as shown in Example 7–20.

Students sometimes have difficulty deciding whether to use

$$z = \frac{\overline{X} - \mu}{\sigma/\sqrt{n}} \quad \text{or} \quad z = \frac{X - \mu}{\sigma}$$

The formula

$$z = \frac{\overline{X} - \mu}{\sigma/\sqrt{n}}$$

should be used to gain information about a sample mean, as shown in this section. The formula

$$z = \frac{X - \mu}{\sigma}$$

is used to gain information about an individual data value obtained from the population. Notice that the first formula contains $\overline{X}$, the symbol for the sample mean, while the second formula contains X, the symbol for an individual data value. The next example illustrates the uses of the two formulas.

Example 7–21

The average number of pounds of meat a person consumes a year is 218.4 pounds. Assume that the standard deviation is 25 pounds and the distribution is approximately normal.

Source: Michael D. Shook and Robert L. Shook, *The Book of Odds* (New York: Penguin Putnam Inc., 1991), 164.

a. Find the probability that a person selected at random consumes less than 224 pounds per year.

b. If a sample of 40 individuals is selected, find the probability that the mean of the sample will be less than 224 pounds per year.

Solution

a. Since the question asks about an individual person, the formula $z = (X - \mu)/\sigma$ is used.

The distribution is shown in Figure 7–42.

Figure 7–42

Area under the Curve for Part *a* of Example 7–21

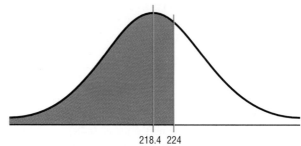

218.4 224

Distribution of individual data values for the population

The *z* value is

$$z = \frac{X - \mu}{\sigma} = \frac{224 - 218.4}{25} = 0.22$$

The area between 0 and 0.22 is 0.0871; this area must be added to 0.5000 to get the total area to the left of $z = 0.22$.

$$0.0871 + 0.5000 = 0.5871$$

Hence, the probability of selecting an individual who consumes less than 224 pounds of meat per year is 0.5871, or 58.71% (i.e., $P(X < 224) = 0.5871$).

b. Since the question concerns the mean of a sample with a size of 40, the formula $z = (\overline{X} - \mu)/(\sigma/\sqrt{n})$ is used.

The area is shown in Figure 7–43.

Figure 7–43

Area under the Curve for Part *b* of Example 7–21

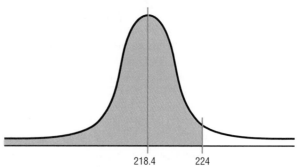

218.4   224

Distribution of means for all samples of size 40 taken from the population

The z value is

$$z = \frac{\overline{X} - \mu}{\sigma/\sqrt{n}} = \frac{224 - 218.4}{25/\sqrt{40}} = 1.42$$

The area between $z = 0$ and $z = 1.42$ is 0.4222; this value must be added to 0.5000 to get the total area.

$$0.4222 + 0.5000 = 0.9222$$

Hence, the probability that the mean of a sample of 40 individuals is less than 224 pounds per year is 0.9222, or 92.22%. That is, $P(\overline{X} < 224) = 0.9222$.

Comparing the two probabilities, one can see that the probability of selecting an individual who consumes less than 224 pounds of meat per year is 58.71%, but the probability of selecting a sample of 40 people with a mean consumption of meat that is less than 224 pounds per year is 92.22%. This rather large difference is due to the fact that the distribution of sample means is much less variable than the distribution of individual data values.

Finite Population Correction Factor (Optional)

The formula for the standard error of the mean, $\sigma/\sqrt{n}$, is accurate when the samples are drawn with replacement or are drawn without replacement from a very large or infinite population. Since sampling with replacement is for the most part unrealistic, a *correction factor* is necessary for computing the standard error of the mean for samples drawn without replacement from a finite population. Compute the correction factor by using the following formula:

$$\sqrt{\frac{N - n}{N - 1}}$$

where N is the population size and n is the sample size.

This correction factor is necessary if relatively large samples are taken from a small population, because the sample mean will then more accurately estimate the population mean and there will be less error in the estimation. Therefore, the standard error of the mean must be multiplied by the correction factor to adjust it for large samples taken from a small population. That is,

$$\sigma_{\overline{X}} = \frac{\sigma}{\sqrt{n}} \cdot \sqrt{\frac{N - n}{N - 1}}$$

Finally, the formula for the z value becomes

$$z = \frac{\overline{X} - \mu}{\dfrac{\sigma}{\sqrt{n}} \cdot \sqrt{\dfrac{N - n}{N - 1}}}$$

When the population is large and the sample is small, the correction factor is generally not used, since it will be very close to 1.00.

The formulas and their uses are summarized in Table 7–1.

Table 7–1   Summary of Formulas and Their Uses

Formula	Use
1. $z = \dfrac{X - \mu}{\sigma}$	Used to gain information about an individual data value when the variable is normally distributed.
2. $z = \dfrac{\overline{X} - \mu}{\sigma/\sqrt{n}}$	Used to gain information when applying the central limit theorem about a sample mean when the variable is normally distributed or when the sample size is 30 or more.

Exercises

7–93. If samples of a specific size are selected from a population and the means are computed, what is this distribution of means called?

7–94. Why do most of the sample means differ somewhat from the population mean? What is this difference called?

7–95. What is the mean of the sample means?

7–96. What is the standard deviation of the sample means called? What is the formula for this standard deviation?

7–97. What does the central limit theorem say about the shape of the distribution of sample means?

7–98. What formula is used to gain information about an individual data value when the variable is normally distributed?

7–99. What formula is used to gain information about a sample mean when the variable is normally distributed or when the sample size is 30 or more?

For Exercises 7–100 through 7–117, assume that the sample is taken from a large population and the correction factor can be ignored.

7–100. A survey found that Americans generate an average of 17.2 pounds of glass garbage each year. Assume the standard deviation of the distribution is 2.5 pounds. Find the probability that the mean of a sample of 55 families will be between 17 and 18 pounds.

Source: Michael D. Shook and Robert L. Shook, *The Book of Odds* (New York: Penguin Putnam Inc., 1991), 14.

7–101. The mean serum cholesterol of a large population of overweight adults is 220 mg/dl and the standard deviation is 16.3 mg/dl. If a sample of 30 adults is selected, find the probability that the mean will be between 220 and 222 mg/dl.

7–102. For a certain large group of individuals, the mean hemoglobin level in the blood is 21.0 grams per milliliter (g/ml). The standard deviation is 2 g/ml. If a sample of 25

individuals is selected, find the probability that the mean will be greater than 21.3 g/ml. Assume the variable is normally distributed.

7–103. The mean weight of 18-year-old females is 126 pounds, and the standard deviation is 15.7. If a sample of 25 females is selected, find the probability that the mean of the sample will be greater than 128.3 pounds. Assume the variable is normally distributed.

7–104. The mean grade point average of the engineering majors at a large university is 3.23, with a standard deviation of 0.72. In a class of 48 students, find the probability that the mean grade point average of the students is less than 3.15.

7–105. The average price of a pound of sliced bacon is $2.02. Assume the standard deviation is $0.08. If a random sample of 40 one-pound packages is selected, find the probability that the mean of the sample will be less than $2.00.

Source: *Statistical Abstract of the United States 1994.* (Washington, DC: U. S. Bureau of the Census).

7–106. The average hourly wage of fast-food workers employed by a nationwide chain is $5.55. The standard deviation is $1.15. If a sample of 50 workers is selected, find the probability that the mean of the sample will be between $5.25 and $5.90.

7–107. The mean score on a dexterity test for 12-year-olds is 30. The standard deviation is 5. If a psychologist administers the test to a class of 22 students, find the probability that the mean of the sample will be between 27 and 31. Assume the variable is normally distributed.

7–108. A recent study of the life span of portable radios found the average to be 3.1 years, with a standard deviation of 0.9 year. If the number of radios owned by the students in one dormitory is 47, find the probability that the mean lifetime of these radios will be less than 2.7 years.

7–109. The average age of lawyers is 43.6 years, with a standard deviation of 5.1 years. If a law firm employs 50 lawyers, find the probability that the average age of the group is greater than 44.2 years old.

7–110. The average annual precipitation for Des Moines is 30.83 inches, with a standard deviation of 5 inches. If a random sample of 10 years is selected, find the probability that the mean will be between 32 and 33 inches. Assume the variable is normally distributed.

7–111. Procter & Gamble reported that an American family of 4 washes an average of one ton (2000 pounds) of clothes each year. If the standard deviation of the distribution is 187.5 pounds, find the probability that the mean of a randomly selected sample of 50 families of four will be between 1980 and 1990 pounds.

Source: Lewis H. Lapham, Michael Pollan, and Eric Etheridge, *The Harper's Index Book* (New York: Henry Holt & Co., 1987).

7–112. The average annual salary in Pennsylvania was $24,393 in 1992. Assume that salaries were normally distributed for a certain group of wage earners, and the standard deviation of this group was $4,362.

a. Find the probability that a randomly selected individual earned less than $26,000.

b. Find the probability that for a randomly selected sample of 25 individuals, the mean salary was less than $26,000.

c. Why is the probability for Part *b* higher than the probability for Part *a*?

Associated Press, December 23, 1992.

7–113. The average time it takes a group of adults to complete a certain achievement test is 46.2 minutes. The standard deviation is 8 minutes. Assume the variable is normally distributed.

a. Find the probability that a randomly selected adult will complete the test in less than 43 minutes.

b. Find the probability that if 50 randomly selected adults take the test, the mean time it takes the group to complete the test will be less than 43 minutes.

c. Does it seem reasonable that an adult would finish the test in less than 43 minutes? Explain.

d. Does it seem reasonable that the mean of the 50 adults could be less than 43 minutes?

7–114. Assume that the mean systolic blood pressure of normal adults is 120 millimeters of mercury (mm Hg) and the standard deviation is 5.6. Assume the variable is normally distributed.

a. If an individual is selected, find the probability that the individual's pressure will be between 120 and 121.8 mm Hg.

b. If a sample of 30 adults is randomly selected, find the probability that the sample mean will be between 120 and 121.8 mm Hg.

c. Why is the answer to Part *a* so much smaller than the answer to Part *b*?

7–115. The average cholesterol content of a certain brand of eggs is 215 milligrams and the standard deviation is 15 milligrams. Assume the variable is normally distributed.

a. If a single egg is selected, find the probability that the cholesterol content will be more than 220 milligrams.

b. If a sample of 25 eggs is selected, find the probability that the mean of the sample will be larger than 220 milligrams.

Source: *Living Fit* (Englewood Cliffs, NJ: Best Foods, CPC International, Inc., 1991).

7–116. At a large publishing company, the mean age of proofreaders is 36.2 years, and the standard deviation is 3.7 years. Assume the variable is normally distributed.

a. If a proofreader from the company is randomly selected, find the probability that his or her age will be between 36 and 37.5 years.

b. If a random sample of 15 proofreaders is selected, find the probability that the mean age of the proofreaders in the sample will be between 36 and 37.5 years.

7–117. The average labor cost for car repairs for a large chain of car repair shops is $48.25. The standard deviation is $4.20. Assume the variable is normally distributed.

a. If a store is selected at random, find the probability that the labor cost will range between $46 and $48.

b. If 20 stores are selected at random, find the probability that the mean of the sample will be between $46 and $48.

c. Which answer is larger? Explain why.

For Exercises 7–118 and 7–119, check to see whether the correction factor should be used. If so, be sure to include it in the calculations.

*7–118. In a study of the life expectancy of 500 people in a certain geographic region, the mean age at death was 72.0 years and the standard deviation was 5.3 years. If a sample of 50 people from this region is selected, find the probability that the mean life expectancy will be less than 70 years.

*7–119. A study of 800 homeowners in a certain area showed that the average value of the homes was $82,000 and the standard deviation was $5,000. If 50 homes are for sale, find the probability that the mean of the values of these homes is greater than $83,500.

*7–120. The average breaking strength of a certain brand of steel cable is 2000 pounds, with a standard deviation of 100 pounds. A sample of 20 cables is selected and tested. Find the sample mean that will cut off the upper 95% of all of the samples of size 20 taken from the population. Assume the variable is normally distributed.

*7–121. The standard deviation of a variable is 15. If a sample of 100 individuals is selected, compute the standard error of the mean. What size sample is necessary to double the standard error of the mean?

*7–122. In Exercise 7–121, what size sample is needed to cut the standard error of the mean in half?

7–6

The Normal Approximation to the Binomial Distribution

The normal distribution is often used to solve problems that involve the binomial distribution since when n is large (say, 100), the calculations are too difficult to do by hand using the binomial distribution. Recall from Chapter 6 that a binomial distribution has the following characteristics:

1. There must be a fixed number of trials.

2. The outcome of each trial must be independent.

3. Each experiment can have only two outcomes or be reduced to two outcomes.

4. The probability of a success must remain the same for each trial.

Also, recall that a binomial distribution is determined by n (the number of trials) and p (the probability of a success). When p is approximately 0.5, and as n increases, the shape of the binomial distribution becomes similar to the normal distribution. The larger n and the closer p is to 0.5, the more similar the shape of the binomial distribution is to the normal distribution.

Objective 7. Use the normal approximation to compute probabilities for a binomial variable.

But when p is close to 0 or 1 and n is relatively small, the normal approximation is inaccurate. As a rule of thumb, statisticians generally agree that the normal approximation should be used only when $n \cdot p$ and $n \cdot q$ are both greater than or equal to 5. (*Note:* $q = 1 - p$.) For example, if p is 0.3 and n is 10, then $np = (10)(0.3) = 3$, and the normal distribution should not be used as an approximation. On the other hand, if $p = 0.5$ and $n = 10$, then $np = (10)(0.5) = 5$ and $nq = (10)(0.5) = 5$, and the normal distribution can be used as an approximation. See Figure 7–44.

In addition to the previous condition of $np \geq 5$ and $nq \geq 5$, a correction for continuity may be used in the normal approximation.

A **correction for continuity** is a correction employed when a continuous distribution is used to approximate a discrete distribution.

The continuity correction means that for any specific value of X, say 8, the boundaries of X in the binomial distribution (in this case, 7.5 to 8.5) must be used. (See Chapter 1, Section 1–3.) Hence, when one employs the normal distribution to approximate the binomial, the boundaries of any specific value X must be used as they are shown in the binomial distribution. For example, for $P(X = 8)$, the correction is $P(7.5 < X < 8.5)$. For $P(X \leq 7)$, the correction is $P(X < 7.5)$. For $P(X \geq 3)$, the correction is $P(X > 2.5)$.

Students sometimes have difficulty deciding whether to add 0.5 or subtract 0.5 from the data value for the correction factor. Table 7–2 summarizes the different situations.

Figure 7–44

Comparison of the
Binomial Distribution and
the Normal Distribution

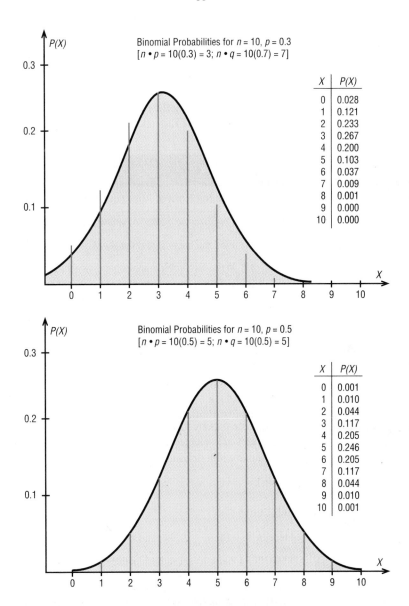

Table 7–2	Summary of the Normal Approximation to the Binomial Distribution
Binomial	**Normal**
When finding	Use
1. $P(X = a)$	$P(a - 0.5 < X < a + 0.5)$
2. $P(X \geq a)$	$P(X > a - 0.5)$
3. $P(X > a)$	$P(X > a + 0.5)$
4. $P(X \leq a)$	$P(X < a + 0.5)$
5. $P(X < a)$	$P(X < a - 0.5)$
For all cases, $\mu = n \cdot p$, $\sigma = \sqrt{n \cdot p \cdot q}$, $n \cdot p \geq 5$, and $n \cdot q \geq 5$.	

The formulas for the mean and standard deviation for the binomial distribution are necessary for calculations. They are

$$\mu = n \cdot p \text{ and } \sigma = \sqrt{n \cdot p \cdot q}$$

The steps for using the normal distribution to approximate the binomial distribution are shown in the next Procedure Table.

Procedure Table

Procedure for the Normal Approximation to the Binomial Distribution

STEP 1   Check to see whether the normal approximation can be used.

STEP 2   Find the mean μ and the standard deviation σ.

STEP 3   Write the problem in probability notation, using X.

STEP 4   Rewrite the problem by using the continuity correction factor, and show the corresponding area under the normal distribution.

STEP 5   Find the corresponding z values.

STEP 6   Find the solution.

Example 7–22

A magazine reported that 6% of American drivers read the newspaper while driving. If 300 drivers are selected at random, find the probability that exactly 25 say they read the newspaper while driving.

Source: USA Snapshot, *USA Today,* June 26, 1995.

Solution

Here, $p = 0.06$, $q = 0.94$, and $n = 300$.

STEP 1   Check to see whether the normal approximation can be used.

$$np = (300)(0.06) = 18 \quad nq = (300)(0.94) = 282$$

Since $np \geq 5$ and $nq \geq 5$, the normal distribution can be used.

STEP 2   Find the mean and standard deviation.

$$\mu = np = (300)(0.06) = 18$$
$$\sigma = \sqrt{npq} = \sqrt{(300)(0.06)(0.94)} = \sqrt{16.92} = 4.11$$

STEP 3   Write the problem in probability notation: $P(X = 25)$.

STEP 4   Rewrite the problem by using the continuity correction factor. See approximation number 1 in Table 7–2. $P(25 - 0.5 < X < 25 + 0.5) = P(24.5 < X < 25.5)$. Show the corresponding area under the normal distribution curve (see Figure 7–45).

Figure 7–45

Area under the Curve
and X Values for
Example 7–22

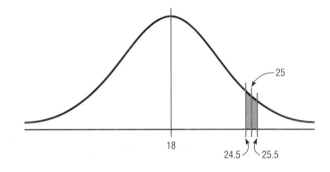

STEP 5 Find the corresponding z values. Since 25 represents any value between 24.5 and 25.5, find both z values.

$$z_1 = \frac{25.5 - 18}{4.11} = 1.82 \qquad z_2 = \frac{24.5 - 18}{4.11} = 1.58$$

STEP 6 Find the solution. Find the corresponding areas in the table: The area for $z = 1.82$ is 0.4656, and the area for $z = 1.58$ is 0.4429. Subtract the areas to get the approximate value: $0.4656 - 0.4429 = 0.0227$, or 2.27%.

Hence, the probability that exactly 25 people read the newspaper while driving is 2.27%.

Example 7–23

Of the members of a bowling league, 10% are widowed. If 200 bowling league members are selected at random, find the probability that 10 or more will be widowed.

Solution

Here, $p = 0.10$, $q = 0.90$, and $n = 200$.

STEP 1 Since np is $(200)(0.10) = 20$ and nq is $(200)(0.90) = 180$, the normal approximation can be used.

STEP 2 $\mu = np = (200)(0.10) = 20$

$\sigma = \sqrt{npq} = \sqrt{(200)(0.10)(0.90)} = \sqrt{18} = 4.24$

STEP 3 $P(X \geq 10)$.

STEP 4 See approximation number 2 in Table 7–2. $P(X > 10 - 0.5) = P(X > 9.5)$. The desired area is shown in Figure 7–46.

Figure 7–46

Area under the Curve and
X Value for Example 7–23

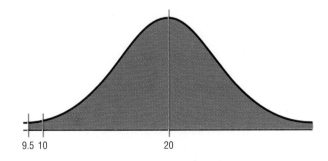

STEP 5   Since the problem is to find the probability of 10 or more positive responses, the normal distribution graph is as shown in Figure 7–46. Hence, the area between 9.5 and 20 must be added to 0.5000 to get the correct approximation.
The z value is

$$z = \frac{9.5 - 20}{4.24} = -2.48$$

STEP 6   The area between 20 and 9.5 is 0.4934. Thus, the probability of getting 10 or more responses is $0.5000 + 0.4934 = 0.9934$, or 99.34%.

It can be concluded, then, that the probability of 10 or more widowed people in a random sample of 200 bowling league members is 99.34%.

Example 7–24

If a baseball player's batting average is 0.320 (32%), find the probability that the player will get at most 26 hits in 100 times at bat.

Solution

Here, $p = 0.32$, $q = 0.68$, and $n = 100$.

STEP 1   Since $np = (100)(0.320) = 32$ and $nq = (100)(0.680) = 68$, the normal distribution can be used to approximate the binomial distribution.

STEP 2 $\mu = np = (100)(0.320) = 32$
$\sigma = \sqrt{npq} = \sqrt{(100)(0.32)(0.68)} = \sqrt{21.76} = 4.66$

STEP 3 $P(X \leq 26)$.

STEP 4   See approximation number 4 in Table 7–2. $P(X < 26 + 0.5) = P(X < 26.5)$. The desired area is shown in Figure 7–47.

Figure 7–47

Area under the Curve for Example 7–24

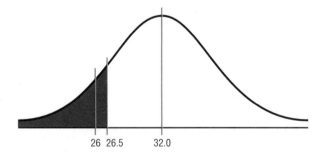

26 26.5   32.0

STEP 5   The z value is

$$z = \frac{26.5 - 32}{4.66} = -1.18$$

STEP 6   The area between the mean and 26.5 is 0.3810. Since the area in the left tail is desired, 0.3810 must be subtracted from 0.5000. So the probability is $0.5000 - 0.3810 = 0.1190$, or 11.9%.

The closeness of the normal approximation is shown in the next example.

Example 7–25

When $n = 10$ and $p = 0.5$, use the binomial distribution table (Table B in Appendix C) to find the probability that $X = 6$. Then use the normal approximation to find the probability that $X = 6$.

Solution

From Table B, for $n = 10$, $p = 0.5$, and $X = 6$, the probability is 0.205.

For the normal approximation,

$$\mu = np = (10)(0.5) = 5$$
$$\sigma = \sqrt{npq} = \sqrt{(10)(0.5)(0.5)} = 1.58$$

Now, $X = 6$ is represented by the boundaries 5.5 and 6.5. So the z values are

$$z_1 = \frac{6.5 - 5}{1.58} = 0.95 \qquad z_2 = \frac{5.5 - 5}{1.58} = 0.32$$

The corresponding area for 0.95 is 0.3289, and the corresponding area for 0.32 is 0.1255.

The solution is $0.3289 - 0.1255 = 0.2034$, which is very close to the binomial table value of 0.205. The desired area is shown in Figure 7–48.

Figure 7–48

Area under the Curve for Example 7–25

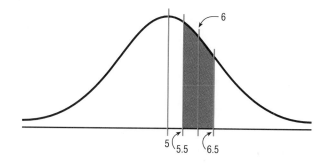

The normal approximation also can be used to approximate other distributions, such as the Poisson distribution (see Table C in Appendix C).

Exercises

7–123. Explain why the normal distribution can be used as an approximation to the binomial distribution. What conditions must be met to use the normal distribution to approximate the binomial distribution? Why is a correction for continuity necessary?

7–124. (ans) Use the normal approximation to the binomial to find the probabilities for the specific value(s) of X.

a. $n = 30, p = 0.5, X = 18$
b. $n = 50, p = 0.8, X = 44$
c. $n = 100, p = 0.1, X = 12$
d. $n = 10, p = 0.5, X \geq 7$
e. $n = 20, p = 0.7, X \leq 12$
f. $n = 50, p = 0.6, X \leq 40$

7–125. Check each binomial distribution to see whether it can be approximated by the normal distribution (i.e., are $np \geq 5$ and $nq \geq 5$?).

a. $n = 20, p = 0.5$ d. $n = 50, p = 0.2$
b. $n = 10, p = 0.6$ e. $n = 30, p = 0.8$
c. $n = 40, p = 0.9$ f. $n = 20, p = 0.85$

7–126. Of all 3- to 5-year-old children, 56% are enrolled in school. If a sample of 500 such children is randomly selected, find the probability that at least 250 will be enrolled in school.

Source: *Statistical Abstract of the United States 1994* (Washington, DC: U.S. Bureau of the Census).

7–127. Two out of five adult smokers acquired the habit by age 14. If 400 smokers are randomly selected, find the probability that 170 or more acquired the habit by age 14.

Source: *Harper's Index* 289, no. 1735 (December 1994).

7–128. A theater owner has found that 5% of his patrons do not show up for the performance that they purchased tickets for. If the theatre has 100 seats, find the probability that six or more patrons will not show up for the sold-out performance.

7–129. In a survey, 15% of Americans said they believe that they will eventually get cancer. In a random sample of 80 Americans used in the survey, find these probabilities.
 a. At least six people in the sample believe they will get cancer.
 b. Fewer than five people in the sample believe they will get cancer.

Source: *Harper's Index* (December 1994), p. 13.

7–130. If the probability that a newborn child will be female is 50%, find the probability that in 100 births, 55 or more will be females.

7–131. *Prevention* magazine reports that 18% of drivers use a car phone while driving. Find the probability that in a random sample of 100 drivers, exactly 18 will use a car phone while driving.

Source: USA Snapshot, *USA Today,* June 26, 1995.

7–132. A contractor states that 90% of all homes sold have burglar alarm systems. If a contractor sells 250 homes, find the probability that fewer than 5 of them will not have burglar alarm systems.

7–133. In 1993, only 3% of elementary and secondary schools in the United States did not have microcomputers. If a random sample of 180 elementary and secondary schools is selected, find the probability that in 1993 6 or fewer schools did not have microcomputers.

Source: *Statistical Abstract of the United States 1994.* (Washington, DC: U. S. Bureau of the Census).

7–134. Last year, 17% of American workers carpooled to work. If a random sample of 30 workers is selected, find the probability that exactly 5 people will carpool to work.

Source: USA Snapshot, *USA Today,* July 6, 1995.

7–135. A political candidate estimates that 30% of the voters in his party favor his proposed tax reform bill. If there are 400 people at a rally, find the probability that at least 100 favor his tax bill.

*7–136. Recall that for use of the normal distribution as an approximation to the binomial distribution, the conditions $np \geq 5$ and $nq \geq 5$ must be met. For each given probability, compute the minimum sample size needed for use of the normal approximation.
 a. $p = 0.1$ *d.* $p = 0.8$
 b. $p = 0.3$ *e.* $p = 0.9$
 c. $p = 0.5$

7–7

Summary

The normal distribution can be used to describe a variety of variables, such as heights, weights, and temperatures. The normal distribution is bell-shaped, unimodal, symmetric, and continuous; its mean, median, and mode are equal. Since each variable has its own distribution with mean μ and standard deviation σ, mathematicians use the standard normal distribution, which has a mean of 0 and a standard deviation of 1. Other approximately normally distributed variables can be transformed into the standard normal distribution with the formula $z = (X - \mu)/\sigma$.

The normal distribution can also be used to describe a sampling distribution of sample means. These samples must be of the same size and randomly selected with replacement from the population. The means of the samples will differ somewhat from the population mean, since samples are generally not perfect representations of the population from which they came. The mean of the sample means will be equal to the population mean, and the standard deviation of the sample means will be equal to the population standard deviation divided by the square root of the sample size. The central limit theorem states that as the size of the samples increases, the distribution of sample means will be approximately normal.

The normal distribution can be used to approximate other distributions, such as the binomial distribution. For the normal distribution to be used as an approximation, the conditions $np \geq 5$ and $nq \geq 5$ must be met. Also, a correction for continuity may be used for more accurate results.

Important Terms

central limit theorem   277

correction for continuity   284

negatively or left skewed distribution   249

normal distribution   250

positively or right skewed distribution   249

sampling distribution of sample means   275

sampling error   275

standard error of the mean   277

standard normal distribution   252

symmetrical distribution   249

z value   252

Important Formulas

Formula for the z value (or standard score):

$$z = \frac{X - \mu}{\sigma}$$

Formula for finding a specific data value:

$$X = z \cdot \sigma + \mu$$

Formula for the mean of the sample means:

$$\mu_{\bar{X}} = \mu$$

Formula for the standard error of the mean:

$$\sigma_{\bar{X}} = \frac{\sigma}{\sqrt{n}}$$

Formula for the z value for the central limit theorem:

$$z = \frac{\bar{X} - \mu}{\sigma/\sqrt{n}}$$

Formulas for the mean and standard deviation for the binomial distribution:

$$\mu = n \cdot p \qquad \sigma = \sqrt{n \cdot p \cdot q}$$

Review Exercises

7–137. Find the area under the standard normal distribution curve for each.

a. Between $z = 0$ and $z = 1.95$

b. Between $z = 0$ and $z = 0.37$

c. Between $z = 1.32$ and $z = 1.82$

d. Between $z = -1.05$ and $z = 2.05$

e. Between $z = -0.03$ and $z = 0.53$

f. Between $z = +1.10$ and $z = -1.80$

g. To the right of $z = 1.99$

h. To the right of $z = -1.36$

i. To the left of $z = -2.09$

j. To the left of $z = 1.68$

7–138. Using the standard normal distribution, find each probability.

a. $P(0 < z < 2.07)$

b. $P(-1.83 < z < 0)$

c. $P(-1.59 < z < +2.01)$

d. $P(1.33 < z < 1.88)$

e. $P(-2.56 < z < 0.37)$

f. $P(z > 1.66)$

g. $P(z < -2.03)$

h. $P(z > -1.19)$

i. $P(z < 1.93)$

j. $P(z > -1.77)$

7–139. The average reaction time for a four-year-old child to respond to a noise is 1.5 seconds, with a standard deviation of 0.3 second. Find the probability that it will take the child the following time to react to the noise. Assume the variable is normally distributed.

a. Between 1.0 and 1.3 seconds

b. More than 2.2 seconds

c. Less than 1.1 seconds

7–140. The average diastolic blood pressure of a certain age group of people is 85 mm Hg. The standard deviation is 6. If an individual from this age group is selected, find the probability that his or her pressure will be the following. Assume the variable is normally distributed.

a. Greater than 90

b. Below 80

c. Between 85 and 95

d. Between 88 and 92

7–141. The average number of miles a mail carrier walks on a typical route in a certain city is 5.6. The standard deviation is 0.8 mile. Find the probability that a carrier chosen at random walks the following miles. Assume the variable is normally distributed.

a. Between 5 and 6 miles

b. Less than 4 miles

c. More than 6.3 miles

7–142. The average number of years a person lives after being diagnosed with a certain disease is 4 years. The standard deviation is 6 months. Find the probability that an individual diagnosed with the disease will live the following numbers of years. Assume the variable is normally distributed.

a. More than 5 years
b. Less than 2 years
c. Between 3.5 and 5.2 years
d. Between 4.2 and 4.8 years

7–143. Americans spend on average $617 per year on their insured vehicles. Assume the variable is normally distributed. If the standard deviation of the distribution is $52, find the probability that a randomly selected insured-vehicle owner will spend the following amounts on his or her vehicle per year.

a. Between $600 and $700
b. Less than $575
c. More than $624

Source: *Statistical Abstract of the United States 1994* (Washington, DC: U.S. Bureau of the Census).

7–144. The average weight of an airline passenger's suitcase is 45 pounds. The standard deviation is 2 pounds. If 15% of the suitcases are overweight, find the maximum weight allowed by the airline. Assume the variable is normally distributed.

7–145. An educational study to be conducted requires a test score in the middle 40% range. If $\mu = 100$ and $\sigma = 15$, find the highest and lowest acceptable test scores that would enable a candidate to participate in the study. Assume the variable is normally distributed.

7–146. The average cost of XYZ brand running shoes is $83 per pair, with a standard deviation of $8. If 9 pairs of running shoes are selected, find the probability that the mean cost of a pair of shoes will be less than $80. Assume the variable is normally distributed.

7–147. Americans spend on average 12.2 minutes in the shower. If the standard deviation of the variable is 2.3 minutes and the variable is normally distributed, find the probability that the mean time of a sample of 12 Americans who shower will be less than 11 minutes.

Source: USA Snapshot, *USA Today,* July 11, 1995.

7–148. The probability of winning on a slot machine is 5%. If a person plays the machine 500 times, find the probability of winning 30 times. Use the normal approximation to the binomial distribution.

7–149. Of the total population of older Americans, 18% live in Florida. For a randomly selected sample of 200 older Americans, find the probability that more than 40 live in Florida.

Source: Elizabeth Vierck, *Fact Book on Aging* (Santa Barbara, CA: ABC-CLIO 1990).

7–150. In a large university, 30% of the incoming freshmen elect to enroll in a Personal Finance course offered by the university. Find the probability that of 800 randomly selected incoming freshmen, at least 260 have elected to enroll in the course.

7–151. Of the total population of the United States, 20% live in the Northeast. If 200 residents of the United States are selected at random, find the probability that at least 50 live in the Northeast.

Source: *Statistical Abstract of the United States 1994* (Washington, DC: U.S. Bureau of the Census).

Statistics Today

What Is Normal? Revisited

Many of the variables measured in medical tests—blood pressure, triglyceride level, etc.—are approximately normally distributed for the majority of the population in the United States. Thus, researchers can find the mean and standard deviation of these variables. Then, using these two measures along with the z values, they can find normal intervals for healthy individuals. For example, 95% of the systolic blood pressures of healthy individuals fall within two standard deviations of the mean. If an individual's pressure is outside the determined normal range (either above or below), the physician will look for a possible cause and prescribe treatment if necessary.

Chapter Quiz

Determine whether each statement is true or false. If the statement is false, explain why.

1. The total area under the normal distribution is infinite.

2. The standard normal distribution is a continuous distribution.

3. All variables that are approximately normally distributed can be transformed into standard normal variables.

4. The z value corresponding to a number below the mean is always negative.

5. The area under the standard normal distribution to the left of $z = 0$ is negative.

6. The central limit theorem applies to means of samples selected from different populations.

Select the best answer.

7. The mean of the standard normal distribution is
   a. 0
   b. 1
   c. 100
   d. variable

8. Approximately what percentage of normally distributed data values will fall within one standard deviation above or below the mean?
   a. 68%
   b. 95%
   c. 99.7%
   d. variable

9. Which is not a property of the standard normal distribution?
   a. It's symmetric about the mean.
   b. It's uniform.
   c. It's bell-shaped.
   d. It's unimodal.

10. When a distribution is positively skewed, the relationship of the mean, median, and mode from left to right will be
    a. Mean, median, mode
    b. Mode, median, mean
    c. Median, mode, mean
    d. Mean, mode, median

11. The standard deviation of all possible sample means equals
    a. The population standard deviation
    b. The population standard deviation divided by the population mean
    c. The population standard deviation divided by the square root of the sample size
    d. The square root of the population standard deviation

Complete the following statements with the best answer.

12. When using the standard normal distribution, $P(z < 0)$ = _____.

13. The difference between a sample mean and a population mean is due to _____.

14. The mean of the sample means equals _____.

15. The standard deviation of all possible sample means is called _____.

16. The normal distribution can be used to approximate the binomial distribution when $n \cdot p$ and $n \cdot q$ are both greater than or equal to _____.

17. The correction factor for the central limit theorem should be used when the sample size is greater than _____ the size of the population.

18. Find the area under the standard normal distribution for each.
    a. Between 0 and 1.50
    b. Between 0 and -1.25
    c. Between 1.56 and 1.96
    d. Between -1.20 and -2.25
    e. Between -0.06 and 0.73
    f. Between 1.10 and -1.80
    g. To the right of $z = 1.75$
    h. To the right of $z = -1.28$
    i. To the left of $z = -2.12$
    j. To the left of $z = 1.36$

19. Using the standard normal distribution, find each probability.
    a. $P(0 < z < 2.16)$
    b. $P(-1.87 < z < 0)$
    c. $P(-1.63 < z < 2.17)$
    d. $P(1.72 < z < 1.98)$
    e. $P(-2.17 < z < 0.71)$
    f. $P(z > 1.77)$
    g. $P(z < -2.37)$
    h. $P(z > -1.73)$
    i. $P(z < 2.03)$
    j. $P(z > -1.02)$

20. The time it takes for a certain pain reliever to begin to reduce symptoms is 30 minutes, with a standard deviation of 4 minutes. Assuming the variable is normally distributed, find the probability that it will take the medication
    a. Between 34 and 35 minutes to begin to work
    b. More than 35 minutes to begin to work
    c. Less than 25 minutes to begin to work
    d. Between 35 and 40 minutes to begin to work

21. The average height of a certain age group of people is 53 inches. The standard deviation is 4 inches. If the variable is normally distributed, find the probability that a selected individual's height will be
    a. Greater than 59 inches
    b. Less than 45 inches
    c. Between 50 and 55 inches
    d. Between 58 and 62 inches

22. The average number of gallons of lemonade consumed by the football team during a game is 20, with a standard deviation of 3 gallons. Assume the variable is normally distributed. When a game is played, find the probability of using
    a. Between 20 and 25 gallons
    b. Less than 19 gallons
    c. More than 21 gallons
    d. Between 26 and 28 gallons

23. The average number of years a person takes to complete a graduate degree program is 3. The standard deviation is 4 months. Assume the variable is normally distributed. If an individual enrolls in the program, find the probability that it will take
   a. More than 4 years to complete the program
   b. Less than 3 years to complete the program
   c. Between 3.8 and 4.5 years to complete the program
   d. Between 2.5 and 3.1 years to complete the program

24. On the daily run of an express bus, the average number of passengers is 48. The standard deviation is 3. Assume the variable is normally distributed. Find the probability that the bus will have
   a. Between 36 and 40 passengers
   b. Fewer than 42 passengers
   c. More than 48 passengers
   d. Between 43 and 47 passengers

25. The average thickness of books on a library shelf is 8.3 centimeters. The standard deviation is 0.6 centimeter. If 20% of the books are oversized, find the minimum thickness of the oversized books on the library shelf. Assume the variable is normally distributed.

26. Membership in an elite organization requires a test score in the upper 30% range. If $\mu = 115$ and $\sigma = 12$, find the lowest acceptable score that would enable a candidate to apply for membership. Assume the variable is normally distributed.

27. The average repair cost of a microwave oven is $55, with a standard deviation of $8. The costs are normally distributed. If 12 ovens are repaired, find the probability that the mean of the repair bills will be greater than $60.

28. The average electric bill in a residential area is $72 for the month of April. The standard deviation is $6. If the amounts of the electric bills are normally distributed, find the probability that the mean of the bill for 15 residents will be less than $75.

29. The probability of winning on a slot machine is 8%. If a person plays the machine 300 times, find the probability of winning at most 20 times. Use the normal approximation to the binomial distribution.

30. If 10% of the people in a certain factory are members of a union, find the probability that in a sample of 2000, fewer than 180 people are union members.

31. In a corporation, 30% of the employees contributed to a retirement program offered by the company. Find the probability that of 300 randomly selected employees, at least 105 are enrolled in the program.

32. A company that installs swimming pools for residential customers sells a pool package to 18% of the customers its representatives visit. If the sales reps visit 180 customers, find the probability of their making at least 40 sales.

Critical Thinking Challenges

Sometimes a researcher must decide whether or not a variable is normally distributed. There are several ways to do this. One simple but very subjective method uses special graph paper, which is called *normal probability paper*. For the distribution of systolic blood pressure readings given in Chapter 3 of the textbook, the following method can be used:

1. Make a table, as shown.

Boundaries	Frequency	Cumulative frequency	Cumulative percent frequency
89.5–104.5	24		
104.5–119.5	62		
119.5–134.5	72		
134.5–149.5	26		
149.5–164.5	12		
164.5–179.5	4		
	200		

2. Find the cumulative frequencies for each class and place the results in the third column.

3. Find the cumulative percents for each class by dividing each cumulative frequency by 200 (the total frequencies) and multiplying by 100%. (For the first class, it would be 24/200 × 100% = 12%.) Place these values in the last column.

4. Using the normal probability paper, label the x axis with the class boundaries as shown and plot the percents.

5. If the points fall approximately in a straight line, it can be concluded that the distribution is normal. Do you feel that this distribution is approximately normal? Explain your answer.

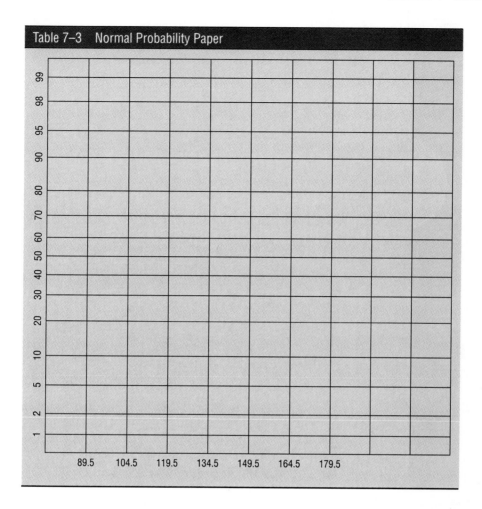

Table 7–3   Normal Probability Paper

6. To find an approximation of the mean or median, draw a horizontal line from the 50% point on the *y* axis over to the curve and then a vertical line down to the *x* axis. Compare this approximation of the mean with the computed mean.

7. To find an approximation of the standard deviation, locate the values on the *x* axis that correspond to the 16% and 84% values on the *y* axis. Subtract these two values and divide the result by 2. Compare this approximate standard deviation to the computed standard deviation.

8. Explain why the method used in Step 7 works.

Data Analysis

1. Select a variable (interval or ratio) and collect 30 data values.
   a. Construct a frequency distribution for the variable.
   b. Use the procedure described in the critical thinking challenge "When Is a Distribution Normal?" to graph the distribution on normal probability paper.
   c. Can you conclude that the data are approximately normally distributed? Explain your answer.

2. Repeat exercise 1 using the data from Data Set X in Appendix D. Use all the data.

3. Repeat exercise 1 using the data from Data Set XI in Appendix D.

chapter

8

Confidence Intervals and Sample Size

Objectives

After completing this chapter, you should be able to

1. Find the confidence interval for the mean when σ is known or $n \geq 30$.

2. Determine the minimum sample size for finding a confidence interval for the mean.

3. Find the confidence interval for the mean when σ is unknown and $n < 30$.

4. Find the confidence interval for a proportion.

5. Determine the minimum sample size for finding a confidence interval for a proportion.

6. Find a confidence interval for a variance and a standard deviation.

Would You Change the Channel?

In the article shown on the next page, a survey by the Roper Organization found that 45% of the people who were offended by a television program would change the channel, while 15% would turn off their television sets. The article further states that the margin of error is 3 percentage points, and 4000 adults were interviewed.

Several questions arise:

1. How do these estimates compare with the true population percentages?

2. What is meant by a margin of error of 3 percentage points?

3. Is the sample of 4000 large enough to represent the population of all adults who watch television in the United States?

After reading this chapter, you will be able to answer these questions, since this chapter explains how statisticians can use statistics to make estimates of parameters.

TV Survey Eyes On-Off Habits

NEW YORK (AP)—You're watching TV and the show offends you. Do you sit there and take it or do you get up and go?

A survey released yesterday on the public's attitudes toward television said 12 percent of the respondents do nothing, 45 percent change channels and 15 percent turn off their sets.

National Association of Broadcasters and the Network Television Association, an alliance of the Big Three broadcast networks, co-sponsored the survey, which has been taken every two years since 1959.

The Roper Organization polled 4,000 adults nationwide in personal interviews between November and December last year. The results had a margin of sampling error of plus or minus 3 percentage points.

The study, "America's Watching," said that even in an age of channel "grazing" about two in three viewers regularly make a special effort to watch shows they particularly like.

And despite increasing cable alternatives, about three-quarters of these "appointment" viewers with cable said that ABC, CBS and NBC still offer most of their favorites.

Viewers find TV news highly credible, the survey also said. In the event of conflicting news reports, 56 percent of respondents said they would believe television's account above others.

A hefty 69 percent said they get most of their news from TV. Newspapers were second, with 43 percent. Radio, word-of-mouth and magazines fell far behind.

A majority of the respondents reported having seen something on television that they found "either personally offensive or morally objectionable" in the past few weeks. But only 42 percent could specify whether it was profanity, sex or violence.

Source: Associated Press. Reprinted with permission.

8–1

Introduction

One aspect of inferential statistics is **estimation,** which is the process of estimating the value of a parameter from information obtained from a sample. For example, *The Book of Odds,* by Michael D. Shook and Robert L. Shook (New York: Penguin Putnam, Inc., 1991), contains the following statements:

> *"One out of 4 Americans is currently dieting." (Calorie Control Council. Used with permission.)*
> *"Seventy-two percent of Americans have flown in commercial airlines." ("The Bristol Meyers Report: Medicine in the Next Century." Used with permission.)*
> *"The average kindergarten student has seen more than 5000 hours of television." (U.S. Department of Education.)*
> *"The average school nurse makes $32,786 a year." (National Association of School Nurses. Used with permission.)*
> *"The average amount of life insurance is $108,000 per household with life insurance." (American Council of Life Insurance. Used with permission.)*

Since the populations from which these values were obtained are large, these values are only *estimates* of the true parameters and are derived from data collected from samples.

The statistical procedures for estimating the population mean, proportion, variance, and standard deviation will be explained in this chapter.

An important question in estimation is that of sample size. How large should the sample be in order to make an accurate estimate? This question is not easy to answer since the size of the sample depends on several factors, such as the accuracy desired and the probability of making a correct estimate. The question of sample size will be explained in this chapter also.

8–2

Confidence Intervals for the Mean (σ Known or $n \geq 30$) and Sample Size

Objective 1. Find the confidence interval for the mean when σ is known or $n \geq 30$.

Suppose a college president wishes to estimate the average age of the students attending classes this semester. The president could select a random sample of 100 students and find the average age of these students, say 22.3 years. From the sample mean, the president could infer that the average age of all the students is 22.3 years. This type of estimate is called a *point estimate*.

A **point estimate** is a specific numerical value estimate of a parameter. The best point estimate of the population mean μ is the sample mean $\overline{X}$.

One might ask why other measures of central tendency, such as the median and mode, are not used to estimate the population mean. The reason is that the means of samples vary less than other statistics (such as medians and modes) when many samples are selected from the same population. Therefore, the sample mean is the best estimate of the population mean.

Sample measures (i.e., statistics) are used to estimate population measures (i.e., parameters). These statistics are called **estimators.** As previously stated, the sample mean is a better estimator of the population mean than the sample median or sample mode.

A good estimator should satisfy the three properties described next.

> ### Three Properties of a Good Estimator
>
> 1. The estimator should be an **unbiased estimator.** That is, the expected value or the mean of the estimates obtained from samples of a given size is equal to the parameter being estimated.
> 2. The estimator should be consistent. For a **consistent estimator,** as sample size increases, the value of the estimator approaches the value of the parameter estimated.
> 3. The estimator should be a **relatively efficient estimator.** That is, of all the statistics that can be used to estimate a parameter, the relatively efficient estimator has the smallest variance.

Confidence Intervals

As stated in Chapter 7, the sample mean will be, for the most part, somewhat different from the population mean due to sampling error. Therefore, one might ask a second question: How good is a point estimate? The answer is that there is no way of knowing how close the point estimate is to the population mean.

This answer places some doubt on the accuracy of point estimates. For this reason, statisticians prefer another type of estimate called an *interval estimate*.

An **interval estimate** of a parameter is an interval or a range of values used to estimate the parameter. This estimate may or may not contain the value of the parameter being estimated.

In an interval estimate, the parameter is specified as being between two values. For example, an interval estimate for the average age of all students might be $26.9 < \mu < 27.7$, or 27.3 ± 0.4 years.

Either the interval contains the parameter or it does not. A degree of confidence (usually a percent) can be assigned before an interval estimate is made. For instance, one may wish to be 95% confident that the interval contains the true population mean. Another question then arises. Why 95%? Why not 99% or 99.5%?

If one desires to be more confident, such as 99% or 99.5% confident, then the interval must be larger. For example, a 99% confidence interval for the mean age of

Historical Note

Point and interval estimates were known as long ago as the late 1700s. However, it wasn't until 1937 that a mathematician, J. Neyman, formulated practical applications for them.

college students might be $26.7 < \mu < 27.9$, or 27.3 ± 0.6. Hence, a trade-off occurs. To be more confident that the interval contains the true population mean, one must make the interval wider.

The **confidence level** of an interval estimate of a parameter is the probability that the interval estimate will contain the parameter.

A **confidence interval** is a specific interval estimate of a parameter determined by using data obtained from a sample and by using the specific confidence level of the estimate.

Intervals constructed in this way are called *confidence intervals*. Three common confidence intervals are used: the 90%, the 95%, and the 99% confidence intervals.

The algebraic derivation of the formula for determining a confidence interval for a mean will be shown later. A brief intuitive explanation will be given first.

The central limit theorem states that when the sample size is large, approximately 95% of the sample means will fall within ± 1.96 standard errors of the population mean. That is,

$$\mu \pm 1.96\left(\frac{\sigma}{\sqrt{n}}\right)$$

Now, if a specific sample mean is selected, say $\overline{X}$, there is a 95% probability that it falls within the range of $\mu \pm 1.96\ (\sigma/\sqrt{n})$. Likewise, there is a 95% probability that the interval specified by

$$\overline{X} \pm 1.96\left(\frac{\sigma}{\sqrt{n}}\right)$$

will contain μ, as will be shown later. Stated another way,

$$\overline{X} - 1.96\left(\frac{\sigma}{\sqrt{n}}\right) < \mu < \overline{X} + 1.96\left(\frac{\sigma}{\sqrt{n}}\right)$$

Hence, one can be 95% confident that the population mean is contained within that interval when the values of the variable are normally distributed in the population.

The value used for the 95% confidence interval, 1.96, is obtained from Table E in Appendix C. For a 99% confidence interval, the value 2.58 is used instead of 1.96 in the formula. This value is also obtained from Table E and is based on the standard normal distribution. Since other confidence intervals are used in statistics, the symbol $z_{\alpha/2}$ (read "zee sub alpha over two") is used in the general formula for confidence intervals. The Greek letter α (alpha) represents the total area in both of the tails of the standard normal distribution curve. $\alpha/2$ represents the area in each one of the tails. More will be said after Examples 8–1 and 8–2 about finding other values for $z_{\alpha/2}$.

The relationship between α and the confidence level is that the stated confidence level is the percentage equivalent to the decimal value of $1 - \alpha$, and vice versa. When the 95% confidence interval is to be found, $\alpha = 0.05$, since $1 - 0.05 = 0.95$, or 95%. When $\alpha = 0.01$, then $1 - \alpha = 1 - 0.01 = 0.99$, and the 99% confidence interval is being calculated.

Formula for the Confidence Interval of the Mean for a Specific α

$$\overline{X} - z_{\alpha/2}\left(\frac{\sigma}{\sqrt{n}}\right) < \mu < \overline{X} + z_{\alpha/2}\left(\frac{\sigma}{\sqrt{n}}\right)$$

For a 95% confidence interval, $z_{\alpha/2} = 1.96$; and for a 99% confidence interval, $z_{\alpha/2} = 2.58$.

The term $z_{\alpha/2}\,(\sigma/\sqrt{n})$ is called the *maximum error of estimate*. For a specific value, say $\alpha = 0.05$, 95% of the sample means will fall within this error value on either side of the population mean, as previously explained. See Figure 8–1.

Figure 8–1

95% Confidence Interval

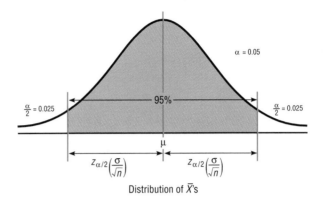

Distribution of $\bar{X}$'s

The **maximum error of estimate** is the maximum likely difference between the point estimate of a parameter and the actual value of the parameter.

A more detailed explanation of the error of estimate follows the examples illustrating the computation of confidence intervals.

Rounding Rule for a Confidence Interval for a Mean   When computing a confidence interval for a population mean using *raw data,* round off to one more decimal place than the number of decimal places in the original data. When computing a confidence interval for a population mean using a sample mean and a standard deviation, round off to the same number of decimal places as given for the mean.

Example 8–1

The president of a large university wishes to estimate the average age of the students presently enrolled. From past studies, the standard deviation is known to be 2 years. A sample of 50 students is selected, and the mean is found to be 23.2 years. Find the 95% confidence interval of the population mean.

Solution

Since the 95% confidence interval is desired, $z_{\alpha/2} = 1.96$. Hence, substituting in the formula

$$\bar{X} - z_{\alpha/2}\left(\frac{\sigma}{\sqrt{n}}\right) < \mu < \bar{X} + z_{\alpha/2}\left(\frac{\sigma}{\sqrt{n}}\right)$$

one gets

$$23.2 - 1.96\left(\frac{2}{\sqrt{50}}\right) < \mu < 23.2 + 1.96\left(\frac{2}{\sqrt{50}}\right)$$

$$23.2 - 0.6 < \mu < 23.2 + 0.6$$

$$22.6 < \mu < 23.8$$

or 23.2 ± 0.6 years. Hence, the president can say, with 95% confidence, that the average age of the students is between 22.6 and 23.8 years, based on 50 students.

Example 8-2

A certain medication is known to increase the pulse rate of its users. The standard deviation of the pulse rate is known to be 5 beats per minute. A sample of 30 users had an average pulse rate of 104 beats per minute. Find the 99% confidence interval of the true mean.

Solution

Since the 99% confidence interval is desired, $z_{\alpha/2} = 2.58$. Hence, substituting in the formula

$$\overline{X} - z_{\alpha/2}\left(\frac{\sigma}{\sqrt{n}}\right) < \mu < \overline{X} + z_{\alpha/2}\left(\frac{\sigma}{\sqrt{n}}\right)$$

one gets $104 - 2.58\left(\frac{5}{\sqrt{30}}\right) < \mu < 104 + 2.58\left(\frac{5}{\sqrt{30}}\right)$

$$104 - 2.4 < \mu < 104 + 2.4$$

$$101.6 < \mu < 106.4$$

rounded to $102 < \mu < 106$ or 104 ± 2

Hence one can be 99% confident that the mean pulse rate of all users of this medication is between 102 and 106 beats per minute, based on a sample of 30 users.

Another way of looking at a confidence interval is shown in Figure 8–2. According to the central limit theorem, approximately 95% of the sample means fall within 1.96 standard deviations of the population mean if the sample size is 30 or more or if σ is known when n is less than 30 when the population is normally distributed. If it were possible to build a confidence interval about each sample mean, as was done in the previous examples for μ, 95% of these intervals would contain the population mean, as shown in Figure 8–3. Hence, one can be 95% confident that an interval built around a specific sample mean would contain the population mean.

Figure 8–2

95% Confidence Interval
for Sample Means

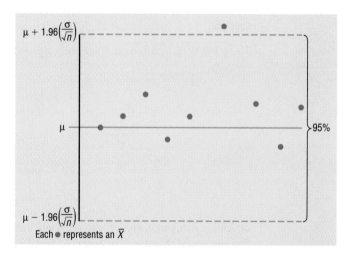

Figure 8–3

95% Confidence Intervals
for Each Sample Mean

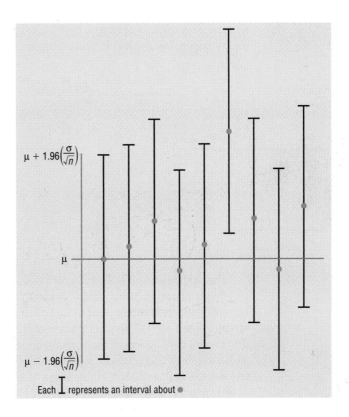

Each ⊥ represents an interval about ●

If one desires to be 99% confident, the confidence intervals must be enlarged so that 99 out of every 100 intervals contain the population mean.

Since other confidence intervals (besides 90%, 95%, and 99%) are sometimes used in statistics, an explanation of how to find the values for $z_{\alpha/2}$ is necessary. As stated previously, the Greek letter α represents the total of the areas in both tails of the normal distribution. The value for α is found by subtracting the decimal equivalent for the desired confidence level from 1. For example, if one wanted to find the 98% confidence interval, one would change 98% to 0.98 and find $\alpha = 1 - 0.98$, or 0.02. Then $\alpha/2$ is obtained by dividing α by 2. So $\alpha/2$ is 0.02/2, or 0.01. Finally, $z_{0.01}$ is the z value that will give an area of 0.01 in the right tail of the standard normal distribution curve. See Figure 8–4.

Figure 8–4

Finding $\alpha/2$ for a 98%
Confidence Interval

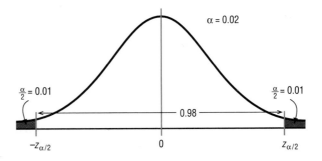

Once $\alpha/2$ is determined, the corresponding $z_{\alpha/2}$ value can be found by using the procedure shown in Chapter 7 (see Example 7–17), which is reviewed here. To get the $z_{\alpha/2}$ value for a 98% confidence interval, subtract 0.01 from 0.5000 to get 0.4900. Next, locate the area that is closest to 0.4900 (in this case, 0.4901) in Table E, and then find the corresponding z value. In this example, it is 2.33. See Figure 8–5.

Figure 8–5

Finding $z_{\alpha/2}$ for a 98% Confidence Interval

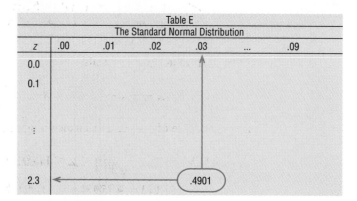

For confidence intervals, only the positive z value is used in the formula.

When the original variable is normally distributed and α is known, the standard normal distribution can be used to find confidence intervals regardless of the size of the sample. When $n \geq 30$, the distribution of means will be approximately normal even if the original distribution of the variable departs from normality. Also, if $n \geq 30$ (some authors use $n > 30$), s can be substituted for σ in the formula for confidence intervals; and the standard normal distribution can be used to find confidence intervals for means, as shown in the next example.

Example 8–3

The following data represent a sample of the assets (in millions of dollars) of 30 credit unions in Southwestern Pennsylvania. Find the 90% confidence interval of the mean.

$12.23	$16.56	$ 4.39
2.89	1.24	2.17
13.19	9.16	1.42
73.25	1.91	14.64
11.59	6.69	1.06
8.74	3.17	18.13
7.92	4.78	16.85
40.22	2.42	21.58
5.01	1.47	12.24
2.27	12.77	2.76

Source: *Pittsburgh Post Gazette,* June 7, 1998.

Solution:

STEP 1   Find the mean and standard deviation for the data. Use the formulas shown in Chapter 3 or your calculator. The mean $\overline{X} = 11.091$. The standard deviation $s = 14.405$.

STEP 2   Find $\alpha/2$. Since the 90% confidence interval is to be used, $\alpha = 1 - 0.90 = 0.10$, and

$$\frac{\alpha}{2} = \frac{0.10}{2} = 0.05$$

STEP 3   Find $z_{\alpha/2}$. Subtract 0.05 from 0.5000 to get 0.4500. The corresponding z value obtained from Table E is 1.65. (*Note:* This value is found by using the z value for an area between 0.4495 and 0.4505. A more precise z value obtained mathematically is 1.645 and is sometimes used; however, 1.65 will be used in this textbook.)

STEP 4   Substitute in the formula

$$\overline{X} - z_{\alpha/2}\left(\frac{s}{\sqrt{n}}\right) < \mu < \overline{X} + z_{\alpha/2}\left(\frac{s}{\sqrt{n}}\right)$$

(s is used in place of σ when σ is unknown, since $n \geq 30$.)

$$11.091 - 1.65\left(\frac{14.405}{\sqrt{30}}\right) < \mu < 11.091 + 1.65\left(\frac{14.405}{\sqrt{30}}\right)$$

$$11.091 - 4.339 < \mu < 11.091 + 4.339$$

$$6.752 < \mu < 15.430$$

Hence, one can be 90% confident that the population mean of the assets of all credit unions is between $6.752 million and $15.430 million, based on a sample of 30 credit unions.

Comment to Computer and Statistical Calculator Users

This chapter and subsequent chapters include examples using raw data. If you are using computer or calculator programs to find the solutions, the answers you get may vary somewhat from the ones given in the textbook. This is due to the fact that computers and calculators do not round the answers in the intermediate steps and can use 12 or more decimal places for computation. Also, they use more exact values than those given in the tables at the back of this book. These discrepancies are part and parcel of statistics.

Sample Size

Objective 2. Determine the minimum sample size for finding a confidence interval for the mean.

Sample size determination is closely related to statistical estimation. Quite often, one asks, "How large a sample is necessary to make an accurate estimate?" The answer is not simple, since it depends on three things: the maximum error of estimate, the population standard deviation, and the degree of confidence. For example, how close to the true mean does one want to be (2 units, 5 units, etc.), and how confident does one wish to be (90%, 95%, 99%, etc.)? For the purpose of this chapter, it will be assumed that the population standard deviation of the variable is known or has been estimated from a previous study.

The formula for sample size is derived from the maximum error of estimate formula,

$$E = z_{\alpha/2}\left(\frac{\sigma}{\sqrt{n}}\right)$$

and this formula is solved for n as follows:

$$E\sqrt{n} = z_{\alpha/2}(\sigma)$$

$$\sqrt{n} = \frac{z_{\alpha/2} \cdot \sigma}{E}$$

Hence, $n = \left(\frac{z_{\alpha/2} \cdot \sigma}{E}\right)^2$

> ### Formula for the Minimum Sample Size Needed for an Interval Estimate of the Population Mean
>
> $$n = \left(\frac{z_{\alpha/2} \cdot \sigma}{E} \right)^2$$
>
> where E is the maximum error of estimate. If necessary, round the answer up to obtain a whole number. That is, if there is any fraction or decimal portion in the answer, use the next whole number for sample size, n.

Example 8–4

The college president asks the statistics teacher to estimate the average age of the students at their college. How large a sample is necessary? The statistics teacher would like to be 99% confident that the estimate should be accurate within one year. From a previous study, the standard deviation of the ages is known to be 3 years.

Solution

Since $\alpha = 0.01$ (or $1 - 0.99$), $z_{\alpha/2} = 2.58$, and $E = 1$, substituting in the formula, one gets

$$n = \left(\frac{z_{\alpha/2} \cdot \sigma}{E} \right)^2 = \left[\frac{(2.58)(3)}{1} \right]^2 = 59.9$$

which is rounded up to 60. Therefore, in order to be 99% confident that the estimate is within 1 year of the true mean age, the teacher needs a sample size of at least 60 students.

Notice that when one is finding the sample size, the size of the population is irrelevant when the population is large or infinite or when sampling is done with replacement. In other cases, an adjustment is made in the formula for computing sample size. This adjustment is beyond the scope of this book.

The formula for determining sample size requires the use of the population standard deviation. What then happens when σ is unknown? In this case, an attempt is made to estimate σ. One such way is to use the standard deviation, s, obtained from a sample taken previously as an estimate for σ. The standard deviation can also be estimated by dividing the range by 4.

Sometimes, interval estimates rather than point estimates are reported. For instance, one may read a statement such as "On the basis of a sample of 200 families, the survey estimates that an American family of two spends an average of $84 per week for groceries. One can be 95% confident that this estimate is accurate within $3 of the true mean." This statement means that the 95% confidence interval of the true mean is

$$\$84 - \$3 < \mu < \$84 + \$3$$
$$\$81 < \mu < \$87$$

The algebraic derivation of the formula for a confidence interval is shown next. As explained in Chapter 7, the sampling distribution of the mean is approximately normal when large samples ($n \geq 30$) are taken from a population. Also,

$$z = \frac{\bar{X} - \mu}{\sigma/\sqrt{n}}$$

Furthermore, there is a probability of $1 - \alpha$ that a z will have a value between $-z_{\alpha/2}$ and $+z_{\alpha/2}$. Hence,

$$-z_{\alpha/2} < \frac{\overline{X} - \mu}{\sigma/\sqrt{n}} < z_{\alpha/2}$$

Using algebra,

$$-z_{\alpha/2} \cdot \frac{\sigma}{\sqrt{n}} < \overline{X} - \mu < z_{\alpha/2} \cdot \frac{\sigma}{\sqrt{n}}$$

Subtracting $\overline{X}$ from both sides and from the middle, one gets

$$-\overline{X} - z_{\alpha/2} \cdot \frac{\sigma}{\sqrt{n}} < -\mu < -\overline{X} + z_{\alpha/2} \cdot \frac{\sigma}{\sqrt{n}}$$

Multiplying by -1, one gets

$$\overline{X} + z_{\alpha/2} \cdot \frac{\sigma}{\sqrt{n}} > \mu > \overline{X} - z_{\alpha/2} \cdot \frac{\sigma}{\sqrt{n}}$$

Reversing the inequality, one gets the formula for the confidence interval:

$$\overline{X} - z_{\alpha/2} \cdot \frac{\sigma}{\sqrt{n}} < \mu < \overline{X} + z_{\alpha/2} \cdot \frac{\sigma}{\sqrt{n}}$$

Exercises

8–1. What is the difference between a point estimate and an interval estimate of a parameter? Which is better? Why?

8–2. What information is necessary to calculate a confidence interval?

8–3. What is the maximum error of estimate?

8–4. What is meant by the 95% confidence interval of the mean?

8–5. What are three properties of a good estimator?

8–6. What statistic best estimates μ?

8–7. What is necessary to determine sample size?

8–8. When one is determining the sample size for a confidence interval, is the size of the population relevant?

8–9. Find each.
a. $z_{\alpha/2}$ for the 99% confidence interval
b. $z_{\alpha/2}$ for the 98% confidence interval
c. $z_{\alpha/2}$ for the 95% confidence interval
d. $z_{\alpha/2}$ for the 90% confidence interval
e. $z_{\alpha/2}$ for the 94% confidence interval

8–10. Find the 95% confidence interval for the mean paid attendance at the Major League All Star games. A random sample of the paid attendances is shown.

47,596	68,751	5,838
69,831	28,843	53,107
31,391	48,829	50,706
62,892	55,105	63,974
56,674	38,362	51,549
31,938	31,851	56,088
34,906	38,359	72,086
34,009	50,850	43,801
46,127	49,926	54,960
32,785	48,321	49,671

Source: *Time Almanac,* Boston, MA, 1999.

8–11. A sample of the reading scores of 35 fifth-graders has a mean of 82. The standard deviation of the sample is 15.
a. Find the 95% confidence interval of the mean reading scores of all fifth-graders.
b. Find the 99% confidence interval of the mean reading scores of all fifth-graders.
c. Which interval is larger? Explain why.

8–12. Find the 90% confidence interval of the population mean for the incomes of Western Pennsylvania Credit Unions. A random sample of 50 credit unions is shown. The data are in thousands of dollars.

$ 84	14	31	72	26
49	252	104	31	8
3	18	72	23	55
133	16	29	225	138
85	24	391	72	158
4340	346	19	5	846
461	254	125	61	123
60	29	10	366	47
28	254	6	77	21
97	6	17	8	82

Source: *Pittsburgh Post Gazette,* June 7, 1998.

8–13. A study of 40 English composition professors showed that they spent, on average, 12.6 minutes correcting a student's term paper.

a. Find the 90% confidence interval of the mean time for all composition papers when $\sigma = 2.5$ minutes.

b. If a professor stated that he spent, on average, 30 minutes correcting a term paper, what would be your reaction?

8–14. A study of 40 bowlers showed that their average score was 186. The standard deviation of the population is 6.

a. Find the 95% confidence interval of the mean score for all bowlers.

b. Find the 95% confidence interval of the mean score if a sample of 100 bowlers is used instead of a sample of 40.

c. Which interval is smaller? Explain why.

8–15. A study found that 8- to 12-year-olds spend an average of $18.50 per trip to a mall. If a sample of 49 children was used, find the 90% confidence interval of the mean. Assume the standard deviation of the sample is $1.56.

Source: *USA Today,* July 25, 1995.

8–16. A study found that the average time it took a person to find a new job was 5.9 months. If a sample of 36 job seekers was surveyed, find the 95% confidence interval of the mean. Assume the standard deviation of the sample is 0.8 month.

Source: *The Book of Odds* by Michael D. Shook and Robert Shook (New York: Penguin Putnam, Inc., 1991), 46.

8–17. Find the 90% confidence interval for the mean number of local jobs for top corporations in Southwestern Pennsylvania. A sample of 40 selected corporations is shown.

7,685	3,100	725	850
11,778	7,300	3,472	540
11,370	5,400	1,570	160
9,953	3,114	2,600	2,821
6,200	3,483	8,954	8
1,000	1,650	1,200	390
1,999	400	3,473	600
1,270	873	400	713
11,960	1,195	2,290	175
887	1,703	4,236	1,400

Source: *Pittsburgh Tribune Review,* July 6, 1997.

8–18. A random sample of 48 days taken at a large hospital shows that an average of 38 patients were treated in the emergency room per day. The standard deviation of the population is 4.

a. Find the 99% confidence interval of the mean number of ER patients treated each day at the hospital.

b. Find the 99% confidence interval of the mean number of ER patients treated each day if the standard deviation were 8 instead of 4.

c. Why is the confidence interval for part *b* wider than the one for part *a*?

8–19. Noise levels at various area urban hospitals were measured in decibels. The mean of the noise levels in 84 corridors was 61.2 decibels, and the standard deviation was 7.9. Find the 95% confidence interval of the true mean.

Source: M. Bayo, A. Garcia, and A. Garcia, "Noise Levels in an Urban Hospital and Workers' Subjective Responses," *Archives of Environmental Health* 50, no. 3 (May–June 1995).

8–20. A researcher is interested in estimating the average salary of police officers in a large city. She wants to be 95% confident that her estimate is correct. If the standard deviation is $1050, how large a sample is needed to get the desired information and to be accurate within $200?

8–21. A university dean wishes to estimate the average number of hours his part-time instructors teach per week. The standard deviation from a previous study is 2.6 hours. How large a sample must be selected if he wants to be 99% confident of finding whether the true mean differs from the sample mean by 1 hour?

8–22. In the hospital study cited in Exercise 8–19, the mean noise level in the 171 ward areas was 58.0 decibels, and the standard deviation was 4.8. Find the 90% confidence interval of the true mean.

Source: M. Bayo, A. Garcia, and A. Garcia, "Noise Levels in an Urban Hospital and Workers' Subjective Responses," *Archives of Environmental Health* 50, no. 3 (May–June 1995).

8–23. An insurance company is trying to estimate the average number of sick days that full-time food-service workers use per year. A pilot study found the standard deviation to be 2.5 days. How large a sample must be selected if the company wants to be 95% confident of getting an interval that contains the true mean with a maximum error of 1 day?

8–24. A restaurant owner wishes to find the 99% confidence interval of the true mean cost of a dry martini. How large should the sample be if she wishes to be

accurate within $0.10? A previous study showed that the standard deviation of the price was $0.12.

8–25. A health care professional wishes to estimate the birth weights of infants. How large a sample must she select if she desires to be 90% confident that the true mean is within 6 ounces of the sample mean? The standard deviation of the birth weights is known to be 8 ounces.

Technology Step by Step

MINITAB
Step by Step

Finding a z Confidence Interval for the Mean

Example MT8–1

Find the 95% confidence interval when $\sigma = 11$ using the following sample:

| 43 | 52 | 18 | 20 | 25 | 45 | 43 | 21 | 42 | 32 | 24 | 32 | 19 | 25 | 26 | 44 |
| 42 | 41 | 41 | 53 | 22 | 25 | 23 | 21 | 27 | 33 | 36 | 47 | 19 | 20 |

1. Enter the data into C1 of a MINITAB worksheet.
2. Select **Stat**>**Basic Statistics**>**1-Sample Z**.
3. Select C1 for the variable.
4. Click in the box for **Sigma** and enter **11.** If the standard deviation for the population were unknown, you would calculate the standard deviation for the sample and use *s*.

1-Sample Z Dialog Box

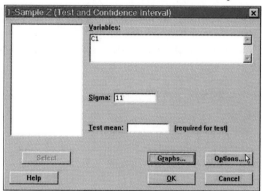

5. Click the [**Options**] button. In the dialog box make sure the **Confidence Level** is 95.0 (in Release 12 the Confidence Level is set in the 1-Sample Z dialog box, without clicking Options). You may need to click inside the textbox before you type the new level.
6. Click [**OK**].

The results will be displayed in the session window.

Session Window with Z-Interval

```
Results for: Worksheet 1

One-Sample Z: C1
The assumed sigma = 11

Variable          N      Mean     StDev    SE Mean        95.0% CI
C1               30     32.03     11.01       2.01  (   28.10,    35.97)
```

TI-83
Step by Step

For confidence intervals, the calculator will accept either raw data or summary statistics.

Finding a z Confidence Interval for the Mean (Data)

1. Enter the data into L_1.
2. Press **STAT** and move the cursor to TESTS.
3. Press 7.
4. Select Data, and press **ENTER** and move cursor to σ.
5. Enter the values for σ. Make sure L_1 is selected and Freq is 1.
6. Type in the correct confidence level.
7. Select Calculate and press **ENTER.**

Example TI8–1

Find the 99% confidence interval for the mean when $\sigma = 6$ using the following data:

| 27 | 16 | 9 | 14 | 32 | 15 | 16 | 18 | 16 | 13 |

Following the steps listed above, we arrive at the screens shown here. The confidence interval is $12.713 < \mu < 22.487$. The values of the sample mean and standard deviation are also given.

Input

Output

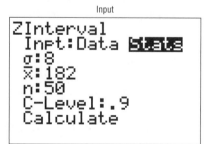

Finding a z Confidence Interval for the Mean (Statistics)

1. Press **STAT** and move the cursor to TESTS.
2. Press **7.**
3. Select Stats and press **ENTER.** Move cursor to σ.
4. Enter the value for the population standard deviation, σ.
5. Enter the sample mean, $\overline{X}$.
6. Enter the sample size.
7. Type in the correct confidence level.
8. Select Calculate and press **ENTER.**

Example TI8–2

Find the 90% confidence interval of the mean when $\overline{X} = 182$, $\sigma = 8$, and $n = 50$.

Input

Output

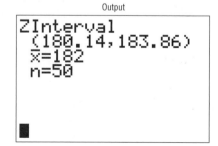

As shown, the 90% confidence interval of the mean is $180.14 < \mu < 183.86$.

Excel
Step by Step

Example XL8–1

Find a 95% confidence interval for the mean based on the following 30-number sample data set using the sample standard deviation:

43	52	18	20	25	45	43	21	42	32	24	32
19	25	26	44	42	41	41	53	22	25	23	
21	27	33	36	47	19	20					

1. Enter the data in column A.
2. Use the function STDEV to find the standard deviation.
3. Select a blank cell, and select the function CONFIDENCE.
4. Enter the sample size and standard deviation for the data set.
5. Enter alpha (here, 0.05) and click [OK].

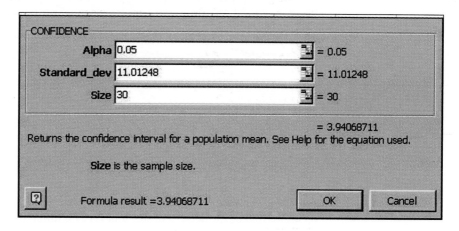

The number returned by the function CONFIDENCE specifies the interval around the mean as [mean − CONFIDENCE, mean + CONFIDENCE]. That is, the width of the interval is twice the value returned by this function. Using the AVERAGE function, we find the mean of data set to be 32.0333. So, the 95% confidence interval of the population mean is

$$32.0333 - 3.9407 < \mu < 32.0333 + 3.9407$$
$$28.0926 < \mu < 35.9740$$

Confidence Intervals for the Mean (σ Unknown and $n < 30$)

Objective 3. Find the confidence interval for the mean when σ is unknown and $n < 30$.

When σ is known and the variable is normally distributed or when σ is unknown and $n \geq 30$, the standard normal distribution is used to find confidence intervals for the mean. However, in many situations, the population standard deviation is not known and the sample size is less than 30. In such situations, the standard deviation from the sample can be used in place of the population standard deviation for confidence intervals. But a somewhat different distribution, called the *t* **distribution,** must be used when the sample size is less than 30 and the variable is normally or approximately normally distributed.

Some important characteristics of the *t* distribution are described next.

Historical Note

The *t* distribution was formulated in 1908 by an Irish brewing employee named W. S. Gosset. Gosset was involved in researching new methods of manufacturing ale. Because brewing employees were not allowed to publish results, Gosset published his finding using the pseudonym Student; hence, the *t* distribution is sometimes called the Student's distribution.

Characteristics of the *t* Distribution

The *t* distribution shares some characteristics of the normal distribution and differs from it in others. The *t* distribution is similar to the standard normal distribution in the following ways.

1. It is bell-shaped.
2. It is symmetrical about the mean.
3. The mean, median, and mode are equal to 0 and are located at the center of the distribution.
4. The curve never touches the *x* axis.

The *t* distribution differs from the standard normal distribution in the following ways.

1. The variance is greater than 1.
2. The *t* distribution is actually a family of curves based on the concept of *degrees of freedom*, which is related to sample size.
3. As the sample size increases, the *t* distribution approaches the standard normal distribution. See Figure 8–6.

Figure 8–6

The *t* Family of Curves

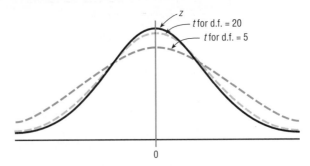

Many statistical distributions use the concept of degrees of freedom, and the formulas for finding the degrees of freedom vary for different statistical tests. The **degrees of freedom** are the number of values that are free to vary after a sample statistic has been computed, and they tell the researcher which specific curve to use when a distribution consists of a family of curves.

For example, if the mean of 5 values is 10, then 4 of the 5 values are free to vary. But once 4 values are selected, the fifth value must be a specific number to get a sum of 50, since $50 \div 5 = 10$. Hence, the degrees of freedom are $5 - 1 = 4$, and this value tells the researcher which *t* curve to use.

The symbol d.f. will be used for degrees of freedom. The degrees of freedom for a confidence interval for the mean are found by subtracting 1 from the sample size. That is, d.f. $= n - 1$. *Note:* For some statistical tests used later in this book, the degrees of freedom are not equal to $n - 1$.

The formula for finding a confidence interval about the mean using the *t* distribution is given next.

Formula for a Specific Confidence Interval for the Mean When σ Is Unknown and $n < 30$

$$\bar{X} - t_{\alpha/2}\left(\frac{s}{\sqrt{n}}\right) < \mu < \bar{X} + t_{\alpha/2}\left(\frac{s}{\sqrt{n}}\right)$$

The degrees of freedom are $n - 1$.

The values for $t_{\alpha/2}$ are found in Table F in Appendix C. The top row of Table F, labeled "Confidence Intervals," is used to get these values. The other two rows, labeled "One Tail" and "Two Tails," will be explained in the next chapter and should not be used here.

The next example shows how to find the value in Table F for $t_{\alpha/2}$.

Example 8–5

Find the $t_{\alpha/2}$ value for a 95% confidence interval when the sample size is 22.

Solution

d.f. = $22 - 1$, or 21. Find 21 in the left column and 95% in the row labeled "Confidence Intervals." The intersection where the two meet gives the value for $t_{\alpha/2}$, which is 2.080. See Figure 8–7.

Figure 8–7

Finding $t_{\alpha/2}$ for Example 8–5

Table F							
The *t* Distribution							
	Confidence Intervals	50%	80%	90%	95%	98%	99%
d.f.	One Tail α	0.25	0.10	0.05	0.025	0.01	0.005
	Two Tails α	0.50	0.20	0.10	0.05	0.02	0.01
1							
2							
3							
⋮							
21					2.080	2.518	2.831
⋮							
z ∞		.674	1.282^a	1.645^b	1.960	2.326^c	2.576^d

Note: At the bottom of Table F where d.f. = ∞, the $z_{\alpha/2}$ values can be found for specific confidence intervals. The reason is that as the degrees of freedom increase, the *t* distribution approaches the standard normal distribution.

The next two examples show how to find the confidence interval when one is using the *t* distribution.

Example 8–6

Ten randomly selected automobiles were stopped, and the tread depth of the right front tire was measured. The mean was 0.32 inch, and the standard deviation was 0.08 inch. Find the 95% confidence interval of the mean depth. Assume that the variable is approximately normally distributed.

Historical Note

Gosset derived the *t* distribution by selecting small random samples of measurements taken from a population of incarcerated criminals. For the measures he used the lengths of one of their fingers.

Solution

Since σ is unknown and *s* must replace it, the *t* distribution (Table F) must be used for 95%. Hence, with 9 degrees of freedom, $t_{\alpha/2} = 2.262$.

The 95% confidence interval of the population mean is found by substituting in the formula

$$\overline{X} - t_{\alpha/2}\left(\frac{s}{\sqrt{n}}\right) < \mu < \overline{X} + t_{\alpha/2}\left(\frac{s}{\sqrt{n}}\right)$$

Hence, $0.32 - (2.262)\left(\frac{0.08}{\sqrt{10}}\right) < \mu < 0.32 + (2.262)\left(\frac{0.08}{\sqrt{10}}\right)$

$$0.32 - 0.057 < \mu < 0.32 + 0.057$$

$$0.26 < \mu < 0.38$$

Therefore, one can be 95% confident that the population mean tread depth of all right front tires is between 0.26 and 0.38 inch based on a sample of 10 tires.

| Example 8–7 | The data represent a sample of the number of home fires started by candles for the past several years. (Data are from the National Fire Protection Association.) Find the 99% confidence interval for the mean number of home fires started by candles each year. |

| 5460 | 5900 | 6090 | 6310 | 7160 | 8440 | 9930 |

Solution

STEP 1 Find the mean and standard deviation for the data.
Use the formulas in Chapter 3 or your calculator.
The mean $\bar{X} = 7041.4$
The standard deviation $s = 1610.3$

STEP 2 Find $t_{\alpha/2}$ in Table F. Use the 99% confidence interval with d.f. = 6. It is 3.707.

STEP 3 Substitute in the formula and solve

$$\bar{X} - t_{\alpha/2}\left(\frac{s}{\sqrt{n}}\right) < \mu < \bar{X} + t_{\alpha/2} + \left(\frac{s}{\sqrt{n}}\right)$$

$$7041.4 - 3.707\left(\frac{1610.3}{\sqrt{7}}\right) < \mu < 7041.4 + 3.707\left(\frac{1610.3}{\sqrt{7}}\right)$$

$$7041.4 - 2256.2 < \mu < 7041.4 + 2256.2$$

$$4785.2 < \mu < 9297.6$$

One can be 99% confident that the population mean of home fires started by candles each year is between 4785.2 and 9297.6, based on a sample of home fires occurring over a period of 7 years.

Students sometimes have difficulty deciding whether to use $z_{\alpha/2}$ or $t_{\alpha/2}$ values when finding confidence intervals for the mean. As stated previously, when the σ is known, $z_{\alpha/2}$ values can be used *no matter what the sample size is,* as long as the variable is normally distributed or $n \geq 30$. When σ is unknown and $n \geq 30$, s can be used in the formula and $z_{\alpha/2}$ values can be used. Finally, when σ is unknown and $n < 30$, s is used in the formula and $t_{\alpha/2}$ values are used, as long as the variable is approximately normally distributed. These rules are summarized in Figure 8–8.

Figure 8–8

When to Use the z or t Distribution

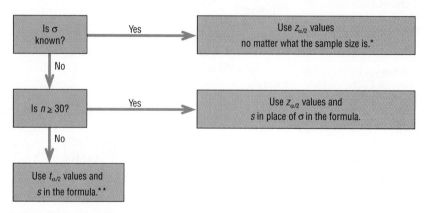

*Variable must be normally distibuted when $n < 30$.
**Variable must be approximately normally distibuted.

It should be pointed out that some statisticians have a different point of view. They use $z_{\alpha/2}$ values when σ is known and $t_{\alpha/2}$ values when σ is unknown. In these circumstances, a t table that contains t values for sample sizes greater than or equal to 30 would be needed. The procedure shown in Figure 8–8 is the one used throughout this textbook.

Exercises

8–26. What are the properties of the t distribution?

8–27. Who developed the t distribution?

8–28. What is meant by degrees of freedom?

8–29. When should the t distribution be used to find a confidence interval for the mean?

8–30. (ans) Find the values for each.
  a. $t_{\alpha/2}$ and $n = 18$ for the 99% confidence interval for the mean
  b. $t_{\alpha/2}$ and $n = 23$ for the 95% confidence interval for the mean
  c. $t_{\alpha/2}$ and $n = 15$ for the 98% confidence interval for the mean
  d. $t_{\alpha/2}$ and $n = 10$ for the 90% confidence interval for the mean
  e. $t_{\alpha/2}$ and $n = 20$ for the 95% confidence interval for the mean

For Exercises 8–31 through 8–46, assume that all variables are approximately normally distributed.

8–31. The average hemoglobin reading for a sample of 20 teachers was 16 grams per 100 milliliters, with a sample standard deviation of 2 grams. Find the 99% confidence interval of the true mean.

8–32. A meteorologist who sampled 15 cold weather fronts found that the average speed at which they traveled across a certain state was 18 miles per hour. The standard deviation of the sample was 2 miles per hour. Find the 95% confidence interval of the mean.

8–33. A state representative wishes to estimate the mean number of women representatives per state legislature. A random sample of 17 states is selected, and the number of women representatives is shown below. Based on the sample, what is the point estimate of the mean? Find the 90% confidence interval of the mean population. (*Note:* The population mean is actually 31.72 or about 32.) Compare this value to the point estimate and the confidence interval. There is something unusual about the data. Describe it and state how it would affect the confidence interval.

5	33	35	37	24
31	16	45	19	13
18	29	15	39	18
58	132			

8–34. A sample of 20 tuna showed that they swim an average of 8.6 miles per hour. The standard deviation for the sample was 1.6. Find the 90% confidence interval of the true mean.

8–35. A sample of 6 adult elephants had an average weight of 12,200 pounds, with a sample standard deviation of 200 pounds. Find the 95% confidence interval of the true mean.

8–36. The daily salaries of substitute teachers for 8 local school districts is shown. What is the point estimate for the mean? Find the 90% confidence interval of the mean for the salaries of substitute teachers in the region.

$60	$56	$60	$55	$70	$55	$60	$55

Source: *Pittsburgh Tribune Review,* November 30, 1997.

8–37. A recent study of 28 city residents showed that the mean of the time they had lived at their present address was 9.3 years. The standard deviation of the sample was 2 years. Find the 90% confidence interval of the true mean.

8–38. An automobile shop manager timed six employees and found that the average time it took them to change a water pump was 18 minutes. The standard deviation of the sample was 3 minutes. Find the 99% confidence interval of the true mean.

8–39. A recent study of 25 students showed that they spent an average of $18.53 for gasoline per week. The standard deviation of the sample was $3.00. Find the 95% confidence interval of the true mean.

8–40. For a group of 10 men subjected to a stress situation, the mean number of heartbeats per minute was 126, and the standard deviation was 4. Find the 95% confidence interval of the true mean.

8–41. For the stress test described in Exercise 8–40, six women had an average heart rate of 115 beats per minute. The standard deviation of the sample was 6 beats. Find the 95% confidence interval of the true mean for the women.

8–42. For a sample of 24 operating rooms taken in the hospital study mentioned in Exercise 8–19, the mean noise level was 41.6 decibels, and the standard deviation was 7.5. Find the 95% confidence interval of the true mean of the noise levels in the operating rooms.

Source: M. Bayo, A. Garcia, and A. Garcia, "Noise Levels in an Urban Hospital and Workers' Subjective Responses," *Archives of Environmental Health* 50, no. 3 (May–June 1995).

8–43. Find the 98% confidence interval of the mean taxable assessed values for the City of Pittsburgh. A random sample of 14 wards is shown. Data are in millions of dollars. Note: Based on all the wards, the population mean is $65.70. Does the confidence interval actually include the population mean?

$22	$24	$120	$382	$50	$38	$297
29	23	70	56	17	51	38

Source: *Pittsburgh Tribune Review,* May 22, 1999.

8–44. For a group of 20 students taking a final exam, the mean heart rate was 96 beats per minute, and the standard deviation was 5. Find the 95% confidence interval of the true mean.

8–45. The average yearly income for 28 married couples living in city C is $58,219. The standard deviation of the sample is $56. Find the 95% confidence interval of the true mean.

*8–46. A *one-sided confidence* interval can be found for a mean by using

$$\mu > \bar{X} - t_\alpha \frac{s}{\sqrt{n}} \quad \text{or} \quad \mu < \bar{X} + t_\alpha \frac{s}{\sqrt{n}}$$

where t_α is the value found under the row labeled one tail. Find two one-sided 95% confidence intervals of the population mean for the data shown and interpret the

answers. The data represent the daily revenues from 20 parking meters in a small municipality.

$2.60	$1.05	$2.45	$2.90
1.30	3.10	2.35	2.00
2.40	2.35	2.40	1.95
2.80	2.50	2.10	1.75
1.00	2.75	1.80	1.95

THE MAN IN THE STREET
(SUBJECT TO A SAMPLING ERROR OF PLUS OR MINUS THREE PERCENTAGE POINTS)

Source: Drawing by Nurit Karlin: © 1988 *The New Yorker Magazine,* Inc.

Technology Step by Step

MINITAB
Step by Step

Finding a *t* Confidence Interval for the Mean

Example MT8–2

Find the 95% confidence interval when $\sigma = 11$ using the following sample:

625	675	535	406	512	680	483	522	619	575

1. Type the data into **C1** of a MINITAB worksheet.
2. Select **Stat>Basic Statistics>1-Sample t**.

1-Sample t Dialog Box

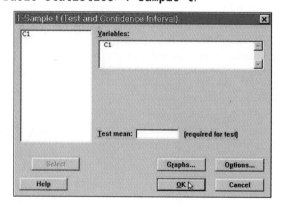

3. Double-click **C1** for the variable.
4. Click on **Options** and be sure the **Confidence Level** is 95.0 (in Release 12 the Confidence Level is set in the 1-Sample t dialogue box, without clicking Options). You may need to click inside the textbox to change it. Click [OK].
5. Click [OK].

In the session window you will see the results. The 95% confidence interval estimate for μ is between 500.4 and 626.0. The sample size, mean, standard deviation, and standard error of the mean are also shown.

Session Window with
t-interval

One-Sample T: C1

Variable	N	Mean	StDev	SE Mean	95.0% CI
C1	10	563.2	87.9	27.8	(500.4, 626.0)

TI-83
Step by Step

Finding a *t* Confidence Interval for the Mean (Data)

1. Enter the data into L_1.
2. Press **STAT** and move the cursor to TESTS.
3. Press **8.**
4. Select Data and press **ENTER.** Move cursor to c-level.
5. Type in the correct confidence level. Make sure L_1 is selected and Freq is 1.
6. Select Calculate and press **ENTER.**

Example TI8–3

Find the 95% confidence interval for the following data:

$$62 \quad 81 \quad 86 \quad 79 \quad 73 \quad 88 \quad 90 \quad 98 \quad 78 \quad 93 \quad 87 \quad 82$$
$$78 \quad 59 \quad 63 \quad 97 \quad 93 \quad 84$$

Input

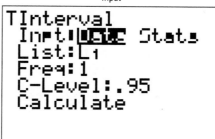

Output

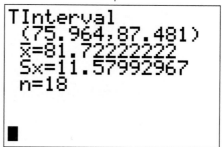

As shown on the screen, the 95% confidence interval of the mean is $75.964 < \mu < 87.481$. The values of the sample mean and standard deviation are also given.

Finding a *t* Confidence Interval for the Mean (Stats)

1. Press **STAT** and move the cursor to TESTS.

2. Press **8** for TInterval.

3. Select Stats and press **ENTER.** Move cursor to $\overline{X}$.

4. Enter the value for the sample mean, $\overline{X}$.

5. Enter the value for the sample standard deviation, s_x.

6. Enter the sample size, *n*.

7. Type in the correct confidence interval.

8. Select Calculate and press **ENTER.**

Example TI8–4

As in Example 8–6, find the 95% confidence interval when $\overline{X} = 0.32$, $s = 0.08$, and $n = 10$.

Input	Output
TInterval Inpt:Data **Stats** x̄:.32 Sx:.08 n:10 C-Level:.95 Calculate	TInterval (.26277,.37723) x̄=.32 Sx=.08 n=10

As shown, the 95% confidence interval is $0.26277 < \mu < 0.37723$.

8–4

Confidence Intervals and Sample Size for Proportions

Objective 4. Find the confidence interval for a proportion.

A *USA Today* Snapshots feature (July 2, 1993) stated that 12% of the pleasure boats in the United States were named *Serenity.* The parameter 12% is called a **proportion.** It means that of all the pleasure boats in the United States, 12 out of every 100 are named *Serenity.* A proportion represents a part of a whole. It can be expressed as a fraction, decimal, or percentage. In this case, $12\% = 0.12 = \frac{12}{100}$ or $\frac{3}{25}$. Proportions can also represent probabilities. In this case, if a pleasure boat is selected at random, the probability that it is called *Serenity* is 0.12.

Proportions can be obtained from samples or populations. The following symbols will be used.

Symbols Used in Proportion Notation

$$p = \text{symbol for the population proportion}$$

$$\hat{p} \text{ (read "}p\text{ hat")} = \text{symbol for the sample proportion}$$

For a sample proportion,

$$\hat{p} = \frac{X}{n} \quad \text{and} \quad \hat{q} = \frac{n - X}{n} \quad \text{or} \quad 1 - \hat{p}$$

where $X =$ number of sample units that possess the characteristics of interest and $n =$ sample size.

For example, in a study, 200 people were asked if they were satisfied with their job or profession; 162 said that they were. In this case, $n = 200$, $X = 162$, and $\hat{p} = X/n = 162/200 = 0.81$. It can be said that for this sample, 0.81 or 81% of those surveyed were satisfied with their job or profession. The sample proportion is $\hat{p} = 0.81$.

The proportion of people who did not respond favorably when asked if they were satisfied with their job or profession constituted $\hat{q}$, where $\hat{q} = (n - X)/n$. For this survey, $\hat{q} = (200 - 162)/200 = 38/200$, or 0.19, or 19%.

When $\hat{p}$ and $\hat{q}$ are given in decimals or fractions, $\hat{p} + \hat{q} = 1$. When $\hat{p}$ and $\hat{q}$ are given in percentages, $\hat{p} + \hat{q} = 100\%$. It follows, then, that $\hat{q} = 1 - \hat{p}$, or $\hat{p} = 1 - \hat{q}$, when $\hat{p}$ and $\hat{q}$ are in decimal or fraction form. For the sample survey on job satisfaction, $\hat{q}$ can also be found by using $\hat{q} = 1 - \hat{p}$, or $1 - 0.81 = 0.19$.

Similar reasoning applies to population proportions; i.e., $p = 1 - q$, $q = 1 - p$, and $p + q = 1$, when p and q are expressed in decimal or fraction form. When p and q are expressed as percentages, $p + q = 100\%$, $p = 100\% - q$, and $q = 100\% - p$.

Example 8–8

In a recent survey of 150 households, 54 had central air conditioning. Find $\hat{p}$ and $\hat{q}$, where $\hat{p}$ is the proportion of households that have central air conditioning.

Solution

Since $X = 54$ and $n = 150$,

$$\hat{p} = \frac{X}{n} = \frac{54}{150} = 0.36 = 36\%$$

$$\hat{q} = \frac{n - X}{n} = \frac{150 - 54}{150} = \frac{96}{150} = 0.64 = 64\%$$

One can also find $\hat{q}$ by using the formula $\hat{q} = 1 - \hat{p}$. In this case, $\hat{q} = 1 - 0.36 = 0.64$.

As with means, the statistician, given the sample proportion, tries to estimate the population proportion. Point and interval estimates for a population proportion can be made by using the sample proportion. For a point estimate of p (the population proportion), $\hat{p}$ (the sample proportion) is used. On the basis of the three properties of a good estimator, $\hat{p}$ is unbiased, consistent, and relatively efficient. But as with means, one is not able to decide how good the point estimate of p is. Therefore, statisticians also use an interval estimate for a proportion, and they can assign a probability that the interval will contain the population proportion.

Confidence Intervals

To construct a confidence interval about a proportion, one must use the maximum error of estimate, which is

$$E = z_{\alpha/2}\sqrt{\frac{\hat{p}\hat{q}}{n}}$$

Confidence intervals about proportions must meet the criteria that $np \geq 5$ and $nq \geq 5$.

Formula for a Specific Confidence Interval for a Proportion
$$\hat{p} - (z_{\alpha/2})\sqrt{\frac{\hat{p}\hat{q}}{n}} < p < \hat{p} + (z_{\alpha/2})\sqrt{\frac{\hat{p}\hat{q}}{n}}$$
when np and nq are each greater than or equal to 5.

Rounding Rule for a Confidence Interval for a Proportion   Round off to three decimal places.

Example 8–9

A sample of 500 nursing applications included 60 from men. Find the 90% confidence interval of the true proportion of men who applied to the nursing program.

Solution

Since $\alpha = 1 - 0.90 = 0.10$, and $z_{\alpha/2} = 1.65$, substituting in the formula

$$\hat{p} - (z_{\alpha/2})\sqrt{\frac{\hat{p}\hat{q}}{n}} < p < \hat{p} + (z_{\alpha/2})\sqrt{\frac{\hat{p}\hat{q}}{n}}$$

when $\hat{p} = 60/500 = 0.12$ and $\hat{q} = 1 - 0.12 = 0.88$, one gets

$$0.12 - (1.65)\sqrt{\frac{(0.12)(0.88)}{500}} < p < 0.12 + (1.65)\sqrt{\frac{(0.12)(0.88)}{500}}$$

$$0.12 - 0.024 < p < 0.12 + 0.024$$

$$0.096 < p < 0.144$$

or

$$9.6\% < p < 14.4\%$$

Hence, one can be 90% confident that the percentage of men who applied is between 9.6% and 14.4%.

When a specific percentage is given, the percentage becomes $\hat{p}$ when it is changed to a decimal. For example, if the problem states that 12% of the applicants were men, then $\hat{p} = 0.12$.

Example 8–10

A survey of 200,000 boat owners found that 12% of the pleasure boats were named *Serenity*. Find the 95% confidence interval of the true proportion of boats named *Serenity*. (Source: *USA Today* Snapshot, July 2, 1993)

Solution

From the Snapshot, $\hat{p} = 0.12$ (i.e., 12%), and $n = 200,000$. Since $z_{\alpha/2} = 1.96$, substituting in the formula

$$\hat{p} - (z_{\alpha/2})\sqrt{\frac{\hat{p}\hat{q}}{n}} < p < \hat{p} + (z_{\alpha/2})\sqrt{\frac{\hat{p}\hat{q}}{n}} \qquad \text{yields}$$

$$0.12 - (1.96)\sqrt{\frac{(0.12)(0.88)}{200,000}} < p < 0.12 + (1.96)\sqrt{\frac{(0.12)(0.88)}{200,000}}$$

$$0.119 < p < 0.121$$

Hence, one can say with 95% confidence that the true percentage of boats named *Serenity* is between 11.9% and 12.1%.

Sample Size for Proportions

In order to find the sample size needed to determine a confidence interval about a proportion, use the following formula.

Objective 5. Determine the minimum sample size for finding a confidence interval for a proportion.

Formula for Minimum Sample Size Needed for Interval Estimate of a Population Proportion

$$n = \hat{p}\hat{q}\left(\frac{z_{\alpha/2}}{E}\right)^2$$

If necessary, round up to obtain a whole number.

This formula can be found by solving the maximum error of estimate value for n:

$$E = z_{\alpha/2}\sqrt{\frac{\hat{p}\hat{q}}{n}}$$

There are two situations to consider. First, if some approximation of $\hat{p}$ is known (e.g., from a previous study), that value can be used in the formula.

Second, if no approximation of $\hat{p}$ is known, one should use $\hat{p} = 0.5$. This value will give a sample size sufficiently large to guarantee an accurate prediction, given the confidence interval and the error of estimate. The reason is that when $\hat{p}$ and $\hat{q}$ are each 0.5, the product $\hat{p} \cdot \hat{q}$ is at maximum, as shown here.

$\hat{p}$	$\hat{q}$	$\hat{p}\hat{q}$
0.1	0.9	0.09
0.2	0.8	0.16
0.3	0.7	0.21
0.4	0.6	0.24
0.5	**0.5**	**0.25**
0.6	0.4	0.24
0.7	0.3	0.21
0.8	0.2	0.16
0.9	0.1	0.09

Example 8–11

A researcher wishes to estimate, with 95% confidence, the proportion of people who own a home computer. A previous study shows that 40% of those interviewed had a computer at home. The researcher wishes to be accurate within 2% of the true proportion. Find the minimum sample size necessary.

Solution

Since $z_{\alpha/2} = 1.96$, $E = 0.02$, $\hat{p} = 0.40$, and $\hat{q} = 0.60$, then

$$n = \hat{p}\hat{q}\left(\frac{z_{\alpha/2}}{E}\right)^2 = (0.40)(0.60)\left(\frac{1.96}{0.02}\right)^2 = 2304.96$$

which, when rounded up, is 2305 people to interview.

Example 8–12

The same researcher wishes to estimate the proportion of executives who own a car phone. She wants to be 90% confident and be accurate within 5% of the true proportion. Find the minimum sample size necessary.

Solution

Since no prior knowledge of $\hat{p}$ is known, statisticians assign the values $\hat{p} = 0.5$ and $\hat{q} = 0.5$. The sample size obtained by using these values will be large enough to ensure the specified degree of confidence. Hence,

$$n = \hat{p}\hat{q}\left(\frac{z_{\alpha/2}}{E}\right)^2 = (0.5)(0.5)\left(\frac{1.65}{0.05}\right)^2 = 272.25$$

which, when rounded up, is 273 executives to ask.

In determining the sample size, the size of the population is irrelevant. Only the degree of confidence and the maximum error are necessary to make the determination.

Exercises

8–47. In each case, find $\hat{p}$ and $\hat{q}$.
a. $n = 80$ and $X = 40$
b. $n = 200$ and $X = 90$
c. $n = 130$ and $X = 60$
d. $n = 60$ and $X = 35$
e. $n = 95$ and $X = 43$

8–48. (ans) Find $\hat{p}$ and $\hat{q}$ for each percentage. (Use each percentage for $\hat{p}$.)
a. 12%
b. 29%
c. 65%
d. 53%
e. 67%

8–49. A U.S. Travel Data Center survey conducted for *Better Homes and Gardens* of 1500 adults found that 39% said that they would take more vacations this year than last year. Find the 95% confidence interval of the true proportion of adults who said that they will travel more this year.

Source: *USA Today,* April 20, 1995.

8–50. A recent study of 100 people in Miami found 27 were obese. Find the 90% confidence interval of the population proportion of individuals living in Miami who are obese.

Source: Based on information from the Center for Disease Control and Prevention, *USA Today,* March 4, 1997.

8–51. An employment counselor found that in a sample of 100 unemployed workers, 65% were not interested in returning to work. Find the 95% confidence interval of the true proportion of workers who do not wish to return to work.

8–52. A *Today*/CNN/Gallup Poll of 1015 adults found that 13% approved of the job Congress was doing in 1995. Find the 95% confidence interval of the true proportion of adults who feel this way.

Source: *USA Today,* March 31, 1995.

8–53. A survey found that out of 200 workers, 168 said they were interrupted three or more times an hour by phone messages, faxes, etc. Find the 90% confidence interval of the population proportion of workers who are interrupted three or more times an hour.

Source: Based on information from *USA Today* Snapshot, August 6, 1997.

8–54. A survey of 120 female freshmen showed that 18 did not wish to work after marriage. Find the 95% confidence interval of the true proportion of females who do not wish to work after marriage.

8–55. A study by the University of Michigan found that one in five 13- and 14-year-olds is a sometime smoker. To see how the smoking rate of the students at a large school district compared to the national rate, the superintendent surveyed 200 13- and 14-year-old students and found that 23% said they were sometime smokers. Find the 99% confidence interval of the true proportion and compare this with the University of Michigan's study.

Source: *USA Today,* July 20, 1995.

8–56. A survey of 80 recent fatal traffic accidents showed that 46 were alcohol-related. Find the 95% confidence interval of the true proportion of fatal alcohol-related accidents.

8–57. A survey of 90 families showed that 40 owned at least one gun. Find the 95% confidence interval of the true proportion of families who own at least one gun.

8–58. In a certain state, a survey of 500 workers showed that 45% belonged to a union. Find the 90% confidence interval of the true proportion of workers who belong to a union.

8–59. For a certain age group, a study of 100 people who died showed that 25% had died of cancer. Find the proportion of individuals who die of cancer in that age group. Use the 98% confidence interval.

8–60. A researcher wishes to be 95% confident that her estimate of the true proportion of individuals who travel overseas is within 0.04 of the true proportion. Find the sample size necessary. In a prior study, a sample of 200 people showed that 80 traveled overseas last year.

8–61. A medical researcher wishes to determine the percentage of females who take vitamins. He wishes to be 99% confident that the estimate is within 2 percentage points of the true proportion. A recent study of 180 females showed that 25% took vitamins.
a. How large should the sample size be?
b. If no estimate of the sample proportion is available, how large should the sample be?

8–62. A recent study indicated that 29% of the 100 women over 55 in the study were widows.
a. How large a sample must one take to be 90% confident that the estimate is within 0.05 of the true proportion of women over 55 who are widows?
b. If no estimate of the sample proportion is available, how large should the sample be?

8–63. A researcher wishes to estimate the proportion of adult males who are under 5 feet 5 inches tall. She wants to

be 90% confident that her estimate is within 5% of the true proportion.
a. How large a sample should be taken if in a sample of 300 males, 30 were under 5 feet 5 inches tall?
b. If no estimate of the sample proportion is available, how large should the sample be?

8–64. An educator desires to estimate, within 0.03, the true proportion of high school students who study at least 1 hour each school night. He wants to be 98% confident.
a. How large a sample is necessary? Previously, he conducted a study and found that 60% of the 250 students surveyed spent at least 1 hour each school night studying.
b. If no estimate of the sample proportion is available, how large should the sample be?

*8–65. If a sample of 600 people is chosen and the researcher desires to have a maximum error of estimate of 4% on the specific proportion who favor gun control, find the degree of confidence. A recent study showed that 50% were in favor of some form of gun control.

*8–66. In a study, 68% of 1015 adults said they believe the Republicans favor the rich. If the margin of error was 3 percentage points, what was the confidence interval used for the proportion?
Source: *USA Today,* March 31, 1995.

Technology Step by Step

MINITAB
Step by Step

Finding a Confidence Interval for a Proportion

MINITAB will calculate a confidence interval given the statistics from a sample or given the raw data. These instructions use the sample information in Example 8–9.

Example MT8–3

There were 500 nursing applications in a sample, including 60 from men. Find the 90% confidence interval of the actual proportion of male applicants.

1. Select **Stat>Basic Statistics>1 Proportion.**

1 Proportion Dialog Box

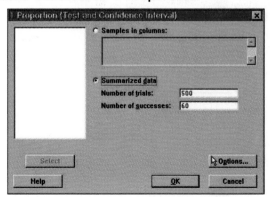

2. Click on the button for **Summarized data.**

3. Click in the box for Number of trials and enter 500.

4. In the Number of successes box enter 60.

5. Click on [Options].

1 Proportion Options Dialog Box

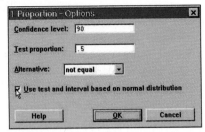

6. Type **90.0** for the confidence level.

7. Check the box for Use test and interval based on normal distribution.

8. Click [OK] twice.

The results for the confidence interval will be displayed in the session window. The population proportion is most likely between 10% and 14%. The Z- and P-values shown are used in hypothesis testing, covered in the next chapter.

Confidence Interval for 1 Proportion

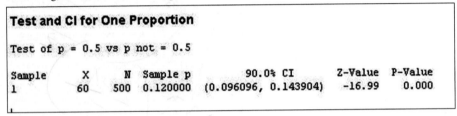

Test and CI for One Proportion

Test of p = 0.5 vs p not = 0.5

Sample	X	N	Sample p	90.0% CI	Z-Value	P-Value
1	60	500	0.120000	(0.096096, 0.143904)	-16.99	0.000

TI-83
Step by Step

Finding a Confidence Interval for a Proportion

1. Press **STAT** and move the cursor to TESTS.

2. Press **A (ALPHA, MATH)** for 1-PropZInt.

3. Enter the value for X.

4. Enter the value for n.

5. Type in the correct confidence interval.

6. Select Calculate and press ENTER.

Example TI8–5

Find the 95% confidence interval of p when X = 60 and n = 500, as in Example 8–9.

Input   Output

The 95% confidence level for p is $0.09152 < p < 0.14848$. $\hat{p}$ is also given.

8–5

Confidence Intervals for Variances and Standard Deviations

Objective 6. Find a confidence interval for a variance and a standard deviation.

In the previous section, confidence intervals were calculated for means and proportions. This section will explain how to find confidence intervals for variances and standard deviations. In statistics, the variance and standard deviation of a variable are as important as the mean. For example, when products that fit together (such as pipes) are manufactured, it is important to keep the variations of the diameters of the products as small as possible; otherwise, they will not fit together properly and have to be scrapped. In the manufacture of medicines, the variance and standard deviation of the medication in the pills play an important role in making sure patients receive the proper dosage. For these reasons, confidence intervals for variances and standard deviations are necessary.

In order to calculate these confidence intervals, a new statistical distribution is needed. It is called the **chi-square distribution.**

The chi-square variable is similar to the t variable in that its distribution is a family of curves based on the number of degrees of freedom. The symbol for chi-square is χ^2 (Greek letter chi, pronounced "ki"). Several of the distributions are shown in Figure 8–9, along with the corresponding degrees of freedom. The chi-square distribution is obtained from the values of $(n-1)s^2/\sigma^2$ when random samples are selected from a normally distributed population whose variance is σ^2.

Figure 8–9

The Chi-Square Family of Curves

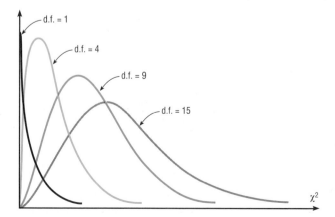

Historical Note

The χ^2 distribution with 2 degrees of freedom was formulated by a mathematician named Hershel in 1869 while he was studying the accuracy of shooting arrows at a target. Many other mathematicians have since contributed to its development.

A chi-square variable cannot be negative, and the distributions are positively skewed. At about 100 degrees of freedom, the chi-square distribution becomes somewhat symmetrical. The area under each chi-square distribution is equal to 1.00 or 100%.

Table G in Appendix C gives the values for the chi-square distribution. These values are used in the denominators of the formulas for confidence intervals. Two different values are used in the formula. One value is found on the left side of the table and the other is on the right. For example, to find the table values corresponding to the 95% confidence interval, one must first change 95% to a decimal and subtract it from 1 ($1 - 0.95 = 0.05$). Then divide the answer by 2 ($\alpha/2 = 0.05/2 = 0.025$). This is the column on the right side of the table, used to get the values for χ^2_{right}. To get the value for χ^2_{left}, subtract the value of $\alpha/2$ from 1 ($1 - 0.05/2 = 0.975$). Finally, find the appropriate row corresponding to the degrees of freedom, $n - 1$. A similar procedure is used to find the values for a 90% or 99% confidence interval.

Example 8–13

Find the values for χ^2_{right} and χ^2_{left} for a 90% confidence interval when $n = 25$.

Solution

To find χ^2_{right}, subtract $1 - 0.90 = 0.10$ and divide by 2 to get 0.05.

To find χ^2_{left}, subtract $1 - 0.05$ to get 0.95. Hence, use the 0.95 and 0.05 columns and the row corresponding to 24 d.f. See Figure 8–10.

Figure 8–10

χ^2 Table for Example 8–13

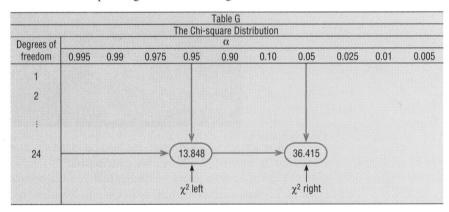

The answers are

$$\chi^2_{\text{right}} = 36.415$$
$$\chi^2_{\text{left}} = 13.848$$

The best estimates for σ^2 and σ are s^2 and s respectively.

In order to find confidence intervals for variances and standard deviations, one must assume that the variable is normally distributed.

The formulas for the confidence intervals are shown next.

Formula for the confidence interval for a variance

$$\frac{(n-1)s^2}{\chi^2_{\text{right}}} < \sigma^2 < \frac{(n-1)s^2}{\chi^2_{\text{left}}}$$

d.f. $= n - 1$

Formula for the confidence interval for a standard deviation

$$\sqrt{\frac{(n-1)s^2}{\chi^2_{\text{right}}}} < \sigma < \sqrt{\frac{(n-1)s^2}{\chi^2_{\text{left}}}}$$

d.f. $= n - 1$

Recall that s^2 is the symbol for the sample variance and s is the symbol for the sample standard deviation. If the problem gives the sample standard deviation (s), be sure to *square* it when using the formula. But if the problem gives the sample variance (s^2), do not square it when using the formula, since the variance is already in square units.

Rounding Rule for a Confidence Interval for a Variance or Standard Deviation   When computing a confidence interval for a population variance or standard deviation using raw data, round off to one more decimal place than the number of decimal places in the original data.

When computing a confidence interval for a population variance or standard deviation using a sample variance or standard deviation, round off to the same number of decimal places as given for the sample variance or standard deviation.

The next example shows how to find a confidence interval for a variance and standard deviation.

Example 8–14

Find the 95% confidence interval for the variance and standard deviation of the nicotine content of cigarettes manufactured if a sample of 20 cigarettes has a standard deviation of 1.6 milligrams.

Solution

Since $\alpha = 0.05$, the two critical values for the 0.025 and 0.975 levels for 19 degrees of freedom are 32.852 and 8.907. The 95% confidence interval for the variance is found by substituting in the formula:

$$\frac{(n-1)s^2}{\chi^2_{\text{right}}} < \sigma^2 < \frac{(n-1)s^2}{\chi^2_{\text{left}}}$$

$$\frac{(20-1)(1.6)^2}{32.852} < \sigma^2 < \frac{(20-1)(1.6)^2}{8.907}$$

$$1.5 < \sigma^2 < 5.5$$

Hence, one can be 95% confident that the true variance for the nicotine content is between 1.5 and 5.5.

For the standard deviation, the confidence interval is

$$\sqrt{1.5} < \sigma < \sqrt{5.5}$$

$$1.2 < \sigma < 2.3$$

Hence, one can be 95% confident that the true standard deviation for the nicotine content of all cigarettes manufactured is between 1.2 and 2.3 milligrams based on a sample of 20 cigarettes.

Example 8–15

Find the 90% confidence interval for the variance and standard deviation for the price of an adult single-day ski lift ticket. The data represent a selected sample of nationwide ski resorts. Assume the variable is normally distributed.

$59	$54	$53	$52	$51
39	49	46	49	48

Source: *USA Today,* October 6, 1998.

Solution

STEP 1   Find the variance for the data. Use the formulas in Chapter 3 or your calculator.
The variance $s^2 = 28.2$

STEP 2   Find χ^2_{right} and χ^2_{left} from Table G in Appendix C. Since $\alpha = 0.10$, the two critical values are 3.325 and 16.919 using d.f. = 9 and 0.95 and 0.05.

STEP 3   Substitute in the formula and solve.

$$\frac{(n-1)s^2}{\chi^2_{\text{right}}} < \sigma^2 < \frac{(n-1)s^2}{\chi^2_{\text{left}}}$$

$$\frac{(10-1)(28.2)}{16.919} < \sigma^2 < \frac{(10-1)(28.2)}{3.325}$$

$$15.0 < \sigma^2 < 76.3$$

For the standard deviation

$$\sqrt{15} < \sigma < \sqrt{76.3}$$

$$3.87 < \sigma < 8.73$$

Hence one can be 90% confident that the standard deviation for the price of all single-day ski lift tickets of the population is between $3.87 and $8.73 based on a sample of 10 nationwide ski resorts. (Two decimal places are used since the data are in dollars and cents.)

....

Note: If you are using the standard deviation instead (as in Example 8–14) of the variance, be sure to square the standard deviation when substituting in the formula.

Exercises

8–67. What distribution must be used when computing confidence intervals for variances and standard deviations?

8–68. What assumption must be made when computing confidence intervals for variances and standard deviations?

8–69. Using Table G, find the values for χ^2_{left} and χ^2_{right}.
a. $\alpha = 0.05$ $n = 16$
b. $\alpha = 0.10$ $n = 5$
c. $\alpha = 0.01$ $n = 23$
d. $\alpha = 0.05$ $n = 29$
e. $\alpha = 0.10$ $n = 14$

8–70. Find the 95% confidence interval for the variance and standard deviation for the lifetime of batteries if a sample of 20 batteries has a standard deviation of 1.7 months. Assume the variable is normally distributed.

8–71. Find the 90% confidence interval for the variance and standard deviation for the time it takes an inspector to check a bus for safety if a sample of 27 buses has a standard deviation of 6.8 minutes. Assume the variable is normally distributed.

8–72. Find the 99% confidence interval for the variance and standard deviation of the weights of 25 gallon containers of motor oil if a sample of 14 containers has a variance of 3.2. The weights are given in ounces. Assume the variable is normally distributed.

8–73. The sugar content (in grams) for a random sample of 4-ounce containers of applesauce is shown. Find the 99% confidence interval for the population variance and standard deviation. Assume the variable is normally distributed.

18.6	19.5	20.2	20.4	19.3
21.0	20.3	19.6	20.7	18.9
22.1	19.7	20.8	18.9	20.7
21.6	19.5	20.1	20.3	19.9

8–74. Find the 90% confidence interval for the variance and standard deviation of the ages of seniors at Oak Park College if a sample of 24 students has a standard deviation of 2.3 years. Assume the variable is normally distributed.

8–75. Find the 98% confidence interval for the variance and standard deviation for the time it takes a telephone company to transfer a call to the correct office. A sample of 15 calls has a standard deviation of 1.6 minutes. Assume the variable is normally distributed.

8–76. A random sample of stock prices per share is shown. Find the 90% confidence interval for the variance and standard deviation for the prices. Assume the variable is normally distributed.

$26.69	$13.88	$28.37	$12.00
75.37	7.50	47.50	43.00
3.81	53.81	13.62	45.12
6.94	28.25	28.00	60.50
40.25	10.87	46.12	14.75

Source: *Pittsburgh Tribune Review,* July 6, 1997.

8–77. A service station advertises that customers will have to wait no more than 30 minutes for an oil change. A sample of 28 oil changes has a standard deviation of 5.2 minutes. Find the 95% confidence interval of the population standard deviation of the time spent waiting for an oil change.

8–78. Find the 95% confidence interval for the variance and standard deviation of the ounces of coffee that a machine dispenses in 12-ounce cups. Assume the variable is normally distributed. The data are given here.

12.03, 12.10, 12.02, 11.98, 12.00, 12.05, 11.97, 11.99.

*8–79. A confidence interval for a standard deviation for large samples taken from a normally distributed population can be approximated by

$$s - z_{\alpha/2} \cdot \frac{s}{\sqrt{2n}} < \sigma < s + z_{\alpha/2} \cdot \frac{s}{\sqrt{2n}}$$

Find the 95% confidence interval for the population standard deviation of calculator batteries. A sample of 200 calculator batteries has a standard deviation of 18 months.

<table>
<tr><td>**8–6**</td></tr>
<tr><td>**Summary**</td></tr>
</table>

An important aspect of inferential statistics is estimation. Estimations of parameters of populations are accomplished by selecting a random sample from that population and choosing and computing a statistic that is the best estimator of the parameter. A good estimator must be unbiased, consistent, and relatively efficient. The best estimators of μ and p are $\overline{X}$ and $\hat{p}$, respectively. The best estimators of σ^2 and σ are s^2 and s, respectively.

There are two types of estimates of a parameter: point estimates and interval estimates. A point estimate is a specific value. For example, if a researcher wishes to estimate the average length of a certain adult fish, a sample of the fish is selected and measured. The mean of this sample is computed—e.g., 3.2 centimeters. From this sample mean, the researcher estimates the population mean to be 3.2 centimeters.

The problem with point estimates is that the accuracy of the estimate cannot be determined. For this reason, statisticians prefer to use the interval estimate. By computing an interval about the sample value, statisticians can be 95% or 99% (or some other percentage) confident that their estimate contains the true parameter. The confidence level is determined by the researcher. The higher the confidence level, the wider the interval of the estimate must be. For example, a 95% confidence interval of the true mean length of a certain species of fish might be

$$3.17 < \mu < 3.23$$

whereas the 99% confidence interval might be

$$3.15 < \mu < 3.25$$

When the confidence interval of the mean is computed, the z or t values are used, depending on whether or not the population standard deviation is known and depending on the size of the sample. If σ is known or $n \geq 30$, the z values can be used. If σ is not known, the t values must be used when the sample size is less than 30, and the population is normally distributed.

Closely related to computing confidence intervals is determining the sample size to make an estimate of the mean. The following information is needed to determine the minimum sample size necessary.

1. The degree of confidence must be stated.
2. The population standard deviation must be known or be able to be estimated.
3. The maximum error of estimate must be stated.

Confidence intervals and sample sizes can also be computed for proportions, using the normal distribution, and confidence intervals for variances and standard deviations can be computed using the chi-square distribution.

Important Terms

chi-square distribution   325

confidence interval   300

confidence level   300

consistent estimator   299

degrees of freedom   312

estimation   298

estimator   299

interval estimate   299

maximum error of estimate   301

point estimate   299

proportion   318

relatively efficient estimator   299

t distribution   311

unbiased estimator   299

Important Formulas

Formula for the confidence interval of the mean when σ is known (when $n \geq 30$, s can be used if σ is unknown):

$$\bar{X} - z_{\alpha/2}\left(\frac{\sigma}{\sqrt{n}}\right) < \mu < \bar{X} + z_{\alpha/2}\left(\frac{\sigma}{\sqrt{n}}\right)$$

Formula for the sample size for means:

$$n = \left(\frac{z_{\alpha/2} \cdot \sigma}{E}\right)^2$$

where E is the maximum error.

Formula for the confidence interval of the mean when σ is unknown and $n < 30$:

$$\bar{X} - t_{\alpha/2}\left(\frac{s}{\sqrt{n}}\right) < \mu < \bar{X} + t_{\alpha/2}\left(\frac{s}{\sqrt{n}}\right)$$

Formula for the confidence interval for a proportion:

$$\hat{p} - (z_{\alpha/2})\sqrt{\frac{\hat{p}\hat{q}}{n}} < p < \hat{p} + (z_{\alpha/2})\sqrt{\frac{\hat{p}\hat{q}}{n}}$$

where $\hat{p} = X/n$ and $\hat{q} = 1 - \hat{p}$.

Formula for the sample size for proportions:

$$n = \hat{p}\hat{q}\left(\frac{z_{\alpha/2}}{E}\right)^2$$

Formula for the confidence interval for a variance:

$$\frac{(n-1)s^2}{\chi^2_{\text{right}}} < \sigma^2 < \frac{(n-1)s^2}{\chi^2_{\text{left}}}$$

Formula for confidence interval for a standard deviation:

$$\sqrt{\frac{(n-1)s^2}{\chi^2_{\text{right}}}} < \sigma < \sqrt{\frac{(n-1)s^2}{\chi^2_{\text{left}}}}$$

Review Exercises

8–80. *TV Guide* reported that Americans have their television sets turned on an average of 54 hours per week. A researcher surveyed 50 local households and found they had their sets on an average of 47.3 hours. The standard deviation of the sample was 6.2 hours. Find the 90% confidence interval of the true mean. How does this compare with the *TV Guide* findings?

Source: *TV Guide* 43, no. 30 (1995), p. 26.

8–81. A study of 36 members of the Central Park Walkers showed that they could walk at an average rate of 2.6 miles per hour. The sample standard deviation is 0.4. Find the 95% confidence interval for the mean for all walkers.

8–82. The average weight of 60 randomly selected compact automobiles was 2627 pounds. The sample standard deviation was 400 pounds. Find the 99% confidence interval of a true mean weight of the automobiles.

8–83. A U.S. Travel Data Center survey reported that Americans stayed an average of 7.5 nights when they went on vacation. The sample size was 1500. Find the 95% confidence interval of the true mean. Assume the standard deviation was 0.8 day.

Source: *USA Today,* April 20, 1995.

8–84. A recent study of 25 commuters showed that they spent an average of $18.23 on public transportation per week. The standard deviation of the sample was $3.00. Find the 95% confidence interval of the true mean. Assume the variable is normally distributed.

8–85. For a certain urban area, in a sample of five months, an average of 28 mail carriers were bitten by dogs each month. The standard deviation of the sample was 3. Find the 90% confidence interval of the true mean number of mail carriers who are bitten by dogs each month. Assume the variable is normally distributed.

8–86. A researcher is interested in estimating the average salary of teachers in a large urban school district. She wants to be 95% confident that her estimate is correct. If the standard deviation is $1050, how large a sample is needed to be accurate within $200?

8–87. A researcher wishes to estimate, within $25, the true average amount of postage a community college spends each year. If she wishes to be 90% confident, how large a sample is necessary? The standard deviation is known to be $80.

8–88. A U.S. Travel Data Center's survey of 1500 adults found that 42% of respondents stated that they favor historical sites as vacations. Find the 95% confidence interval of the true proportion of all adults who favor visiting historical sites as vacations.

Source: *USA Today,* April 20, 1995.

8–89. In a study of 200 accidents that required treatment in an emergency room, 40% occurred at home. Find the 90% confidence interval of the true proportion of accidents that occur at home.

8–90. In a recent study of 100 people, 85 said that they were dissatisfied with their local elected officials. Find the 90% confidence interval of the true proportion of individuals who are dissatisfied with their local elected officials.

8–91. A nutritionist wishes to determine, within 2%, the true proportion of adults who snack before bedtime. If she wishes to be 95% confident that her estimate contains the population proportion, how large a sample will she need? A previous study found that 18% of the 100 people surveyed said they did snack before bedtime.

8–92. A survey of 200 adults showed that 15% played basketball for regular exercise. If a researcher desires to find the 99% confidence interval of the true proportion of adults who play basketball and be within 1% of the true population, how large a sample should be selected?

8–93. The standard deviation of the diameter of 28 oranges was 0.34 inch. Find the 99% confidence interval of the true standard deviation of the diameters of the oranges.

8–94. A random sample of 22 lawn mowers was selected, and the motors were tested to see how many miles per gallon of gasoline each one obtained. The variance of the measurements was 2.6. Find the 95% confidence interval of the true variance.

8–95. A random sample of 15 snowmobiles was selected, and the lifetimes (in months) of the batteries was measured. The variance of the sample was 8.6. Find the 90% confidence interval of the true variance.

8–96. The heights of 28 police officers from a large-city police force were measured. The standard deviation of the sample was 1.83 inches. Find the 95% confidence interval of the standard deviation of the heights of the officers.

Statistics Today

Would You Change the Channel? Revisited

The estimates given in the article are point estimates. However, since the margin of error is stated to be 3 percentage points, an interval estimate can easily be obtained. For example, if 45% of the people changed the channel, then the confidence interval of the true percentages of people who changed channels would be $42\% < p < 48\%$. The article fails to state whether a 90%, 95%, or some other percentage was used for the confidence interval.

Using the formula given in Section 8–4, a minimum sample size of 1068 would be needed to obtain a 95% confidence interval for *p,* as shown next. Use $\hat{p}$ and $\hat{q}$ as 0.5, since no value is known for $\hat{p}$.

$$n = \hat{p}\hat{q}\left(\frac{z_{\alpha/2}}{E}\right)^2$$

$$n = (0.5)(0.5)\left(\frac{1.96}{0.03}\right)^2 = 1067.1$$

$$n = 1068$$

Data Analysis

The Data Bank is found in Appendix D, or on the World Wide Web by following links from www.mhhe.com/math/stat/bluman/

1. From the Data Bank choose a variable, find the mean, and construct the 95% and 99% confidence intervals of the population mean. Use a sample of at least 30 subjects. Find the mean of the population and determine whether it falls within the confidence interval.

2. Repeat Exercise 1 using a different variable and a sample of 15.

3. Repeat Exercise 1 using a proportion. For example, construct a confidence interval for the proportion of individuals who did not complete high school.

4. From Data Set III in Appendix D, select a sample of 30 values and construct the 95% and 99% confidence

intervals of the mean length in miles of major North American rivers. Find the mean of all the values and determine if the confidence intervals contain the mean.

5. From Data Set VI in Appendix D, select a sample of 20 values and find the 90% confidence interval of the mean of the number of acres. Find the mean of all the values and determine if the confidence interval contains the mean.

6. From Data Set XIV in Appendix D, select a sample of 20 weights for the Pittsburgh Steelers and find the proportion of players who weigh over 250 pounds. Construct a 95% confidence interval for this proportion. Find the proportion of all Steelers who weigh more than 250 pounds. Does the confidence interval contain this value?

Chapter Quiz

Determine whether each statement is true or false. If the statement is false, explain why.

1. Interval estimates are preferred over point estimates since a confidence level can be specified.

2. For a specific confidence interval, the larger the sample size, the smaller the maximum error of estimate will be.

3. An estimator is consistent if, as the sample size decreases, the value of the estimator approaches the value of the parameter estimated.

4. In order to determine the sample size needed to estimate a parameter, one must know the maximum error of estimate.

Select the best answer.

5. When a 99% confidence interval is calculated instead of a 95% confidence interval with *n* being the same, the maximum error of estimate will be
   a. Smaller

   b. Larger
   c. The same
   d. It cannot be determined

6. The best point estimate of the population mean is
   a. The sample mean
   b. The sample median
   c. The sample mode
   d. The sample midrange

7. When the population standard deviation is unknown and sample size is less than 30, what table value should be used in computing a confidence interval for a mean?
   a. z
   b. t
   c. chi-square
   d. None of the above

Complete the following statements with the best answer.

8. A good estimator should be _____, _____, and _____.

9. The maximum difference between the point estimate of a parameter and the actual value of the parameter is called _____.

10. The statement "The average height of an adult male is 5 feet 10 inches" is an example of a(n) _____ estimate.

11. The three confidence intervals used most often are the _____ %, _____ %, and _____ %.

12. A random sample of 49 shoppers showed that they spend an average of $23.45 per visit at the Saturday Mornings Bookstore. The standard deviation of the sample was $2.80. Find the 90% confidence interval of the true mean.

13. An irate patient complained that the cost of a doctor's visit was too high. She randomly surveyed 20 other patients and found that the mean amount of money they spent on each doctor's visit was $44.80. The standard deviation of the sample was $3.53. Find the 95% confidence interval of the population mean. Assume the variable is normally distributed.

14. The average weight of 40 randomly selected school buses was 4150 pounds. The standard deviation was 480 pounds. Find the 99% confidence interval of the true mean weight of the buses.

15. In a study of 10 insurance sales reps from a certain large city, the average age of the group was 48.6, and the standard deviation was 4.1 years. Find the 95% confidence interval of the population mean age of all insurance sales reps in that city.

16. In a hospital, a sample of 8 weeks was selected, and it was found that an average of 438 patients were treated in the emergency room each week. The standard deviation was 16. Find the 99% confidence interval of the true mean. Assume the variable is normally distributed.

17. For a certain urban area, it was found that in a sample of 4 months, an average of 31 burglaries occurred each month. The standard deviation was 4. Find the 90% confidence interval of the true mean number of burglaries each month.

18. A university dean wishes to estimate the average number of hours freshmen study each week. The standard deviation from a previous study is 2.6 hours. How large a sample must be selected if he wants to be 99% confident of finding whether the true mean differs from the sample mean by 0.5 hour?

19. A researcher wishes to estimate within $300 the true average amount of money a county spends on road repairs each year. If she wants to be 90% confident, how large a sample is necessary? The standard deviation is known to be $900.

20. A recent study of 75 workers found that 53 people rode the bus to work each day. Find the 95% confidence interval of the proportion of all workers who rode the bus to work.

21. In a study of 150 accidents that required treatment in an emergency room, 36% involved children under 6 years of age. Find the 90% confidence interval of the true proportion of accidents that involve children under the age of 6.

22. A survey of 90 families showed that 40 owned at least one television set. Find the 95% confidence interval of the true proportion of families who own at least one television set.

23. A nutritionist wishes to determine, within 3%, the true proportion of adults who do not eat any lunch. If he wishes to be 95% confident that his estimate contains the population proportion, how large a sample will be necessary? A previous study found that 15% of the 125 people surveyed said they did not eat lunch.

24. A sample of 25 novels has a standard deviation of nine pages. Find the 95% confidence interval of the population standard deviation.

25. Find the 90% confidence interval for the variance and standard deviation for the time it takes a state police inspector to check a truck for safety if a sample of 27 trucks has a standard deviation of 6.8 minutes. Assume the variable is normally distributed.

26. A sample of 20 automobiles has a pollution by-product release standard deviation of 2.3 ounces when one gallon of gasoline is used. Find the 90% confidence interval of the population standard deviation.

Critical Thinking Challenges

A confidence interval for a median can be found by using the formulas

$$U = \frac{n+1}{2} + \frac{z_{\alpha/2}\sqrt{n}}{2} \quad \text{(round up)}$$

$$L = n - U + 1$$

to define positions in the set of ordered data values.

Suppose a data set has 30 values, and one wants to find the 95% confidence interval for the median. Substituting in the formulas, one gets

$$U = \frac{30+1}{2} + \frac{1.96\sqrt{30}}{2} = 21 \quad \text{(rounded up)}$$

$$L = 30 - 21 + 1 = 10$$

when $n = 30$ and $z_{\alpha/2} = 1.96$.

Arrange the data in order from smallest to largest, and then select the 10th and 21st values of the data array; hence, $X_{10} <$ med $< X_{21}$.

Find the 90% confidence interval for the median for the data in Exercise 8–12.

Data Projects

Use MINITAB, the TI-83, or a computer program of your choice to complete the following exercises.

1. Select several variables, such as the number of points a football team scored in each game of a specific season, the number of passes completed, or the number of yards gained. Using confidence intervals for the mean, determine the 90%, 95%, and 99% confidence intervals. (Use z or t, whichever is relevant.) Decide which you think is most appropriate. When this is completed, write a summary of your findings by answering the following questions.

   a. What was the purpose of the study?
   b. What was the population?
   c. How was the sample selected?
   d. What were the results obtained using confidence intervals?
   e. Did you use z or t? Why?

2. Using the same data or different data, construct a confidence interval for a proportion. For example, you might want to find the proportion of passes completed by the quarterback or the proportion of passes that were intercepted. Write a short paragraph summarizing the results.

You may use the following websites to obtain raw data:

   http://www.mhhe.com/math/stat/bluman/
   http://lib.stat.cmu.edu/DASL
   http://www.oecd.org/statlist.htm
   http://www.statcan.ca/english/

chapter

9 Hypothesis Testing

Objectives

After completing this chapter, you should be able to

1. Understand the definitions used in hypothesis testing.

2. State the null and alternative hypotheses.

3. Find critical values for the z test.

4. State the five steps used in hypothesis testing.

5. Test means for large samples using the z test.

6. Test means for small samples using the t test.

7. Test proportions using the z test.

8. Test variances or standard deviations using the chi-square test.

9. Test hypotheses using confidence intervals.

10. Explain the relationship between type I and type II errors and the power of a test.

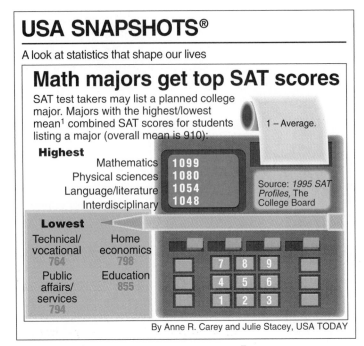

USA SNAPSHOTS®

A look at statistics that shape our lives

Math majors get top SAT scores

SAT test takers may list a planned college major. Majors with the highest/lowest mean[1] combined SAT scores for students listing a major (overall mean is 910):

1 – Average.

Highest

Mathematics	1099
Physical sciences	1080
Language/literature	1054
Interdisciplinary	1048

Source: *1995 SAT Profiles,* The College Board

Lowest

Technical/ vocational 764	Home economics 798
Public affairs/ services 794	Education 855

By Anne R. Carey and Julie Stacey, USA TODAY

Source: *USA Today,* October 12, 1995. Copyright *USA Today.* Used with permission.

How Much Better Is Better?

Suppose a school superintendent reads the *USA Today* Snapshot shown here, which states that the overall mean score for the SAT test is 910. Furthermore, suppose that, for a sample of students, the average of the SAT scores in the superintendent's school district is 960. Can the superintendent conclude that the students in his school district scored higher than average? At first glance, you might be inclined to say yes, since 960 is higher than 910. But recall that the means of samples vary about the population mean when samples are selected from a specific population. So the question arises, "Is there a real difference in the means, or is the difference simply due to chance (i.e., sampling error)?" In this chapter, you will learn how to answer that question using statistics that explain hypothesis testing.

9–1

Introduction

Researchers are interested in answering many types of questions. For example, a scientist might want to know whether the earth is warming up. A physician might want to know whether a new medication will lower a person's blood pressure. An educator might wish to see whether a new teaching technique is better than a traditional one. A retail merchant might want to know whether the public prefers a certain color in a new line of fashion. Automobile manufacturers are interested in determining whether seat belts will reduce the severity of injuries caused by accidents. These types of questions can be addressed through statistical **hypothesis testing,** which is a decision-making process for evaluating claims about a population. In hypothesis testing, the researcher must define the population under study, state the particular hypotheses that will be investigated, give the significance level, select a sample from the population, collect the data, perform the calculations required for the statistical test, and reach a conclusion.

Hypotheses concerning parameters such as means and proportions can be investigated. There are two specific statistical tests used for hypotheses concerning means: the *z test* and the *t test.* This chapter will explain in detail the hypothesis-testing procedure along with the *z* test and the *t* test. In addition, a hypothesis-testing procedure for testing a single variance or standard deviation using the chi-square distribution is explained in Section 9–6.

The three methods used to test hypotheses are

1. The traditional method
2. The *P*-value method
3. The confidence interval method

The *traditional method* will be explained first. It has been used since the hypothesis-testing method was formulated. A newer method, called the *P-value method,* has become popular with the advent of modern computers and high-powered statistical calculators. It will be explained at the end of Section 9–3. The third method, the *confidence interval method,* is explained in Section 9–7 and illustrates the relationship between hypothesis testing and confidence intervals.

9–2

Steps in Hypothesis Testing—Traditional Method

Objective 1. Understand the definitions used in hypothesis testing.

Every hypothesis-testing situation begins with the statement of a hypothesis.

A **statistical hypothesis** is a conjecture about a population parameter. This conjecture may or may not be true.

There are two types of statistical hypotheses for each situation: the null hypothesis and the alternative hypothesis.

The **null hypothesis,** symbolized by H_0, is a statistical hypothesis that states that there is no difference between a parameter and a specific value, or that there is no difference between two parameters.

The **alternative hypothesis,** symbolized by H_1, is a statistical hypothesis that states the existence of a difference between a parameter and a specific value, or states that there is a difference between two parameters.

As an illustration of how hypotheses should be stated, three different statistical studies will be used as examples.

Situation A   A medical researcher is interested in finding out whether a new medication will have any undesirable side effects. The researcher is particularly concerned with the pulse rate of the patients who take the medication. Will the pulse rate increase, decrease, or remain unchanged after a patient takes the medication?

Objective 2. State the null and alternative hypotheses.

Since the researcher knows that the mean pulse rate for the population under study is 82 beats per minute, the hypotheses for this situation are

$$H_0: \mu = 82 \qquad \text{and} \qquad H_1: \mu \neq 82$$

The null hypothesis specifies that the mean will remain unchanged, and the alternative hypothesis states that it will be different. This test is called a *two-tailed test* (a term that will be formally defined later in this section), since the possible side effects of the medicine could be to raise or lower the pulse rate.

Situation B   A chemist invents an additive to increase the life of an automobile battery. If the mean lifetime of the automobile battery is 36 months, then his hypotheses are

$$H_0: \mu \leq 36 \qquad \text{and} \qquad H_1: \mu > 36$$

In this situation, the chemist is interested only in increasing the lifetime of the batteries, so his alternative hypothesis is that the mean is greater than 36 months. The null hypothesis is that the mean is less than or equal to 36 months. This test is called right-tailed, since the interest is in an increase only.

Situation C   A contractor wishes to lower heating bills by using a special type of insulation in houses. If the average of the monthly heating bills is $78, her hypotheses about heating costs with the use of insulation are

$$H_0: \mu \geq \$78 \qquad \text{and} \qquad H_1: \mu < \$78$$

This test is a left-tailed test, since the contractor is interested only in lowering heating costs.

In order to state hypotheses correctly, researchers must translate the *conjecture* or *claim* from words into mathematical symbols. The basic symbols used are as follows:

Equal	$=$	Less than	$<$
Not equal	$\neq$	Greater than or equal to	$\geq$
Greater than	$>$	Less than or equal to	$\leq$

The null and alternative hypotheses are stated together, and the null hypothesis contains the equal sign, as shown (where k represents a specified number).

Two-tailed test	Right-tailed test	Left-tailed test
$H_0: \mu = k$	$H_0: \mu \leq k$	$H_0: \mu \geq k$
$H_1: \mu \neq k$	$H_1: \mu > k$	$H_1: \mu < k$

The formal definitions of the different types of tests are given later in this section.

Table 9–1 shows some common phrases that are used in hypotheses conjectures and the corresponding symbols. This table should be helpful in translating verbal conjectures into mathematical symbols.

Table 9–1   Hypothesis-Testing Common Phrases

>	<
Is greater than	Is less than
Is above	Is below
Is higher than	Is lower than
Is longer than	Is shorter than
Is bigger than	Is smaller than
Is increased	Is decreased or reduced from

$\geq$	$\leq$
Is greater than or equal to	Is less than or equal to
Is at least	Is at most
Is not less than	Is not more than

=	$\neq$
Is equal to	Is not equal to
Is exactly the same as	Is different from
Has not changed from	Has changed from
Is the same as	Is not the same as

Example 9–1

State the null and alternative hypotheses for each conjecture.

a. A researcher thinks that if expectant mothers use vitamin pills, the birth weight of the babies will increase. The average birth weights of the population is 8.6 pounds.

b. An engineer hypothesizes that the mean number of defects can be decreased in a manufacturing process of compact discs by using robots instead of humans for certain tasks. The mean number of defective discs per 1000 is 18.

c. A psychologist feels that playing soft music during a test will change the results of the test. The psychologist is not sure whether the grades will be higher or lower. In the past, the mean of the scores was 73.

Solution

a. H_0: $\mu \leq 8.6$ and H_1: $\mu > 8.6$.   c. H_0: $\mu = 73$ and H_1: $\mu \neq 73$.
b. H_0: $\mu \geq 18$ and H_1: $\mu < 18$.

After stating the hypothesis, the researcher's next step is to design the study. The researcher selects the correct *statistical test,* chooses an appropriate *level of significance,* and formulates a plan for conducting the study. In situation A, for instance, the researcher will select a sample of patients who will be given the drug. After allowing a suitable period of time for the drug to be absorbed, the researcher will measure each person's pulse rate.

Recall that when samples of a specific size are selected from a population, the means of these samples will vary about the population mean and the distribution of the sample means will be approximately normal when the sample size is 30 or more. (See Section 7–5). So even if the null hypothesis is true, the mean of the pulse rates of the sample of patients will not, in most cases, be exactly equal to the population mean of 82 beats per minute. There are two possibilities. Either the null hypothesis is true, and the difference between the sample mean and the population mean is due to chance *or* the null hypothesis is false, and the sample came from a population whose mean is not 82 beats per minute but some other value that is not known. These situations are shown in Figure 9–1.

Figure 9–1

Situations in
Hypothesis Testing

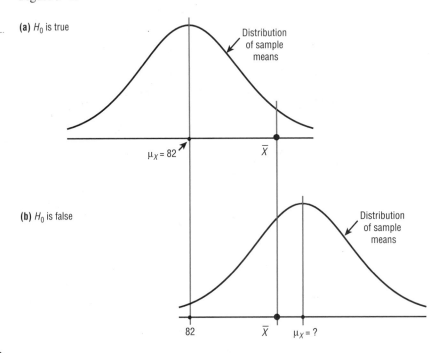

(a) H_0 is true

Distribution of sample means

$\mu_X = 82$ $\overline{X}$

(b) H_0 is false

Distribution of sample means

82    $\overline{X}$ $\mu_X = ?$

The farther away the sample mean is from the population mean, the more evidence there would be for rejecting the null hypothesis. The probability that the sample came from a population whose mean is 82 decreases as the distance or absolute value of the difference between the means increases.

If the mean pulse rate of the sample were, say, 83, the researcher would probably conclude that this difference was due to chance and would not reject the null hypothesis. But if the sample mean were, say, 90, then in all likelihood the researcher would conclude that the medication increased the pulse rate of the users and would reject the null hypothesis. The question is, "Where does the researcher draw the line?" This decision is not made on feelings or intuition; it is made statistically. That is, the difference must be significant and in all likelihood not due to chance. Here is where the concepts of statistical test and level of significance are used.

A **statistical test** uses the data obtained from a sample to make a decision about whether or not the null hypothesis should be rejected.

The numerical value obtained from a statistical test is called the **test value.**

In this type of statistical test, the mean is computed for the data obtained from the sample and is compared with the population mean. Then a decision is made to reject or not reject the null hypothesis on the basis of the value obtained from the statistical test. If the difference is significant, the null hypothesis is rejected. If it is not, then the null hypothesis is not rejected.

In the hypothesis-testing situation, there are four possible outcomes. In reality, the null hypothesis may or may not be true, and a decision is made to reject or not reject it on the basis of the data obtained from a sample. The four possible outcomes are shown in Figure 9–2. Notice that there are two possibilities for a correct decision and two possibilities for an incorrect decision.

Figure 9–2

Possible Outcomes of a Hypothesis Test

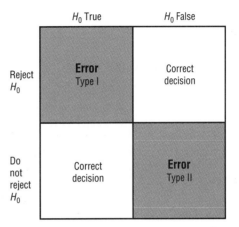

If a null hypothesis is true and it is rejected, then a *type I error* is made. In situation A, for instance, the medication might not significantly change the pulse rate of all the users in the population; but it might change the rate, by chance, of the subjects in the sample. In this case, the researcher will reject the null hypothesis when it is really true, thus committing a type I error.

On the other hand, the medication might not change the pulse rate of the subjects in the sample, but when it is given to the general population, it might cause a significant increase or decrease in the pulse rate of the users. The researcher, on the basis of the data obtained from the sample, will not reject the null hypothesis, thus committing a *type II error*.

In situation B, the additive might not significantly increase the lifetimes of automobile batteries in the population, but it might increase the lifetimes of the batteries in the sample. In this case, the null hypothesis would be rejected when it is really true. This would be a type I error. On the other hand, the additive might not work on the batteries selected for the sample, but if it were to be used in the general population of batteries, it might significantly increase their lifetimes. The researcher, on the basis of information obtained from the sample, would not reject the null hypothesis, thus committing a type II error.

A **type I error** occurs if one rejects the null hypothesis when it is true.

A **type II error** occurs if one does not reject the null hypothesis when it is false.

The hypothesis-testing situation can be likened to a jury trial. In a jury trial, there are four possible outcomes. Either the defendant is guilty or innocent, and he or she will be convicted or acquitted. See Figure 9–3.

Figure 9–3

Hypothesis Testing and a
Jury Trial

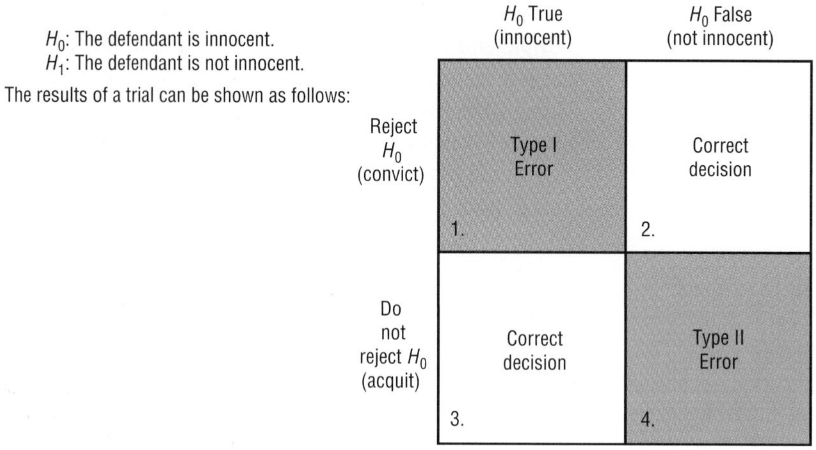

H_0: The defendant is innocent.
H_1: The defendant is not innocent.

The results of a trial can be shown as follows:

	H_0 True (innocent)	H_0 False (not innocent)
Reject H_0 (convict)	Type I Error 1.	Correct decision 2.
Do not reject H_0 (acquit)	Correct decision 3.	Type II Error 4.

Now the hypotheses are

H_0: The defendant is innocent

H_1: The defendant is not innocent (i.e., guilty)

Next, the evidence is presented in court by the prosecutor, and based on this evidence, the jury decides the verdict, innocent or guilty.

If the defendant is convicted but he or she did not commit the crime, then a type I error has been committed. See block 1 of Figure 9–3. On the other hand, if the defendant is convicted and he or she has committed the crime, then a correct decision has been made. See block 2.

If the defendant is acquitted and he or she did not commit the crime, a correct decision has been made by the jury. See block 3. However, if the defendant is acquitted and he or she did commit the crime, then a type II error has been made. See block 4.

The decision of the jury does not prove that the defendant did or did not commit the crime. The decision is based on the evidence presented. If the evidence is strong enough, the defendant will be convicted in most cases. If the evidence is weak, the defendant will be acquitted in most cases. Nothing is proven absolutely. Likewise, the decision to reject or not reject the null hypothesis does not prove anything. *The only way to prove anything statistically is to use the entire population,* which, in most cases, is not possible. The decision, then, is made on the basis of probabilities. That is, when there is a large difference between the mean obtained from the sample and the hypothesized mean, the null hypothesis is probably not true. The question is, "How large a difference is necessary to reject the null hypothesis?" Here is where the level of significance is used.

The **level of significance** is the maximum probability of committing a type I error. This probability is symbolized by α (Greek letter **alpha**). That is, P(type I error) = α.

The probability of a type II error is symbolized by β (Greek letter **beta**). That is, P (type II error) = β. In most hypothesis-testing situations, β cannot easily be computed; however, α and β are related in that decreasing one increases the other.

Statisticians generally agree on using three arbitrary significance levels: the 0.10, 0.05, and 0.01 level. That is, if the null hypothesis is rejected, the probability of a type I error will be 10%, 5%, or 1%, depending on which level of significance is used. Here is another way of putting it: When $\alpha = 0.10$, there is a 10% chance of rejecting a true null hypothesis; when $\alpha = 0.05$, there is a 5% chance of rejecting a true null hypothesis; and when $\alpha = 0.01$, there is a 1% chance of rejecting a true null hypothesis.

In a hypothesis-testing situation, the researcher decides what level of significance to use. It does not have to be the 0.10, 0.05, or 0.01 level. It can be any level, depending on the seriousness of the type I error. After a significance level is chosen, a *critical value* is selected from a table for the appropriate test. If a *z* test is used, for example, the *z* table (Table E in Appendix C) is consulted to find the critical value. The critical value determines the critical and noncritical regions.

The **critical value(s)** separates the critical region from the noncritical region. The symbol for critical value is C.V.

The **critical** or **rejection region** is the range of values of the test value that indicates that there is a significant difference and that the null hypothesis should be rejected.

The **noncritical** or **nonrejection region** is the range of values of the test value that indicates that the difference was probably due to chance and that the null hypothesis should not be rejected.

The critical value can be on the right side of the mean or on the left side of the mean for a one-tailed test. Its location depends on the inequality sign of the alternative hypothesis. For example, in situation B, where the chemist is interested in increasing the average lifetime of automobile batteries, the alternative hypothesis is H_1: $\mu > 36$. Since the inequality sign is $>$, the null hypothesis will be rejected only when the sample mean is significantly greater than 36. Hence, the critical value must be on the right side of the mean. Therefore, this test is called a *one-tailed right test.*

A **one-tailed test** indicates that the null hypothesis should be rejected when the test value is in the critical region on one side of the mean. A one-tailed test is either **right-tailed** or **left-tailed,** depending on the direction of the inequality of the alternative hypothesis.

To obtain the critical value, the researcher must choose an alpha level. In situation B, suppose the researcher chose $\alpha = 0.01$. Then, the researcher must find a *z* value such that 1% of the area falls to the right of the *z* value and 99% falls to the left of the *z* value, as shown in Figure 9–4(a).

Objective 3. Find critical values for the *z* test.

Next, the researcher must find the value in Table E closest to 0.4900. Note that because the table gives the area between 0 and the *z*, 0.5000 must be subtracted from 0.9900 to get 0.4900. The critical *z* value is 2.33, since that value gives the area closest to 0.4900, as shown in Figure 9–4(b).

Figure 9–4

Finding the Critical Value for $\alpha = 0.01$ (Right-Tailed Test)

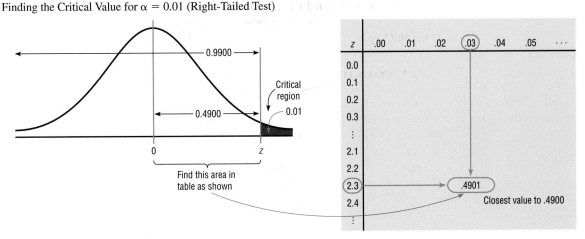

(a) The critical region

(b) The critical value from Table E

The critical and noncritical regions and the critical value are shown in Figure 9–5.

Figure 9–5

Critical and Noncritical
Regions for $\alpha = 0.01$
(Right-Tailed Test)

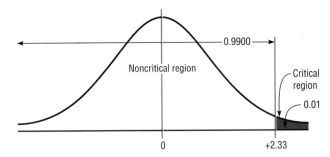

Now, move on to situation C, where the contractor is interested in lowering the heating bills. The alternative hypothesis is H_1: $\mu < \$78$. Hence, the critical value falls to the left of the mean. This test is thus a left-tailed test. At $\alpha = 0.01$, the critical value is -2.33, as shown in Figure 9–6.

Figure 9–6

Critical and Noncritical
Regions for $\alpha = 0.01$
(Left-Tailed Test)

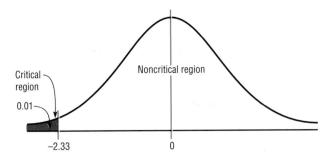

When a researcher conducts a two-tailed test, as in situation A, the null hypothesis can be rejected when there is a significant difference in either direction, above or below the mean.

In a **two-tailed test,** the null hypothesis should be rejected when the test value is in either of the two critical regions.

For a two-tailed test, then, the critical region must be split into two equal parts. If $\alpha = 0.01$, then half of the area, or 0.005, must be to the right of the mean and half must be to the left of the mean, as shown in Figure 9–7.

Figure 9–7

Finding the Critical
Values for $\alpha = 0.01$
(Two-Tailed Test)

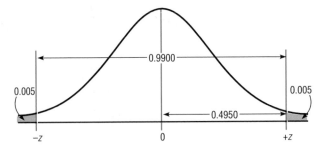

In this case, the area to be found in Table E is 0.4950. The critical values are $+2.58$ and -2.58, as shown in Figure 9–8.

Figure 9–8

Critical and Noncritical
Regions for $\alpha = 0.01$
(Two-Tailed Test)

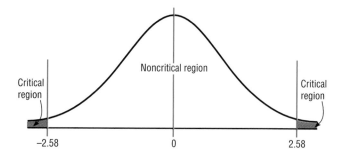

A similar procedure is used for other values of α.

Figure 9–9 with rejection regions shaded shows the critical values (C.V.) for the three situations discussed in this section for values of $\alpha = 0.10$, $\alpha = 0.05$, and $\alpha = 0.01$. The procedure for finding critical values is outlined next (where k is a specified number).

Figure 9–9

Summary of Hypothesis
Testing and Critical
Values

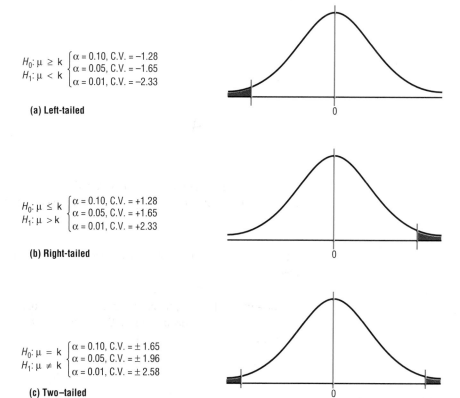

$H_0: \mu \geq k$ $\begin{cases} \alpha = 0.10, \text{C.V.} = -1.28 \\ \alpha = 0.05, \text{C.V.} = -1.65 \\ \alpha = 0.01, \text{C.V.} = -2.33 \end{cases}$
$H_1: \mu < k$

(a) Left-tailed

$H_0: \mu \leq k$ $\begin{cases} \alpha = 0.10, \text{C.V.} = +1.28 \\ \alpha = 0.05, \text{C.V.} = +1.65 \\ \alpha = 0.01, \text{C.V.} = +2.33 \end{cases}$
$H_1: \mu > k$

(b) Right-tailed

$H_0: \mu = k$ $\begin{cases} \alpha = 0.10, \text{C.V.} = \pm 1.65 \\ \alpha = 0.05, \text{C.V.} = \pm 1.96 \\ \alpha = 0.01, \text{C.V.} = \pm 2.58 \end{cases}$
$H_1: \mu \neq k$

(c) Two–tailed

Procedure Table

Finding the Critical Values for Specific α Values, Using Table E

STEP 1   Draw the figure and indicate the appropriate area.

a. If the test is left-tailed, the critical region, with an area equal to α, will be on the left side of the mean.

(continued)

Procedure Table (concluded)

       b. If the test is right-tailed, the critical region, with an area equal to α, will be on the right side of the mean.

       c. If the test is two-tailed, α must be divided by 2; half of the area will be to the right of the mean, and the other half will be to the left of the mean.

STEP 2   For a one-tailed test, subtract the area (equivalent to α) in the critical region from 0.5000, since Table E gives the area under the standard normal distribution curve between 0 and any z to the right of 0. For a two-tailed test, subtract the area (equivalent to α/2) from 0.5000.

STEP 3   Find the area in Table E corresponding to the value obtained in step 2. If the exact value cannot be found in the table, use the closest value.

STEP 4   Find the z value that corresponds to the area. This will be the critical value.

STEP 5   Determine the sign of the critical value for a one-tailed test.
       a. If the test is left-tailed, the critical value will be negative.
       b. If the test is right-tailed, the critical value will be positive.
       For a two-tailed test, one value will be positive and the other negative.

Example 9–2

Using Table E in Appendix C, find the critical value(s) for each situation and draw the appropriate figure, showing the critical region.

a. A left-tailed test with $\alpha = 0.10$.

b. A two-tailed test with $\alpha = 0.02$.

c. A right-tailed test with $\alpha = 0.005$.

Solution *a*

STEP 1   Draw the figure and indicate the appropriate area. Since this is a left-tailed test, the area of 0.10 is located in the left tail, as shown in Figure 9–10.

STEP 2   Subtract 0.10 from 0.5000 to get 0.4000.

STEP 3   In Table E, find the area that is closest to 0.4000; in this case, it is 0.3997.

STEP 4   Find the z value that corresponds to this area. It is 1.28.

STEP 5   Determine the sign of the critical value (i.e., the z value). Since this is a left-tailed test, the sign of the critical value is negative. Hence, the critical value is -1.28. See Figure 9–10.

Figure 9–10

Critical Value and Critical Region for Part *a* of Example 9–2

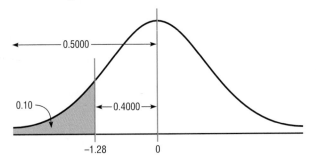

Solution *b*

STEP 1   Draw the figure and indicate the appropriate area. In this case, there are two areas equivalent to α/2, or 0.02/2 = 0.01.

STEP 2   Subtract 0.01 from 0.5000 to get 0.4900.

STEP 3   Find the area in Table E closest to 0.4900. In this case, it is 0.4901.

STEP 4   Find the z value that corresponds to this area. It is 2.33.

STEP 5   Determine the sign of the critical value. Since this test is a two-tailed test, there are two critical values: one is positive and the other is negative. They are $+2.33$ and -2.33. See Figure 9–11.

Figure 9–11

Critical Values and Critical Regions for Part b of Example 9–2

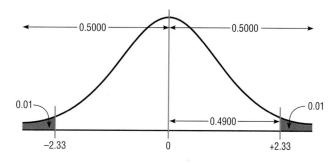

Solution c

STEP 1   Draw the figure and indicate the appropriate area. Since this is a right-tailed test, the area 0.005 is located in the right tail, as shown in Figure 9–12.

Figure 9–12

Critical Value and Critical Region for Part c of Example 9–2

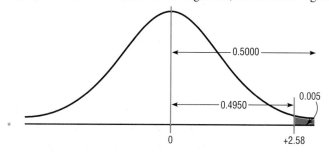

STEP 2   Subtract 0.005 from 0.5000 to get 0.4950.

STEP 3   Find the area in Table E closest to 0.4950. In this case it is 0.4951.

STEP 4   Find the z value that corresponds to this area. It is 2.58.

STEP 5   Determine the sign of the critical value. Since this is a right-tailed test, the sign is positive; hence, the critical value is $+2.58$.

Objective 4. State the five steps used in hypothesis testing.

In hypothesis testing, the following steps are recommended.

1. State the hypotheses. Be sure to state both the null and the alternative hypotheses.
2. Design the study. This step includes selecting the correct statistical test, choosing a level of significance, and formulating a plan to carry out the study. The plan should include information such as the definition of the population, the way the sample will be selected, and the methods that will be used to collect the data.
3. Conduct the study and collect the data.
4. Evaluate the data. The data should be tabulated in this step, and the statistical test should be conducted. Finally, decide whether to reject or not reject the null hypothesis.
5. Summarize the results.

For the purposes of this chapter, a simplified version of the hypothesis-testing procedure will be used, since designing the study and collecting the data will be omitted. The steps are summarized in the Procedure Table.

Procedure Table

Solving Hypothesis-Testing Problems (Traditional Method)

STEP 1   State the hypotheses, and identify the claim.

STEP 2   Find the critical value(s) from the appropriate table in Appendix C.

STEP 3   Compute the test value.

STEP 4   Make the decision to reject or not reject the null hypothesis.

STEP 5   Summarize the results.

Exercises

9–1. Define *null* and *alternative hypotheses,* and give an example of each.

9–2. What is meant by a type I error? A type II error? How are they related?

9–3. What is meant by a statistical test?

9–4. Explain the difference between a one-tailed and a two-tailed test.

9–5. What is meant by the critical region? The noncritical region?

9–6. What symbols are used to represent the null hypothesis and the alternative hypothesis?

9–7. What symbols are used to represent the probabilities of type I and type II errors?

9–8. Explain what is meant by a significant difference.

9–9. When should a one-tailed test be used? A two-tailed test?

9–10. List the steps in hypothesis testing.

9–11. In hypothesis testing, why can't the hypothesis be proved true?

9–12. (ans) Using the z table (Table E), find the critical value (or values) for each.
  a. $\alpha = 0.01$, two-tailed test
  b. $\alpha = 0.05$, right-tailed test
  c. $\alpha = 0.005$, left-tailed test
  d. $\alpha = 0.10$, left-tailed test
  e. $\alpha = 0.05$, two-tailed test
  f. $\alpha = 0.04$, right-tailed test
  g. $\alpha = 0.01$, left-tailed test
  h. $\alpha = 0.10$, two-tailed test
  i. $\alpha = 0.02$, right-tailed test
  j. $\alpha = 0.02$, two-tailed test

9–13. For each conjecture, state the null and alternative hypotheses.
  a. The average age of taxi drivers in New York City is 36.3 years.
  b. The average income of nurses is $36,250.
  c. The average age of disc jockeys is greater than 27.6 years.
  d. The average pulse rate of female joggers is less than 72 beats per minute.
  e. The average bowling score of people who enrolled in a basic bowling class is less than 100.
  f. The average cost of a VCR is $297.75.
  g. The average electric bill for residents of White Pine Estates exceeds $52.98 per month.
  h. The average number of calories of brand A's low-calorie meals is at most 300.
  i. The average weight loss of people who use brand A's low-calorie meals for six weeks is at least 3.6 pounds.

z Test for a Mean

Objective 5. Test means for large samples using the *z* test.

In this chapter, two statistical tests will be explained: the *z* test, used to test for the mean of a large sample, and the *t* test, used for the mean of a small sample. This section explains the *z* test, and Section 9–4 explains the *t* test.

Many hypotheses are tested using a statistical test based on the following general formula:

$$\text{test value} = \frac{(\text{observed value}) - (\text{expected value})}{\text{standard error}}$$

The observed value is the statistic (such as the mean) that is computed from the sample data. The expected value is the parameter (such as the mean) that one would expect to obtain if the null hypothesis were true—in other words, the hypothesized value. The denominator is the standard error of the statistic being tested (in this case, the standard error of the mean).

The *z* test is defined formally as follows.

The **z test** is a statistical test for the mean of a population. It can be used when $n \geq 30$, or when the population is normally distributed and σ is known.

The formula for the *z* test is

$$z = \frac{\bar{X} - \mu}{\sigma/\sqrt{n}}$$

where

$\bar{X}$ = sample mean
μ = hypothesized population mean
σ = population deviation
n = sample size

For the *z* test, the observed value is the value of the sample mean. The expected value is the value of the population mean, assuming that the null hypothesis is true. The denominator $\sigma/\sqrt{n}$ is the standard error of the mean.

The formula for the *z* test is the same formula shown in Chapter 7 for the situation where one is using a distribution of sample means. Recall that the central limit theorem allows one to use the standard normal distribution to approximate the distribution of sample means when $n \geq 30$. If σ is unknown, *s* can be used when $n \geq 30$.

Note: The student's first encounter with hypothesis testing can be somewhat challenging and confusing, since there are many new concepts being introduced at the same time. *In order to understand all the concepts, the student must carefully follow each step in the examples and try each exercise that is assigned.* Only after careful study and patience will these concepts become clear.

As stated in the previous section, there are five steps for solving *hypothesis-testing* problems:

STEP 1   State the hypotheses and identify the claim.

STEP 2   Find the critical value(s).

STEP 3   Compute the test value.

STEP 4   Make the decision to reject or not reject the null hypothesis.

STEP 5   Summarize the results.

In this study on margarine consumption, many conclusions are stated. State hypotheses that may have been used to test these conclusions. Define the population and the sample used. Do you think the sample would be representative of all adults? Explain your answer. Comment on the sample size.

Stick Margarine May Boost Risk of Heart Attacks

By Nancy Hellmich
USA TODAY

The bad news about margarine continues to mount.

Eating stick margarine increases the risk of heart attack, according to new findings from the Harvard Nurses' Health Study, an ongoing analysis of the diets of 90,000 nurses.

This new report adds to growing evidence that trans fatty acids—the fats that form when liquid vegetable oils are processed or hydrogenated—raise blood cholesterol in much the same way that saturated fat does.

Other culprits besides margarine: any solid vegetable shortening, some fried foods at chains such as McDonald's and Burger King and processed foods made with partially hydrogenated vegetable oils (check the ingredient list).

The new study reported in Saturday's *Lancet* shows:

• Women who frequently use margarine have more than a 50% higher risk of heart disease than those who infrequently use margarine.

• Those who eat a couple of cookies a day (cookies often contain partially hydrogenated oils) are at a 50% higher risk.

• Women who eat lots of foods high in trans fatty acids have a 70% higher risk of heart disease than those who don't.

In recent years, many people have switched to margarine from butter, which is high in saturated fat.

But don't go back to butter just because margarine is unhealthy, says Dr. Walter Willett, study author. Instead, consider switching to liquid oils instead of solid margarine or shortening.

He's a strong supporter of olive oil because "there's a strong indication that olive oil is at least safe." And it's a monounsaturated fat that lowers bad cholesterol (LDL) without lowering good cholesterol (HDL).

Also, some tub margarines contain more water and air and therefore less fat and fewer trans fatty acids than stick margarines.

Source: Copyright 1993, *USA TODAY.* Reprinted with permission.

The next example illustrates these steps.

Example 9–3

A researcher reports that the average salary of assistant professors is more than $42,000. A sample of 30 assistant professors has a mean salary of $43,260. At $\alpha = 0.05$, test the claim that assistant professors earn more than $42,000 a year. The standard deviation of the population is $5230.

Solution

STEP 1 State the hypotheses and identify the claim.

$$H_0: \mu \leq \$42,000 \quad \text{and} \quad H_1: \mu > \$42,000 \text{ (claim)}$$

STEP 2 Find the critical value. Since $\alpha = 0.05$ and the test is a right-tailed test, the critical value is $z = +1.65$.

STEP 3   Compute the test value.

$$z = \frac{\overline{X} - \mu}{\sigma/\sqrt{n}} = \frac{\$43,260 - 42,000}{5230/\sqrt{30}} = 1.32$$

STEP 4   Make the decision. Since the test value, +1.32, is less than the critical value, +1.65, and is not in the critical region, the decision is, "Do not reject the null hypothesis." This test is summarized in Figure 9–13.

Figure 9–13

Summary of the *z* Test of Example 9–3

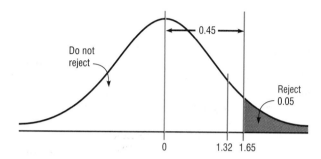

STEP 5   Summarize the results. There is not enough evidence to support the claim that assistant professors earn more on average than $42,000 a year.

Comment: Even though in Example 9–3 the sample mean, $43,260, is higher than the hypothesized population mean of $42,000, it is not *significantly* higher. Hence, the difference may be due to chance. When the null hypothesis is not rejected, there is still a probability of a type II error—i.e, of not rejecting the null hypothesis when it is false.

The probability of a type II error is not easily ascertained. Further explanation about the type II error is given in Section 9–7. For now, it is only necessary to realize that the probability of type II error exists when the decision is not to reject the null hypothesis.

It should also be noted that when the null hypothesis is not rejected, it cannot be accepted as true. There is merely not enough evidence to say that it is false. This guideline may sound a little confusing, but the situation is analogous to a jury trial. The verdict is either guilty or not guilty and is based on the evidence presented. If a person is judged not guilty, it does not mean that the person is proved innocent; it only means that there was not enough evidence to reach the guilty verdict.

Example 9–4

A researcher claims that the average cost of men's athletic shoes is less than $80. He selects a random sample of 36 pairs of shoes from a catalog and finds the following costs. (The costs have been rounded to the nearest dollar.) Is there enough evidence to support the researcher's claim at $\alpha = 0.10$?

$ 60	$70	$75	$55	$80	$55
50	40	80	70	50	95
120	90	75	85	80	60
110	65	80	85	85	45
75	60	90	90	60	95
110	85	45	90	70	70

Solution

STEP 1   State the hypotheses and identify the claim

$$H_0\!: \mu \geq \$80 \qquad \text{and} \qquad H_1\!: \mu < \$80 \text{ (claim)}$$

STEP 2 Find the critical value. Since $\alpha = 0.10$ and the test is a left tailed test, the critical value is -1.28

STEP 3 Compute the test value. Since the exercise gives raw data, it is necessary to find the mean and standard deviation of the data. Using the formulas in Chapter 3 or your calculator, $\overline{X} = 75.0$ and $s = 19.2$. Substitute in the formula

$$z = \frac{\overline{X} - \mu}{s/\sqrt{n}} = \frac{75 - 80}{19.2/\sqrt{36}} = -1.56$$

STEP 4 Make the decision. Since the test value, -1.56, falls in the critical region, the decision is to reject the null hypothesis. See Figure 9–14.

Figure 9–14

Critical and Test Values for Example 9–4

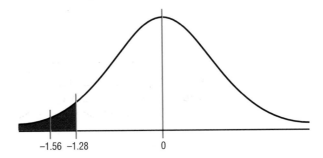

STEP 5 Summarize the results. There is enough evidence to support the claim that the average cost of men's athletic shoes is less than $80.

Comment: In Example 9–4, the difference is said to be significant. However, when the null hypothesis is rejected, there is always a chance of a type I error. In this case, the probability of a type I error is at most 0.10, or 10%.

Example 9–5

The Medical Rehabilitation Education Foundation reports that the average cost of rehabilitation for stroke victims is $24,672. To see if the average cost of rehabilitation is different at a particular hospital, a researcher selected a random sample of 35 stroke victims at the hospital and found that the average cost of their rehabilitation is $25,226. The standard deviation of the population is $3,251. At $\alpha = 0.01$, can it be concluded that the average cost of stroke rehabilitation at a particular hospital is different from $24,672?

Source: Snapshot, *USA Today,* September 18, 1995.

Solution

STEP 1 State the hypotheses and identify the claim.

$$H_0: \mu = \$24{,}672 \qquad \text{and} \qquad H_1: \mu \neq \$24{,}672 \text{ (claim)}$$

STEP 2 Find the critical values. Since $\alpha = 0.01$ and the test is a two-tailed test, the critical values are $+2.58$ and -2.58.

STEP 3 Compute the test value.

$$z = \frac{\overline{X} - \mu}{\sigma/\sqrt{n}} = \frac{25{,}226 - 24{,}672}{3{,}251/\sqrt{35}} = 1.01$$

STEP 4   Make the decision. Do not reject the null hypothesis, since the test value falls in the noncritical region, as shown in Figure 9–15.

Figure 9–15

Critical and Test Values
for Example 9–5

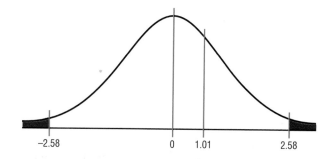

-2.58   0   1.01   2.58

STEP 5   Summarize the results. There is not enough evidence to support the claim that the average cost of rehabilitation at the particular hospital is different from $24,672.

As with confidence intervals, the central limit theorem states that when the population standard deviation σ is unknown, the sample standard deviation s can be used in the formula as long as the sample size is 30 or more. The formula for the z test in this case is

$$z = \frac{\overline{X} - \mu}{s/\sqrt{n}}$$

When n is less than 30 and σ is unknown, the t test must be used. The t test will be explained in the next section.

Students sometimes have difficulty summarizing the results of a hypothesis test. Figure 9–16 shows the four possible outcomes and the summary statement for each situation.

Figure 9–16

Outcomes of a
Hypothesis-Testing
Situation

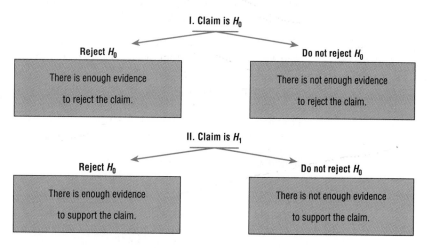

First of all, the claim can be either the null or alternative hypothesis, and one should identify which it is. Second, after the study is completed, the null hypothesis is either rejected or not rejected. From these two facts, the decision can be identified in the appropriate block of Figure 9–16.

For example, suppose a researcher claims that the mean weight of an adult animal of a particular species is 42 pounds. In this case, the claim would be the null hypothesis,

H_0: $\mu = 42$, since the researcher is asserting that the parameter is a specific value. If the null hypothesis is rejected, the conclusion would be that there is enough evidence to reject the claim that the mean weight of the adult animal is 42 pounds. See Figure 9–17(a).

Figure 9–17

Outcomes of a Hypothesis-Testing Situation for Two Specific Cases

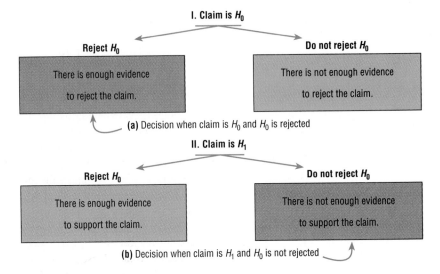

On the other hand, suppose the researcher claims that the mean weight of the adult animals is not 42 pounds. The claim would be the alternative hypothesis, H_1: $\mu \neq 42$. Furthermore, suppose that the null hypothesis is not rejected. The conclusion, then, would be that there is not enough evidence to support the claim that the mean weight of the adult animals is not 42 pounds. See Figure 9–17(b).

Again, remember that nothing is being proven true or false. The statistician is only stating that there is or is not enough evidence to say that a claim is *probably* true or false. As noted previously, the only way to prove something would be to use the entire population under study, and usually this cannot be done, especially when the population is large.

P-Value Method for Hypothesis Testing

Statisticians usually test hypotheses at the common α levels of 0.05 or 0.01 and sometimes at 0.10. Recall that the choice of the level depends on the seriousness of the type I error. Besides listing an α value, many computer statistical packages give a *P*-value for hypothesis tests.

The **P-value** (or probability value) is the probability of getting a sample statistic (such as the mean) or a more extreme sample statistic in the direction of the alternative hypothesis when the null hypothesis is true.

In other words, the *P*-value is the actual area under the standard normal distribution curve (or other curve, depending on what statistical test is being used) representing the probability of a particular sample statistic or a more extreme sample statistic occurring if the null hypothesis is true.

For example, suppose that a null hypothesis is H_0: $\mu \leq 50$ and the mean of a sample is $\overline{X} = 52$. If the computer printed a *P*-value of 0.0356 for a statistical test, then the probability of getting a sample mean of 52 or greater is 0.0356 if the true population mean is 50 (for the given sample size and standard deviation). The relationship between the *P*-value and the α value can be explained in this manner. For $P = 0.0356$, the null hypothesis would be rejected at $\alpha = 0.05$ but not at $\alpha = 0.01$. See Figure 9–18.

Figure 9–18

Comparison of α Values and *P*-Values

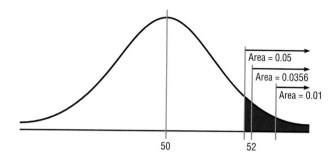

When the hypothesis test is two-tailed, the area in one tail must be doubled. For a two-tailed test, if α is 0.05 and the area in one tail is 0.0356, the *P*-value would be 2(0.0356) = 0.0712. That is, the null hypothesis should not be rejected at $\alpha = 0.05$, since 0.0712 is greater than 0.05. In summary, then, if the *P*-value falls in the critical region, reject the null hypothesis. If the *P*-value falls in the noncritical region, do not reject the null hypothesis.

The *P*-values for the *z* test can be found by using Table E in Appendix C. First find the area under the standard normal distribution curve corresponding to the *z* test value then subtract this area from 0.5000 to get the *P*-value for a right-tailed or a left-tailed test. To get the *P*-value for a two-tailed test, double this area after subtracting. This procedure is shown in step 3 of Examples 9–6 and 9–7.

The *P*-value method for testing hypotheses differs from the traditional method somewhat. The steps for the *P*-value method are summarized next.

Procedure Table

Solving Hypothesis-Testing Problems (*P*-Value Method)

STEP 1   State the hypotheses and identify the claim.

STEP 2   Compute the test value.

STEP 3   Find the *P*-value.

STEP 4   Make the decision.

STEP 5   Summarize the results.

The next two examples show how to use the *P*-value method to test hypotheses.

Example 9–6

A researcher wishes to test the claim that the average age of lifeguards in Ocean City is greater than 24 years. She selects a sample of 36 guards and finds the mean of the sample to be 24.7 years, with a standard deviation of 2 years. Is there evidence to support the claim at $\alpha = 0.05$? Find the *P*-value.

Solution

STEP 1   State the hypotheses and identify the claim.

$$H_0: \mu \leq 24 \quad \text{and} \quad H_1: \mu > 24 \text{ (claim)}$$

STEP 2   Compute the test value.

$$z = \frac{24.7 - 24}{2/\sqrt{36}} = 2.10$$

STEP 3   Find the *P*-value. Using Table E in Appendix C, find the corresponding area under the normal distribution for $z = 2.10$. It is 0.4821. Subtract this value for the area from 0.5000 to find the area in the right tail.

$$0.5000 - 0.4821 = 0.0179$$

Hence, the *P*-value is 0.0179.

STEP 4   Make the decision. Since the *P*-value is less than 0.05, the decision is to reject the null hypothesis. See Figure 9–19.

Figure 9–19

P-Value and α Value for Example 9–6

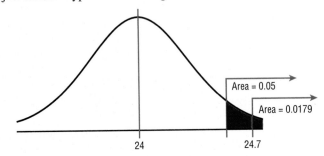

STEP 5   Summarize the results. There is enough evidence to support the claim that the average age of lifeguards in Ocean City is greater than 24 years.

Note: Had the researcher chosen $\alpha = 0.01$, the null hypothesis would not have been rejected, since the *P*-value (0.0179) is greater than 0.01.

Example 9–7

A researcher claims that the average wind speed in a certain city is 8 miles per hour. A sample of 32 days has an average wind speed of 8.2 miles per hour. The standard deviation of the sample is 0.6 mile per hour. At $\alpha = 0.05$, is there enough evidence to reject the claim? Use the *P*-value method.

Solution

STEP 1   State the hypotheses and identify the claim.

$$H_0\colon \mu = 8 \text{ (claim)} \qquad \text{and} \qquad H_1\colon \mu \neq 8$$

STEP 2   Compute the test value.

$$z = \frac{8.2 - 8}{0.6/\sqrt{32}} = 1.89$$

STEP 3   Find the *P*-value. Using Table E, find the corresponding area for $z = 1.89$. It is 0.4706. Subtract the value from 0.5000.

$$0.5000 - 0.4706 = 0.0294$$

Since this is a two-tailed test, the area 0.0294 must be doubled to get the *P*-value.

$$2 (0.0294) = 0.0588$$

STEP 4   Make the decision. The decision is not to reject the null hypothesis, since the *P*-value is greater than 0.05. See Figure 9–20.

Figure 9–20

P-Values and α Values for
Example 9–7

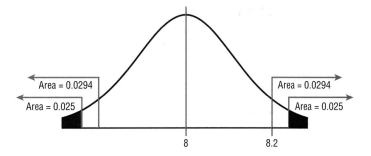

Area = 0.0294

Area = 0.025

Area = 0.0294

Area = 0.025

8   8.2

STEP 5   Summarize the results. There is not enough evidence to reject the claim that
the average wind speed is 8 miles per hour.

In Examples 9–6 and 9–7, the *P*-value and the α value were shown on the normal
distribution curve in order to illustrate the relationship between the two values; however,
it is not necessary to draw the normal distribution curve in order to make the decision
whether or not to reject the null hypothesis. One can use the following rule:

Decision Rule When Using a *P*-Value

If *P*-value $\leq \alpha$, reject the null hypothesis.
If *P*-value $> \alpha$, do not reject the null hypothesis.

In Example 9–6, *P*-value = 0.0179 and $\alpha = 0.05$. Since *P*-value $\leq \alpha$, the null hypothesis was rejected. In Example 9–7, *P*-value = 0.0588 and $\alpha = 0.05$. Since *P*-value
$> \alpha$ the null hypothesis was not rejected.

The *P*-values given on calculators and computers are slightly different from those
found with Table E. This is due to the fact that *z* values and the values in Table E have
been rounded. *Also, most calculators and computers give the exact P-value for two-tailed
tests, so it should not be doubled (as it should when the area found in Table E is used).*

A clear distinction between the α value and the *P*-value should be made. The α
value is chosen by the researcher before the statistical test is conducted. The *P*-value is
computed after the sample mean has been found.

There are two schools of thought on *P*-values. Some researchers do not choose an
α value but report the *P*-value and allow the reader to decide whether the null hypothesis should be rejected.

In this case, the following guidelines can be used, but be advised that these guidelines are not written in stone, and some statisticians may have other opinions.

If *P*-value ≤ 0.01, reject the null hypothesis. The difference is highly significant.

If *P*-value > 0.01 but *P*-value ≤ 0.05, reject the null hypothesis. The difference is
significant.

If *P*-value > 0.05 but *P*-value ≤ 0.10, consider the consequences of type I error before rejecting the null hypothesis.

If *P*-value > 0.10, do not reject the null hypothesis. The difference is not significant.

Others decide on the α value in advance and use the *P*-value to make the decision,
as shown in the previous two examples. A note of caution is needed here: If a researcher
selects $\alpha = 0.01$ and the *P*-value is 0.03, the researcher may decide to change the α value
from 0.01 to 0.05 so that the null hypothesis will be rejected. This, of course, should not
be done. If the α level is selected in advance, it should be used when making the decision.

One additional note on hypothesis testing is that the researcher should distinguish between *statistical significance* and *practical significance*. When the null hypothesis is rejected at a specific significance level, it can be concluded that the difference is probably not due to chance and thus is statistically significant. However, the results may not have any practical significance. For example, suppose that a new fuel additive increases the miles per gallon a car can get by $\frac{1}{4}$ mile for a sample of 1000 automobiles. The results may be statistically significant at the 0.05 level, but it would hardly be worthwhile to market the product for such a small increase. Hence, there is no practical significance to the results. It is up to the researcher to use common sense when interpreting the results of a statistical test.

Exercises

For Exercises 9–14 through 9–27, perform each of the following steps.
a. State the hypotheses and identify the claim.
b. Find the critical value(s).
c. Compute the test value.
d. Make the decision.
e. Summarize the results.

Use diagrams to show the critical region (or regions), and use the traditional method of hypothesis testing unless otherwise specified.

9–14. A researcher feels that the average budget of major city public libraries is $25 million. A random sample of 40 libraries is selected and their annual budgets (in millions of dollars) are shown. At $\alpha = 0.05$, is there enough evidence to reject the researcher's claim?

16.7	17.6	26.5	6.3	16.5
11.9	23.7	14.3	94.0	4.7
11.6	26.5	5.6	58.6	3.2
14.2	3.5	10.9	11.8	15.2
30.1	19.7	11.7	38.8	36.3
4.8	7.9	14.2	18.0	24.5
69.2	8.5	19.2	5.0	15.3
41.0	27.1	10.3	3.7	13.6

Source: *Time Almanac*, 1999, p. 711.

9–15. A survey claims that the average cost of a hotel room in Atlanta is $69.21. In order to test the claim, a researcher selects a sample of 30 hotel rooms and finds that the average cost is $68.43. The standard deviation of the population is $3.72. At $\alpha = 0.05$, is there enough evidence to reject the claim?

Source: *USA Today*, August 11, 1995.

9–16. The manager of a large factory believes that the average hourly wage of the employees is below $9.78 per hour. A sample of 18 employees has a mean hourly wage of $9.60. The standard deviation of all salaries is $1.42.

Assume the variable is normally distributed. At $\alpha = 0.10$, is there enough evidence to support the manager's claim?

9–17. A researcher estimates that the average revenue of the largest businesses in the United States is greater than $24 billion. A sample of 50 companies is selected and the revenues (in billions of dollars) are shown. At $\alpha = 0.05$, is there enough evidence to support the researcher's claim?

178	122	91	44	35
61	56	46	20	32
30	28	28	20	27
29	16	16	19	15
41	38	36	15	25
31	30	19	19	19
24	16	15	15	19
25	25	18	14	15
24	23	17	17	22
22	21	20	17	20

Source: *Time Almanac 1999*, pp. 825–826.

9–18. A maker of frozen meals claims that the average caloric content of its meals is 800, and the standard deviation is 25. A researcher tested 12 meals and found that the average number of calories was 873. Is there enough evidence to reject the claim at $\alpha = 0.02$? Assume the variable is normally distributed.

9–19. A report in *USA Today* stated that the average age of commercial jets in the United States is 14 years. An executive of a large airline company selects a sample of 36 planes and finds the average age of the planes is 11.8 years. The standard deviation of the sample is 2.7 years. At $\alpha = 0.01$, can it be concluded that the average age of the planes in his company is less than the national average?

Source: *USA Today*, July 7, 1995.

9–20. A diet clinic states that there is an average loss of 24 pounds for patients who stay on the program for 20 weeks. The standard deviation is 5 pounds. The clinic tries a new diet, reducing the salt intake to see whether that strategy will produce a greater weight loss. A group of 40 volunteers loses an average of 16.3 pounds each over 20 weeks. Should the clinic change to the new diet? Use $\alpha = 0.05$.

9–21. A statistician claims that the average age of people who purchase lottery tickets is 70. A sample of 30 is selected, and their ages are recorded. At $\alpha = 0.05$ is there enough evidence to reject the statistician's claim?

49	80	24	61	79	68
63	72	46	65	76	71
90	56	70	71	71	67
52	82	74	39	49	69
22	56	70	74	62	45

9–22. A survey found that women over the age of 55 consume an average of 1660 calories a day. In order to see if the number of calories consumed by women over age 55 in assisted-living residences is the same, a researcher sampled 43 women over the age of 55 in a large assisted-living facility and found the mean number of calories consumed was 1446. The standard deviation of the sample is 56 calories. At $\alpha = 0.10$, test the claim that there is no difference between the number of calories consumed by the residents and that consumed by other women over 55.

9–23. A manufacturer states that the average lifetime of its lightbulbs is 3 years, or 36 months. The standard deviation is 8 months. Fifty bulbs are selected, and the average lifetime is found to be 32 months. Should the manufacturer's statement be rejected at $\alpha = 0.01$?

9–24. A real estate agent claims that the average price of a home sold in Beaver County, Pennsylvania, is $60,000. A random sample of 36 homes sold in the county is selected and the prices are shown. Is there enough evidence to reject the agent's claim at $\alpha = 0.05$?

$ 9,500	$ 54,000	$ 99,000	$ 94,000	$ 80,000
29,000	121,500	184,750	15,000	164,450
6,000	13,000	188,400	121,000	308,000
42,000	7,500	32,900	126,900	25,225
95,000	92,000	38,000	60,000	211,000
15,000	28,000	53,500	27,000	21,000
76,000	85,000	25,225	40,000	97,000
284,000				

Source: *Pittsburgh Tribune Review,* June 13, 1999.

9–25. The average serum cholesterol level in a certain group of patients is 240 milligrams. The standard deviation is 18 milligrams. A new medication is designed to lower the cholesterol level if taken for one month. A sample of 40

people used the medication for 30 days, after which their average cholesterol level was 229 milligrams. At $\alpha = 0.01$, does the medication lower the cholesterol level of the patients?

9–26. The state's education secretary claims that the average cost of one year's tuition for all private high schools in the state is $2350. A sample of 30 private high schools is selected, and the average tuition is $2315. The standard deviation for the population is $38. At $\alpha = 0.05$, is there enough evidence to reject the claim that the average cost of tuition is equal to $2350?

9–27. To see if young men ages 8–17 spend more or less than the national average of $24.44 per shopping trip to a local mall, the manager surveyed 33 young men and found the average amount spent per visit was $22.97. The standard deviation of the sample was $3.70. At $\alpha = 0.02$, can it be concluded that the average amount spent at a local mall is not equal to the national average of $24.44?

Source: *USA Today,* July 25, 1995.

9–28. What is meant by a *P*-value?

9–29. State whether or not the null hypothesis should be rejected on the basis of the given *P*-value.
  a. *P*-value = 0.258, $\alpha = 0.05$, one-tailed test
  b. *P*-value = 0.0684, $\alpha = 0.10$, two-tailed test
  c. *P*-value = 0.0153, $\alpha = 0.01$, one-tailed test
  d. *P*-value = 0.0232, $\alpha = 0.05$, two-tailed test
  e. *P*-value = 0.002, $\alpha = 0.01$, one-tailed test

9–30. A college professor claims that the average cost of a paperback textbook is greater than $27.50. A sample of 50 books has an average cost of $29.30. The standard deviation of the sample is $5.00. Find the *P*-value for the test. On the basis of the *P*-value, should the null hypothesis be rejected at $\alpha = 0.05$?

9–31. A study found that the average stopping distance of a school bus traveling 50 miles per hour was 264 feet (Snapshot, *USA Today,* March 12, 1992). A group of automotive engineers decided to conduct a study of its school buses and found that for 20 buses, the average stopping distance of buses traveling 50 miles per hour was 262.3 feet. The standard deviation of the population was 3 feet. Test the claim that the average stopping distance of the company's buses is actually less than 264 feet. Find the *P*-value. On the basis of the *P*-value, should the null hypothesis be rejected at $\alpha = 0.01$? Assume that the variable is normally distributed.

9–32. A store manager hypothesizes that the average number of pages a person copies on the store's copy machine is less than 40. A sample of 50 customers' orders is selected. At $\alpha = 0.01$ is there enough evidence to support the claim? Use the *P*-value hypothesis testing method.

2	2	2	5	32
5	29	8	2	49
21	1	24	72	70
21	85	61	8	42
3	15	27	113	36
37	5	3	58	82
9	2	1	6	9
80	9	51	2	122
21	49	36	43	61
3	17	17	4	1

9–33. A manufacturer states that the average lifetime of its television sets is more than 84 months. The standard deviation of the population is 10 months. One hundred sets are randomly selected and tested. The average lifetime of the sample is 85.1 months. Test the claim that the average lifetime of the sets is more than 84 months, and find the P-value. On the basis of the P-value, should the null hypothesis be rejected at $\alpha = 0.01$?

9–34. A special cable has a breaking strength of 800 pounds. The standard deviation of the population is 12 pounds. A researcher selects a sample of 20 cables and finds that the average breaking strength is 793 pounds. Can one reject the claim that the breaking strength is 800 pounds? Find the P-value. Should the null hypothesis be rejected at $\alpha = 0.01$? Assume that the variable is normally distributed.

9–35. The average hourly wage last year for members of the hospital clerical staff in a large city was $6.32. The standard deviation of the population was $0.54. This year a sample of 50 workers had an average hourly wage of $6.51. Test the claim, at $\alpha = 0.05$, that the average has not changed by finding the P-value for the test.

9–36. Ten years ago, the average acreage of farms in a certain geographic region was 65 acres. The standard deviation of the population was 7 acres. A recent study consisting of 22 farms showed that the average was 63.2 acres per farm. Test the claim, at $\alpha = 0.10$, that the average has not changed by finding the P-value for the test. Assume that σ has not changed and the variable is normally distributed.

9–37. A car dealer recommends that transmissions should be serviced at 30,000 miles. In order to see whether her customers are adhering to this recommendation, the dealer selects a sample of 40 customers and finds that the average mileage of the automobiles serviced is 30,456. The standard deviation of the sample is 1684 miles. By finding the P-value, determine whether the owners are having their transmissions serviced at 30,000 miles for $\alpha = 0.10$. Do you think the α value of 0.10 is an appropriate significance level?

9–38. A motorist claims that the South Boro Police issue an average of 60 speeding tickets per day. The following data show the number of speeding tickets issued each day for a period of one month. Assume σ is 13.42. Is there enough evidence to reject the motorist's claim at $\alpha = 0.05$? Use the P-value method.

72	45	36	68	69	71	57	60
83	26	60	72	58	87	48	59
60	56	64	68	42	57	57	
58	63	49	73	75	42	63	

9–39. A manager states that in his factory, the average number of days per year missed by the employees due to illness is less than the national average of 10. The following data show the number of days missed by 40 employees last year. Is there sufficient evidence to believe the manager's statement at $\alpha = 0.05$? (Use s to estimate σ.) Use the P-value method.

0	6	12	3	3	5	4	1
3	9	6	0	7	6	3	4
7	4	7	1	0	8	12	3
2	5	10	5	15	3	2	5
3	11	8	2	2	4	1	9

*9–40. Suppose a statistician chose to test a hypothesis at $\alpha = 0.01$. The critical value for a right-tailed test is $+2.33$. If the test value was 1.97, what would the decision be? What would happen if, after seeing the test value, he decided to choose $\alpha = 0.05$? What would the decision be? Explain the contradiction, if there is one.

*9–41. The president of a company states that the average hourly wage of her employees is $8.65. A sample of 50 employees has the distribution shown. At $\alpha = 0.05$, is the president's statement believable? (Use s to approximate σ.)

Class	Frequency
8.35–8.43	2
8.44–8.52	6
8.53–8.61	12
8.62–8.70	18
8.71–8.79	10
8.80–8.88	2

MINITAB
Step by Step

Hypothesis Test for the Mean, *z* Distribution

These instructions show how MINITAB can be used to calculate the test statistic and its *P*-value. Example 9–4 is used as an example. Since the *P*-value approach is used, we do not require a critical value from the table.

Example MT9–1

1. Enter the data from Example 9–4 on shoe costs into a column of MINITAB. Name the column **ShoeCost.**

2. In this example sigma is unknown. The standard deviation for the sample will be calculated and used as an estimate for the population standard deviation.
   a. Select **Calc>Column Statistics.**
   b. Check the button for Standard deviation. You can only do one of these at a time.
   c. Use ShoeCost for the Input variable.
   d. Store the result in s. It will also be displayed in the session window: s = 19.161.
   e. Click [OK].

Calc>Column Statistics

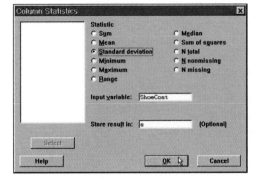

To calculate the test statistic and *P*-value:

3. Select **Stat>Basic Statistics>1 Sample Z.**

4. Choose the ShoeCost variable.

5. Click in the text box for Sigma. Type **s** (since *s* equals the sample standard deviation, per step 2 above).

6. Click the button for Test mean and enter the hypothesized value of 80 in the text box.

7. Click on [Options]. Change the Alternative to less than (in Release 12 change the Alternative in the 1-Sample Z dialogue box, without clicking Options). This setting is crucial for calculating the *P*-value.

8. Click [OK].

1-Sample z Dialog Box

Since the P-value of 0.059 is less than α, reject the null hypothesis. Summarize: There is enough evidence in the sample to conclude the mean cost is less than \$80. Since the "less than" alternative was used, the P-value indicates the probability to the left of the test statistic (-1.57).

TI-83
Step by Step

For most hypothesis tests and confidence intervals, the calculator will accept raw data in lists denoted by Data or summary statistics denoted by Stats. In hypothesis testing, the calculator gives P-values instead of critical values.

Hypothesis Test for the Mean, z Distribution (Data)

1. Enter the data into L_1.
2. Press **STAT** and move the cursor to TESTS.
3. Press l for z-Test.
4. Select Data and press **ENTER** then press ▼.
5. Type in the value for μ.
6. Type in the value for σ. Make sure L_1 is selected and Freq is 1.
7. Select the correct alternative hypothesis—either $\mu = \mu_0$, $\mu < \mu_0$ or $\mu > \mu_0$, where μ_0 represents the hypothesized mean.
8. Select Calculate and press **ENTER**.

Example TI9–1

Test the claim from Example 9–4, H_1: $\mu < 80$ for the following data:

60	70	75	55	80	55
50	40	80	70	50	95
120	90	75	85	80	60
110	65	80	85	85	45
75	60	90	90	60	95
110	85	45	90	70	70

Use $\alpha = 0.10$.

First run 1-Var Stats to find $s_x = 19.16097224$ and use this value as an approximation for σ since $n \geq 30$.
Press **VARS** to access the Vars menu.
Press **5** for Statistics.
Press **3** for s_x.
The alternative hypothesis is $\mu < 80$. The test value is -1.565682556. The P-value is 0.0587114841. The decision is to reject, H_0 since $0.0587114841 < 0.10$.

Input

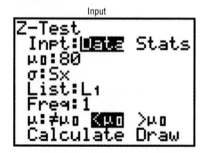

Output

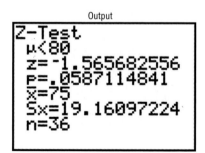

Hypothesis Test for the Mean, *z* Distribution (Stats)

1. Select `Stat`. Move the cursor to `TESTS` and press **1** for `z-Test`.

2. Select Stats and press **ENTER**, then press ▼.

3. Type the values for μ, σ, $\overline{X}$, and n.

4. Select the correct alternative hypothesis.

5. Select `Calculate` and press **ENTER**.

Example TI9–2

Test the claim from Example 9–7, H_0: $\mu = 8$ when $s = 0.6$, $\overline{X} = 8.2$, and $n = 32$. Use $\alpha = 0.05$. Use $s = 0.6$ to approximate σ since $n \geq 30$.

The alternative hypothesis is H_1: $\mu \neq 8$. The *z* test value is 1.885618083. The *P*-value is 0.0593463074. The decision is not to reject H_0 since $0.0593463074 > 0.05$.

<table>
<tr><td>Input</td><td>Output</td></tr>
<tr><td>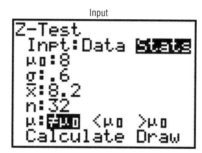</td><td>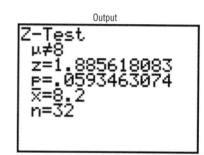</td></tr>
</table>

9–4

t Test for a Mean

Objective 6. Test means for small samples using the *t* test.

When the population standard deviation is unknown and the sample size is less than 30, the *z* test is inappropriate for testing hypotheses involving means. A different test, called the *t* test, is used. The *t* test is used when σ is unknown and $n < 30$. (Some authors use $n \leq 30$.)

As stated in Chapter 8, the *t* distribution is similar to the standard normal distribution in the following ways.

1. It is bell-shaped.

2. It is symmetrical about the mean.

3. The mean, median, and mode are equal to 0 and are located at the center of the distribution.

4. The curve never touches the *x* axis.

The *t* distribution differs from the standard normal distribution in the following ways.

1. The variance is greater than 1.

2. The *t* distribution is a family of curves based on the *degrees of freedom,* which is a number related to sample size. (Recall that the symbol for degrees of freedom is d.f. See Section 8–3 for an explanation of degrees of freedom.)

3. As the sample size increases, the *t* distribution approaches the normal distribution.

The t test is defined next.

The **t test** is a statistical test for the mean of a population and is used when the population is normally or approximately normally distributed, σ is unknown, and $n < 30$.
  The formula for the t test is

$$t = \frac{\bar{X} - \mu}{s/\sqrt{n}}$$

The degrees of freedom are d.f. $= n - 1$.

The formula for the t test is similar to the formula for the z test. But since the population standard deviation σ is unknown, the sample standard deviation s is used instead.

The critical values for the t test are given in Table F in Appendix C. For a one-tailed test, find the α level by looking at the top row of the table and finding the appropriate column. Find the degrees of freedom by looking down the left-hand column. Notice that degrees of freedom are given for values from 1 through 28. When the degrees of freedom are 29 or more, the row with ∞ (infinity) is used. Note that the values in this row are the same as the values for the z distribution, since as the sample size increases, the t distribution approaches the z distribution. When the sample size is 30 or more, statisticians generally agree that the two distributions can be considered identical, since the difference between their values is relatively small.

Example 9–8

Find the critical t value for $\alpha = 0.05$ with d.f. $= 16$ for a right-tailed t test.

Solution

Find the 0.05 column in the top row and 16 in the left-hand column. Where the row and column meet, the appropriate critical value is found; it is $+1.746$. See Figure 9–21.

Figure 9–21

Finding the Critical Value for the t Test in Table F (Example 9–8)

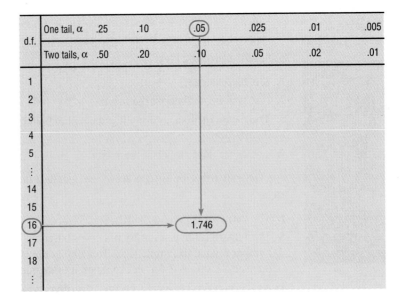

Example 9–9

Find the critical *t* value for $\alpha = 0.01$ with d.f. = 22 for a left-tailed test.

Solution

Find the 0.01 column in the row labeled "One tail" and find 22 in the left column. The critical value is -2.508 since the test is a one-tailed left test.

Example 9–10

Find the critical values for $\alpha = 0.10$ with d.f. = 18 for a two-tailed *t* test.

Solution

Find the 0.10 column in the row labeled "Two tails" and find 18 in the column labeled "d.f." The critical values are $+1.734$ and -1.734.

Example 9–11

Find the critical value for $\alpha = 0.05$ with d.f. = 28 for a right-tailed *t* test.

Solution

Find the 0.05 column in the "One tail" row and 28 in the left column. The critical value is $+1.701$.

When testing hypotheses by using the *t* test (traditional method), follow the same procedure as for the *z* test, except use Table F.

STEP 1   State the hypotheses and identify the claim.

STEP 2   Find the critical value(s) from Table F.

STEP 3   Compute the test value.

STEP 4   Make the decision to reject or not reject the null hypothesis.

STEP 5   Summarize the results.

Remember that the t test should be used when the population is approximately normally distributed, the population standard deviation is unknown, and the sample size is less than 30.

The next three examples illustrate the application of the *t* test.

Example 9–12

A job placement director claims that the average starting salary for nurses is \$24,000. A sample of 10 nurses has a mean of \$23,450 and a standard deviation of \$400. Is there enough evidence to reject the director's claim at $\alpha = 0.05$?

Solution

STEP 1 H_0: $\mu = \$24,000$ (claim) and H_1: $\mu \neq \$24,000$.

STEP 2   The critical values are $+2.262$ and -2.262 for $\alpha = 0.05$ and d.f. = 9.

STEP 3   The test value is

$$t = \frac{\overline{X} - \mu}{s/\sqrt{n}} = \frac{23,450 - 24,000}{400/\sqrt{10}} = -4.35$$

STEP 4 Reject the null hypothesis, since $-4.35 < -2.262$, as shown in Figure 9–22.

Figure 9–22

Summary of the t Test of Example 9–12

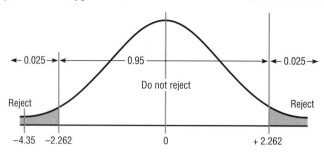

STEP 5 There is enough evidence to reject the claim that the starting salary of nurses is $24,000.

Example 9–13

An educator claims that the average salary of substitute teachers in school districts in Allegheny County, Pennsylvania, is less than $60 per day. A random sample of 8 school districts is selected, and the daily salaries are shown. Is there enough evidence to support the educator's claim at $\alpha = 0.10$?

$$\$60 \quad \$56 \quad \$60 \quad \$55 \quad \$70 \quad \$55 \quad \$60 \quad \$55$$

Source: The *Pittsburgh Tribune Review,* November 30, 1997.

Solution

STEP 1 $H_0: \mu \geq \$60$ and $H_1: \mu < \$60$ (claim).

STEP 2 At $\alpha = 0.10$ and d.f. $= 7$, the critical value is -1.415.

STEP 3 In order to compute the test value, the mean and standard deviation must be found. Using either the formulas in Chapter 3 or your calculator, $\bar{X} = \$58.88$, and $s = 5.08$:

$$t = \frac{\bar{X} - \mu}{s/\sqrt{n}} = \frac{58.88 - 60}{5.08/\sqrt{8}} = -0.624$$

STEP 4 Do not reject the null hypothesis since -0.624 falls in the noncritical region. See Figure 9–23.

Figure 9–23

Critical Value and Test Value for Example 9–13

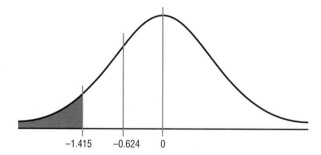

STEP 5 There is not enough evidence to support the educator's claim that the average salary of substitute teachers in Allegheny County is less than $60 per day.

The *P*-values for the *t* test can be found by using Table F; however, specific *P*-values for *t* tests cannot be obtained from the table since only selected values of α (e.g., 0.01, 0.05) are given. To find specific *P*-values for *t* tests, one would need a table similar to Table E for each degree of freedom. Since this is not practical, only *intervals* can be found for *P*-values. The next three examples show how to use Table F to determine intervals for *P*-values for the *t* test.

Example 9–14

Find the *P*-value when the *t* test value is 2.056, the sample size is 11, and the test is right-tailed.

Solution

To get the *P*-value, look across the row with 10 degrees of freedom (d.f. $= n - 1$) in Table F and find the two values that 2.056 falls between. They are 1.812 and 2.228. Since this is a right-tailed test, look up to the row labeled, "One Tail α" and find the two α values corresponding to 1.812 and 2.228. They are 0.05 and 0.025. See Figure 9–24. Hence, the *P*-value would be contained in the interval $0.025 < P\text{-value} < 0.05$. This means that the *P*-value is between 0.025 and 0.05. If α were 0.05, one would reject the null hypothesis since the *P*-value is less than 0.05. But if α were 0.01, one would not reject the null hypothesis since the *P*-value is greater than 0.01. (Actually, it is greater than 0.025.)

Figure 9–24

Finding the *P*-Value for Example 9–14

Confidence intervals		50%	80%	90%	95%	98%	99%
One tail, α		0.25	0.10	0.05	0.025	0.01	0.005
d.f.	Two tails, α	0.50	0.20	0.10	0.05	0.02	0.01
1		1.000	3.078	6.314	12.706	31.821	63.657
2		.816	1.886	2.920	4.303	6.965	9.925
3		.765	1.638	2.353	3.182	4.541	5.841
4		.741	1.533	2.132	2.776	3.747	4.604
5		.727	1.476	2.015	2.571	3.365	4.032
6		.718	1.440	1.943	2.447	3.143	3.707
7		.711	1.415	1.895	2.365	2.998	3.499
8		.706	1.397	1.860	2.306	2.896	3.355
9		.703	1.383	1.833	2.262	2.821	3.250
10		.700	1.372	1.812	2.228	2.764	3.169
11		.697	1.363	1.796	2.201	2.718	3.106
12		.695	1.356	1.782	2.179	2.681	3.055
13		.694	1.350	1.771	2.160	2.650	3.012
14		.692	1.345	1.761	2.145	2.624	2.977
15		.691	1.341	1.753	2.131	2.602	2.947
$\vdots$		$\vdots$	$\vdots$	$\vdots$	$\vdots$	$\vdots$	$\vdots$
(*z*) ∞		.674	1.282	1.645	1.960	2.326	2.576

*2.056 falls between 1.812 and 2.228

Example 9–15

Find the *P*-value when the *t* test value is 2.983, the sample size is 6, and the test is two-tailed.

Solution

To get the P-value, look across the row with d.f. $= 5$ and find the two values that 2.983 falls between. They are 2.571 and 3.365. Then look up to the row labeled "Two tails α" to find the corresponding α values.

In this case, they are 0.05 and 0.02. Hence the P-value is contained in the interval 0.02 $< P$-value < 0.05. This means that the P-value is between 0.02 and 0.05. In this case, if α $= 0.05$, the null hypothesis can be rejected since P-value < 0.05, but if $\alpha = 0.01$, the null hypothesis cannot be rejected since P-value > 0.01 (actually P-value > 0.02).

Note: **Since many students will be using calculators or computer programs that give the specific P-value for the t test and other tests presented later in this textbook, these specific values, in addition to the intervals, will be given for the answers to the examples and exercises.**

The P-value obtained from a calculator for Example 9–14 is 0.033. The P-value obtained from a calculator for Example 9–15 is 0.031.

To test hypotheses using the P-value method, follow the same steps as explained in Section 9–3. These steps are repeated here.

STEP 1 State the hypotheses and identify the claim.

STEP 2 Compute the test value.

STEP 3 Find the P-value.

STEP 4 Make the decision.

STEP 5 Summarize the results.

This method is shown in the next example.

Example 9–16

A physician claims that joggers' maximal volume oxygen uptake is greater than the average of all adults. A sample of 15 joggers has a mean of 40.6 milliliters per kilogram (ml/kg) and a standard deviation of 6 ml/kg. If the average of all adults is 36.7 ml/kg, is there enough evidence to support the physician's claim at $\alpha = 0.05$?

Solution

STEP 1 State the hypotheses and identify the claim.

$$H_0: \mu \leq 36.7 \text{ and } H_1: \mu > 36.7 \text{ (claim)}$$

STEP 2 Compute the test value. The test value is

$$t = \frac{\bar{X} - \mu}{s/\sqrt{n}} = \frac{40.6 - 36.7}{6/\sqrt{15}} = 2.517$$

STEP 3 Find the P-value. Looking across the row with d.f. $= 14$ in Table F, 2.517 falls between 2.145 and 2.624 corresponding to $\alpha = 0.025$ and $\alpha = 0.01$ since this is a right-tailed test. Hence, P-value > 0.01 and P-value < 0.025 or $0.01 < P$-value < 0.025. That is, the P-value is somewhere between 0.01 and 0.025. (The P-value obtained from a calculator is 0.012.)

STEP 4 Reject the null hypothesis since *P*-value < 0.05 (i.e., *P*-value $< \alpha$).

STEP 5 There is enough evidence to support the claim that the joggers' maximal volume oxygen uptake is greater than 36.7 ml/kg.

Students sometimes have difficulty deciding whether to use the *z* test or *t* test. The rules are the same as those pertaining to confidence intervals.

1. If σ is known, use the *z* test. The variable must be normally distributed if $n < 30$.
2. If σ is unknown but $n \geq 30$, use the *z* test and use *s* in place of σ in the formula.
3. If σ is unknown and $n < 30$, use the *t* test. (The population must be approximately normally distributed.)

These rules are summarized in Figure 9–25.

Figure 9–25

Using the *z* or *t* Test

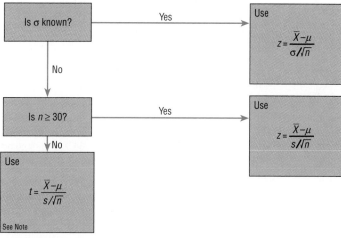

Note: With d.f. = $n - 1$, and the population must be approximately normally distributed.

Exercises

9–42. In what ways is the *t* distribution similar to the standard normal distribution? In what ways is the *t* distribution different from the standard normal distribution?

9–43. What are the degrees of freedom for the *t* test?

9–44. How does the formula for the *t* test differ from the formula for the *z* test?

9–45. Find the critical value (or values) for the *t* test for each.

a. $n = 10$, $\alpha = 0.05$, right-tailed
b. $n = 18$, $\alpha = 0.10$, two-tailed
c. $n = 6$, $\alpha = 0.01$, left-tailed
d. $n = 9$, $\alpha = 0.025$, right-tailed
e. $n = 15$, $\alpha = 0.05$, two-tailed
f. $n = 23$, $\alpha = 0.005$, left-tailed
g. $n = 28$, $\alpha = 0.01$, two-tailed
h. $n = 17$, $\alpha = 0.02$, two-tailed

9–46. (ans) Using Table F, find the *P*-value interval for each test value.

a. $t = 2.321$, $n = 15$, right-tailed
b. $t = 1.945$, $n = 28$, two-tailed
c. $t = -1.267$, $n = 8$, left-tailed
d. $t = 1.562$, $n = 17$, two-tailed
e. $t = 3.025$, $n = 24$, right-tailed
f. $t = -1.145$, $n = 5$, left-tailed
g. $t = 2.179$, $n = 13$, two-tailed
h. $t = 0.665$, $n = 10$, right-tailed

For Exercises 9–47 through 9–60, perform each of the following steps.

a. State the hypotheses and identify the claim.
b. Find the critical value(s).
c. Find the test value.
d. Make the decision.
e. Summarize the results.

Use the traditional method of hypothesis testing unless otherwise specified.

Assume that the population is approximately normally distributed.

9–47. The average amount of rainfall during the summer months for the northeast part of the United States is 11.52 inches. A researcher selects a random sample of 10 cities in the northeast and finds that the average amount of rainfall for 1995 was 7.42 inches. The standard deviation of the sample is 1.3 inches. At $\alpha = 0.05$, can it be concluded that for 1995 the mean rainfall was below 11.52 inches?

Source: Based on information in *USA Today,* September 14, 1995.

9–48. A state executive claims that the average number of acres in Western Pennsylvania State Parks is less than 2000 acres. A random sample of five parks is selected and the number of acres is shown. At $\alpha = 0.01$ is there enough evidence to support the claim?

959	1187	493	6249	541

Source: *Pittsburgh Tribune Review,* May 25, 1997.

9–49. A rental agency claims that the average rent that small-business establishments pay in Eagle City is $800. A sample of 10 establishments shows an average rental rate of $863. The standard deviation of the sample is $20. At $\alpha = 0.05$, is there evidence to reject the agency's claim?

9–50. A large hospital instituted a fitness program to reduce absenteeism. The director reported that the average number of working hours lost due to illness per employee is 48 hours per year. After one year, a sample of 18 employees showed an average of 41 hours of work lost, with a standard deviation of 5. Did the program reduce absenteeism? Use $\alpha = 0.10$.

9–51. A researcher estimates that the average height of the buildings of 30 or more stories in a large city is at least 700 feet. A random sample of 10 buildings is selected and the heights in feet are shown. At $\alpha = 0.025$ is there enough evidence to reject the claim?

485	511	841	725	615
520	535	635	616	582

Source: *Pittsburgh Tribune Review,* January 27, 1997.

9–52. Cushman and Wakefield reported that the average annual rent for office space in Tampa was $17.63 per square foot. A real estate agent selected a random sample of 15 rental properties (offices) and found the mean rent was $18.72 per square foot, and the standard deviation was $3.64. At $\alpha = 0.05$, test the claim that there is no difference in the rents.

Source: Snapshot, *USA Today,* September 13, 1995.

9–53. A rental agent states that the average monthly rent for a studio apartment in Shadyside is $750. A sample of 12 renters shows that the mean is $732 and the standard deviation is $17. Is there enough evidence to reject the agent's claim? Use $\alpha = 0.01$.

9–54. A manufacturer claims that a new snowmaking machine can save money for ski resort owners. He states that in a sample of 10 tons, the cost of making a ton of snow was $5.75. The average of the old machine was $6.62. The standard deviation of the sample was $1.05. At $\alpha = 0.10$, does the new machine save money?

9–55. An officer states that the average fine levied by the safety office against companies is at most $350. A company owner suspects it is higher. She samples 12 companies and finds that the average is $358. The standard deviation of the sample is $16. Test the claim that the average is higher than $350, at $\alpha = 0.05$.

9–56. A person claims that the average price of a car wash in West Newton is $3.00. A sample of five car washes had an average price of $3.70 and a standard deviation of $0.30. At $\alpha = 0.05$, test the claim.

9–57. A recent survey stated that the average single-person household received at least 37 telephone calls per month. To test the claim, a researcher surveyed 29 single-person households and found that the average number of calls was 34.9. The standard deviation of the sample was 6. At $\alpha = 0.05$, can the claim be rejected? Use the *P*-value method.

9–58. A student suspected the average cost of a Saturday night date was no longer $30.00. To test her hypothesis, she randomly selected 16 men from the dormitory and asked them how much they spent on a date last Saturday. She found that the average cost was $31.17. The standard deviation of the sample was $5.51. At $\alpha = 0.05$, is there enough evidence to support her claim? Use the *P*-value method.

9–59. From past experience, a teacher believes that the average score on a real estate exam is 75. A sample of 20 students' exam scores is as follows:

$$80, 68, 72, 73, 76, 81, 71, 71, 65, 50,$$
$$63, 71, 70, 70, 76, 75, 69, 70, 72, 74$$

Test the claim that the students' average is still 75. Use $\alpha = 0.01$. Use the *P*-value method.

9–60. A new laboratory technician read a report that the average number of students using the computer laboratory per hour was 16. To test this hypothesis, he selected a day at random and kept track of the number of students who used the lab over an eight-hour period. The results were as follows:

$$20, 24, 18, 16, 16, 19, 21, 23$$

At $\alpha = 0.05$, test the claim that the average is actually 16. Use the *P*-value method.

Technology Step by Step

MINITAB
Step by Step

Hypothesis Test for the Mean, *t* Distribution

Example MT9–2

Test the claim H_0: $\mu = 2$, when $\alpha = 0.10$ using the following data:

$$6 \quad 8 \quad 3 \quad 2 \quad 0 \quad 0 \quad 1 \quad 5 \quad 4 \quad 3 \quad 3 \quad 2$$

1. Enter the data into column C1 of a MINITAB worksheet.
2. Select **Stat>Basic Statistics>1 Sample t.**
3. Chose C1 as the variable.
4. Click inside the textbox for Test mean and enter the hypothesized value of **2.**
5. Click [Options]. The Alternative should be not equal.
6. Click [OK].

1-Sample *t* Session
Window

```
One-Sample T: C1

Test of mu = 2 vs mu not = 2

Variable            N       Mean     StDev    SE Mean
C1                 12       3.083     2.392     0.690

Variable                  90.0% CI                T        P
C1             (    1.843,    4.323)             1.57    0.145
```

The results in the session window are shown. Since the *P*-value is greater than the significance level (0.145 > 0.1), do not reject H_0. Summary: There is not enough evidence in the sample to conclude that the population mean does not equal 2.

TI-83
Step by Step

Hypothesis Test for the Mean, *t* Distribution (Data)

1. Enter the data into L₁.
2. Press **STAT** and move the cursor to TESTS.
3. Press **2** for T-Test, and select Data. Press **ENTER,** then press ▼.
4. Type in the value for μ and make sure L₁ is selected and Freq is 1.
5. Select the correct alternative hypothesis.
6. Select Calculate and press **ENTER.**

Example TI9–3

Test the claim from Example 9–13, H_1: $\mu < 60$ at $\alpha = 0.10$, using the data values 60, 56, 60, 55, 70, 55, 60, and 55.

    The alternative hypothesis is H_1: $\mu < 60$. The test value is -0.6259753964. The *P*-value is 0.2755939019, which is not significant at $\alpha = 0.10$, since $0.2755939019 > 0.10$. The decision is not to reject H_0. The values for $\overline{X}$, *s,* and *n* are also given.

Input
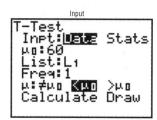

Output
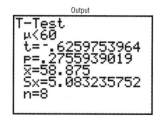

Hypothesis Test for the Mean, *t* Distribution (Stats)

1. Select **STAT** and move the cursor to TESTS. Then press **2** for T-Test.
2. Move the cursor to Stats. Press **ENTER,** then press ▼.
3. Type the values for μ, $\bar{X}$, s, and n.
4. Select the correct alternative hypothesis.
5. Select Calculate and press **ENTER.**

Example TI9–4

Test the claim from Example 9–16, H_1: $\mu > 36.7$ at $\alpha = 0.05$ when $\bar{X} = 40.6$, $s = 6$, and $n = 15$.

Input

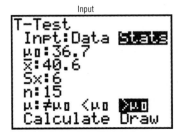

Output

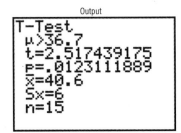

The test value is 2.517439175, and the *P*-value is 0.0123111889. The decision is to reject H_0, since $0.0123111889 < 0.05$.

9–5

z Test for a Proportion

Objective 7. Test proportions using the *z* test.

Many hypothesis-testing situations involve proportions. Recall from Chapter 8 that a *proportion* is the same as a percentage of the population.

The following data were obtained from *The Book of Odds* by Michael D. Shook and Robert L. Shook (New York: Penguin Putnam, Inc. 1991):

- 59% of consumers purchase gifts for their fathers
- 85% of people over 21 said they have entered a sweepstakes
- 51% of Americans buy generic products
- 35% of Americans go out for dinner once a week

A hypothesis test involving a population proportion can be considered as a binomial experiment when there are only two outcomes and the probability of a success does not change from trial to trial. Recall from Section 6–4 in Chapter 6 that the mean is $\mu = np$ and the standard deviation is $\sigma = \sqrt{npq}$ for the binomial distribution.

Since the normal distribution can be used to approximate the binomial distribution when $np \geq 5$ and $nq \geq 5$, the standard normal distribution can be used to test hypotheses for proportions.

Formula for the *z* Test for Proportions

$$z = \frac{\hat{p} - p}{\sqrt{pq/n}}$$

where

$\hat{p} = \dfrac{X}{n}$ (sample proportion)

p = population proportion

n = sample size

The formula is derived from the normal approximation to the binomial and follows the general formula

$$\text{test value} = \frac{(\text{observed value}) - (\text{expected value})}{\text{standard error}}$$

We obtain $\hat{p}$ from the sample (i.e., observed value), p is the expected value (i.e., hypothesized population proportion), and $\sqrt{pq/n}$ is the standard error.

The formula $z = \dfrac{\hat{p} - p}{\sqrt{pq/n}}$ can be derived from the formula $z = \dfrac{X - \mu}{\sigma}$ by substituting

$\mu = np$ and $\sigma = \sqrt{npq}$ and then dividing numerator and denominator by n. Some algebra is used. See Exercise 9–78.

The steps for hypothesis testing are the same as those shown in Section 9–3. Table E is used to find critical values and P-values.

The next three examples show the traditional method of hypothesis testing. The last example shows the P-value method.

Sometimes it is necessary to find $\hat{p}$, as shown in Examples 9–17, 9–19, and 9–20, and sometimes $\hat{p}$ is given in the exercise. See Example 9–18.

Example 9–17

An educator estimates that the dropout rate for seniors at high schools in Ohio is 15%. Last year, 38 seniors from a random sample of 200 Ohio seniors withdrew. At $\alpha = 0.05$, is there enough evidence to reject the educator's claim?

Solution

STEP 1   State the hypotheses and identify the claim.

$$H_0: p = 0.15 \text{ (claim)} \qquad \text{and} \qquad H_1: p \neq 0.15$$

STEP 2   Find the critical value(s). Since $\alpha = 0.05$ and the test is two-tailed, the critical values are ± 1.96.

STEP 3   Compute the test value. First, it is necessary to find $\hat{p}$.

$$\hat{p} = \frac{X}{n} = \frac{38}{200} = 0.19 \quad \text{and} \quad p = 0.15 \quad \text{and} \quad q = 1 - 0.15 = 0.85$$

Substitute in the formula and solve.

$$z = \frac{\hat{p} - p}{\sqrt{pq/n}} = \frac{0.19 - 0.15}{\sqrt{(0.15)(0.85)/200}} = 1.58$$

STEP 4   Make the decision. Do not reject the null hypothesis, since the test value falls outside the critical region, as shown in Figure 9–26.

Figure 9–26

Critical and Test Values
for Example 9–17

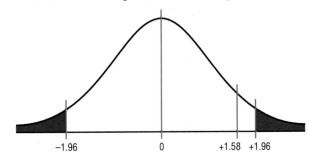

−1.96   0   +1.58 +1.96

STEP 5 Summarize the results. There is not enough evidence to reject the claim that the dropout rate for seniors in high schools in Ohio is 15%.

Example 9–18

A telephone company representative estimates that 40% of its customers have call-waiting service. To test this hypothesis, she selected a sample of 100 customers and found that 37% had call waiting. At $\alpha = 0.01$, is there enough evidence to reject the claim?

Solution

STEP 1 State the hypotheses and identify the claim.

$$H_0: p = 0.40 \text{ (claim)} \quad \text{and} \quad H_1: p \neq 0.40$$

STEP 2 Find the critical value(s). Since $\alpha = 0.01$ and this test is two-tailed, the critical values are ± 2.58.

STEP 3 Compute the test value. It is not necessary to find $\hat{p}$ since it is given in the exercise; $\hat{p} = 0.37$. Substitute in the formula and solve.

$$p = 0.40 \quad \text{and} \quad q = 1 - 0.40 = 0.60$$

$$z = \frac{\hat{p} - p}{\sqrt{pq/n}} = \frac{0.37 - 0.40}{\sqrt{(0.40)(0.60)/100}} = -0.612$$

STEP 4 Make the decision. Do not reject the null hypothesis, since the test value falls in the noncritical region, as shown in Figure 9–27.

Figure 9–27

Critical and Test Values for Example 9–18

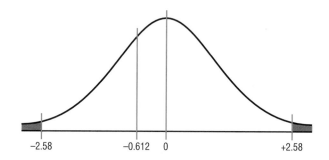

STEP 5 Summarize the results. There is not enough evidence to reject the claim that 40% of the telephone company's customers have call waiting.

Example 9–19

A statistician read that at least 77% of the population oppose replacing $1 bills with $1 coins. To see if this claim is valid, the statistician selected a sample of 80 people and found that 55 were opposed to replacing the $1 bills. At $\alpha = 0.01$, test the claim that at least 77% of the population are opposed to the change.

Source: *USA Today,* June 14, 1995.

Solution

STEP 1 State the hypotheses and identify the claim.

$$H_0: p \geq 0.77 \text{ (claim)} \quad \text{and} \quad H_1: p < 0.77$$

STEP 2 Find the critical value(s). Since $\alpha = 0.01$ and the test is left-tailed, the critical value is -2.33.

STEP 3 Compute the test value.

$$\hat{p} = \frac{X}{n} = \frac{55}{80} = 0.6875$$

$$p = 0.77 \quad \text{and} \quad q = 1 - 0.77 = 0.23$$

$$z = \frac{\hat{p} - p}{\sqrt{pq/n}} = \frac{0.6875 - 0.77}{\sqrt{(0.77)(0.23)/80}} = -1.75$$

STEP 4 Do not reject the null hypothesis, since the test value does not fall in the critical region, as shown in Figure 9–28.

Figure 9–28

Critical and Test Values for Example 9–19

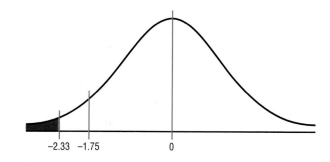

−2.33 −1.75   0

STEP 5 There is not enough evidence to reject the claim that at least 77% of the population oppose replacing \$1 bills with \$1 coins.

Example 9–20

An attorney claims that more than 25% of all lawyers advertise. A sample of 200 lawyers in a certain city showed that 63 had used some form of advertising. At $\alpha = 0.05$, is there enough evidence to support the attorney's claim? Use the *P*-value method.

Interesting Fact

Many people have what is called "white-coat hypertension." This condition causes the blood pressure of people who have normal pressure to rise when they visit their physician. However, some people have the reverse effect. Their pressure is above normal in everyday life but actually drops to normal level at the doctor's office. Researchers suspect that for these people, a visit to a physician is a vacation from their fast-paced, high-pressure lives.
Source: Reprinted with permission from *Psychology Today* magazine. Copyright © 1996. (Sussex Publishers, Inc.)

Solution

STEP 1 State the hypotheses and identify the claim.

$$H_0: p \leq 0.25 \qquad \text{and} \qquad H_1: p > 0.25 \text{ (claim)}$$

STEP 2 Compute the test value

$$\hat{p} = \frac{X}{n} = \frac{63}{200} = 0.315$$

$$P = 0.25 \quad \text{and} \quad q = 1 - 0.25 = 0.75$$

$$z = \frac{\hat{p} - p}{\sqrt{pq/n}} = \frac{0.315 - 0.25}{\sqrt{(0.25)(0.75)/200}} = 2.12$$

STEP 3 Find the *P*-value. The area under the curve for $z = 2.12$ is 0.4830. Subtracting the area from 0.5000, one gets $0.5000 - 0.4830 = 0.0170$. The *P*-value is 0.0170.

STEP 4 Reject the null hypothesis, since $0.0170 < 0.05$ (i.e., *P*-value < 0.05). See Figure 9–29.

Figure 9–29

P-Value and α Value for
Example 9–20

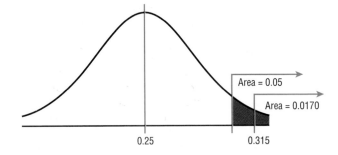

Area = 0.05

Area = 0.0170

0.25   0.315

STEP 5   There is enough evidence to support the attorney's claim that more than 25%
of the lawyers use some form of advertising.

Exercises

9–61. Give three examples of proportions.

9–62. Why is a proportion considered a binomial variable?

9–63. When one is testing hypotheses using proportions, what are the necessary requirements?

9–64. What are the mean and the standard deviation of a proportion?

For Exercises 9–65 through 9–75, perform each of the following steps.

a. State the hypotheses and identify the claim.
b. Find the critical value(s)
c. Compute the test value.
d. Make the decision.
e. Summarize the results.

Use the traditional method of hypothesis testing unless otherwise specified.

9–65. A toy manufacturer claims that at least 23% of the 14-year-old residents of a certain city own a skateboard. A sample of forty 14-year-olds shows that seven own a skateboard. Is there enough evidence to reject the manufacturer's claim at $\alpha = 0.05$?

9–66. From past records, a hospital found that 37% of all full-term babies born in the hospital weighed more than 7 pounds 2 ounces. This year a sample of 100 babies showed that 23 weighed over 7 pounds 2 ounces. At $\alpha = 0.01$, is there enough evidence to say the percentage has changed?

9–67. A study on crime suggested that at least 40% of all arsonists were under 21 years of age. Checking local crime statistics, a researcher found that 30 out of 80 arson suspects were under 21. At $\alpha = 0.10$, should the crime study's statement be rejected?

9–68. A recent study found that, at most, 32% of people who have been in a plane crash have died. In a sample of 100 people who were in a plane crash, 38 died. Should the study's claim be rejected? Use $\alpha = 0.05$.

9–69. An item in *USA Today* reported that 63% of Americans owned an answering machine. A survey of 143 employees at a large school showed that 85 owned an answering machine. At $\alpha = 0.05$, test the claim that the percentage is the same as stated in *USA Today*.

Source: *USA Today,* May 22, 1995.

9–70. The *Statistical Abstract* reported that 17% of adults attended a musical play in the past year. To test this claim, a researcher surveyed 90 people and found that 22 had attended a musical play in the past year. At $\alpha = 0.05$, test the claim that this figure is correct.

Source: *Statistical Abstract of the United States 1994.*

9–71. A recent study claimed that at least 15% of all eighth-grade students are overweight. In a sample of 80 students, 9 were found to be overweight. At $\alpha = 0.05$, is there enough evidence to reject the claim?

9–72. At a large university, a study found that no more than 25% of the students who commute travel more than 14 miles to campus. At $\alpha = 0.10$, test the findings that if in a sample of 100 students, 30 drove more than 14 miles. Use the *P*-value method.

9–73. A telephone company wants to advertise that more than 30% of all its customers have more than two telephones. To support this ad, the company selects a sample of 200 customers and finds that 72 have more than two telephones. Does the evidence support the ad? Use $\alpha = 0.05$. Use the *P*-value method.

9–74. Experts claim that 10% of murders are committed by women. Is there enough evidence to reject the claim if in a sample of 67 murders, 10 were committed by women? Use $\alpha = 0.01$. Use the *P*-value method.

9–75. Researchers suspect that 18% of all high school students smoke at least one pack of cigarettes a day. At Wilson High School, with an enrollment of 300 students, a study found that 50 students smoked at least one pack of cigarettes a day. At $\alpha = 0.05$, test the claim that 18% of all high school students smoke at least one pack of cigarettes a day. Use the *P*-value method.

When *np* or *nq* is not 5 or more, the binomial table (Table B in Appendix C) must be used to find critical values in hypothesis tests involving proportions.

*9–76. A coin is tossed nine times and three heads appear. Can one conclude that the coin is not balanced? Use $\alpha = 0.10$. (*Hint:* Use the binomial table and find $2P(X \le 3)$ with $p = 0.5$ and $n = 9$.)

*9–77. In the past, 20% of all airline passengers flew first class. In a sample of 15 passengers, 5 flew first class. At $\alpha = 0.10$, can one conclude that the proportions have changed?

9–78. Show that $z = \dfrac{\hat{p} - p}{\sqrt{pq/n}}$ can be derived from

$z = \dfrac{X - \mu}{\sigma}$ by substituting $\mu = np$ and $\sigma = \sqrt{npq}$ and

dividing numerator and denominator by *n*.

Technology Step by Step

TI-83 Step by Step

z-Test for a Proportion

1. Press **STAT** and move the cursor to TESTS.
2. Press **5** for 1-PropZTest.
3. Enter the hypothesized proportion *p*, *X*, and *n*.
4. Select the correct alternative hypothesis.
5. Select Calculate and press **ENTER**.

Example TI9–5

Test the claim $H_1: p \ne 0.15$ when $X = 38$ and $n = 200$ at $\alpha = 0.05$.

Input

```
1-PropZTest
 P0:.15
 x:38
 n:200
 prop≠P0 <P0 >P0
 Calculate Draw
```

Output

```
1-PropZTest
 prop≠.15
 z=1.584236069
 P=.1131400131
 p̂=.19
 n=200
```

The test value is 1.584236069. The *P*-value is 0.1131400131. The decision is not to reject H_0. The values for $\hat{p}$ and *n* are also given.

9–6

χ^2 Test for a Variance or Standard Deviation

Objective 8. Test variances or standard deviations using the chi-square test.

In Chapter 8, the chi-square distribution was used to construct a confidence interval for a single variance or standard deviation. This distribution is also used to test a claim about a single variance or standard deviation.

In order to find the area under the chi-square distribution, use Table G in Appendix C. There are three cases to consider:

1. Finding the chi-square critical value for a specific α when the hypothesis test is right-tailed.
2. Finding the chi-square critical value for a specific α when the hypothesis test is left-tailed.
3. Finding the chi-square critical values for a specific α when the hypothesis test is two-tailed.

Example 9–21

Find the critical chi-square value for 15 degrees of freedom when $\alpha = 0.05$ and the test is right-tailed.

Solution

The distribution is shown in Figure 9–30.

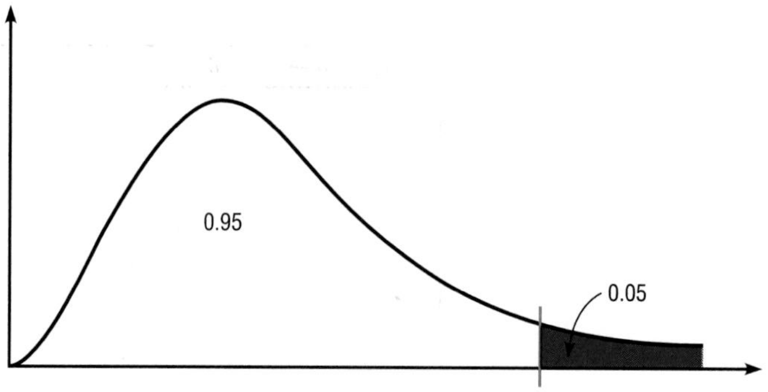
0.95
0.05

Figure 9–30
Chi-Square Distribution for Example 9–21

Find the α value at the top of Table G and find the corresponding degrees of freedom in the left column. The critical value is located where the two columns meet—in this case, 24.996. See Figure 9–31.

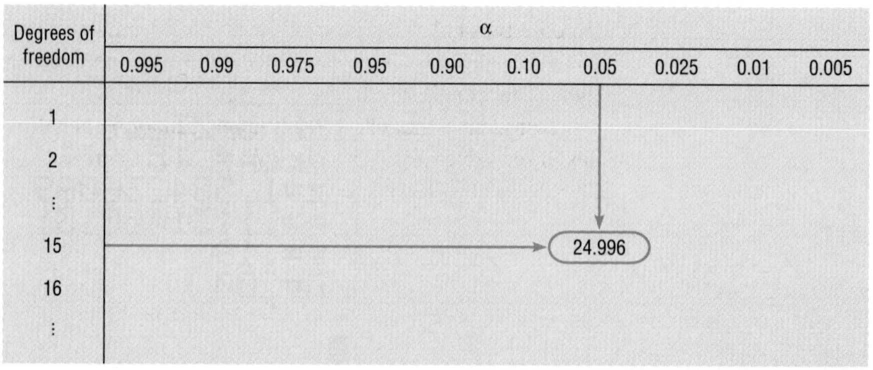
Degrees of freedom
α
0.995
0.99
0.975
0.95
0.90
0.10
0.05
0.025
0.01
0.005
1
2
⋮
15
24.996
16
⋮

Figure 9–31
Locating the Critical Value in Table G for Example 9–21

Example 9–22

Find the critical chi-square value for 10 degrees of freedom when $\alpha = 0.05$ and the test is left-tailed.

Solution

This distribution is shown in Figure 9–32.

Figure 9–32

Chi-Square Distribution for Example 9–22

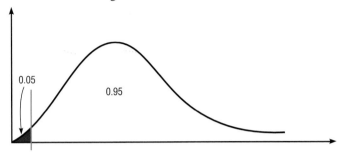

When the test is one-tailed left, the α value must be subtracted from 1, i.e., $1 - 0.05 = 0.95$. The left side of the table is used, because the chi-square table gives the area to the right of the critical value, and the chi-square statistic cannot be negative. The table is set up so that it gives the values for the area to the right of the critical value. In this case, 95% of the area will be to the right of the value.

For 0.95 and 10 degrees of freedom, the critical value is 3.940. See Figure 9–33.

Figure 9–33

Locating the Critical Value in Table G for Example 9–22

Degrees of freedom	α									
	0.995	0.99	0.975	0.95	0.90	0.10	0.05	0.025	0.01	0.005
1										
2										
⋮										
10				3.940						
⋮										

Example 9–23

Find the critical chi-square values for 22 degrees of freedom when $\alpha = 0.05$ and a two-tailed test is conducted.

Solution

When a two-tailed test is conducted, the area must be split, as shown in Figure 9–34. Note that the area to the right of the larger value is 0.025 (0.05/2), and the area to the right of the smaller value is 0.975 (1.00 − 0.05/2).

Remember that chi-square values cannot be negative. Hence, one must use α values in the table of 0.025 and 0.975. With 22 degrees of freedom, the critical values are 36.781 and 10.982, respectively.

Figure 9–34

Chi-Square Distribution for Example 9–23

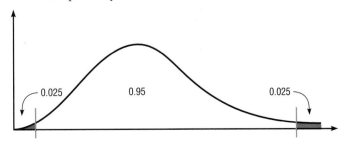

After the degrees of freedom reach 30, Table G gives values only for multiples of 10 (40, 50, 60, etc.). When the exact degrees of freedom one is seeking are not specified in the table, the closest smaller value should be used. For example, if the given degrees of freedom are 36, use the table value for 30 degrees of freedom. This guideline keeps the type I error equal to or below the α value.

When one is testing a claim about a single variance, there are three possible test situations: right-tailed test, left-tailed test, and two-tailed test.

If a researcher believes the variance of a population to be greater than some specific value, say 225, then the researcher states the hypotheses as

$$H_0: \sigma^2 \leq 225 \qquad \text{and} \qquad H_1: \sigma^2 > 225$$

and conducts a right-tailed test.

If the researcher believes the variance of a population to be less than 225, then the researcher states the hypotheses as

$$H_0: \sigma^2 \geq 225 \qquad \text{and} \qquad H_1: \sigma^2 < 225$$

and conducts a left-tailed test.

Finally, if a researcher does not wish to specify a direction, he or she states the hypotheses as

$$H_0: \sigma^2 = 225 \qquad \text{and} \qquad H_1: \sigma^2 \neq 225$$

and conducts a two-tailed test.

Formula for the Chi-Square Test for a Single Variance

$$\chi^2 = \frac{(n-1)s^2}{\sigma^2}$$

with degrees of freedom equal to $n - 1$ and where
n = sample size
s^2 = sample variance
σ^2 = population variance

One might ask, "Why is it important to test variances?" There are several reasons. First, in any situation where consistency is required, such as in manufacturing, one would like to have the smallest variation possible in the products. For example, when bolts are manufactured, the variation in diameters due to the process must be kept to a minimum, or the nuts will not fit them properly. In education, consistency is required on a test. That is, if the same students take the same test several times, they should get approximately the same grades, and the variance of each of the students' grades should be small. On the other hand, if the test is to be used to judge learning, the overall standard deviation of all of the grades should be large so that one can differentiate those who have learned the subject from those who have not learned it.

Three assumptions are made for the chi-square test, as outlined next.

Assumptions for the Chi-Square Test for a Single Variance

1. The sample must be randomly selected from the population.
2. The population must be normally distributed for the variable under study.
3. The observations must be independent of each other.

The traditional method for hypothesis-testing follows the same five steps listed in the preceding sections. They are repeated here.

STEP 1   State the hypotheses and identify the claim.

STEP 2   Find the critical value(s).

STEP 3   Compute the test value.

STEP 4   Make the decision.

STEP 5   Summarize the results.

The next three examples illustrate the traditional hypothesis-testing procedure for variances.

Example 9–24

An instructor wishes to see whether the variation in scores of the 23 students in her class is less than the variance of the population. The variance of the class is 198. Is there enough evidence to support the claim that the variation of the students is less than the population variance ($\sigma^2 = 225$) at $\alpha = 0.05$? Assume that the scores are normally distributed.

Solution

STEP 1   State the hypotheses and identify the claim.

$$H_0: \sigma^2 \geq 225 \quad \text{and} \quad H_1: \sigma^2 < 225 \text{ (claim)}$$

STEP 2   Find the critical value. Since this test is left-tailed and $\alpha = 0.05$, use the value $1 - 0.05 = 0.95$. The degrees of freedom are $n - 1 = 23 - 1 = 22$. Hence, the critical value is 12.338. Note that the critical region is on the left, as shown in Figure 9–35.

Figure 9–35

Critical Value for Example 9–24

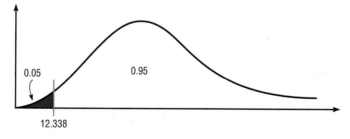

STEP 3   Compute the test value.

$$\chi^2 = \frac{(n-1)s^2}{\sigma^2} = \frac{(23-1)(198)}{225} = 19.36$$

STEP 4   Make the decision. Since the test value 19.36 falls in the noncritical region, as shown in Figure 9–36, the decision is to not reject the null hypothesis.

Figure 9–36

Critical and Test Values for Example 9–24

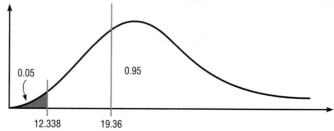

STEP 5   Summarize the results. There is not enough evidence to support the claim that the variation in test scores of the instructor's students is less than the variation in scores of the population.

Example 9–25

A hospital administrator believes that the standard deviation of the number of people using out-patient surgery per day is greater than eight. A random sample of 15 days is selected. The data are shown. At $\alpha = 0.10$ is there enough evidence to support the administrator's claim? Assume the variable is normally distributed.

25	30	5	15	18
42	16	9	10	12
12	38	8	14	27

Solution

STEP 1   State the hypotheses and identify the claim.

$$H_0: \sigma^2 \le 64 \qquad \text{and} \qquad H_1: \sigma^2 > 64 \text{ (claim)}$$

Since the standard deviation is given, it should be squared to get the variance.

STEP 2   Find the critical value. Since this test is right-tailed with d.f. of $15 - 1 = 14$ and $\alpha = 0.10$, the critical value is 21.064.

STEP 3   Compute the test value. Since raw data are given, the standard deviation of the sample must be found using the formula in Chapter 3 or your calculator. It is $s = 11.2$.

$$\chi^2 = \frac{(n-1)s^2}{\sigma^2} = \frac{(15-1)(11.2)^2}{8} = 27.44$$

STEP 4   Make the decision. The decision is to reject the null hypothesis since the test value, 27.44, is greater than the critical value 21.064, and falls in the critical region. See Figure 9–37.

Figure 9–37

Critical and Test Value for Example 9–25

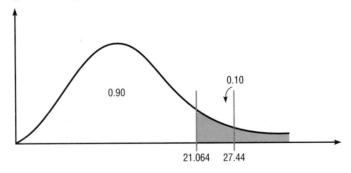

STEP 5   Summarize the results. There is enough evidence to support the claim that the standard deviation is greater than 8.

Example 9–26

A cigarette manufacturer wishes to test the claim that the variance of the nicotine content of its cigarettes is 0.644. Nicotine content is measured in milligrams, and assume that it is normally distributed. A sample of 20 cigarettes has a standard deviation of 1.00 milligram. At $\alpha = 0.05$, is there enough evidence to reject the manufacturer's claim?

Solution

STEP 1   State the hypotheses and identify the claim.

$$H_0: \sigma^2 = 0.644 \text{ (claim)} \qquad \text{and} \qquad H_1: \sigma^2 \neq 0.644$$

STEP 2   Find the critical values. Since this test is a two-tailed test at $\alpha = 0.05$, the critical values for 0.025 and 0.975 must be found. The degrees of freedom are 19; hence, the critical values are 32.852 and 8.907. The critical or rejection regions are shown in Figure 9–38.

Figure 9–38

Critical Values for
Example 9–26

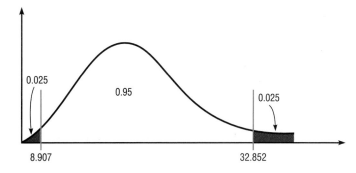

STEP 3   Compute the test value.

$$\chi^2 = \frac{(n-1)s^2}{\sigma^2} = \frac{(20-1)(1.0)^2}{0.644} = 29.5$$

Since the standard deviation s is given in the problem, it must be squared for the formula.

STEP 4   Make the decision. Do not reject the null hypothesis, since the test value falls between the critical values ($8.907 < 29.5 < 32.852$) and in the noncritical region, as shown in Figure 9–39.

Figure 9–39

Critical and Test Values
for Example 9–26

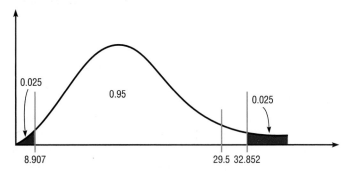

STEP 5   Summarize the results. There is not enough evidence to reject the manufacturer's claim that the variance of the nicotine content of the cigarettes is equal to 0.644.

Approximate P-values for the chi-square test can be found using Table G in the Appendix. The procedure is somewhat more complicated than the previous procedures for finding P-values for the z and t tests since the chi-square distribution is not exactly symmetrical and χ^2 values cannot be negative. As we did for the t test, we will determine an *interval* for the *P*-value based on the table. The next three examples show the procedure.

Find the *P*-value when $\chi^2 = 19.274$, $n = 8$, and the test is right-tailed.

Solution

To get the *P*-value, look across the row with d.f. = 7 in Table G and find the two values that 19.274 falls between. They are 18.475 and 20.278. Look up to the top row and find the α values corresponding to 18.475 and 20.278. They are 0.01 and 0.005, respectively. See Figure 9–40. Hence the *P*-value is contained in the interval $0.005 < P\text{-value} < 0.01$. (The *P*-value obtained from a calculator is 0.007.)

Figure 9–40

P-Value Interval for
Example 9–27

Degrees of freedom	0.995	0.99	0.975	0.95	0.90	0.10	0.05	0.025	0.01	0.005
						α				
1	—	—	0.001	0.004	0.016	2.706	3.841	5.024	6.635	7.879
2	0.010	0.020	0.051	0.103	0.211	4.605	5.991	7.378	9.210	10.597
3	0.072	0.115	0.216	0.352	0.584	6.251	7.815	9.348	11.345	12.838
4	0.207	0.297	0.484	0.711	1.064	7.779	9.488	11.143	13.277	14.860
5	0.412	0.554	0.831	1.145	1.610	9.236	11.071	12.833	15.086	16.750
6	0.676	0.872	1.237	1.635	2.204	10.645	12.592	14.449	16.812	18.548
7	0.989	1.239	1.690	2.167	2.833	12.017	11.067	16.013	18.475	20.278
8	1.344	1.646	2.180	2.733	3.490	13.362	15.507	17.535	20.090	21.955
9	1.735	2.088	2.700	3.325	4.168	14.684	16.919	19.023	21.666	23.589
10	2.156	2.558	3.247	3.940	4.865	15.987	18.307	20.483	23.209	25.188
⋮	⋮	⋮	⋮	⋮	⋮	⋮	⋮	⋮	⋮	⋮
100	67.328	70.065	74.222	77.929	82.358	118.498	124.342	129.561	135.807	140.169

* 19.274 falls between 18.475 and 20.278

Find the *P*-value when $\chi^2 = 3.823$, $n = 13$, and the test is left-tailed.

Solution

To get the *P*-value look across the row with d.f. = 12 and find the two values that 3.823 falls between. They are 3.571 and 4.404. Look up to the top row and find the values corresponding to 3.571 and 4.404. They are 0.99 and 0.975, respectively. When the χ^2 test value falls on the left side, each of the values must be subtracted from one to get the interval that *P*-value falls between.

$$1 - 0.99 = 0.01 \qquad \text{and} \qquad 1 - 0.975 = 0.025$$

Hence the *P*-value falls in the interval

$$0.01 < P\text{-value} < 0.025$$

(The *P*-value obtained from a calculator is 0.014.)

When the χ^2 test is two-tailed, both interval values must be doubled. If a two-tailed test was being used in Example 9–28 then the interval would be 2 (0.01) < *P*-value < 2 (0.025) or 0.02 < *P*-value < 0.05.

The *P*-value method for hypothesis testing for a variance or standard deviation follows the same steps shown in the preceding sections.

STEP 1   State the hypotheses and identify the claim.

STEP 2   Compute the test value.

STEP 3   Find the *P*-value.

STEP 4   Make the decision.

STEP 5   Summarize the results.

The next example shows the *P*-value method for variances or standard deviations.

Example 9–29

A researcher knows from past studies that the standard deviation of the times it takes to inspect a car is 16.8 minutes. A sample of 24 cars is selected and inspected. The standard deviation was 12.5 minutes. At $\alpha = 0.05$, can it be concluded that the standard deviation has changed? Use the *P*-value method.

Solution

STEP 1   State the hypotheses and identify the claim

$$H_0: \sigma = 16.8 \qquad \text{and} \qquad H_1: \sigma \neq 16.8 \text{ (claim)}$$

STEP 2   Compute the test value.

$$\chi^2 = \frac{(n-1)s^2}{\sigma^2} = \frac{(24-1)(12.5)^2}{(16.8)^2} = 12.733$$

STEP 3   Find the *P*-value. Using Table G with d.f. = 23, the value 12.733 falls between 11.689 and 13.091, corresponding to 0.975 and 0.95. Since these values are found on the left side of the distribution, each value must be subtracted from one. Hence $1 - 0.975 = 0.025$ and $1 - 0.95 = 0.05$. Since this is a two-tailed test, the area must be doubled to obtain the *P*-value interval. Hence $0.05 < P\text{-value} < 0.10$, or somewhere between 0.05 and 0.10. (The *P*-value obtained from a calculator is 0.085.)

STEP 4   Make the decision. Since $\alpha = 0.05$ and the *P*-value is between 0.05 and 0.10, the decision is do not reject the null hypothesis since $P\text{-value} > \alpha$.

STEP 5   Summarize the results. There is not enough evidence to support the claim that the standard deviation has changed.

Exercises

9–79. Using Table G, find the critical value(s) for each, show the critical and noncritical regions, and state the appropriate null and alternative hypotheses. Use $\sigma^2 = 225$.
a. $\alpha = 0.05, n = 18$, right-tailed
b. $\alpha = 0.10, n = 23$, left-tailed
c. $\alpha = 0.05, n = 15$, two-tailed
d. $\alpha = 0.10, n = 8$, two-tailed
e. $\alpha = 0.01, n = 17$, right-tailed
f. $\alpha = 0.025, n = 20$, left-tailed
g. $\alpha = 0.01, n = 13$, two-tailed
h. $\alpha = 0.025, n = 29$, left-tailed

9–80. (ans) Using Table G, find the *P*-value interval for each χ^2 test value.
a. $\chi^2 = 29.321, n = 16$, right-tailed
b. $\chi^2 = 10.215, n = 25$, left-tailed

c. $\chi^2 = 24.672$, $n = 11$, two-tailed
d. $\chi^2 = 23.722$, $n = 9$, right-tailed
e. $\chi^2 = 13.974$, $n = 28$, two-tailed
f. $\chi^2 = 10.571$, $n = 19$, left-tailed
g. $\chi^2 = 12.144$, $n = 6$, two-tailed
h. $\chi^2 = 8.201$, $n = 23$, two-tailed

For Exercises 9–81 through 9–87, assume that the variables are normally or approximately normally distributed.
    Use the traditional method of hypothesis testing unless otherwise specified.

9–81. A nutritionist claims that the standard deviation of the number of calories in one tablespoon of the major brands of pancake syrup is 60. A sample of major brands of syrup is selected and the number of calories is shown. At $\alpha = 0.10$ can the claim be rejected?

53	210	100	200	100	220
210	100	240	200	100	210
100	210	100	210	100	60

Source: Based on information from *The Complete Book of Food Counts* by Corrine T. Netzer, Dell Publishers, New York, 1997.

9–82. Eighteen 5-gallon containers of fruit punch are selected, and each is tested to determine its weight in ounces. The variance of the sample is 6.5. Test the claim that the population variance is greater than 6.2, at $\alpha = 0.01$.

9–83. A company claims that the variance of the sugar content of its yogurt is less than or equal to 25. (The sugar content is measured in milligrams per ounce.) A sample of 20 servings is selected, and the sugar content is measured. The variance of the sample is found to be 36. At $\alpha = 0.10$, is there enough evidence to reject the claim?

9–84. A researcher suggests that the variances of the ages of freshmen at Oak Park College is greater than 1.6, as past studies have shown. A sample of 50 freshmen is selected, and the variance is found to be 2.3 years. At $\alpha = 0.05$, is the variation of the freshmen's ages greater than 1.6?

9–85. The manager of a large company claims that the standard deviation of the time (in minutes) that it takes a telephone call to be transferred to the correct office in her company is 1.2 minutes or less. A sample of 15 calls is selected and the calls timed. The standard deviation of the sample is 1.8 minutes. At $\alpha = 0.01$, test the claim that the standard deviation is less than or equal to 1.2 minutes. Use the *P*-value method.

9–86. A machine fills 12-ounce bottles with soda. In order for the machine to function properly, the standard deviation of the sample must be less than or equal to 0.03 ounce. A sample of 8 bottles is selected, and the number of ounces of soda in each bottle is as given below. At $\alpha = 0.05$, can we reject the claim that the machine is functioning properly? Use the *P*-value method.

12.03, 12.10, 12.02, 11.98,
12.00, 12.05, 11.97, 11.99

9–87. A construction company purchases steel cables. A sample of 12 cables is selected, and the breaking strength (in pounds) of each is found. The data are shown below. The lot is rejected if the standard deviation of the sample is greater than 2 pounds. At $\alpha = 0.01$, should the lot be rejected?

2001, 1998, 2002, 2000, 1998, 1999,
1997, 2005, 2003, 2001, 1999, 2006

Technology Step by Step

TI-83
Step by Step

P-Values for χ^2

To find a *P*-value for a χ^2 test statistic value on the right side of the χ^2 distribution
Example: $\chi^2 = 27.44$, d.f. $= 14$.

1. Press **2nd DISTR** then **7** to get χ^2cdf (.

2. Enter **27.44, 2nd [EE] 99, 14)**.

The *P*-value is 0.0168672575. This is the *P*-value for Example 9–25.
To find a *P*-value for a χ^2 test statistic value on the left side of the χ^2 distribution.
Example: $\chi^2 = 19.36$, d.f. $= 22$.

1. Press **2nd DISTR** then **7** to get χ^2cdf (.

2. Enter **0, 19.36, 22)**.

The *P*-value is 0.3769943702. This is the *P*-value for Example 9–24.
Be sure to double the *P*-value for a two-tailed test when using the above procedures.

```
X²cdf(27.44,1ε99
,14)
         .0168672575
X²cdf(0,19.36,22
)
         .3769943702
```

9–7

Additional Topics Regarding Hypothesis Testing

In hypothesis testing, there are several other concepts that might be of interest to students in elementary statistics. These topics include the relationship between hypothesis testing and confidence intervals, and some additional information about the type II error.

Confidence Intervals and Hypothesis Testing

Objective 9. Test hypotheses using confidence intervals.

There is a relationship between confidence intervals and hypothesis testing. When the null hypothesis is rejected in a hypothesis-testing situation, the confidence interval for the mean using the same level of significance *will not* contain the hypothesized mean. Likewise, when the null hypothesis is not rejected, the confidence interval computed using the same level of significance *will* contain the hypothesized mean. The next two examples show this concept for two-tailed tests.

Example 9–30

Sugar is packed in 5-pound bags. An inspector suspects the bags may not contain 5 pounds. A sample of 50 bags produces a mean of 4.6 pounds and a standard deviation of 0.7 pound. Is there enough evidence to conclude that the bags do not contain 5 pounds as stated at $\alpha = 0.05$? Also, find the 95% confidence interval of the true mean.

Solution

H_0: $\mu = 5$, and H_1: $\mu \neq 5$ (claim). The critical values are $+1.96$ and -1.96. The test value is

$$z = \frac{\overline{X} - \mu}{s/\sqrt{n}} = \frac{4.6 - 5.0}{0.7/\sqrt{50}} = \frac{-0.4}{0.099} = -4.04$$

Since $-4.04 < -1.96$, the null hypothesis is rejected. There is enough evidence to support the claim that the bags do not weigh 5 pounds.

The 95% confidence for the mean is given by

$$\overline{X} - z_{\alpha/2} \cdot \frac{s}{\sqrt{n}} < \mu < \overline{X} + z_{\alpha/2} \cdot \frac{s}{\sqrt{n}}$$

$$4.6 - (1.96)\left(\frac{0.7}{\sqrt{50}}\right) < \mu < 4.6 + (1.96)\left(\frac{0.7}{\sqrt{50}}\right)$$

$$4.4 < \mu < 4.8$$

Notice that the 95% confidence interval of μ does *not* contain the hypothesized value $\mu = 5$. Hence, there is agreement between the hypothesis test and the confidence interval.

Example 9–31

A researcher claims that adult hogs fed a special diet will have an average weight of 200 pounds. A sample of 10 hogs has an average weight of 198.2 pounds and a standard deviation of 3.3 pounds. At $\alpha = 0.05$, can the claim be rejected? Also, find the 95% confidence interval of the true mean.

Solution

H_0: $\mu = 200$ pounds (claim) and H_1: $\mu \neq 200$ pounds. The t test must be used since σ is unknown and $n < 30$. The critical values at $\alpha = 0.05$ with 9 degrees of freedom are $+2.262$ and -2.262. The test value is

$$t = \frac{\overline{X} - \mu}{s/\sqrt{n}} = \frac{198.2 - 200}{3.3/\sqrt{10}} = \frac{-1.8}{1.0436} = -1.72$$

Thus, the null hypothesis is not rejected. There is not enough evidence to reject the claim that the weight of the adult hogs is 200 pounds.

The 95% confidence interval of the mean is

$$\overline{X} - t_{\alpha/2} \cdot \frac{s}{\sqrt{n}} < \mu < \overline{X} + t_{\alpha/2} \cdot \frac{s}{\sqrt{n}}$$

$$198.2 - (2.262)\left(\frac{3.3}{\sqrt{10}}\right) < \mu < 198.2 + (2.262)\left(\frac{3.3}{\sqrt{10}}\right)$$

$$198.2 - 2.361 < \mu < 198.2 + 2.361$$

$$195.8 < \mu < 200.6$$

The 95% confidence interval does contain the hypothesized mean $\mu = 200$.

....

In summary, then, when the null hypothesis is rejected, the confidence interval computed at the same significance level will not contain the value of the mean that is stated in the null hypothesis. On the other hand, when the null hypothesis is not rejected, the confidence interval computed at the same significance level will contain the value of the mean stated in the null hypothesis. These results are true for other hypothesis-testing situations and are not limited to means tests.

The relationship between confidence intervals and hypothesis testing presented here is valid for two-tailed tests. The relationship between one-tailed hypothesis tests and one-sided or one-tailed confidence intervals is also valid; however, this technique is beyond the scope of this textbook.

Type II Error and the Power of a Test

Recall that in hypothesis testing, there are two possibilities: Either the null hypothesis (H_0) is true, or it is false. Furthermore, on the basis of the statistical test, the null hypothesis is either rejected or not rejected. These results give rise to four possibilities, as shown in Figure 9–41. This figure is similar to Figure 9–2.

Figure 9–41

Possibilities in Hypothesis Testing

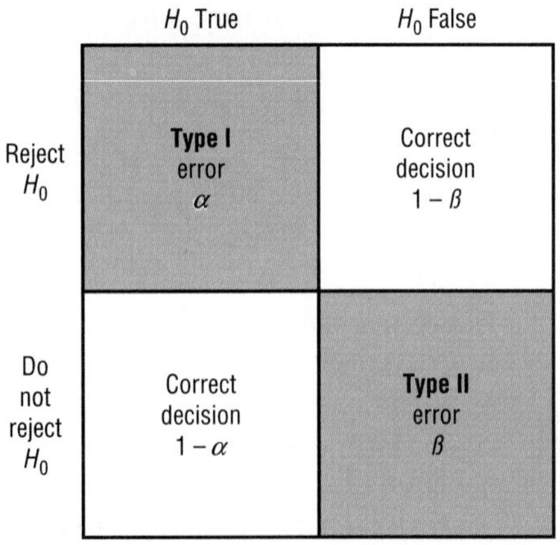

As stated previously, there are two types of errors: type I and type II. A type I error can occur only when the null hypothesis is rejected. By choosing a level of significance, say 0.05 or 0.01, the researcher can determine the probability of committing a type I error. For example, suppose that the null hypothesis was H_0: $\mu \le 50$, and it was rejected. At the 0.05 level, the researcher has only a 5% chance of being wrong, i.e., of rejecting a true null hypothesis.

On the other hand, if the null hypothesis is not rejected, then either it is true or a type II error has been committed. A type II error occurs when the null hypothesis is indeed false, but it is not rejected. The probability of committing a type II error is denoted as β.

The value of β is not easy to compute. It depends on several things, including the value of α, the size of the sample, the population standard deviation, and the actual difference between the hypothesized value of the parameter being tested and the true parameter. The researcher has control over two of these factors, namely, the selection of α and the size of the sample. The standard deviation of the population is sometimes known or can be estimated. The major problem, then, is knowing the actual difference between the hypothesized parameter and the true parameter. If this difference were known, then the value of the parameter would be known; and if the parameter were known, then there would be no need to do any hypothesis testing. Hence, the value of β cannot be computed. But this does not mean that it should be ignored. What the researcher usually does is try to minimize the size of β or maximize what is called the **power of a test.**

The power of a statistical test measures the sensitivity of the test to detect a real difference in parameters if one actually exists. The power of a test is a probability and, like all probabilities, can have values ranging from zero to one. The higher the power, the more sensitive the test is to detecting a real difference between parameters if there is a difference. In other words, the closer the power of a test is to one, the better the test is for rejecting the null hypothesis if the null hypothesis is, in fact, false.

The power of a test is equal to $1 - \beta$; that is, 1 minus the probability of committing a type II error. The power of the test is shown in the upper right-hand block of Figure 9–41. If somehow it were known that $\beta = 0.04$, then the power of a test would be $1 - 0.04 = 0.96$ or 96%. In this case, the probability of rejecting the null hypothesis when it is false is 96%.

As stated previously, the power of a test depends on the probability of committing a type II error, and since β is not easily computed, the power of a test cannot be easily computed. (See the Critical Thinking Challenge on page 396.)

However, there are some guidelines that can be used when conducting a statistical study concerning the power of a test. When conducting a statistical study, use the test that has the highest power for the data. There are times when the researcher has a choice of two or more statistical tests to test his or her hypotheses. The tests with the highest power should be used. It is important, however, to remember that statistical tests have assumptions that need to be considered.

If these assumptions cannot be met, then another test with lower power should be used. The power of a test can be increased by increasing the value of α. For example, instead of using $\alpha = 0.01$, use $\alpha = 0.05$. Recall that as α increases, β decreases. So if β is decreased, then $1 - \beta$ will increase, thus increasing the power of the test.

Another way to increase the power of a test is to select a larger sample size. A larger sample size would make the standard error of the mean smaller and consequently reduce β. (The derivation is omitted.)

These two methods should not be used at the whim of the researcher. Before α can be increased, the researcher must consider the consequences of committing a type I error. If these consequences are more serious than the consequences of committing a type II error, then α should not be increased.

Likewise, there are consequences of increasing sample size. These consequences might include an increase in the amount of money required to do the study and an increase in the time needed to tabulate the data. When these consequences result, increasing the sample size may not be practical.

There are several other methods a researcher can use to increase the power of a statistical test, but these methods are beyond the scope of this book.

One final comment is necessary. When the researcher fails to reject the null hypothesis, this does not mean that there is not enough evidence to support alternative hypotheses. It may be that the null hypothesis is false, but the statistical test has too low a power to detect the real difference; hence, one can conclude only that in this study, there is not enough evidence to reject the null hypothesis.

The relationship among α, β, and the power of a test can be analyzed in more detail than the explanation given here. However, it is hoped that this explanation will show the student that there is no magic formula or statistical test that can guarantee foolproof results when a decision is made about the validity of H_0. Whether the decision is to reject H_0 or not to reject H_0, there is in either case a chance of being wrong. The goal, then, is to try to keep the probabilities of type I and type II errors as small as possible.

Exercises

9–88. Explain how confidence intervals are related to hypothesis testing.

9–89. A ski-shop manager claims that the average of the sales for her shop is $1800 a day during the winter months. Ten winter days are selected at random, and the mean of the sales is $1830. The standard deviation of the population is $200. Can one reject the claim at $\alpha = 0.05$? Find the 95% confidence interval of the mean. Does the confidence interval interpretation agree with the hypothesis test results? Explain. Assume that the variable is normally distributed.

9–90. Charter bus records show that in past years, the buses carried an average of 42 people per trip to Niagara Falls. The standard deviation of the population in the past was found to be 8. This year, the average of 10 trips showed a mean of 48 people booked. Can one reject the claim, at $\alpha = 0.10$, that the average is still the same? Find the 90% confidence interval of the mean. Does the confidence interval interpretation agree with the hypothesis-testing results? Explain. Assume that the variable is normally distributed.

9–91. The sales manager of a rental agency claims that the monthly maintenance fee for a condominium in the Lakewood region is $86. Past surveys showed that the standard deviation of the population is $6. A sample of 15 owners shows that they pay an average of $84. Test the manager's claim at $\alpha = 0.01$. Find the 99% confidence interval of the mean. Does the confidence interval interpretation agree with the results of the hypothesis test? Explain. Assume that the variable is normally distributed.

9–92. The average time it takes a person in a one-person canoe to complete a certain river course is 47 minutes. Because of rapid currents in the spring, a group of 10 people traverse the course in 42 minutes. The standard deviation, known from previous trips, is 7 minutes. Test the claim that this group's time was different because of the strong currents. Use $\alpha = 0.10$. Find the 90% confidence level of the true mean. Does the confidence interval interpretation agree with the results of the hypothesis test? Explain. Assume that the variable is normally distributed.

9–93. From past studies the average time college freshmen spend studying is 22 hours per week. The standard deviation is 4 hours. This year, 60 students were surveyed, and the average time that they spent studying was 20.8 hours. Test the claim that the time students spend studying has changed. Use $\alpha = 0.01$. It is believed that the standard deviation is unchanged. Find the 99% confidence interval of the mean. Do the results agree? Explain.

9–94. A survey taken several years ago found that the average time a person spent reading the local daily newspaper was 10.8 minutes. The standard deviation of the population was 3 minutes. In order to see whether the average time had changed since the newspaper's format was revised, the newspaper editor surveyed 36 individuals.

The average time that the 36 people spent reading the paper was 12.2 minutes. At $\alpha = 0.02$, is there a change in the average time an individual spends reading the newspaper? Find the 98% confidence interval of the mean. Do the results agree? Explain.

9–95. What is meant by the power of a test?

9–96. How is the power of a test related to the type II error?

9–97. How can the power of a test be increased?

Summary

This chapter introduces the basic concepts of hypothesis testing. A statistical hypothesis is a conjecture about a population. There are two types of statistical hypotheses: the null and the alternative hypotheses. The null hypothesis states that there is no difference, and the alternative hypothesis specifies a difference. To test the null hypothesis, researchers use a statistical test. Many test values are computed by using

$$\text{test value} = \frac{(\text{observed value}) - (\text{expected value})}{\text{standard error}}$$

Two common statistical tests are the z test and the t test. The z test is used when the population standard deviation is known and the variable is normally distributed or when σ is not known and the sample size is greater than or equal to 30. In this case, s is used in place of σ regardless of the shape of the distribution. When the population standard deviation is not known and the variable is normally distributed, the sample standard deviation is used, but a t test should be conducted when the sample size is less than 30. The z test is also used to test proportions when $np \geq 5$ and $nq \geq 5$.

Researchers compute a test value from the sample data in order to decide whether the null hypothesis should or should not be rejected. Statistical tests can be one-tailed or two-tailed, depending on the hypotheses.

The null hypothesis is rejected when the difference between the population parameter and the sample statistic is said to be significant. The difference is significant when the test value falls in the critical region of the distribution. The critical region is determined by α, the level of significance of the test. The level is the probability of committing a type I error. This error occurs when the null hypothesis is rejected when it is true. Three generally agreed-upon significance levels are 0.10, 0.05, and 0.01. A second kind of error, the type II error, can occur when the null hypothesis is not rejected when it is false.

Finally, one can test a single variance by using a chi-square test.

All hypothesis-testing situations using the traditional method should include the following steps:

1. State the null and alternative hypotheses and identify the claim.
2. State an alpha level and find the critical value(s).
3. Compute the test value.
4. Make the decision to reject or not reject the null hypothesis.
5. Summarize the results.

All hypothesis-testing situations using the P-value method should include the following steps:

1. State the hypotheses and identify the claim.
2. Compute the test value.
3. Find the P-value.
4. Make the decision.
5. Summarize the results.

Important Terms

α (alpha)   342

alternative
hypothesis   338

chi-square test   380

β (beta)   342

critical or rejection
region   343

critical value   343

hypothesis testing   337

left-tailed test   343

level of significance   342

noncritical or nonrejection
region   343

null hypothesis   338

one-tailed test   343

power of a test   389

P-value   354

right-tailed test   343

statistical
hypothesis   337

statistical test   340

test value   340

t test   364

two-tailed test   344

type I error   341

type II error   341

z test   349

Important Formulas

Formula for the z test for means:

$$z = \frac{\bar{X} - \mu}{\sigma/\sqrt{n}} \quad \text{for any value } n$$

$$z = \frac{\bar{X} - \mu}{s/\sqrt{n}} \quad \text{for} \quad n \geq 30$$

Formula for the t test for means:

$$t = \frac{\bar{X} - \mu}{s/\sqrt{n}} \quad \text{for} \quad n < 30$$

Formula for the z test for proportions:

$$z = \frac{X - \mu}{\sigma} \quad \text{or} \quad z = \frac{\hat{p} - p}{\sqrt{pq/n}}$$

Formula for the chi-square test for variance or standard deviation:

$$\chi^2 = \frac{(n - 1)s^2}{\sigma^2}$$

Review Exercises

For exercises 9–98 through 9–117, perform each of the following steps.

a. State the hypotheses and identify the claim.
b. Find the critical value(s).
c. Compute the test value.
d. Make the decision.
e. Summarize the results.

Use the traditional method of hypothesis testing unless otherwise specified.

9–98.   The Medical Rehabilitation Education Foundation found that the average cost of cardiac rehabilitation is $16,411. An administrator at Pine Valley Rehabilitation Center sampled 40 cardiac patients and found the mean cost was $14,706 and the standard deviation was $2,016. At $\alpha = 0.05$, can it be concluded that the average cost is different from $16,411?

Source: Snapshot, *USA Today,* September 18, 1995.

9–99.   A meteorologist claims that the average of the highest temperatures in the United States is 98°. A random sample of 50 cities is selected and the highest temperatures are recorded. The data are shown. At $\alpha = 0.05$ can the claim be rejected?

97	94	96	105	99
96	80	95	101	97
101	87	88	97	94
98	95	88	94	94
99	99	98	96	96
97	98	99	92	97
99	108	97	98	114
91	96	102	99	102
100	93	88	102	99
98	80	95	101	61

Source: *The World Almanac & Book of Facts 1999,* p. 222.

9–100.   A recent study stated that if a person smoked, the average of the number of cigarettes he or she smoked was 14 per day. To test the claim, a researcher selected a random sample of 40 smokers and found that the mean number of cigarettes smoked per day was 18. The standard deviation of the sample was 6. At $\alpha = 0.05$, is the number of cigarettes a person smokes per day actually different from 14?

9–101.   A high school counselor wishes to test the theory that the average age of the dropouts in her school district is

16.3 years. She samples 32 recent dropouts and finds that their mean age is 16.9 years. At $\alpha = 0.01$, is the theory refutable? The standard deviation of the population is 0.3. Use the P-value method.

9–102. In a certain city, a researcher wishes to determine whether the average age of its citizens is really 61.2 years, as the mayor claims. A sample of 22 residents has an average age of 59.8 years. The standard deviation of the sample is 1.5 years. At $\alpha = 0.01$, is the average age of the residents really different from 61.2 years? Assume that the variable is approximately normally distributed.

9–103. The average temperature during the summer months for the northeastern part of the United States is 67.0°. A sample of 10 cities had an average temperature of 69.6° for the summer of 1995. The standard deviation of the sample is 1.1°. At $\alpha = 0.10$, can it be concluded that the summer of 1995 was warmer than average?

Source: *USA Today*, September 14, 1995.

9–104. A recent study claimed that the average age of murder victims in a small city was less than or equal to 23.2 years. A sample of 18 recent victims had a mean of 22.6 years and a standard deviation of 2 years. At $\alpha = 0.05$, is the average age higher than originally believed? Assume that the variable is approximately normally distributed. Use the P-value method.

9–105. A park manager suggests that the average attendance for the 10 most popular national parks last year was 6 million. The number of visits for the 10 parks this year is shown. At $\alpha = 0.02$ has the average attendance changed? The data are in millions of people.

4.7	17.2	14.0	6.1	6.1
4.9	9.3	9.4	6.4	6.1

Source: *USA Today*, August 29, 1997.

9–106. The financial aid director of a college believes that at least 30% of the students are receiving some sort of financial aid. To see whether his belief is correct, the director selects a sample of 60 students and finds that 15 are receiving financial aid. At $\alpha = 0.05$, test the claim that at least 30% of the students are receiving financial aid.

9–107. A dietitian read in a survey that at least 60% of adults eat eggs for breakfast at least four times a week. To test this claim, she selected a random sample of 100 adults and asked them how many days a week they ate eggs. In her sample, 54% responded that they ate eggs at least four times a week. At $\alpha = 0.10$, do her results refute the survey?

9–108. A contractor desires to build new homes with fireplaces. He read in a survey that 80% of all home buyers want a fireplace. To test this claim, he selected a sample of 30 home buyers and found that 20 wanted a fireplace. Use $\alpha = 0.02$ to test the claim.

9–109. A radio manufacturer claims that 65% of teenagers 13–16 years old have their own portable radios. A researcher wishes to test the claim and selects a random sample of 80 teenagers. She finds that 57 have their own portable radios. At $\alpha = 0.05$, should the claim be rejected? Use the P-value method.

9–110. A football coach claims that the average weight of all the opposing teams' members is 225 pounds. For a test of the claim, a sample of 50 players is taken from all the opposing teams. The mean is found to be 230 pounds and the standard deviation is 15 pounds. At $\alpha = 0.01$, test the coach's claim. Find the P-value and make the decision.

9–111. An advertisement claims that Fasto Stomach Calm will provide relief from indigestion in less than 10 minutes. For a test of the claim, 35 individuals were given the product; the average time until relief was 9.25 minutes. From past studies, the standard deviation is known to be 2 minutes. Can one conclude that the claim is justified? Find the P-value and let $\alpha = 0.05$.

9–112. A film editor feels that the standard deviation for the number of minutes in a video is 3.4 minutes. A sample of 24 videos has a standard deviation of 4.2 minutes. At $\alpha = 0.05$, is the sample standard deviation different from what the editor hypothesized?

9–113. The standard deviation of the fuel consumption of a certain automobile is hypothesized to be greater than or equal to 4.3 miles per gallon. A sample of 20 automobiles produced a standard deviation of 2.6 miles per gallon. Is the standard deviation really less than previously thought? Use $\alpha = 0.05$. Use the P-value method.

9–114. A real estate agent claims that the standard deviation of the rental rates of apartments in a certain county is $95. A random sample of rates is shown. At $\alpha = 0.02$ can the claim be refuted?

$400	$345	$325	$395	$400	$300
375	435	495	525	290	460
425	250	200	525	375	390

Source: *Pittsburgh Tribune Review*, June 13, 1999.

9–115. A manufacturer claims that the standard deviation of the drying time of a certain type of paint is 18 minutes. A sample of five test panels produced a standard deviation of 21 minutes. Test the claim at $\alpha = 0.05$.

9–116. In order to see whether people are keeping their car tires inflated to the correct level, 35 pounds per square inch (psi), a tire company manager selects a sample of 36 tires and checks the pressure. The mean of the sample is 33.5 psi, and the standard deviation is 3. Are the tires properly inflated? Use $\alpha = 0.10$. Find the 90% confidence interval of the mean. Do the results agree? Explain.

9–117. A biologist knows that the average length of a leaf of a certain full-grown plant is 4 inches. The standard deviation of the population is 0.6 inch. A sample of 20 leaves of that type of plant given a new type of plant food had an average length of 4.2 inches. Is there reason to believe that the new food is responsible for a change in the growth of the leaves? Use $\alpha = 0.01$. Find the 99% confidence interval of the mean. Do the results concur? Explain. Assume that the variable is approximately normally distributed.

Statistics Today

How Much Better Is Better? Revisited

Now that you have learned the techniques of hypothesis testing presented in this chapter, you realize that the difference between the sample mean and the population mean must be *significant* before one can conclude that the students really scored above average. The superintendent should follow the steps in the hypothesis-testing procedure and be able to reject the null hypothesis before announcing that his students scored higher than average.

Data Analysis

The Data Bank is found in Appendix D, or on the World Wide Web by following links from **www.mhhe.com/math/stats/bluman/**

1. From the Data Bank, select a random sample of at least 30 individuals, and test one or more of the following hypotheses by using the z test. Use $\alpha = 0.05$.
   a. For serum cholesterol, H_0: $\mu = 220$ milligram percent (mg%).
   b. For systolic pressure, H_0: $\mu = 120$ millimeters of mercury (mm Hg).
   c. For IQ, H_0: $\mu = 100$.
   d. For sodium level, H_0: $\mu = 140$ milliequivalents per liter (mEq/l).

2. Select a random sample of 15 individuals and test one or more of the hypotheses in Exercise 1 by using the t test. Use $\alpha = 0.05$.

3. Select a random sample of at least 30 individuals, and using the z test for proportions, test one or more of the following hypotheses. Use $\alpha = 0.05$.
   a. For educational level, H_0: $p = 0.50$ for level 2.
   b. For smoking status, H_0: $p = 0.20$ for level 1.
   c. For exercise level, H_0: $p = 0.10$ for level 1.
   d. For gender, H_0: $p = 0.50$ for males.

4. Select a sample of 20 individuals and test the hypothesis H_0: $\sigma^2 = 225$ for IQ level. Use $\alpha = 0.05$.

5. Using the data from Data Set X in Appendix D, test the hypothesis that H_0: $\mu = 300$. Use the z test. Use $\alpha = 0.05$.

6. Using the data from Data Set XVI, select a sample of 10 hospitals and test H_0: $\mu \geq 250$ for the number of beds. Use $\alpha = 0.05$.

7. Using the data obtained in the previous problem, test the hypothesis H_0: $\sigma \geq 150$. Use $\alpha = 0.05$.

Quiz

Determine whether each statement is true or false. If the statement is false, explain why.

1. No error is committed when the null hypothesis is rejected when it is false.

2. When one is conducting the t test, the population must be approximately normally distributed.

3. The test value separates the critical region from the noncritical region.

4. The values of a chi-square test cannot be negative.

5. The chi-square test for variances is always one-tailed.

Select the best answer.

6. When the value of α is increased, the probability of committing a type I error is
   a. Decreased
   b. Increased
   c. The same
   d. None of the above.

7. If one wishes to test the claim that the mean of the population is 100, the appropriate null hypothesis is

a. $\overline{X} = 100$
b. $\mu \geq 100$
c. $\mu \leq 100$
d. $\mu = 100$

8. The degrees of freedom for the chi-square test for variances or standard deviations are
   a. One
   b. n
   c. $n - 1$
   d. None of the above.

9. For the z test, if σ is unknown and $n \geq 30$, one can substitute for σ
   a. n
   b. s
   c. χ^2
   d. t

Complete the following statements with the best answer.

10. Rejecting the null hypothesis when it is true is called a _____ error.

11. The probability of a type II error is referred to as _____.

12. A conjecture about a population parameter is called a _____.

13. To test the hypothesis H_0: $\mu \leq 87$, one would use a _____ tailed test.

14. The degrees of freedom for the t test are _____.

For the following exercises where applicable:

   a. **State the hypotheses**
   b. **Find the critical value(s)**
   c. **Compute the test value**
   d. **Make the decision**
   e. **Summarize the results**

Use the traditional method of hypothesis testing unless otherwise specified.

15. A sociologist wishes to see if it is true that for a certain group of professional women, the average age at which they have their first child is 28.6. A random sample of 36 women is selected and their ages at the birth of their first children are recorded. At $\alpha = 0.05$, does the evidence refute the sociologist's assertion?

32	28	26	33	35	34
29	24	22	25	26	28
28	34	33	32	30	29
30	27	33	34	28	25
24	33	25	37	35	33
34	36	38	27	29	26

16. A real estate agent believes that the average closing cost of purchasing a new home is $6500 over the purchase price. She selects 40 new home sales at random and finds that the average closing costs are $6600. The standard deviation of the population is $120. Test her belief at $\alpha = 0.05$.

17. A recent study stated that if a person chewed gum, the average number of sticks of gum he or she chewed daily was eight. To test the claim, a researcher selected a random sample of 36 gum chewers and found the mean number of sticks of gum chewed per day was nine. The standard deviation was 1. At $\alpha = 0.05$, is the number of sticks of gum a person chews per day actually greater than eight?

18. A high school counselor wishes to see if the average number of dropouts in his school is 21. He reviews the last 17 years and finds that the number of dropouts each year is as shown. At $\alpha = 0.01$, is the hypothesis refutable?

12	18	24	16	21	20	18	19
19	22	25	16	18	19	19	20	23

19. In a New York modeling agency, a researcher wishes to see if the average height of female models is really less than 67 inches, as the chief claims. A sample of 20 models has an average height of 65.8 inches. The standard deviation of the sample is 1.7 inches. At $\alpha = 0.05$, is the average height of the models really less than 67 inches? Use the P-value method.

20. A taxi company claims that its drivers average at least 12.4 years of experience. In a study of 15 taxi drivers, the average experience was 11.2 years. The standard deviation was 2. At $\alpha = 0.10$, is the number of years' experience of the taxi drivers really less than the taxi company claimed?

21. A recent study in a small city stated that the average age of robbery victims was 63.5. A sample of 20 recent victims had a mean of 63.7 years and a standard deviation of 1.9. At $\alpha = 0.05$, is the average age higher than originally believed? Use the P-value method.

22. A magazine article stated that the average age of women who are getting married for the first time is 26 years. A researcher decided to test this hypothesis at $\alpha = 0.02$. She selected a sample of 25 women who were recently married for the first time and found the average was 25.1. The standard deviation was 3 years. Should the null hypothesis be rejected on the basis of the sample?

23. The president of a local college reported that at least 25% of the students have received scholarships. To see if this is true, the registrar selected a sample of 100 students and found that 22 received scholarships. At $\alpha = 0.05$, test the claim that at least 25% of the students have scholarships.

24. A dietitian read in a survey that at least 55% of adults do not eat breakfast at least three days a week. To verify this, she selected a random sample of 80 adults and asked them how many days a week they skipped breakfast. A total of 50% responded that they skipped breakfast at least three days a week. At $\alpha = 0.10$, test the claim.

25. A contractor wanted to build new homes with two-car garages. He read in a survey that 70% of all home buyers want a two-car garage. To verify the figure, he selected a sample of 30 home buyers and found that 23 wanted a two-car garage. At $\alpha = 0.02$, test the claim.

26. A magazine claims that 75% of all teenage boys have their own radios. A researcher wished to test the claim and selected a random sample of 60 teenage boys. She found that 54 had their own radios. At $\alpha = 0.01$, should the claim be rejected?

27. Find the *P*-value for the *z* test in Exercise 15.

28. Find the *P*-value for the *z* test in Exercise 16.

29. A copyeditor thinks the standard deviation for the number of pages in a romance novel is greater than six.

A sample of 25 novels has a standard deviation of nine pages. At $\alpha = 0.05$, is it higher as the editor hypothesized?

30. It has been hypothesized that the standard deviation of the germination time of radish seeds is eight days. The standard deviation of a sample of 60 radish plants' germination times was six days. At $\alpha = 0.01$, test the claim.

31. The standard deviation of the pollution by-products released in the burning of one gallon of gas is 2.3 ounces. A sample of 20 automobiles tested produced a standard deviation of 1.9 ounces. Is the standard deviation really less than previously thought? Use $\alpha = 0.05$.

32. A manufacturer claims that the standard deviation of the strength of wrapping cord is nine pounds. A sample of 10 wrapping cords produced a standard deviation of 11 pounds. At $\alpha = 0.05$, test the claim. Use the *P*-value method.

33. Find the 90% confidence interval of the mean in Exercise 15. Is μ contained in the interval?

34. Find the 95% confidence interval for the mean in Exercise 16. Is μ contained in the interval?

Critical Thinking Challenge

The power of a test $(1 - \beta)$ can be calculated when a specific value of the mean is hypothesized in the alternative hypothesis; for example, let H_0: $\mu = 50$ and let H_1: $\mu = 52$. In order to find the power of a test, it is necessary to find the value of β. This can be done by the following steps:

1. For a specific value of α find the corresponding value of $\overline{X}$ using $z = \dfrac{\overline{X} - \mu}{\sigma/\sqrt{n}}$, where μ is the hypothesized value given in H_0. Use a right-tailed test.

2. Using the value of $\overline{X}$ found in step 1 and the value of μ in the alternative hypothesis find the area corresponding to z in the formula $z = \dfrac{\overline{X} - \mu}{\sigma/\sqrt{n}}$.

3. Subtract this area from 0.5000. This is the value of β.

4. Subtract the value of β from one. This will give you the power of a test.

Figure 9–42 shows the steps.

Figure 9–42

Relationship Among α, β, and the Power of a Test

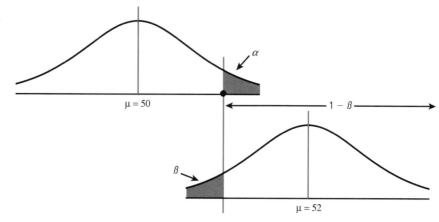

a. Find the power of a test using the hypotheses given previously and $\alpha = 0.05$, $\sigma = 3$, and $n = 30$.

b. Select several other values for μ in H_1 and compute the power of the test. Generalize the results.

🌐 Data Projects

Use MINITAB, the TI-83, or a computer program of your choice to complete the following exercises.

1. Choose a variable such as the number of miles students live from the college or the number of daily admissions to an ice-skating rink for a one-month period. Before collecting the data, decide what a likely average might be, then complete the following:
   a. Write a brief statement of the purpose of the study.
   b. Define the population.
   c. State the hypotheses for the study.
   d. Select an α value.
   e. State how the sample was selected.
   f. Show the raw data.
   g. Decide which statistical test is appropriate and compute the test statistic (z or t). Why is the test appropriate?
   h. Find the critical value(s).
   i. State the decision.
   j. Summarize the results.

2. Decide on a question that could be answered with a "yes" or "no" or "not sure (undecided)." For example, "Are you satisfied with the variety of food the cafeteria serves?" Before collecting the data, hypothesize what you expect the proportion of students who will respond "yes" will be. Then complete the following:
   a. Write a brief statement of the purpose of the study.
   b. Define the population.
   c. State the hypotheses for the study.
   d. Select an α value.
   e. State how the sample was selected.
   f. Show the raw data.
   g. Decide which statistical test is appropriate and compute the test statistic (z or t). Why is the test appropriate?
   h. Find the critical value(s).
   i. State the decision.
   j. Summarize the results.

Do you think the project supported your initial hypothesis? Explain your answer.

You may use the following websites to obtain raw data:

    http://www.mhhe.com/math/stat/bluman/
    http://lib.stat.cmu.edu/DASL
    http://www.oecd.org/statlist.htm
    http://www.statcan.ca/english/

chapter

10

Testing the Difference between Two Means, Two Variances, and Two Proportions

Objectives

After completing this chapter, you should be able to

1. Test the difference between two large sample means using the z test.

2. Test the difference between two variances or standard deviations.

3. Test the difference between two means for small independent samples.

4. Test the difference between two means for small dependent samples.

5. Test the difference between two proportions.

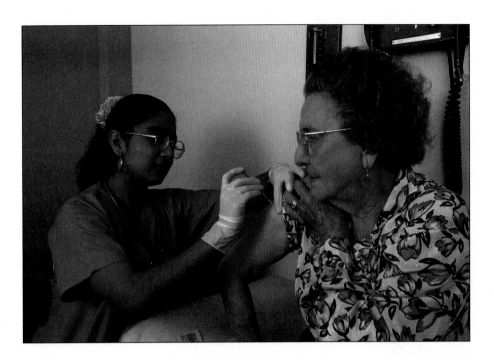

Statistics Today

To Vaccinate or Not to Vaccinate? Small or Large?

Influenza is a serious disease among the elderly, especially those living in nursing homes. Those residents are more susceptible to influenza than elderly persons living in the community because the former are usually older and more debilitated, and they live in a closed environment where they are exposed more so than community residents to the virus if it is introduced into the home. Three researchers decided to investigate the use of vaccine and its value in determining outbreaks of influenza in small nursing homes.

These researchers surveyed 83 licensed homes in seven counties in Michigan. Part of the study consisted of comparing the number of people being vaccinated in small nursing homes (100 or fewer beds) with the number in larger nursing homes (more than 100 beds). Unlike the statistical methods presented in the previous chapter, these researchers used the techniques explained in this chapter to compare two sample proportions to see if there was a significant difference in the vaccination rates of patients in small nursing homes as compared to those in larger nursing homes.

Source: Nancy Arden, Arnold S. Monto, and Suzanne E. Ohmit, "Vaccine Use and the Risk of Outbreaks in a Sample of Nursing Homes during an Influenza Epidemic," *American Journal of Public Health* 85, no. 3 (March 1995), pp. 399–401.

10–1

Introduction

The basic concepts of hypothesis testing were explained in Chapter 9. With the z and t tests, a sample mean, variance, or proportion can be compared to a specific population mean, variance, or proportion in order to determine whether the null hypothesis should be rejected.

There are, however, many instances when researchers wish to compare two sample means using experimental and control groups. For example, the average lifetimes of two different brands of bus tires might be compared to see whether there is any difference in tread wear. Two different brands of fertilizer might be tested to see whether one is better than the other for growing plants. Or two brands of cough syrup might be tested to see whether one brand is more effective than the other.

In the comparison of two means, the same basic steps for hypothesis testing shown in Chapter 9 are used, and the z and t tests are also used. When comparing two means using the t test, the researcher must decide if the two samples are *independent* or *dependent*. The concepts of independent and dependent samples will be explained in Sections 10–4 and 10–5.

Furthermore, when the samples are independent, there are two different formulas that can be used depending on whether or not the variances are equal. To determine if the variances are equal, use the F test shown in Section 10–3. Finally, the z test can be used to compare two proportions, as shown in Section 10–6.

10–2

Testing the Difference between Two Means: Large Samples

Objective 1. Test the difference between two large sample means using the z test.

Suppose a researcher wishes to determine whether there is a difference in the average age of nursing students who enroll in a nursing program at a community college and those who enroll in a nursing program at a university. In this case, the researcher is not interested in the average age of all beginning nursing students; instead, he is interested in *comparing* the means of the two groups. His research question is: Does the mean age of nursing students who enroll at a community college differ from the mean age of nursing students who enroll at a university? Here, the hypotheses are:

$$H_0: \mu_1 = \mu_2$$
$$H_1: \mu_1 \neq \mu_2$$

where

μ_1 = mean age of all beginning nursing students at the community college

μ_2 = mean age of all beginning nursing students at the university

Another way of stating the hypotheses for this situation is

$$H_0: \mu_1 - \mu_2 = 0$$
$$H_1: \mu_1 - \mu_2 \neq 0$$

If there is no difference in population means, subtracting them will give a difference of zero. If they are different, subtracting will give a number other than zero. Both methods of stating hypotheses are correct; however, the first method will be used in this book.

Assumptions for the Test to Determine the Difference between Two Means

1. The samples must be independent of each other. That is, there can be no relationship between the subjects in each sample.
2. The populations from which the samples were obtained must be normally distributed, and the standard deviations of the variable must be known, or the sample sizes must be greater than or equal to 30.

The theory behind testing the difference between two means is based on selecting pairs of samples and comparing the means of the pairs. The population means need not be known.

All possible pairs of samples are taken from populations. The means for each pair of samples are computed and then subtracted, and the differences are plotted. If both populations have the same mean, then most of the differences will be zero or close to zero. Occasionally, there will be a few large differences due to chance alone, some positive and some negative. If the differences are plotted, the curve will be shaped like the normal distribution and have a mean of zero, as shown in Figure 10–1.

Figure 10–1

Differences of Means of Pairs of Samples

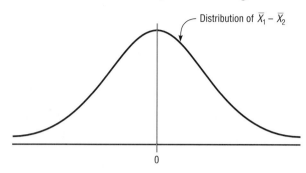

Distribution of $\bar{X}_1 - \bar{X}_2$

0

The variance of the difference $\bar{X}_1 - \bar{X}_2$ is equal to the sum of the individual variances of $\bar{X}_1$ and $\bar{X}_2$. That is,

$$\sigma^2_{\bar{X}_1 - \bar{X}_2} = \sigma^2_{\bar{X}_1} + \sigma^2_{\bar{X}_2}$$

where $\qquad \sigma^2_{\bar{X}_1} = \dfrac{\sigma^2_1}{n_1} \qquad$ and $\qquad \sigma^2_{\bar{X}_2} = \dfrac{\sigma^2_2}{n_2}$

So the standard deviation of $\bar{X}_1 - \bar{X}_2$ is

$$\sqrt{\frac{\sigma^2_1}{n_1} + \frac{\sigma^2_2}{n_2}}$$

Formula for the z Test for Comparing Two Means from Independent Populations

$$z = \frac{(\bar{X}_1 - \bar{X}_2) - (\mu_1 - \mu_2)}{\sqrt{\dfrac{\sigma^2_1}{n_1} + \dfrac{\sigma^2_2}{n_2}}}$$

This formula is based on the general format of

$$\text{test value} = \frac{(\text{observed value}) - (\text{expected value})}{\text{standard error}}$$

where $\bar{X}_1 - \bar{X}_2$ is the observed difference, and the expected difference $\mu_1 - \mu_2$ is zero when the null hypothesis is $\mu_1 = \mu_2$, since that is equivalent to $\mu_1 - \mu_2 = 0$. Finally, the standard error of difference is

$$\sqrt{\frac{\sigma^2_1}{n_1} + \frac{\sigma^2_2}{n_2}}$$

In the comparison of two sample means, the difference may be due to chance, in which case the null hypothesis will not be rejected, and the researcher can assume that the means of the populations are basically the same. The difference in this case is not significant. See Figure 10–2(a). On the other hand, if the difference is significant, the null hypothesis is rejected, and the researcher can conclude that the population means are different. See Figure 10–2(b).

Figure 10–2

Hypothesis-Testing Situations in the Comparison of Means

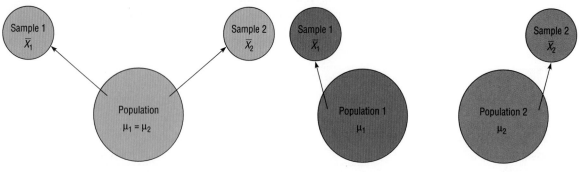

(a) Difference is not significant

Do not reject $H_0: \mu_1 = \mu_2$ since $\overline{X}_1 - \overline{X}_2$ is not significant.

(b) Difference is significant

Reject $H_0: \mu_1 = \mu_2$ since $\overline{X}_1 - \overline{X}_2$ is significant.

These tests can also be one-tailed, using the following hypotheses.

Right-tailed		Left-tailed	
$H_0: \mu_1 \leq \mu_2$	or	$H_0: \mu_1 \geq \mu_2$	or
$H_1: \mu_1 > \mu_2$	$H_0: \mu_1 - \mu_2 \leq 0$	$H_1: \mu_1 < \mu_2$	$H_0: \mu_1 - \mu_2 \geq 0$
	$H_1: \mu_1 - \mu_2 > 0$		$H_1: \mu_1 - \mu_2 < 0$

The same critical values used in Section 9–3 are used here. They can be obtained from Table E in Appendix C.

If σ_1^2 and σ_2^2 are not known, the researcher can use the variances obtained from each sample, s_1^2 and s_2^2, but both sample sizes must be 30 or more. The formula then is

$$z = \frac{(\overline{X}_1 - \overline{X}_2) - (\mu_1 - \mu_2)}{\sqrt{\dfrac{s_1^2}{n_1} + \dfrac{s_2^2}{n_2}}}$$

provided that $n_1 \geq 30$ and $n_2 \geq 30$.

When one or both sample sizes are less than 30 and σ_1 and σ_2 are unknown, the t test must be used, as shown in Section 10–4.

The basic format for hypothesis testing using the traditional method is reviewed here.

STEP 1 State the hypotheses and identify the claim.

STEP 2 Find the critical value(s).

STEP 3 Compute the test value.

STEP 4 Make the decision.

STEP 5 Summarize the results.

Example 10–1

A survey found that the average hotel room rate in New Orleans is $88.42 and the average room rate in Phoenix is $80.61. Assume that the data were obtained from two samples of 50 hotels each and that the standard deviations were $5.62 and $4.83 respectively. At $\alpha = 0.05$, can it be concluded that there is a significant difference in the rates?

Source: *USA Today,* January 10, 1995.

Solution

STEP 1   State the hypotheses and identify the claim.

$$H_0: \mu_1 = \mu_2 \quad \text{and} \quad H_1: \mu_1 \neq \mu_2 \text{ (claim)}$$

STEP 2   Find the critical values. Since $\alpha = 0.05$, the critical values are $+1.96$ and -1.96.

STEP 3   Compute the test value.

$$z = \frac{(\bar{X}_1 - \bar{X}_2) - (\mu_1 - \mu_2)}{\sqrt{\dfrac{s_1^2}{n_1} + \dfrac{s_2^2}{n_2}}} = \frac{(88.42 - 80.61) - 0}{\sqrt{\dfrac{5.62^2}{50} + \dfrac{4.83^2}{50}}} = 7.45$$

STEP 4   Make the decision. Reject the null hypothesis at $\alpha = 0.05$, since $7.45 > 1.96$. See Figure 10–3.

Figure 10–3

Critical and Test Values for Example 10–1

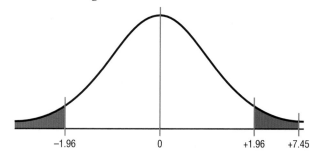

STEP 5   Summarize the results. There is enough evidence to support the claim that the means are not equal. Hence, there is a significant difference in the rates.

The P-values for this test can be determined using the same procedure as shown in Section 9–3. For example, if the test value for a two-tailed test is 1.40, then the P-value obtained from Table E is 0.1616. This value is obtained by looking up the area for $z = 1.40$, which is 0.4192. Then 0.4192 is subtracted from 0.5000 to get 0.0808. Finally, this value is doubled to get 0.1616 since the test is two-tailed. If $\alpha = 0.05$, the decision would be do not reject the null hypothesis, since P-value $< \alpha$.

The P-value method for hypothesis testing for this chapter also follows the same format as stated in Chapter 9. The steps are reviewed here.

STEP 1   State the hypotheses and identify the claim.

STEP 2   Compute the test value.

STEP 3   Find the P-value.

STEP 4   Make the decision.

STEP 5   Summarize the results.

The next example illustrates these steps.

Example 10–2

A researcher hypothesizes that the average number of sports colleges offer for males is greater than the average number of sports colleges offer for females. A sample of the number of sports offered by colleges is shown. At $\alpha = 0.10$, is there enough evidence to support the claim?

Males					Females				
6	11	11	8	15	6	8	11	13	8
6	14	8	12	18	7	5	13	14	6
6	9	5	6	9	6	5	5	7	6
6	9	18	7	6	10	7	6	5	5
15	6	11	5	5	16	10	7	8	5
9	9	5	5	8	7	5	5	6	5
8	9	6	11	6	9	18	13	7	10
9	5	11	5	8	7	8	5	7	6
7	7	5	10	7	11	4	6	8	7
10	7	10	8	11	14	12	5	8	5

Source: *USA Today,* March 4, 1997.

Solution

STEP 1 State the hypotheses and identify the claim:

$$H_0: \mu_1 \le \mu_2$$
$$H_1: \mu_1 > \mu_2 \text{ (claim)}$$

STEP 2 Compute the test value. Using a calculator or the formulas in Chapter 3, find the mean and standard deviation for each data set.

For the males $\overline{X}_1 = 8.6$ and $s_1 = 3.3$

For the females $\overline{X}_2 = 7.9$ and $s_2 = 3.3$

Substitute in the formula:

$$z = \frac{(\overline{X}_1 - \overline{X}_2) - (\mu_1 - \mu_2)}{\sqrt{\dfrac{s_1^2}{n_1} + \dfrac{s_2^2}{n_2}}} = \frac{(8.6 - 7.9) - 0}{\sqrt{\dfrac{3.3^2}{50} + \dfrac{3.3^2}{50}}} = 1.06*$$

STEP 3 Find the *P*-value. For $z = 1.06$, the area is 0.3554, and $0.5000 - 0.3554 = 0.1446$ or a *P*-value of 0.1446.

STEP 4 Make the decision. Since the *P*-value is larger than α (i.e., $0.1446 > 0.10$), the decision is do not reject the null hypothesis. See Figure 10–4.

STEP 5 Summarize the results. There is not enough evidence to support the claim that colleges offer more sports for males than they do for females.

Figure 10–4

P-Value and α Value for Example 10–2

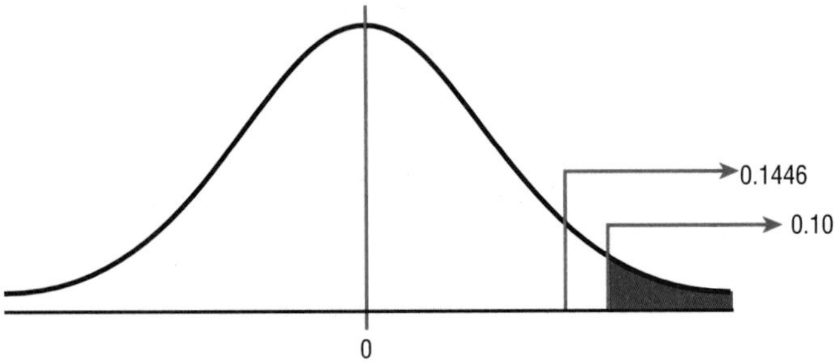

→0.1446

→ 0.10

0

*Note: Calculator results may differ due to rounding.

Sometimes, the researcher is interested in testing a specific difference in means other than zero. For example, he or she might hypothesize that the nursing students at a community college are, on the average, 3.2 years older than those at a university. In this case, the hypotheses are

$$H_0: \mu_1 - \mu_2 \leq 3.2 \qquad \text{and} \qquad H_1: \mu_1 - \mu_2 > 3.2$$

The formula for the z test is still

$$z = \frac{(\overline{X}_1 - \overline{X}_2) - (\mu_1 - \mu_2)}{\sqrt{\dfrac{\sigma_1^2}{n_1} + \dfrac{\sigma_2^2}{n_2}}}$$

where $\mu_1 - \mu_2$ is the hypothesized difference or expected value. In this case, $\mu_1 - \mu_2 = 3.2$.

Confidence intervals for the difference between two means can also be found. When hypothesizing a difference of 0, if the confidence interval contains 0, the null hypothesis is not rejected. If the confidence interval does not contain 0, the null hypothesis is rejected.

Confidence intervals for the difference between two means can be found by using the following formula:

Formula for Confidence Interval for Difference between Two Means: Large Samples

$$(\overline{X}_1 - \overline{X}_2) - z_{\alpha/2} \sqrt{\frac{\sigma_1^2}{n_1} + \frac{\sigma_2^2}{n_2}} < \mu_1 - \mu_2$$

$$< (\overline{X}_1 - \overline{X}_2) + z_{\alpha/2} \sqrt{\frac{\sigma_1^2}{n_1} + \frac{\sigma_2^2}{n_2}}$$

When $n_1 \geq 30$ and $n_2 \geq 30$, s_1^2 and s_2^2 can be used in place of σ_1^2 and σ_2^2.

Example 10–3

Find the 95% confidence interval for the difference between the means for the data in Example 10–1.

Solution

Substitute in the formula using $z_{\alpha/2} = 1.96$.

$$(\overline{X}_1 - \overline{X}_2) - z_{\alpha/2} \sqrt{\frac{\sigma_1^2}{n_1} + \frac{\sigma_2^2}{n_2}} < \mu_1 - \mu_2$$

$$< (\overline{X}_1 - \overline{X}_2) + z_{\alpha/2} \sqrt{\frac{\sigma_1^2}{n_1} + \frac{\sigma_2^2}{n_2}}$$

$$(88.42 - 80.61) - 1.96 \sqrt{\frac{5.62^2}{50} + \frac{4.83^2}{50}} < \mu_1 - \mu_2$$

$$< (88.42 - 80.61) + 1.96 \sqrt{\frac{5.62^2}{50} + \frac{4.83^2}{50}}$$

$$7.81 - 2.05 < \mu_1 - \mu_2 < 7.81 + 2.05$$

$$5.76 < \mu_1 - \mu_2 < 9.86$$

Since the confidence interval does not contain zero, the decision is to reject the null hypothesis, which agrees with the previous result.

Exercises

10–1. Explain the difference between testing a single mean and testing the difference between two means.

10–2. When a researcher selects all possible pairs of samples from a population in order to find the difference between the means of each pair, what will be the shape of the distribution of the differences when the original distributions are normally distributed? What will be the mean of the distribution? What will be the standard deviation of the distribution?

10–3. What two assumptions must be met when one is using the z test to test differences between two means? When can the sample standard deviations, s_1 and s_2, be used in place of the population standard deviations σ_1 and σ_2?

10–4. Show two different ways to state that the means of two populations are equal.

For Exercises 10–5 through 10–17, perform each of the following steps.
a. State the hypotheses and identify the claim.
b. Find the critical value(s).
c. Compute the test value.
d. Make the decision.
e. Summarize the results.

Use the traditional method of hypothesis testing unless otherwise specified.

10–5. A researcher wishes to see if the average length of the major rivers in the United States is the same as the average length of the major rivers in Europe. The data in miles of a sample of rivers are shown below. At $\alpha = 0.01$, is there enough evidence to reject the claim?

United States			Europe		
729	560	434	481	724	820
329	332	360	532	357	505
450	2315	865	1776	1122	496
330	410	1036	1224	634	230
329	800	447	1420	326	626
600	1310	652	877	580	210
1243	605	360	447	567	252
525	926	722	824	932	600
850	310	430	634	1124	1575
532	375	1979	565	405	2290
710	545	259	675	454	
300	470	425			

Source: *The World Almanac and Book of Facts 1999*, pp. 458–459.

10–6. A study was conducted to see if there was a difference between spouses and significant others in coping skills when living with or caring for a person with multiple sclerosis. These skills were measured by questionnaire responses. The results of the two groups are given below on one factor, ambivalence. At $\alpha = 0.10$, is there a difference in the means of the two groups?

Spouses	Significant others
$\overline{X}_1 = 2$	$\overline{X}_2 = 1.7$
$s_1 = 0.6$	$s_2 = 0.7$
$n_1 = 120$	$n_2 = 34$

Source: Elsie E. Gulick, "Coping among Spouses or Significant Others of Persons with Multiple Sclerosis," *Nursing Research* 44, no. 4 (July/August 1995), p. 224.

10–7. A medical researcher wishes to see whether the pulse rates of smokers are higher than the pulse rates of nonsmokers. Samples of 100 smokers and 100 nonsmokers are selected. The results are shown here. Can the researcher conclude, at $\alpha = 0.05$, that smokers have higher pulse rates than nonsmokers?

Smokers	Nonsmokers
$\overline{X}_1 = 90$	$\overline{X}_2 = 88$
$s_1 = 5$	$s_2 = 6$
$n_1 = 100$	$n_2 = 100$

10–8. A statistician claims that the average score on a standardized test of students who major in psychology is greater than that of students who major in mathematics. The results of the test, given to 50 students in each group, are shown here. Is there enough evidence to support the statistician's claim at $\alpha = 0.01$?

Psychology	Mathematics
$\overline{X}_1 = 118$	$\overline{X}_2 = 115$
$\sigma_1 = 15$	$\sigma_2 = 15$
$n_1 = 50$	$n_2 = 50$

10–9. Using data from the "Noise Levels in an Urban Hospital" study cited in Exercise 8–19, test the claim that the noise level in the corridors is higher than that in the clinics. Use $\alpha = 0.02$. The data are shown here.

Corridors	Clinics
$\overline{X}_1 = 61.2$ dBA	$\overline{X}_2 = 59.4$ dBA
$s_1 = 7.9$	$s_2 = 7.9$
$n_1 = 84$	$n_2 = 34$

Source: M. Bayo, A. Garcia, and A. Garcia, "Noise Levels in an Urban Hospital and Workers' Subjective Responses," *Archives of Environmental Health* 50, no. 3 (May–June 1995).

10–10. A real estate agent compares the selling prices of homes in two suburbs of Seattle to see whether there is a difference in price. The results of the study are shown here. Is there evidence to reject the claim that the average cost of a home in both locations is the same? Use $\alpha = 0.01$.

Suburb 1	Suburb 2
$\overline{X}_1 = \$63,255$	$\overline{X}_2 = \$59,102$
$s_1 = \$5602$	$s_2 = \$4731$
$n_1 = 35$	$n_2 = 40$

10–11. In a study of women science majors, the following data were obtained on two groups, those who left their profession within a few months after graduation (leavers) and those who remained in their profession after they graduated (stayers). Test the claim that those who stayed had a higher science grade point average than those who left. Use $\alpha = 0.05$.

Leavers	Stayers
$\overline{X}_1 = 3.16$	$\overline{X}_2 = 3.28$
$s_1 = 0.52$	$s_2 = 0.46$
$n_1 = 103$	$n_2 = 225$

Source: Paula Rayman and Belle Brett, "Women Science Majors: What Makes a Difference in Persistence after Graduation?" *The Journal of Higher Education* 66, no. 4 (July–August 1995), pp. 388–414.

10–12. A college admissions officer believes that students enrolling from Bakersfield School District have higher scores on an entrance achievement test than those who enroll from West River School District. A sample of 30 students from each district is selected, and the scores are shown below. Is there enough evidence to support the officer's belief at $\alpha = 0.05$?

Bakersfield					West River				
656	682	753	609	636	626	630	710	699	730
594	617	698	642	580	531	630	587	598	504
636	721	649	532	691	643	540	543	703	515
642	701	631	597	715	582	606	547	525	613
683	614	626	698	634	637	570	609	624	527
642	632	584	705	749	624	623	673	615	518

10–13. A school administrator hypothesizes that colleges spend more for male sports than they do for female sports. A sample of two different colleges is selected, and the annual expenses per student at each school are shown. At $\alpha = 0.01$, is there enough evidence to support the claim?

Males

$ 7,040	6,576	1,664	12,919	8,605
22,220	3,377	10,128	7,723	2,063
8,033	9,463	7,656	11,456	12,244
6,670	12,371	9,626	5,472	16,175
8,383	623	6,797	10,160	8,725
14,029	13,763	8,811	11,480	9,544
15,048	5,544	10,652	11,267	10,126
8,796	13,351	7,120	9,505	9,571
7,551	5,811	9,119	9,732	5,286
5,254	7,550	11,015	12,403	12,703

Females

$10,333	6,407	10,082	5,933	3,991
7,435	8,324	6,989	16,249	5,922
7,654	8,411	11,324	10,248	6,030
9,331	6,869	6,502	11,041	11,597
5,468	7,874	9,277	10,127	13,371
7,055	6,909	8,903	6,925	7,058
12,745	12,016	9,883	14,698	9,907
8,917	9,110	5,232	6,959	5,832
7,054	7,235	11,248	8,478	6,502
7,300	993	6,815	9,959	10,353

Source: *USA Today,* March 4, 1997.

10–14. Is there a difference in average miles traveled for each of two taxi companies during a randomly selected week? The data are shown below. Use $\alpha = 0.05$. Assume that the populations are normally distributed. Use the *P*-value method.

Moonview Cab Company	Starlight Taxi Company
$\overline{X}_1 = 837$	$\overline{X}_2 = 753$
$\sigma_1 = 30$	$\sigma_2 = 40$
$n_1 = 35$	$n_2 = 40$

10–15. In the study cited in Exercise 10–11, the researchers collected the data shown here on a self-esteem questionnaire. At $\alpha = 0.05$, can it be concluded that there is a difference in the self-esteem scores of the two groups? Use the *P*-value method.

Leavers	Stayers
$\overline{X}_1 = 3.05$	$\overline{X}_2 = 2.96$
$s_1 = 0.75$	$s_2 = 0.75$
$n_1 = 103$	$n_2 = 225$

Source: Paula Rayman and Belle Brett, "Women Science Majors: What Makes a Difference in Persistence after Graduation?" *The Journal of Higher Education* 66, no. 4 (July–August 1995), pp. 388–414.

10–16. The dean of students wants to see whether there is a significant difference in ages of resident students and commuting students. She selects a sample of 50 students from each group. The ages are shown here. At $\alpha = 0.05$, decide if there is enough evidence to reject the claim of no difference in the ages of the two groups. Use the standard deviations from the samples, and the P-value method.

Resident students

22	25	27	23	26	28	26	24
25	20	26	24	27	26	18	19
18	30	26	18	18	19	32	23
19	19	18	29	19	22	18	22
26	19	19	21	23	18	20	18
22	21	19	21	21	22	18	20
19	23						

Commuter students

18	20	19	18	22	25	24	35
23	18	23	22	28	25	20	24
26	30	22	22	22	21	18	20
19	26	35	19	19	18	19	32
29	23	21	19	36	27	27	20
20	21	18	19	23	20	19	19
20	25						

10–17. Two groups of students are given a problem-solving test, and the results are compared. Find the 90% confidence interval of the true difference in means.

Mathematics majors	Computer science majors
$\bar{X}_1 = 83.6$	$\bar{X}_2 = 79.2$
$s_1 = 4.3$	$s_2 = 3.8$
$n_1 = 36$	$n_2 = 36$

10–18. A study of teenagers found the following information on the length of time (in minutes) each talked on the telephone. Find the 95% confidence interval of the true differences in means.

Boys	Girls
$\bar{X}_1 = 21$	$\bar{X}_2 = 18$
$\sigma_1 = 2.1$	$\sigma_2 = 3.2$
$n_1 = 50$	$n_2 = 50$

10–19. Two brands of cigarettes are selected and their nicotine content compared. The data are shown here. Find the 99% confidence interval of the true difference in the means.

Brand A	Brand B
$\bar{X}_1 = 28.6$ milligrams	$\bar{X}_2 = 32.9$ milligrams
$\sigma_1 = 5.1$ milligrams	$\sigma_2 = 4.4$ milligrams
$n_1 = 30$	$n_2 = 40$

10–20. Two brands of batteries are tested and their voltage compared. The data follow. Find the 95% confidence interval of the true difference in the means. Assume that both variables are normally distributed.

Brand X	Brand Y
$\bar{X}_1 = 9.2$ volts	$\bar{X}_2 = 8.8$ volts
$\sigma_1 = 0.3$ volt	$\sigma_2 = 0.1$ volt
$n_1 = 27$	$n_2 = 30$

*10–21. A researcher claims that students in a private school have exam scores that are at most 8 points higher than those of students in public schools. Random samples of 60 students from each type of school are selected and given an exam. The results are shown below. At $\alpha = 0.05$, test the claim.

Private school	Public school
$\bar{X}_1 = 110$	$\bar{X}_2 = 104$
$s_1 = 15$	$s_2 = 15$
$n_1 = 60$	$n_2 = 60$

Technology Step by Step

MINITAB
Step by Step

Testing the Difference between Two Means
(Large Independent Samples)

MINITAB uses a t test to test for differences between means (and to compute confidence intervals) for two populations when the population standard deviations are unknown, regardless of sample size. The results will match those obtained when the formula presented in Section 10–2 is used.

Example MT10–1

Test for a difference between means using the college athletics data from Example 10–2, at $\alpha = 0.10$.

1. Enter the data for Example 10–2 into C1 and C2. Name the columns **MaleSports** and **FemaleSports.**
2. Select `Stat>Basic Statistics>2-Sample t.`
3. Click the button for `Samples in different columns`. There is one sample in each column.
4. Click in the box for `First:` and double-click C1 `MaleSports` in the list.
5. Click in the box for `Second:` and double-click C2 `FemaleSports` in the list. Do not check the box for `Assume equal variances`. The completed dialog box is shown.

2-Sample t Dialog Box

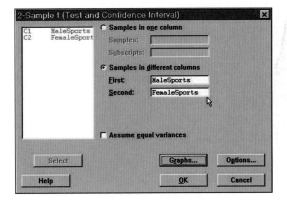

6. Click [`Options`].
   a. Type in **90** for the `Confidence level`.
   b. Type **0** for the `Test mean`.
   c. Select `greater than` for the `Alternative`.
7. Click [`OK`] twice.

Since the resulting *P*-value is greater than the significance level ($0.172 > 0.1$), do not reject the null hypothesis.

TI-83
Step by Step

z Test for the Difference between Two Means (Data)

1. Enter the data into L_1 and L_2.
2. Press **STAT** and move the cursor to `TESTS`.
3. Press **3** for `2-SampZTest`.
4. Select Data and press **ENTER** then press ▼.
5. Enter the values for σ_1 and σ_2. Make sure L_1 and L_2 are selected and `Freq1` and `Freq2` $= 1$.
6. Select the appropriate alternative hypothesis.
7. Select `Calculate` and press **ENTER.**

Example TI10–1

Test the claim from Example 10–2, using $\sigma_1 = 3.3$ and $\sigma_2 = 3.3$ at $\alpha = 0.10$.

Input	Input
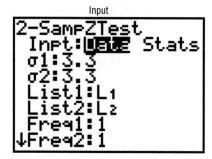	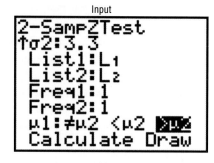

Output	Output
	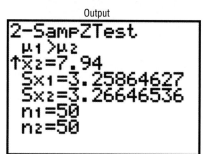

The test value is 0.939 and the *P*-value is 0.17. The decision is not to reject the null hypothesis at $\alpha = 0.10$, since $0.17 > 0.10$. The calculator also gives the values of the sample means and standard deviations. Any differences between the values of the test statistic and *P*-value displayed here and those obtained by other means are probably due to rounding.

z Test for the Difference between Two Means (Stats)

1. Press **STAT** and move the cursor to TESTS.
2. Press **3** for `2-SampZTest`.
3. Select `Stats` and press **ENTER,** then press ▼.
4. Enter the values for σ_1, σ_2, $\bar{X}_1$, n_1, $\bar{X}_2$, and n_2.
5. Select the appropriate alternative hypothesis.
6. Select `Calculate` and press **ENTER.**

Example TI10–2

Test the claim from Example 10–1, H_1: $\mu_1 \neq \mu_2$ at $\alpha = 0.05$ when

A	B
$\bar{X}_1 = 88.42$	$\bar{X}_2 = 80.61$
$\sigma_1 = 5.62$	$\sigma_2 = 4.83$
$n_1 = 50$	$n_2 = 50$

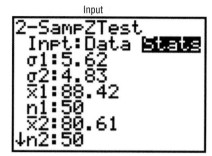

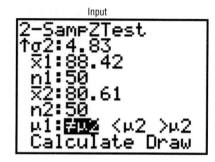

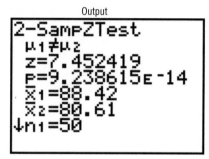

The test value is 7.452419, and the *P*-value is 0.00000000000009. The decision is to reject the null hypothesis, since 0.00000000000009 < 0.05.

Excel
Step by Step

z Test for the Difference between Two Means

Excel has a two-sample *z* test in its Data Analysis tools. To perform a *z* test for the difference between the means of two populations given two independent samples:

1. Enter the first sample data set in column A.

2. Enter the second sample data set in column B.

3. If the population variances are not known, but $n \geq 30$ for both samples, use the formulas =VAR(A1:An) and =VAR(B1:Bn), where An and Bn are the last cells with data in each column, to find the variances of the sample data sets.

4. Select Tools, Data Analysis and choose z-Test: Two Sample for Means.

5. Enter the ranges for the data in columns A and B and enter **0** for Hypothesized Mean Difference.

6. If the population variances are known, enter them for Variable 1 and Variable 2. Otherwise, use the sample variances obtained in step 3.

7. Specify the confidence level, Alpha.

8. Specify a location for output, and click [OK].

Example XL10–1

Test the claim that the two population means are equal using the sample data provided below, at $\alpha = 0.05$. Assume the population variances are $\sigma_A^2 = 10.067$ and $\sigma_B^2 = 7.067$.

Set A	10	2	15	18	13	15	16	14	18	12	15	15	14	18	16
Set B	5	8	10	9	9	11	12	16	8	8	9	10	11	7	6

Two-Sample *z* Test
Dialog Box

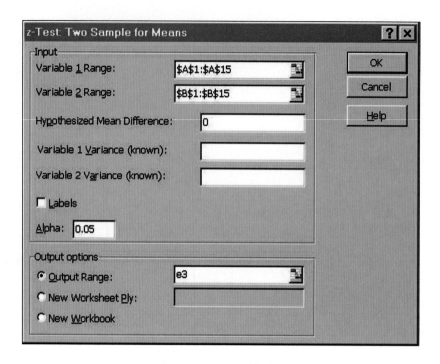

The two-sample *z* test dialog box is shown here (before the variances are entered); the results appear in the table that Excel generates. Note that the *P*-value and critical *z* value are provided for both the one-tailed test and the two-tailed test. The *P*-values are here expressed in scientific notation: $7.09045E\text{-}06 = 7.09045 \times 10^{-6} = 0.00000709045$. Because this value is less than 0.05, we reject the null hypothesis and conclude that the population means are not equal.

z-Test: Two Sample for Means		
	Variable 1	*Variable 2*
Mean	14.06666667	9.266666667
Known Variance	10.067	7.067
Observations	15	15
Hypothesized Mean Difference	0	
z	4.491149228	
P(Z<=z) one-tail	3.54522E-06	
z Critical one-tail	1.644853	
P(Z<=z) two-tail	7.09045E-06	
z Critical two-tail	1.959961082	

10–3

Testing the Difference between Two Variances

In addition to comparing two means, statisticians are also interested in comparing two variances or standard deviations. For example, is the variation in the temperatures for a certain month for two cities different?

In another situation, a researcher may be interested in comparing the variance of the cholesterol of men with the variance of the cholesterol of women. For the comparison

Objective 2. Test the difference between two variances or standard deviations.

of two variances or standard deviations, an **F test** is used. The F test should not be confused with the chi-square test, which compares a single sample variance to a specific population variance, as shown in Chapter 9.

If two independent samples are selected from two normally distributed populations in which the variances are equal ($\sigma_1^2 = \sigma_2^2$) and if the variances s_1^2 and s_2^2 are compared as $\dfrac{s_1^2}{s_2^2}$, the sampling distribution of the variances is called the **F distribution.**

Characteristics of the F Distribution

1. The values of F cannot be negative, because variances are always positive or zero.
2. The distribution is positively skewed.
3. The mean value of F is approximately equal to 1.
4. The F distribution is a family of curves based on the degrees of freedom of the variance of the numerator and the degrees of freedom of the variance of the denominator.

Figure 10–5 shows the shapes of several curves for the F distribution.

Figure 10–5

The F Family of Curves

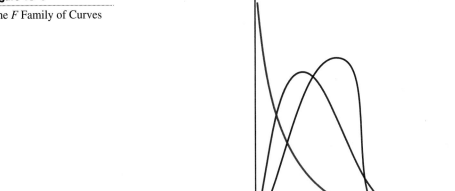

Formula for the F Test

$$F = \frac{s_1^2}{s_2^2}$$

where s_1^2 is the larger of the two variances.

The F test has two terms for the degrees of freedoms: that of the numerator, $n_1 - 1$, and that of the denominator, $n_2 - 1$, where n_1 is the sample size from which the larger variance was obtained.

When one is finding the F test value, *the larger of the variances is placed in the numerator of the F formula;* this is not necessarily the variance of the larger of the two sample sizes.

Table H in Appendix C gives the F critical values for $\alpha = 0.005, 0.01, 0.025, 0.05,$ and 0.10 (each α value involves a separate table in Table H). These are one-tailed values; if a two-tailed test is being conducted, then the $\alpha/2$ value must be used. For example, if a two-tailed test with $\alpha = 0.05$ is being conducted, then the $0.05/2 = 0.025$ table of Table H should be used.

Example 10–4

Find the critical value for a right-tailed F test when $\alpha = 0.05$, the degrees of freedom for the numerator (abbreviated d.f.N.) are 15, and the degrees of freedom for the denominator (d.f.D.) are 21.

Solution

Since this test is right-tailed with $\alpha = 0.05$, use the 0.05 table. The d.f.N. is listed across the top, and the d.f.D. is listed in the left column. The critical value is found where the row and column intersect in the table. In this case, it is 2.18. See Figure 10–6.

Figure 10–6

Finding the Critical
Value in Table H for
Example 10–4

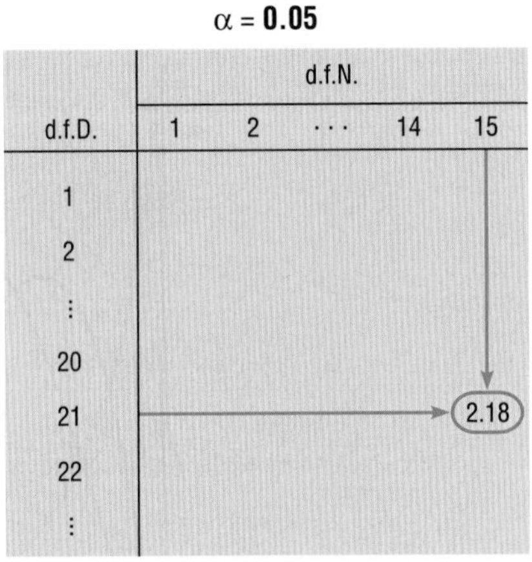

As noted previously, when the F test is used, the larger variance is always placed in the numerator of the formula. When one is conducting a two-tailed test, α is split; and even though there are two values, only the right one is used. The reason is that the F test value is always greater than or equal to 1.

Example 10–5

Find the critical value for a two-tailed F test with $\alpha = 0.05$ when the sample size from which the variance for the numerator was obtained was 21 and the sample size from which the variance for the denominator was obtained was 12.

Solution

Since this is a two-tailed test with $\alpha = 0.05$, the $0.05/2 = 0.025$ table must be used. Here, d.f.N. $= 21 - 1 = 20$, and d.f.D. $= 12 - 1 = 11$; hence, the critical value is 3.23. See Figure 10–7.

Figure 10–7

Finding the Critical
Value in Table H for
Example 10–5

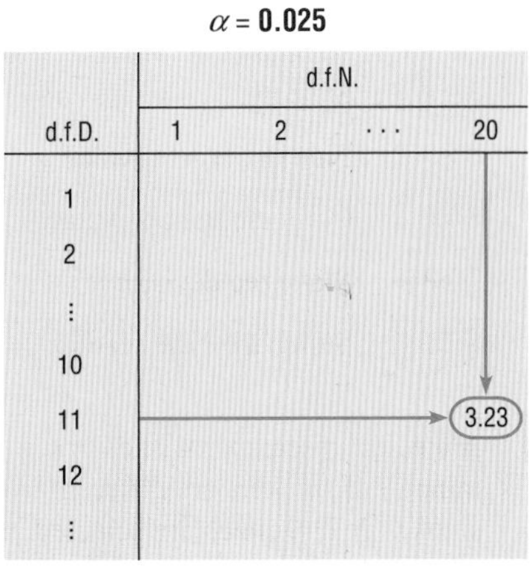

$\alpha = 0.025$

When the degrees of freedom values cannot be found in the table, the closest value on the smaller side should be used. For example, if d.f.N. = 14, this value is between the given table values of 12 and 15; therefore, 12 should be used to be on the safe side.

When one is testing the equality of two variances, the following hypotheses are used.

Right-tailed	Left-tailed	Two-tailed
H_0: $\sigma_1^2 \leq \sigma_2^2$	H_0: $\sigma_1^2 \geq \sigma_2^2$	H_0: $\sigma_1^2 = \sigma_2^2$
H_1: $\sigma_1^2 > \sigma_2^2$	H_1: $\sigma_1^2 < \sigma_2^2$	H_1: $\sigma_1^2 \neq \sigma_2^2$

There are four key points to keep in mind when using the F test.

Notes for the Use of the F Test

1. The larger variance should always be designated as s_1^2 and be placed in the numerator of the formula.

$$F = \frac{s_1^2}{s_2^2}$$

2. For a two-tailed test, the α value must be divided by 2 and the critical value be placed on the right side of the F curve.

3. If the standard deviations instead of the variances are given in the problem, they must be squared for the formula for the F test.

4. When the degrees of freedom cannot be found in Table H, the closest value on the smaller side should be used.

Assumptions for Testing the Difference between Two Variances

1. The populations from which the samples were obtained must be normally distributed. (*Note:* The test should not be used when the distributions depart from normality.)

2. The samples must be independent of each other.

Remember also that in tests of hypotheses using the traditional method, the following five steps should be taken.

STEP 1 State the hypotheses and identify the claim.

STEP 2 Find the critical value.

STEP 3 Compute the test value.

STEP 4 Make the decision.

STEP 5 Summarize the results.

Example 10–6

A medical researcher wishes to see whether the variances of the heart rates (in beats per minute) of smokers are different from the variances of heart rates of people who do not smoke. Two samples are selected, and the data are as shown. Using $\alpha = 0.05$, is there enough evidence to support the claim?

Smokers	Nonsmokers
$n_1 = 26$	$n_2 = 18$
$s_1^2 = 36$	$s_2^2 = 10$

Solution

STEP 1 State the hypotheses and identify the claim.

$$H_0: \sigma_1^2 = \sigma_2^2 \qquad \text{and} \qquad H_1: \sigma_1^2 \neq \sigma_2^2 \, (claim)$$

STEP 2 Find the critical value. Use the 0.025 table in Table H since $\alpha = 0.05$ and this is a two-tailed test. Here, d.f.N. $= 26 - 1 = 25$, and d.f.D. $= 18 - 1 = 17$. The critical value is 2.56 (d.f.N. $= 24$ was used). See Figure 10–8.

Figure 10–8

Critical Value for Example 10–6

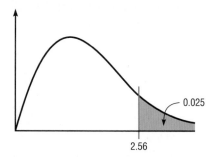

2.56

0.025

STEP 3 Compute the test value.

$$F = \frac{s_1^2}{s_2^2} = \frac{36}{10} = 3.6$$

STEP 4 Make the decision. Reject the null hypothesis, since $3.6 > 2.56$.

STEP 5 Summarize the results. There is enough evidence to support the claim that the variances of the heart rates of smokers and nonsmokers are different.

Example 10–7

An instructor hypothesizes that the standard deviation of the final exam grades in her statistics class is larger for the male students than it is for the female students. The data from the final exam for the last semester are as shown. Is there enough evidence to support her claim, using $\alpha = 0.01$?

Males	Females
$n_1 = 16$	$n_2 = 18$
$s_1 = 4.2$	$s_2 = 2.3$

Solution

STEP 1   State the hypotheses and identify the claim.

$$H_0: \sigma_1^2 \leq \sigma_2^2 \quad \text{and} \quad H_1: \sigma_1^2 > \sigma_2^2 \text{ (claim)}$$

STEP 2   Find the critical value. Here, d.f.N. $= 16 - 1 = 15$, and d.f.D. $= 18 - 1 = 17$. From the 0.01 table, the critical value is 3.31.

STEP 3   Compute the test value.

$$F = \frac{s_1^2}{s_2^2} = \frac{4.2^2}{2.3^2} = 3.33$$

STEP 4   Make the decision. Reject the null hypothesis, since $3.33 > 3.31$.

STEP 5   Summarize the results. There is enough evidence to support the claim that the standard deviation of the final exam grades for the male students is larger than that for the female students.

....

Finding P-values for the F test statistic is somewhat more complicated since it requires looking through all the F tables (Table H in Appendix C) using the specific d.f.N. and d.f.D. values. For example, suppose that a certain test has an $F = 3.58$ and d.f.N. $= 5$ and d.f.D. $= 10$. In order to find the P-value interval for $F = 3.58$ one must first find the corresponding F values for d.f.N. $= 5$ and d.f.D. $= 10$ for $\alpha = 0.005$ on page C–18, 0.01 on page C–19, 0.025 on page C–20, and 0.10 on page C–22 in Table H. Then make a table as shown.

α	0.10	0.05	0.025	0.01	0.005
F	2.52	3.33	4.24	5.64	6.87
Reference page	C–22	C–21	C–20	C–19	C–18

Now locate the two F values that the test value 3.58 falls between. In this case, 3.58 falls between 3.33 and 4.24 corresponding to 0.05 and 0.025. Hence, the P-value for a right-tailed test for $F = 3.58$ falls between 0.025 and 0.05 (i.e., $0.025 < P\text{-value} < 0.05$). For a right-tailed test, then, one would reject the null hypothesis at $\alpha = 0.05$ but not at $\alpha = 0.01$. The P-value obtained from a calculator is 0.0481. Remember that for a two-tailed test the values found in the Table H for α must be doubled. In this case, $0.05 < P\text{-value} < 0.10$ for an $F = 3.58$.

Once you understand the concept, you can dispense with making a table as shown and find the P-value directly from Table H.

....

Example 10–8

The CEO of an airport hypothesizes that the variance for the number of passengers for American airports is greater than the variance for the number of passengers for foreign airports. At $\alpha = 0.10$ is there enough evidence to support the hypothesis? The data in millions of passengers per year is shown for selected airports. Use the P-value method. Assume the variable is normally distributed.

American airports		Foreign airports	
36.8	73.5	60.7	51.2
72.4	61.2	42.7	38.6
60.5	40.1		

Source: Airports Council International

Solution

STEP 1 State the hypotheses and identify the claim.

$$H_0: \sigma_1^2 \le \sigma_2^2 \qquad \text{and} \qquad H_1: \sigma_1^2 > \sigma_2^2 \text{ (claim)}$$

STEP 2 Compute the test value. Using the formula in Chapter 3 or a calculator, find the variance for each group.

$$s_1^2 = 246.38 \qquad \text{and} \qquad s_2^2 = 95.87$$

Substitute in the formula and solve.

$$F = \frac{s_1^2}{s_2^2} = \frac{246.38}{95.87} = 2.57$$

STEP 3 Find the P-value in Table H using d.f.N. $= 5$ and d.f.D. $= 3$.

α	0.10	0.05	0.025	0.01	0.005
F	5.31	9.01	14.88	28.24	45.39

Since 2.57 is less than 5.31, the P-value is greater than 0.10. (The P-value obtained from a calculator is 0.234.)

STEP 4 Make the decision. The decision is do not reject the null hypothesis since P-value > 0.10.

STEP 5 Summarize the results. There is not enough evidence to support the claim that the variance in the number of passengers for American airports is greater than the variance for the number of passengers for foreign airports.

··

If the exact degrees of freedom is not specified in Table H, the closest smaller value should be used. For example, if $\alpha = 0.05$ (right-tailed test) and d.f.N. $= 18$ and d.f.D. $= 20$, use the column d.f.N. $= 15$ and the row d.f.D. $= 20$ to get an $F = 2.20$.

Note: It is not necessary to place the larger variance in the numerator when performing the F test. Critical values for left-tailed hypotheses tests can be found by interchanging the degrees of freedom and taking the reciprocal of the value found in Table H.

Also, one should use caution when performing the F test since on rare occasions, the data can run contrary to the hypotheses. For example, if the hypotheses are H_0: $\sigma_1^2 \le \sigma_2^2$ and H_1: $\sigma_1^2 > \sigma_2^2$, but if $s_1^2 < s_2^2$, then the F test should not be performed and one would not reject the null hypothesis.

Exercises

10–22. When one is computing the F test value, what condition is placed on the variance that is in the numerator?

10–23. Why is the critical region always on the right side in the use of the F test?

10–24. What is one application of the F test?

10–25. What are the two different degrees of freedom associated with the F distribution?

10–26. What are the characteristics of the F distribution?

10–27. Using Table H, find the critical value for each.

a. sample 1: $s_1^2 = 128$, $n_1 = 23$
   sample 2: $s_2^2 = 162$, $n_2 = 16$
   two-tailed, $\alpha = 0.01$
b. sample 1: $s_1^2 = 37$, $n_1 = 14$
   sample 2: $s_2^2 = 89$, $n_2 = 25$
   right-tailed, $\alpha = 0.01$
c. sample 1: $s_1^2 = 232$, $n_1 = 30$
   sample 2: $s_2^2 = 387$, $n_2 = 46$
   two-tailed, $\alpha = 0.05$
d. sample 1: $s_1^2 = 164$, $n_1 = 21$
   sample 2: $s_2^2 = 53$, $n_2 = 17$
   two-tailed, $\alpha = 0.10$
e. sample 1: $s_1^2 = 92.8$, $n_1 = 11$
   sample 2: $s_2^2 = 43.6$, $n_2 = 11$
   right-tailed, $\alpha = 0.05$

10–28. (ans) Using Table H, find the P-value interval for each F test value.
a. $F = 2.97$, d.f.N. = 9, d.f.D. = 14, right-tailed
b. $F = 3.32$, d.f.N. = 6, d.f.D. = 12, two-tailed
c. $F = 2.28$, d.f.N. = 12, d.f.D. = 20, right-tailed
d. $F = 3.51$, d.f.N. = 12, d.f.D. = 21, right-tailed
e. $F = 4.07$, d.f.N. = 6, d.f.D. = 10, two-tailed
f. $F = 1.65$, d.f.N. = 19, d.f.D. = 28, right-tailed
g. $F = 1.77$, d.f.N. = 28, d.f.D. = 28, right-tailed
h. $F = 7.29$, d.f.N. = 5, d.f.D. = 8, two-tailed

For Exercises 10–29 through 10–42, perform the following steps. Assume that all variables are normally distributed.
a. State the hypotheses and identify the claim.
b. Find the critical value.
c. Compute the test value.
d. Make the decision.
e. Summarize the results.

Use the traditional method of hypothesis testing unless otherwise specified.

10–29. An instructor claims that when a composition course is taught in conjunction with a word-processing course, the variance in the final grades will be larger than when the composition course is taught without the word-processing component. Two groups are randomly selected. The variance of the exams of the group that also had word-processing instruction is 103, and the variance of the exams of the students who did not have the word-processing component is 73. Each sample consists of 20 students. At $\alpha = 0.05$, can the instructor's claim be supported?

10–30. A consumer advocate claims that there is no difference in the variance of the number of hours that two companies' batteries will last. A sample of 10 batteries is selected from company X, and the variance of hours is 24. A sample of 10 batteries from company Y has a variance of 40. At $\alpha = 0.10$, test the claim that there is no difference in the variance of the life of the batteries.

10–31. A tax collector wishes to see if the variance of the values of the tax-exempt properties is different for two large cities. The values of the tax-exempt properties for two samples are shown. The data are given in millions of dollars. At $\alpha = 0.05$ is there enough evidence to support the tax collector's claim that the variances are different?

City A				City B			
$113	22	14	8	$ 82	11	5	15
25	23	23	30	295	50	12	9
44	11	19	7	12	68	81	2
31	19	5	2	20	16	4	5

10–32. In the hospital study cited in Exercise 8–19, it was found that the standard deviation of the sound levels from 20 areas designated as "casualty doors" was 4.1 dBA and the standard deviation of 24 areas designated as operating theaters was 7.5 dBA. At $\alpha = 0.05$, can one substantiate the claim that there is a difference in the standard deviations?

Source: M. Bayo, A. Garcia, and A. Garcia, "Noise Levels in an Urban Hospital and Workers' Subjective Responses," *Archives of Environmental Health* 50, no. 3 (May–June 1995), p. 249.

10–33. A nurse claims that the variations of the lengths of newborn males is different from the variations of the lengths of newborn females. A sample of 15 newborn males is selected, and the standard deviation is 1.3 inches. The standard deviation of a sample of 15 newborn females is 0.9 inch. At $\alpha = 0.10$, can the nurse conclude that the variation of the lengths is different?

10–34. A researcher wishes to see if the variance in the number of vehicles passing through the toll booths during a fiscal year on the Pennsylvania Turnpike is different from the variance in the number of vehicles passing through the toll booths on the expressways in Pennsylvania during the same year. The data are shown. At $\alpha = 0.05$, can it be concluded that the variances are different?

PA Turnpike	PA Expressways
3,694,560	2,774,251
719,934	204,369
3,768,285	456,123
1,838,271	1,068,107
3,358,175	3,534,092
6,718,905	235,752
3,469,431	499,043
4,420,553	2,016,046
920,264	253,956
1,005,469	826,710
1,112,722	133,619
2,855,109	3,453,745

Source: *Pittsburgh Post-Gazette*, June 28, 1998.

10–35. A researcher claims that the variation of blood pressure of overweight individuals is greater than the variation of blood pressure of normal-weight individuals.

The standard deviation of the pressures of 28 overweight people was found to be 6.2 mm Hg, and the standard deviation of the pressures of 25 normal-weight people was 2.7 mm Hg. At $\alpha = 0.01$, can the researcher conclude that the blood pressures of overweight individuals are more variable than those of individuals who are of normal weight?

10–36. An educator claims that the variation in the number of years of teaching experience of senior high school teachers is greater than the variation in the number of years of teaching experience of elementary school teachers. Two groups are randomly selected. The variance of the number of years of teaching experience of 18 elementary teachers is 1.9, and the variance of the years of experience of 26 senior high school teachers is 2.8. At $\alpha = 0.10$, can the educator conclude that the variation in the years of teaching experience of senior high school teachers is greater than the variation of the elementary school teachers?

10–37. A researcher wishes to test the variation in the number of pounds lost by men who follow two popular liquid diets. Ten men follow diet A for four months, and the standard deviation of the weight loss is 6.3 pounds. Twelve men follow diet B for four months, and the standard deviation of the weight loss is 4.8 pounds. At $\alpha = 0.05$, can the researcher substantiate the claim that the variation in pounds lost following diet A is greater than the variation in pounds lost following diet B?

10–38. A researcher wishes to see if the variance of the areas in square miles for counties in Indiana is less than the variance of the areas for counties in Iowa. A random sample of counties is selected, and the data are shown. At $\alpha = 0.01$ can it be concluded that the variance of the areas for counties in Indiana is less than the variance of the areas for counties in Iowa?

Indiana				Iowa			
406	393	396	485	640	580	431	416
431	430	369	408	443	569	779	381
305	215	489	293	717	568	714	731
373	148	306	509	571	577	503	501
560	384	320	407	568	434	615	402

Source: *The World Almanac and Book of Facts 1999*, p. 423.

10–39. Test the claim that the variance of heights of tall buildings in Denver is equal to the variance in heights of tall buildings in Detroit at $\alpha = 0.10$. The data are given in feet.

Denver			Detroit		
714	698	544	620	472	430
504	438	408	562	448	420
404			534	436	

Source: *The World Almanac and Book of Facts 1999*, p. 625.

10–40. A researcher claims that the variation in the salaries of elementary school teachers is greater than the variation in the salaries of secondary school teachers. A sample of the salaries of 30 elementary school teachers has a variance of $8324, and a sample of the salaries of 30 secondary school teachers has a variance of $2862. At $\alpha = 0.05$, can the researcher conclude that the variation in the elementary school teachers' salaries is greater than the variation in the secondary teachers' salaries? Use the *P*-value method.

10–41. The weights in ounces of a sample of running shoes for men and women are shown below. Calculate the variances for each sample and test the claim that the variances are equal at $\alpha = 0.05$. Use the *P*-value method.

Men			Women		
11.9	10.4	12.6	10.6	10.2	8.8
12.3	11.1	14.7	9.6	9.5	9.5
9.2	10.8	12.9	10.1	11.2	9.3
11.2	11.7	13.3	9.4	10.3	9.5
13.8	12.8	14.5	9.8	10.3	11.0

Source: *Consumer Reports*, May 1995, p. 316.

10–42. Upright vacuum cleaners have either a hard body type or a soft body type. Shown below are the weights in pounds of a sample of each type. At $\alpha = 0.05$, can the claim that there is a difference in the variances of the weights of the two types be substantiated?

Hard body types				Soft body types			
21	17	17	20	24	13	11	13
16	17	15	20	12	15		
23	16	17	17				
13	15	16	18				
18							

Source: *Consumer Reports*, January 1995, p. 46.

Technology Step by Step

MINITAB
Step by Step

Test for the Difference between Two Variances

This item is new to Release 13 of MINITAB.

Example MT10–2

Test the hypothesis that the two population variances are equal based on the sample data provided below, using the *P*-value method.

Sample 1	12	15	18	13	10	9	6	3	11	5
Sample 2	18	20	13	7	9	4				

1. Enter the data into two columns of MINITAB.
2. Select **Stat>Basic Statistics>2-Variances**
3. Click the button for `Samples in different columns`.
4. Click in the box labeled `First` then double-click `C1`.
5. Double-click `C2`. The dialog box should look like the one shown here.

Dialog Box for
2-Variances

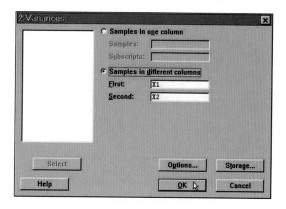

In the graph window and the session window the P-value for the F test is 0.399. We would fail to reject the hypothesis that the population variances are equal at any standard level of significance.

TI-83
Step by Step

F Test for the Difference between Two Variances (Data)

1. Enter the data into L_1 and L_2.
2. Press STAT and move the cursor to TESTS.
3. Press **D (ALPHA X^{-1})** for `2-SampFTest`.
4. Select `Data` and press **ENTER,** then press ▼.
5. Enter the values for σ_1 and σ_2. Make sure L_1 and L_2 are selected and `Freq1` and `Freq2 = 1`.
6. Select the appropriate alternative hypothesis.
7. Select `Calculate` and press ENTER.

Example TI10–3

Test the claim from Example 10–8 that the variance in number of passengers for American airports is greater than that for foreign airports, at $\alpha = 0.10$. Here are the sample data:

American Airports	Foreign Airports
36.8	60.7
73.5	51.2
72.4	42.7
61.2	38.6
60.5	
40.1	

Input

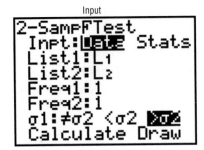

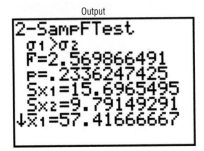

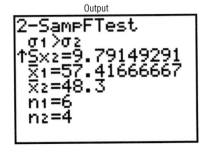

The test value is 2.569866491, and the *P*-value is 0.2336247425. The decision is not to reject the null hypothesis, since 0.2336247425 > 0.10. The calculator also gives various other statistics.

F Test for the Difference between Two Variances (Stats)

1. Press **STAT** and move the cursor to TESTS.
2. Press **D (ALPHA X⁻¹)** for 2-SampFTest.
3. Select Stats and press **ENTER,** then press ▼.
4. Enter the sample 1 standard deviation in Sx1 and sample size in n1.
5. Enter the sample 2 standard deviation in Sx2 and sample size in n2.
6. Select the appropriate alternative hypothesis.
7. Select Calculate and press ENTER.

Example TI10–4

Test the claim from Example 10–7, H_1: $\sigma_1 > \sigma_2$ at $\alpha = 0.01$, when

	A	B
	$s_1 = 4.2$	$s_2 = 2.3$
	$n_1 = 16$	$n_2 = 18$

Input

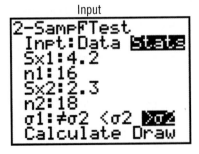

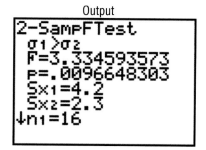

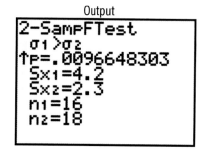

As shown on the output screen, the test value is 3.334593573, and the *P*-value is 0.0096648303. The decision is to reject the null hypothesis, since 0.0096648303 < 0.01.

Excel
Step by Step

F Test for the Difference between Two Variances

Excel has a two-sample *F* test in its Data Analysis tools. To perform an *F* test for the difference between the variances of two populations given two independent samples:

1. Enter the first sample data set in column A.

2. Enter the second sample data set in column B.

3. Select `Tools, Data Analysis` and choose `F-Test Two-Sample for Variances`.

4. Enter the ranges for the data in columns A and B.

5. Specify the confidence level, `Alpha`.

6. Specify a location for output, and click [`OK`].

Example XL10–2

Test the hypothesis that the two population variances are equal using the sample data provided below, at $\alpha = 0.05$.

Set A	63	73	80	60	86	83	70	72	82
Set B	86	93	64	82	81	75	88	63	63

The results appear in the table that Excel generates, shown here. For this example, the output shows that the null hypothesis cannot be rejected at an alpha level of 0.05.

F-Test Two-Sample for Variances		
	Variable 1	Variable 2
Mean	74.33333333	77.22222222
Variance	82.75	132.9444444
Observations	9	9
df	8	8
F	0.622440451	
P(F<=f) one-tail	0.258814151	
F Critical one-tail	0.290858004	

10–4

Testing the Difference between Two Means: Small Independent Samples

Objective 3. Test the difference between two means for small independent samples.

In Section 10–2, the z test was used to test the difference between two means when the population standard deviations were known and the variables were normally or approximately normally distributed, or when both sample sizes were greater than or equal to 30. In many situations, however, these conditions cannot be met—that is, the population standard deviations are not known, and one or both sample sizes are less than 30. In these cases, a t test is used to test the difference between means when the two samples are independent and when the samples are taken from two normally or approximately normally distributed populations. Samples are **independent** when they are not related.

There are actually two different options for the use of t tests. *One option is used when the variances of the populations are not equal, and the other option is used when the variances are equal.* To determine whether two sample variances are equal, the researcher can use an F test, as shown in the previous section.

Note, however, that not all statisticians are in agreement about using the F test before using the t test. Some believe that conducting the F and t tests at the same level of significance will change the overall level of significance of the t test. Their reasons are beyond the scope of this textbook.

Formulas for the t Tests—For Testing the Difference between Two Means—Small Samples

Variances are assumed to be unequal:

$$t = \frac{(\overline{X}_1 - \overline{X}_2) - (\mu_1 - \mu_2)}{\sqrt{\dfrac{s_1^2}{n_1} + \dfrac{s_2^2}{n_2}}}$$

where the degrees of freedom are equal to the smaller of $n_1 - 1$ or $n_2 - 1$.

Variances are assumed to be equal:

$$t = \frac{(\overline{X}_1 - \overline{X}_2) - (\mu_1 - \mu_2)}{\sqrt{\dfrac{(n_1 - 1)s_1^2 + (n_2 - 1)s_2^2}{n_1 + n_2 - 2}}\sqrt{\dfrac{1}{n_1} + \dfrac{1}{n_2}}}$$

where the degrees of freedom are equal to $n_1 + n_2 - 2$.

When the variances are unequal, the first formula

$$t = \frac{(\overline{X}_1 - \overline{X}_2) - (\mu_1 - \mu_2)}{\sqrt{\dfrac{s_1^2}{n_1} + \dfrac{s_2^2}{n_2}}}$$

follows the format of

$$\text{test value} = \frac{(\text{observed value}) - (\text{expected value})}{\text{standard error}}$$

where $\overline{X}_1 - \overline{X}_2$ is the observed difference between sample means and where the expected value $\mu_1 - \mu_2$ is equal to zero when no difference between population means is hypothesized. The denominator $\sqrt{(s_1^2/n_1) + (s_2^2/n_2)}$ is the standard error of the difference between two means. Since mathematical derivation of the standard error is somewhat complicated, it will be omitted here.

When the variances are assumed to be equal, the second formula

$$t = \frac{(\bar{X}_1 - \bar{X}_2) - (\mu_1 - \mu_2)}{\sqrt{\dfrac{(n_1 - 1)s_1^2 + (n_2 - 1)s_2^2}{n_1 + n_2 - 2}} \sqrt{\dfrac{1}{n_1} + \dfrac{1}{n_2}}}$$

also follows the format of

$$\text{test value} = \frac{(\text{observed value}) - (\text{expected value})}{\text{standard error}}$$

For the numerator, the terms are the same as in the first formula. However, a note of explanation is needed for the denominator of the second test statistic. Since both populations are assumed to have the same variance, the standard error is computed with what is called a pooled estimate of the variance. A **pooled estimate of the variance** is a weighted average of the variance using the two sample variances and the *degrees of freedom* of each variance as the weights. Again, since the algebraic derivation of the standard error is somewhat complicated, it is omitted.

In summary, then, to use the *t* test, first use the *F* test to determine whether or not the variances are equal. Then use the appropriate *t* test formula. This procedure involves two five-step processes.

Example 10–9

The average size of a farm in Indiana County, Pennsylvania, is 191 acres. The average size of a farm in Greene County, Pennsylvania, is 199 acres. Assume the data were obtained from two samples with standard deviations of 38 acres and 12 acres, respectively, and sample sizes of 8 and 10, respectively. Can it be concluded at $\alpha = 0.05$ that the average size of the farms in the two counties is different? Assume the populations are normally distributed.

Source: *Pittsburgh Tribune-Review,* August 28, 1994.

Solution

The approach here will be to use the *F* test to determine whether or not the variances are equal. The null hypothesis is that the variances are equal.

STEP 1   State the hypotheses and identify the claim.

$$H_0: \sigma_1^2 = \sigma_2^2 \text{ (claim)} \qquad \text{and} \qquad H_1: \sigma_1^2 \neq \sigma_2^2$$

STEP 2   Find the critical value. The critical value for the *F* test found in Table H (Appendix C) for $\alpha = 0.05$ is 4.20, since there are 7 and 9 degrees of freedom. (*Note:* Use the 0.025 table.)

STEP 3   Compute the test value.

$$F = \frac{s_1^2}{s_2^2} = \frac{38^2}{12^2} = 10.03$$

STEP 4   Make the decision. Reject the null hypothesis since 10.03 falls in the critical region. See Figure 10–9.

Figure 10–9

Critical and *F* Test Values
for Example 10–9

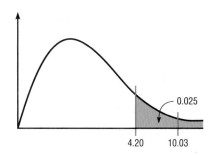

0.025

4.20   10.03

STEP 5   Summarize the results. It can be concluded that the variances are not equal. Since the variances are not equal, the first formula will be used to test the equality of the means.

STEP 1   State the hypotheses and identify the claim for the means.

$$H_0: \mu_1 = \mu_2 \quad \text{and} \quad H_1: \mu_1 \neq \mu_2 \text{ (claim)}$$

STEP 2   Find the critical values. Since the test is two-tailed, since $\alpha = 0.05$, and since the variances are unequal, the degrees of freedom are the smaller of $n_1 - 1$ or $n_2 - 1$. In this case, the degrees of freedom are $8 - 1 = 7$. Hence, from Table F, the critical values are $+2.365$ and -2.365.

STEP 3   Compute the test value. Since the variances are unequal, use the first formula.

$$t = \frac{(\bar{X}_1 - \bar{X}_2) - (\mu_1 - \mu_2)}{\sqrt{\dfrac{s_1^2}{n_1} + \dfrac{s_2^2}{n_2}}} = \frac{(191 - 199) - 0}{\sqrt{\dfrac{38^2}{8} + \dfrac{12^2}{10}}} = -0.57$$

STEP 4   Make the decision. Do not reject the null hypothesis, since $-0.57 > -2.365$. See Figure 10–10.

Figure 10–10

Critical and Test Values
for Example 10–9

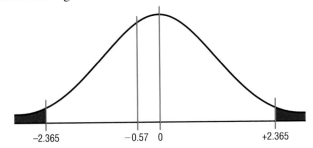

STEP 5   Summarize the results. There is not enough evidence to support the claim that the average size of the farms is different.

Example 10–10   A researcher wishes to determine whether the salaries of professional nurses employed by private hospitals are higher than those of nurses employed by government-owned hospitals. She selects a sample of nurses from each type of hospital and calculates the means and standard deviations of their salaries. At $\alpha = 0.01$, can she conclude that the private hospitals pay more than the government hospitals? Assume that the populations are approximately normally distributed. Use the P-value method.

Private	Government
$\bar{X}_1 = \$26{,}800$	$\bar{X}_2 = \$25{,}400$
$s_1 = \$600$	$s_2 = \$450$
$n_1 = 10$	$n_2 = 8$

Solution

The F test will be used to determine whether or not the variances are equal. The null hypothesis is that the variances are equal.

STEP 1   State the hypotheses and identify the claim.

$$H_0: \sigma_1^2 = \sigma_2^2 \text{ (claim)} \quad \text{and} \quad H_1: \sigma_1^2 \neq \sigma_2^2$$

STEP 2   Compute the test value.

$$F = \frac{s_1^2}{s_2^2} = \frac{600^2}{450^2} = 1.78$$

STEP 3   Find the P-value using Table H using d.f.N. = 9 and d.f.D. = 7. Since 1.78 < 2.72, P-value > 0.20. (The P-value obtained from a calculator is 0.460.)

STEP 4   Make the decision. Do not reject the null hypothesis since P-value > 0.01 (the α value).

STEP 5   Summarize the results. There is not enough evidence to reject the claim that the variances are equal; therefore, the second formula is used to test the difference between the two means, as shown next.

STEP 1   State the hypotheses and identify the claim.

$$H_0: \mu_1 \le \mu_2 \qquad \text{and} \qquad H_1: \mu_1 > \mu_2 \text{ (claim)}$$

STEP 2   Compute the test value. Use the second formula, since the variances are assumed to be equal.

$$t = \frac{(\bar{X}_1 - \bar{X}_2) - (\mu_1 - \mu_2)}{\sqrt{\dfrac{(n_1 - 1)s_1^2 + (n_2 - 1)s_2^2}{n_1 + n_2 - 2}}\sqrt{\dfrac{1}{n_1} + \dfrac{1}{n_2}}}$$

$$= \frac{(26{,}800 - 25{,}400) - 0}{\sqrt{\dfrac{(10 - 1)(600)^2 + (8 - 1)(450)^2}{10 + 8 - 2}}\sqrt{\dfrac{1}{10} + \dfrac{1}{8}}}$$

$$= 5.47$$

STEP 3   Find the P-value using Table F. The P-value for $t = 5.47$ with d.f. = 16 (i.e., 10 + 8 − 2) is P-value < 0.005. (The P-value obtained from a calculator is 0.00002.)

STEP 4   Make the decision. Since P-value < 0.01 (the α value), the decision is to reject the null hypothesis.

STEP 5   Summarize the results. There is enough evidence to support the claim that the salaries paid to nurses employed by private hospitals are higher than those paid to nurses employed by government-owned hospitals.

··

When raw data are given in the exercises, use your calculator or the formulas in Chapter 3 to find the means and variances for the data sets. Then follow the procedures shown in this section to test the hypotheses.

Confidence intervals can also be found for the difference between two means with the following formulas:

Confidence Intervals for the Difference of Two Means: Small Independent Samples

Variances unequal

$$(\bar{X}_1 - \bar{X}_2) - t_{\alpha/2}\sqrt{\frac{s_1^2}{n_1} + \frac{s_2^2}{n_2}} < \mu_1 - \mu_2$$

$$< (\bar{X}_1 - \bar{X}_2) + t_{\alpha/2}\sqrt{\frac{s_1^2}{n_1} + \frac{s_2^2}{n_2}}$$

d.f. = the smaller value of $n_1 - 1$ or $n_2 - 1$ *continued*

Confidence Intervals for the Difference of Two Means: Small Independent Samples (concluded)

Variances equal

$$(\bar{X}_1 - \bar{X}_2) - t_{\alpha/2} \sqrt{\frac{(n_1 - 1)s_1^2 + (n_2 - 1)s_2^2}{n_1 + n_2 - 2}} \cdot \sqrt{\frac{1}{n_1} + \frac{1}{n_2}}$$

$$< \mu_1 - \mu_2 < (\bar{X}_1 - \bar{X}_2) + t_{\alpha/2} \sqrt{\frac{(n_1 - 1)s_1^2 + (n_2 - 1)s_2^2}{n_1 + n_2 - 2}} \cdot \sqrt{\frac{1}{n_1} + \frac{1}{n_2}}$$

d.f. $= n_1 + n_2 - 2$

Remember that when one is testing the difference between two means from independent samples, two different statistical test formulas can be used. One formula is used when the variances are equal, the other when the variances are not equal. As shown in Section 10–3, some statisticians use an F test to determine whether or not the two variances are equal.

Example 10–11

Find the 95% confidence interval for the data in Example 10–9.

Solution

Substitute in the formula

$$(\bar{X}_1 - \bar{X}_2) - t_{\alpha/2} \sqrt{\frac{s_1^2}{n_1} + \frac{s_2^2}{n_2}} < \mu_1 - \mu_2$$

$$< (\bar{X}_1 - \bar{X}_2) + t_{\alpha/2} \sqrt{\frac{s_1^2}{n_1} + \frac{s_2^2}{n_2}}$$

$$(191 - 199) - 2.365 \sqrt{\frac{38^2}{8} + \frac{12^2}{10}} < \mu_1 - \mu_2$$

$$< (191 - 199) + 2.365 \sqrt{\frac{38^2}{8} + \frac{12^2}{10}}$$

$$-41.02 < \mu_1 - \mu_2 < 25.02$$

Since 0 is contained in the interval, the decision is not to reject the null hypothesis, H_0: $\mu_1 = \mu_2$.

Exercises

For Exercises 10–43 through 10–53, perform each of the following steps. Assume that all variables are normally or approximately normally distributed. Be sure to test for equality of variance first.

a.   State the hypotheses and identify the claim.
b.   Find the critical value(s).
c.   Compute the test value.
d.   Make the decision.
e.   Summarize the results.

Use the traditional method of hypothesis testing unless otherwise specified.

10–43. A real estate agent wishes to determine whether tax assessors and real estate appraisers agree on the values of homes. A random sample of the two groups appraised 10 homes. The data are shown here. Is there a significant difference in the values of the homes for each group? Let $\alpha = 0.05$. Find the 95% confidence interval for the difference of the means.

Real estate appraisers	Tax assessors
$\bar{X}_1 = \$83,256$	$\bar{X}_2 = \$88,354$
$s_1 = \$3256$	$s_2 = \$2341$
$n_1 = 10$	$n_2 = 10$

10–44. A researcher suggests that male nurses earn more than female nurses. A survey of 16 male nurses and 20 female nurses reports the following data. Is there enough evidence to support the claim that male nurses earn more than female nurses? Use $\alpha = 0.05$.

Male	Female
$\bar{X}_1 = \$23,800$	$\bar{X}_2 = \$23,750$
$s_1 = \$300$	$s_2 = \$250$
$n_1 = 16$	$n_2 = 20$

10–45. An instructor thinks that math majors can write and debug computer programs faster than business majors. A sample of 12 math majors took an average of 36 minutes to write a specific program and debug it; a sample of 18 business majors took an average of 39 minutes. The standard deviations were 4 minutes and 9 minutes, respectively. At $\alpha = 0.10$, is there evidence to support the claim?

10–46. A researcher estimates that high school girls miss more days of school than high school boys. A sample of 16 girls showed that they missed an average of 3.9 days of school per school year; a sample of 22 boys showed that they missed an average of 3.6 days of school per year. The standard deviations are 0.6 and 0.8 respectively. At $\alpha = 0.01$, is there enough evidence to support the researcher's claim?

10–47. A health care worker wishes to see if the average number of family day care homes per county is greater than the average number of day care centers per county. The number of centers for a selected sample of counties is shown. At $\alpha = 0.01$ can it be concluded that the average number of family day care homes is greater than the average number of day care centers?

Number of family day care homes			Number of day care centers		
25	57	34	5	28	37
42	21	44	16	16	48

Source: *Pittsburgh Tribune Review,* April 19, 1998.

10–48. A researcher wishes to test the claim that on average more juveniles than adults are classified as missing persons. Records for the last five years are shown. At $\alpha = 0.10$ is there enough evidence to support the claim?

Juveniles	65,513	65,934	64,213	61,954	59,167
Adults	31,364	34,478	36,937	35,946	38,209

Source: *USA Today,* March 24, 1999.

10–49. The local branch of the Internal Revenue Service spent an average of 21 minutes helping each of 10 people prepare their tax returns. The standard deviation was 5.6 minutes. A volunteer tax preparer spent an average of 27 minutes helping 14 people prepare their taxes. The standard deviation was 4.3 minutes. At $\alpha = 0.02$, is there a difference in the average time spent by the two services? Find the 98% confidence interval for the two means.

10–50. The average price of seven ABC dishwashers was $815, and the average price of nine XYZ dishwashers was $845. The standard deviations were $19 and $9, respectively. At $\alpha = 0.05$, can one conclude that the XYZ dishwashers cost more?

10–51. The average monthly premium paid by 12 administrators for hospitalization insurance is $56. The standard deviation is $3. The average monthly premium paid by 27 nurses is $63. The standard deviation is $5.75. At $\alpha = 0.05$, do the nurses pay more for hospitalization insurance? Use the *P*-value method.

10–52. Health Care Knowledge Systems reported that an insured woman spends on average 2.3 days in the hospital for a routine childbirth, while an uninsured woman spends on average 1.9 days. Assume two samples of 16 women each were used and the standard deviations are both equal to 0.6 of a day. At $\alpha = 0.01$, test the claim that the means are equal. Find the 99% confidence interval for the differences of the means. Use the *P*-value method.

Source: Michael D. Shook and Robert L. Shook, *The Book of Odds* (New York: Penguin Putnam, Inc., 1991).

10–53. The times (in minutes) it took six white mice to learn to run a simple maze and the times it took six brown mice to learn to run the same maze are given here. At $\alpha = 0.05$, does the color of the mice make a difference in their learning rate? Find the 95% confidence interval for the difference of the means. Use the *P*-value method.

White mice	18	24	20	13	15	12
Brown mice	25	16	19	14	16	10

Technology Step by Step

MINITAB
Step by Step

Testing the Difference between Two Means (Small Independent Samples)

Example MT10–3

Using the sample data from Example MT10–2 (repeated below), is there evidence to support the claim that the population means are equal?

Sample 1	12	15	18	13	10	9	6	3	11	5
Sample 2	18	20	13	7	9	4				

1. Select **Stat>Basic Statistics>2-Sample t.**
2. Click the button for **Samples in different columns.** There is one sample in each column.
3. Click in the box for **First:** and double-click **C1**.
4. Double-click **C2** in the list.
5. Check the box for **Assume equal variances**, since the *F* test in Example MT10–2 showed that we can make that assumption. The pooled standard deviation formula will be used to calculate the test statistic and *P*-value.

Two-Sample t Dialog Box

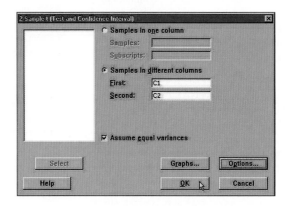

6. The session window is shown. The *P*-value for the difference is 0.560. Do not reject the null hypothesis.

```
F-Test (normal distribution)
Test Statistic: 0.541
P-Value     : 0.399
```

Two-Sample T-Test and CI: C1, C2

```
Two-sample T for C1 vs C2
N        Mean     StDev     SE Mean
C1   10      10.20     4.64       1.5
C2    6      11.83     6.31       2.6

Difference = mu C1 - mu C2
Estimate for difference: -1.63
95% CI for difference: (-7.50, 4.23)
T-Test of difference = 0 (vs not =): T-Value = -0.60 P-Value = 0.560
DF = 14
```

TI-83 Step by Step

t Test for the Difference between Two Means (Data)

1. Enter the data into L_1 and L_2.
2. Press **STAT** and move the cursor to **TESTS.**
3. Press **4** for **2-SampTTest.**
4. Select **Data,** press **ENTER,** then press ▼. Make sure L_1 and L_2 are selected and **Freq1** and **Freq2** = 1.

5. Select the correct alternative hypothesis.

6. Select either No or Yes for Pooled. If the variances are assumed equal, select Yes. If the variances are assumed unequal, select No.

7. Select Calculate and press ENTER.

Example TI10–5

Test the claim H_0: $\mu_1 = \mu_2$ at $\alpha = 0.05$. Assume the variances are equal.

	A					B			
32	38	37	36	36	30	36	35	36	31
34	39	36	37	42	34	37	33	32	

Input

Input

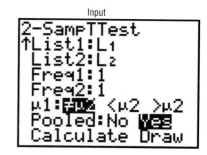

Output

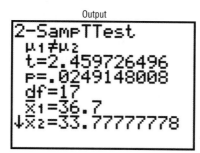

Output

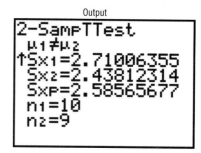

As shown, the test value is 2.459726496, and the *P*-value is 0.0249148008. The decision is to reject the null hypothesis, since $0.0249148008 < 0.05$. The calculator also gives the values for d.f., $\overline{X}_1$, $\overline{X}_2$, s_1, s_2, n_1, and n_2.

t Test for the Difference between Two Means (Stats)

1. Press **STAT** and move the cursor to TESTS.

2. Press **4** for 2-SampTTest.

3. Select Stats; press **ENTER**, then press ▼.

4. Enter the values for $\overline{X}_1$, s_1, n_1, $\overline{X}_2$, s_2, and n_2.

5. Select the correct alternative hypothesis.

6. Select Yes if population variances are assumed equal; otherwise, select No.

7. Select Calculate and press **ENTER.**

Example TI10–6

Test the claim from Example 10–9, H_1: $\mu_1 \neq \mu_2$ at $\alpha = 0.05$. Assume the population variances are not equal. The statistics for the sample data are:

A	B
$\overline{X}_1 = 191$	$\overline{X}_2 = 199$
$s_1 = 38$	$s_2 = 12$
$n_1 = 8$	$n_2 = 10$

Input

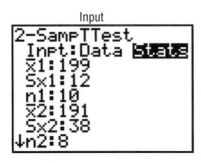

Input

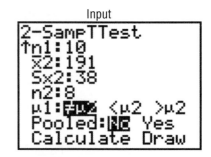

Output

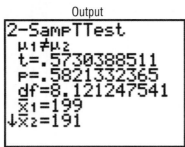

Output

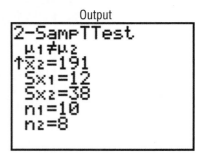

As shown, the test value is -0.5730388511, and the *P*-value is 0.5821332365. The decision is not to reject the null hypothesis, since $0.5821332365 > 0.05$. (*Note:* The degrees of freedom shown by the calculator are computed differently from those in the textbook.)

Excel
Step by Step

Testing the Difference between Two Means (Small Independent Samples)

Excel has a two-sample *t* test in its Data Analysis tools. To perform the *t* test for the difference between means, see the following example.

Example XL10–3

Test the hypothesis that there is no difference between population means based on the following sample data. Assume the population variances are not equal. Use $\alpha = 0.05$.

Set A	32	38	37	36	36	34	39	36	37	42
Set B	30	36	35	36	31	34	37	33	32	

1. Enter the 10-number data set A in column A.

2. Enter the 9-number data set B in column B.

3. Select `Tools`, `Data Analysis` and choose `t-Test: Two-Sample Assuming Unequal Variances`.

4. Enter the data ranges, hypothesized mean difference (here, zero), and α.

5. Select a location for output and click [`OK`].

Two-Sample t Test
in Excel

t-Test: Two-Sample Assuming Unequal Variances   ? X

Input

Variable 1 Range:	A1:A10
Variable 2 Range:	B1:B9
Hypothesized Mean Difference:	0

☐ Labels

Alpha: 0.05

OK

Cancel

Help

Output options

⦿ Output Range:	D9
○ New Worksheet Ply:	
○ New Workbook	

t-Test: Two-Sample Assuming Unequal Variances	Variable 1	Variable 2
Mean	36.7	33.77777778
Variance	7.344444444	5.944444444
Observations	10	9
Hypothesized Mean Difference	0	
df	17	
t Stat	2.474205364	
P(T<=t) one-tail	0.012095	
t Critical one-tail	1.739606432	
P(T<=t) two-tail	0.024189999	
t Critical two-tail	2.109818524	

The output reports both one- and two-tailed *P*-values.

Another test in Data Analysis is used if the sample variances are equal.

10–5

Testing the Difference between Two Means: Small Dependent Samples

Objective 4. Test the difference between two means for small dependent samples.

In the previous section, the *t* test was used to compare two sample means when the samples were independent. In this section, a different version of the *t* test is explained. This version is used when the samples are dependent. Samples are considered to be **dependent samples** when the subjects are paired or matched in some way.

For example, suppose a medical researcher wants to see whether a drug will affect the reaction time of its users. To test this hypothesis, the researcher must pretest the subjects in the sample first. That is, they are given a test to ascertain their normal reaction times. Then after taking the drug, the subjects are tested again, using a posttest. Finally, the means of the two tests are compared to see whether there is a difference. Since the same subjects are used in both cases, the samples are *related;* subjects scoring high on the pretest will generally score high on the posttest, even after consuming the drug. Likewise, those scoring lower on the pretest will tend to score lower on the posttest. In order to take this effect into account, the researcher employs a *t* test using the differences

Speaking of   STATISTICS

On the basis of the conclusions in this study, state several hypotheses that may have been used to support these conclusions. Comment on how you think these hypotheses were tested. For example, how would one determine whether standardized tests are biased against girls?

Public School Teachers Show Bias for Boys

By Katy Kelly
USA TODAY

Public schools favor boys, says a report out today.

"There is an illusion that schools are treating boys and girls equally," says Alice McKee, American Association of University Women, which commissioned the report.

Based on a review of recent research, it suggests:

• Teachers ask academic questions of boys 80% more often than of girls.
• School curricula generally ignore or stereotype females.
• Most standardized tests are biased against girls.

• Preschool boys get more individual attention, even hugs, from teachers than girls do.

Bias is not intentional: "When teachers are made aware they're doing this . . . they are willing and eager to change," McKee says.

The report lists 40 recommendations, including requiring course work on gender issues for teacher certification.

New classroom materials are needed too, McKee says: "Boys and girls should be able to study women Nobel Prize winners in addition to Betsy Ross sewing the flag."

between the pretest values and the posttest values. This way makes sure only the gain or loss in values is compared.

Here are some other examples of dependent samples. A researcher may want to design an SAT preparation course to help students raise their test scores the second time they take the SAT exam. Hence, the differences between the two exams are compared. A medical specialist may want to see whether a new counseling program will help subjects lose weight. Therefore, the preweights of the subjects will be compared with the postweights.

Besides samples in which the same subjects are used in a pre–post situation, there are other cases where the samples are considered dependent. For example, students might be matched or paired according to some variable that is pertinent to the study; then one student is assigned to one group and the other student is assigned to a second group. For instance, in a study involving learning, students can be selected and paired according to their IQs. That is, two students with the same IQ will be paired. Then one will be assigned to one sample group (which might receive instruction by computers), and the other student will be assigned to another sample group (which might receive instruction by the lecture-discussion method). These assignments will be done randomly. Since a student's IQ is important to learning, it is a variable that should be controlled. By matching subjects on IQ, the researcher can eliminate the variable's influence, for the most part. Matching, then, helps to reduce type II error by eliminating extraneous variables.

Two notes of caution should be mentioned. First, when subjects are matched according to one variable, the matching process does not eliminate the influence of other variables. Matching students according to IQ does not account for their mathematical ability or their familiarity with computers. Since all variables influencing a study cannot be controlled, it is up to the researcher to determine which variables should be used in matching. Second, when the same subjects are used for a pre–post study, sometimes the knowledge that they are participating in a study can influence the results. For example, if people are placed in a special program, they may be more highly motivated to succeed simply because they have been selected to participate; the program itself may have little effect on their success.

When the samples are dependent, a special t test for dependent means is used. This test employs the difference in values of the matched pairs. The hypotheses are as follows, where μ_D is the symbol for the expected mean of the differences of the matched pairs.

Two-tailed	**Left-tailed**	**Right-tailed**
$H_0: \mu_D = 0$	$H_0: \mu_D \geq 0$	$H_0: \mu_D \leq 0$
$H_1: \mu_D \neq 0$	$H_1: \mu_D < 0$	$H_1: \mu_D > 0$

The general procedure for finding the test value involves several steps.
First, find the differences of the values of the pairs of data.

$$D = X_1 - X_2$$

Second, find the mean ($\bar{D}$) of the differences, using the formula

$$\bar{D} = \frac{\Sigma D}{n}$$

where n is the number of data pairs.
Third, find the standard deviation (s_D) of the differences, using the formula

$$s_D = \sqrt{\frac{\Sigma D^2 - \dfrac{(\Sigma D)^2}{n}}{n - 1}}$$

Fourth, find the estimated standard error ($s_{\bar{D}}$) of the differences, which is

$$s_{\bar{D}} = \frac{s_D}{\sqrt{n}}$$

Finally, find the test value, using the formula

$$t = \frac{\bar{D} - \mu_D}{s_D / \sqrt{n}} \qquad \text{with d.f.} = n - 1$$

The formula in the final step follows the basic format of

$$\text{test value} = \frac{(\text{observed value}) - (\text{expected value})}{\text{standard error}}$$

where the observed value is the mean of the differences. The expected value μ_D is zero if the hypothesis is $\mu_D = 0$. The standard error of the difference is the standard deviation of the difference divided by the square root of the sample size. Both populations must be normally or approximately normally distributed. Example 10–12 illustrates the hypothesis-testing procedure in detail.

Example 10–12

A physical education director claims by taking a special vitamin, a weight lifter can increase his strength. Eight athletes are selected and given a test of strength, using the standard bench press. After two weeks of regular training, supplemented with the vitamin, they are tested again. Test the effectiveness of the vitamin regimen at $\alpha = 0.05$. Each value in the data that follow represents the maximum number of pounds the athlete can bench press. Assume that the variable is approximately normally distributed.

Athlete	1	2	3	4	5	6	7	8
Before (X_1)	210	230	182	205	262	253	219	216
After (X_2)	219	236	179	204	270	250	222	216

Solution

STEP 1 State the hypotheses and identify the claim. In order for the vitamin to be effective, the "before" weights must be significantly less than the "after" weights; hence, the mean of the differences must be less than zero.

$$H_0: \mu_D \geq 0 \quad \text{and} \quad H_1: \mu_D < 0 \text{ (claim)}$$

STEP 2 Find the critical value. The degrees of freedom are $n - 1$. In this case, d.f. $= 8 - 1 = 7$. The critical value for a left-tailed test with $\alpha = 0.05$ is -1.895.

STEP 3 Compute the test value.

a. Make a table.

		A	**B**
Before (X_1)	After (X_2)	$D = (X_1 - X_2)$	$D^2 = (X_1 - X_2)^2$
210	219		
230	236		
182	179		
205	204		
262	270		
253	250		
219	222		
216	216		

b. Find the differences and place the results in column A.

$$210 - 219 = -9$$
$$230 - 236 = -6$$
$$182 - 179 = +3$$
$$205 - 204 = +1$$
$$262 - 270 = -8$$
$$253 - 250 = +3$$
$$219 - 222 = -3$$
$$216 - 216 = \underline{\quad 0}$$
$$\Sigma D = -19$$

c. Find the mean of the differences.

$$\bar{D} = \frac{\Sigma D}{n} = \frac{-19}{8} = -2.375$$

d. Square the differences and place the results in column B.

$$(-9)^2 = 81$$
$$(-6)^2 = 36$$
$$(+3)^2 = 9$$
$$(+1)^2 = 1$$
$$(-8)^2 = 64$$
$$(+3)^2 = 9$$
$$(-3)^2 = 9$$
$$0^2 = 0$$
$$\Sigma D^2 = 209$$

The completed table is shown next.

Before (X_1)	After (X_2)	**A** $D = (X_1 - X_2)$	**B** $D^2 = (X_1 - X_2)^2$
210	219	−9	81
230	236	−6	36
182	179	+3	9
205	204	+1	1
262	270	−8	64
253	250	+3	9
219	222	−3	9
216	216	0	0
		$\Sigma D = -19$	$\Sigma D^2 = 209$

e. Find the standard deviation of the differences.

$$s_D = \sqrt{\frac{\Sigma D^2 - \dfrac{(\Sigma D)^2}{n}}{n-1}} = \sqrt{\frac{209 - \dfrac{(-19)^2}{8}}{8-1}} = 4.84$$

f. Find the test value.

$$t = \frac{\bar{D} - \mu_D}{s_D/\sqrt{n}} = \frac{-2.375 - 0}{4.84/\sqrt{8}} = -1.388$$

STEP 4   Make the decision. The decision is not to reject the null hypothesis at $\alpha = 0.05$, since $-1.388 > -1.895$, as shown in Figure 10–11.

Figure 10–11

Critical and Test Values
for Example 10–12

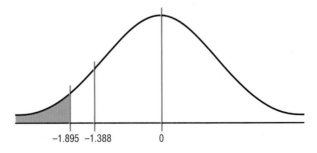

$$-1.895\ \ -1.388 \qquad 0$$

STEP 5   Summarize the results. There is not enough evidence to support the claim that the vitamin increases the strength of weight lifters.

The formulas for this *t* test are summarized next.

Formulas for the *t* Test for Dependent Samples

$$t = \frac{\bar{D} - \mu_D}{s_D/\sqrt{n}}$$

with d.f. = $n - 1$ and where

$$\bar{D} = \frac{\Sigma D}{n} \quad \text{and} \quad s_D = \sqrt{\frac{\Sigma D^2 - \frac{(\Sigma D)^2}{n}}{n - 1}}$$

Example 10–13

A dietitian wishes to see if a person's cholesterol level will change if the diet is supplemented by a certain mineral. Six subjects were pretested and then took the mineral supplement for a six-week period. The results are shown in the table. (Cholesterol level is measured in milligrams per deciliter.) Can it be concluded that the cholesterol level has been changed at $\alpha = 0.10$? Assume the variable is approximately normally distributed.

Subject	1	2	3	4	5	6
Before (X_1)	210	235	208	190	172	244
After (X_2)	190	170	210	188	173	228

Solution

STEP 1 State the hypotheses and identify the claim. If the diet is effective, the "before" cholesterol levels should be different from the "after" levels.

$$H_0: \mu_D = 0 \quad \text{and} \quad H_1: \mu_D \neq 0 \text{ (claim)}$$

STEP 2 Find the critical value. The degrees of freedom are 5. At $\alpha = 0.10$, the critical values are ± 2.015.

STEP 3 Compute the test value.

a. Make a table.

Before (X_1)	After (X_2)	A $D = (X_1 - X_2)$	B $D^2 = (X_1 - X_2)^2$
210	190		
235	170		
208	210		
190	188		
172	173		
244	228		

b. Find the differences and place the results in column A.

$$210 - 190 = 20$$
$$235 - 170 = 65$$
$$208 - 210 = -2$$
$$190 - 188 = 2$$
$$172 - 173 = -1$$
$$244 - 228 = \underline{16}$$
$$\Sigma D = 100$$

c. Find the mean of the differences.

$$\bar{D} = \frac{\Sigma D}{n} = \frac{100}{6} = 16.7$$

d. Square the differences and place the results in column B.

$$(20)^2 = 400$$
$$(65)^2 = 4225$$
$$(-2)^2 = 4$$
$$(2)^2 = 4$$
$$(-1)^2 = 1$$
$$(16)^2 = \underline{256}$$
$$\Sigma D^2 = 4890$$

Then complete the table as shown.

Before (X_1)	After (X_2)	A $D = (X_1 - X_2)$	B $D^2 = (X_1 - X_2)^2$
210	190	20	400
235	170	65	4225
208	210	-2	4
190	188	2	4
172	173	-1	1
244	228	16	256
		$\Sigma D = 100$	$\Sigma D^2 = 4890$

e. Find the standard deviation of the differences.

$$s_D = \sqrt{\frac{\Sigma D^2 - \frac{(\Sigma D)^2}{n}}{n-1}} = \sqrt{\frac{4890 - \frac{(100)^2}{6}}{5}} = 25.4$$

f. Find the test value.

$$t = \frac{\bar{D} - \mu_D}{s_D/\sqrt{n}} = \frac{16.7 - 0}{25.4/\sqrt{6}} = 1.610$$

STEP 4   Make the decision. The decision is not to reject the null hypothesis, since the test value 1.610 is in the noncritical region, as shown in Figure 10–12.

Figure 10–12

Critical and Test Values for Example 10–13

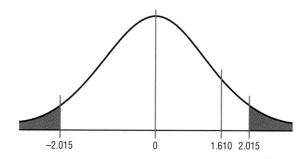

-2.015   0   1.610 2.015

STEP 5   Summarize the results. There is not enough evidence to support the claim that the mineral changes a person's cholesterol level.

The steps for this t test are summarized in the Procedure Table.

Procedure Table

Testing the Difference between Means for Dependent Samples

STEP 1   State the hypotheses and identify the claim.

STEP 2   Find the critical value(s).

STEP 3   Compute the test value.

a. Make a table, as shown.

X_1	X_2	**A** $D = (X_1 - X_2)$	**B** $D^2 = (X_1 - X_2)^2$
$\vdots$	$\vdots$		
		$\Sigma D = $ _____	$\Sigma D^2 = $ _____

b. Find the differences and place the results in column A.

$$D = (X_1 - X_2)$$

c. Find the mean of the differences.

$$\bar{D} = \frac{\Sigma D}{n}$$

d. Square the differences and place the results in column B. Complete the table.

$$D^2 = (X_1 - X_2)^2$$

e. Find the standard deviation of the differences.

$$s_D = \sqrt{\frac{\Sigma D^2 - \dfrac{(\Sigma D)^2}{n}}{n-1}}$$

f. Find the test value.

$$t = \frac{\bar{D} - \mu_D}{s_D/\sqrt{n}} \qquad \text{with d.f.} = n - 1$$

STEP 4   Make the decision.

STEP 5   Summarize the results.

The P-values for the t test are found in Table F. For a two-tailed test with d.f. $= 5$ and $t = 1.610$, the P-value is found between 1.476 and 2.015; hence, $0.10 < P\text{-value} < 0.20$. Thus, the null hypothesis cannot be rejected at $\alpha = 0.10$.

If a specific difference is hypothesized, the following formula should be used:

$$t = \frac{\bar{D} - \mu_D}{s_D/\sqrt{n}}$$

where μ_D is the hypothesized difference.

For example, if a dietician claims that people on a specific diet will lose an average of 3 pounds in a week, the hypotheses are

$$H_0\text{: } \mu_D = 3 \qquad \text{and} \qquad H_1\text{: } \mu_D \neq 3$$

The value 3 will be substituted in the test statistic formula for μ_D.

Confidence intervals can be found for the mean differences with the following formula.

Confidence Interval for the Mean Difference

$$\bar{D} - t_{\alpha/2}\frac{s_D}{\sqrt{n}} < \mu_D < \bar{D} + t_{\alpha/2}\frac{s_D}{\sqrt{n}}$$

$$\text{d.f.} = n - 1$$

Example 10–14

Find the 90% confidence interval for the data in Example 10–13.

Solution

Substitute in the formula

$$\bar{D} - t_{\alpha/2}\frac{s_D}{\sqrt{n}} < \mu_D < \bar{D} + t_{\alpha/2}\frac{s_D}{\sqrt{n}}$$

$$16.7 - 2.015 \cdot \frac{25.4}{\sqrt{6}} < \mu_D < 16.7 + 2.015 \cdot \frac{25.4}{\sqrt{6}}$$

$$16.7 - 20.89 < \mu_D < 16.7 + 20.89$$

$$-4.19 < \mu_D < 37.59$$

Since 0 is contained in the interval, the decision is not to reject the null hypothesis, $H_0: \mu_D = 0$.

Exercises

10–54. Explain the difference between independent and dependent samples.

10–55. Classify each as independent or dependent samples.
a. Heights of identical twins
b. Test scores of the same students in English and psychology
c. The effectiveness of two different brands of aspirin
d. Effects of a drug on reaction time, measured by a "before" and an "after" test
e. The effectiveness of two different diets on two different groups of individuals

For Exercises 10–56 through 10–64, perform each of the following steps. Assume that all variables are normally or approximately normally distributed.
a. State the hypotheses and identify the claim.
b. Find the critical value(s).
c. Compute the test value.
d. Make the decision.
e. Summarize the results.

Use the traditional method of hypothesis testing unless otherwise specified.

10–56. A program for reducing the number of days missed by food handlers in a certain restaurant chain was conducted. The owners hypothesized that after the program, the workers would miss fewer days of work due to illness. The table shows the number of days 10 workers missed per month before and after completing the program. Is there enough evidence to support the claim, at $\alpha = 0.05$, that the food handlers missed fewer days after the program?

Before	2	3	6	7	4	5	3	1	0	0
After	1	4	3	8	3	3	1	0	1	0

10–57. As an aid for improving students' study habits, nine students were randomly selected to attend a seminar on the importance of education in life. The table shows the number of hours each student studied per week before and after the seminar. At $\alpha = 0.10$, did attending the seminar increase the number of hours the students studied per week?

Before	9	12	6	15	3	18	10	13	7
After	9	17	9	20	2	21	15	22	6

10–58. A doctor is interested in determining whether a film about exercise will change 10 people's attitudes about

Here is a study that involves two groups of children. What was the sample size for each group? State a possible null and alternative hypothesis for this study. What statistical test was probably used? What was the decision? Explain your answer.

Pedaling a solution for couch-potato kids

By Nanci Hellmich
USA Today

Getting kids to exercise instead of vegging out in front of the TV has long been a problem for parents. But researchers have figured out how to get kids moving while watching TV.

Obesity researchers at St. Luke's-Roosevelt Hospital Center in New York City wanted to know what would happen if kids had to ride a stationary bike to keep the TV working. So they designed and built TV-cycles.

They randomly assigned 10 overweight, sedentary children ages 8 to 12, to a bicycle that required the child to pedal for TV time or a bicycle that was in front of the TV but not necessary for its operation.

They found that:

➤ The children who had to pedal to watch TV biked an average of an hour a week compared with eight minutes for the others.

➤ The treatment group watched one hour of TV a week, while the other children watched for an average of more than 20 hours.

➤ The treatment group significantly decreased overall body fat.

"This was a non-nagging approach" to get kids to exercise, says lead investigator David Allison, associate professor of medical psychology at Columbia University College of Physicians and Surgeons. "We told parents to just let the bicycle do the work."

One problem: It was tough on parents not to be able to watch TV, and they had to find ways to occupy their kids.

"This is an example of changing the environment to promote activity. We need to think of how we can take this concept and make it work for more people," Allison says.

The study was discussed Sunday at the Experimental Biology meeting in Washington, D.C.

Source: *USA Today,* April 19, 1999. Used with permission.

---

exercise. The results of his questionnaire are shown below. A higher numerical value shows a more favorable attitude toward exercise. Is there enough evidence to support the claim, at $\alpha = 0.05$, that there was a change in attitude? Find the 95% confidence interval for the difference of the two means.

Before	12	11	14	9	8	6	8	5	4	7
After	13	12	10	9	8	8	7	6	5	5

10–59. Suppose a researcher wishes to test the effects of a new diet designed to reduce the blood sodium level of 10 patients. The patients' sodium levels are checked before they are placed on the diet and after two weeks on the diet. At $\alpha = 0.05$, is there enough evidence to conclude that the diet is effective in lowering the sodium content of their blood?

Patient	Before	After
1	146	135
2	138	133
3	152	147
4	163	156
5	136	138
6	147	141
7	148	139
8	141	132
9	143	138
10	142	131

10–60. An office manager wishes to see whether the typing speed of 10 secretaries can be increased by changing over to computers. The number of words typed per minute is given here. At $\alpha = 0.10$, is there enough evidence to conclude that by using computers, the secretaries can type more words per minute?

Secretary	Typewriter	Computer
1	63	68
2	72	80
3	85	95
4	97	93
5	82	80
6	101	106
7	73	82
8	62	78
9	58	65
10	75	83

10–61. A composition teacher wishes to see whether a new grammar program will reduce the number of grammatical errors her students make when writing a two-page essay. The data are shown here. At $\alpha = 0.025$, can it be concluded that the number of errors has been reduced?

Student	1	2	3	4	5	6
Errors before	12	9	0	5	4	3
Errors after	9	6	1	3	2	3

10–62. A sports-shoe manufacturer claims that joggers who wear its brand of shoe will jog faster than those who don't. A sample of eight joggers is taken, and they agree to test the claim on a 1-mile track. The rates (in minutes) of the joggers while wearing the manufacturer's shoe and while wearing any other brand of shoe are shown here. Test the claim at $\alpha = 0.025$.

Runner	1	2	3	4	5	6	7	8
Manufacturer's brand	8.2	6.3	9.2	8.6	6.8	8.7	8.0	6.9
Other brand	7.1	6.8	9.8	8.0	5.8	8.0	7.4	8.0

10–63. A researcher wanted to compare the pulse rates of identical twins to see whether there was any difference. Eight sets of twins were selected. The rates are given in the table as number of beats per minute. At $\alpha = 0.01$, is there a significant difference in the average pulse rates of twins? Find the 99% confidence interval for the difference of the two. Use the P-value method.

Twin A	87	92	78	83	88	90	84	93
Twin B	83	95	79	83	86	93	80	86

10–64. A reporter hypothesizes that the average assessed values of land in a large city have changed during a five-year period. A random sample of wards is selected, and the data (in millions of dollars) are shown below. At $\alpha = 0.05$ can it be concluded that the average taxable assessed values have changed? Use the P-value method.

Ward	1994	1999
A	184	161
B	414	382
C	22	22
D	99	109
E	116	120
F	49	52
G	24	28
H	50	50
I	282	297
J	25	40
K	141	148
L	45	56
M	12	20
N	37	38
O	9	9
P	17	19

Source: *Pittsburgh Tribune Review,* May 2, 1999.

*10–65. Instead of finding the mean of the differences between X_1 and X_2 by subtracting $X_1 - X_2$, one can find it by finding the means of X_1 and X_2 and then subtracting the means. Show that these two procedures will yield the same results.

Technology Step by Step

MINITAB
Step by Step

Testing the Difference between Two Means (Small Dependent Samples)

Example MT10–4

1. Enter the data from Example 10–12 into C1 and C2. Name the columns **Before** and **After.**

2. Select **Stat>BasicStatistics>Pairedt.**

3. Double-click C1 Before. (First sample:)

4. Double-click C2 After (Second sample:)

Paired t Dialog Box

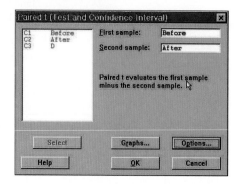

The second sample will be subtracted from the first. The differences are not stored or displayed.

5. Click [Options]. Change the alternative hypothesis to less than. This effects the *P*-value. Click [OK].

6. Click [OK].

Paired T-Test and CI: Before, After

```
Paired T for Before–After

                N       Mean      StDev    SE Mean
Before          8      222.13     25.92       9.16
After           8      224.50     27.91       9.87
Difference      8       -2.38      4.84       1.71

95% upper bound for mean difference: 0.87
T-Test of mean difference = 0 (vs < 0): T-Value = -1.39 P-Value = 0.104
```

The *P*-value is 0.104 for a test statistic of $t = -1.39$. The sample difference of -2.38 in the strength measurement is not statistically significant, so the claim that vitamins increase strength is not supported by the evidence.

TI–83
Step by Step

t Test for the Difference between Two Means (Small Dependent Samples)

To solve this type of problem, first enter the differences between the matched pairs in column L_1. Then, use the one-sample t test on that column.

1. Enter the differences into L_1 (one way to do this is to enter the sample data sets in L_2 and L_3, then enter **L3-L2 [STO] L1** at the home screen).

2. Press **STAT** and move the cursor to TESTS.

3. Press **2** for T-Test.

4. Select Data and press **ENTER**, then press ▼. Make sure L1 is selected and Freq = 1.

5. Enter **0** for the value of μ_0.

6. Select the correct alternative hypothesis.

7. Select Calculate and press **ENTER**.

Example TI10–7

Test the claim from Example 10–13, H_1: $\mu_D \neq 0$ at $\alpha = 0.10$ when

Before (X_1)	210	235	208	190	172	244
After (X_2)	190	170	210	188	173	228

Subtract $X_1 - X_2$ and enter these values in L_1:

$$20 \quad 65 \quad -2 \quad 2 \quad -1 \quad 16$$

Input

```
T-Test
 Inpt:DATA Stats
 µ0:0
 List:L₁
 Freq:1
 µ:≠µ0 <µ0 >µ0
 Calculate Draw
```

Output

```
T-Test
 µ≠0
 t=1.607891603
 p=.1687705833
 x̄=16.66666667
 Sx=25.39028686
 n=6
```

The test value is 1.607891603 and the *P*-value is 0.1687705833. The decision is not to reject the null hypothesis at $\alpha = 0.10$, since $0.1687705833 > 0.10$.

Excel
Step by Step

t Test for the Difference between Two Means (Small Dependent Samples)

Example XL10–4

Test the hypothesis that there is no difference in population means, based on the following sample paired data. Use $\alpha = 0.05$.

Set A	33	35	28	29	32	34	30	34
Set B	27	29	36	34	30	29	28	24

1. Enter the eight-number data set A in column A.
2. Enter the eight-number data set B in column B.
3. Select `Tools`, `Data Analysis` and choose `t-Test: Paired Two Sample for Means`.
4. Enter the data ranges and hypothesized mean difference (here, zero), and α.
5. Select a location for output and click [OK].

Dialog Box for Paired
Data t Test

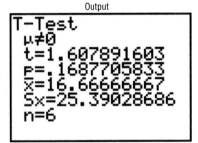

The screen shows a P-value of 0.3253988 for the two-tailed case. This is greater than the confidence level, 0.05, so we fail to reject the null hypothesis.

t-Test: Paired Two Sample for Means

	Variable 1	Variable 2
Mean	31.875	29.625
Variance	6.696428571	14.55357143
Observations	8	8
Pearson Correlation	-0.757913399	
Hypothesized Mean Difference	0	
df	7	
t Stat	1.057517468	
P(T<=t) one-tail	0.1626994	
t Critical one-tail	1.894577508	
P(T<=t) two-tail	0.3253988	
t Critical two-tail	2.36462256	

10–6

Testing the Difference between Proportions

Objective 5. Test the difference between two proportions.

The z test with some modifications can be used to test the equality of two proportions. For example, a researcher might ask: Is the proportion of men who exercise regularly less than the proportion of women who exercise regularly? Is there a difference in the percentage of students who own a personal computer and the percentage of nonstudents who own one? Is there a difference in the proportion of college graduates who pay cash for purchases and the proportion of noncollege graduates who pay cash?

Recall from Chapter 8 that the symbol $\hat{p}$ ("p hat") is the sample proportion used to estimate the population proportion, denoted by p. For example, if in a sample of 30 college students, 9 are on probation, then the sample proportion is $\hat{p} = \frac{9}{30}$, or 0.3. The population proportion p is the number of all students who are on probation divided by the number of students who attend the college. The formula for $\hat{p}$ is

$$\hat{p} = \frac{X}{n}$$

where

$X = $ number of units that possess the characteristic of interest

$n = $ sample size

When one is testing the difference between two population proportions, p_1 and p_2, the hypotheses can be stated as follows, if no difference between the proportions is hypothesized.

$$H_0: p_1 = p_2 \qquad \text{or} \qquad H_0: p_1 - p_2 = 0$$
$$H_1: p_1 \neq p_2 \qquad\qquad H_1: p_1 - p_2 \neq 0$$

Similar statements using $\geq$ and $<$ or $\leq$ and $>$ can be formed for one-tailed tests.

For two proportions, $\hat{p}_1 = X_1/n_1$ is used to estimate p_1 and $\hat{p}_2 = X_2/n_2$ is used to estimate p_2. The standard error of difference is

$$\sigma_{(\hat{p}_1 - \hat{p}_2)} = \sqrt{\sigma_{p_1}^2 + \sigma_{p_2}^2} = \sqrt{\frac{p_1 q_1}{n_1} + \frac{p_2 q_2}{n_2}}$$

where $\sigma_{p_1}^2$ and $\sigma_{p_2}^2$ are the variances of the proportions, $q_1 = 1 - p_1$, $q_2 = 1 - p_2$, and n_1 and n_2 are the respective sample sizes.

Since p_1 and p_2 are unknown, a weighted estimate of p can be computed by using the formula

$$\bar{p} = \frac{n_1 \hat{p}_1 + n_2 \hat{p}_2}{n_1 + n_2}$$

and $\bar{q} = 1 - \bar{p}$. This weighted estimate is based on the hypothesis that $p_1 = p_2$. Hence, $\bar{p}$ is a better estimate than either $\hat{p}_1$ or $\hat{p}_2$, since it is a combined average using both $\hat{p}_1$ and $\hat{p}_2$.

Since $\hat{p}_1 = X_1/n_1$ and $\hat{p}_2 = X_2/n_2$, $\bar{p}$ can be simplified to

$$\bar{p} = \frac{X_1 + X_2}{n_1 + n_2}$$

Finally, the standard error of difference in terms of the weighted estimate is

$$\sigma_{(\hat{p}_1 - \hat{p}_2)} = \sqrt{\bar{p}\bar{q}\left(\frac{1}{n_1} + \frac{1}{n_2}\right)}$$

The formula for the test value is shown next.

Formula for the z Test for Comparing Two Proportions

$$z = \frac{(\hat{p}_1 - \hat{p}_2) - (p_1 - p_2)}{\sqrt{\bar{p}\bar{q}\left(\frac{1}{n_1} + \frac{1}{n_2}\right)}}$$

where

$$\bar{p} = \frac{X_1 + X_2}{n_1 + n_2} \qquad \hat{p}_1 = \frac{X_1}{n_1}$$

$$\bar{q} = 1 - \bar{p} \qquad \hat{p}_2 = \frac{X_2}{n_2}$$

This formula follows the format

$$\text{test value} = \frac{(\text{observed value}) - (\text{expected value})}{\text{standard error}}$$

There are two requirements for use of the z test: (1) The samples must be independent of each other and (2) $n_1 p_1$ and $n_1 q_1$ must be 5 or more and $n_2 p_2$ and $n_2 q_2$ must be 5 or more.

Example 10–15

In the nursing home study mentioned in the chapter-opening "Statistics Today," the researchers found that 12 out of 34 small nursing homes had a resident vaccination rate of less than 80%, while 17 out of 24 large nursing homes had a vaccination rate of less than 80%. At $\alpha = 0.05$, test the claim that there is no difference in the proportions of the small and large nursing homes with a resident vaccination rate of less than 80%.

Source: Nancy Arden, Arnold S. Monto, and Suzanne E. Ohmit, "Vaccine Use and the Risk of Outbreaks in a Sample of Nursing Homes during an Influenza Epidemic," *American Journal of Public Health* 85, no. 3 (March 1995), pp. 399–401.

Solution

Let $\hat{p}_1$ be the proportion of the small nursing homes with a vaccination rate of less than 80% and $\hat{p}_2$ be the proportion of the large nursing homes with a vaccination rate of less than 80%. Then

$$\hat{p}_1 = \frac{X_1}{n_1} = \frac{12}{34} = 0.35 \qquad \text{and} \qquad \hat{p}_2 = \frac{X_2}{n_2} = \frac{17}{24} = 0.71$$

$$\bar{p} = \frac{X_1 + X_2}{n_1 + n_2} = \frac{12 + 17}{34 + 24} = \frac{29}{58} = 0.5$$

$$\bar{q} = 1 - \bar{p} = 1 - 0.5 = 0.5$$

Now, follow the steps in hypothesis testing.

STEP 1   State the hypotheses and identify the claim.

$$H_0: p_1 = p_2 \text{ (claim)} \qquad \text{and} \qquad H_1: p_1 \neq p_2$$

STEP 2   Find the critical values. Since $\alpha = 0.05$, the critical values are $+1.96$ and -1.96.

STEP 3   Compute the test value.

$$z = \frac{(\hat{p}_1 - \hat{p}_2) - (p_1 - p_2)}{\sqrt{\bar{p}\bar{q}\left(\dfrac{1}{n_1} + \dfrac{1}{n_2}\right)}}$$

$$= \frac{(0.35 - 0.71) - 0}{\sqrt{(0.5)(0.5)\left(\dfrac{1}{34} + \dfrac{1}{24}\right)}} = \frac{-0.36}{0.1333} = -2.7$$

STEP 4   Make the decision. Reject the null hypothesis, since $-2.7 < -1.96$. See Figure 10–13.

Figure 10–13

Critical and Test Values for Example 10–15

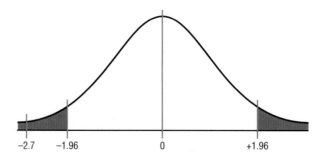

STEP 5   Summarize the results. There is enough evidence to reject the claim that there is no difference in the proportions of small and large nursing homes with a resident vaccination rate of less than 80%.

Example 10–16

A sample of 50 randomly selected men with high triglyceride levels consumed 2 tablespoons of oat bran daily for six weeks. After six weeks, 60% of the men had lowered their triglyceride level. A sample of 80 men consumed 2 tablespoons of wheat bran for six weeks. After six weeks, 25% had lower triglyceride levels. Is there a significant difference in the two proportions, at the 0.01 significance level?

Solution

Since the statistics are given in percentages, $\hat{p}_1 = 60\%$, or 0.60, and $\hat{p}_2 = 25\%$, or 0.25. In order to compute $\bar{p}$, one must find X_1 and X_2.

Since $\hat{p}_1 = X_1/n_1$, $X_1 = \hat{p}_1 \cdot n_1$, and since $\hat{p}_2 = X_2/n_2$, $X_2 = \hat{p}_2 \cdot n_2$.

$$X_1 = (0.60)(50) = 30 \qquad X_2 = (0.25)(80) = 20$$

$$\bar{p} = \frac{X_1 + X_2}{n_1 + n_2} = \frac{30 + 20}{50 + 80} = \frac{50}{130} = 0.385$$

$$\bar{q} = 1 - \bar{p} = 1 - 0.385 = 0.615$$

STEP 1   State the hypotheses and identify the claim.

$$H_0: p_1 = p_2 \qquad \text{and} \qquad H_1: p_1 \neq p_2 \text{ (claim)}$$

STEP 2   Find the critical values. Since $\alpha = 0.01$, the critical values are $+2.58$ and -2.58.

STEP 3   Compute the test value.

$$z = \frac{(\hat{p}_1 - \hat{p}_2) - (p_1 - p_2)}{\sqrt{\bar{p}\bar{q}\left(\dfrac{1}{n_1} + \dfrac{1}{n_2}\right)}}$$

$$= \frac{(0.60 - 0.25) - 0}{\sqrt{(0.385)(0.615)\left(\dfrac{1}{50} + \dfrac{1}{80}\right)}} = 3.99$$

STEP 4   Make the decision. Reject the null hypothesis, since $3.99 > 2.58$. See Figure 10–14.

Figure 10–14

Critical and Test Values for Example 10–16

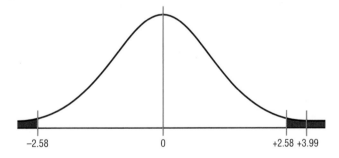

-2.58 \qquad 0 \qquad $+2.58$ $+3.99$

STEP 5   Summarize the results. There is enough evidence to support the claim that there is a difference in proportions.

The P-value for the difference of proportions can be found from Table E, as shown in Section 9–3. For Example 10–16, 3.99 is beyond 3.09; hence, the null hypothesis can be rejected since the P-value is less than 0.001.

The formula for the confidence interval for the difference between two proportions is shown next.

Confidence Interval for the Difference between Two Proportions

$$(\hat{p}_1 - \hat{p}_2) - z_{\alpha/2}\sqrt{\frac{\hat{p}_1\hat{q}_1}{n_1} + \frac{\hat{p}_2\hat{q}_2}{n_2}} < p_1 - p_2 < (\hat{p}_1 - \hat{p}_2) + z_{\alpha/2}\sqrt{\frac{\hat{p}_1\hat{q}_1}{n_1} + \frac{\hat{p}_2\hat{q}_2}{n_2}}$$

Example 10–17

Find the 95% confidence interval for the difference of proportions for the data in Example 10–15.

Solution

$$\hat{p}_1 = \frac{12}{34} = 0.35 \qquad \hat{q}_1 = 0.65$$

$$\hat{p}_2 = \frac{17}{24} = 0.71 \qquad \hat{q}_2 = 0.29$$

Substitute in the formula.

$$(\hat{p}_1 - \hat{p}_2) - z_{\alpha/2}\sqrt{\frac{\hat{p}_1\hat{q}_1}{n_1} + \frac{\hat{p}_2\hat{q}_2}{n_2}} < p_1 - p_2$$

$$< (\hat{p}_1 - \hat{p}_2) + z_{\alpha/2}\sqrt{\frac{\hat{p}_1\hat{q}_1}{n_1} + \frac{\hat{p}_2\hat{q}_2}{n_2}}$$

$$(0.35 - 0.71) - 1.96\sqrt{\frac{(0.35)(0.65)}{34} + \frac{(0.71)(0.29)}{24}}$$

$$< p_1 - p_2 < (0.35 - 0.71) + 1.96\sqrt{\frac{(0.35)(0.65)}{34} + \frac{(0.71)(0.29)}{24}}$$

$$-0.36 - 0.242 < p_1 - p_2 < -0.36 + 0.242$$

$$-0.602 < p_1 - p_2 < -0.118$$

Since 0 is not contained in the interval, the decision is to reject the null hypothesis, H_0: $p_1 = p_2$.

Exercises

10–66. Find the proportions $\hat{p}$ and $\hat{q}$ for each.
a. $n = 48$, $X = 34$
b. $n = 75$, $X = 28$
c. $n = 100$, $X = 50$
d. $n = 24$, $X = 6$
e. $n = 144$, $X = 12$

10–67. Find each X given $\hat{p}$.
a. $\hat{p} = 0.16$, $n = 100$
b. $\hat{p} = 0.08$, $n = 50$
c. $\hat{p} = 6\%$, $n = 80$
d. $\hat{p} = 52\%$, $n = 200$
e. $\hat{p} = 20\%$, $n = 150$

10–68. Find $\bar{p}$ and $\bar{q}$ for each.
a. $X_1 = 60$, $n_1 = 100$, $X_2 = 40$, $n_2 = 100$
b. $X_1 = 22$, $n_1 = 50$, $X_2 = 18$, $n_2 = 30$
c. $X_1 = 18$, $n_1 = 60$, $X_2 = 20$, $n_2 = 80$
d. $X_1 = 5$, $n_1 = 32$, $X_2 = 12$, $n_2 = 48$
e. $X_1 = 12$, $n_1 = 75$, $X_2 = 15$, $n_2 = 50$

For Exercises 10–69 through 10–80, perform the following steps.

a.   State the hypotheses and identify the claim.
b.   Find the critical value(s).
c.   Compute the test value.
d.   Make the decision.
e.   Summarize the results.

Use the traditional method of hypothesis testing unless otherwise specified.

10–69. A sample of 150 people from a certain industrial community showed that 80 people suffered from a lung disease. A sample of 100 people from a rural community showed that 30 suffered from the same lung disease. At $\alpha = 0.05$, is there a difference between the proportion of people who suffer from the disease in the two communities?

10–70. A recent survey showed that in a sample of 80 surgeons, 15 smoked; in a sample of 50 general practitioners, 5 smoked. At $\alpha = 0.05$, is the proportion of surgeons who smoke higher than the proportion of general practitioners who smoke?

10–71. In a sample of 100 store customers, 43 used a MasterCard. In another sample of 100, 58 used a Visa card. At $\alpha = 0.05$, is there a difference in the proportion of people who use each type of credit card?

10–72. In Cleveland, a sample of 73 mail carriers showed that 10 had been bitten by an animal during one week. In Philadelphia, in a sample of 80 mail carriers, 16 had received animal bites. Is there a significant difference in the proportions? Use $\alpha = 0.05$. Find the 95% confidence interval for the difference of the two proportions.

10–73. A survey found that 83% of the men questioned preferred computer-assisted instruction to lecture, and 75% of the women preferred computer-assisted instruction to lecture. There were 100 individuals in each sample. At $\alpha = 0.05$, test the claim that there is no difference in the proportion of men and the proportion of women who favor computer-assisted instruction over lecture. Find the 95% confidence interval for the difference of the two proportions.

10–74. In a sample of 200 surgeons, 15% thought the government should control health care. In a sample of 200 general practitioners, 21% felt this way. At $\alpha = 0.10$, is there a difference in the proportions? Find the 90% confidence interval for the difference of the two proportions.

10–75. In a sample of 80 Americans, 55% wished that they were rich. In a sample of 90 Europeans, 45% wished that they were rich. At $\alpha = 0.01$, is there a difference in the proportions? Find the 99% confidence interval for the difference of the two proportions.

10–76. In a sample of 200 men, 130 said they used seat belts. In a sample of 300 women, 63 said they used seat belts. Test the claim that men are more safety-conscious than women, at $\alpha = 0.01$. Use the P-value method.

10–77. A survey of 80 homes in a Washington, D.C., suburb showed that 45 were air conditioned. A sample of

120 homes in a Pittsburgh suburb showed that 63 had air conditioning. At $\alpha = 0.05$, is there a difference in the two proportions? Find the 95% confidence interval for the difference of the two proportions.

10–78. A recent study showed that in a sample of 100 people, 30% had visited Disneyland. In another sample of 100 people, 24% had visited Disney World. Are the proportions of people who visited each park different? Use $\alpha = 0.02$. Use the P-value method.

10–79. A sample of 200 teenagers shows that 50 believe that war is inevitable, and a sample of 300 people over 60 shows that 93 believe war is inevitable. Is the proportion of teenagers who believe war is inevitable different from the proportion of people over 60 who do? Use $\alpha = 0.01$. Find the 99% confidence interval for the difference of the two proportions.

10–80. In a sample of 50 high school seniors, eight had their own cars. In a sample of 75 college freshmen, 20 had their own cars. At $\alpha = 0.05$, can it be concluded that a higher proportion of college freshmen have their own cars? Use the P-value method.

10–81. Find the 99% confidence interval for the difference in the population proportions for the data of a study in which 80% of the 150 Republicans surveyed favored the bill for a salary increase and 60% of the 200 Democrats surveyed favored the bill for a salary increase.

10–82. Find the 95% confidence interval for the true difference in proportions for the data of a study in which 40% of the 200 males surveyed opposed the death penalty and 56% of the 100 females surveyed opposed the death penalty.

10–83. If there is a significant difference between p_1 and p_2 and between p_2 and p_3, can one conclude that there is a significant difference between p_1 and p_3?

Technology Step by Step

MINITAB
Step by Step

Testing the Difference between Two Proportions

Example MT10–5

Test for a difference between proportions using the nursing home data from Example 10–15, at $\alpha = 0.05$.

1. Select **Stat>BasicStatistics>2 Proportions.**
2. Click the button for Summarized data.
3. Press **TAB** to move cursor to the First sample box under Trials.

   a. Enter **34 TAB.**

b. Enter **12 TAB.**

c. Enter **24 TAB.**

d. Enter **17 TAB.**

Two Proportions
Dialog Box

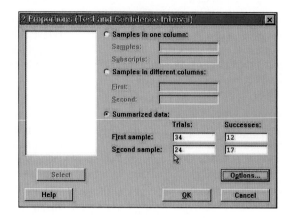

4. Click on [Options].

a. Check the box for Use pooled estimate of p for test.

b. Click [OK] twice.

The results are shown in the session window.

Test and CI for Two Proportions

```
Sample    X     N    Sample p
1         12    34   0.352941
2         17    24   0.708333

Estimate for p(1) - p(2): -0.355392
95% CI for p(1) - p(2): (-0.598025, -0.112759)
Test for p(1) - p(2) = 0 (vs not = 0): Z = -2.67 P-Value = 0.008
```

The *P*-value of the test is 0.008. Reject the null hypothesis. The difference of -0.355 is statistically significant.

TI–83
Step by Step

z Test for the Difference between Two Proportions

1. Press **STAT** and move the cursor to TESTS.

2. Press **6** for 2-PropZTest.

3. Enter the values for X_1, n_1, X_2, and n_2.

4. Select the appropriate alternative hypothesis.

5. Select Calculate and press **ENTER.**

Example TI10–8

Test the claim from Example 10–15, H_1: $p_1 \neq p_2$ at $\alpha = 0.05$ when

$$X_1 = 12 \qquad X_2 = 17$$
$$n_1 = 34 \qquad n_2 = 24$$

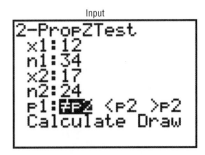

The test value is -2.666053851, and the P-value is 0.0076748288. The decision is to reject the null hypothesis, since $0.0076748288 < 0.05$.

10–7

Summary

Many times, researchers are interested in comparing two population parameters, such as means or proportions. This comparison can be accomplished using special z and t tests. If the samples are independent and the variances are known, the z test is used. The z test is also used when the variances are unknown but both sample sizes are 30 or more. If the variances are not known and one or both sample sizes are less than 30, the t test must be used. For independent samples, a further requirement is that one must determine whether the variances of the populations are equal. The F test is used to determine whether or not the variances are equal. Different formulas are used in each case. If the samples are dependent, the t test for dependent samples is used. Finally, a z test is used to compare two proportions.

Important Terms

Important Formulas

Formula for the z test for comparing two means from independent populations:

$$z = \frac{(\bar{X}_1 - \bar{X}_2) - (\mu_1 - \mu_2)}{\sqrt{\dfrac{\sigma_1^2}{n_1} + \dfrac{\sigma_2^2}{n_2}}}$$

Formula for the confidence interval for difference of two means (large samples):

$$(\bar{X}_1 - \bar{X}_2) - z_{\alpha/2}\sqrt{\frac{\sigma_1^2}{n_1} + \frac{\sigma_2^2}{n_2}} < \mu_1 - \mu_2$$
$$< (\bar{X}_1 - \bar{X}_2) + z_{\alpha/2}\sqrt{\frac{\sigma_1^2}{n_1} + \frac{\sigma_2^2}{n_2}}$$

Formula for the F test for comparing two variances:

$$F = \frac{s_1^2}{s_2^2}$$

Formula for the t test for comparing two means (small independent samples, variances not equal):

$$t = \frac{(\bar{X}_1 - \bar{X}_2) - (\mu_1 - \mu_2)}{\sqrt{\dfrac{s_1^2}{n_1} + \dfrac{s_2^2}{n_2}}}$$

and d.f. = the smaller of $n_1 - 1$ or $n_2 - 1$.

Formula for the t test for comparing two means (independent samples, variances equal):

$$t = \frac{(\bar{X}_1 - \bar{X}_2) - (\mu_1 - \mu_2)}{\sqrt{\dfrac{(n_1 - 1)s_1^2 + (n_2 - 1)s_2^2}{(n_1 + n_2 - 2)}}\sqrt{\dfrac{1}{n_1} + \dfrac{1}{n_2}}}$$

and d.f. = $n_1 + n_2 - 2$.

Formula for the confidence interval for the difference of two means (small independent samples, variances unequal):

$$(\bar{X}_1 - \bar{X}_2) - t_{\alpha/2}\sqrt{\frac{s_1^2}{n_1} + \frac{s_2^2}{n_2}} < \mu_1 - \mu_2$$

$$< (\bar{X}_1 - \bar{X}_2) + t_{\alpha/2}\sqrt{\frac{s_1^2}{n_1} + \frac{s_2^2}{n_2}}$$

and d.f. = smaller of $n_1 - 1$ and $n_2 - 2$.

Formula for the confidence interval for the difference of two means (small independent samples, variances equal):

$$(\bar{X}_1 - \bar{X}_2) - t_{\alpha/2}\sqrt{\frac{(n_1 - 1)s_1^2 + (n_2 - 1)s_2^2}{n_1 + n_2 - 2}} \cdot \sqrt{\frac{1}{n_1} + \frac{1}{n_2}}$$

$$< \mu_1 - \mu_2$$

$$< (\bar{X}_1 - \bar{X}_2) + t_{\alpha/2}\sqrt{\frac{(n_1 - 1)s_1^2 + (n_2 - 1)s_2^2}{n_1 + n_2 - 2}} \cdot \sqrt{\frac{1}{n_1} + \frac{1}{n_2}}$$

and d.f. = $n_1 + n_2 - 2$.

Formula for the t test for comparing two means from dependent samples:

$$t = \frac{\bar{D} - \mu_D}{s_D/\sqrt{n}}$$

where $\bar{D}$ is the mean of the differences,

$$\bar{D} = \frac{\Sigma D}{n}$$

and s_D is the standard deviation of the differences,

$$s_D = \sqrt{\frac{\Sigma D^2 - \dfrac{(\Sigma D)^2}{n}}{n - 1}}$$

Formula for confidence interval for the mean of the difference for dependent samples:

$$\bar{D} - t_{\alpha/2}\frac{s_D}{\sqrt{n}} < \mu_D < \bar{D} + t_{\alpha/2}\frac{s_D}{\sqrt{n}}$$

and d.f. = $n - 1$.

Formula for the z test for comparing two proportions:

$$z = \frac{(\hat{p}_1 - \hat{p}_2) - (p_1 - p_2)}{\sqrt{\bar{p}\bar{q}\left(\dfrac{1}{n_1} + \dfrac{1}{n_2}\right)}}$$

where

$$\bar{p} = \frac{X_1 + X_2}{n_1 + n_2} \qquad \hat{p}_1 = \frac{X_1}{n_1}$$

$$\bar{q} = 1 - \bar{p} \qquad \hat{p}_2 = \frac{X_2}{n_2}$$

Formula for confidence interval for the difference of two proportions:

$$(\hat{p}_1 - \hat{p}_2) - z_{\alpha/2}\sqrt{\frac{\hat{p}_1\hat{q}_1}{n_1} + \frac{\hat{p}_2\hat{q}_2}{n_2}} < p_1 - p_2$$

$$< (\hat{p}_1 - \hat{p}_2) + z_{\alpha/2}\sqrt{\frac{\hat{p}_1\hat{q}_1}{n_1} + \frac{\hat{p}_2\hat{q}_2}{n_2}}$$

Review Exercises

For each problem, perform the following steps. Assume that all variables are normally or approximately normally distributed.

a. State the hypotheses and identify the claim.
b. Find the critical value(s).
c. Compute the test value.
d. Make the decision.
e. Summarize the results.

Use the traditional method of hypothesis testing unless otherwise specified.

10–84. The average annual cost of automobile insurance in 1992 for residents of North Carolina was $541.07, while for residents of Indiana it was $584.17. Test the claim at $\alpha = 0.10$ that there is no difference in the means for both states. Assume samples of 100 residents were used and the standard deviation was $81 for both samples. Find the 90% confidence interval for the difference in the means.

Source: *In Sync* (Erie Insurance, Erie, PA), Fall 1995.

10–85. Two groups of drivers are surveyed to see how many miles per week they drive for pleasure trips. The data are shown. At $\alpha = 0.01$ can it be concluded that single drivers do more driving for pleasure trips on average than married drivers?

Single drivers					Married drivers				
106	110	115	121	132	97	104	138	102	115
119	97	118	122	135	133	120	119	136	96
110	117	116	138	142	139	108	117	145	114
115	114	103	98	99	140	136	113	113	150
108	117	152	147	117	101	114	116	113	135
154	86	115	116	104	115	109	147	106	88
107	133	138	142	140	113	119	99	108	105

10–86. An educator wishes to compare the variances of the amount of money spent per pupil in two states. The data are given below. At $\alpha = 0.05$, is there a significant difference in the variances of the amounts the states spend per pupil?

State 1	State 2
$s_1^2 = \$585$	$s_2^2 = \$261$
$n_1 = 18$	$n_2 = 16$

10–87. In the hospital study cited in Exercise 8–19, the standard deviation of the noise levels of the 11 intensive care units was 4.1 dBA and the standard deviation of the noise levels of 24 nonmedical care areas, such as kitchens and machine rooms, was 13.2 dBA. At $\alpha = 0.10$, is there a significant difference between the standard deviations of these two areas?

Source: M. Bayo, A. Garcia, and A. Garcia, "Noise Levels in an Urban Hospital and Workers' Subjective Responses," *Archives of Environmental Health* 50, no. 3 (May–June 1995), p. 249.

10–88. A researcher wants to compare the variances of the heights (in inches) of major league baseball players with those of players in the minor leagues. A sample of 25 players from each league is selected, and the variances of the heights for each league are 2.25 and 4.85, respectively. At $\alpha = 0.10$, is there a significant difference between the variances of the heights for the two leagues?

10–89. A traffic safety commissioner believes the variation in the number of speeding tickets given on Route 19 is greater than the variation in the number of speeding tickets given on Route 22. Ten weeks are randomly selected; the standard deviation of the number of tickets issued for Route 19 is 6.3, and the standard deviation of the number of tickets issued for Route 22 is 2.8. At $\alpha = 0.05$, can the commissioner conclude that the variance of speeding tickets issued on Route 19 is greater than the variance of speeding tickets issued on Route 22? Use the *P*-value method.

10–90. The variations in the number of absentees per day in two schools are being compared. A sample of 30 days is selected; the standard deviation of the number of absentees

in school A is 4.9, and for school B it is 2.5. At $\alpha = 0.01$, can one conclude that there is a difference in the two standard deviations?

10–91. A researcher claims that the variation in the number of days factory workers miss per year due to illness is greater than the variation in the number of days hospital workers miss per year. A sample of 42 workers from a large hospital has a standard deviation of 2.1 days, and a sample of 65 workers from a large factory has a standard deviation of 3.2 days. Test the claim, at $\alpha = 0.10$.

10–92. The average price of 15 cans of tomato soup from different stores is \$0.73, and the standard deviation is \$0.05. The average price of 24 cans of chicken noodle soup is \$0.91, and the standard deviation is \$0.03. At $\alpha = 0.01$, is there a significant difference in price?

10–93. The average temperatures for a 25-day period for Birmingham, Alabama, and Chicago, Illinois, are shown. Based on the samples, at $\alpha = 0.10$, can it be concluded that it is warmer in Birmingham?

Birmingham					Chicago				
78	82	68	67	68	70	74	73	60	77
75	73	75	64	68	71	72	71	74	76
62	73	77	78	79	71	80	65	70	83
74	72	73	78	68	67	76	75	62	65
73	79	82	71	66	66	65	77	66	64

10–94. A sample of 15 teachers from Rhode Island has an average salary of \$35,270, with a standard deviation of \$3256. A sample of 30 teachers from New York has an average salary of \$29,512, with a standard deviation of \$1432. Is there a significant difference in teachers' salaries between the two states? Use $\alpha = 0.02$. Find the 99% confidence interval for the difference of the two means.

10–95. The average income of 16 families who reside in a large metropolitan city is \$54,356, and the standard deviation is \$8256. The average income of 12 families who reside in a suburb of the same city is \$46,512, with a standard deviation of \$1311. At $\alpha = 0.05$, can one conclude that the income of the families who reside within the city is greater than that of those who reside in the suburb? Use the *P*-value method.

10–96. In an effort to improve the vocabulary of 10 students, a teacher provides a weekly one-hour tutoring session for them. A pretest is given before the sessions and a posttest is given afterward. The results are shown in the table. At $\alpha = 0.01$, can the teacher conclude that the tutoring sessions helped to improve the students' vocabulary?

Before	1	2	3	4	5	6	7	8	9	10
Pretest	83	76	92	64	82	68	70	71	72	63
Posttest	88	82	100	72	81	75	79	68	81	70

10–97. In an effort to increase production of an automobile part, the factory manager decides to play music in the manufacturing area. Eight workers are selected, and the number of items each produced for a specific day is recorded. After one week of music, the same workers are monitored again. The data are given in the following table. At $\alpha = 0.05$, can the manager conclude that the music has increased production?

Worker	1	2	3	4	5	6	7	8
Before	6	8	10	9	5	12	9	7
After	10	12	9	12	8	13	8	10

10–98. St. Petersburg, Russia, has 207 foggy days out of 365 days while Stockholm, Sweden, has 166 foggy days out of 365. At $\alpha = 0.02$, can it be concluded that the proportions of foggy days for the two cities are different? Find the 98% confidence interval for the difference of the two proportions.

Source: Jack Williams, *USA Today, 1995: The Weather Almanac* (New York: Vantage Books, 1994), p. 355.

10–99. In a recent survey of 50 apartment residents, 32 had microwave ovens. In a survey of 60 homeowners, 24 had microwave ovens. At $\alpha = 0.05$, test the claim that the proportions are equal. Find the 95% confidence interval for the difference of the two proportions.

Statistics Today

To Vaccinate or Not to Vaccinate? Small or Large? Revisited

Using a z test to compare two proportions, the researchers found that the proportion of residents in smaller nursing homes who were vaccinated (80.8%) was statistically greater than that of residents in large nursing homes who were vaccinated (68.7%). Using statistical methods presented in later chapters, they also found that the larger size of the nursing home and the lower frequency of vaccination were significant predictions of influenza outbreaks in nursing homes.

Data Analysis

The Data Bank is found in Appendix D, or on the World Wide Web by following links from **www.mhhe.com/math/stat/bluman/.**

1. From the Data Bank, select a variable and compare the mean of the variable for a random sample of at least 30 men with the mean of the variable for the random sample of at least 30 women. Use a z test.

2. Repeat the experiment in Exercise 1 using a different variable and two samples of size 15. Compare the means by using a t test. Assume that the variances are equal.

3. Compare the proportion of men who are smokers with the proportion of women who are smokers. Use the data in the Data Bank. Choose random samples of size 30 or more. Use the z test for proportions.

4. Using the data from Data Set XIV, test the hypothesis that the means of the weights of the players for two professional football teams are equal. Use an α value of your choice. Be sure to include the five steps of hypothesis testing. Use a z test.

5. For the same data used in the previous exercise, test the equality of the variances of the weights.

6. Using the data from Data Set XV, test the hypothesis that the means of the sizes of earthquakes of the two hemispheres are equal. Select an α value and use a t test.

Quiz

Determine whether each statement is true or false. If the statement is false, explain why.

1. When one is testing the difference between two means for small samples, it is not important to distinguish whether or not the samples are independent of each other.

2. If the same diet is given to two groups of randomly selected individuals, the samples are considered to be dependent.

3. When computing the F test value, one always places the larger variance in the numerator of the fraction.

4. Tests for variances are always two-tailed.

Select the best answer.

5. To test the equality of two variances, one would use a(n) ———— test.

   a. z   *c.* chi-square
   b. t   *d.* F

6. To test the equality of two proportions, one would use a(n) ———— test.

   a. z   *c.* chi-square
   b. t   *d.* F

7. The mean value of the F is approximately equal to

   a. 0   *c.* 1
   b. 0.5   *d.* It cannot be determined.

8. What test can be used to test the difference between two small sample means?

   a. z   *c.* chi-square
   b. t   *d.* F

Complete the following statements with the best answer.

9. If one hypothesizes that there is no difference between means, this is represented as H_0: ————.

10. When one is testing the difference between two means, a ———— estimate of the variances is used when the variances are equal.

11. When the t test is used for testing the equality of two means, the populations must be ————.

12. The values of F cannot be ————.

13. The formula for the F test for variances is ————.

For each of the following problems, perform the following steps.

   a. State the hypotheses.
   b. Find the critical value(s).
   c. Compute the test value.
   d. Make the decision.
   e. Summarize the results.

Use the traditional method of hypothesis testing unless otherwise specified.

14. A researcher wishes to see if there is a difference in the cholesterol levels of two groups of men. A random sample of 30 men between the ages of 25 and 40 is selected and tested. The average level is 223. A second sample of 25 men between the ages of 41 and 56 is selected and tested. The average of this group is 229. The population standard deviation for both groups is 6. At $\alpha = 0.01$, is there a difference in the cholesterol levels between the two groups? Find the 99% confidence interval for the difference of the two means.

15. The data shown are the rental fees for two random samples of apartments in a large city. At $\alpha = 0.10$ can it be concluded that the average rental fees for apartments in the East is greater than the average rental fee in the West?

East					West				
$495	390	540	445	420	$525	400	310	375	750
410	550	499	500	550	390	795	554	450	370
389	350	450	530	350	385	395	425	500	550
375	690	325	350	799	380	400	450	365	425
475	295	350	485	625	375	360	425	400	475
275	450	440	425	675	400	475	430	410	450
625	390	485	550	650	425	450	620	500	400
685	385	450	550	425	295	350	300	360	400

Source: *Pittsburgh Post-Gazette,* July 11, 1999.

16. A politician wishes to compare the variances of the amount of money spent for road repair in two different counties. The data are given here. At $\alpha = 0.05$, is there a significant difference in the variances of the amounts spent in the two counties? Use the *P*-value method.

County A	County B
$s_1 = \$11,596$	$s_2 = \$14,837$
$n_1 = 15$	$n_2 = 18$

17. A researcher wants to compare the variances of the heights (in inches) of four-year college basketball players with those of players in junior colleges. A sample of 30 players from each type of school is selected, and the variances of the heights for each type are 2.43 and 3.15, respectively. At $\alpha = 0.10$, is there a significant difference between the variances of the heights in the two types of schools?

18. The data shown are based on a survey taken in February and July and indicate the number of hours per day of household television usage. At $\alpha = 0.05$ test the claim that there is no difference in the standard deviations of the number of hours televisions are used.

February			July		
7.6	9.3	8.2	7.4	10.3	9.4
7.4	7.9	6.8	4.6	7.3	7.1
7.5	7.1	6.4	6.8	7.7	8.2
4.3	10.6	9.8	5.4	6.2	7.1

19. The variances of the amount of fat in two different types of ground beef are compared. Eight samples of the first type, Super Lean, have a variance of 18.2 grams; 12 of the second type, Ultimate Lean, have a variance of 9.4 grams. At $\alpha = 0.10$, can it be concluded that there is a difference in the variances of the two types of ground beef?

20. It is hypothesized that the variations of the number of days high school teachers miss per year due to illness

are greater than the variations of the number of days nurses miss per year. A sample of 56 high school teachers has a standard deviation of 3.4 days, while a sample of 70 nurses has a standard deviation of 2.8. Test the hypothesis at $\alpha = 0.10$.

21. The variations in the number of retail thefts per day in two shopping malls are being compared. A sample of 21 days is selected. The standard deviation of the number of retail thefts in mall A is 6.8, and for mall B, it is 5.3. At $\alpha = 0.05$, can it be concluded that there is a difference in the two standard deviations?

22. The average price of a sample of 12 bottles of diet salad dressing taken from different stores is $1.43. The standard deviation is $0.09. The average price of a sample of 16 low-calorie frozen desserts is $1.03. The standard deviation is $0.10. At $\alpha = 0.01$, is there a significant difference in price? Find the 99% confidence interval of the difference in the means.

23. The data shown represent the number of accidents people had when using jet skis and other types of wet bikes. At $\alpha = 0.05$ can it be concluded that the average number of accidents per year has increased during the last five years?

1987–1991			1992–1996		
376	650	844	1650	2236	3002
1162	1513		4028	4010	

Source: *USA Today*, August 27, 1997.

24. A sample of 12 chemists from Washington state shows an average salary of $39,420 with a standard deviation of $1659, while a sample of 26 chemists from New Mexico has an average salary of $30,215 with a standard deviation of $4116. Is there a significant difference between the two states in chemists' salaries at $\alpha = 0.02$? Find the 98% confidence interval of the difference in the means.

25. The average income of 15 families who reside in a large metropolitan East Coast city is $62,456. The standard deviation is $9652. The average income of 11 families who reside in a rural area of the Midwest is

$60,213, with a standard deviation of $2009. At $\alpha = 0.05$, can it be concluded that the families who live in the cities have a higher income than those who live in the rural areas? Use the *P*-value method.

26. In an effort to improve the mathematical skills of 10 students, a teacher provides a weekly one-hour tutoring session for the students. A pretest is given before the sessions, and a posttest is given after. The results are shown here. At $\alpha = 0.01$, can it be concluded that the sessions help to improve the students' mathematical skills?

Student	1	2	3	4	5	6	7	8	9	10
Pretest	82	76	91	62	81	67	71	69	80	85
Posttest	88	80	98	80	80	73	74	78	85	93

27. In order to increase egg production, a farmer decided to increase the amount of time the lights in his hen house were on. Ten hens were selected, and the number of eggs each produced was recorded. After one week of lengthened light time, the same hens were monitored again. The data are given here. At $\alpha = 0.05$, can it be concluded that the increased light time increased egg production?

Hen	1	2	3	4	5	6	7	8	9	10
Before	4	3	8	7	6	4	9	7	6	5
After	6	5	9	7	4	5	10	6	9	6

28. In a sample of 80 workers from a factory in city A, it was found that 5% were unable to read, while in a sample of 50 workers in city B, 8% were unable to read. Can it be concluded that there is a difference in the proportions of nonreaders in the two cities? Use $\alpha = 0.10$. Find the 90% confidence interval for the difference of the two proportions.

29. In a recent survey of 45 apartment residents, 28 had phone answering machines. In a survey of 55 homeowners, 20 had phone answering machines. At $\alpha = 0.05$, test the claim that the proportions are equal. Find the 95% confidence interval for the difference of the two proportions.

Critical Thinking Challenges

1. In the article at the top of the next page, researchers for Japan Airlines are trying to reduce flight fatigue by masking cabin noise. No data or statistics are given for the results of the study. Design a statistical study to see if the noise-canceling system reduced flight fatigue in airline passengers by answering the following questions:

    a. How could airline fatigue be measured?

    b. How could a population be defined?

    c. How could a sample be selected?

    d. Suggest other features that might influence flight fatigue (duration of the flights, time of day, etc.). How might these be controlled?

    e. What statistical tests might be used to analyze the data?

f. Find some information on jet lag in books and periodicals in the library and write a brief summary of these findings.

2. In the article at the bottom of this page, researchers concluded that physical exercise can keep the brain sharp into old age. After reading the study, answer the following questions:

a. Do you think the conclusions derived from studying rats would be valid for humans?

b. What could be a possible hypothesis for a study such as this?

c. What statistical test could be used to test the hypothesis?

d. Cite several reasons why the study might be controversial.

e. What factors other than exercise might influence the results of the study?

A Dull Roar

—Charles N. Barnard

One culprit causing flight fatigue is cabin noise—which comes not only from jet engines but also from the rush of air over the airplane fuselage. On the theory that you can't escape this racket but maybe you can disguise it, Japan Airlines offers a **noise-canceling system** through special battery-powered headphones produced by Sony. The system generates a 250 Hz noise of its own, which masks and flattens out other sounds between 60 and 2,000 Hz. Passengers can use the headphones in the usual way for movies and audio channels, or to lull themselves to sleep with "white noise."

Does this help with jet lag? Well, a good long sleep always speeds up *my* lag!

Source: "A Dull Roar," *Modern Maturity* 38, no. 1 (January/February 1995), p. 20. Used with permission.

Building Biceps Could Boost Brainpower, Too

By Ellen Hale
Gannett News Service

Exercise can keep the brain sharp into old age and might help prevent Alzheimer's disease and other mental disorders that accompany aging, says a new study that provides some of the first direct evidence linking physical activity and mental ability.

The study, reported in the journal *Nature*, is the first to show that growth factors in the brain—compounds responsible for the brain's health—can be controlled by exercise.

Combined with previous research that shows exercisers live longer and score higher on tests of mental function, the new findings add hard proof of the importance of physical activity in the aging process.

"Here's another argument for getting active and staying active," says Dr. Carl Cotman of the University of California at Irvine.

Cotman's research was on rodents, but the effects of exercise are nearly identical in humans and rats, and rats have "surprisingly similar" exercise habits, Cotman says.

In his study, which promises to be controversial, rats were permitted to choose how much they wanted to exercise, and each had its own activity habits—just like humans. Some were "couch" rats, Cotman says, rarely getting on the treadmill; others were "runaholics," with one obsessively logging five miles every night on the wheel. "Those little feet must have been paddling away like crazy," Cotman says.

The rats that exercised had much higher levels of BDNF (brain-derived neurotrophic factor), the most widely distributed growth factor in the brain and one reported to decline with the onset of Alzheimer's.

Cotman predicts there is a minimum level of exercise that provides the maximum benefit. The rat that ran five miles nightly, for example, did not raise its level of growth factor much more than those that ran a mile or two.

Source: Ellen Hale, "Building Biceps Could Boost Brainpower, Too," *USA Today*, January 12, 1995, Copyright 1995. *USA Today*. Reprinted with permission.

Data Projects

Where appropriate, use MINITAB, the TI-83, or a computer program of your choice to complete the following exercises.

1. Choose a variable for which you would like to determine if there is a difference in the averages for two groups. Make sure that the samples are independent. For example, you may wish to see if men see more movies or spend more money on lunch than women. Select a sample of data values (10 to 50) and complete the following:
   a. Write a brief statement as to the purpose of the study.
   b. Define the population.
   c. State the hypotheses for the study.
   d. Select an α value.
   e. State how the sample was selected.
   f. Show the raw data.
   g. Decide which statistical test is appropriate and compute the test statistic (z or t). Why is the test appropriate?
   h. Find the critical value(s).
   i. State the decision.
   j. Summarize the results.

2. Choose a variable that will permit using dependent samples. For example, you might wish to see if a person's weight has changed after a diet. Select a sample of data (10 to 50) value pairs (e.g., before and after), and then complete the following:
   a. Write a brief statement as to the purpose of the study.
   b. Define the population.
   c. State the hypotheses for the study.
   d. Select an α value.
   e. State how the sample was selected.
   f. Show the raw data.
   g. Decide which statistical test is appropriate and compute the test statistic (z or t). Why is the test appropriate?
   h. Find the critical value(s).
   i. State the decision.
   j. Summarize the results.

3. Choose a variable that will enable you to compare proportions of two groups. For example, you might want to see if the proportion of freshmen who buy used books is lower than (or higher than or the same as) the proportion of sophomores who buy used books. After collecting 30 or more responses from the two groups, complete the following:
   a. Write a brief statement as to the purpose of the study.
   b. Define the population.
   c. State the hypotheses for the study.
   d. Select an α value.
   e. State how the sample was selected.
   f. Show the raw data.
   g. Decide which statistical test is appropriate and compute the test statistic (z or t). Why is the test appropriate?
   h. Find the critical value(s).
   i. State the decision.
   j. Summarize the results.

You may use the following websites to obtain raw data:

http://www.mhhe.com/math/stat/bluman/
http://lib.stat.cmu.edu/DASL
http://www.oecd.org/statlist.htm
http://www.statcan.ca/english/

Hypothesis-Testing Summary 1

1. Comparison of a sample mean with a specific population mean.

Example: $H_0: \mu = 100$

a. Use the z test when σ is known:

$$z = \frac{\bar{X} - \mu}{\sigma/\sqrt{n}}$$

b. Use the t test when σ is unknown:

$$t = \frac{\bar{X} - \mu}{s/\sqrt{n}} \quad \text{with} \quad \text{d.f.} = n - 1$$

2. Comparison of a sample variance or standard deviation with a specific population variance or standard deviation.

Example: $H_0: \sigma^2 = 225$

Use the chi-square test:

$$\chi^2 = \frac{(n - 1)s^2}{\sigma^2} \quad \text{with} \quad \text{d.f.} = n - 1$$

3. Comparison of two sample means.

Example: $H_0: \mu_1 = \mu_2$

a. Use the z test when the population variances are known:

$$z = \frac{(\bar{X}_1 - \bar{X}_2) - (\mu_1 - \mu_2)}{\sqrt{\dfrac{\sigma_1^2}{n_1} + \dfrac{\sigma_2^2}{n_2}}}$$

b. Use the t test for independent samples when the population variances are unknown and the sample variances are unequal:

$$t = \frac{(\bar{X}_1 - \bar{X}_2) - (\mu_1 - \mu_2)}{\sqrt{\dfrac{s_1^2}{n_1} + \dfrac{s_2^2}{n_2}}}$$

with d.f. = the smaller of $n_1 - 1$ or $n_2 - 1$.

c. Use the t test for independent samples when the population variances are unknown and assumed to be equal:

$$t = \frac{(\bar{X}_1 - \bar{X}_2) - (\mu_1 - \mu_2)}{\sqrt{\dfrac{(n_1 - 1)s_1^2 + (n_2 - 1)s_2^2}{n_1 + n_2 - 2}}\sqrt{\dfrac{1}{n_1} + \dfrac{1}{n_2}}}$$

with d.f. = $n_1 + n_2 - 2$.

d. Use the t test for means for dependent samples:

Example: $H_0: \mu_D = 0$

$$t = \frac{\bar{D} - \mu_D}{s_D/\sqrt{n}} \quad \text{with} \quad \text{d.f.} = n - 1$$

where n = number of pairs.

4. Comparison of a sample proportion with a specific population proportion.

Example: $H_0: P = 0.32$

Use the z test:

$$z = \frac{X - \mu}{\sigma} \quad \text{or} \quad z = \frac{\hat{p} - p}{\sqrt{pq/n}}$$

5. Comparison of two sample proportions.

Example: $H_0: p_1 = p_2$

Use the z test:

$$z = \frac{(\hat{p}_1 - \hat{p}_2) - (p_1 - p_2)}{\sqrt{\bar{p}\bar{q}\left(\dfrac{1}{n_1} + \dfrac{1}{n_2}\right)}}$$

where

$$\bar{p} = \frac{X_1 + X_2}{n_1 + n_2} \qquad \hat{p}_1 = \frac{X_1}{n_1}$$

$$\bar{q} = 1 - \bar{p} \qquad \hat{p}_2 = \frac{X_2}{n_2}$$

6. Comparison of two sample variances or standard deviations.

Example: $H_0: \sigma_1^2 = \sigma_2^2$

Use the F test:

$$F = \frac{s_1^2}{s_2^2}$$

where

s_1^2 = larger variance   d.f.N. = $n_1 - 1$

s_2^2 = smaller variance   d.f.D. = $n_2 - 1$

chapter

11 Correlation and Regression

Objectives

After completion this chapter, you
should be able to

1. Draw a scatter plot for a set
of ordered pairs.

2. Compute the correlation
coefficient.

3. Test the hypothesis
$H_0: \rho = 0$.

4. Compute the equation of the
regression line.

5. Compute the coefficient of
determination.

6. Compute the standard error
of estimate.

7. Find a prediction interval.

8. Be familiar with the concept
of multiple regression.

Statistics Today

Do Dust Storms Affect Respiratory Health?

Southeast Washington state has a long history of seasonal dust storms. Several researchers decided to see what effect, if any, these storms had on the respiratory health of the people living in the area. They undertook (among other things) to see if there was a relationship between the amount of dust and sand particles in the air when the storms occur and the number of hospital emergency room visits for respiratory disorders at three community hospitals in southeast Washington. Using methods of correlation and regression, which are explained in this chapter, they were able to determine the effect of these dust storms on local residents.

Source: B. Hefflin, B. Jalaludin, N. Cobb, C. Johnson, L. Jecha, and R. Etzel, "Surveillance for Dust Storms and Respiratory Diseases in Washington State, 1991," *Archives of Environmental Health* 49, no. 3 (May–June 1994), pp. 170–74. Reprinted with permission of the Helen Dwight Reid Education Foundation. Published by Neldrof Publications, 1319 18th St. N.W., Washington, D.C. 20036-1802. Copyright 1994.

11–1

Introduction

In the previous chapters, two areas of inferential statistics, hypothesis testing and confidence intervals, were explained. Another area of inferential statistics involves determining whether a relationship between two or more numerical or quantitative variables exists. For example, a businessperson may want to know whether the volume of sales for a given month is related to the amount of advertising the firm does that month. Educators are interested in determining whether the number of hours a student studies is related to the student's score on a particular exam. Medical researchers are interested in questions such as "Is caffeine related to heart damage?" or "Is there a relationship between a person's age and his or her blood pressure?" A zoologist may want to know whether the birth weight of a certain animal is related to its life span. These are only a few of the many questions that can be answered by using the techniques of correlation and regression analysis. **Correlation** is a statistical method used to determine whether a relationship between variables exists. **Regression** is a statistical method used to describe the nature of the relationship between variables—that is, positive or negative, linear or nonlinear.

The purpose of this chapter is to answer the following questions statistically:

1. Are two or more variables related?

2. If so, what is the strength of the relationship?

3. What type of relationship exists?

4. What kind of predictions can be made from the relationship?

To answer the first two questions, statisticians use a measure to determine whether two or more variables are related and also to determine the strength of the relationship between or among the variables. This measure is called a *correlation coefficient.* For example, there are many variables that contribute to heart disease, among them lack of exercise, smoking, heredity, age, stress, and diet. Of these variables, some are more important than others; therefore, a physician who wants to help a patient must know which factors are most important.

To answer the third question, one must ascertain what type of relationship exists. There are two types of relationships: simple and multiple. In a **simple relationship,** there are only two variables under study. For example, a manager may wish to see whether the number of years the salespeople have been working for the company has anything to do with the amount of sales they make. This type of study involves a simple relationship, since there are only two variables, years of experience and amount of sales.

In **multiple relationships,** many variables are under study. For example, an educator may wish to investigate the relationship between a student's success in college and factors such as the number of hours devoted to studying, the student's GPA, and the student's high school background. This type of study involves several variables.

Simple relationships can also be positive or negative. A **positive relationship** exists when both variables increase or decrease at the same time. For instance, a person's height and weight are related; and the relationship is positive, since the taller a person is, generally, the more the person weighs. In a **negative relationship,** as one variable increases, the other variable decreases, and vice versa. For example, if one measures the strength of people over 60 years of age, one will find that as age increases, strength generally decreases. The word *generally* is used here because there are exceptions.

Finally, the fourth question asks what type of predictions can be made. Predictions are made in all areas and on a daily basis. Examples include weather forecasting, stock market analyses, sales predictions, crop predictions, gasoline predictions, and sports predictions. Some predictions are more accurate than others, due to the strength of the relationship. That is, the stronger the relationship is between variables, the more accurate the prediction is.

11–2

Scatter Plots

Objective 1. Draw a scatter plot for a set of ordered pairs.

In simple correlation and regression studies, the researcher collects data on two numerical or quantitative variables to see whether a relationship exists between the variables. For example, if a researcher wishes to see whether there is a relationship between number of hours studied and test scores on an exam, he must select a random sample of students, determine the hours each studied, and obtain their grades on the exam. A table can be made for the data, as shown here.

Student	Hours studied, x	Grade, y (%)
A	6	82
B	2	63
C	1	57
D	5	88
E	2	68
F	3	75

The two variables for this study are called the independent variable and the dependent variable. The **independent variable** is the variable in regression that can be controlled or manipulated. In this case, "number of hours studied" is the independent variable and is designated as the x variable. The **dependent variable** is the variable in regression that cannot be controlled or manipulated. The grade the student received on the exam is the dependent variable, designated as the y variable. The reason for this distinction between the variables is that one assumes that the grade the student earns *depends* on the number of hours the student studied. Also, one assumes that, to some extent, the student can regulate or *control* the number of hours he or she studies for the exam.

The determination of the x and y variables is not always clear-cut and sometimes is an arbitrary decision. For example, if a researcher studies the effects of age on a person's blood pressure, the researcher can generally assume that age affects blood pressure. Hence, the variable "age" can be called the *independent variable* and the variable "blood pressure" can be called the *dependent variable*. On the other hand, if a researcher is studying the attitudes of husbands on a certain issue and the attitudes of their wives on the same issue, it is difficult to say which variable is the independent variable and which is the dependent variable. In this study, the researcher can arbitrarily designate the variables as independent and dependent.

The independent and dependent variables can be plotted on a graph called a *scatter plot*. The independent variable, x, is plotted on the horizontal axis and the dependent variable, y, is plotted on the vertical axis.

A **scatter plot** is a graph of the ordered pairs (x, y) of numbers consisting of the independent variable, x, and the dependent variable, y.

The scatter plot is a visual way to describe the nature of the relationship between the independent and dependent variables. The scales of the variables can be different and the coordinates of the axes are determined by the smallest and largest data values of the variables.

The procedure for drawing a scatter plot is shown in the next three examples.

Example 11–1

Construct a scatter plot for the data obtained in a study of age and systolic blood pressure of six randomly selected subjects. The data are shown in the following table.

Subject	Age, x	Pressure, y
A	43	128
B	48	120
C	56	135
D	61	143
E	67	141
F	70	152

Solution

STEP 1   Draw and label the *x* and *y* axes.

STEP 2   Plot each point on the graph, as shown in Figure 11–1.

Figure 11–1

Scatter Plot for
Example 11–1

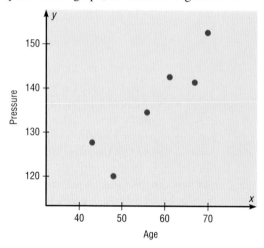

Example 11–2

Construct a scatter plot for the data obtained in a study on the number of absences and the final grades of seven randomly selected students from a statistics class. The data are shown here.

Student	Number of absences, *x*	Final grade, *y* (%)
A	6	82
B	2	86
C	15	43
D	9	74
E	12	58
F	5	90
G	8	78

Solution

STEP 1   Draw and label the *x* and *y* axes.

STEP 2   Plot each point on the graph, as shown in Figure 11–2.

Figure 11–2

Scatter Plot for
Example 11–2

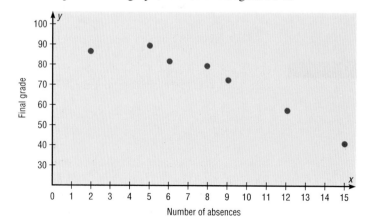

Example 11–3

Construct a scatter plot for the data obtained in a study on the number of hours nine people exercise each week and the amount of milk (in ounces) each person consumes per week. The data follow.

Subject	Hours, x	Amount, y
A	3	48
B	0	8
C	2	32
D	5	64
E	8	10
F	5	32
G	10	56
H	2	72
I	1	48

Solution

STEP 1   Draw and label the x and y axes.

STEP 2   Plot each point on the graph, as shown in Figure 11–3.

Figure 11–3

Scatter Plot for
Example 11–3

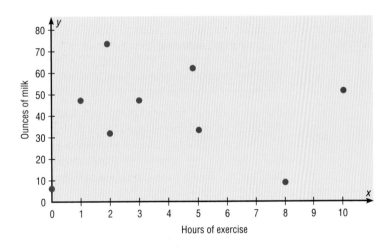

After the plot is drawn, it should be analyzed to determine which type of relationship, if any, exists. For example, the plot shown in Figure 11–1 suggests a positive relationship, since as a person's age increases, blood pressure tends to increase also. The plot of the data shown in Figure 11–2 suggests a negative relationship, since as the number of absences increases, the final grade decreases. Finally, the plot of the data shown in Figure 11–3 shows no specific type of relationship, since no pattern is discernible.

Note that the data shown in Figures 11–1 and 11–2 also suggest a linear relationship, since the points seem to fit a straight line, although not perfectly. Sometimes a scatter plot, such as the one in Figure 11–4, will show a curvilinear relationship between the data. In this situation, the methods shown in this section and the next cannot be used. Methods for curvilinear relationships are beyond the scope of this book.

Figure 11–4

Scatter Plot Suggesting a
Curvilinear Relationship

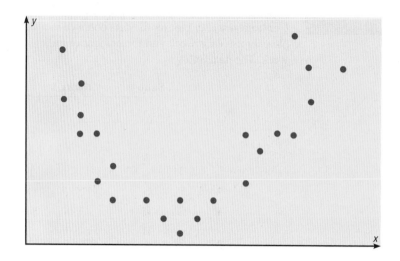

<table>
<tr><td>

11–3

Correlation

Correlation Coefficient

Objective 2. Compute the
correlation coefficient.

</td></tr>
</table>

As stated in Section 11–1, statisticians use a measure called the *correlation coefficient* to determine the strength of the relationship between two variables. There are several types of correlation coefficients. The one explained in this section is called the **Pearson product moment correlation coefficient** (PPMC), named after statistician Karl Pearson, who pioneered the research in this area.

The **correlation coefficient** computed from the sample data measures the strength and direction of a linear relationship between two variables. The symbol for the sample correlation coefficient is r. The symbol for the population correlation coefficient is ρ (Greek letter rho).

The *range of the correlation coefficient* is from -1 to $+1$. If there is a *strong positive linear relationship* between the variables, the value of r will be close to $+1$. If there is a *strong negative linear relationship* between the variables, the value of r will be close to -1. When there is no linear relationship between the variables or only a weak relationship, the value of r will be close to 0. See Figure 11–5.

Figure 11–5

Range of Values for the
Correlation Coefficient

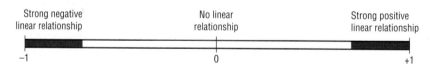

Strong negative linear relationship	No linear relationship	Strong positive linear relationship
-1	0	$+1$

There are several ways to compute the value of the correlation coefficient. One method is to use the formula shown next.

Formula for the Correlation Coefficient r

$$r = \frac{n(\Sigma\, xy) - (\Sigma\, x)(\Sigma\, y)}{\sqrt{[n(\Sigma\, x^2) - (\Sigma x)^2][n(\Sigma\, y^2) - (\Sigma y)^2]}}$$

where n is the number of data pairs.

Rounding Rule for the Correlation Coefficient   Round the value of r to three decimal places.

The formula looks somewhat complicated, but using a table to compute the values, as shown in the next example, makes it somewhat easier to determine the value of r.

Example 11–4

Compute the value of the correlation coefficient for the data obtained in the study of age and blood pressure given in Example 11–1.

Solution

STEP 1   Make a table, as shown here.

Subject	Age, x	Pressure, y	xy	x^2	y^2
A	43	128			
B	48	120			
C	56	135			
D	61	143			
E	67	141			
F	70	152			

STEP 2   Find the values of xy, x^2, and y^2 and place these values in the corresponding columns of the table.

The completed table is shown next.

Subject	Age, x	Pressure, y	xy	x^2	y^2
A	43	128	5,504	1,849	16,384
B	48	120	5,760	2,304	14,400
C	56	135	7,560	3,136	18,225
D	61	143	8,723	3,721	20,449
E	67	141	9,447	4,489	19,881
F	70	152	10,640	4,900	23,104
	$\Sigma x = 345$	$\Sigma y = 819$	$\Sigma xy = 47{,}634$	$\Sigma x^2 = 20{,}399$	$\Sigma y^2 = 112{,}443$

STEP 3   Substitute in the formula and solve for r.

$$r = \frac{n(\Sigma xy) - (\Sigma x)(\Sigma y)}{\sqrt{[n(\Sigma x^2) - (\Sigma x)^2][n(\Sigma y^2) - (\Sigma y)^2]}}$$

$$= \frac{(6)(47{,}634) - (345)(819)}{\sqrt{[(6)(20{,}399) - (345)^2][(6)(112{,}443) - (819)^2]}} = 0.897$$

The correlation coefficient suggests a strong positive relationship between age and blood pressure.

Example 11–5

Compute the value of the correlation coefficient for the data obtained in the study of the number of absences and the final grade of the seven students in the statistics class given in Example 11–2.

Solution

STEP 1   Make a table.

STEP 2   Find the values of xy, x^2, and y^2 and place these values in the corresponding columns of the table.

Student	Number of absences, x	Final grade, y ($\%$)	xy	x^2	y^2
A	6	82	492	36	6,724
B	2	86	172	4	7,396
C	15	43	645	225	1,849
D	9	74	666	81	5,476
E	12	58	696	144	3,364
F	5	90	450	25	8,100
G	8	78	624	64	6,084
	$\Sigma x = 57$	$\Sigma y = 511$	$\Sigma xy = 3745$	$\Sigma x^2 = 579$	$\Sigma y^2 = 38{,}993$

STEP 3   Substitute in the formula and solve for r.

$$r = \frac{n(\Sigma xy) - (\Sigma x)(\Sigma y)}{\sqrt{[n(\Sigma x^2) - (\Sigma x)^2][n(\Sigma y^2) - (\Sigma y)^2]}}$$

$$= \frac{(7)(3745) - (57)(511)}{\sqrt{[(7)(579) - (57)^2][(7)(38{,}993) - (511)^2]}} = -0.944$$

The value of r suggests a strong negative relationship between a student's final grade and the number of absences a student has. That is, the more absences a student has the lower is his or her grade.

Example 11–6

Compute the value of the correlation coefficient for the data given in Example 11–3 for the number of hours a person exercises and the amount of milk a person consumes per week.

Solution

STEP 1   Make a table.

STEP 2   Find the values of xy, x^2, and y^2 and place these values in the corresponding columns of the table.

Subject	Hours, x	Amount, y	xy	x^2	y^2
A	3	48	144	9	2,304
B	0	8	0	0	64
C	2	32	64	4	1,024
D	5	64	320	25	4,096
E	8	10	80	64	100
F	5	32	160	25	1,024
G	10	56	560	100	3,136
H	2	72	144	4	5,184
I	1	48	48	1	2,304
	$\Sigma x = 36$	$\Sigma y = 370$	$\Sigma xy = 1520$	$\Sigma x^2 = 232$	$\Sigma y^2 = 19{,}236$

STEP 3   Substitute in the formula and solve for r.

$$r = \frac{n(\Sigma xy) - (\Sigma x)(\Sigma y)}{\sqrt{[n(\Sigma x^2) - (\Sigma x)^2][n(\Sigma y^2) - (\Sigma y)^2]}}$$

$$= \frac{(9)(1520) - (36)(370)}{\sqrt{[(9)(232) - (36)^2][(9)(19{,}236) - (370)^2]}} = 0.067$$

The value of r indicates a very weak positive relationship between the variables.

Figure 11–7

Test Value for
Example 11–7

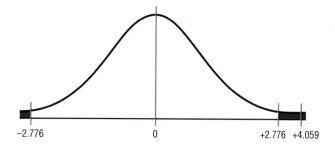

$$-2.776 \qquad\qquad 0 \qquad\qquad +2.776 \quad +4.059$$

STEP 5   Summarize the results. There is a significant relationship between the
variables of age and blood pressure.

The second method that can be used to test the significance of r is the P-value
method. The method is the same as that shown in Chapters 9 and 10. It uses the following steps.

STEP 1   State the hypotheses.

STEP 2   Find the test value. (In this case, use the t test.)

STEP 3   Find the P-value. (In this case, use Table F.)

STEP 4   Make the decision.

STEP 5   Summarize the results.

Referring to Example 11–7, the t value obtained in step 3 is 4.059 and d.f. = 4. Using Table F with d.f. = 4 and the row "Two Tails," the value 4.059 falls between 3.747
and 4.604; hence, $0.01 < P$-value < 0.02. (The P-value obtained from a calculator is
0.015.) That is, the P-value falls between 0.01 and 0.02. The decision then is to reject
the null hypothesis since P-value < 0.05.

The third method of testing the significance of r is to use Table I in Appendix C.
This table shows the values of the correlation coefficient that are significant for a specific α level and a specific number of degrees of freedom. For example, for 7 degrees of
freedom and $\alpha = 0.05$, the table gives a critical value of 0.666. Any value of r greater
than $+0.666$ or less than -0.666 will be significant, and the null hypothesis will be rejected. See Figure 11–8. When Table I is used, one need not compute the t test value.
Table I is for two-tailed tests only.

Figure 11–8

Finding the Critical Value
from Table I

d.f.	$\alpha = 0.05$	$\alpha = 0.01$
1		
2		
3		
4		
5		
6		
7	0.666	

| **Example 11–8** | Using Table I, test the significance of the correlation coefficient, $r = 0.067$, obtained in Example 11–6, at $\alpha = 0.01$. |

Solution

$$H_0: \rho = 0 \quad \text{and} \quad H_1: \rho \neq 0$$

Since the sample size is 9, there are 7 degrees of freedom. When $\alpha = 0.01$ and with 7 degrees of freedom, the value obtained from Table I is 0.798. For a significant relationship, a value of r greater than $+0.798$ or less than -0.798 is needed. Since $r = 0.067$, the null hypothesis is not rejected. Hence, there is not enough evidence to say that there is a significant linear relationship between the variables. See Figure 11–9.

Figure 11–9

Rejection and Nonrejection Regions for Example 11–8

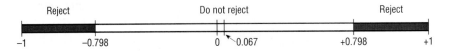

Correlation and Causation

Researchers must understand the nature of the linear relationship between the independent variable x and the dependent variable y. When a hypothesis test indicates that a significant linear relationship exists between the variables, researchers must consider the possibilities outlined next.

Possible Relationships between Variables

When the null hypothesis has been rejected for a specific α value, any of the following five possibilities can exist.

1. *There is a direct cause-and-effect relationship between the variables.* That is, x causes y. For example, water causes plants to grow, poison causes death, and heat causes ice to melt.

2. *There is a reverse cause-and-effect relationship between the variables.* That is, y causes x. For example, suppose a researcher believes excessive coffee consumption causes nervousness, but the researcher fails to consider that the reverse situation may occur. That is, it may be that an extremely nervous person craves coffee to calm his or her nerves.

continued

Possible Relationships between Variables *(concluded)*

3. *The relationship between the variables may be caused by a third variable.* For example, if a statistician correlated the number of deaths due to drowning and the number of cans of soft drink consumed during the summer, he or she would probably find a significant relationship. However, the soft drink is not necessarily responsible for the deaths, since both variables may be related to heat and humidity.

4. *There may be a complexity of interrelationships among many variables.* For example, a researcher may find a significant relationship between students' high school grades and college grades. But there probably are many other variables involved, such as IQ, hours of study, influence of parents, motivation, age, and instructors.

5. *The relationship may be coincidental.* For example, a researcher may be able to find a significant relationship between the increase in the number of people who are exercising and the increase in the number of people who are committing crimes. But common sense dictates that any relationship between these two values must be due to coincidence.

Thus, when the null hypothesis is rejected, the researcher must consider all possibilities and select the appropriate one as determined by the study. Remember, correlation does not necessarily imply causation.

Exercises

11–1. What is meant by the statement that two variables are related?

11–2. How is a linear relationship between two variables measured in statistics? Explain.

11–3. What is the symbol for the sample correlation coefficient? The population correlation coefficient?

11–4. What is the range of values for the correlation coefficient?

11–5. What is meant when the relationship between the two variables is positive? Negative?

11–6. Give examples of two variables that are positively correlated and two that are negatively correlated.

11–7. Give an example of a correlation study and identify the independent and dependent variables.

11–8. What is the diagram of the independent and dependent variables called? Why is drawing this diagram important?

11–9. What is the name of the correlation coefficient used in this section?

11–10. What statistical test is used to test the significance of the correlation coefficient?

11–11. When two variables are correlated, can the researcher be sure that one variable causes the other? Why or why not?

For Exercises 11–12 through 11–27, perform the following steps.

 a. Draw the scatter plot for the variables.
 b. Compute the value of the correlation coefficient.
 c. State the hypotheses.
 d. Test the significance of the correlation coefficient at $\alpha = 0.05$, using Table I.
 e. Give a brief explanation of the type of relationship.

11–12. A manager wishes to find out whether there is a relationship between the number of radio ads aired per week and the amount of sales (in thousands of dollars) of a product. The data for the sample follow. (The information in this exercise will be used for Exercises 11–40 and 11–64.)

No. of ads, x	2	5	8	8	10	12
Sales, y	$2	$4	$7	$6	$9	$10

11–13. A researcher wishes to determine if a person's age is related to the number of hours he or she exercises per week. The data for the sample are shown here. (The information in this exercise will be used for Exercises 11–41, 11–64, and 11–81.)

Age, x	18	26	32	38	52	59
Hours, y	10	5	2	3	1.5	1

11–14. A study was conducted to determine the relationship between a person's monthly income and the number of meals that person eats away from home per

Speaking of **STATISTICS**

In correlation and regression studies, it is difficult to control all variables. This study shows some of the consequences when researchers overlook certain aspects in studies. Suggest ways that the extraneous variables might be controlled in future studies.

Coffee Not Disease Culprit, Study Says

NEW YORK (AP)—Two new studies suggest that coffee drinking, even up to 5 1/2 cups per day, does not increase the risk of heart disease, and other studies that claim to have found increased risks might have missed the true culprits, a researcher says.

"It might not be the coffee cup in one hand, it might be the cigarette or coffee roll in the other," said Dr. Peter W.F. Wilson, the author of one of the new studies.

He noted in a telephone interview Thursday that many coffee drinkers, particularly heavy coffee drinkers, are smokers. And one of the new studies found that coffee drinkers had excess fat in their diets.

The findings of the new studies conflict sharply with a study reported in November 1985 by Johns Hopkins University scientists in Baltimore.

The Hopkins scientists found that coffee drinkers who consumed five or more cups of coffee per day had three times the heart-disease risk of non-coffee drinkers.

The reason for the discrepancy appears to be that many of the coffee drinkers in the Hopkins study also smoked—and it was the smoking that increased their heart-disease risk, said Wilson.

Wilson, director of laboratories for the Framingham Heart Study in Framingham, Mass., said Thursday at a conference sponsored by the American Heart Association in Charleston, S.C., that he had examined the coffee intake of 3,937 participants in the Framingham study during 1956–66 and an additional 2,277 during the years 1972–1982.

In contrast to the subjects in the Hopkins study, most of these coffee drinkers consumed two or three cups per day, Wilson said. Only 10 percent drank six or more cups per day.

He then looked at blood cholesterol levels and heart and blood vessel disease in the two groups. "We ran these analyses for coronary heart disease, heart attack, sudden death and stroke and in absolutely every analysis, we found no link with coffee," Wilson said.

He found that coffee consumption was linked to a significant decrease in total blood cholesterol in men, and to a moderate increase in total cholesterol in women.

Source: Associated Press. Reprinted with permission.

month. The data from the sample are shown here. (The information in this exercise will be used for Exercises 11–42, 11–64, and 11–82.)

Income, x	$500	$1200	$1500	$945	$850	$400	$540
Meals, y	8	12	16	10	9	3	7

11–15. The director of an alumni association for a small college wants to determine whether there is any type of relationship between the amount of an alumnus's contribution and the years the alumnus has been out of school. The data follow. (The information in this exercise will be used for Exercises 11–43, 11–65, and 11–83.)

Years, x	1	5	3	10	7	6
Contribution, y	$500	$100	$300	$50	$75	$80

11–16. A store manager wishes to find out whether there is a relationship between the age of her employees and the number of sick days they take each year. The data for the sample follow. (The information in this exercise will be used for Exercises 11–44, 11–65, and 11–84.)

Age, x	18	26	39	48	53	58
Days, y	16	12	9	5	6	2

11–17. An educator wants to see how strong the relationship is between a student's score on a test and his or her grade point average. The data obtained from the sample follow. (The information in this exercise will be used for Exercises 11–45 and 11–65.)

Test score, x	98	105	100	100	106	95	116	112
GPA, y	2.1	2.4	3.2	2.7	2.2	2.3	3.8	3.4

11–18. An English instructor is interested in finding the strength of a relationship between the final exam grades of students enrolled in Composition I and Composition II classes. The data are given below in percentages. (The information in this exercise will be used for Exercises 11–46 and 11–66.)

Comp I, x	83	97	80	95	73	78	91	86
Comp II, y	78	95	83	97	78	72	90	80

11–19. An insurance company wants to determine the strength of the relationship between the number of hours a person works per week and the number of injuries or accidents that person has over a period of one week. The data follow. (The information in this exercise will be used for Exercises 11–47 and 11–66.)

Hours worked, x	40	32	36	44	41
No. of accidents, y	1	0	3	8	5

11–20. An emergency service wishes to see whether a relationship exists between the outside temperature and the number of emergency calls it receives for a seven-hour period. The data follow. (The information in this exercise will be used for Exercises 11–48 and 11–66.)

Temperature, x	68	74	82	88	93	99	101
No. of calls, y	7	4	8	10	11	9	13

11–21. A researcher wishes to see if there is a relationship between the number of calories a sandwich has and the sodium content of the sandwich. Several different types of sandwiches are used. The data follow. (The information in this exercise will be used for Exercise 11–49.)

Calories, x	419	419	386	240	354	231	174
Sodium (mg), y	495	645	1202	581	990	1159	787

Source: *Consumer Reports* 60, no. 9 (September 1995), p. 588.

11–22. A medical researcher wishes to determine how the dosage (in milligrams) of a drug affects the heart rate of the patient. The data for seven patients are given here. (The information in this exercise will be used for Exercise 11–50.)

Drug dosage, x	0.125	0.20	0.25	0.30	0.35	0.40	0.50
Heart rate, y	95	90	93	92	88	80	82

11–23. A researcher wishes to determine if there is a relationship between the energy efficiency rating (EER) of an air conditioner and its cost. The EER is determined by the Appliance Manufacturers' Association. The data follow. (The information in this exercise will be used for Exercise 11–51.)

EER, x	9.5	10	10	10	10.3	9.1	8.6
Cost, y	$370	360	400	400	420	350	310

Source: *Consumer Reports* 60, no. 6 (June 1995), p. 406.

11–24. In the air conditioner study cited in the previous exercise, the researcher also wishes to see if there is a relationship between the air conditioner's moisture removal capacity (in pints per hour) and its weight (in pounds). The data are shown here. (The information in this exercise will be used for Exercise 11–52.)

Moisture removal capacity, x	1.9	1.6	2.0	1.6	2.0	2.0	2.1
Weight, y	57	60	72	55	69	55	48

11–25. A psychologist selects six families with two children each, a boy and a girl. She is interested in comparing the IQs of each gender to determine whether there is a relationship between them. The data are given here. (The information in this exercise will be used for Exercise 11–53.)

IQ, females, x	107	95	116	109	101	98
IQ, males, y	107	102	112	104	105	103

11–26. A researcher wishes to determine whether there is a relationship between the age (in years) of grocery store cash registers and monthly maintenance cost. The data follow. (The information in this exercise will be used for Exercise 11–54.)

Age, x	2	4	3	1	2	6	4
Cost, y	$75	$90	$70	$65	$83	$90	$87

11–27. A study was conducted at a large hospital to determine whether there was a relationship between the number of years nurses were employed and the number of nurses who voluntarily resign. The results are shown below. (The information in this exercise will be used for Exercise 11–28 and 11–55.)

Years employed, x	5	6	3	7	8	2
Resignations, y	7	4	7	1	2	8

*11–28. One of the formulas for computing r is

$$r = \frac{\Sigma\,(x - \bar{x})(y - \bar{y})}{(n - 1)(s_x)(s_y)}$$

Using the data in Exercise 11–27, compute r with this formula. Compare the results.

*11–29. Compute r for the following data set. Explain the reason for this value of r. Now, interchange the values of x and y and compute r again. Compare this value with the previous one. Explain the results of the comparison.

x	1	2	3	4	5
y	3	5	7	9	11

x	−3	−2	−1	0	1	2	3
y	9	4	1	0	1	4	9

*11–30. Compute r for the following data and test the hypothesis H_0: $\rho = 0$. Draw the scatter plot; then explain the results.

Regression

Objective 4. Compute the equation of the regression line.

In studying relationships between two variables, collect the data and then construct a scatter plot. The purpose of the scatter plot, as indicated previously, is to determine the nature of the relationship. The possibilities include a positive linear relationship, a negative linear relationship, a curvilinear relationship, or no discernible relationship. After the scatter plot is drawn, the next steps are to compute the value of the correlation coefficient and to test the significance of the relationship. If the value of the correlation coefficient is significant, the next step is to determine the equation of the **regression line,** which is the data's line of best fit. (*Note:* Determining the regression line when r is not significant and then making predictions using the regression line is meaningless.) The purpose of the regression line is to enable the researcher to see the trend and make predictions on the basis of the data.

Line of Best Fit

Figure 11–10 shows a scatter plot for the data of two variables. It shows that several lines can be drawn on the graph near the points. Given a scatter plot, one must be able to draw the *line of best fit. Best fit* means that the sum of the squares of the vertical distances from each point to the line is at a minimum. The reason one needs a line of best fit is that the values of y will be predicted from the values of x; hence, the closer the points are to the line, the better the fit and the prediction will be. See Figure 11–11.

Figure 11–10

Scatter Plot with Three Lines Fit to the Data

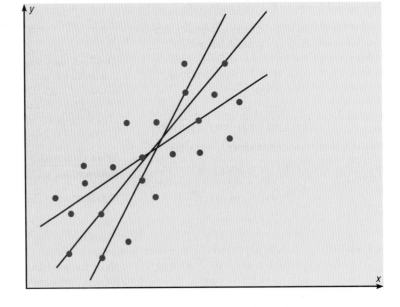

Figure 11–12 shows the relationship between the values of the correlation coefficient and the variability of the scores about the regression line. The closer the points fit the regression line, the higher the absolute value of r is and the closer it will be to $+1$ or -1. When all the points fall exactly on the line, r will equal $+1$ or -1, and this indicates

a perfect linear relationship between the variables. Figure 11–12(a) shows r as positive and approximately equal to 0.50; Figure 11–12(b) shows r approximately equal to 0.90; and Figure 11–12(c) shows r equal to +1. For positive values of r, the line slopes upward from left to right.

Figure 11–11

Line of Best Fit for a Set of Data Points

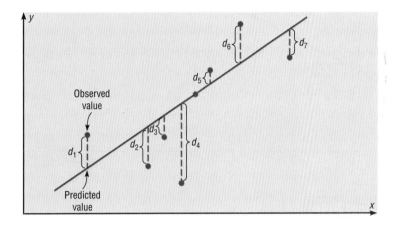

Figure 11–12

Relationship between the Correlation Coefficient and the Line of Best Fit

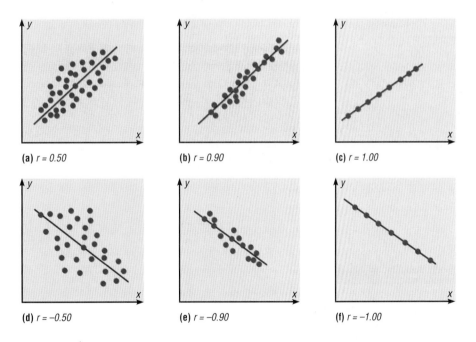

(a) $r = 0.50$ (b) $r = 0.90$ (c) $r = 1.00$

(d) $r = -0.50$ (e) $r = -0.90$ (f) $r = -1.00$

Figure 11–12(d) shows r as negative and approximately equal to -0.50; Figure 11–12(e) shows r approximately equal to -0.90; and Figure 11–12(f) shows r equal to -1. For negative values of r, the line slopes downward from left to right.

Determination of the Regression Line Equation

In algebra, the equation of a line is usually given as $y = mx + b$, where m is the slope of the line and b is the y intercept. (Students who need an algebraic review of the properties of a line should refer to Appendix A Section A–3 before studying this section.) In statistics, the equation of the regression line is written as $y' = a + bx$, where a is the y' intercept and b is the slope of the line. See Figure 11–13.

Figure 11–13

A Line as Represented in Algebra and in Statistics

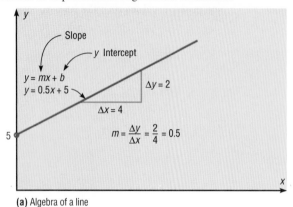

(a) Algebra of a line

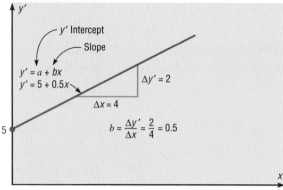

(b) Statistical notation for a regression line

Historical Note

In 1795, Adrien Marie Legendre (1752–1833) measured the meridian arc on the earth's surface from Barcelona, Spain, to Dunkirk, England. This measure was used as the basis for the measure of the meter. Legendre developed the least-squares method around the year 1805.

There are several methods for finding the equation of the regression line. Two formulas are given here. *These formulas use the same values that are used in computing the value of the correlation coefficient.* The mathematical development of these formulas is beyond the scope of this book.

Formulas for the Regression Line $y' = a + bx$

$$a = \frac{(\Sigma y)(\Sigma x^2) - (\Sigma x)(\Sigma xy)}{n(\Sigma x^2) - (\Sigma x)^2}$$

$$b = \frac{n(\Sigma xy) - (\Sigma x)(\Sigma y)}{n(\Sigma x^2) - (\Sigma x)^2}$$

where a is the y' intercept and b is the slope of the line.

Rounding Rule for the Intercept and Slope   Round the values of a and b to three decimal places.

Example 11–9

Find the equation of the regression line for the data in Example 11–4, and graph the line on the scatter plot of the data.

Solution

The values needed for the equation are $n = 6$, $\Sigma x = 345$, $\Sigma y = 819$, $\Sigma xy = 47,634$, and $\Sigma x^2 = 20,399$. Substituting in the formulas, one gets

$$a = \frac{(\Sigma y)(\Sigma x^2) - (\Sigma x)(\Sigma xy)}{n(\Sigma x^2) - (\Sigma x)^2} = \frac{(819)(20,399) - (345)(47,634)}{(6)(20,399) - (345)^2} = 81.048$$

$$b = \frac{n(\Sigma xy) - (\Sigma x)(\Sigma y)}{n(\Sigma x^2) - (\Sigma x)^2} = \frac{(6)(47,634) - (345)(819)}{(6)(20,399) - (345)^2} = 0.964$$

Hence, the equation of the regression line, $y' = a + bx$, is

$$y' = 81.048 + 0.964x$$

The graph of the line is shown in Figure 11–14.

Figure 11–14

Regression Line for
Example 11–9

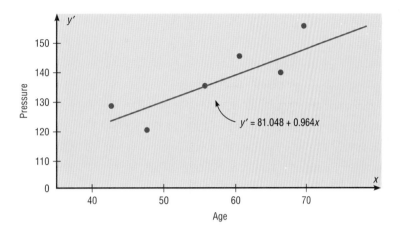

Note: When drawing the scatter plot and the regression line, it is sometimes desirable to *truncate* the graph (see Chapter 2). The reason is to show the line drawn in the range of the independent and dependent variables. For example, the regression line in Figure 11–14 is drawn between the x values of approximately 43 and 82 and the y' values of approximately 120 and 152. The range of the x values in the original data shown in Example 11–4 is $70 - 43 = 27$ and the range of the y' values is $152 - 120 = 32$. Notice that the x axis has been truncated; the distance between 0 and 40 is not shown in the proper scale compared to the distance between 40 and 50, 50 and 60, etc. The y' axis has been similarly truncated.

The important thing to remember is that when the x axis and sometimes the y' axis have been truncated, do not use the y' intercept value, a, to graph the line. To be on the safe side when graphing the regression line, use a value for x selected from the range of x values.

Example 11–10

Find the equation of the regression line for the data in Example 11–5, and graph the line on the scatter plot.

Solution

The values needed for the equation are $n = 7$, $\Sigma x = 57$, $\Sigma y = 511$, $\Sigma xy = 3745$, and $\Sigma x^2 = 579$. Substituting in the formulas, one gets

$$a = \frac{(\Sigma y)(\Sigma x^2) - (\Sigma x)(\Sigma xy)}{n(\Sigma x^2) - (\Sigma x)^2} = \frac{(511)(579) - (57)(3745)}{(7)(579) - (57)^2} = 102.493$$

$$b = \frac{n(\Sigma xy) - (\Sigma x)(\Sigma y)}{n(\Sigma x^2) - (\Sigma x)^2} = \frac{(7)(3745) - (57)(511)}{(7)(579) - (57)^2} = -3.622$$

Hence, the equation of the regression line, $y' = a + bx$, is

$$y' = 102.493 - 3.622x$$

The graph of the line is shown in Figure 11–15.

Figure 11–15

Regression Line for
Example 11–10

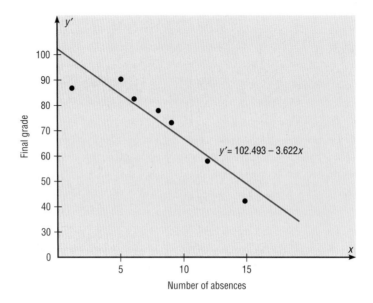

The sign of the correlation coefficient and the sign of the slope of the regression line will always be the same. That is, if r is positive, then b will be positive; if r is negative, then b will be negative. The reason is that the numerators of the formulas are the same and determine the signs of r and b, and the denominators are always positive.

The regression line can be used to make predictions for the dependent variable. The method for making predictions is shown in the next example.

Example 11–11

Using the equation of the regression line found in Example 11–9, predict the blood pressure for a person who is 50 years old.

Solution

Substituting 50 for x in the regression line $y' = 81.048 + 0.964x$ gives

$$y' = 81.048 + (0.964)(50) = 129.248 \text{ (rounded to 129)}$$

In other words, the predicted systolic blood pressure for a 50-year-old person is 129.

The value obtained in Example 11–11 is a point prediction, and with point predictions, no degree of accuracy or confidence can be determined. More information on prediction is given in Section 11–5, including prediction intervals.

When r is not significantly different from 0, the best predictor of y is the mean of the data values of y. For valid predictions, the value of the correlation coefficient must be significant. Also, two other assumptions must be met.

Assumptions for Valid Predictions in Regression

1. For any specific value of the independent variable x, the value of the dependent variable y must be normally distributed about the regression line. See Figure 11–16(a).
2. The standard deviation of each of the dependent variables must be the same for each value of the independent variable. See Figure 11–16(b).

Figure 11–16

Assumptions for Predictions

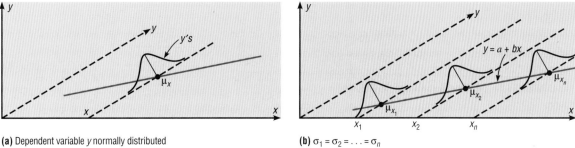

(a) Dependent variable y normally distributed

(b) $\sigma_1 = \sigma_2 = \ldots = \sigma_n$

When predictions are made beyond the bounds of the data, they must be interpreted cautiously. For example, in 1979, some experts predicted that the United States would run out of oil by the year 2003. This prediction was based on the current consumption and on known oil reserves at that time. However, since then, the automobile industry has produced many new fuel-efficient vehicles. Also, there are many as yet undiscovered oil fields. Finally, science may someday discover a way to run a car on something as unlikely but as common as peanut oil. In addition, the price of a gallon of gasoline was predicted to reach $10 a few years later. Fortunately this has not come to pass. *Remember that when predictions are made, they are based on present conditions or on the premise that present trends will continue.* This assumption may or may not prove true in the future.

The steps for finding the value of the correlation coefficient and the regression line equation are summarized in the Procedure Table.

Procedure Table

Finding the Correlation Coefficient and the Regression Line Equation

STEP 1   Make a table, as shown in Step 2.

STEP 2   Find the values of xy, x^2, and y^2. Place them in the appropriate columns and sum each column.

x	y	xy	x^2	y^2
.	.	.	.	.
.	.	.	.	.
.	.	.	.	.
$\Sigma x =$	$\Sigma y =$	$\Sigma xy =$	$\Sigma x^2 =$	$\Sigma y^2 =$

STEP 3   Substitute in the formula to find the value of r.

$$r = \frac{n(\Sigma xy) - (\Sigma x)(\Sigma y)}{\sqrt{[n(\Sigma x^2) - (\Sigma x)^2][n(\Sigma y^2) - (\Sigma y)^2]}}$$

STEP 4   When r is significant, substitute in the formulas to find the values of a and b for the regression line equation $y' = a + bx$.

$$a = \frac{(\Sigma y)(\Sigma x^2) - (\Sigma x)(\Sigma xy)}{n(\Sigma x^2) - (\Sigma x)^2} \qquad b = \frac{n(\Sigma xy) - (\Sigma x)(\Sigma y)}{n(\Sigma x^2) - (\Sigma x)^2}$$

Exercises

11–31. What two things should be done before one performs a regression analysis?

11–32. What are the assumptions for regression analysis?

11–33. What is the general form for the regression line used in statistics?

11–34. What is the symbol for the slope? For the y intercept?

11–35. What is meant by the *line of best fit*?

11–36. When all the points fall on the regression line, what is the value of the correlation coefficient?

11–37. What is the relationship of the sign of the correlation coefficient and the sign of the slope of the regression line?

11–38. As the value of the correlation coefficient increases from 0 to 1, or decreases from 0 to -1, how do the points of the scatter plot fit the regression line?

11–39. How is the value of the correlation coefficient related to the accuracy of the predicted value for a specific value of x?

11–40. For the data in Exercise 11–12, find the equation of the regression line, and predict y' when $x = 7$ ads.

11–41. For the data in Exercise 11–13, find the equation of the regression line, and predict y' when $x = 35$ years.

11–42. For the data in Exercise 11–14, find the equation of the regression line, and predict y' when $x = \$1100$.

11–43. For the data in Exercise 11–15, find the equation of the regression line, and predict y' when $x = 4$ years.

11–44. For the data in Exercise 11–16, find the equation of the regression line, and predict y' when $x = 47$ years.

11–45. For the data in Exercise 11–17, find the equation of the regression line, and predict y' when $x = 104$.

11–46. For the data in Exercise 11–18, find the equation of the regression line, and predict y' when $x = 88$.

11–47. For the data in Exercise 11–19, find the equation of the regression line, and predict y' when $x = 35$ hours.

11–48. For the data in Exercise 11–20, find the equation of the regression line, and predict y' when $x = 80°$.

11–49. For the data in Exercise 11–21, find the equation of the regression line, and predict y' when $x = 400$ calories.

11–50. For the data in Exercise 11–22, find the equation of the regression line, and predict y' when $x = 0.27$ milligram.

11–51. For the data in Exercise 11–23, find the equation of the regression line, and predict y' when $x = 9$ EER.

11–52. For the data in Exercise 11–24, find the equation of the regression line, and predict y' when $x = 1.8$ pints.

11–53. For the data in Exercise 11–25, find the equation of the regression line, and predict y' when $x = 104$.

11–54. For the data in Exercise 11–26, find the equation of the regression line, and predict y' when $x = 4$ years.

11–55. For the data in Exercise 11–27, find the equation of the regression line, and predict y' when $x = 4$ years.

"Explain that to me."

Source: Reprinted with special permission of King Features Syndicate, Inc.

For Exercises 11–56 through 11–61, do a complete regression analysis by performing the following steps.
a. Draw a scatter plot.
b. Compute the correlation coefficient.
c. State the hypotheses.
d. Test the hypotheses at $\alpha = 0.05$. Use Table I.
e. Determine the regression line equation.
f. Plot the regression line on the scatter plot.
g. Summarize the results.

11–56. The following data were obtained for the years 1993 through 1998 and indicate the number of fireworks (in millions) used and the related injuries. Predict the number of injuries if 100 million fireworks are used during a given year.

Fireworks in use, x	67.6	87.1	117	115	118	113
Related injuries, y	12,100	12,600	12,500	10,900	7,800	7,000

Source: National Council of Fireworks Safety, American Pyrotechnic Assoc.

11–57. The following data were obtained from a survey of the number of years people smoked and the percentage of lung damage they sustained. Predict the percentage of lung damage for a person who has smoked for 30 years.

Years, x	22	14	31	36	9	41	19
Damage, y	20	14	54	63	17	71	23

11–58. A researcher wishes to determine whether a person's age is related to the number of hours he or she jogs per week. The data for the sample follow.

Age, x	34	22	48	56	62
Hours, y	5.5	7	3.5	3	1

11–59. The following data were obtained from a sample of counties in Southwestern Pennsylvania and indicate the number (in thousands) of tons of bituminous coal produced in each county and the number of employees working in coal production in each county. Predict the number of employees needed to produce 500 thousand tons of coal. The data are given here.

Tons, x	227	5410	5328	147	729	8095
No. of employees, y	110	731	1031	20	118	1162

Tons, x	635	6157
No. of employees, y	103	752

11–60. A statistics instructor is interested in finding the strength of a relationship between the final exam grades of students enrolled in Statistics I and in Statistics II. The data are given here in percentages.

Statistics I, x	87	92	68	72	95	78	83	98
Statistics II, y	83	88	70	74	90	74	83	99

11–61. An educator wants to see how the number of absences a student in her class has affects the student's final grade. The data obtained from a sample follow.

No. of absences, x	10	12	2	0	8	5
Final grade, y	70	65	96	94	75	82

For Exercises 11–62 and 11–63 do a complete regression analysis and test the significance of r at $\alpha = 0.05$ using the P-value method.

11–62. A physician wishes to know whether there is a relationship between a father's weight (in pounds) and his newborn son's weight (in pounds). The data are given here.

Father's weight, x	176	160	187	210	196	142	205	215
Son's weight, y	6.6	8.2	9.2	7.1	8.8	9.3	7.4	8.6

11–63. A car dealership owner wishes to determine if there is a relationship between the number of years of experience a salesperson has and the number of cars sold per month. The data are given here.

Years of experience, x	3	9	2	5	1
No. of cars sold per month, y	5	14	12	21	8

*11–64. For Exercises 11–12, 11–13, and 11–14, find the mean of the x and y variables, then substitute the mean of the x variable into the corresponding regression line equations found in Exercises 11–40, 11–41, and 11–42 and find y'. Compare the value of y' with $\bar{y}$ for each exercise. Generalize the results.

*11–65. The y intercept value, a, can also be found by using the following equation:

$$a = \bar{y} - b\bar{x}$$

Verify this result by using the data in Exercises 11–15 and 11–43, 11–16 and 11–44, and 11–17 and 11–45.

*11–66. The value of the correlation coefficient can also be found by using the formula

$$r = \frac{bs_x}{s_y}$$

Where s_x is the standard deviation of the x value, and s_y is the standard deviation of the y values. Verify this result for Exercises 11–18, 11–19, and 11–20.

Technology Step by Step

MINITAB
Step by Step

Correlation and Regression

These instructions use data from the age and blood pressure study discussed in Examples 11–1, 11–4, 11–7, and 11–9.

Example MT11–1

1. Enter the data into two columns of the worksheet. Name the columns **Age** and **Pressure.**
2. Select **Stat>Basic Statistics>Correlation.**
3. Double-click `Pressure` and double-click `Age`. The dependent variable should be first.
4. Check the box for `Display p-values`.

Correlation Dialog Box

5. Click [OK].

The correlation coefficient will be displayed in the session window. It is $r = 0.897$. To make a scatterplot:

6. Select **GRAPH>PLOT**.

7. Double-click on Pressure for the [Y] variable and Age for the Predictor [X] variable.

8. The Display type should be Symbol. If not, click the drop down list arrow and change it.

9. Click [OK]. A graph window will open showing the scatter plot.

Plot Dialog Box

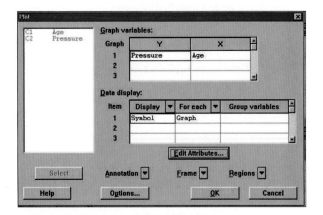

To determine the equation of the least-squares regression line:

10. Select **Stat>Regression>Regression.**

11. Double-click Pressure in the variable list to select it for the Response variable, Y.

12. Double-click Age in the variable list to select it for the Predictors variable, X.

13. Click on [Storage] then check the boxes for Residuals and Fits.

14. Click [OK] twice.

The session window will contain the equation of the regression line and additional analysis. Note that the *P*-value shown to the right of the Predictor "Age," 0.015, indicates the significance of *r*. See the discussion after Example 11–7.

Regression Analysis: Pressure versus Age
```
The regression equation is
Pressure = 81.0 + 0.964 Age
```

Predictor	Coef	SE Coef	T	P
Constant	81.05	13.88	5.84	0.004
Age	0.9644	0.2381	4.05	0.015

```
S = 5.641      R-Sq = 80.4%      R-Sq (adj) = 75.5%
```

Analysis of Variance

Source	DF	SS	MS	F	P
Regression	1	522.21	522.21	16.41	0.015
Residual Error	4	127.29	31.82		
Total	5	649.50			

TI-83
Step by Step

Correlation and Regression

To graph a scatter plot:

1. Enter the *x* values in L_1 and the *y* values in L_2.

2. Make sure the Window values are appropriate. Select an Xmin slightly less than the smallest *x* data value and an Xmax slightly larger than the largest *x* data value. Do the same for Ymin and Ymax. Also, you may need to change the Xscl and Yscl values, depending on the data.

3. Press **2nd [STAT PLOT] 1** for Plot 1. The other *y* functions should be turned off.

4. Move the cursor to On and press **ENTER** on the Plot 1 menu.

5. Move the cursor to the graphic that looks like a scatter plot next to Type (first graph), and press **ENTER.** Make sure the *X* list is L_1, and the *Y* list is L_2.

6. Press GRAPH.

Example TI11–1

Draw a scatter plot for the data from Example 11–1.

x	43	48	56	61	67	70
y	128	120	135	143	141	152

Input

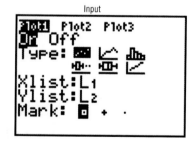

Input

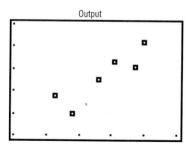

Output

The input and output screens are shown.

To find the equation of the regression line:

1. Press **STAT** and move the cursor to `Calc`.

2. Press **8** for `LinReg(a+bx)` then **ENTER.** The values for *a* and *b* will be displayed.

In order to have the calculator compute and display the correlation coefficient and coefficient of determination as well as the equation of the line, you must set the diagnostics display mode to on. Follow these steps:

1. Press **2nd [CATALOG].**

2. Use the arrow keys to scroll down to `Diagnostic On`.

3. Press **ENTER** to copy the command to the home screen.

4. Press **ENTER** to execute the command.

Example TI11–2

Find the equation of the regression line for the data in the previous example, as shown in Example 11–9.

The input and output screens are shown.

The equation of the regression line is $y' = 81.04808549 + 0.964381122x$.

To plot the regression line on the scatter plot:

1. Press **Y=** and **CLEAR** to clear any previous equations.

2. Press **VARS** and then **5** for `Statistics`.

3. Move the cursor to **EQ** and press **1** for `RegEQ`. The line will be in the Y= screen.

4. Press **GRAPH.**

Example TI11–3

Draw the regression line found in the previous example on the scatter plot.

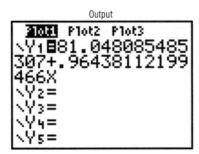

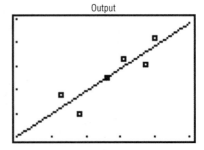

The output screens are shown.

To test the significance of b and ρ:

1. Press **STAT** and move the cursor to TESTS.

2. Press **E (ALPHA SIN)** for LinRegTTest. Make sure the Xlist is L_1, the YList is L_2, and the Freq is 1.

3. Select the appropriate alternative hypothesis.

4. Move the cursor to Calculate and press **ENTER.**

Example TI11–4

Test the hypothesis from Examples 11–4 and 11–7, H_0: $\rho = 0$ for the data in Example 11–1. Use $\alpha = 0.05$.

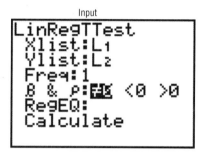

Input
```
LinRegTTest
 Xlist:L₁
 Ylist:L₂
 Freq:1
 β & ρ:≠0  <0  >0
 RegEQ:
 Calculate
```

Output
```
LinRegTTest
 y=a+bx
 β≠0 and ρ≠0
 t=4.050983638
 P=.0154631742
 df=4
↓a=81.04808549
```

Output
```
LinRegTTest
 y=a+bx
 β≠0 and ρ≠0
↑b=.964381122
 s=5.641090817
 r²=.8040221364
 r=.8966728145
```

In this case, the t test value is 4.050983638. The P-value is 0.0154631742, which is significant. The decision is to reject the null hypothesis at $\alpha = 0.05$, since $0.0154631742 < 0.05$; $r = 0.8966728145$, $r^2 = 0.8040221364$.

Excel
Step by Step

Scatter Plots

Creating scatter plots in Excel is straightforward using the Chart Wizard.

1. Click on the Chart Wizard icon (it looks like a colorful histogram).

2. Select chart type XY (Scatter) under the Standard Types tab. Click on [Next >].

3. Enter the data range, and specify whether the data for each variable are stored in columns (as we have done in our examples) or rows. Click on [Next >].

4. The next dialog box enables you to set various options for displaying the plot. In most cases, the defaults will be okay. After entering the desired options (note that there are several tabs for this screen), click on [Next >].

5. Use this final dialog box to specify where the chart will be located. Click on [Finish].

Correlation Coefficient

The CORREL function returns the correlation coefficient.

1. Enter the data in columns A and B.

2. Select a blank cell, then click on the f_x button.

3. Under `Function category`, select `Statistical`. From the `Function name` list, select `CORREL`.

4. Enter the data range (**A1:AN,** where N is the number of sample data pairs) for the first variable in `Array1`. Enter the data range for the second variable in `Array2`. The correlation coefficient will be displayed in the selected cell.

11–5

Coefficient of Determination and Standard Error of Estimate

The previous sections stated that if the correlation coefficient is significant, the equation of the regression line can be determined. Also, for various values of the independent variable x, the corresponding values of the dependent variable y can be predicted. Several other measures are associated with the correlation and regression techniques. They include the coefficient of determination, the standard error of estimate, and the prediction interval. But before these concepts can be explained, the different types of variation associated with the regression model must be defined.

Types of Variation for the Regression Model

Consider the following hypothetical regression model.

x	1	2	3	4	5
y	10	8	12	16	20

The equation of the regression line is $y' = 4.8 + 2.8x$, and $r = 0.919$. The sample y values are 10, 8, 12, 16, and 20. The predicted values, designated by y', for each x can be found by substituting each x value into the regression equation and finding y'. For example, when $x = 1$,

$$y' = 4.8 + 2.8x = 4.8 + (2.8)(1) = 7.6$$

Now, for each x, there are an observed y value and a predicted y' value. For example, when $x = 1$, $y = 10$, and $y' = 7.6$. Recall that the closer the observed values are to the predicted values, the better the fit is and the closer r is to $+1$ or -1.

The *total variation* $\Sigma (y - \bar{y})^2$ is the sum of the squares of the vertical distances each point is from the mean. The total variation can be divided into two parts: that which is attributed to the relationship of x and y and that which is due to chance. The variation obtained from the relationship (i.e., from the predicted y' values) is $\Sigma (y' - \bar{y})^2$ and is called the *explained variation*. Most of the variations can be explained by the relationship. The closer the value r is to $+1$ or -1, the better the points fit the line, and the closer $\Sigma (y' - \bar{y})^2$ is to $\Sigma (y - \bar{y})^2$. In fact, if all points fall on the regression line, $\Sigma (y' - \bar{y})^2$ will equal $\Sigma (y - \bar{y})^2$, since y' would be equal to y in each case.

On the other hand, the variation due to chance, found by $\Sigma (y - y')^2$, is called the *unexplained variation*. This variation cannot be attributed to the relationships. When the unexplained variation is small, the value of r is close to $+1$ or -1. If all points fall on the regression line, the unexplained variation, $\Sigma (y - y')^2$, will be 0. Hence, the *total variation* is equal to the sum of the explained variation and the unexplained variation. That is,

$$\Sigma (y - \bar{y})^2 = \Sigma (y' - \bar{y})^2 + \Sigma (y - y')^2$$

These values are shown in Figure 11–17. For a single point, the differences are called *deviations*. For the hypothetical regression model given earlier, for $x = 1$ and $y = 10$, one gets $y' = 7.6$ and $\bar{y} = 13.2$.

Figure 11–17

Deviations for the
Regression Equation

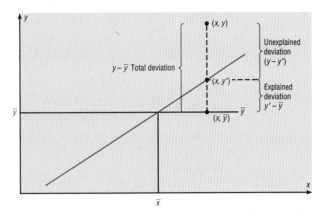

The procedure for finding the three types of variation is illustrated next.

STEP 1   Find the predicted y' values.

$$\text{for } x = 1 \quad y' = 4.8 + 2.8x = 4.8 + (2.8)(1) = 7.6$$
$$\text{for } x = 2 \quad y' = 4.8 + (2.8)(2) = 10.4$$
$$\text{for } x = 3 \quad y' = 4.8 + (2.8)(3) = 13.2$$
$$\text{for } x = 4 \quad y' = 4.8 + (2.8)(4) = 16$$
$$\text{for } x = 5 \quad y' = 4.8 + (2.8)(5) = 18.8$$

Hence, the values for this example are as follows:

x	y	y'
1	10	7.6
2	8	10.4
3	12	13.2
4	16	16
5	20	18.8

STEP 2   Find the mean of the y values.

$$\bar{y} = \frac{10 + 8 + 12 + 16 + 20}{5} = 13.2$$

STEP 3   Find the total variation, $\Sigma (y - \bar{y})^2$.

$$(10 - 13.2)^2 = 10.24$$
$$(8 - 13.2)^2 = 27.04$$
$$(12 - 13.2)^2 = 1.44$$
$$(16 - 13.2)^2 = 7.84$$
$$(20 - 13.2)^2 = \underline{46.24}$$
$$\Sigma (y - \bar{y})^2 = 92.8$$

STEP 4   Find the explained variation, $\Sigma (y' - \bar{y})^2$.

$$(7.6 - 13.2)^2 = 31.36$$
$$(10.4 - 13.2)^2 = 7.84$$
$$(13.2 - 13.2)^2 = 0.00$$
$$(16 - 13.2)^2 = 7.84$$
$$(18.8 - 13.2)^2 = \underline{31.36}$$
$$\Sigma (y' - \bar{y})^2 = 78.4$$

STEP 5 Find the unexplained variation, $\Sigma\,(y - y')^2$.

$$(10 - 7.6)^2 = 5.76$$
$$(8 - 10.4)^2 = 5.76$$
$$(12 - 13.2)^2 = 1.44$$
$$(16 - 16)^2 = 0.00$$
$$(20 - 18.8)^2 = \underline{1.44}$$
$$\Sigma\,(y - y')^2 = 14.4$$

Notice that

total variation = explained variation + unexplained variation

$$92.8 \quad = \quad 78.4 \quad + \quad 14.4$$

Coefficient of Determination

Objective 5. Compute the coefficient of determination.

The *coefficient of determination* is the ratio of the explained variation to the total variation and is denoted by r^2. That is,

$$r^2 = \frac{\text{explained variation}}{\text{total variation}}$$

For the example, $r^2 = 78.4/92.8 = 0.845$. The term r^2 is usually expressed as a percentage. So in this case, 84.5% of the total variation is explained by the regression line using the independent variable.

Another way to arrive at the value for r^2 is to square the correlation coefficient. In this case, $r = 0.919$, and $r^2 = 0.845$, which is the same value found by using the variation ratio.

The **coefficient of determination** is a measure of the variation of the dependent variable that is explained by the regression line and the independent variable. The symbol for the coefficient of determination is r^2.

Of course, it is usually easier to find the coefficient of determination by squaring r and converting it to a percentage. Therefore, if $r = 0.90$, then $r^2 = 0.81$, which is equivalent to 81%. This result means that 81% of the variation in the dependent variable is accounted for by the variations in the independent variable. The rest of the variation, 0.19 or 19%, is unexplained. This value is called the *coefficient of nondetermination* and is found by subtracting the coefficient of determination from 1. As the value of r approaches 0, r^2 decreases more rapidly. For example, if $r = 0.6$, then $r^2 = 0.36$, which means that only 36% of the variation in the dependent variable can be attributed to the variation in the independent variable.

Formula for the Coefficient of Nondetermination

$$1.00 - r^2$$

Standard Error of Estimate

Objective 6. Compute the standard error of estimate.

When a y' value is predicted for a specific x value, the prediction is a point prediction. However, a prediction interval about the y' value can be constructed, just as a confidence interval was constructed for an estimate of the population mean. The prediction interval uses a statistic called the *standard error of estimate*.

The **standard error of estimate,** denoted by s_{est} is the standard deviation of the observed y values about the predicted y' values. The formula for the standard error of estimate is

$$s_{est} = \sqrt{\frac{\Sigma(y - y')^2}{n - 2}}$$

The standard error of estimate is similar to the standard deviation, but the mean is not used. As can be seen from the formula, the standard error of estimate is the square root of the unexplained variation—i.e., the variation due to the difference of the observed values and the expected values—divided by $n - 2$. So the closer the observed values are to the predicted values, the smaller the standard error of estimate will be.

The next example shows how to compute the standard error of estimate.

Example 11–12

A researcher collects the following data and determines that there is a significant relationship between the age of a copy machine and its monthly maintenance cost. The regression equation is $y' = 55.57 + 8.13x$ Find the standard error of estimate.

Machine	Age, x (years)	Monthly cost, y
A	1	$ 62
B	2	78
C	3	70
D	4	90
E	4	93
F	6	103

Solution

STEP 1 Make a table, as shown.

x	y	y'	$y - y'$	$(y - y')^2$
1	62			
2	78			
3	70			
4	90			
4	93			
6	103			

STEP 2 Using the regression line equation, $y' = 55.57 + 8.13x$, compute the predicted values y' for each x and place the results in the column labeled y'.

$$x = 1 \quad y' = 55.57 + (8.13)(1) = 63.70$$
$$x = 2 \quad y' = 55.57 + (8.13)(2) = 71.83$$
$$x = 3 \quad y' = 55.57 + (8.13)(3) = 79.96$$
$$x = 4 \quad y' = 55.57 + (8.13)(4) = 88.09$$
$$x = 6 \quad y' = 55.57 + (8.13)(6) = 104.35$$

STEP 3 For each y, subtract y' and place the answer in the column labeled $y - y'$.

$$62 - 63.70 = -1.70 \qquad 90 - (88.09) = 1.91$$
$$78 - 71.83 = 6.17 \qquad 93 - (88.09) = 4.91$$
$$70 - 79.96 = -9.96 \qquad 103 - (104.35) = -1.35$$

STEP 4 Square the numbers found in step 3 and place the squares in the column labeled $(y - y')^2$.

STEP 5 Find the sum of the numbers in the last column. The completed table follows.

x	y	y'	y − y'	(y − y')²
1	62	63.70	−1.70	2.89
2	78	71.83	6.17	38.0689
3	70	79.96	−9.96	99.2016
4	90	88.09	1.91	3.6481
4	93	88.09	4.91	24.1081
6	103	104.35	−1.35	1.8225
				169.7392

STEP 6 Substitute in the formula and find s_{est}.

$$s_{est} = \sqrt{\frac{\Sigma (y - y')^2}{n - 2}} = \sqrt{\frac{169.7392}{6 - 2}} = 6.51$$

In this case, the standard deviation of observed values about the predicted values is 6.51.

The standard error of estimate can also be found by using the formula

$$s_{est} = \sqrt{\frac{\Sigma y^2 - a \Sigma y - b \Sigma xy}{n - 2}}$$

Example 11–13

Find the standard error of estimate for the data for Example 11–12 by using the preceding formula. The equation of the regression line is $y' = 55.57 + 8.13x$.

Solution

STEP 1 Make a table.

STEP 2 Find the product of x and y values and place the results in the third column.

STEP 3 Square the y values and place the results in the fourth column.

STEP 4 Find the sums of the second, third, and fourth columns. The completed table is shown here.

x	y	xy	y²
1	62	62	3,844
2	78	156	6,084
3	70	210	4,900
4	90	360	8,100
4	93	372	8,649
6	103	618	10,609
	$\Sigma y = 496$	$\Sigma xy = 1778$	$\Sigma y^2 = 42{,}186$

STEP 5 From the regression equation, $y' = 55.57 + 8.13x$, $a = 55.57$, and $b = 8.13$.

STEP 6 Substitute in the formula and solve for s_{est}.

$$s_{est} = \sqrt{\frac{\Sigma y^2 - a \Sigma y - b \Sigma xy}{n - 2}}$$

$$= \sqrt{\frac{42,186 - (55.57)(496) - (8.13)(1778)}{6 - 2}} = 6.48$$

This value is close to the value found in Example 11–12. The difference is due to rounding.

Prediction Interval

Objective 7. Find a prediction interval.

The standard error of estimate can be used for constructing a **prediction interval** (similar to a confidence interval) about a y' value.

When a specific value x is substituted into the regression equation, one gets y' which is a point estimate for y. For example, if the regression line equation for the age of a machine and the monthly maintenance cost is $y' = 55.57 + 8.13x$ (Example 11–12), then the predicted maintenance cost for a 3-year-old machine would be $y' = 5.57 + 8.13(3)$ or $79.96. Since this is a point estimate, one has no ideas how accurate it is. But one can construct a prediction interval about the estimate. By selecting an α value, one can achieve a $1 - \alpha$ confidence that the interval contains the actual mean of the y values that correspond to the given value of x.

The reason is that there are possible sources of prediction errors in finding the regression line equation. One source occurs when finding the standard error of estimate, s_{est}. Two others are errors made in estimating the slope and the y intercept, since the equation of the regression line will change somewhat if different random samples are used when calculating the equation.

Formula for the Prediction Interval about a Value y'

$$y' - t_{\alpha/2}s_{est}\sqrt{1 + \frac{1}{n} + \frac{n(x - \overline{X})^2}{n\Sigma x^2 - (\Sigma x)^2}} < y < y'$$

$$+ t_{\alpha/2}s_{est}\sqrt{1 + \frac{1}{n} + \frac{n(x - \overline{X})^2}{n\Sigma x^2 - (\Sigma x)^2}}$$

with d.f. $= n - 2$

Example 11–14

For the data in Example 11–12, find the 95% prediction interval for the monthly maintenance cost of a machine that is 3 years old.

Solution

STEP 1   Find Σx, Σx^2, and $\overline{X}$.

$$\Sigma x = 20 \qquad \Sigma x^2 = 82 \qquad \overline{X} = \frac{20}{6} = 3.3$$

STEP 2   Find y' for $x = 3$

$$y' = 55.57 + 8.13x$$
$$y' = 55.57 + 8.13(3) = 79.96$$

STEP 3   Find s_{est}

$$s_{est} = 6.48, \text{ as shown in Example 11–13.}$$

STEP 4   Substitute in the formula and solve : $t_{\alpha/2} = 2.776$, d.f. $= 6 - 2 = 4$ for 95%

$$y' - t_{\alpha/2}s_{est} \sqrt{1 + \frac{1}{n} + \frac{n(x - \overline{X})^2}{n \sum x^2 - (\sum x)^2}} < y < y'$$

$$+ t_{\alpha/2}s_{est} \sqrt{1 + \frac{1}{n} + \frac{n(x - \overline{X})^2}{n \sum x^2 - (\sum x)^2}}$$

$$79.96 - (2.776)(6.48) \sqrt{1 + \frac{1}{6} + \frac{6(3 - 3.3)^2}{6(82) - (20)^2}} < y < 79.96$$

$$+ (2.776)(6.48) \sqrt{1 + \frac{1}{6} + \frac{6(3 - 3.3)^2}{6(82) - (20)^2}}$$

$$79.96 - (2.776)(6.48)(1.08) < y < 79.96 + (2.776)(6.48)(1.08)$$

$$79.96 - 19.43 < y < 79.96 + 19.43$$

$$60.53 < y < 99.39$$

Hence, one can be 95% confident that the interval $60.53 < y < 99.39$ contains the actual value of y.

Exercises

11–67. What is meant by the *explained variation*? How is it computed?

11–68. What is meant by the *unexplained variation*? How is it computed?

11–69. What is meant by the *total variation*? How is it computed?

11–70. Define *coefficient of determination*.

11–71. How is the coefficient of determination found?

11–72. Define *coefficient of nondetermination*.

11–73. How is the coefficient of nondetermination found?

For Exercises 11–74 through 11–79, find the coefficients of determination and nondetermination and explain the meaning of each.

11–74. $r = 0.81$

11–75. $r = 0.70$

11–76. $r = 0.45$

11–77. $r = 0.37$

11–78. $r = 0.15$

11–79. $r = 0.05$

11–80. Define *standard error of estimate* for regression. When can the standard error of estimate be used to construct a prediction interval about a value y'?

11–81. Compute the standard error of estimate for Exercise 11–13. The regression line equation was found in Exercise 11–41.

11–82. Compute the standard error of estimate for Exercise 11–14. The regression line equation was found in Exercise 11–42.

11–83. Compute the standard error of estimate for Exercise 11–15. The regression line equation was found in Exercise 11–43.

11–84. Compute the standard error of estimate for Exercise 11–16. The regression line equation was found in Exercise 11–44.

11–85. For the data in Exercises 11–13, 11–41, and 11–81, find the 90% prediction interval when $x = 20$ years.

11–86. For the data in Exercises 11–14, 11–42, and 11–82, find the 95% prediction interval when $x = \$1100$.

11–87. For the data in Exercises 11–15, 11–43, and 11–83, find the 90% prediction interval when $x = 4$ years.

11–88. For the data in Exercises 11–16, 11–44, and 11–84, find the 98% prediction interval when $x = 47$ years.

11–6

Multiple Regression (Optional)

Objective 8. Be familiar with the concept of multiple regression.

The previous sections explained the concepts of simple linear regression and correlation. In simple linear regression, the regression equation contains one independent variable, x, and one dependent variable, y', and is written as

$$y' = a + bx$$

where a is the y' intercept and b is the slope of the regression line.

In **multiple regression,** there are several independent variables and one dependent variable, and the equation is

$$y' = a + b_1x_1 + b_2x_2 + \cdots + b_kx_k$$

where $x_1, x_2, \ldots, x_k$ are the independent variables.

For example, suppose a nursing instructor wishes to see whether there is a relationship between a student's grade point average, age, and score on the state board nursing examination. The two independent variables are GPA (denoted by x_1) and age (denoted by x_2). The instructor will collect the data for all three variables for a sample of nursing students. Rather than conduct two separate simple regression studies, one using the GPA and state board scores and another using ages and state board scores, the instructor can conduct one study using multiple regression analysis with two independent variables—GPA and ages—and one dependent variable—state board scores.

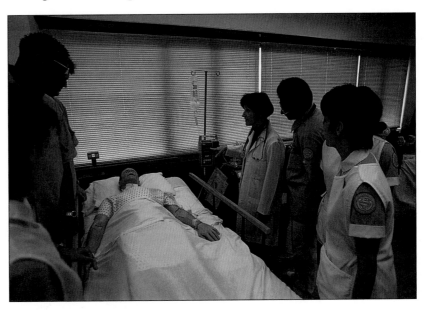

A multiple regression correlation, R, can also be computed to determine if a significant relationship exists between the independent variables and the dependent variable. Multiple regression analysis is used when a statistician thinks there are several independent variables contributing to the variation of the dependent variable. This analysis then can be used to increase the accuracy of predictions for the dependent variable over one independent variable alone.

Two other examples for multiple regression analysis are when a store manager wants to see whether the amount spent on advertising and the amount of floor space used for a display affect the amount of sales of a product, and when a sociologist wants to see whether the amount of time children spend watching television and playing video games is related to their weight. Multiple regression analysis can also be

This study indicates that as girls get older, their use of computers declines. Identify the independent and dependent variables. What type of relationship is this?

Computer Use Declines as Girls Age

By Dottie Enrico
USA TODAY

NEW YORK—First, there was math anxiety. Now a study has discovered a form of "PC avoidance" that plagues young girls with personal computers.

A nine-month poll of 1,200 families, conducted earlier this year by FIND/SVP and education consultants Gruenwald Associates, has turned up a surprising fact: Young girls actually spend more time using home PCs than young boys until they reach grades 4–7.

After that usage drops off.

One possible explanation: Software designers, who are mostly male, aren't creating computer games that appeal to girls as they grow.

The survey found that between kindergarten and 3rd grade, girls spend an average 4.8 hours each week on their home PCs compared with 3.8 hours a week for boys.

By the time they enter high school, girls are spending 5.7 hours a week at their keyboards while boys are spending 7.7 hours.

But as they grow, the study showed, girls use their PC skills for word processing and to build learning skills; boys use them mostly for games.

Source: *USA Today*, October 23, 1995. Copyright 1995. *USA TODAY.* Used with permission.

conducted using more than two independent variables, denoted by $x_1, x_2, x_3, \ldots, x_m$. Since these computations are quite complicated and for the most part would be done on a computer, this chapter will show the computations for two independent variables only.

For example, the nursing instructor wishes to see whether a student's grade point average and age are related to the student's score on the state board nursing examination. She selects five students and obtains the following data.

Student	GPA, x_1	Age, x_2	State board score, y
A	3.2	22	550
B	2.7	27	570
C	2.5	24	525
D	3.4	28	670
E	2.2	23	490

The multiple regression equation obtained from the data is

$$y' = -44.572 + 87.679x_1 + 14.519x_2$$

If a student has a GPA of 3.0 and is 25 years old, her predicted state board score can be computed by substituting these values in the equation for x_1 and x_2 respectively, as shown.

$$y' = -44.572 + 87.679(3.0) + 14.519(25)$$
$$= 581.44 \text{ or } 581$$

Hence, if a student has a GPA of 3.0 and is 25 years old, the student's predicted state board score is 581.

The Multiple Regression Equation

A multiple regression equation with two independent variables (x_1 and x_2) and one dependent variable would have the form

$$y' = a + b_1x_1 + b_2x_2$$

A multiple regression with three independent variables (x_1, x_2, and x_3) and one dependent variable would have the form

$$y' = a + b_1x_1 + b_2x_2 + b_3x_3$$

General Form of the Multiple Regression Equation

The general form of the multiple regression equation with k independent variables is

$$y' = a + b_1x_1 + b_2x_2 + \cdots + b_kx_k$$

The x's are the independent variables. The value for a is more or less an intercept, although a multiple regression equation with two independent variables constitutes a plane rather than a line. The b's are called *partial regression* coefficients. Each b represents the amount of change in y' for one unit of change in the corresponding x value when the other x values are held constant. In the example just shown, the regression equation was $y' = -44.572 + 87.679x_1 + 14.519x_2$. In this case, for each unit of change in the student's GPA, there is a change of 87.679 units in the state board score with the student's age (x_2) being held constant. And for each unit of change in x_2 (the student's age), there is a change of 14.519 units in the state board score with the GPA held constant.

Assumptions for Multiple Regression

The assumptions for multiple regression are similar to those for simple regression.

1. For any specific value of the independent variable, the values of the y variable are normally distributed. (This is called the *normality* assumption.)
2. The variances (or standard deviations) for the y variables are the same for each value of the independent variable. (This is called the *equal variance* assumption.)
3. There is a linear relationship between the dependent variable and the independent variables. (This is called the *linearity* assumption.)
4. The independent variables are not correlated. (This is called the *nonmulticolinearity* assumption.)
5. The values for the y variables are independent. (This is called the *independence* assumption.)

In multiple regression, as in simple regression, the strength of the relationship between the independent variables and the dependent variable is measured by a correlation coefficient. This **multiple correlation coefficient** is symbolized by R. The value of R can range from 0 to $+1$; R can never be negative. The closer to $+1$, the stronger the relationship; the closer to 0, the weaker the relationship. The value of R takes into account all the independent variables and can be computed using the values of the individual correlation coefficients. The formula for the multiple correlation coefficient when there are two independent variables is shown next.

Formula for the Multiple Correlation Coefficient

The formula for R is

$$R = \sqrt{\frac{r_{yx_1}^2 + r_{yx_2}^2 - 2r_{yx_1} \cdot r_{yx_2} \cdot r_{x_1x_2}}{1 - r_{x_1x_2}^2}}$$

where r_{yx_1} is the value of the correlation coefficient for the variables y and x_1; r_{yx_2} is the value of the correlation coefficient for the variables y and x_2; and $r_{x_1x_2}$ is the value of the correlation coefficient for the variables x_1 and x_2.

In this case, R is 0.989, as shown in the next example. The multiple correlation coefficient is always higher than the individual correlation coefficients. For this specific example, the multiple correlation coefficient is higher than the two individual correlation coefficients computed by using grade point average and state board scores ($r_{yx_1} = 0.845$) or age and state board scores ($r_{yx_2} = 0.791$). *Note:* $r_{x_1x_2} = 0.371$.

Example 11–15

For the data regarding state board scores, find the value of R.

Solution

The values of the correlation coefficients are

$$r_{yx_1} = 0.845$$
$$r_{yx_2} = 0.791$$
$$r_{x_1x_2} = 0.371$$

Substituting in the formula, one gets

$$R = \sqrt{\frac{r_{yx_1}^2 + r_{yx_2}^2 - 2r_{yx_1} \cdot r_{yx_2} \cdot r_{x_1x_2}}{1 - r_{x_1x_2}^2}}$$

$$R = \sqrt{\frac{(0.845)^2 + (0.791)^2 - 2(0.845)(0.791)(0.371)}{1 - 0.371^2}}$$

$$R = \sqrt{\frac{0.8437569}{0.862359}} = \sqrt{0.9784288} = 0.989$$

Hence, the correlation between a student's grade point average and age with the student's score on the nursing state board examination is 0.989. In this case, there is a strong relationship among the variables; the value of R is close to 1.00.

As with simple regression, R^2 is the *coefficient of multiple determination,* and it is the amount of variation explained by the regression model. The expression $1 - R^2$ represents the amount of unexplained variation, called the *error* or *residual variation.* Since $R = 0.989$, $R^2 = 0.978$ and $1 - R^2 = 1 - 0.978 = 0.022$.

Testing the Significance of R

An F test is used to test the significance of R. The hypotheses are

$$H_0: \rho = 0 \quad \text{and} \quad H_1: \rho \neq 0$$

where ρ represents the population correlation coefficient for multiple correlation.

F test for Significance of R

The formula for the F test is

$$F = \frac{R^2/k}{(1 - R^2)/(n - k - 1)}$$

where n is the number of data groups $(x_1, x_2, \ldots, y)$ and k is the number of independent variables.

The degrees of freedom are d.f.N $= n - k$ and d.f.D $= n - k - 1$.

Example 11–16

Test the significance of the R obtained in Example 11–15 at $\alpha = 0.05$.

Solution

$$F = \frac{R^2/k}{(1 - R^2)/(n - k - 1)}$$

$$= \frac{0.978/2}{(1 - 0.978)/(5 - 2 - 1)} = \frac{0.489}{0.011} = 44.45$$

The critical value obtained from Table H with $\alpha = 0.05$, d.f.N $= 3$, and d.f.D $= 5 - 2 - 1 = 2$ is 19.16. Hence, the decision is to reject the null hypothesis and conclude that there is a significant relationship among the student's GPA, age, and score on the nursing state board examination.

Adjusted R^2

Since the value of R^2 is dependent on n (the number of data pairs) and k (the number of variables), statisticians also calculate what is called an **adjusted R^2**, denoted by R^2_{adj}. This is based on the number of degrees of freedom.

Formula for the Adjusted R^2

The formula for the adjusted R^2 is

$$R^2_{adj} = 1 - \left[\frac{(1 - R^2)(n - 1)}{n - k - 1} \right]$$

The adjusted R^2 is smaller than R^2 and takes into account the fact that when n and k are approximately equal, the value of R may be artificially high, due to sampling error rather than a true relationship among the variables. This occurs because the chance variations of all the variables are used in conjunction with each other to derive the regression equation. Even if the individual correlation coefficients for each independent variable and the dependent variable were all zero, the multiple correlation coefficient due to sampling error could be higher than zero.

Hence, both R^2 and R^2_{adj} are usually reported in a multiple regression analysis.

Example 11–17

Calculate the adjusted R^2 for the data in the previous example. The value for R is 0.989.

Solution

$$R_{adj}^2 = 1 - \left[\frac{(1 - R^2)(n - 1)}{n - k - 1} \right]$$

$$= 1 - \left[\frac{(1 - 0.989^2)(5 - 1)}{5 - 2 - 1} \right]$$

$$= 1 - 0.043758$$

$$= 0.956$$

In this case, when the number of data pairs and the number of independent variables are accounted for, the adjusted multiple coefficient of determination is 0.956.

Exercises

11–89. Explain the similarities and differences between simple linear regression and multiple regression.

11–90. What is the general form of the multiple regression equation? What does the a represent? What do the b's represent?

11–91. Why would a researcher prefer to conduct a multiple regression study rather than separate regression studies using one independent variable and the dependent variable?

11–92. What are the assumptions for multiple regression?

11–93. How do the values of the individual correlation coefficients compare to the value of the multiple correlation coefficient?

11–94. A researcher has determined that a significant relationship exists among an employee's age (x_1), grade point average (x_2), and income (y). The multiple regression equation is $y' = -34,127 + 132x_1 + 20,805x_2$. Predict the income of a person who is 32 years old and has a GPA of 3.4.

11–95. A manufacturer found that a significant relationship exists among the number of hours an assembly line employee works per shift (x_1), the total number of items produced (x_2), and the number of defective items produced (y). The multiple regression equation is $y' = 9.6 + 2.2x_1 - 1.08x_2$. Predict the number of defective items produced by an employee who has worked nine hours and produced 24 items.

11–96. A real estate agent found that there is a significant relationship among the number of acres on a farm (x_1), the number of rooms in the farmhouse (x_2), and the selling price in thousands of dollars (y) of farms in a specific area. The regression equation is $y' = 44.9 - 0.0266x_1 + 7.56x_2$. Predict the selling price of a farm that has 371 acres and a farmhouse with six rooms.

11–97. An educator has found a significant relationship among a college graduate's IQ (x_1), score on the verbal section of the SAT exam (x_2), and income for the first year following graduation from college (y). Predict the income of a college graduate whose I.Q. is 120 and verbal SAT score is 650. The regression equation is $y' = 5000 + 97x_1 + 35x_2$.

11–98. A medical researcher found a significant relationship among a person's age (x_1), cholesterol level (x_2), sodium level of the blood (x_3), and systolic blood pressure (y). The regression equation is $y' = 97.7 + 0.691x_1 + 219x_2 - 299 x_3$. Predict the blood pressure of a person who is 35 years old and has a cholesterol level of 194 milligrams per deciliter (mg/dl) and a sodium blood level of 142 milliequivalents per liter (mEq/1).

11–99. Explain the meaning of the multiple correlation coefficient, R.

11–100. What is the range of values R can assume?

11–101. Define R^2 and R_{adj}^2.

11–102. What are the hypotheses used to test the significance of R?

11–103. What is the test used to test the significance of R?

11–104. What is the meaning of the adjusted R^2? Why is it computed?

Technology Step by Step

**MINITAB
Step by Step**

Multiple Regression (Two or More Independent Variables)

These instructions use data from the nursing examination example discussed at the beginning of Section 11–6.

Example MT11–2

1. Enter the data for the example into three columns of MINITAB. Name the columns **GPA, AGE,** and **SCORE.**
2. To determine the correlation coefficients select **Stat>Basic Statistics>Correlation.**
3. Double-click on each variable. Make sure the Display p-values box is not checked.

Dialog Box
for Correlation

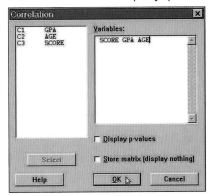

4. Click [OK].
5. Click **Stat>Regression>Regression.**
6. Double-click on SCORE, the response variable.
7. Double-click each predictor variable: GPA, then AGE.

Multiple Regression
Dialog Box

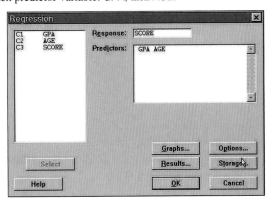

8. Click on [Storage]. Check the box for Residuals and the box for Fits. Click [OK].
9. Click [OK].

Correlations: SCORE, AGE, GPA

	SCORE	AGE
AGE	0.791	
GPA	0.845	0.371

Cell Contents: Pearson correlation

Regression Analysis: SCORE versus GPA, AGE
```
The regression equation is
SCORE = - 44.8 + 87.6 GPA + 14.5 AGE

Predictor             Coef        SE Coef          T          P
Constant            -44.81          69.25      -0.65      0.584
GPA                  87.64          15.24       5.75      0.029
AGE                 14.533          2.914       4.99      0.038
S = 14.01      R-Sq = 97.9%       R-Sq(adj) = 95.7%

Analysis of Variance
Source                  DF            SS         MS          F          P
Regression               2       18027.5     9013.7      45.93      0.021
Residual Error           2         392.5      196.3
Total                    4       18420.0

Source        DF       Seq SS
GPA            1      13145.2
AGE            1       4882.3
```

The session window shows the correlation coefficient for each pair of variables. The multiple correlation coefficient is significant at 0.021. Ninety-six percent of the variation from the mean is explained by regression. The regression equation is: SCORE = −44.8 + 87.6 GPA + 14.5 AGE.

11–7

Summary

Many relationships among variables exist in the real world. One way to determine whether a relationship exists is to use the statistical techniques known as correlation and regression. The strength and direction of the relationship is measured by the value of the correlation coefficient. It can assume values between and including −1 and +1. The closer the value of the correlation coefficient is to +1 or −1, the stronger the relationship is between the variables. A value of +1 or −1 indicates a perfect relationship. A positive relationship between two variables means that for small values of the independent variable, the values of the dependent variable will be small, and that for large values of the independent variable, the values of the dependent variable will be large. A negative relationship between two variables means that for small values of the independent variable, the values of the dependent variable will be large, and that for large values of the independent variable, the values of the dependent variable will be small.

Relationships can be linear or curvilinear. To determine the shape, one draws a scatter plot of the variables. If the relationship is linear, the data can be approximated by a straight line, called the *regression line* or the *line of best fit.* The closer the value of *r* is to +1 or −1, the closer the points will fit the line.

In addition, relationships can be multiple. That is, there can be two or more independent variables and one dependent variable. A coefficient of correlation and a regression equation can be found for multiple relationships, just as they can be found for simple relationships.

The coefficient of determination is a better indicator of the strength of a relationship than the correlation coefficient. It is better because it identifies the percentage of variation of the dependent variable that is directly attributable to the variation of the independent variable. The coefficient of determination is obtained by squaring the correlation coefficient and converting the result to a percentage.

Another statistic used in correlation and regression is the standard error of estimate, which is an estimate of the standard deviation of the *y* values about the predicted *y'* values. The standard error of estimate can be used to construct a prediction interval about a specific value point estimate *y'* of the mean of the *y* values for a given value of *x*.

"At this point in my report, I'll ask all of you to follow me to the conference room directly below us!"

Source: Cartoon by Bradford Veley, Marquette, Michigan. Reprinted with permission.

Finally, remember that a significant relationship between two variables does not necessarily mean that one variable is a direct cause of the other variable. In some cases this is true, but other possibilities that should be considered include a complex relationship involving other (perhaps unknown) variables, a third variable interacting with both variables, or a relationship due solely to chance.

Important Terms

adjusted R^2 501

coefficient of determination   492

correlation   463

correlation coefficient   468

dependent variable   465

independent variable   465

multiple correlation coefficient   499

multiple regression   497

multiple relationship   464

negative relationship   464

Pearson product moment correlation coefficient   468

population correlation coefficient   471

positive relationship   464

prediction interval   495

regression   464

regression line   478

scatter plot   465

simple relationship   464

standard error of estimate   493

Important Formulas

Formula for the correlation coefficient:

$$r = \frac{n(\Sigma\, xy) - (\Sigma\, x)(\Sigma\, y)}{\sqrt{[n(\Sigma\, x^2) - (\Sigma\, x)^2][n(\Sigma\, y^2) - (\Sigma\, y)^2]}}$$

Formula for the t test for the correlation coefficient:

$$t = r\,\sqrt{\frac{n-2}{1-r^2}} \qquad d.f. = n - 2$$

The regression line equation: $y' = a + bx$, where

$$a = \frac{(\Sigma\, y)(\Sigma\, x^2) - (\Sigma\, x)(\Sigma\, xy)}{n(\Sigma\, x^2) - (\Sigma\, x)^2}$$

$$b = \frac{n(\Sigma\, xy) - (\Sigma\, x)(\Sigma\, y)}{n(\Sigma\, x^2) - (\Sigma\, x)^2}$$

Formula for the standard error of estimate:

$$s_{est} = \sqrt{\frac{\Sigma\, (y - y')^2}{n - 2}}$$

or

$$s_{est} = \sqrt{\frac{\Sigma\, y^2 - a\,\Sigma\, y - b\,\Sigma\, xy}{n - 2}}$$

Formula for the prediction interval for a value y':

$$y' - t_{\alpha/2}s_{est}\sqrt{1 + \frac{1}{n} + \frac{n(x - \bar{X})^2}{n\,\Sigma\, x^2 - (\Sigma\, x)^2}} < y$$

$$< y' + t_{\alpha/2}s_{est}\sqrt{1 + \frac{1}{n} + \frac{n(x - \bar{X})^2}{n\,\Sigma\, x^2 - (\Sigma\, x)^2}}$$

d.f. $= n - 2$

Formula for the multiple correlation coefficient:

$$R = \sqrt{\frac{r_{yx_1}^2 + r_{yx_2}^2 - 2r_{yx_1} \cdot r_{yx_2} \cdot r_{x_1x_2}}{1 - r_{x_1x_2}^2}}$$

Formula for the F test for the multiple correlation coefficient:

$$F = \frac{R^2/k}{(1 - R^2)/(n - k - 1)}$$

with d.f.N $= n - k$ and d.f.D $= n - k - 1$.

Formula for the adjusted R^2:

$$R_{adj}^2 = 1 - \left[\frac{(1 - R^2)(n - 1)}{n - k - 1}\right]$$

Review Exercises

For Exercises 11–105 through 11–111, do a complete regression analysis by performing the following steps.

a. Draw the scatter plot.

b. Compute the value of the correlation coefficient.

c. Test the significance of the correlation coefficient at $\alpha = 0.01$, using Table I.

d. Determine the regression line equation.

e. Plot the regression line on the scatter plot.

f. Predict y' for a specific value of x.

11–105. A study is done to see whether there is a relationship between a student's grade point average and the number of hours the student watches television each week. The data are shown here. If there is a significant relationship, predict the GPA of a student who watches television 9 hours per week.

Hours, x	6	10	8	15	5	6	12
GPA, y	2.4	4	3.2	1.6	3.7	3	3.5

11–106. A researcher wishes to determine if there is a relationship between the number of day care centers and the number of group day care homes for counties in Pennsylvania. If there is a significant relationship predict the number of group care homes a county has if the county has 20 day care centers.

Day care centers, x	5	28	37	16	16	48
Group day care homes, y	2	7	4	10	6	9

Source: State Department of Public Welfare

11–107. A study is done to see whether there is a relationship between a mother's age and the number of children she has. The data are shown here. If there is a

significant relationship, predict the number of children of a mother whose age is 34.

Mother's age, x	18	22	29	20	27	32	33	36
No. of children, y	2	1	3	1	2	4	3	5

11–108. A study is conducted to determine the relationship between a driver's age and the number of accidents he or she has over a one-year period. The data are shown here. (This information will be used for Exercise 11–112.) If there is a significant relationship, predict the number of accidents of a driver who is 28.

Driver's age, x	16	24	18	17	23	27	32
No. of accidents, y	3	2	5	2	0	1	1

11–109. A researcher desires to know whether the typing speed of a secretary (in words per minute) is related to the time (in hours) that it takes the secretary to learn to use a new word-processing program. The data follow.

Speed, x	Time, y
48	7
74	4
52	8
79	3.5
83	2
56	6
85	2.3
63	5
88	2.1
74	4.5
90	1.9
92	1.5

If there is a significant relationship, predict the time it will take the average secretary who has a typing speed of 72 words per minute to learn the word-processing program. (This information will be used for Exercises 11–113 and 11–115.)

11–110. A study was conducted with vegetarians to see whether the number of grams of protein each ate per day was related to diastolic blood pressure. The data are given here. (This information will be used for Exercises 11–114 and 11–116.) If there is a significant relationship, predict the diastolic pressure of a vegetarian who consumes 8 grams of protein per day.

Grams, x	4	6.5	5	5.5	8	10	9	8.2	10.5
Pressure, y	73	79	83	82	84	92	88	86	95

11–111. A study was conducted to determine whether there is a relationship between strength and speed. A sample of 20-year-old men was selected. Each was asked to do push-ups and to run a specific course. The number of push-ups and time it took to run the course (in seconds) are given in the table. If there is a significant relationship, predict the running time for a person who can do 18 push-ups.

Push-ups, x	5	8	10	10	11	13	15	18	23
Time, y	61	65	43	56	62	73	48	49	50

11–112. For Exercise 11–108, find the standard error of estimate.

11–113. For Exercise 11–109, find the standard error of estimate.

11–114. For Exercise 11–110, find the standard error of estimate.

11–115. For Exercise 11–109, find the 90% prediction interval for time when the speed is 72 words per minute.

11–116. For Exercise 11–110, find the 95% prediction interval for pressure when the number of grams is 8.

11–117. (Opt.) A study found a significant relationship among a person's years of experience on a particular job (x_1), the number of work days missed per month (x_2), and the person's age (y). The regression equation is $y' = 12.8 + 2.09x_1 + 0.423x_2$. Predict a person's age if he or she has been employed for four years and has missed two work days a month.

11–118. (Opt.) Find R when $r_{yx_1} = 0.681$ and $r_{yx_2} = 0.872$ and $r_{x_1x_2} = 0.746$.

11–119. (Opt.) Find R^2_{adj} when $R = 0.873$, $n = 10$, and $k = 3$.

Statistics Today

Do Dust Storms Affect Respiratory Health? Revisited

The researchers correlated the dust pollutant levels in the atmosphere and the number of daily emergency room visits for several respiratory disorders, such as bronchitis, sinusitis, asthma, and pneumonia. Using the Pearson correlation coefficient, they found overall a significant but low correlation, $r = 0.13$, for bronchitis visits only. However, they found a much higher correlation value for sinusitis, P-value $= 0.08$, when pollutant levels exceeded maximums set by the Environmental Protection Agency (EPA). In addition, they found statistically significant correlation coefficients, $r = 0.94$, for sinusitis visits and $r = 0.74$ for upper-respiratory-tract infection visits two days after the dust pollutants exceeded the maximum levels set by the EPA.

Data Analysis

The Data Bank is found in Appendix D, or on the World Wide Web by following links from www.mhhe.com/math/stat/bluman/

1. From the Data Bank, choose two variables that might be related: e.g., IQ and educational level; age and cholesterol level; exercise and weight; or weight and systolic pressure. Do a complete correlation and regression analysis by performing the following steps. Select a random sample of at least 10 subjects.

  1. Draw a scatter plot.
  2. Compute the correlation coefficient.
  3. Test the hypothesis H_0: $\rho = 0$.
  4. Find the regression line equation.
  5. Summarize the results.

2. Repeat exercise 1 using Data Set XI in Appendix D. Let $x =$ the number of miles and $y =$ the number of vehicles operated. If you are using a TI-83 calculator or a computer program, use all the data values; otherwise, select a sample of 10 values.

3. Repeat exercise 1 using the data from Data Set XI in Appendix D and let $x =$ the number of miles and $y =$ the number of stations. If you are using a TI-83 calculator or a computer program, use all the data values; otherwise, select a sample of 10 values.

Quiz

Determine whether each statement is true or false. If the statement is false, explain why.

1. A negative relationship between two variables means that for the most part, as the x variable increases, the y variable increases.

2. A correlation coefficient of -1 implies a perfect linear relationship between the variables.

3. Even if the correlation coefficient is high or low, it may not be significant.

4. When the correlation coefficient is significant, one can assume x causes y.

5. It is not possible to have a significant correlation by chance alone.

6. In multiple regression, there are several dependent variables and one independent variable.

Select the best answer.

7. The strength of the relationship between two variables is determined by the value of
   a. r
   b. a
   c. x
   d. s_{est}

8. To test the significance of r, a(n) _____ test is used.
   a. t
   b. F
   c. χ^2
   d. None of the above

9. The test of significance for r has _____ degrees of freedom.
   a. 1
   b. n
   c. $n - 1$
   d. $n - 2$

10. The equation of the regression line used in statistics is
    a. $x = a + by$
    b. $y = bx + a$
    c. $y' = a + bx$
    d. $x = ay + b$

11. The coefficient of determination is _____.
    a. r
    b. r^2
    c. a
    d. b

Complete the following statements with the best answer.

12. A statistical graph of two variables is called a _____.

13. The x variable is called the _____ variable.

14. The range of r is from _____ to _____.

15. The sign of r and _____ will always be the same.

16. The regression line is called the _____.

17. If all the points fall on a straight line, the value of r will be _____ or _____.

For Problems 18−21, do a complete regression analysis.
   a. Draw the scatter plot.
   b. Compute the value of the correlation coefficient.
   c. Test the significance of the correlation coefficient at $\alpha = 0.05$.
   d. Determine the regression line equation.
   e. Plot the regression line on the scatter plot.
   f. Predict y' for a specific value of x

18. The relationship between a father's age and the number of children he has is studied. The data are shown here. If there is a significant relationship, predict the number of children of a father whose age is 35.

Father's age, x	19	21	27	20	25	31	32	38
No. of children, y	1	2	3	0	1	4	3	4

19. A study is conducted to determine the relationship between a driver's age and the number of accidents he or she has over a one-year period. The data are shown here. If there is a significant relationship, predict the number of accidents of a driver who is 64.

Driver's age, x	63	65	60	62	66	67	59
No. of accidents, y	2	3	1	0	3	1	4

20. A researcher desires to know if the age of a child is related to the number of cavities he or she has. The data are shown here. If there is a significant relationship, predict the number of cavities for a child of 11.

Age of child, x	6	8	9	10	12	14
No. of cavities, y	2	1	3	4	6	5

21. A study is conducted with a group of dieters to see if the number of grams of fat each consumes per day is related to cholesterol level. The data are shown here. If there is a significant relationship, predict the cholesterol level of a dieter who consumes 8.5 grams of fat per day.

Fat grams, x	6.8	5.5	8.2	10	8.6
Cholesterol level, y	183	201	193	283	222
Fat grams, x	9.1	8.6	10.4		
Cholesterol level, y	250	190	218		

22. For Problem 20, find the standard error of the estimate.

23. For Problem 21, find the standard error of the estimate.

24. For Problem 20, find the 90% prediction interval of the number of cavities for a 7-year-old.

25. For Problem 21, find the 95% prediction interval of the cholesterol level of a person who consumes 10 grams of fat.

26. (Opt.) A study was conducted and a significant relationship was found among the number of hours a teenager watches television per day (x_1), the number of hours the teenager talks on the telephone per day (x_2), and the teenager's weight (y). The regression equation

is $y' = 98.7 + 3.82x_1 + 6.51x_2$. Predict a teenager's weight if she averages 3 hours of TV and 1.5 hours on the phone per day.

27. (Opt.) Find R when $r_{yx_1} = 0.561$ and $r_{yx_2} = 0.714$ and $r_{x_1x_2} = 0.625$.

28. (Opt.) Find R^2_{adj} when $R = 0.774$, $n = 8$, and $k = 2$.

Critical Thinking Challenges

When the points in a scatter plot show a curvilinear trend rather than a linear trend, statisticians have methods of fitting curves rather than straight lines to the data, thus obtaining a better fit and a better prediction model. One type of curve that can be used is the logarithmic regression curve. The data shown are the number of items of a new product sold over a period of 15 months at a certain store. Notice that sales rise during the beginning months and then level off later on.

Month, x	1	3	6	8	10	12	15
No. of items sold, y	10	12	15	19	20	21	21

1. Draw the scatter plot for the data.

2. Find the equation of the regression line.

3. Describe how the line fits the data.

4. Using the log key on your calculator, transform the x values into log x values.

5. Using the log x values instead of the x values, find the equation of the a and b for the regression line.

6. Next, plot the curve $y = a + b \log x$ on the graph.

7. Compare the line $y = a + bx$ with the curve $y = a + b \log x$ and decide which one fits the data better.

8. Compute r using the x and y values and then compute r using the log x and y values. Which is higher?

9. In your opinion, which (the line or the logarithmic curve) would be a better predictor for the data? Why?

Data Projects

Where appropriate, use MINITAB, the TI-83, or a computer program of your choice to complete the following exercises.

1. Select two variables that might be related, such as the age of a person and the number of cigarettes the person smokes, or the number of credits a student has and the number of hours the student watches television. Sample at least 10 people.
   a. Write a brief statement as to the purpose of the study.
   b. Define the population.
   c. State how the sample was selected.
   d. Show the raw data.
   e. Draw a scatter plot for the data values.
   f. Write a statement analyzing the scatter plot.
   g. Compute the value of the correlation coefficient.
   h. Test the significance of r. (State the hypotheses, select α, find the critical values, make the decision, and analyze the results.)
   i. Find the equation of the regression line and draw it on the scatter plot. (*Note:* Even if r is not significant, complete this step.)
   j. Summarize the overall results.

2. For the data in exercise 1, use MINITAB to answer the following.
   a. Does a linear correlation exist between x and y?
   b. If so, find the regression equation.
   c. Explain how good a model the regression equation is by finding the coefficient of determination and coefficient of correlation and interpreting the strength of these values.
   d. Find the prediction interval for y. Use the α value that you selected in exercise 1.

You may use the following websites to obtain raw data:

   http://www.mhhe.com/math/stat/bluman/
   http://lib.stat.cmu.edu/DASL
   http://www.oecd.org/statlist.htm
   http://www.statcan.ca/english/

chapter

12 Other Chi-Square Tests

Objectives

After completing this chapter, you should be able to

1. Test a distribution for goodness of fit using chi-square.

2. Test two variables for independence using chi-square.

3. Test proportions for homogeneity using chi-square.

Statistics Today

Statistics and Heredity

An Austrian monk, Gregor Mendel (1822–1884), studied genetics, and his principles are the foundation for modern genetics. Mendel used his spare time to grow a variety of peas at the monastery. One of his many experiments involved crossbreeding peas that had smooth yellow seeds with peas that had wrinkled green seeds. He noticed that the results occurred with regularity. That is, some of the offspring had smooth yellow seeds, some had smooth green seeds, some had wrinkled yellow seeds, and some had wrinkled green seeds. Furthermore, after several experiments, the percentages of each type seemed to remain approximately the same. Mendel formulated his theory based on the assumption of dominant and recessive traits and tried to predict the results. He then crossbred his peas and examined 556 seeds over the next generation.

Finally, he compared the actual results with the theoretical results to see if his theory was correct. In order to do this, he used a "simple" chi-square test, which is explained in this chapter.

Source: J. Hodges, Jr., D. Krech, and R. Crutchfield, *Stat Lab, An Empirical Introduction to Statistics* (New York: McGraw-Hill, 1975), pp. 228–229. Used with permission.

12–1

Introduction

The chi-square distribution was used in Chapters 8 and 9 to find a confidence interval for a variance or standard deviation and to test a hypothesis about a single variance or standard deviation.

It can also be used for tests concerning *frequency distributions,* such as "If a sample of buyers is given a choice of automobile colors, will each color be selected with the same frequency?" The chi-square distribution can be used to test the *independence* of two variables. For example, "Are senators' opinions on gun control independent of party affiliations?" That is, do the Republicans feel one way and the Democrats feel differently, or do they have the same opinion?

Finally, the chi-square distribution can be used to test the *homogeneity of proportions.* For example, is the proportion of high school seniors who attend college immediately after graduating the same for the northern, southern, eastern, and western parts of the United States?

This chapter explains the chi-square distribution and its applications. In addition to the applications mentioned here, the chi-square has many other uses in statistics.

12-2

Test for Goodness of Fit

Objective 1. Test a distribution for goodness of fit using chi-square.

Historical Note

Karl Pearson (1857–1936) first used the chi-square distribution as a goodness-of-fit test for data. He developed many types of descriptive graphs and gave them unusual names such as stigmograms, topograms, stereograms, and radiograms.

In addition to being used to test a single variance, the chi-square statistic can be used to see whether a frequency distribution fits a specific pattern. For example, in order to meet customer demands, a manufacturer of running shoes may wish to see whether buyers show a preference for a specific style. A traffic engineer may wish to see whether accidents occur more often on some days than on others, so that he can increase police patrols accordingly. An emergency service may want to see whether it receives more calls at certain times of the day than others, so that it can provide adequate staffing.

When one is testing to see whether a frequency distribution fits a specific pattern, the chi-square **goodness-of-fit test** is used. For example, suppose a market analyst wished to see whether consumers have any preference among five flavors of a new fruit soda. A sample of 100 people provided the following data:

Cherry	Strawberry	Orange	Lime	Grape
32	28	16	14	10

If there were no preference, one would expect that each flavor would be selected with equal frequency. In this case, the equal frequency is 100/5 = 20. That is, *approximately* 20 people would select each flavor.

Since the frequencies for each flavor were obtained from a sample, these actual frequencies are called the **observed frequencies.** The frequencies obtained by calculation (as if there were no preference) are called the **expected frequencies.** A completed table for the test follows:

Frequency	Cherry	Strawberry	Orange	Lime	Grape
Observed	32	28	16	14	10
Expected	20	20	20	20	20

The observed frequencies will almost always differ from the expected frequencies due to sampling error; i.e., the values differ from sample to sample. But the question is: Are these differences significant (a preference exists), or are they due to chance? The chi-square goodness-of-fit test will enable the researcher to determine the answer.

Before computing the test value, one must state the hypotheses. The null hypothesis should be a statement indicating that there is no difference or no change. For this example, the hypotheses are as follows:

H_0: Consumers show no preference for flavors of the fruit soda.

H_1: Consumers show a preference.

In the goodness-of-fit test, the degrees of freedom are equal to the number of categories minus 1. For this example, there are five categories (cherry, strawberry, orange, lime, and grape); hence, the degrees of freedom are 5 − 1 = 4. This is because the number of subjects in each of the first four categories is free to vary. But in order for the sum to be 100—the total number of subjects—the number of subjects in the last category is fixed.

Formula for the Chi-Square Goodness-of-Fit Test

$$\chi^2 = \Sigma \frac{(O - E)^2}{E}$$

with degrees of freedom equal to the number of categories minus 1, and where

O = observed frequency
E = expected frequency

Two assumptions are needed for the goodness-of-fit test. These assumptions are given next.

Assumptions for the Chi-Square Goodness-of-Fit Test

1. The data are obtained from a random sample.
2. The expected frequency for each category must be 5 or more.

This test is a right-tailed test, since when the $(O - E)$ values are squared, the answer will be positive or zero. This formula is explained in the next example.

Example 12–1

Is there enough evidence to reject the claim that there is no preference in the selection of fruit soda flavors, using the data shown previously? Let $\alpha = 0.05$.

Solution

STEP 1   State the hypotheses and identify the claim.

H_0: Consumers show no preference for flavors (claim).
H_1: Consumers show a preference.

STEP 2   Find the critical value. The degrees of freedom are $5 - 1 = 4$, and $\alpha = 0.05$. Hence, the critical value from Table G in Appendix C is 9.488.

STEP 3   Compute the test value by subtracting the expected value from the corresponding observed value, squaring the result and dividing by the expected value, and finding the sum. The expected value for each category is 20, as shown previously.

$$\chi^2 = \Sigma \frac{(O - E)^2}{E}$$

$$= \frac{(32 - 20)^2}{20} + \frac{(28 - 20)^2}{20} + \frac{(16 - 20)^2}{20} + \frac{(14 - 20)^2}{20} + \frac{(10 - 20)^2}{20}$$

$$= 18.0$$

STEP 4   Make the decision. The decision is to reject the null hypothesis, since $18.0 > 9.488$, as shown in Figure 12–1.

Figure 12–1

Critical and Test Values
for Example 12–1

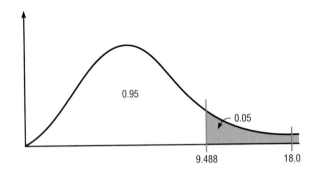

0.95

0.05

9.488   18.0

STEP 5 Summarize the results. There is enough evidence to reject the claim that consumers show no preference for the flavors.

To get some idea of why this test is called the goodness-of-fit test, examine graphs of the observed values and expected values. See Figure 12–2. From the graphs, one can see whether the observed values and expected values are close together or far apart.

Figure 12–2

Graphs of the Observed
and Expected Values for
Soda Flavors

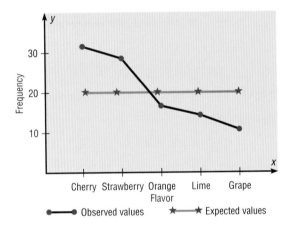

When the observed values and expected values are close together, the chi-square test value will be small. Then, the decision will be not to reject the null hypothesis—hence, there is "a good fit." See Figure 12–3(a). When the observed values and the expected values are far apart, the chi-square test value will be large. Then, the null hypothesis will be rejected—hence, there is "not a good fit." See Figure 12–3(b).

Figure 12–3

Results of the
Goodness-of-Fit Test

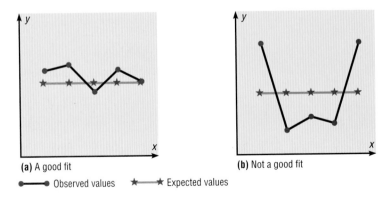

(a) A good fit   (b) Not a good fit

The steps for the chi-square goodness-of-fit test are summarized in the Procedure Table.

Procedure Table

The Chi-Square Goodness-of-Fit Test

STEP 1 State the hypotheses and identify the claim.

STEP 2 Find the critical value. The test is always right-tailed.

STEP 3 Compute the test value.

Find the sum of the $\dfrac{(O - E)^2}{E}$ values.

STEP 4 Make the decision.

STEP 5 Summarize the results.

When there is perfect agreement between the observed and the expected values, $\chi^2 = 0$. Also, χ^2 can never be negative. Finally, the test is right-tailed because "H_0: Good fit" and "H_1: Not a good fit" means that χ^2 will be small in the first case and χ^2 will be large in the second case.

Example 12–2

The Russel Reynold Association surveyed retired senior executives who had returned to work. They found that after returning to work: 38% were employed by another organization, 32% were self-employed, 23% were either freelancing or consulting, and 7% had formed their own companies. In order to see if these percentages are consistent with those of Allegheny County residents, a local researcher surveyed 300 retired executives who had returned to work and found that 122 were working for another company, 85 were self-employed, 76 were either freelancing or consulting, and 17 had formed their own companies. At $\alpha = 0.10$, test the claim that the percentages are the same for those people in Allegheny County.

Source: Michael L. Shook and Robert D. Shook, *The Book of Odds* (New York: Penguin Putnam Inc., 1991).

Solution

STEP 1 State the hypotheses and identify the claim.

H_0: The retired executives who returned to work are distributed as follows: 38% are employed by another organization, 32% are self-employed, 23% are either freelancing or consulting, and 7% have formed their own companies (claim).

H_1: The distribution is not the same as stated in the null hypothesis.

STEP 2 Find the critical value. Since $\alpha = 0.10$ and the degrees of freedom are $4 - 1 = 3$, the critical value is 6.251.

STEP 3 Compute the test value. The expected values are computed as follows:

$$0.38 \times 300 = 114 \qquad 0.23 \times 300 = 69$$
$$0.32 \times 300 = 96 \qquad 0.07 \times 300 = 21$$

$$\chi^2 = \Sigma \frac{(O - E)^2}{E}$$

$$= \frac{(122 - 114)^2}{114} + \frac{(85 - 96)^2}{96} + \frac{(76 - 69)^2}{69} + \frac{(17 - 21)^2}{21}$$

$$= 3.2939$$

STEP 4   Make the decision. Since $3.2939 < 6.251$, the decision is not to reject the null hypothesis. See Figure 12–4.

Figure 12–4

Critical and Test Values for Example 12–2

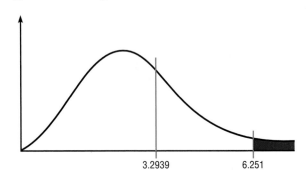

3.2939   6.251

STEP 5   Summarize the results. There is not enough evidence to reject the claim. It can be concluded that the percentages are not significantly different than those given in the null hypothesis.

Example 12–3

The advisor of an ecology club at a large college believes that the group consists of 10% freshmen, 20% sophomores, 40% juniors, and 30% seniors. The membership for the club this year consisted of 14 freshmen, 19 sophomores, 51 juniors, and 16 seniors. At $\alpha = 0.10$, test the advisor's conjecture.

Solution

STEP 1   State the hypotheses and identify the claim.

H_0: The club consists of 10% freshmen, 20% sophomores, 40% juniors, and 30% seniors (claim).

H_1: The distribution is not the same as stated in the null hypothesis.

STEP 2   Find the critical value. Since $\alpha = 0.10$ and the degrees of freedom are $4 - 1 = 3$, the critical value is 6.251.

STEP 3   Compute the test value. The expected values are computed as follows:

$$0.10 \times 100 = 10 \qquad 0.40 \times 100 = 40$$
$$0.20 \times 100 = 20 \qquad 0.30 \times 100 = 30$$

$$\chi^2 = \Sigma \frac{(O - E)^2}{E}$$

$$= \frac{(14 - 10)^2}{10} + \frac{(19 - 20)^2}{20} + \frac{(51 - 40)^2}{40} + \frac{(16 - 30)^2}{30}$$

$$= 11.208$$

STEP 4   Reject the null hypothesis, since $11.208 > 6.251$, as shown in Figure 12–5.

Figure 12–5

Critical and Test Values
for Example 12–3

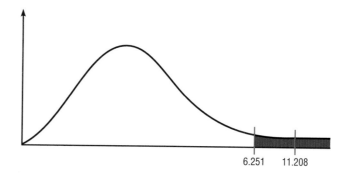

6.251   11.208

STEP 5   Summarize the results. There is enough evidence to reject the advisor's
claim.

The *P*-value method of hypothesis testing can also be used for the chi-square tests
explained in this chapter. The *P*-values for chi-square are found in Table G in Appendix
C. The method used to find the *P*-value for a chi-square test value is the same as the
method shown in Section 9–6. The *P*-value for $\chi^2 = 3.2939$ with d.f. = 3 (for the data
in Example 12–2) is greater than 0.10 since 6.251 is the value in Table G for $\alpha = 0.10$.
(The *P*-value obtained from a calculator is 0.348.) Hence *P*-value > 0.10. The decision
is do not reject the null hypothesis, which is consistent with the decision made in Example 12–2 using the traditional method of hypothesis testing.

For use of the chi-square goodness-of-fit test, statisticians have determined that the
expected frequencies should be at least 5, as stated in the assumptions. The reasoning is
as follows: The chi-square distribution is continuous, whereas the goodness-of-fit test is
discrete. However, the continuous distribution is a good approximation and can be used
when the expected value for each class is at least 5. If an expected frequency of a class
is less than 5, then that class can be combined with another class so that the expected
frequency is 5 or more.

**Test of Normality
(Optional)**

The chi-square goodness-of-fit test can be used to test a variable to see if it is normally
distributed. The null hypotheses are

H_0:   The variable is normally distributed.

H_1:   The variable is not normally distributed.

The procedure is somewhat complicated. It involves finding the expected frequencies for each class of a frequency distribution using the standard normal distribution.
Then the actual frequencies (i.e., observed frequencies) are compared to the expected
frequencies using the chi-square goodness-of-fit test. If the observed frequencies are
close in value to the expected frequencies, the chi-square test value will be small, and
the null hypothesis cannot be rejected. In this case, it can be concluded that the variable
is approximately normally distributed.

On the other hand, if there is a large difference between the observed frequencies
and the expected frequencies, the chi-square test value will be larger, and the null hypothesis can be rejected. In this case, it can be concluded that the variable is not normally
distributed. Example 12–4 illustrates the procedure for the chi-square test of normality.
In order to find the areas in the examples, you might want to review Section 7–3.

The next example shows how to do the calculations.

Example 12–4

Use chi-square to determine if the variable shown in the frequency distribution is normally distributed. Use $\alpha = 0.05$.

Boundaries	Frequency
89.5–104.5	24
104.5–119.5	62
119.5–134.5	72
134.5–149.5	26
149.5–164.5	12
164.5–179.5	4
	200

Solution

H_0: The variable is normally distributed.

H_1: The variable is not normally distributed.

First, find the mean and standard deviation of the variable. Then find the area under the standard normal distribution using z values and Table E for each class. Find the expected frequencies for each class by multiplying the area by 200. Finally, find the chi-square test value by using the formula $\chi^2 = \Sigma \dfrac{(O - E)^2}{E}$.

Boundaries	f	X_m	$f \cdot X_m$	$f \cdot X_m^2$
89.5–104.5	24	97	2328	225,816
104.5–119.5	62	112	6944	777,728
119.5–134.5	72	127	9144	1,161,288
134.5–149.5	26	142	3692	524,264
149.5–164.5	12	157	1884	295,788
164.5–179.5	4	172	688	118,336
	200		24,680	3,103,220

$$\overline{X} = \frac{24,680}{200} = 123.4$$

$$s = \sqrt{\frac{3,103,220 - 24,680^2/200}{199}} = \sqrt{290} = 17.03$$

The area to the left of $x = 104.5$ is found as follows:

$$z = \frac{104.5 - 123.4}{17.03} = -1.11$$

The area for $z < -1.11$ is $0.5000 - 0.3665 = 0.1335$.

The area between 104.5 and 119.5 is found as follows:

$$z = \frac{119.5 - 123.4}{17.03} = -0.23$$

The area for $-1.11 < z < -0.23$ is $0.3665 - 0.0910 = 0.2755$.

The area between 119.5 and 134.5 is found as follows:

$$z = \frac{134.5 - 123.4}{17.03} = 0.65$$

The area for $-0.23 < z < 0.65$ is $0.2422 + 0.0910 = 0.3332$.

The area between 134.5 and 149.5 is found as follows:

$$z = \frac{149.5 - 123.4}{17.03} = 1.53$$

The area for $0.65 < z < 1.53$ is $0.4370 - 0.2422 = 0.1948$.

The area between 149.5 and 164.5 is found as follows:

$$z - \frac{164.5 - 123.4}{17.03} = 2.41$$

The area for $1.53 < z < 2.41$ is $0.4920 - 0.4370 = 0.0550$.

The area to the right of $x = 164.5$ is found as follows:

$$z = \frac{164.5 - 123.4}{17.03} = 2.41$$

$0.5000 - 0.4920 = 0.0080$.

The expected frequencies are found by

$$0.1335 \cdot 200 = 26.7$$
$$0.2755 \cdot 200 = 55.1$$
$$0.3332 \cdot 200 = 66.64$$
$$0.1948 \cdot 200 = 38.96$$
$$0.0550 \cdot 200 = 11.0$$
$$0.0080 \cdot 200 = 1.6$$

Note: Since the expected frequency for the last category is less than five, it can be combined with the previous category.

The χ^2 is found by

O	24	62	72	26	16
E	26.7	55.1	66.64	38.96	12.6

$$\chi^2 = \frac{(24 - 26.7)^2}{26.7} + \frac{(62 - 55.1)^2}{55.1} + \frac{(72 - 66.64)^2}{66.64} + \frac{(26 - 38.96)^2}{38.96}$$
$$+ \frac{(16 - 12.6)^2}{12.6}$$
$$= 6.797$$

The C.V. with d.f. = 4 and $\alpha = 0.05$ is 9.488, so the null hypothesis is not rejected. Hence, the distribution can be considered approximately normal.

Exercises

12–1. How does the goodness-of-fit test differ from the chi-square variance test?

12–2. How are the degrees of freedom computed for the goodness-of-fit test?

12–3. How are the expected values computed for the goodness-of-fit test?

12–4. When the expected frequencies are less than 5 for a specific class, what should be done so that one can use the goodness-of-fit test?

For Exercises 12–5 through 12–19, perform the following steps
 a. State the hypotheses and identify the claim.
 b. Find the critical value.
 c. Compute the test value.
 d. Make the decision.
 e. Summarize the results.

Use the traditional method of hypothesis testing unless otherwise specified.

12–5. A staff member of an emergency medical service wishes to determine whether the number of accidents is equally distributed during the week. A week was selected at random, and the following data were obtained. Is there evidence to reject the hypothesis that the number of accidents is equally distributed throughout the week, at $\alpha = 0.05$?

Day	Mon.	Tues.	Wed.	Thurs.	Fri.	Sat.	Sun.
No. of accidents	28	32	15	14	38	43	19

12–6. A children's raincoat manufacturer wants to know whether customers prefer any specific color over other colors in children's raincoats. He selects a random sample of 50 raincoats sold and notes the colors. The data are shown here. At $\alpha = 0.10$ is there a color preference for the raincoats?

Color	Yellow	Red	Green	Blue
No. sold	17	13	8	12

12–7. The chef at the Slippery Rock Country Club wishes to see whether there is any preference in the flavors of sherbet served for dessert at the club. A random sample of sales is selected, and the data are shown here. At $\alpha = 0.01$ are the flavors selected with equal frequency?

Flavor	Lemon	Orange	Raspberry	Lime
No. sold	12	24	19	9

12–8. A bank manager wishes to see whether there is any preference in the times that customers use the bank. Six hours are selected, and the number of customers visiting the bank during each hour are as shown here. At $\alpha = 0.05$, do the customers show a preference for specific times?

Time	10:00	11:00	12:00	1:00	2:00	3:00
No. of customers	26	33	42	36	24	19

12–9. The American Red Cross reports that 42% of Americans have type O blood, 44% have type A blood, 10% have type B blood, and 4% have type AB blood. A county medical examiner hypothesizes the distribution of blood types is the same in his county as it is nationally. A random sample of 200 people is selected, and the following data are tallied. At $\alpha = 0.10$, test the examiner's hypothesis.

Type	A	O	B	AB
Frequency	58	65	55	22

Source: Robert D. Shook and Michael L. Shook, *The Book of Odds* (New York: Penguin Putnam, Inc., 1991), p. 161.

12–10. The chair of the history department of a college hypothesizes that the final grades are distributed as 40% A's, 30% B's, 20% C's, 5% D's, and 5% F's. At the end of the semester, the following numbers of grades were earned. For $\alpha = 0.05$, is the grade distribution for the department different from that expected?

Grade	A	B	C	D	F
Number	45	52	39	8	6

12–11. *USA Today* reported that 21% of loans granted by credit unions were for home mortgages, 39% were for automobile purchases, 20% were for credit card and other unsecured loans, 12% were for real estate other than home loans, and 8% were for other miscellaneous needs. In order to see if her credit union customers had similar needs, a manager surveyed a random sample of 100 loans and found that 25 were for home mortgages, 44 for automobile purchases, 19 for credit card and unsecured loans, 8 for real estate other than home loans, and 4 for miscellaneous needs. At $\alpha = 0.05$, is the distribution the same as reported in the newspaper?

Source: *USA Today*, July 21, 1995.

12–12. A *USA Today*/CNN/Gallup poll shows that 74% of respondents felt that other motorists were driving more aggressively than they did five years ago, 23% felt that other motorists were driving the same way they did five years ago, and 3% felt other motorists were driving less aggressively than they were driving five years ago. A sample survey of 180 senior drivers found that 125 felt that other motorists were driving more aggressively than they did five years ago, 36 felt that other motorists were driving about the same as they did five years ago, and 19 felt that other motorists were driving less aggressively than they did five years ago. At $\alpha = 0.10$ test the claim that senior drivers feel the same way as those who were surveyed in the *USA Today*/CNN/Gallup poll.

Source: Based on information in *USA Today*, August 29, 1997.

12–13. A *USA Today* Snapshot states that 53% of adult shoppers prefer to pay cash for purchases, 30% use checks, 16% use credit cards, and 1% have no preference. The owner of a large store randomly selected 800 shoppers and asked their payment preferences. The results were that 400 paid cash, 210 paid by check, 170 paid with a credit card, and 20 had no preference. At $\alpha = 0.01$, test the claim that the owner's customers have the same preferences as those surveyed.

Source: *USA Today,* July 19, 1995.

12–14. The owner of a sporting-goods store wishes to see whether his customers show any preference for the month in which they purchase hunting rifles. The sales of rifles for the end of last year are shown here. At $\alpha = 0.05$, test the claim that there is no preference for the month in which the customers purchase guns.

Month	Sept.	Oct.	Nov.	Dec.
No. sold	18	23	28	15

12–15. The dean of students of a college wishes to test the claim that the distribution of students is as follows: 40% business (BU), 25% computer science (CS), 15% science (SC), 10% social science (SS), 5% liberal arts (LA), and 5% general studies (GS). Last semester, the program enrollment was distributed as shown here. At $\alpha = 0.10$, is the distribution of students the same as hypothesized?

Major	BU	CS	SC	SS	LA	GS
Number	72	53	32	20	16	7

12–16. A quality control engineer for a manufacturing plant wishes to determine whether the number of defective items manufactured during the week is approximately the same on each day. A week is selected at random, and the number of defective items produced each day is shown here. At $\alpha = 0.05$, test the claim that the defective items are produced with the same frequency each day. Use the *P*-value method.

Day	Mon.	Tues.	Wed.	Thurs.	Fri.
Number	32	16	23	19	40

12–17. A software department manager believes that 50% of her customers purchase word-processing programs, 25% purchase spreadsheet programs, and 25% purchase database programming. A sample of purchases shows the following distribution. At $\alpha = 0.05$, is her assumption correct? Use the *P*-value method.

Program	Word processing	Spreadsheet	Database
No. of purchases	38	23	19

*12–18. Three coins are tossed 72 times and the number of heads is as shown. At $\alpha = 0.05$, test the null hypothesis that the coins are balanced and randomly tossed. (*Hint:* Use the binomial distribution.)

No. of heads	0	1	2	3
Frequency	3	10	17	42

*12–19. Select a three-digit state lottery number over a period of 50 days. Count the number of times each digit, 0 through 9, occurs. Test the claim, at $\alpha = 0.05$, that the digits occur at random.

Technology Step by Step

MINITAB
Step by Step

Chi-Square Test for Goodness of Fit

These instructions are based on Example 12–1 and the data on flavor preferences presented at the beginning of Section 12–2. Test the claim that the sample data do not indicate any preference in the population at large. Let $\alpha = 0.05$.

Example MT12–1

1. Enter the observed counts into C1. Name the column **Observed.**

2. Enter the expected counts into C2. Name the column **Expected.**

3. Select **Calc>Calculator.**

   a. Type **Chi-square** in the box for Store result in variable:.

   b. In the Expression: box, enter the formula as shown: **SUM ((Observed-Expected)**2/Expected).**

   c. Click [OK].

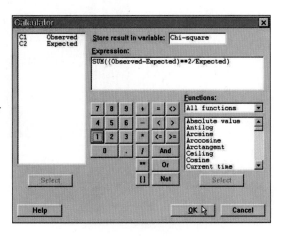

The chi-square test statistic (18) will be displayed in the first row of column C3. To calculate the *P*-value:

4. Select **Calc>Probability Distributions.**

5. Click on **Chi-Square.**

    a. Click the button for **Inverse Cumulative probability.**

    b. Click in the box for **Degrees of freedom** and type **4** (one less than the number of categories.)

    c. Click the button for **Input constant** then click in the box and type **18.**

    d. Click [**OK**].

Subtract the result from 1. The *P*-value = $1 - 0.9988 = 0.0012$. Reject the null hypothesis.

Excel
Step by Step

Chi-Square Test

Excel has several functions for χ^2 calculations in its Statistical functions category. We can use CHITEST to solve Example 12–4:

Example XL12–1

Starting with the table of observed and expected frequencies:

O	24	62	72	26	16
E	26.7	55.1	66.64	38.96	12.6

1. Enter the data set *O* in row 1.

2. Enter the data set *E* in row 2.

3. Select a blank cell, and enter the formula **=CHITEST(B1:F1,B2:F2).**

CHITEST Dialog Box and Worksheet

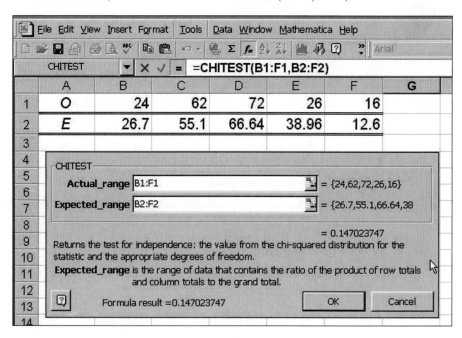

This test gives a *P*-value directly. Here the *P*-value 0.1470 is greater than the confidence level of 0.05 specified in the example, so we do not reject the null hypothesis.

**Tests Using
Contingency Tables**

When data can be tabulated in table form in terms of frequencies, several types of hypotheses can be tested using the chi-square test.

Two such tests are the independence of variables test and the homogeneity of proportions test. The test of independence of variables is used to determine whether two variables are independent of or related to each other when a single sample is selected. The test of homogeneity of proportions is used to determine whether the proportions for a variable are equal when several samples are selected from different populations. Both tests use the chi-square distribution and a contingency table, and the test value is found the same way. The independence test will be explained first.

Test for Independence

Objective 2. Test two
variables for independence
using chi-square.

The chi-square **independence test** can be used to test the independence of two variables. For example, suppose a new postoperative procedure is administered to a number of patients in a large hospital. One can ask the question, "Do the doctors feel differently about this procedure from the nurses, or do they feel basically the same way?" Note that the question is not whether or not they prefer the procedure but whether there is a difference of opinion between the two groups.

To answer this question, a researcher selects a sample of nurses and doctors and tabulates the data in table form, as shown.

Group	Prefer new procedure	Prefer old procedure	No preference
Nurses	100	80	20
Doctors	50	120	30

As the survey indicates, 100 nurses prefer the new procedure, 80 prefer the old procedure, and 20 have no preference; 50 doctors prefer the new procedure, 120 like the old procedure, and 30 have no preference. Since the main question is whether there is a difference in opinion, the null hypothesis is stated as follows:

H_0: The opinion about the procedure is *independent* of the profession.

The alternative hypothesis is stated as follows:

H_1: The opinion about the procedure is *dependent* on the profession.

If the null hypothesis is not rejected, the test means that both professions feel basically the same way about the procedure, and the differences are due to chance. If the null hypothesis is rejected, the test means that one group feels differently about the procedure from the other. Remember that rejection does *not* mean that one group favors the procedure and the other does not. Perhaps both groups favor it or both dislike it, but in different proportions.

In order to test the null hypothesis using the chi-square independence test, one must compute the expected frequencies, assuming that the null hypothesis is true. These frequencies are computed by using the observed frequencies given in the table.

When data are arranged in table form for the chi-square independence test, the table is called a **contingency table.** The table is made up of R rows and C columns. The table here has two rows and three columns.

Group	Prefer new procedure	Prefer old procedure	No preference
Nurses	100	80	20
Doctors	50	120	30

Note that row and column headings do not count in determining the number of rows and columns.

Interesting Fact

You're never too old—or too young—to be your best. George Foreman won the world heavyweight boxing championship at 46. William Pitt was 24 when he became prime minister of Great Britain. Benjamin Franklin was a newspaper columnist at age 16 and a framer of the Constitution when he was 81. Source: *Prime,* Fall 1996, p. 107

A contingency table is designated as an $R \times C$ (rows times columns) table. In this case, $R = 2$ and $C = 3$; hence, this table is a 2×3 contingency table. Each block in the table is called a *cell* and is designated by its row and column position. For example, the cell with a frequency of 80 is designated as $C_{1,2}$, or row 1, column 2. The cells are shown below.

	Column 1	Column 2	Column 3
Row 1	$C_{1,1}$	$C_{1,2}$	$C_{1,3}$
Row 2	$C_{2,1}$	$C_{2,2}$	$C_{2,3}$

The degrees of freedom for any contingency table are (rows $-$ 1) times (columns $-$ 1); that is, d.f. $= (R - 1)(C - 1)$. In this case, $(2 - 1)(3 - 1) = (1)(2) = 2$. The reason for this formula for d.f. is that all the expected values except one are free to vary in each row and in each column.

Using the previous table, one can compute the expected frequencies for each block (or cell) as shown next.

a. Find the sum of each row and each column, and find the grand total, as shown.

Group	Prefer new procedure	Prefer old procedure	No preference	
Nurses	100	80	20	Row 1 sum 200
Doctors	+ 50	+ 120	+ 30	Row 2 sum 200
	150	200	50	400
	Column 1 sum	Column 2 sum	Column 3 sum	Grand total

b. For each cell, multiply the corresponding row sum by the column sum and divide by the grand total, to get the expected value:

$$\text{expected value} = \frac{\text{row sum} \times \text{column sum}}{\text{grand total}}$$

For example, for $C_{1,2}$, the expected value, denoted by $E_{1,2}$, is (refer to the previous tables)

$$E_{1,2} = \frac{(200)(200)}{400} = 100$$

For each cell, the expected values are computed as follows:

$$E_{1,1} = \frac{(200)(150)}{400} = 75 \qquad E_{1,2} = \frac{(200)(200)}{400} = 100 \qquad E_{1,3} = \frac{(200)(50)}{400} = 25$$

$$E_{2,1} = \frac{(200)(150)}{400} = 75 \qquad E_{2,2} = \frac{(200)(200)}{400} = 100 \qquad E_{2,3} = \frac{(200)(50)}{400} = 25$$

The expected values can now be placed in the corresponding cells along with the observed values, as shown.

Group	Prefer new procedure	Prefer old procedure	No preference	
Nurses	100 (75)	80 (100)	20 (25)	200
Doctors	50 (75)	120 (100)	30 (25)	200
	150	200	50	400

The rationale for the computation of the expected frequencies for a contingency table uses proportions. For $C_{1,1}$ a total of 150 out of 400 people prefer the new procedure. And since there are 200 nurses, one would expect, if the null hypothesis were true, (150/400)(200), or 75, of the nurses to be in favor of the new procedure.

The formula for the test value for the independence test is the same as the one used for the goodness-of-fit test. It is

$$\chi^2 = \Sigma \frac{(O - E)^2}{E}$$

For the previous example, compute the $(O - E)^2/E$ values for each cell, and then find the sum.

$$\chi^2 = \Sigma \frac{(O - E)^2}{E}$$
$$= \frac{(100 - 75)^2}{75} + \frac{(80 - 100)^2}{100} + \frac{(20 - 25)^2}{25} + \frac{(50 - 75)^2}{75}$$
$$+ \frac{(120 - 100)^2}{100} + \frac{(30 - 25)^2}{25}$$
$$= 26.67$$

The final steps are to make the decision and summarize the results. This test is always a right-tailed test, and the degrees of freedom are $(R - 1)(C - 1) = (2 - 1)(3 - 1) = 2$. If $\alpha = 0.05$, the critical value from Table G is 5.991. Hence, the decision is to reject the null hypothesis, since $26.67 > 5.991$. See Figure 12–6.

Figure 12–6

Critical and Test Values
for the Postoperative
Procedures Example

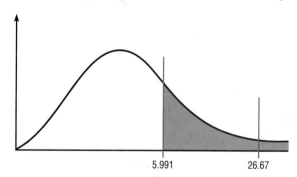

The conclusion is that there is enough evidence to support the claim that opinion is related to (dependent on) profession—i.e., that the doctors and nurses differ in their opinions about the procedure.

Two more examples illustrate the procedure for the chi-square test of independence.

Example 12–5

A sociologist wishes to see whether the number of years of college a person has completed is related to his or her place of residence. A sample of 88 people is selected and classified as shown.

Location	No college	Four-year degree	Advanced degree	Total
Urban	15	12	8	35
Suburban	8	15	9	32
Rural	6	8	7	21
Total	29	35	24	88

At $\alpha = 0.05$, can the sociologist conclude that a person's location is dependent on the number of years of college?

Solution

STEP 1 State the hypotheses and identify the claim.

> H_0: A person's place of residence is independent of the number of years of college completed.
>
> H_1: A person's place of residence is dependent on the number of years of college completed (claim).

STEP 2 Find the critical value. The critical value is 9.488, since the degrees of freedom are $(3 - 1)(3 - 1) = (2)(2) = 4$.

STEP 3 Compute the test value. To compute the test value, one must first compute the expected values.

$$E_{1,1} = \frac{(35)(29)}{88} = 11.53 \quad E_{1,2} = \frac{(35)(35)}{88} = 13.92 \quad E_{1,3} = \frac{(35)(24)}{88} = 9.55$$

$$E_{2,1} = \frac{(32)(29)}{88} = 10.55 \quad E_{2,2} = \frac{(32)(35)}{88} = 12.73 \quad E_{2,3} = \frac{(32)(24)}{88} = 8.73$$

$$E_{3,1} = \frac{(21)(29)}{88} = 6.92 \quad E_{3,2} = \frac{(21)(35)}{88} = 8.35 \quad E_{3,3} = \frac{(21)(24)}{88} = 5.73$$

The completed table is as shown.

Location	No college	Four-year degree	Advanced degree	Total
Urban	15 (11.53)	12 (13.92)	8 (9.55)	35
Suburban	8 (10.55)	15 (12.73)	9 (8.73)	32
Rural	6 (6.92)	8 (8.35)	7 (5.73)	21
	29	35	24	88

Then, the chi-square test value is

$$\chi^2 = \Sigma \frac{(O - E)^2}{E}$$

$$= \frac{(15 - 11.53)^2}{11.53} + \frac{(12 - 13.92)^2}{13.92} + \frac{(8 - 9.55)^2}{9.55}$$

$$+ \frac{(8 - 10.55)^2}{10.55} + \frac{(15 - 12.73)^2}{12.73} + \frac{(9 - 8.73)^2}{8.73}$$

$$+ \frac{(6 - 6.92)^2}{6.92} + \frac{(8 - 8.35)^2}{8.35} + \frac{(7 - 5.73)^2}{5.73}$$

$$= 3.01$$

STEP 4 Make the decision. The decision is not to reject the null hypothesis since $3.01 < 9.488$. See Figure 12–7.

Figure 12–7

Critical and Test Values
for Example 12–5

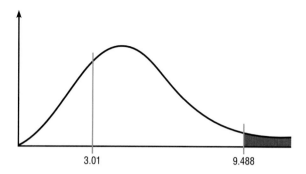

3.01   9.488

STEP 5   Summarize the results. There is not enough evidence to support the claim
that a person's place of residence is dependent on the number of years of
college completed.

Example 12–6

A researcher wishes to determine whether there is a relationship between the gender of
an individual and the amount of alcohol consumed. A sample of 68 people is selected,
and the following data are obtained.

Gender	Alcohol consumption			Total
	Low	**Moderate**	**High**	**Total**
Male	10	9	8	27
Female	13	16	12	41
Total	23	25	20	68

At $\alpha = 0.10$, can the researcher conclude that alcohol consumption is related to gender?

Solution

STEP 1   State the hypotheses and identify the claim.

H_0: The amount of alcohol that a person consumes is independent of the
individual's gender.

H_1: The amount of alcohol that a person consumes is dependent on the
individual's gender (claim).

STEP 2   Find the critical value. The critical value is 4.605, since the degrees of
freedom are $(2 - 1)(3 - 1) = 2$.

STEP 3   Compute the test value. First, compute the expected values.

$$E_{1,1} = \frac{(27)(23)}{68} = 9.13 \quad E_{1,2} = \frac{(27)(25)}{68} = 9.93 \quad E_{1,3} = \frac{(27)(20)}{68} = 7.94$$

$$E_{2,1} = \frac{(41)(23)}{68} = 13.87 \quad E_{2,2} = \frac{(41)(25)}{68} = 15.07 \quad E_{2,3} = \frac{(41)(20)}{68} = 12.06$$

The completed table is shown next.

Gender	Alcohol consumption			Total
	Low	**Moderate**	**High**	**Total**
Male	10 (9.13)	9 (9.93)	8 (7.94)	27
Female	13 (13.87)	16 (15.07)	12 (12.06)	41
	23	25	20	68

Then, the test value is

$$\chi^2 = \Sigma \frac{(O - E)^2}{E}$$

$$= \frac{(10 - 9.13)^2}{9.13} + \frac{(9 - 9.93)^2}{9.93} + \frac{(8 - 7.94)^2}{7.94}$$

$$+ \frac{(13 - 13.87)^2}{13.87} + \frac{(16 - 15.07)^2}{15.07} + \frac{(12 - 12.06)^2}{12.06}$$

$$= 0.283$$

STEP 4   Make the decision. The decision is not to reject the null hypothesis, since $0.283 < 4.605$. See Figure 12–8.

Figure 12–8

Critical and Test Values for Example 12–6

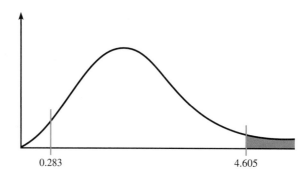

STEP 5   Summarize the results. There is not enough evidence to support the claim that the amount of alcohol a person consumes is dependent on the individual's gender.

Test for Homogeneity of Proportions

Objective 3. Test proportions for homogeneity using chi-square.

The second chi-square test that uses a contingency table is called the **homogeneity of proportions test.** In this situation, samples are selected from several different populations and the researcher is interested in determining whether the proportions of elements that have a common characteristic are the same for each population. The sample sizes are specified in advance, making either the row totals or column totals in the contingency table known before the samples are selected. For example, a researcher may select a sample of 50 freshmen, 50 sophomores, 50 juniors, and 50 seniors, and then find the proportion of students who are smokers in each level. The researcher will then compare the proportions for each group to see if they are equal. The hypotheses in this case would be

H_0: $p_1 = p_2 = p_3 = p_4$
H_1: At least one proportion is different from the others.

If one does not reject the null hypothesis, it can be assumed that the proportions are equal and the differences in them are due to chance. Hence, the proportion of students who smoke is the same for grade levels freshmen through senior. When the null hypothesis is rejected, it can be assumed that the proportions are not all equal. The computational procedure is the same as that for the test of independence as shown in the next example.

Example 12–7

A researcher selected a sample of 150 seniors from each of three area high schools and asked each senior, "Do you drive to school in a car owned by either you or your parents?" The data are shown in the table. At $\alpha = 0.05$, test the claim that the proportion of students who drive their own or their parents' cars is the same at all three schools.

	School 1	School 2	School 3	Total
Yes	18	22	16	56
No	32	28	34	94
	50	50	50	150

Solution

STEP 1   State the hypotheses.

H_0: $p_1 = p_2 = p_3$
H_1: At least one proportion is different from the others.

STEP 2   Find the critical value. The formula for the degrees of freedom is the same as before: (rows $-$ 1) (columns $-$ 1) = (2 $-$ 1) (3 $-$ 1) = 1(2) = 2. The critical value is 5.991.

STEP 3   Compute the test value. First, compute the expected values.

$$E_{1,1} = \frac{(56)(50)}{150} = 18.67 \qquad E_{2,1} = \frac{(94)(50)}{150} = 31.33$$

$$E_{1,2} = \frac{(56)(50)}{150} = 18.67 \qquad E_{2,2} = \frac{(94)(50)}{150} = 31.33$$

$$E_{1,3} = \frac{(56)(50)}{150} = 18.67 \qquad E_{2,3} = \frac{(94)(50)}{150} = 31.33$$

The completed table is shown here.

	School 1	School 2	School 3	Total
Yes	18 (18.67)	22 (18.67)	16 (18.67)	56
No	32 (31.33)	28 (31.33)	34 (31.33)	94
	50	50	50	150

The test value is

$$\chi^2 = \Sigma \frac{(O - E)^2}{E}$$

$$= \frac{(18 - 18.67)^2}{18.67} + \frac{(22 - 18.67)^2}{18.67} + \frac{(16 - 18.67)^2}{18.67}$$

$$+ \frac{(32 - 31.33)^2}{31.33} + \frac{(28 - 31.33)^2}{31.33} + \frac{(34 - 31.33)^2}{31.33}$$

$$= 1.596$$

STEP 4   Make the decision. The decision is not to reject the null hypothesis, since 1.596 < 5.991.

STEP 5   Summarize the results. There is not enough evidence to reject the null hypothesis that the proportions of high school students who drive their own or their parents' cars to school are equal for each school.

When the degrees of freedom for a contingency table are equal to 1—i.e., the table is a 2×2 table—some statisticians suggest using the *Yates correction for continuity*. The formula for the test is then

$$\chi^2 = \Sigma \frac{(|O - E| - 0.5)^2}{E}$$

Since the chi-square test is already conservative, most statisticians agree that the Yates correction is not necessary. (See Exercise 12–53.)

The steps for the chi-square independence and homogeneity tests are summarized in the Procedure Table.

Procedure Table

The Chi-Square Independence and Homogeneity Tests

STEP 1 State the hypotheses and identify the claim.

STEP 2 Find the critical value in the right tail. Use Table G.

STEP 3 Compute the test value. To compute the test value, first find the expected values. For each cell of the contingency table, use the formula

$$E = \frac{(\text{row sum})(\text{column sum})}{\text{grand total}}$$

to get the expected value. To find the test value, use the formula

$$\chi^2 = \Sigma \frac{(O - E)^2}{E}$$

STEP 4 Make the decision.

STEP 5 Summarize the results.

The assumptions for the two chi-square tests are given next.

Assumptions for the Chi-Square Independence and Homogeneity Tests

1. The data are obtained from a random sample.
2. The expected value in each cell must be 5 or more.

If the expected values are not 5 or more, combine categories.

Exercises

12–20. How is the chi-square independence test similar to the goodness-of-fit test? How is it different?

12–21. How are the degrees of freedom computed for the independence test?

12–22. When the observed frequencies are close to the expected frequencies, what is the value of chi-square?

12–23. Generally, how would the null and alternative hypotheses be stated for the chi-square independence test?

12–24. What is the name of the table used in the independence test?

12–25. How are the expected values computed for each cell in the table?

Can you find a mistake in this USA SNAPSHOT?

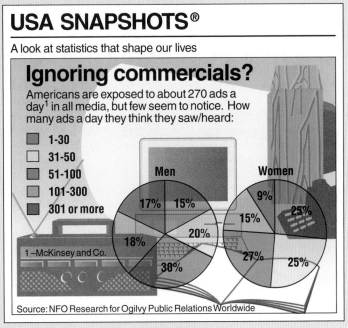

Source: Copyright 1999. *USA Today.* Reprinted with permission.

12–26. Explain how the chi-square independence test differs from the chi-square homogeneity of proportions test.

12–27. How are the null and alternative hypotheses stated for the test of homogeneity of proportions?

For Exercises 12–28 through 12–51, perform the following steps.

a. State the hypotheses and identify the claim.
b. Find the critical value.
c. Compute the test value.
d. Make the decision.
e. Summarize the results.

Use the traditional method of hypothesis testing unless otherwise specified.

12–28. A study is being conducted to determine whether there is a relationship between jogging and blood pressure. A random sample of 210 subjects is selected, and they are classified as shown in the table. At $\alpha = 0.05$, test the claim that jogging and blood pressure are not related.

Jogging status	Blood pressure		
	Low	**Moderate**	**High**
Joggers	34	57	21
Nonjoggers	15	63	20

12–29. A researcher wishes to see whether the age of an individual is related to coffee consumption. A sample of 152 people is selected, and they are classified as shown in the table. At $\alpha = 0.01$, is there a relationship between coffee consumption and age?

Age	Coffee consumption		
	Low	**Moderate**	**High**
21–30	18	16	12
31–40	9	15	27
41–50	5	12	10
51 and over	13	9	6

12–30. A survey of the 164 state representatives is conducted to see whether their opinions on a bill are related to their party affiliation. The following data are obtained. At $\alpha = 0.01$, can the researcher conclude that opinions are related to party affiliations?

Party	Opinion		
	Approve	**Disapprove**	**No opinion**
Republican	27	15	13
Democrat	43	18	12
Independent	9	15	12

This study involves three groups: smokers, ex-smokers, and nonsmokers. Suggest how a chi-square independence test could be used to arrive at the conclusions. What hypothesis could be used in this study? Do you agree with the results of the study? Explain your answer.

At Least They're Checking for Radon

Smokers have a tough time kicking the habit—and it's not just cigarettes. They are also reluctant to give up other unhealthy acts.

University of Rhode Island researchers studied the readiness of 19,000 smokers, ex-smokers, and non-smokers to begin practicing 10 healthful behaviors. Smokers proved the least willing to wear seat belts, cut fat intake, exercise, eat more fiber, watch their weight, stay out of the sun, and use sunscreen.

People who had never smoked were most receptive to change, with ex-smokers usually falling in between. Only three behaviors—getting a Pap smear, going for a mammogram, and checking home radon levels—were unaffected by smoking status.

Health psychologists have been searching for a "gateway behavior"—a health-promoting practice that, once adopted, would lead folks to begin other healthy habits. "A lot of people think exercise might be that behavior," notes Joseph S. Rossi, Ph.D., research director at Rhode Island's Cancer Prevention

Research Center. But giving up smoking may be an even better candidate, he told the Society of Behavioral Medicine.

Why does tobacco use coincide with so many other unhealthy practices? Perhaps cigarettes are so harmful that smokers consider it pointless to take up jogging or eat more broccoli, says Rossi. On the other hand he admits that it's hard to see why they wouldn't want to buckle up.

Source: Reprinted with permission from *Psychology Today* magazine. Copyright © 1995. (Sussex Publishers, Inc.)

12–31. An automobile manufacturer wishes to determine whether the age of the purchaser is related to the price of the car purchased. A sample of 222 drivers shows the following data. At $\alpha = 0.05$ is the purchase price of the car independent of the age of the driver?

	Selling price		
Age	Under $20,000	$20,001–$30,000	$30,001–$40,000
21–30	16	25	3
31–40	44	23	15
41–50	31	15	18
51 and over	9	11	12

12–32. A researcher wishes to determine if on-line service or Internet use is independent of the type of user. A sample of 300 computer users shows the following data. At $\alpha = 0.10$, test the claim that usage is independent of the user.

	Service usage		
	Increase	Same	Decrease
Business	79	21	0
Consumer	122	63	15

Source: *USA Today* Snapshot, August 31, 1995.

12–33. 300 men and 210 women were asked about how many ads in all media they think they saw or heard during one day. The results are shown.

	Number				
	1–30	31–50	51–100	101–300	301 or more
Men	45	60	90	54	51
Women	50	50	54	30	26

At $\alpha = 0.01$ is the number of ads people feel that they see or hear related to the gender to the person?

Source: Based on information from *USA Today* Snapshots, February 23, 1999.

12–34. An instructor wishes to see if the way people obtain information is independent of their educational background. A survey of 400 high school and college graduates yielded the following information. At $\alpha = 0.05$, test the claim that the way people obtain information is independent of their educational background.

	Television	Newspapers	Other sources
High school	159	90	51
College	27	42	31

Source: *USA Today* Snapshot, 1993.

12–35. A university official wishes to determine whether the instructor's degree is related to the students' opinion of the quality of instruction received. A sample of students' evaluations of various instructors is selected; the data are shown here. At $\alpha = 0.10$, can the official conclude that the degree of the instructor is related to students' opinions about that instructor's effectiveness in the classroom?

Rating	Bachelor's	Master's	Doctorate
Excellent	14	9	4
Average	16	5	7
Poor	3	12	16

Header spanning Bachelor's, Master's, Doctorate: **Degree**

12–36. A researcher wishes to determine whether the marital status of a student is related to his or her grade in a statistics course. The data below were obtained from a random sample of 142 students. At $\alpha = 0.05$, is the marital status independent of the grade received in the course?

Marital status	A	B	C	D or F
Single	27	32	16	10
Married	14	19	16	8

Header spanning A, B, C, D or F: **Grade**

12–37. A study is being conducted to determine whether the age of the customer is related to the type of movie he or she rents. A sample of renters gives the data shown here. At $\alpha = 0.10$, is the type of movie selected related to the customer's age?

Age	Documentary	Comedy	Mystery
12–20	14	9	8
21–29	15	14	9
30–38	9	21	39
39–47	7	22	17
48 and over	6	38	12

Header spanning Documentary, Comedy, Mystery: **Type of movie**

12–38. A study was conducted to determine whether the preference for a two-wheel drive or a four-wheel drive vehicle is related to the gender of the purchaser. A sample of 90 buyers was selected, and the data are shown here. At $\alpha = 0.05$, test the claim that vehicle preference is independent of gender.

Gender	Two-wheel drive	Four-wheel drive
Male	23	43
Female	18	6

12–39. A survey at a ballpark shows the following selection of snacks purchased. At $\alpha = 0.10$, is the snack chosen independent of the gender of the consumer?

Gender	Hot dog	Peanuts	Popcorn
Male	12	21	19
Female	13	8	25

Header spanning Hot dog, Peanuts, Popcorn: **Snack**

12–40. To test the effectiveness of a new drug, a researcher gives one group of individuals the new drug and another group a placebo. The results of the study are shown here. At $\alpha = 0.10$, can the researcher conclude that the drug is effective? Use the *P*-value method.

Medication	Effective	Not effective
Drug	32	9
Placebo	12	18

12–41. A book publisher wishes to determine whether there is a difference in the type of book selected by males and females for recreational reading. A random sample provides the data given here. At $\alpha = 0.05$, test the claim that the type of book selected is independent of the gender of the individual. Use the *P*-value method.

Gender	Mystery	Romance	Self-help
Male	243	201	191
Female	135	149	202

Header spanning Mystery, Romance, Self-help: **Type of Book**

12–42. According to a recent survey, 32% of Americans say they are "very likely" to become organ donors. A researcher surveys 50 drivers in each of three neighborhoods to determine the percentage of those willing to donate their organs. The results are shown here. At $\alpha = 0.01$, test the claim that the proportions of those who will donate their organs are equal in all three neighborhoods.

	Neighborhood A	Neighborhood B	Neighborhood C
Will donate	28	14	21
Will not donate	22	36	29
Total	50	50	50

Source: Lewis H. Lapham, et al., *The Harper's Index Book* (New York: Henry Holt & Co., 1987), p. 27.

12–43. According to a recent survey, 64% of Americans between the ages of 6 and 17 cannot pass a basic fitness test. A physical education instructor wishes to determine if the percentages of such students in different schools in his school district are the same. He administers a basic fitness test to 120 students in each of four schools. The results are shown here. At $\alpha = 0.05$, test the claim that the proportions who pass the test are equal.

	Southside	West End	East Hills	Jefferson
Passed	49	38	46	34
Failed	71	82	74	86
Total	120	120	120	120

Source: Lewis H. Lapham, et al., *The Harper's Index Book* (New York: Henry Holt & Co., 1987), p. 57.

12–44. An advertising firm has decided to ask 92 customers at each of three local shopping malls if they are willing to take part in a market research survey. According to previous studies, 38% of Americans refuse to take part in such surveys. The results are shown here. At $\alpha = 0.01$, test the claim that the proportions of those who are willing to participate are equal.

	Mall A	Mall B	Mall C
Will participate	52	45	36
Will not participate	40	47	56
Total	92	92	92

Source: Lewis H. Lapham, et al., *The Harper's Index Book* (New York: Henry Holt & Co., 1987), p. 41.

12–45. An insurance firm wished to see if the proportion of drivers who admit to driving after drinking varies according to the age of the driver. The firm surveyed 86 drivers in each of four age groups to see if they admitted to driving after drinking. The results are shown here. At $\alpha = 0.05$, test the claim that the proportions of those who said yes are equal for the age groups.

	Ages 21–29	30–39	40–49	50 and over
Yes	32	28	26	21
No	54	58	60	65
Total	86	86	86	86

12–46. According to a recent survey, 59% of Americans aged 8 to 17 would prefer that their mother work outside the home, regardless of what she does now. A school district psychologist decided to select three samples of 60 students each in elementary, middle, and high school to see how the students in her district felt about the issue. At $\alpha = 0.10$, test the claim that the proportions of the students who prefer that their mother have a job are equal.

	Elementary	Middle	High
Prefers mother work	29	38	51
Prefers mother not work	31	22	9
Total	60	60	60

Source: Daniel Weiss, *100% American* (New York: Poseidon Press, 1988), p. 59.

12–47. A researcher surveyed 100 randomly selected lawyers in each of four areas of the country and asked them if they had performed *pro bono* work for 25 or fewer hours in the last year. The results are shown here. At $\alpha = 0.10$, is there enough evidence to reject the claim that the proportions of those who accepted *pro bono* work for 25 hours or less are the same in each area?

	North	South	East	West
Yes	43	39	22	28
No	57	61	78	72
Total	100	100	100	100

Source: Daniel Weiss, *100% American* (New York: Poseidon Press, 1988), p. 59.

12–48. On average, 79% of American fathers are in the delivery room when their children are born. A physician's assistant surveyed 300 first-time fathers to determine if they had been in the delivery room when their children were born. The results are shown here. At $\alpha = 0.05$, is there enough evidence to reject the claim that the proportions of those who were in the delivery room at the time of birth are the same?

	Hospital A	Hospital B	Hospital C	Hospital D
Present	66	60	57	56
Not present	9	15	18	19
Total	75	75	75	75

Source: Daniel Weiss, *100% American* (New York: Poseidon Press, 1988), p. 79.

12–49. A children's playground equipment manufacturer read in a survey that 55% of all American playground injuries occur on the monkey bars. The manufacturer wishes to investigate playground injuries in four different parts of the country to determine if the proportions of accidents on the monkey bars are equal. The results are shown here. At $\alpha = 0.05$, test the claim that the proportions are equal. Use the *P*-value method.

Accidents	North	South	East	West
On monkey bars	15	18	13	16
Not on monkey bars	15	12	17	14
Total	30	30	30	30

Source: Michael D. Shook and Robert L. Shook, *The Book of Odds* (New York: Penguin Putnam Inc., 1991), p. 96.

12–50. According to the American Automobile Association, 31 million Americans travel over the Thanksgiving holiday. To determine whether to stay open or not, a national restaurant chain surveyed 125 customers at each of four locations to see if they would be traveling over the holiday. The results are shown here. At $\alpha = 0.10$, test the claim that the proportions of Americans who will travel over the Thanksgiving holiday are equal. Use the P-value method.

	Loca-tion A	Loca-tion B	Loca-tion C	Loca-tion D
Will travel	37	52	46	49
Will not travel	88	73	79	76
Total	125	125	125	125

Source: Michael D. Shook and Robert L. Shook, *The Book of Odds* (New York: Penguin Putnam Inc., 1991), p. 67.

12–51. The vice president of a large supermarket chain wished to determine if his customers made a list before going grocery shopping. He surveyed 288 customers in three stores. The results are shown here. At $\alpha = 0.10$, test the claim that the proportions of the customers in the three stores who made a list before going shopping are equal.

	Store A	Store B	Store C
Made list	77	74	68
No list	19	22	28
Total	96	96	96

Source: Daniel Weiss, *100% American* (New York: Poseidon Press, 1988), p. 82.

*12–52. For a 2 × 2 table, *a, b, c,* and *d* are the observed values for each cell, as shown.

a	b
c	d

The chi-square test value can be computed as

$$\chi^2 = \frac{n(ad - bc)^2}{(a + b)(a + c)(c + d)(b + d)}$$

where $n = a + b + c + d$. Compute the χ^2 test value by using the above formula and the formula $\Sigma (O - E)^2/E$, and compare the results for the following table.

12	15
9	23

*12–53. For the contingency table shown in Exercise 12–52, compute the chi-square test value by using Yates's correction for continuity.

*12–54. When the chi-square test value is significant, and there is a relationship between the variables, the strength of this relationship can be measured by using the *contingency coefficient.* The formula for the contingency coefficient is

$$C = \sqrt{\frac{\chi^2}{\chi^2 + n}}$$

where χ^2 is the test value and *n* is the sum of frequencies of the cells. The contingency coefficient will always be less than 1. Compute the contingency coefficient for Exercises 12–28 and 12–40.

Technology Step by Step

MINITAB Step-by-Step

Chi-Square Test for Independence

Example MT12–2

1. Enter the observed frequencies for Example 12–5 into three columns of MINITAB. Name the columns but not the rows. Exclude totals. The complete worksheet is shown.

Worksheet

C1	C2	C3
NoCollege	FourYear	Advanced
15	12	8
8	15	9
6	8	7

2. Select **Stat>Tables>Chi-Square Test.**
3. Drag the mouse over the three columns in the list.
4. Click **[Select]**.

Table Chi-Square Test
Dialog Box

The three columns will be placed in the Columns box as a sequence, C1 through C3.

5. Click [OK].

Chi-Square Test: NoCollege, FourYear, Advanced

```
Expected counts are printed below observed counts
         NoColleg  FourYear  Advanced   Total
    1          15        12         8      35
            11.53     13.92      9.55

    2           8        15         9      32
            10.55     12.73      8.73

    3           6         8         7      21
             6.92      8.35      5.73

Total          29        35        24      88

Chi-sq =   1.041 +   0.265 +   0.250 +
           0.614 +   0.406 +   0.009 +
           0.122 +   0.015 +   0.283 = 3.006
DF = 4,  P-Value = 0.557
```

The chi-square test statistic 3.006 has a *P*-value of 0.557. Do not reject the null hypothesis. The sample data do not support a relationship between level of education and place of residence.

**TI-83
Step by Step**

Chi-Square Test for Independence

1. Press **MATRIX** and move the cursor to Edit, then press **ENTER**.

2. Enter the number of rows and columns. Then press **ENTER**.

3. Enter the values in the matrix as they appear in the contingency table.

4. Press **STAT** and move the cursor to TESTS. Press **C** (**ALPHA PRGM**) for χ^2-Test. Make sure the observed matrix is [A] and the expected measure is [B].

5. Move the cursor to Calculate and press **ENTER**.

Example TI12–1

Using the data shown below from Example 12–6, test the claim of independence at $\alpha = 0.10$.

10	9	8
13	16	12

Input

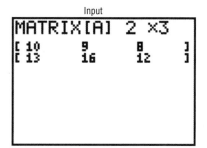

Input

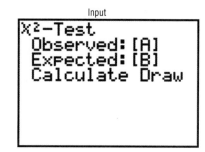

Output

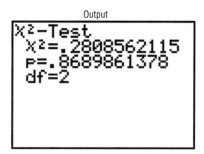

The test value is 0.2808562115. The *P*-value is 0.8689861378. The decision is to not reject the null hypothesis, since this value is greater than 0.10. You can find the expected values by pressing **MATRIX,** moving the cursor to [B], and pressing **ENTER** twice.

12–4

Summary

Three uses of the chi-square distribution were explained in this chapter. It can be used as goodness-of-fit test, in order to determine whether the frequencies of a distribution are the same as the hypothesized frequencies. For example, is the number of defective parts produced by a factory the same each day? This test is always a right-tailed test.

The test of independence is used to determine whether two variables are related or are independent. This test uses a contingency table and is always a right-tailed test. An example of its use is a test to determine whether the attitudes of urban residents about the recycling of trash differ from the attitudes of rural residents.

Finally, the homogeneity of proportions test is used to determine if several proportions are all equal when samples are selected from different populations.

The chi-square distribution is also used for other types of statistical hypothesis tests, such as the Kruskal-Wallis test, which is explained in Chapter 14.

Important Terms

contingency table   523

expected
frequency   512

goodness-of-fit test   512

homogeneity of
proportions test   528

independence test   523

observed
frequency   512

Important Formulas

Formula for the chi-square test for goodness of fit:

$$\chi^2 = \Sigma \frac{(O - E)^2}{E}$$

with degrees of freedom equal to the number of categories minus 1 and where

$$O = \text{observed frequency}$$
$$E = \text{expected frequency}$$

Formula for the chi-square independence and homogeneity of proportions tests:

$$\chi^2 = \Sigma \frac{(O - E)^2}{E}$$

with degrees of freedom equal to (rows − 1)(columns − 1). Formula for the expected value for each cell:

$$E = \frac{\text{(row sum)(column sum)}}{\text{grand total}}$$

Review Exercises

For Exercises 12–55 through 12–64, follow these steps.
a. State the hypotheses and identify the claim.
b. Find the critical value(s).
c. Compute the test value.
d. Make the decision.
e. Summarize the results.

Use the traditional method of hypothesis testing unless otherwise specified.

12–55. A company owner wishes to determine whether the number of sales of a product is equally distributed over five regions. A month is selected at random, and the number of sales is recorded. The data are as shown here. At $\alpha = 0.05$, test the claim that the number of items sold in each region is the same.

Region	NE	SE	MW	NW	SW
Sales	236	324	182	221	365

12–56. An ad is placed in newspapers in four counties asking for volunteers to test a new medication for reducing blood pressure. The number of inquiries received in each area is as shown here. At $\alpha = 0.01$, test the claim that all the ads produced the same number of responses.

County	1	2	3	4
No. of inquiries	87	62	56	93

12–57. The federal government has proposed labeling tires by fuel efficiency to save fuel and cut emissions. A survey was taken to see who would use these labels. At $\alpha = 0.10$, is the gender of the individual related to whether or not a person would use these labels? The data from a sample are shown here.

Gender	Yes	No	Undecided
Men	114	30	6
Women	136	16	8

Source: *USA Today* Snapshot, September 11, 1995.

12–58. A survey at a county fair shows the following selection of condiments for the hamburgers that are purchased. At $\alpha = 0.10$, is the condiment chosen independent of the gender of the individual?

	Condiment		
Gender	Relish	Catsup	Mustard
Men	15	18	10
Women	25	14	8

12–59. A survey was taken on how a lump-sum pension would be invested by 45-year-olds and 65-year-olds. The data are shown here. At $\alpha = 0.05$, is there a relationship between the age of the investor and the way the money would be invested?

	Large company stock funds	Small company stock funds	Inter-national stock funds	CDs or money market funds	Bonds
Age 45	20	10	10	15	45
Age 65	42	24	24	6	24

Source: *USA Today,* September 11, 1995.

12–60. A car manufacturer wishes to determine whether the type of car purchased is related to the individual's gender. The data obtained from a sample are shown here. At $\alpha = 0.01$, is the gender of the purchaser related to the type of car purchased?

	Type of vehicle purchased			
Gender of purchaser	Sedan	Compact	Station wagon	SUV
Male	33	27	23	17
Female	21	34	41	18

12–61. A guidance counselor wishes to determine if the proportions of high school girls in his school district who have jobs are equal to the national average of 36%. He surveys 80 female students, ages 16 through 18, to determine if they work or not. The results are shown below. At $\alpha = 0.01$, test the claim that the proportions of girls who work are equal. Use the P-value method.

	16-year-olds	17-year-olds	18-year-olds
Work	45	31	38
Don't work	35	49	42
Total	80	80	80

Source: Michael D. Shook and Robert L. Shook, *The Book of Odds* (New York: Penguin Putnam Inc., 1991), p. 196.

12–62. The risk of injury is higher for males as compared to females (57% versus 43%). A hospital emergency room supervisor wishes to determine if the proportions of injuries to males in his hospital are the same for each of four months. He surveys 100 injuries treated in his ER for each month. The results are shown here. At $\alpha = 0.05$, can he reject the claim that the proportions of injuries for males are equal for each of the four months?

	May	June	July	August
Male	51	47	58	63
Female	49	53	42	37
Total	100	100	100	100

Source: Michael D. Shook and Robert L. Shook, *The Book of Odds* (New York: Penguin Putnam Inc., 1991), p. 98.

12–63. A researcher surveyed 50 randomly selected subjects in four cities and asked if they felt that their anger was the most difficult behavior to control. The results are shown below. At $\alpha = 0.10$, is there enough evidence to reject the claim that the proportion of those who felt this way in each city is the same?

	City A	City B	City C	City D
Yes	12	15	10	21
No	38	35	40	29
Total	50	50	50	50

12–64. In order to see if the proportion of transactions that result in a glitch is the same for three stores, a researcher sampled 200 transactions at each store and asked if the purchaser had any problems. The results are shown here. At $\alpha = 0.01$, test the claim that the proportions are equal.

	Store 1	Store 2	Store 3
Yes	87	56	43
No	113	144	157
Total	200	200	200

Statistics Today

Statistics and Heredity—*Revisited*

Using probability, Mendel predicted the following:

	Smooth		Wrinkled	
	Yellow	Green	Yellow	Green
Expected	0.5625	0.1875	0.1875	0.0625

The observed results were

	Smooth		Wrinkled	
	Yellow	Green	Yellow	Green
Observed	0.5666	0.1942	0.1816	0.0556

Using chi-square tests on the data, Mendel found that his predictions were accurate in most cases (i.e., a good fit), thus supporting his theory. He reported many highly successful experiments. Mendel's genetic theory is simple but useful in predicting the results of hybridization.

A Fly in the Ointment

Although Mendel's theory is basically correct, an English statistician, R. A. Fisher, examined Mendel's data some 50 years later. He found that the observed (actual) results agreed too closely with the expected (theoretical) results and concluded that the data had in some way been falsified. The results are too good to be true. Several explanations have been proposed, ranging from deliberate misinterpretation to an assistant's error, but no one can be sure why this happened.

Data Analysis

The Data Bank is located in Appendix D, or on the World Wide Web by following links from www.mhhe.com/math/stat/bluman

1. Test the hypothesis that the marital status of individuals is equally distributed among four groups. Use the chi-square goodness-of-fit-test. Choose a sample of at least 50 individuals.

2. Use the chi-square test of independence to test the hypothesis that smoking is independent of the gender of the individual. Use a sample size of at least 50 individuals.

3. Using the data from Data Set IV in Appendix D, classify the heights of the buildings in New York City as follows: 500–599, 600–699, 700–799, 800–899, 900–999, 1000 and over. Then use the chi-square goodness-of-fit test to determine if the number of buildings in New York City is evenly distributed according to height. Use an α value of your choice.

4. Using the data from Data Set XII in Appendix D, select a weather variable, such as flash floods, and use the chi-square goodness-of-fit test to see if the number of deaths per year is evenly distributed over the years. Use an α value of your choice.

5. Using the data from Data Set XV in Appendix D, set up a contingency table as follows:

	Eastern Hemisphere	Western Hemisphere
2.0–4.9		
5.0–7.0		

Then count the number of Richter scale measurements in each category. Next use the chi-square test of independence to see if there is a relationship between the magnitudes of the earthquake and the hemisphere in which they occur. Use an α value of your choice.

Quiz

Determine whether each statement is true or false. If the statement is false, explain why.

1. The chi-square test of independence is always two-tailed.

2. The test values for the chi-square goodness-of-fit test and the independence test are computed using the same formula.

3. When the null hypothesis is rejected in the goodness-of-fit test, it means there is a close agreement between the observed and expected frequencies.

Select the best answer.

4. The values of the chi-square variable cannot be
   a. Positive   c. Negative
   b. 0   d. None of the above

5. The null hypothesis for the chi-square test of independence is that the variables are
   a. Dependent   c. Related
   b. Independent   d. Always 0

6. The degrees of freedom for the goodness-of-fit test are
   a. 0   c. Sample size—1
   b. 1   d. Number of categories—1

Complete the following statements with the best answer.

7. The degrees of freedom for a 4 × 3 contingency table are _____ .

8. An important assumption for the chi-square test is that the observations must be _____ .

9. The chi-square goodness-of-fit test is always _____ tailed.

10. In the chi-square independence test, the expected frequency for each class must always be _____ .

For Problems 11 through 17, follow these steps.
   a. State the hypotheses.
   b. Find the critical value.
   c. Compute the test value.
   d. Make the decision.
   e. Summarize the results.

Use the traditional method of hypothesis testing unless otherwise specified.

11. A company owner wishes to determine if the number of advertisements of a product are equally distributed over five geographic locations. A month is selected at random, and the number of ads is recorded. The data are shown here. At $\alpha = 0.05$, can it be concluded that the number of ads is the same in each region?

Region	NE	SE	NW	MW	SW
Sales	215	287	201	193	306

12. An ad is placed in five campus buildings asking for volunteers to test a new weight-loss medication. The number of inquiries received in each building is shown here. At $\alpha = 0.01$, can it be concluded that all the ads produced the same number of responses?

Building	A	B	C	D	E
No. of inquiries	73	82	49	51	68

13. A survey found that 62% of the respondents stated that they never watched the home shopping channels on cable television, 23% stated that they watched the channels rarely, 11% stated that they watched them occasionally, and 4% stated that they watched them frequently. A group of 200 college students were surveyed, and 105 stated that they never watched the home shopping channels, 72 stated that they watched them rarely, 13 stated that they watched them occasionally, and 10 stated that they watched them frequently. At $\alpha = 0.05$ can it be concluded that the college students differ in their preference for the home shopping channels?

Source: Based on information obtained from *USA Today* Snapshots, February 21, 1997.

14. A recent survey shows the following number of each type of gifts purchased for Mother's Day for a randomly selected week. The data are shown here. At $\alpha = 0.01$, can it be concluded that each gift was purchased with equal frequency?

Gift	Clothing	Flowers	Candy
No. purchased	208	318	423

15. A bookstore manager wishes to determine if there is a difference in the type of novels selected by male and female customers. A random sample of males and females provides the following data. At $\alpha = 0.05$, can it be concluded that the type of novel selected is independent of the gender of the individual?

	Type of novel		
	Science fiction	**Mystery**	**Romance**
Males	482	303	185
Females	291	257	405

16. A pizza-shop owner wishes to determine if the type of pizza a person selects is related to the age of the individual. The data obtained from a sample are shown here. At $\alpha = 0.10$, is the age of the purchaser related to the type of pizza ordered? Use the *P*-value method.

	Type of pizza			
Age	Plain	Pepperoni	Mushroom	Double cheese
10–19	12	21	39	71
20–29	18	76	52	87
30–39	24	50	40	47
40–49	52	30	12	28

17. A survey at a ballpark shows the following selection of pennants sold to fans. The data are presented here. At $\alpha = 0.10$, is the color of the pennant purchased independent of the gender of the individual?

	Blue	Yellow	Red
Men	519	659	876
Women	487	702	787

18. In a survey of children ages 8–11, the following data were obtained as to what their parents should do with the money from a $400 tax credit.

	Keep it for themselves	Give it to their children	Don't know
Girls	162	132	6
Boys	147	147	6

At $\alpha = 0.10$, is there a relationship between the feelings of the children and the gender of the children?

Source: Based on information from *USA Today* Snapshot, March 31, 1999.

Critical Thinking Challenges

1. Use your calculator or the MINITAB random number generator and generate 100 two-digit random numbers. Make a grouped frequency distribution using the chi-square goodness-of-fit test to see if the distribution is random. In order to do this, use an expected frequency of 10 for each class. Can it be concluded that the distribution is random? Explain.

2. Simulate the state lottery by using your calculator or MINITAB to generate 100 three-digit random numbers. Group these numbers 100–199, 200–299, etc. Use the chi-square goodness-of-fit test to see if the numbers are random. The expected frequency for each class should be 10. Explain why.

Data Projects

Where appropriate, use MINITAB, the TI-83, or a computer program of your choice to complete the following exercises.

1. Select a variable and collect some data over a period of a week or several months. For example, you may want to record the number of phone calls you received over seven days, or the number of times you used your credit card each month for the past several months. Using the chi-square goodness-of-fit test, see if the occurrences are equally distributed over the period.

    a. State the purpose of the study.
    b. Define the population.
    c. State how the sample was selected.
    d. State the hypotheses.
    e. Select an α value.
    f. Compute the chi-square test value.
    g. Make the decision.
    h. Write a paragraph summarizing the results.

2. Collect some data on a variable and construct a frequency distribution.

    a. Using the method shown in Section 12–2, decide if the variable you have chosen is approximately normally distributed.

    b. Write a short paper describing your findings and cite some reasons why the variable you selected is or is not approximately normally distributed.

3. Collect some data on a variable that can be divided into groups. For example, you may want to see if there is a difference in the color of cars men own versus the color of cars women own. Using the chi-square independence test, determine if the one variable is independent of the other.

    a. State the purpose of the study.
    b. Define the population.
    c. State how the sample was selected.
    d. State the hypotheses for the study.
    e. Select an α value.
    f. Compute the chi-square test value.
    g. Make the decision.
    h. Summarize the results.

You may use the following websites to obtain raw data:

http://www.mhhe.com/math/stat/bluman/
http://lib.stat.cmu.edu/DASL
http://www.oecd.org/statlist.htm
http://www.statcan.ca/english/

chapter

13 Analysis of Variance

Objectives

After completing this chapter, you should be able to

1. Use the one-way ANOVA technique to determine if there is a significant difference among three or more means.

2. Determine which means differ using the Scheffé or Tukey test if the null hypothesis is rejected in the ANOVA.

3. Use the two-way ANOVA technique to determine if there is a significant difference in the main effects or interaction.

Is Seeing Really Believing?

Many adults look upon the eyewitness testimony of children with skepticism. They believe that young witnesses' testimony is less accurate than the testimony of adults in court cases. Several statistical studies have been done on this subject.

In a preliminary study, three researchers selected fourteen 8-year-olds, fourteen 12-year-olds, and fourteen adults. The researchers showed each group the same video of a crime being committed. The next day, each witness responded to direct and cross-examination questioning. Then the researchers, using statistical methods explained in this chapter, were able to determine if there were differences in the accuracy of the testimony of the three groups on direct examination and on cross-examination. The statistical methods used here differ from the ones explained in Chapter 10 because there are three groups rather than two.

Source: C. Luus, G. Wells, and J. Turtle, "Child Eyewitnesses: Seeing Is Believing," *Journal of Applied Psychology* 80, no. 2, pp. 317–26.

13–1

Introduction

The F test, used to compare two variances shown in Chapter 10, can also be used to compare three or more means. This technique is called *analysis of variance,* or *ANOVA.* It is used to test claims involving three or more means. (*Note:* The F test can also be used to test the equality of two means. But since it is equivalent to the t test in this case, the t test is usually used instead of the F test when there are only two means.) For example, suppose a researcher wishes to see whether the means of the time it takes three groups of students to solve a computer problem using FORTRAN, BASIC, and PASCAL are different. The researcher will use the ANOVA technique for this test. The z and t tests should not be used when three or more means are compared, for reasons given later in this chapter.

For three groups, the F test can only show whether or not a difference exists among the three means. It cannot reveal where the difference lies—i.e., between $\overline{X}_1$ and $\overline{X}_2$, or $\overline{X}_1$ and $\overline{X}_3$, or $\overline{X}_2$ and $\overline{X}_3$. If the F test indicates that there is a difference among the means, other statistical tests are used to find where the difference exists. The most commonly used tests are the Scheffé test and the Tukey test, which are also explained in this chapter.

The analysis of variance that is used to compare three or more means is called a *one-way analysis of variance* since it contains only one variable. In the example above, the variable is the type of computer language used. The analysis of variance can be extended to studies involving two variables, such as type of computer language used and mathematical background of the students. These studies involve a *two-way analysis of variance.* Section 13–4 explains the two-way analysis of variance.

13–2

One-Way Analysis of Variance

Objective 1. Use the one-way ANOVA technique to determine if there is a significant difference among three or more means.

When an *F* test is used to test a hypothesis concerning the means of three or more populations, the technique is called **analysis of variance** (commonly abbreviated as ANOVA). At first glance, one might think that to compare the means of three or more samples, the *t* test can be used, comparing two means at a time. But there are several reasons why the *t* test should not be done.

First, when one is comparing two means at a time, the rest of the means under study are ignored. With the *F* test, all the means are compared simultaneously. Second, when one is comparing two means at a time and making all pairwise comparisons, the probability of rejecting the null hypothesis when it is true is increased, since the more *t* tests that are conducted, the greater is the likelihood of getting significant differences by chance alone. Third, the more means there are to compare, the more *t* tests are needed. For example, for the comparison of 3 means two at a time, 3 *t* tests are required. For the comparison of 5 means two at a time, 10 tests are required. And for the comparison of 10 means two at a time, 45 tests are required.

Assumptions for the *F* Test for Comparing Three or More Means
1. The populations from which the samples were obtained must be normally or approximately normally distributed.
2. The samples must be independent of each other.
3. The variances of the populations must be equal.

Even though one is comparing three or more means in this use of the *F* test, *variances* are used in the test instead of means.

With the *F* test, two different estimates of the population variance are made. The first estimate is called the **between-group variance,** and it involves finding the variance of the means. The second estimate, the **within-group variance,** is made by computing the variance using all the data and is not affected by differences in the means. If there is no difference in the means, the between-group variance estimate will be approximately equal to the within-group variance estimate, and the *F* test value will be approximately equal to 1 and the null hypothesis will not be rejected. However, when the means differ significantly, the between-group variance will be much larger than the within-group variance; the *F* test value will be significantly greater than 1; and the null hypothesis will be rejected. Since variances are compared, this procedure is called *analysis of variance* (ANOVA).

For a test of the difference among three or more means, the following hypotheses should be used:

H_0: $\mu_1 = \mu_2 = \cdots = \mu_n$

H_1: At least one mean is different from the others.

As stated previously, a significant test value means that there is a high probability that this difference in means is not due to chance, but it does not indicate where the difference lies.

The degrees of freedom for this F test are d.f.N. $= k - 1$, where k is the number of groups, and d.f.D. $= N - k$ where N is the sum of the sample sizes of the groups, $N = n_1 + n_2 + \cdots + n_k$. The sample sizes need not be equal. The F test to compare means is always right-tailed.

The next two examples illustrate the computational procedure for the ANOVA technique for comparing three or more means, and the steps are summarized in the Procedure Table shown after the examples.

Example 13–1

A researcher wishes to try three different techniques to lower the blood pressure of individuals diagnosed with high blood pressure. The subjects are randomly assigned to three groups; the first group takes medication, the second group exercises, and the third group follows a special diet. After four weeks, the reduction in each person's blood pressure is recorded. At $\alpha = 0.05$, test the claim that there is no difference among the means. The data follow.

Medication	Exercise	Diet
10	6	5
12	8	9
9	3	12
15	0	8
13	2	4
$\bar{X}_1 = 11.8$	$\bar{X}_2 = 3.8$	$\bar{X}_3 = 7.6$
$s_1^2 = 5.7$	$s_2^2 = 10.2$	$s_3^2 = 10.3$

Solution

STEP 1 State the hypotheses and identify the claim.

H_0: $\mu_1 = \mu_2 = \mu_3$ (claim).
H_1: At least one mean is different from the others.

STEP 2 Find the critical value. Since $k = 3$ and $N = 15$,

d.f.N. $= k - 1 = 3 - 1 = 2$
d.f.D. $= N - k = 15 - 3 = 12$

The critical value is 3.89, obtained from Table H in Appendix C with $\alpha = 0.05$.

STEP 3 Compute the test value, using the procedure outlined here.
   a. Find the mean and variance of each sample (these values are shown below the data).
   b. Find the grand mean. The *grand mean,* denoted by $\bar{X}_{GM}$, is the mean of all values in the samples.

$$\bar{X}_{GM} = \frac{\Sigma X}{N} = \frac{10 + 12 + 9 + \cdots + 4}{15} = \frac{116}{15} = 7.73$$

When samples are equal in size, find $\bar{X}_{GM}$ by summing the $\bar{X}$'s and dividing by k, where $k =$ the number of groups.

c. Find the between-group variance, denoted by s_B^2.

$$s_B^2 = \frac{\sum n_i (\overline{X}_i - \overline{X}_{\text{GM}})^2}{k - 1}$$

$$= \frac{5(11.8 - 7.73)^2 + 5(3.8 - 7.73)^2 + 5(7.6 - 7.73)^2}{3 - 1}$$

$$= \frac{160.13}{2} = 80.07$$

Note: This formula finds the variance among the means using the sample sizes as weights and considers the differences in the means.

d. Find the within-group variance, denoted by s_W^2.

$$s_W^2 = \frac{\sum (n_i - 1)s_i^2}{\sum (n_i - 1)}$$

$$= \frac{(5 - 1)(5.7) + (5 - 1)(10.2) + (5 - 1)(10.3)}{(5 - 1) + (5 - 1) + (5 - 1)}$$

$$= \frac{104.80}{12} = 8.73$$

Note: This formula finds an overall variance by calculating a weighted average of the individual variances. It does not involve using differences of the means.

e. Find the F test value.

$$F = \frac{s_B^2}{s_W^2} = \frac{80.07}{8.73} = 9.17$$

STEP 4   Make the decision. The decision is to reject the null hypothesis, since $9.17 > 3.89$.

STEP 5   Summarize the results. There is enough evidence to reject the claim and conclude that at least one mean is different from the others.

The numerator of the fraction obtained in Step 3, part *c,* of the computational procedure is called the **sum of squares between groups,** denoted by SS_B. The numerator of the fraction obtained in step 3, part *d,* of the computational procedure is called the **sum of squares within groups,** denoted by SS_W. This statistic is also called the *sum of squares for the error.* SS_B is divided by d.f.N. to obtain the between-group variance. SS_W is divided by $N - k$ to obtain the within-group or error variance. These two variances are sometimes called **mean squares,** denoted by MS_B and MS_W. These terms are used to summarize the analysis of variance and are placed in a summary table, as shown in Table 13–1.

Table 13–1   Analysis of Variance Summary Table

Source	Sum of squares	d.f.	Mean square	F
Between	SS_B	$k - 1$	MS_B	
Within (error)	SS_W	$N - k$	MS_W	
Total				

In the table,

SS_B = sum of squares between groups

SS_W = sum of squares within groups

k = number of groups

$N = n_1 + n_2 + \cdots + n_k$ = sum of the sample sizes for the groups

$$MS_B = \frac{SS_B}{k - 1}$$

$$MS_W = \frac{SS_W}{N - k}$$

$$F = \frac{MS_B}{MS_W}$$

The totals are obtained by adding the corresponding columns. For Example 13–1, the ANOVA summary table is shown in Table 13–2.

Table 13–2 Analysis of Variance Summary Table for Example 13–1				
Source	**Sum of squares**	**d.f.**	**Mean square**	**F**
Between	160.13	2	80.07	9.17
Within (error)	104.80	12	8.73	
Total	264.93	14		

Most computer programs will print out an ANOVA summary table.

Example 13–2

A state employee wishes to see if there is a significant difference in the number of employees at the interchanges of three state toll roads. The data are shown next. At $\alpha = 0.05$ can it be concluded that there is a significant difference in the average number of employees at each interchange?

PA Turnpike	Greensburg Bypass/ Mon-Fayette Expressway	Beaver Valley Expressway
7	10	1
14	1	12
32	1	1
19	0	9
10	11	1
11	1	11
$\bar{X}_1 = 15.5$	$\bar{X}_2 = 4.0$	$\bar{X}_3 = 5.8$
$s_1^2 = 81.9$	$s_2^2 = 25.6$	$s_3^2 = 29.0$

Source: Pennsylvania Turnpike Commission.

Solution

STEP 1   State the hypotheses and identify the claim.

H_0: $\mu_1 = \mu_2 = \mu_3$.

H_1: At least one mean is different from the others (claim).

STEP 2   Find the critical value. Since $k = 3$, $N = 18$, and $\alpha = 0.05$,

$$\text{d.f.N.} = k - 1 = 3 - 1 = 2$$
$$\text{d.f.D.} = N - k = 18 - 3 = 15$$

The critical value is 3.68.

STEP 3   Compute the test value.
a. Find the mean and variance of each sample (these values are shown below the data).
b. Find the grand mean.

$$\overline{X}_{GM} = \frac{\Sigma X}{N} = \frac{7 + 14 + 32 + \cdots + 11}{18} = \frac{152}{18} = 8.4$$

c. Find the between-group variance.

$$s_B^2 = \frac{\Sigma n_i (\overline{X}_i - \overline{X}_{GM})^2}{k - 1}$$

$$= \frac{6(15.5 - 8.4)^2 + 6(4 - 8.4)^2 + 6(5.8 - 8.4)^2}{3 - 1}$$

$$= \frac{459.18}{2} = 229.59$$

d. Find the within-group variance.

$$s_W^2 = \frac{\Sigma (n_i - 1)s_i^2}{\Sigma (n_i - 1)}$$

$$= \frac{(6 - 1)(81.9) + (6 - 1)(25.6) + (6 - 1)(29.0)}{(6 - 1) + (6 - 1) + (6 - 1)}$$

$$= \frac{682.5}{15} = 45.5$$

e. Find the *F* test value.

$$F = \frac{s_B^2}{s_W^2} = \frac{229.59}{45.5} = 5.05$$

STEP 4   Make the decision. Since $5.05 > 3.68$, the decision is to reject the null hypothesis.

STEP 5   Summarize the results. There is enough evidence to support the claim that there is a difference among the means. The ANOVA summary table for this example is shown in Table 13–3.

Table 13–3	Analysis of Variance Summary Table for Example 13–2			
Source	**Sum of squares**	**d.f.**	**Mean square**	**F**
Between	459.18	2	229.59	5.05
Within	682.5	15	45.5	
Total	1141.68	17		

The steps for computing the F test value for the ANOVA are summarized in the Procedure Table.

Procedure Table

Finding the F Test Value for the Analysis of Variance

STEP 1 Find the mean and variance of each sample:

$$(\overline{X}_1, s_1^2), (\overline{X}_2, s_2^2), \ldots, (\overline{X}_k, s_k^2).$$

STEP 2 Find the grand mean

$$\overline{X}_{GM} = \frac{\sum X}{N}$$

STEP 3 Find the between-group variance.

$$s_B^2 = \frac{\sum n_i (\overline{X}_i - \overline{X}_{GM})^2}{k - 1}$$

STEP 4 Find the within-group variance.

$$s_W^2 = \frac{\sum (n_i - 1)s_i^2}{\sum (n_i - 1)}$$

STEP 5 Find the F test value.

$$F = \frac{s_B^2}{s_W^2}$$

The degrees of freedom are

$$\text{d.f.N.} = k - 1$$

where k is the number of groups,

$$\text{and d.f.D.} = N - k$$

where N is the sum of the sample sizes of the groups,

$$N = n_1 + n_2 + \cdots + n_k$$

The *P*-values for ANOVA are found by using the procedure shown in Section 10–3. For Example 13–2, find the two α values in the tables for the *F* distribution (Table H) using d.f.N. = 2 and d.f.D. = 15, where *F* = 5.05 falls between. In this case, 5.05 falls between 4.77 and 6.36 corresponding respectively to $\alpha = 0.025$ and $\alpha = 0.01$; hence, $0.01 < P\text{-value} < 0.025$. Since the *P*-value is between 0.01 and 0.025 and since *P*-value < 0.05 (the originally chosen value for α), the decision is to reject the null hypothesis. (The *P*-value obtained from a calculator is 0.021.)

Technology Step by Step

MINITAB
Step by Step

One-Way Analysis of Variance (ANOVA)

These instructions are based on the data on blood pressure from Example 13–1. Test the claim that there is no difference in mean blood pressures among the populations sampled. Let $\alpha = 0.05$.

Example MT13–1

1. Enter the data for Example 13–1 into columns of MINITAB. Name the columns **Medication, Exercise,** and **Diet.**
2. Select `Stat>ANOVA>One-way (Unstacked)`.

One-Way ANOVA
Dialog Box

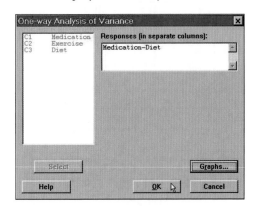

3. Drag the mouse over the three columns in the list box and then click [`Select`].
4. Click [`OK`].

One-way ANOVA: Medication, Exercise, Diet

```
Analysis of Variance
Source     DF        SS       MS       F       P
Factor      2     160.13    80.07    9.17    0.004
Error      12     104.80     8.73
Total      14     264.93
                                  Individual 95% CIs For Mean
                                  Based on Pooled StDev
Level       N      Mean     StDev    --------+---------+---------+--------
Medicati    5    11.800     2.387                            (-------*------)
Exercise    5     3.800     3.194    (-------*------)
Diet        5     7.600     3.209                  (-------*------)
                                     --------+---------+---------+--------
Pooled StDev =     2.955               4.0       8.0      12.0
```

In the session window the ANOVA table will be displayed showing the test statistic *F* = 9.17 whose *P*-value is 0.004. Reject the null hypothesis.

TI-83
Step by Step

One-Way Analysis of Variance (ANOVA)

1. Enter the data into L_1, L_2, L_3, etc.
2. Press **STAT** and move the cursor to TESTS.
3. Press **F (ALPHA COS)** for ANOVA (.
4. Enter each list followed by a comma. End with) and press **ENTER.**

Example TI13–1

Test the claim H_0: $\mu_1 = \mu_2 = \mu_3$ at $\alpha = 0.05$ for the following data from Example 13–1:

Medication	Exercise	Diet
10	6	5
12	8	9
9	3	12
15	0	8
13	2	4

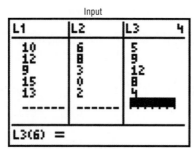

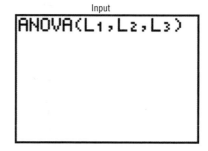

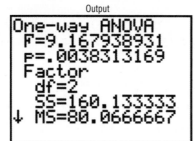

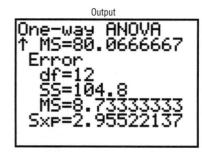

The F test value is 9.167938931. The P-value is 0.0038313169, which is significant at $\alpha = 0.05$. The factor variable has

d.f. = 2

SS = 160.133333

MS = 80.0666667

The error has

d.f. = 12

SS = 104.8

MS = 8.73333333

Excel
Step by Step

One-Way Analysis of Variance (ANOVA)

Example XL13–1

1. Enter this data set in columns A, B, and C.

9	8	12
6	7	15
15	12	18
4	3	9
3	5	10

2. Select Tools, Data Analysis and choose Anova: Single Factor.
3. Enter the data range for the three columns.

Anova: Single Factor
Dialog Box and
Worksheet

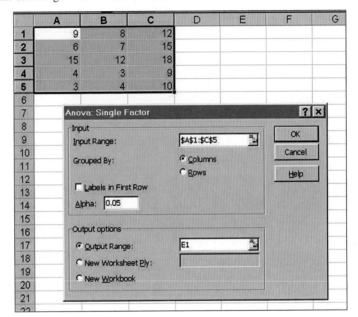

4. Set an alpha level; here, we enter **0.05.**
5. Select a location for output and click [OK].

The results are shown.

Anova: Single Factor

SUMMARY

Groups	Count	Sum	Average	Variance
Column 1	5	37	7.4	23.3
Column 2	5	35	7	11.5
Column 3	5	64	12.8	13.7

ANOVA

Source of Variation	SS	df	MS	F	P-value	F crit
Between Groups	104.9333	2	52.46667	3.245361	0.074708	3.88529
Within Groups	194	12	16.16667			
Total	298.9333	14				

13–3

The Scheffé Test and the Tukey Test

When the null hypothesis is rejected using the F test, the researcher may want to know where the difference among the means is. Several procedures have been developed to determine where the significant differences in the means lie after the ANOVA procedure has been performed. Among the most commonly used tests are the *Scheffé test* and the *Tukey test*.

Scheffé Test

In order to conduct the **Scheffé test,** one must compare the means two at a time, using all possible combinations of means. For example, if there are three means, the following comparisons must be done:

Objective 2. Determine which means differ using the Scheffé or Tukey test if the null hypothesis is rejected in the ANOVA.

$$\overline{X}_1 \text{ versus } \overline{X}_2 \qquad \overline{X}_1 \text{ versus } \overline{X}_3 \qquad \overline{X}_2 \text{ versus } \overline{X}_3$$

Unusual Stats

According to the *British Medical Journal*, the body's circadian rhythms produce drowsiness during the midafternoon, matched only by the 2:00 A.M. to 7:00 A.M. period for sleep-related traffic accidents. Source: *Prime*, Winter 1995, p. 18.

Formula for the Scheffé Test

$$F_S = \frac{(\overline{X}_i - \overline{X}_j)^2}{s_W^2[(1/n_i) + (1/n_j)]}$$

where $\overline{X}_i$ and $\overline{X}_j$ are the means of the samples being compared, n_i and n_j are the respective sample sizes, and s_W^2 is the within-group variance.

To find the critical value F' for the Scheffé test, multiply the critical value for the F test by $k - 1$:

$$F' = (k - 1)(\text{C.V.})$$

There is a significant difference between the two means being compared when F_S is greater than F'. Example 13–3 illustrates the use of the Scheffé test.

Example 13–3

Using the Scheffé test, test each pair of means in Example 13–1 to see whether a specific difference exists, at $\alpha = 0.05$.

Solution

a. For $\overline{X}_1$ versus $\overline{X}_2$,

$$F_S = \frac{(\overline{X}_1 - \overline{X}_2)^2}{s_W^2[(1/n_1) + (1/n_2)]} = \frac{(11.8 - 3.8)^2}{8.73[(1/5) + (1/5)]} = 18.33$$

b. For $\overline{X}_2$ versus $\overline{X}_3$,

$$F_S = \frac{(\overline{X}_2 - \overline{X}_3)^2}{s_W^2[(1/n_2) + (1/n_3)]} = \frac{(3.8 - 7.6)^2}{8.73[(1/5) + (1/5)]} = 4.14$$

c. For $\overline{X}_1$ versus $\overline{X}_3$,

$$F_S = \frac{(\overline{X}_1 - \overline{X}_3)^2}{s_W^2[(1/n_1) + (1/n_3)]} = \frac{(11.8 - 7.6)^2}{8.73[(1/5) + (1/5)]} = 5.05$$

The critical value for the analysis of variance for Example 13–1 was 3.89, found by using Table H with $\alpha = 0.05$, d.f.N $= k - 1 = 2$, and d.f.D. $= N - k = 12$. In this case, it is multiplied by $k - 1$ as shown.

The critical value for F' at $\alpha = 0.05$, with d.f.N. $= 2$ and d.f.D. $= 12$, is

$$F' = (k - 1)(\text{C.V.}) = (3 - 1)(3.89) = 7.78$$

This article contains many facts about coffee and its effects. Which facts are based on descriptive studies and which are based on inferential studies?

Coffee Doesn't Increase Risk of Heart Disease*

By Tim Friend
USA TODAY

In the ongoing debate over coffee's effects on the heart, the latest research scores one in favor of coffee drinkers:

You can safely drink four six-ounce cups of coffee a day.

"Coffee drinkers appear to have no grounds for concern," says Dr. Roy Fried, Kaiser Permanente Medical Center, Kensington, Md.

Previous studies linked coffee to a higher risk of heart disease. The strongest link came from observations that boiled coffee raises bad cholesterol levels, thus raising heart disease risk. But studies also suggested filtered coffee had no effect on cholesterol.

To sort out the confusion, Fried and researchers at the Johns Hopkins Medical Institutions, Baltimore, studied 100 healthy men for eight weeks after the men were taken off coffee and all caffeine. Fried says the first surprise was cholesterol levels dropped.

Then the men were divided into groups: one got four cups of caffeinated coffee a day, another got two cups, a third got four cups of decaf and a fourth got two cups of decaf. A fifth group still got no coffee.

Results, out today in the *Journal of the American Medical Association:*

- Levels of LDL (bad) cholesterol increased by six points in the four-cup-a-day caffeinated group, raising the heart disease risk by 9%.

- Levels of HDL (good) cholesterol increased three points in the same group, lowering heart disease risk 7% to 9%.
- No other group had cholesterol changes except the two-cup-a-day caffeinated group, which had a slight increase in good cholesterol only.

Overall, coffee does raise cholesterol levels, but the effect of LDL was offset by the HDL, Fried says.

Fried says it's still unclear how coffee raises cholesterol.

The coffee used in the study was brewed in automatic-drip makers with paper filters. Fried says the only sweetener used was aspartame; the only lighteners, skim and nonfat powdered dry milk.

What's in the Daily Grind†

The USA drinks more coffee than any other country—an average $1^3/_4$ cups per person per day.

Some coffee facts:

- It contains 393 chemicals, including caffeine, tannins, caramelized sugar and carbon dioxide.
- A 6-oz cup contains about 4 calories.
- Caffeine content is highest in drip-brewed, then percolated, then instant. Even decaf contains a small amount.
- Caffeine is both a stimulant and a diuretic; it's absorbed rapidly, appearing in all tissues and organs within about five minutes.

*Source: Copyright 1997 *USA Today.* Reprinted with permission.
†Source: *The Mount Sinai Schools of Medicine, Complete Book of Nutrition.*

Since only the F test value for part a ($\overline{X}_1$ versus $\overline{X}_2$) is greater than the critical value, 7.78, the only significant difference is between $\overline{X}_1$ and $\overline{X}_2$, that is, between medication and exercise.

On occasion, when the F test value is greater than the critical value, the Scheffé test may not show any significant differences in the pairs of means. This result occurs because the difference may actually lie in the average of two or more means when compared with the other mean. The Scheffé test can be used to make these types of comparisons, but the technique is beyond the scope of this book.

Tukey Test

The **Tukey test** can also be used after the analysis of variance has been completed to make pairwise comparisons between means when the groups have the same sample size. The symbol for the test value in the Tukey test is q.

Formula for the Tukey Test

$$q = \frac{\overline{X}_i - \overline{X}_j}{\sqrt{s_W^2/n}}$$

where $\overline{X}_i$ and $\overline{X}_j$ are the means of the samples being compared, n is the size of the samples, and s_W^2 is the within-group variance.

When the absolute value of q is greater than the critical value for the Tukey test, there is a significant difference between the two means being compared. The procedures for finding q and the critical value from Table N in Appendix C for the Tukey test are shown in the next example.

Example 13–4

Using the Tukey test, test each pair of means in Example 13–1 to see whether a specific difference exists, at $\alpha = 0.05$.

Solution

a. For $\overline{X}_1$ versus $\overline{X}_2$,

$$q = \frac{\overline{X}_1 - \overline{X}_2}{\sqrt{s_W^2/n}} = \frac{11.8 - 3.8}{\sqrt{8.73/5}} = \frac{8}{1.32} = 6.06$$

b. For $\overline{X}_1$ versus $\overline{X}_3$,

$$q = \frac{\overline{X}_1 - \overline{X}_3}{\sqrt{s_W^2/n}} = \frac{11.8 - 7.6}{\sqrt{8.73/5}} = \frac{4.2}{1.32} = 3.18$$

c. For $\overline{X}_2$ versus $\overline{X}_3$,

$$q = \frac{\overline{X}_2 - \overline{X}_3}{\sqrt{s_W^2/n}} = \frac{3.8 - 7.6}{\sqrt{8.73/5}} = \frac{-3.8}{1.32} = -2.88$$

To find the critical value for the Tukey test, use Table N in Appendix C. The number of means, k, is found in the row at the top, and the degrees of freedom for s_W^2 are found in the left column (denoted by v). Since $k = 3$, d.f. $= 12$, and $\alpha = 0.05$, the critical value is 3.77. See Figure 13–1. Hence, the only q value that is greater in absolute value than the critical value is the one for the difference between $\overline{X}_1$ and $\overline{X}_2$. The conclusion, then, is that there is a significant difference in means for medication and exercise. These results agree with the Scheffé analysis.

Figure 13–1

Finding the Critical Value
in Table N for the Tukey
Test (Example 13–4)

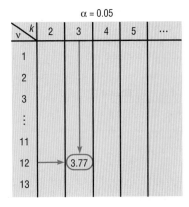

The student might wonder why there are two different tests that can be used after the
ANOVA. Actually, there are several other tests that can be used in addition to the
Scheffé and Tukey tests. It is up to the researcher to select the most appropriate test. The
Scheffé test is the most general, and it can be used when the samples are of different
sizes. Furthermore, the Scheffé test can be used to make comparisons such as the aver-
age of $\overline{X}_1$ and $\overline{X}_2$ compared with $\overline{X}_3$. However, the Tukey test is more powerful than the
Scheffé test when making pairwise comparisons for the means. A rule of thumb for pair-
wise comparisons is to use the Tukey test when the samples are equal in size and the
Scheffé test when the samples differ in size. This rule will be followed in this textbook.

Exercises

13–1. What test is used to compare three or more means?

13–2. State three reasons why multiple t tests cannot be
used to compare three or more means.

13–3. What are the assumptions for ANOVA?

13–4. Define *between-group variance* and *within-group
variance*.

13–5. What is the F test formula for comparing three or
more means?

13–6. State the hypotheses used in the ANOVA test.

13–7. What two tests are used to compare individual
means if the null hypothesis is rejected using the ANOVA
technique?

**If the null hypothesis is rejected in Exercises 13–8
through 13–19, use the Scheffé test when the sample
sizes are unequal to test the differences between the
means, and use the Tukey test when the sample sizes are
equal. Assume that all variables are normally
distributed, that the samples are independent, and that
the population variances are equal. Also, for each
exercise, perform the following steps.**
  a.   State the hypotheses and identify the claim.

  b.   Find the critical value.
  c.   Compute the test value.
  d.   Make the decision.
  e.   Summarize the results, and explain where the
differences in the means are.

**Use the traditional method of hypothesis testing unless
otherwise specified.**

13–8. A school board official wishes to see whether
there is a difference in the average age of teachers,
administrators, and support staff in the local school
district. Employees are randomly selected, and their ages
are recorded as shown in the table. At $\alpha = 0.05$ can the
school board member conclude that the average ages of
the three groups differ?

Teachers	Administrators	Support staff
24	59	34
27	35	29
26	29	35
50	40	31
48	39	40
40	54	45
	56	

13–9. Three brands of microwave ovens are selected, and the number of defects in each is as recorded below. At $\alpha = 0.05$ can one conclude that there is a difference in the means of the number of defects for the three groups?

Brand A	Brand B	Brand C
1	2	0
0	1	0
0	3	1
1	2	5
3	0	3
2	4	2
0	1	0
1	0	0
2	2	1
0	5	2

13–10. The grade point averages of students participating in college sports programs are compared. The data are shown here. At $\alpha = 0.10$, can one conclude that there is a difference in the mean GPA of the three groups?

Football	Basketball	Hockey
3.2	3.8	2.6
2.6	3.1	1.9
2.4	2.6	1.7
2.4	3.9	2.5
1.8	3.3	1.9

13–11. Three different relaxation techniques are given to randomly selected patients in an effort to reduce their stress levels. A special instrument has been designed to

measure the percentage of stress reduction in each person. The data are shown in the table. At $\alpha = 0.05$, can one conclude that there is a difference in the means of the percentages?

Technique I	Technique II	Technique III
3	12	15
10	12	14
5	17	18
1	13	14
13	18	20
3	9	22
4	14	16

13–12. A researcher wishes to see whether there is any difference in the weight gains of athletes following one of three special diets. Athletes are randomly assigned to three groups and placed on the diet for six weeks. The weight gains (in pounds) are shown here. At $\alpha = 0.05$, can the researcher conclude that there is a difference in the diets?

Diet A	Diet B	Diet C
3	10	8
6	12	3
7	11	2
4	14	5
	8	
	6	

A computer printout for this problem is shown next. Use the P-value method and the information in this printout to test the claim.

Computer Printout for Exercise 13–12

```
ANALYSIS OF VARIANCE SOURCE TABLE
Source        df    Sum of Squares   Mean Square      F       P-value

Bet Groups    2        101.095        50.548        7.740     0.00797
W/I Groups    11        71.833         6.530

Total         13       172.929

DESCRIPTIVE STATISTICS
Condit        N              Means          St Dev

diet A        4              5.000          1.826
diet B        6             10.167          2.858
diet C        4              4.500          2.646
```

13–13. A consumer magazine rated dishwashers as excellent, very good, and good. A researcher wishes to see if the average prices for the three groups differ. At

$\alpha = 0.10$, is there a difference in the average prices of the machines rated?

Excellent	Very good	Good
$565	$330	$350
400	840	379
369	510	280
550	470	320
460	380	
400	375	
400	450	
	290	
	319	

Source: *Consumer Reports,* August 1995, p. 536.

13–14. Workers are randomly assigned to four machines on an assembly line. The number of defective parts produced by each worker for one day is recorded. The data are shown here. At $\alpha = 0.05$, test the claim that the mean number of defective parts produced by the workers is the same.

Machine 1	Machine 2	Machine 3	Machine 4
3	8	10	9
2	6	9	15
0	2	8	3
6	0	11	0
4	1	12	2
3	9	15	0
5	7	17	1

13–15. A researcher wishes to see if there is a difference in the weights (in pounds) of four types of lawnmowers. At $\alpha = 0.10$, can one conclude that the weights differ?

Gas (self-propelled)	Gas (push; rear bag)	Electric	Manual
95	73	55	37
101	69	52	24
108	72	51	25
107	71	37	29
97	67	57	22
101	62	54	17
	68	34	17
	71	45	22
		41	20
		53	18
			21

Source: *Consumer Reports,* June 1995, pp. 400, 402.

13–16. A researcher tests the lifetimes (in hours) of three cassette tapes. The data are shown here. At $\alpha = 0.10$, is there a difference in the means?

Tape 1	Tape 2	Tape 3
196	98	94
183	91	106
112	101	85
107	99	102
189	84	101

13–17. A research organization tested microwave ovens. At $\alpha = 0.10$, is there a significant difference in the average prices of the three types of ovens?

Watts		
1000	**900**	**800**
270	240	180
245	135	155
190	160	200
215	230	120
250	250	140
230	200	180
	200	140
	210	130

A computer printout for this exercise is shown on the next page. Use the *P*-value method and the information in this printout to test the claim.

Computer Printout for Exercise 13–17

ANALYSIS OF VARIANCE SOURCE TABLE

Source	df	Sum of Squares	Mean Square	F	P-value
Bet Groups	2	21729.735	10864.867	10.118	0.00102
W/I Groups	19	20402.083	1073.794		
Total	21	42131.818			

DESCRIPTIVE STATISTICS

Condit	N	Means	St Dev
1000	6	233.333	28.23
900	8	203.125	39.36
800	8	155.625	28.21

13–18. The time it takes (in minutes) to treat randomly selected patients in an emergency room for three shifts is recorded here. At $\alpha = 0.05$, can one conclude that there is a significant difference in the mean time it takes to treat patients for the three shifts?

Morning (7–3)	Afternoon (3–11)	Night (11–7)
12	9	6
18	8	15
18	16	8
21	20	9
19	15	5

13–19. The data shown next represent the heights in feet of tall buildings in four large cities. At $\alpha = 0.01$, is there a significant difference among the means of the heights of the buildings in the four cities?

Seattle	Pittsburgh	New Orleans	Montreal
943	841	697	669
740	725	645	640
735	635	531	630
722	616	530	624
609	615	481	612
605	582	478	590
580	535	450	498
574	520	442	428
569	511	439	355
543	485	438	480
543	475	407	479
520	445		450
514	442		450
514	430		449
499	428		425
498	424		425
493	410		440
487			
480			
466			
456			
454			
409			

Source: *The World Almanac and Book of Facts,* 1999, pp. 624–629.

13–4

Two-Way Analysis of Variance

Objective 3. Use the two-way ANOVA technique to determine if there is a significant difference in the main effects or interaction.

The analysis of variance technique shown previously is called a **one-way analysis of variance** since there is only *one independent variable*. The **two-way analysis of variance** is an extension of the one-way analysis of variance; it involves *two independent variables.* The independent variables are also called **factors.**

The two-way analysis of variance is quite complicated, and many aspects of the subject should be considered when one is using a research design involving a two-way ANOVA. For the purposes of this textbook, only a brief introduction to the subject will be given.

In doing a study that involves a two-way analysis of variance, the researcher is able to test the effects of two independent variables or factors on one *dependent variable.* In addition, the interaction effect of the two variables can be tested.

For example, suppose a researcher wishes to test the effects of two different types of plant food and two different types of soil on the growth of certain plants. The two independent variables are the type of plant food and the type of soil, while the dependent variable is the plant growth. Other factors, such as water, temperature, and sunlight, are held constant.

To conduct this experiment, the researcher sets up four groups of plants. See Figure 13–2.

Figure 13–2

Treatment Groups for the
Plant Food–Soil Type
Experiment

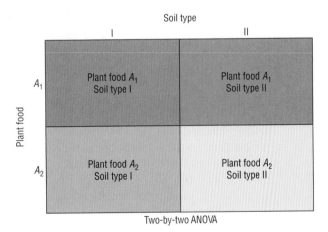

Assume that the plant food type is designated by the letters A_1 and A_2 and the soil type by the Roman numerals I and II. The groups for such a two-way ANOVA are sometimes called **treatment groups.** The four groups are as follows:

Group 1 Plant food A_1, soil type I
Group 2 Plant food A_1, soil type II
Group 3 Plant food A_2, soil type I
Group 4 Plant food A_2, soil type II

The plants are assigned to the groups at random. This design is called a 2×2 (read "two-by-two") design, since each variable consists of two **levels,** i.e., two different treatments.

The two-way ANOVA enables the researcher to test the effects of the plant food and the soil type in a single experiment rather than in separate experiments involving the plant food alone and the soil type alone. Furthermore, the researcher can test an additional hypothesis about the effect of the *interaction* of the two variables, plant food and soil type, on plant growth. For example, is there a difference between the growth of plants using plant food A_1 and soil type II and the growth of plants using plant food A_2 and soil type I? When a difference of this type occurs, the experiment is said to have a significant **interaction effect.** That is, the types of plant food affect the plant growth differently in different soil types.

There are many different kinds of two-way ANOVA designs, depending on the number of levels of each variable. Figure 13–3 shows a few of these designs. As stated previously, the plant food–soil type experiment uses a 2×2 ANOVA.

Figure 13–3

Some Types of Two-Way ANOVA Designs

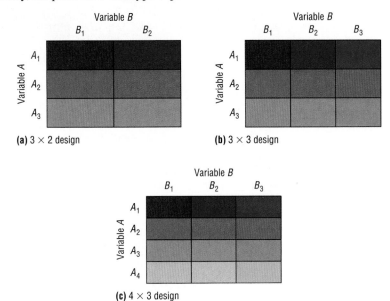

(a) 3×2 design **(b)** 3×3 design

(c) 4×3 design

The design in Figure 13–3(a) is called a 3×2 design, since the factor in the rows has three levels and the factor in the columns has two levels. Figure 13–3(b) is a 3×3 design, since each factor has three levels. Figure 13–3(c) is a 4×3 design.

The two-way ANOVA design has several null hypotheses. There is one for each independent variable and one for the interaction. In the plant food–soil type problem, the hypotheses are as follows:

1. H_0: There is no interaction effect between the type of plant food used and the type of soil used on the plant growth.

   H_1: There is an interaction effect between the food type and the soil type on plant growth.

2. H_0: There is no difference in the means of the heights of the plants grown using different foods.

   H_1: There is a difference in the means of the heights of the plants grown using different foods.

3. H_0: There is no difference in the means of the heights of the plants grown in the different soil types.

H_1: There is a difference in the means of the heights of the plants grown in the different soil types.

The first set of hypotheses concerns the interaction effect; the second and third sets test the effects of the independent variables, which are sometimes called the **main effects.**

As with the one-way ANOVA, a between-group variance estimate is calculated and a within-group variance estimate is calculated. An F test is then performed for each of the independent variables and the interaction. The results of the two-way ANOVA are summarized in a two-way table, as shown in Table 13–4 for the plant experiment.

Table 13–4   ANOVA Summary Table for Plant Food and Soil Type

Source	Sum of squares	d.f.	Mean square	F
Plant food				
Soil type				
Interaction				
Within (error)				
Total				

In general, the two-way **ANOVA summary table** is set up as shown in Table 13–5.

Table 13–5   ANOVA Summary Table

Source	Sum of squares	d.f.	Mean square	F
A	SS_A	$a - 1$	MS_A	F_A
B	SS_B	$b - 1$	MS_B	F_B
$A \times B$	$SS_{A \times B}$	$(a - 1)(b - 1)$	$MS_{A \times B}$	$F_{A \times B}$
Within (error)	SS_W	$ab(n - 1)$	MS_W	
Total				

In the table,

$$SS_A = \text{sum of squares for factor } A$$
$$SS_B = \text{sum of squares for factor } B$$
$$SS_{A \times B} = \text{sum of squares for the interaction}$$
$$SS_W = \text{sum of squares for the error term (within-group)}$$
$$a = \text{number of levels of factor } A$$
$$b = \text{number of levels of factor } B$$
$$n = \text{number of subjects in each group}$$

$$MS_A = \frac{SS_A}{a - 1}$$

$$MS_B = \frac{SS_B}{b - 1}$$

$$MS_{A \times B} = \frac{SS_{A \times B}}{(a - 1)(b - 1)}$$

$$MS_W = \frac{SS_W}{ab(n - 1)}$$

$$F_A = \frac{MS_A}{MS_W}, \text{ with d.f.N.} = a - 1 \text{ and d.f.D.} = ab(n - 1)$$

$$F_B = \frac{MS_B}{MS_W}, \text{ with d.f.N.} = b - 1 \text{ and d.f.D.} = ab(n - 1)$$

$$F_{A \times B} = \frac{MS_{A \times B}}{MS_W}, \text{ with d.f.N.} = (a - 1)(b - 1) \text{ and d.f.D.} = ab(n - 1)$$

The assumptions for the two-way analysis of variance are basically the same as those for the one-way ANOVA, except for sample size.

Assumptions for the Two-Way ANOVA

1. The populations from which the samples were obtained must be normally or approximately normally distributed.
2. The samples must be independent.
3. The variances of the populations from which the samples were selected must be equal.
4. The groups must be equal in sample size.

The computational procedure for the two-way ANOVA is quite lengthy. For this reason, it will be omitted in Example 13–5, and only the two-way ANOVA summary table will be shown. The table used in the example is similar to the one generated by most computer programs. The student should be able to interpret the table and summarize the results.

Example 13–5

A researcher wishes to see whether the type of gasoline used and the type of automobile driven have any effect on gasoline consumption. Two types of gasoline, regular and high-octane, will be used, and two types of automobiles, two-wheel and four-wheel drive, will be used in each group. There will be two automobiles in each group, for a total of eight automobiles used. Using a two-way analysis of variance, the researcher will perform the following steps.

STEP 1 State the hypotheses.

STEP 2 Find the critical value for each F test, using $\alpha = 0.05$.

STEP 3 Complete the summary table to get the test value.

STEP 4 Make the decision.

STEP 5 Summarize the results.

The data (in miles per gallon) are shown here, and the summary table is given in Table 13–6.

	Type of automobile	
Gas	**Two-wheel**	**Four-wheel**
Regular	26.7	28.6
	25.2	29.3
High-octane	32.3	26.1
	32.8	24.2

Table 13–6   ANOVA Summary Table for Example 13–5

Source	SS	d.f.	MS	F
Gasoline, A	3.920			
Automobile, B	9.680			
Interaction ($A \times B$)	54.080			
Within (error)	3.300			
Total	70.980			

Solution

STEP 1   State the hypotheses. The hypotheses for the interaction are as follows:

H_0: There is no interaction effect between the type of gasoline used and the type of automobile a person drives on gasoline consumption.

H_1: There is an interaction effect between the type of gasoline used and the type of automobile a person drives on gasoline consumption.

The hypotheses for the gasoline types are as follows:

H_0: There is no difference between the means of the gasoline consumption for the two types of gasoline.

H_1: There is a difference between the means of the gasoline consumption for the two types of gasoline.

The hypotheses for the types of automobile driven are as follows:

H_0: There is no difference between the means of the gasoline consumption for the two-wheel drive and the four-wheel drive automobiles.

H_1: There is a difference between the means of the gasoline consumption for the two-wheel drive and the four-wheel drive automobiles.

STEP 2   Find the critical values for each F test. In this case, each independent variable, or factor, has two levels. Hence, a 2×2 ANOVA table is used.

Factor A is designated as the gasoline type. It has two levels, regular and high-octane; therefore, $a = 2$. Factor B is designated as the automobile type. It also has two levels; therefore, $b = 2$. The degrees of freedom for each factor are as follows:

$$\text{factor } A:\quad \text{d.f.N.} = a - 1 = 2 - 1 = 1$$

$$\text{factor } B:\quad \text{d.f.N.} = b - 1 = 2 - 1 = 1$$

$$\text{interaction } (A \times B):\quad \text{d.f.N.} = (a - 1)(b - 1)$$

$$= (2 - 1)(2 - 1) = 1 \cdot 1 = 1$$

$$\text{within (error):}\quad \text{d.f.D.} = ab(n - 1)$$

$$= 2 \cdot 2(2 - 1) = 4$$

where n is the number of data values in each group. In this case, $n = 2$.

The critical value for the F_A test is found by using $\alpha = 0.05$, d.f.N. $= 1$, and d.f.D. $= 4$. In this case, $F_A = 7.71$. The critical value for the F_B test is found by using $\alpha = 0.05$, d.f.N. $= 1$, and d.f.D. $= 4$; F_B is also 7.71. Finally, the critical value for the $F_{A \times B}$ test is found by using d.f.N. $= 1$ and d.f.D. $= 4$; it is also 7.71.

Note: If there are different levels of the factors, the critical values will not all be the same. For example, if factor A has three levels and factor b has four levels, and if there are two subjects in each group, then the degrees of freedom are as follows:

d.f.N. $= a - 1 = 3 - 1 = 2$	for factor A
d.f.N. $= b - 1 = 4 - 1 = 3$	for factor B
d.f.N. $= (a - 1)(b - 1) = (3 - 1)(4 - 1) = 2 \cdot 3 = 6$	for factor $A \times B$
d.f.N. $= ab(n - 1) = 3 \cdot 4(2 - 1) = 12$	for the within (error) factor

STEP 3   Complete the ANOVA summary table to get the test values. The mean squares are computed first.

$$\text{MS}_A = \frac{\text{SS}_A}{a - 1} = \frac{3.920}{2 - 1} = 3.920$$

$$\text{MS}_B = \frac{\text{SS}_B}{b - 1} = \frac{9.680}{2 - 1} = 9.680$$

$$\text{MS}_{A \times B} = \frac{\text{SS}_{A \times B}}{(a - 1)(b - 1)} = \frac{54.080}{(2 - 1)(2 - 1)} = 54.080$$

$$\text{MS}_W = \frac{\text{SS}_W}{ab(n - 1)} = \frac{3.300}{4} = 0.825$$

The F values are computed next.

$$F_A = \frac{\text{MS}_A}{\text{MS}_W} = \frac{3.920}{0.825} = 4.752 \qquad \text{d.f.N} = a - 1 = 1 \qquad \text{d.f.D.} = ab(n - 1) = 4$$

$$F_B = \frac{\text{MS}_B}{\text{MS}_W} = \frac{9.680}{0.825} = 11.733 \qquad \text{d.f.N} = b - 1 = 1 \qquad \text{d.f.D.} = ab(n - 1) = 4$$

$$F_{A \times B} = \frac{\text{MS}_{A \times B}}{\text{MS}_W} = \frac{54.080}{0.825} = 65.552 \qquad \text{d.f.N} = (a - 1)(b - 1) = 1 \qquad \text{d.f.D.} = ab(n - 1) = 4$$

The completed ANOVA table is shown in Table 13–7.

Table 13–7 Completed ANOVA Summary Table for Example 13–5

Source	SS	d.f.	MS	F
Gasoline, A	3.920	1	3.920	4.752
Automobile, B	9.680	1	9.680	11.733
Interaction ($A \times B$)	54.080	1	54.080	65.552
Within (error)	3.300	4	0.825	
Total	70.980	7		

STEP 4 Make the decision. Since $F_B = 11.733$ and $F_{A \times B} = 65.552$ are greater than the critical value, 7.71, the null hypotheses concerning the type of automobile driven and the interaction effect should be rejected.

STEP 5 Summarize the results. Since the null hypothesis for the interaction effect was rejected, it can be concluded that the combination of type of gasoline and type of automobile does affect gasoline consumption.

In the preceding analysis, the effect of the type of gasoline used and the effect of the type of automobile driven are called the *main effects*. If there is no significant interaction effect, the main effects can be interpreted independently. However, if there is a significant interaction effect, the main effects must be interpreted cautiously.

To interpret the results of a two-way analysis of variance, researchers suggest drawing a graph, plotting the means of each group, analyzing the graph, and interpreting the results. In Example 13–5, find the means for each group or cell by adding the data values in each cell and dividing by n. The means for each cell are shown in the chart here.

	Type of automobile	
Gas	**Two-wheel**	**Four-wheel**
Regular	$\bar{X} = \dfrac{26.7 + 25.2}{2} = 25.95$	$\bar{X} = \dfrac{28.6 + 29.3}{2} = 28.95$
High-octane	$\bar{X} = \dfrac{32.3 + 32.8}{2} = 32.55$	$\bar{X} = \dfrac{26.1 + 24.2}{2} = 25.15$

The graph of the means for each of the variables is shown in Figure 13–4. In this graph, the lines cross each other. When such an intersection occurs and the interaction is significant, the interaction is said to be **disordinal.** When there is a disordinal interaction, one should not interpret the main effects without considering the interaction effect.

Figure 13–4

Graph of the Means of the
Variables in Example
13–5

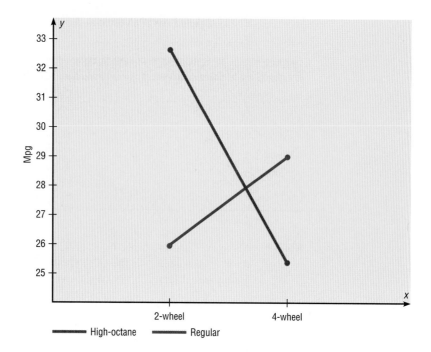

The other type of interaction that can occur is an *ordinal interaction.* Figure 13–5 shows a graph of means in which an ordinal interaction occurs between two variables. The lines do not cross each other, nor are they parallel. If the F test value for the interaction is significant and the lines do not cross each other, then the interaction is said to be **ordinal** and the main effects can be interpreted independently of each other.

Figure 13–5

Graph of Two Variables
Indicating an Ordinal
Interaction

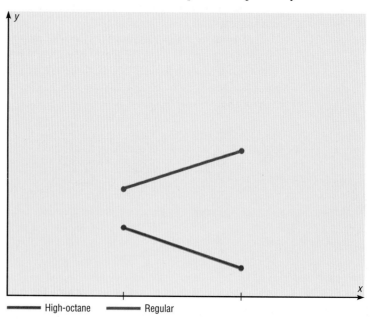

Finally, when there is no significant interaction effect, the lines in the graph will be parallel or approximately parallel. When this situation occurs, the main effects can be interpreted independently of each other because there is no significant interaction.

Figure 13–6 shows the graph of two variables when the interaction effect is not significant; the lines are parallel.

Figure 13–6

Graph of Two Variables Indicating No Interaction

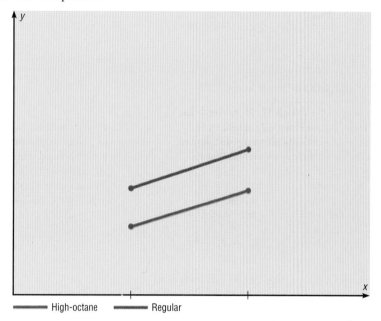

High-octane   Regular

Example 13–5 was an example of a 2×2 two-way analysis of variance, since each independent variable had two levels. For other types of variance problems, such as a 3×2 or a 4×3 ANOVA, interpretation of the results can be quite complicated. Procedures using tests like the Tukey and Scheffé tests for analyzing the cell means exist and are similar to the tests shown for the one-way ANOVA, but they are beyond the scope of this textbook. Many other designs for analysis of variance are available to researchers, such as three-factor designs and repeated-measure designs; they are also beyond the scope of this book.

In summary, the two-way ANOVA is an extension of the one-way ANOVA. The former can be used to test the effects of two independent variables and a possible interaction effect on a dependent variable.

Exercises

13–20. How does the two-way ANOVA differ from the one-way ANOVA?

13–21. Explain what is meant by the *main effects* and the *interaction effect*.

13–22. What is another name for the independent variable?

13–23. How are the values for the mean squares computed?

13–24. How are the F test values computed?

13–25. In the following two-way ANOVA, variable A has three levels and variable B has two levels. There are five data values in each cell. Find each degrees of freedom value.

 a. d.f.N. for factor A

 b. d.f.N. for factor B

 c. d.f.N. for factor $A \times B$

 d. d.f.D. for the within (error) factor

13–26. In the following two-way ANOVA, variable A has six levels and variable B has five levels. There are seven data values in each cell. Find each degrees of freedom value.

 a. d.f.N. for factor A

 b. d.f.N. for factor B

 c. d.f.N. for factor $A \times B$

 d. d.f.D. for the within (error) factor

13–27. What are the two types of interactions that can occur in the two-way ANOVA?

13–28. When can the main effects for the two-way ANOVA be interpreted independently?

13–29. Describe what the graph of the variables would look like for each situation in a two-way ANOVA experiment.
  a. No interaction effect occurs.
  b. An ordinal interaction effect occurs.
  c. A disordinal interaction effect occurs.

For Exercises 13–30 through 13–35, perform the following steps. Assume that all variables are normally or approximately normally distributed, that the samples are independent, and that the population variances are equal.
  a. State the hypotheses.
  b. Find the critical value for each *F* test.
  c. Complete the summary table and find the test value.
  d. Make the decision.
  e. Summarize the results. (Draw a graph of the cell means if necessary.)

13–30. A company wishes to test the effectiveness of its advertising. A product is selected, and two types of ads are written; one is serious and one is humorous. Also the ads are run on both television and radio. Sixteen potential customers are selected and assigned randomly to one of four groups. After seeing or listening to the ad, each customer is asked to rate its effectiveness on a scale of 1 to 20. Various points are assigned for clarity, conciseness, etc. The data are shown here. At $\alpha = 0.01$, analyze the data using a two-way ANOVA.

	Medium	
Type of ad	**Radio**	**Television**
Humorous	6, 10, 11, 9	15, 18, 14, 16
Serious	8, 13, 12, 10	19, 20, 13, 17

ANOVA Summary Table for Exercise 13–30

Source	SS	d.f.	MS	F
Type	10.563			
Medium	175.563			
Interaction	0.063			
Within	66.250			
Total	252.439			

13–31. A medical researcher wishes to test the effects of two diets and the time of day on the sodium level in a person's blood. Eight people are randomly selected and two are randomly assigned to each of the four groups. Analyze the data shown in the tables here using a two-way ANOVA at $\alpha = 0.05$. The sodium content is measured in milliequivalents per liter.

	Diet type			
Time	**I**		**II**	
8:00 A.M.	135	145	138	141
8:00 P.M.	155	162	171	191

ANOVA Summary Table for Exercise 13–31

Source	SS	d.f.	MS	F
Time	1800.0			
Diet	242.0			
Interaction	264.5			
Within	279.0			
Total	2585.5			

Data for Exercise 13–32

	Home Type		
Subcontractor	**I**	**II**	**III**
A	25, 28, 26, 30, 31	30, 32, 35, 29, 31	43, 40, 42, 49, 48
B	15, 18, 22, 21, 17	21, 27, 18, 15, 19	23, 25, 24, 17, 13

Data for Exercise 13–34

	Geographic location			
Type of paint	**North**	**East**	**South**	**West**
Enamel	60, 53, 58, 62, 57	54, 63, 62, 71, 76	80, 82, 62, 88, 71	62, 76, 55, 48, 61
Latex	36, 41, 54, 65, 53	62, 61, 77, 53, 64	68, 72, 71, 82, 86	63, 65, 72, 71, 63

13–32. A contractor wishes to see whether there is a difference in the time (in days) it takes two subcontractors to build three different types of homes. At $\alpha = 0.05$, analyze the data shown here using a two-way ANOVA. See previous page for raw data.

ANOVA Summary Table for Exercise 13–32

Source	SS	d.f.	MS	F
Subcontractor	1672.553			
Home type	444.867			
Interaction	313.267			
Within	328.8			
Total	2759.487			

13–33. Two special training programs in outdoor survival are available for army recruits. One lasts one week and the other lasts two weeks. The officer wishes to test the effectiveness of the programs and see whether there are any gender differences. Six subjects are randomly assigned to each of the programs according to gender. After completing the program, each is given a written test on his or her knowledge of survival skills. The test consists of 100 questions. The scores of the groups are shown here. Use $\alpha = 0.10$ and analyze the data using a two-way ANOVA.

	Duration	
Gender	One week	Two weeks
Female	86, 92, 87, 88, 78, 95	78, 62, 56, 54, 65, 63
Male	52, 67, 53, 42, 68, 71	85, 94, 82, 84, 78, 91

ANOVA Summary Table for Exercise 13–33

Source	SS	d.f.	MS	F
Gender	57.042			
Duration	7.042			
Interaction	3978.375			
Within	1365.5			
Total	5407.959			

13–34. Two types of outdoor paint, enamel and latex, were tested to see how long (in months) each lasted before it began to crack, flake, and peel. They were tested in four geographic locations in the United States to study the effects of climate on the paint. At $\alpha = 0.01$, analyze the data shown using a two-way ANOVA. Each group contained five test panels. See previous page for raw data.

ANOVA Summary Table for Exercise 13–34

Source	SS	d.f.	MS	F
Paint type	12.1			
Location	2501			
Interaction	268.1			
Within	2326.8			
Total	5108			

13–35. A company sells three items: swimming pools, spas, and saunas. The owner decides to see whether the age of the sales representative and the type of item affect monthly sales. At $\alpha = 0.05$, analyze the data shown below using a two-way ANOVA. Sales are given in hundreds of dollars for a randomly selected month, and five salespeople were selected for each group.

ANOVA Summary Table for Exercise 13–35

Source	SS	d.f.	MS	F
Age	168.033			
Product	1762.067			
Interaction	7955.267			
Within	2574			
Total	12,459.367			

Data for Exercise 13–35

Age of salesperson	Product		
	Pool	Spa	Sauna
Over 30	56, 23, 52, 28, 35	43, 25, 16, 27, 32	47, 43, 52, 61, 74
30 or under	16, 14, 18, 27, 31	58, 62, 68, 72, 83	15, 14, 22, 16, 27

Technology Step by Step

MINITAB Step by Step

Two-Way Analysis of Variance (ANOVA)

These instructions are based on the data from Example 13–5. Test for the effects of type of gasoline and type of automobile on gas mileage.

Example MT13–2

1. Enter the data for Example 13–5 into three columns of a worksheet. The data for this analysis has to be "stacked."
   a. All of the gas mileage data is entered in a single column named **MPG**.
   b. The second column should contain codes identifying the gasoline group, named **GasCode** 1 for regular and 2 for high-octane.
   c. The third column will contain codes identifying the type of automobile named **TypeCode** (1 for two-wheel and 2 for four-wheel drive). The first eight rows of data are shown.

Worksheet

↓	C1 MPG	C2 GasCode	C3 TypeCode
1	26.7	1	1
2	25.2	1	1
3	32.3	2	1
4	32.8	2	1
5	28.6	1	2
6	29.3	1	2
7	26.1	2	2
8	24.2	2	2

2. Select **Stat>ANOVA>Two-way.**

Dialog Box for Two-way ANOVA

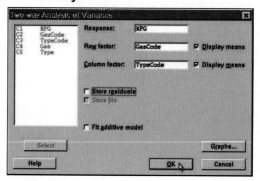

   a. Double-click MPG in the list box.
   b. Double-click GasCode.
   c. Double-click TypeCode.
   d. Click the boxes for Display means.
   e. Click [OK].

The session window will contain the results. There is a significant relationship between the type of vehicle and gas consumption, which are the main effects. Since the interaction is statistically significant, it should also be considered.

Two-way ANOVA: MPG versus GasCode and TypeCode

```
Analysis of Variance for MPG
Source         DF          SS          MS          F          P
GasCode         1       3.920       3.920       4.75      0.095
TypeCode        1       9.680       9.680      11.73      0.027
Interaction     1      54.080      54.080      65.55      0.001
Error           4       3.300       0.825
Total           7      70.980
```

```
                                    Individual 95% CI
GasCode        Mean     ---------+---------+---------+---------+--
1             27.45        (-----------*-----------)
2             28.85                      (-----------*-----------)
                         ---------+---------+---------+---------+--
                             27.00     28.00     29.00     30.00
```

```
                                  Individual 95% CI
   TypeCode          Mean    ---------+---------+---------+---------+--
   1                 29.25                    (----------------*------)
   2                 27.05        (-----------*----------)
                             ---------+---------+---------+---------+--
                                  26.40     27.60     28.80     30.00
```

Excel
Step by Step

Two-Way Analysis of Variance (ANOVA)

The Data Analysis tools include "two-factor" analysis of variance, which is used to solve two-way ANOVA problems. Solve Example 13–5 about the effects of type of gasoline and type of automobile on gas mileage.

Example XL13–2

1. Enter the data from Example 13–5 exactly as shown in the figure.
2. Select `Tools, Data Analysis` and choose `Anova: Two-Factor With Replication`.
3. Enter the data range including labels as shown. Note that there are two rows of data for regular gasoline and two for high-octane.

Worksheet and Dialog Box for Two-Way ANOVA

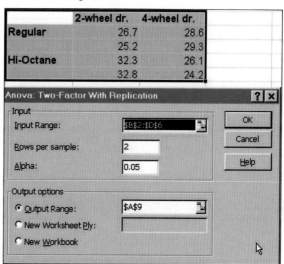

4. Set an alpha level; use 0.05.
5. Select a location for output and click [OK].

The results are shown. Compare these Excel results with those shown in Table 13–7.

ANOVA						
Source of Variation	SS	df	MS	F	P-value	F crit
Sample	3.92	1	3.92	4.752	0.094766	7.7086
Columns	9.68	1	9.68	11.73	0.026648	7.7086
Interaction	54.08	1	54.08	65.55	0.001265	7.7086
Within	3.3	4	0.825			
Total	70.98	7				

Summary

The F test, as shown in Chapter 10, can be used to compare two sample variances to determine whether they are equal. It can also be used to compare three or more means. When three or more means are compared, the technique is called analysis of variance (ANOVA). The ANOVA technique uses two estimates of the population variance. The between-group variance is the variance of the sample means; the within-group variance is the overall variance of all the values. When there is no significant difference among the means, the two estimates will be approximately equal, and the F test value will be close to 1. If there is a significant difference among the means, the between-group variance estimate will be larger than the within-group variance estimate, and a significant test value will result.

If there is a significant difference among means, the researcher may wish to see where this difference lies. Several statistical tests can be used to compare the sample means after the ANOVA technique has been done. The most common are the Scheffé test and the Tukey test. When the sample sizes are the same, the Tukey test can be used. The Scheffé test is more general and can be used when the sample sizes are equal or not equal.

When there is one independent variable, the analysis of variance is called a one-way ANOVA. When there are two independent variables, the analysis of variance is called a two-way ANOVA. The two-way ANOVA enables the researcher to test the effects of two independent variables and a possible interaction effect on one dependent variable.

Important Terms

analysis of variance (ANOVA)   545

ANOVA summary table   547

between-group variance   545

disordinal interaction   567

factors   560

interaction effect   562

level   562

main effect   563

mean square   547

one-way ANOVA   545

ordinal interaction   568

Scheffé test   554

sum of squares between groups   547

sum of squares within groups   547

treatment groups   561

Tukey test   556

two-way ANOVA   560

within-group variance   545

Important Formulas

Formulas for the ANOVA test:

$$\bar{X}_{GM} = \frac{\sum X}{N}$$

$$F = \frac{s_B^2}{s_W^2}$$

where

$$s_B^2 = \frac{\sum n_i(\bar{X}_i - \bar{X}_{GM})^2}{k - 1} \qquad s_W^2 = \frac{\sum (n_i - 1)s_i^2}{\sum (n_i - 1)}$$

d.f.N. $= k - 1$ **$N = n_1 + n_2 + \cdots + n_k$**

d.f.D. $= N - k$ **$k =$ number of groups**

Formulas for the Scheffé test:

$$F_s = \frac{(\bar{X}_i - \bar{X}_j)^2}{s_W^2 [(1/n_i) + (1/n_j)]} \qquad \text{and} \qquad F' = (k - 1)(\text{C.V.})$$

Formula for the Tukey test:

$$q = \frac{\bar{X}_i - \bar{X}_j}{\sqrt{s_W^2/n}}$$

d.f.N. $= k$ **and**   **d.f.D. $=$ degrees of freedom for s_W^2**

Formulas for the two-way ANOVA:

$$MS_A = \frac{SS_A}{a - 1}$$ $$F_A = \frac{MS_A}{MS_W}$$ d.f.N. $= a - 1$
   d.f.D $= ab(n - 1)$

$$MS_B = \frac{SS_B}{b - 1}$$ $$F_B = \frac{MS_B}{MS_W}$$ d.f.N. $= b - 1$
   d.f.D $= ab(n - 1)$

$$MS_{A \times B} = \frac{SS_{A \times B}}{(a - 1)(b - 1)}$$ $$F_{A \times B} = \frac{MS_{A \times B}}{MS_W}$$ d.f.N. $= (a - 1)(b - 1)$
   d.f.D. $= ab(n - 1)$

$$MS_W = \frac{SS_W}{ab(n - 1)}$$

Review Exercises

If the null hypothesis is rejected in Exercises 13–36 through 13–43, use the Scheffé test when the sample sizes are unequal to test the differences between the means, and use the Tukey test when the sample sizes are equal. For these exercises, perform the following steps.

a. State the hypotheses and identify the claim.
b. Find the critical value(s).
c. Compute the test value.
d. Make the decision.
e. Summarize the results, and explain where the differences in means are.

Use the traditional method of hypothesis testing unless otherwise specified.

13–36. The data represent the lengths in feet of three types of bridges in the United States. At $\alpha = 0.01$ test the claim that there is no significant difference in the means of the lengths of the types of bridges.

Simple truss	Segmented concrete	Continuous plate
745	820	630
716	750	573
700	790	525
650	674	510
647	660	480
625	640	460
608	636	451
598	620	450
550	520	450
545	450	425
534	392	420
528	370	360

Source: *World Almanac and Book of Facts* 1999, pp. 629–632.

13–37. The prices of three types of men's shoes are shown here. At $\alpha = 0.05$, can one conclude that there is a significant difference among the mean prices of the three groups?

Dress	Casual	Moccasins
$110	$ 80	$64
95	100	66
95	135	70
265	90	92
59	80	
70		
50		

Source: *Consumer Reports,* January 1995, p. 62.

13–38. The weights in ounces of four types of women's shoes are shown here. At $\alpha = 0.05$, test the claim that there is no difference in the mean weights of the groups.

Dress heels	Dress flats	Casual heels	Casual flats
8	6	11	6
7	6	12	9
7	7	12	7
6			

Source: *Consumer Reports,* January 1995, p. 62.

13–39. A plant owner wants to see whether the average time (in minutes) it takes his employees to commute to work is different for three groups. The data are shown here. At $\alpha = 0.05$, can the owner conclude that there is a significant difference among the means?

Managers	Salespeople	Stock clerks
35	9	15
18	3	6
27	12	27
24	6	22
	14	
	8	
	21	

13–40. The following data represent the number (in thousands) of vehicles that pass through toll booths on three major highways for six randomly selected days. At

$\alpha = 0.05$ is there a significant difference in the means of the number of vehicles that pass through the toll booths for each highway?

Booth 2	Booth 5	Booth 7
719	774	499
838	204	116
718	456	654
920	268	827
112	534	134
469	236	454

13–41. Students are randomly assigned to three reading classes. Each class is taught by a different method. At the end of the course, a comprehensive reading examination is given, and the results are shown here. At $\alpha = 0.05$, is there a significant difference in the means of the examination results?

Class A	Class B	Class C
87	82	97
92	78	90
61	41	83
83	65	92
47	63	91

13–42. Four hospitals are being compared to see whether there is any significant difference in the mean number of operations performed in each. A sample of six days provided the following number of operations performed each day. Test the claim, at $\alpha = 0.05$, that there is no difference in the means.

Hospital A	Hospital B	Hospital C	Hospital D
8	4	5	10
5	9	6	12
6	3	3	13
3	1	7	9
2	0	7	0
7	1	3	1

13–43. Three composition instructors recorded the number of grammatical errors their students made on a term paper. The data are shown here. At $\alpha = 0.01$, is there a significant difference in the average number of errors in the three instructors' classes?

Instructor A	Instructor B	Instructor C
2	6	1
3	7	4
5	12	0
4	4	1
8	9	2
	1	2
	0	

13–44. A teacher wishes to test the math anxiety level of her students in two classes at the beginning of the semester. The classes are Calculus I and Statistics. Furthermore, she wishes to see whether there is a difference owing to the students' ages. Math anxiety is measured by the score on a 100-point anxiety test. Use $\alpha = 0.10$ and a two-way analysis of variance to see whether there is a difference. Five students are randomly assigned to each group. The data are shown here.

	Class	
Age	**Calculus I**	**Statistics**
Under 20	43, 52, 61, 57, 55	19, 20, 31, 36, 24
20 or over	56, 55, 42, 48, 61	63, 78, 67, 71, 75

ANOVA Summary Table for Exercise 13–44

Source	SS	d.f.	MS	F
Age	2376.2			
Class	105.8			
Interaction	2645			
Within	763.2			
Total	5890.2			

13–45. A medical researcher wishes to test the effects of two different diets and two different exercise programs on the glucose level in a person's blood. The glucose level is measured in milligrams per deciliter (mg/dl). Three subjects are randomly assigned to each group. Analyze the data shown here using a two-way ANOVA with $\alpha = 0.05$.

Exercise program	Diet	
	A	**B**
I	62, 64, 66	58, 62, 53
II	65, 68, 72	83, 85, 91

ANOVA Summary Table for Exercise 13–45

Source	SS	d.f.	MS	F
Exercise	816.750			
Diet	102.083			
Interaction	444.083			
Within	108			
Total	1470.916			

Statistics Today

Is Seeing Really Believing? Revisited

In order to see if there were differences in the testimonies of the witnesses in the three age groups, the witnesses responded to 17 questions, 10 on direct examination and 7 on cross-examination. These were then scored for accuracy. An analysis of variance test with age as the independent variable was used to compare the total number of questions answered correctly by the groups. The results showed no significant differences among the age groups for the direct examination questions. However, there was a significant difference among the groups on the cross-examination questions. Further analysis showed the 8-year-olds were significantly less accurate under cross-examination compared to the other two groups. The 12-year-old and adult eyewitnesses did not differ in the accuracy of their cross-examination responses.

Data Analysis

The Data Bank is found in Appendix D, or on the World Wide Web by following links from www.mhhe.com/math/stat/bluman

1. From the Data Bank, select a random sample of subjects, and test the hypothesis that the mean cholesterol levels of the nonsmokers, less-than-one-pack-a-day smokers, and one-pack-plus smokers are equal. Use an ANOVA test. If the null hypothesis is rejected, conduct the Scheffé test to find where the difference is. Summarize the results.

2. Repeat Exercise 2 for the mean IQs of the various educational levels of the subjects.

3. Using the Data Bank, randomly select 12 subjects and randomly assign them to one of the four groups in the following classifications.

	Smoker	Nonsmoker
Male		
Female		

Use one of the following variables—weight, cholesterol, or systolic pressure—as the dependent variable and perform a two-way ANOVA on the data. Use a computer program to generate the ANOVA table.

4. Using the data from Data Set V in Appendix D, test the equality of the means from the life expectancies of the Latin American countries and the former Soviet Union countries. Use ANOVA and select an α value. If the null hypothesis is rejected, use the Sheffé test to find where the difference exists.

Quiz

Determine whether each statement is true or false. If the statement is false, explain why.

1. In analysis of variance, the null hypothesis should be rejected only when there is a significant difference among all pairs of means.

2. The F test does not use the concept of degrees of freedom.

3. When the F test value is close to 1, the null hypothesis should be rejected.

4. The Tukey test is generally more powerful than the Scheffé test for pairwise comparisons.

Select the best answer.

5. The analysis of variance uses the _____ test.
   - *a.* z
   - *b.* t
   - *c.* χ^2
   - *d.* F

6. The null hypothesis in the ANOVA is that all of the means are _____.
   - *a.* Equal
   - *b.* Unequal
   - *c.* Variable
   - *d.* None of the above

7. When one conducts an F test, _____ estimates of the population variance are compared.
   - *a.* Two
   - *b.* Three
   - *c.* Any number of
   - *d.* No

8. If the null hypothesis is rejected in ANOVA, one can use the _____ test to see where the difference in the means is found.
   - *a.* z or t
   - *b.* F or χ^2
   - *c.* Scheffé or Tukey
   - *d.* Any of the above

Complete the following statements with the best answer.

9. When three or more means are compared, one uses the _____ technique.

10. If the null hypothesis is rejected in ANOVA, the _____ test should be used when sample sizes are equal.

11. In a two-way ANOVA, one can test _____ main hypotheses and one interactive hypothesis.

For the following exercises, use the traditional method of hypothesis testing unless otherwise specified.

12. The average number of orders per week from three takeout restaurants is being compared. A week is selected at random, and the number of orders is shown here. Test the claim at $\alpha = 0.05$ that there is no difference in the average number of orders in each restaurant.

Restaurant A	Restaurant B	Restaurant C
467	502	419
318	293	392
384	392	298
427	387	501
504	419	438
309	512	488
356	398	251

13. Four hospitals are being compared to see if there is any significant difference in the mean number of physical therapy sessions held in each. A sample of six days provided the following number of therapy sessions each day. Test the claim at $\alpha = 0.05$ that there is no difference in the means.

Hospital A	Hospital B	Hospital C	Hospital D
7	5	3	11
4	10	7	13
5	4	7	14
2	2	3	9
1	0	6	2
6	1	5	8

14. Three composition instructors recorded the number of spelling errors their students made on a research paper. The data are shown here. At $\alpha = 0.01$, is there a significant difference in the average number of errors in the three classes?

Instructor 1	Instructor 2	Instructor 3
2	4	5
3	6	2
5	8	3
0	4	2
8	9	3
	0	3
	2	

15. A college official wishes to see if the average time (in minutes) it takes people to commute to the college is different for three groups. The data are shown here. At $\alpha = 0.05$, can it be concluded that there is a significant difference among the means?

Students	Faculty	Staff
12	57	15
28	43	12
47	12	28
15	10	35
35	25	49
	19	55
	38	19

16. The bacteria levels (in parts per million) of three pools were tested for a period of five days. The data are shown here. At $\alpha = 0.05$, is there a difference in the means of the bacteria levels of the pools?

Sunset Pool	Blue Spruce Pool	Rainbow Pool
58	98	34
61	87	38
42	98	39
37	83	29
58	76	28

17. A researcher conducted a study of two different diets and two different exercise programs. Three randomly selected subjects were assigned to each group for one month. The values indicate the amount of weight each lost.

Exercise program \\ Diet	A	B
I	5, 6, 4	8, 10, 15
II	3, 4, 8	12, 16, 11

Answer the following questions for the information in the printout shown on next page.
a. What procedure is being used?
b. What are the names of the two variables?
c. How many levels does each variable contain?
d. What are the hypotheses for the study?
e. What are the F values for the hypotheses? State which are significant using the P-values.
f. Based on the answers to part d, which hypotheses can be rejected?

Computer Printout for Problem 17

```
Datafile: NONAME.SST   Procedure: Two-way ANOVA

TABLE OF MEANS:
                            DIET
                     A .....    B .....    Row Mean
EX PROG I .....      5.000     11.000       8.000
       II .....      5.000     13.000       9.000
    Col Mean         5.000     12.000
    Tot Mean         8.500
```

SOURCE TABLE:

Source	df	Sums of Squares	Mean Square	F Ratio	p-value
DIET	1	147.000	147.000	21.000	0.00180
EX PROG	1	3.000	3.000	0.429	0.53106
DIET X EX P	1	3.000	3.000	0.429	0.53106
Within	8	56.000	7.000		
Total	11	209.000			

Critical Thinking Challenges

Shown here are the abstract and two tables from a research study entitled "Adult Children of Alcoholics: Are They at Greater Risk for Negative Health Behaviors?" by Arlene E. Hall. Based on the abstract and the tables, answer the following questions:

1. What was the purpose of the study?

2. How many groups were used in the study?

3. By what means were the data collected?

4. What was the sample size?

5. What type of sampling method was used?

6. How might the population be defined?

7. What may have been the hypothesis for the ANOVA part of the study?

8. Why was the one-way ANOVA procedure used, as opposed to another test, such as the *t* test?

9. What part of the ANOVA table did the conclusion "ACOAs had significantly lower wellness scores (WS) than non-ACOAs" come from?

10. What level of significance was used?

11. In the following excerpts from the article, the researcher states that:

> . . . using the Tukey-HSD procedure revealed a significant difference between ACOAs and non-ACOAs, p = 0.05, but no significant difference was found between ACOAs and Unsures or between non-ACOAs and Unsures.

Using Tables 1 and 2 and the means, explain why the Tukey test would have enabled the researcher to draw this conclusion.

Abstract *The purpose of the study was to examine and compare the health behaviors of adult children of alcoholics (ACOAs) and their non-ACOA peers within a university population. Subjects were 980 undergraduate students from a major university in the East. Three groups (ACOA, non-ACOA, and Unsure) were identified from subjects' responses to three direct questions regarding parental drinking behaviors. A questionnaire was used to collect data for the study. Included were questions related to demographics, parental drinking behaviors, and the College Wellness Check (WS), a health risk appraisal designed especially for college students (Dewey & Cabral, 1986). Analysis of variance procedures revealed that ACOAs had significantly lower wellness scores (WS) than non-ACOAs. Chi-square analyses of the individual variables revealed that ACOAs and non-ACOAs were significantly different on 15 of the 50 variables of the WS. A discriminant analysis procedure revealed the similarities between Unsure subjects and ACOA subjects. The results provide valuable information regarding ACOAs in a nonclinical setting and contribute to our understanding of the influences related to their health risk behaviors.*

Table 1   Means and Standard Deviations for the Wellness Scores (WS) Group by ($N = 945$)

Group	N	$(\bar{X})$	S.D.
ACOAs	143	69.0	13.6
Non-ACOAs	746	73.2	14.5
Unsure	56	70.1	14.0
Total	945	72.3	14.6

Table 2   ANOVA of Group Means for the Wellness Scores (WS)

Source	d.f.	SS	MS	F
Between groups	2	2,403.5	1,201.7	5.9*
Within groups	942	193,237.4	205.1	
Total	944	195,640.8		

*$p < .01$

Source: Arlene E. Hall, "Adult Children of Alcoholics: Are They at Greater Risk for Negative Health Behaviors?" *Journal of Health Education* 12, no. 4 (July–August 1995), pp. 232–238.

Data Projects

Where appropriate, use MINITAB, the TI-83, or a computer program of your choice to complete the following exercises.

Select a variable and collect data for at least three different groups (samples). For example, you could ask students, faculty, and clerical staff how many cups of coffee they drink per day or how many hours they watch television per day. Compare the means using the one-way ANOVA technique, and then complete the following:

a. What is the purpose of the study?
b. Define the population.
c. How were the samples selected?
d. What α value was used?
e. State the hypotheses.
f. What was the F test value?
g. What was the decision?
h. Summarize the results.

You may use the following websites to obtain raw data:

http://www.mhhe.com/math/stat/bluman/
http://lib.stat.cmu.edu/DASL
http://www.oecd.org/statlist.htm
http://www.statcan.ca/english/

Hypothesis-Testing Summary 2*

7. Test of the significance of the correlation coefficient.

    Example: $H_0: \rho = 0$

    Use a t test:

    $$t = r\sqrt{\frac{n-2}{1-r^2}} \text{ with d.f.} = n - 2$$

8. Formula for the F test for the multiple correlation coefficient.

    Example: $H_0: \rho = 0$

    $$F = \frac{R^2/k}{(1-R^2)/(n-k-1)} \text{ with}$$

    d.f.N. $= n - k$ and d.f.D. $= n - k - 1$

9. Comparison of a sample distribution with a specific population.

    Example: H_0: There is no difference between the two distributions.

    Use the chi-square goodness-of-fit test:

    $$\chi^2 = \Sigma \frac{(O-E)^2}{E} \quad \text{with}$$

    d.f. = no. of categories $- 1$

10. Comparison of the independence of two variables.

    Example: H_0: Variable A is independent of variable B.

    Use the chi-square independence test:

    $$\chi^2 = \Sigma \frac{(O-E)^2}{E} \quad \text{with}$$

    d.f. $= (R-1)(C-1)$

11. Test for homogeneity of proportions.

    Example: $H_0: p_1 = p_2 = p_3$

Use the chi-square test:

$$\chi^2 = \Sigma \frac{(O - E)^2}{E} \quad \text{with}$$

$$\text{d.f.} = (R - 1)(C - 1)$$

12. Comparison of three or more sample means.

Example: $H_0: \mu_1 = \mu_2 = \mu_3$

Use the analysis of variance test:

$$F = \frac{s_B^2}{s_W^2}$$

where

$$s_B^2 = \frac{\Sigma\, n_i(\overline{X}_i - \overline{X}_{GM})^2}{k - 1}$$

$$s_W^2 = \frac{\Sigma\, (n_i - 1)s_i^2}{\Sigma\, (n_i - 1)}$$

$$\text{d.f.N.} = k - 1 \qquad N = n_1 + n_2 + \cdots + n_k$$

$$\text{d.f.D.} = N - k \qquad k = \text{number of groups}$$

13. Test when the F value for the ANOVA is significant. Use the Scheffé test to find what pairs of means are significantly different:

$$F_s = \frac{(\overline{X}_i - \overline{X}_j)^2}{s_W^2\,[(1/n_i) + (1/n_j)]} \quad \text{and } F' = (k - 1)(\text{C.V.})$$

Use the Tukey test to find which pairs of means are significantly different:

$$q = \frac{\overline{X}_i - \overline{X}_j}{\sqrt{s_W^2/n}} \text{ with } \begin{array}{l} \text{d.f.N.} = k \\ \text{d.f.D.} = \text{degrees of freedom for } s_W^2 \end{array}$$

Test for the two-way ANOVA.

14. Example: H_0: There is no significant difference for the main effects.

H_0: There is no significant difference for the interaction effect.

$$\text{MS}_A = \frac{\text{SS}_A}{a - 1}$$

$$\text{MS}_B = \frac{\text{SS}_B}{b - 1}$$

$$\text{MS}_{A \times B} = \frac{\text{SS}_{A \times B}}{(a - 1)(b - 1)}$$

$$\text{MS}_W = \frac{\text{SS}_W}{ab(n - 1)}$$

$$F_A = \frac{\text{MS}_A}{\text{MS}_W} \qquad \begin{array}{l} \text{d.f.N.} = a - 1 \\ \text{d.f.D} = ab(n - 1) \end{array}$$

$$F_B = \frac{\text{MS}_B}{\text{MS}_W} \qquad \begin{array}{l} \text{d.f.N.} = (b - 1) \\ \text{d.f.D} = ab(n - 1) \end{array}$$

$$F_{A \times B} = \frac{\text{MS}_{A \times B}}{\text{MS}_W} \qquad \begin{array}{l} \text{d.f.N.} = (a - 1)(b - 1) \\ \text{d.f.D} = ab(n - 1) \end{array}$$

*This summary is a continuation of Hypothesis-Testing Summary 1, at the end of Chapter 10.

chapter

14 Nonparametric Statistics

Objectives

After completing this chapter, you should be able to

1. State the advantages and disadvantages of nonparametric methods.

2. Test hypotheses using the sign test.

3. Test hypotheses using the Wilcoxon rank sum test.

4. Test hypotheses using the signed-rank test.

5. Test hypotheses using the Kruskal-Wallis test.

6. Compute the Spearman rank correlation coefficient.

7. Test hypotheses using the runs test.

Statistics Today

Too Much or Too Little?

Suppose a manufacturer of ketchup wishes to check the bottling machines to see if they are functioning properly. That is, are they dispensing the right amount of ketchup per bottle? A 40-ounce bottle is currently used. Because of the natural variation in the manufacturing process, the amount of ketchup in a bottle will not always be exactly 40 ounces. Some bottles will contain less than 40 ounces, and some will contain more than 40 ounces. To see if the variation is due to chance or to a malfunction in the manufacturing process, a runs test can be used. The runs test is a nonparametric statistical technique. This chapter explains such techniques, which can be used to help the manufacturer determine the answer to the question.

14–1

Introduction

Statistical tests, such as the *z*, *t*, and *F* tests, are called parametric tests. **Parametric tests** are statistical tests for population parameters such as means, variances, and proportions that involve assumptions about the populations from which the samples were selected. One assumption is that these populations are normally distributed. But what if the population in a particular hypothesis-testing situation is *not* normally distributed? Statisticians have developed a branch of statistics known as **nonparametric statistics** or

distribution-free statistics to use when the population from which the samples are selected is not normally distributed. Nonparametric statistics can also be used to test hypotheses that do not involve specific population parameters, such as μ, σ, or p.

For example, a sportswriter may wish to know whether there is a relationship between the rankings of two judges on the diving abilities of 10 Olympic swimmers. In another situation, a sociologist may wish to determine whether men and women enroll at random for a specific drug rehabilitation program. The statistical tests used in these situations are nonparametric or distribution-free tests. The term *nonparametric* is used for both situations.

The nonparametric tests explained in this chapter are the sign test, the Wilcoxon rank sum test, the Wilcoxon signed-rank test, the Kruskal-Wallis test, and the runs test. In addition, the Spearman rank correlation coefficient, a statistic for determining the relationship between ranks, will be explained.

14–2

Advantages and Disadvantages of Nonparametric Methods

As stated previously, nonparametric tests and statistics can be used in place of their parametric counterparts (z, t, and F) when the assumption of normality cannot be met. However, one should not assume that these statistics are a better alternative than the parametric statistics. There are both advantages and disadvantages in the use of nonparametric methods.

Advantages

Objective 1. State the advantages and disadvantages of nonparametric methods.

There are five advantages that nonparametric methods have over parametric methods:

1. They can be used to test population parameters when the variable is not normally distributed.
2. They can be used when the data are nominal or ordinal.
3. They can be used to test hypotheses that do not involve population parameters.
4. In most cases, the computations are easier than those for the parametric counterparts.
5. They are easier to understand.

Disadvantages

There are three disadvantages of nonparametric methods:

1. They are *less sensitive* than their parametric counterparts when the assumptions of the parametric methods are met. Therefore, larger differences are needed before the null hypothesis can be rejected.
2. They tend to use *less information* than the parametric tests. For example, the sign test requires the researcher to determine only whether the data values are above or below the median, not how much above or below the median each value is.
3. They are *less efficient* than their parametric counterparts when the assumptions of the parametric methods are met. That is, larger sample sizes are needed to overcome the loss of information. For example, the nonparametric sign test is about 60% as efficient as its parametric counterpart, the z test. Thus, a sample size of 100 is needed for use of the sign test, compared with a sample size of 60 for use of the z test to obtain the same results.

Since there are both advantages and disadvantages to the nonparametric methods, the researcher should use caution in selecting these methods. If the assumptions can be met, the parametric methods are preferred. However, when parametric assumptions cannot be met, the nonparametric methods are a valuable tool for analyzing the data.

Ranking

Many nonparametric tests involve the **ranking** of data—that is, the positioning of a data value in a data array according to some rating scale. Ranking is an ordinal variable. For example, suppose a judge decides to rate five speakers on an ascending scale of 1 to 10, 1 being the best and 10 being the worst, for categories such as voice, gestures, logical presentation, and platform personality. The ratings are shown in the chart.

Speaker	A	B	C	D	E
Rating	8	6	10	3	1

The rankings are shown in the next chart.

Speaker	E	D	B	A	C
Rating	1	3	6	8	10
Ranking	1	2	3	4	5

Since speaker E received the lowest score, 1 point, he or she is ranked first. Speaker D received the next lower score, 3 points; he or she is ranked second; and so on.

What happens if two or more speakers receive the same number of points? Suppose the judge awards points as follows:

Speaker	A	B	C	D	E
Rating	8	6	10	6	3

The speakers are then ranked as follows:

Speaker	E	D	B	A	C
Rating	3	6	6	8	10
Ranking	1	Tie for 2nd and 3rd		4	5

When there is a tie for two or more places, the average of the ranks must be used. In this case, each would be ranked as

$$\frac{2+3}{2} = \frac{5}{2} = 2.5$$

Hence, the rankings are as follows:

Speaker	E	D	B	A	C
Rating	3	6	6	8	10
Ranking	1	2.5	2.5	4	5

Many times, the data are already ranked, so no additional computations must be done. For example, if the judge does not have to award points but can simply select the speakers who are best, second-best, third-best, and so on, then these ranks can be used directly.

The *P*-values can also be found for nonparametric statistical tests, and the *P*-value method can be used to test hypotheses that use nonparametric tests. For this chapter, the *P*-value method will be limited to some of the nonparametric tests that use the standard normal distribution or the chi-square distribution.

Exercises

14–1. What is meant by *nonparametric statistics*?

14–2. When should nonparametric statistics be used?

14–3. List the advantages and disadvantages of nonparametric statistics.

Ranking is used in many areas of statistics. Here is a ranking of the "healthiest" states. The rankings are based on many factors, as stated in the article. Which factors do you think should be the most important in ranking the states where living is best? Explain your answer.

New Hampshire, Exemplary State of Health

By Karla Price
USA TODAY

If you're in search of healthy living, New Hampshire is the place. If you live in Louisiana, better keep looking.

So says Northwestern National Life Insurance in its state health rankings, out Monday.

New Hampshire ranked healthiest, followed by Minnesota, with Utah and Connecticut tied for third. Louisiana rated 50th, slightly below Mississippi and West Virginia.

The ratings, conducted by Northwestern and a research firm annually since 1990, use 17 measures of

a state population's health. Mortality rates and prevalence of smoking each count 10%; other factors include access to primary care and incidence of disease.

Overall, Americans' health has improved 2.4% since 1990, due to state support for low-income health care, a decline in smoking and lower infant mortality, says Arlene Wheaton, project manager.

New Hampshire ranked high in several categories.

"Our health ranking is based, No. 1, on our low crime rate; second is our quality environment," says Gov. Steve Merrill.

The state isn't without

problems, including more smokers than average and a ranking of 33rd in unemployment.

Louisiana ranked 50th

in high school graduation rates, 48th in access to primary care and rates of heart disease and 46th in violent crime rates.

Ranking Where Living is Best

1 N.H.	14 Maine	* Ariz.	* Ky.
2 Minn.	15 N.J.	28 Del.	* Alaska
3 Utah	16 N.D.	* Calif.	42 Tenn.
* Conn.	* Wash.	30 N.C.	43 N.M.
5 Hawaii	* Md.	* Ill.	* Nev.
6 Vt.	19 R.I.	32 Texas	45 Ala.
7 Mass.	* Ohio	33 Mo.	46 S.C.
8 Iowa	21 Pa.	* Okla.	* Ark.
9 Va.	22 Ind.	* S.D.	48 W.Va.
10 Kan.	* Ore.	36 Ga.	49 Miss.
* Colo.	24 Mich.	* Wyo.	50 La.
* Neb.	25 Idaho	38 N.Y.	* indicates a
13 Wis.	* Mont.	* Fla.	tie.

Source: Northwestern National Life Insurance Co.

Source: *USA Today*, October 18, 1994. *USA TODAY.* Copyright 1994. Used with permission.

For Exercises 14–4 through 14–10, rank each set of data.

14–4. 7, 5, 9, 8, 4, 2, 1

14–5. 21, 65, 31, 41, 61, 41, 72, 34

14–6. 73, 320, 432, 186, 241

14–7. 8, 6, 3, 8, 7, 5, 5, 9, 12, 15, 17, 14

14–8. 22, 25, 28, 28, 18, 32, 37, 41, 41, 43

14–9. 190, 236, 187, 190, 321, 532, 673

14–10. 3.8, 7.9, 3.6, 4.1, 2.5, 7.9, 4.12, 3.21, 4.1

14–3

The Sign Test

Single-Sample Sign Test

Objective 2. Test hypotheses using the sign test.

The simplest nonparametric test, the **sign test** for single samples, is used to test the value of a median for a specific sample. When using the sign test, the researcher hypothesizes the specific value for the median of a population; then he or she selects a sample of data and compares each value with the conjectured median. If the data value is above the conjectured median, it is assigned a + sign. If it is below the conjectured median, it is assigned a − sign. And if it is exactly the same as the conjectured median, it is assigned a 0. Then the number of + and − signs are compared. If the null hypothesis is true, the number of + signs should be approximately equal to the number of − signs. If the null hypothesis is not true, there will be a disproportionate number of + or − signs.

Test Value for the Sign Test

The test value is the smaller number of + or − signs.

For example, if there are 8 positive signs and 3 negative signs, the test value is 3. When the sample size is 25 or less, Table J in Appendix C is used to determine the critical value. For a specific α, if the test value is less than or equal to the critical value obtained from the table, the null hypothesis should be rejected. The values in Table J are obtained from the binomial distribution. The derivation is omitted here.

Example 14–1

A convenience-store owner hypothesizes that the median number of snow cones he sells per day is 40. A random sample of 20 days yields the following data for the number of snow cones sold each day.

18	43	40	16	22
30	29	32	37	36
39	34	39	45	28
36	40	34	39	52

At $\alpha = 0.05$, test the owner's hypothesis.

Solution

STEP 1   State the hypotheses and identify the claim.

$$H_0: \text{median} = 40 \text{ (claim)} \quad \text{and} \quad H_1: \text{median} \neq 40$$

STEP 2   Find the critical value. Compare each value of the data with the median. If the value is greater than the median, replace the value with a + sign. If it is less than the median, replace it with a − sign. And if it is equal to the median, replace it with a 0. The completed table follows.

−	+	0	−	−
−	−	−	−	−
−	−	−	+	−
−	0	−	−	+

Refer to Table J in Appendix C, using $n = 18$ (the total number of + and − signs; omit the zeros) and $\alpha = 0.05$ for a two-tailed test; the critical value is 4. See Figure 14–1.

Figure 14–1

Finding the Critical Value in Table J for Example 14–1.

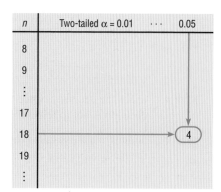

STEP 3   Compute the test value. Count the number of + and − signs obtained in step 2, and use the smaller value as the test value. Since there are 3 + signs and 15 − signs, 3 is the test value.

STEP 4   Make the decision. Compare the test value 3 with the critical value 4. If the test value is less than or equal to the critical value, the null hypothesis is rejected. In this case, the null hypothesis is rejected since $3 < 4$.

STEP 5   Summarize the results. There is enough evidence to reject the claim that the median number of snow cones sold per day is 40.

When the sample size is 26 or more, the normal approximation can be used to find the test value. The formula is given below. The critical value is found in Table E in Appendix C.

Formula for the z Test Value in the Sign Test when $n \geq 26$

$$z = \frac{(X + 0.5) - (n/2)}{\sqrt{n}/2}$$

where

X = smaller number of + or − signs
n = sample size

Example 14–2

Based on past experience, a manufacturer claims that the median lifetime of a rubber washer is at least 8 years. A sample of 50 washers showed that 21 lasted more than 8 years. At $\alpha = 0.05$, is there enough evidence to reject the manufacturer's claim?

Solution

STEP 1   State the hypotheses and identify the claim.

$$H_0: \text{MD} \geq 8 \text{ (claim)} \qquad \text{and} \qquad H_1: \text{MD} < 8$$

STEP 2   Find the critical value. Since $\alpha = 0.05$, and $n = 50$, and since this is a left-tailed test, the critical value is -1.65, obtained from Table E.

STEP 3   Compute the test value.

$$z = \frac{(X + 0.5) - (n/2)}{\sqrt{n}/2} = \frac{(21 + 0.5) - (50/2)}{\sqrt{50}/2}$$

$$= \frac{-3.5}{3.5355} = -0.99$$

STEP 4   Make the decision. Since the test value of -0.99 is greater than -1.65, the decision is not to reject the null hypothesis.

STEP 5   Summarize the results. There is not enough evidence to reject the claim that the median lifetime of the washers is at least 8 years.

In Example 14–2, the sample size was 50 and 21 washers lasted more than 8 years, so $50 - 21$, or 29, washers did not last 8 years. The value of X corresponds to the smaller of the two numbers 21 and 29. In this case, $X = 21$ is used in the formula. The reason is that there would be 21 positive signs, since subtracting 8 years from the value in years of a washer that lasted longer than 8 years would result in a positive answer. When 8 is subtracted from the value in years of a washer that did not last 8 years, the answer would be negative. Assuming that no washer lasted exactly 8 years would result in 21 positive answers and 29 negative answers. Since 21 is the smaller of the two numbers, the value of X is 21.

Suppose a researcher hypothesized that the median age of houses in a certain municipality is 40 years. In a random sample of 100 houses, 68 were older than 40 years. Then the value used for X in the formula would be $100 - 68$, or 32, since it is the smaller of the two numbers 68 and 32. When 40 is subtracted from the age of a house older than 40 years, the resultant answer is positive. When 40 is subtracted from the age of a house that is less than 40 years old, the result is negative. There would be 68 positive signs and 32 negative signs (assuming that no house was exactly 40 years old). Hence, 32 would be used for X, since it is the smaller of the two values.

Paired-Sample Sign Test

The sign test can also be used to test sample means in a comparison of two dependent samples, such as a before and after test. Recall that when dependent samples are taken from normally distributed populations, the t test is used (Section 10–5). When the condition of normality cannot be met, the nonparametric sign test can be used, as shown in the next example.

Example 14–3

A medical researcher believed the number of ear infections in swimmers can be reduced if the swimmers use earplugs. A sample of 10 people was selected, and the number of infections for a four-month period was recorded. During the first two months, the swimmers did not use the earplugs; during the second two months, they did. At the beginning of the second two-month period, each swimmer was examined to make sure that no infections were present. The data are shown here. At $\alpha = 0.05$, can the researcher conclude that using earplugs reduced the number of ear infections?

Swimmer	Before, X_B	After, X_A
A	3	2
B	0	1
C	5	4
D	4	0
E	2	1
F	4	3
G	3	1
H	5	3
I	2	2
J	1	3

Number of ear infections

Solution

STEP 1   State the hypotheses and identify the claim.

H_0: The number of ear infections will not be reduced.

H_1: The number of ear infections will be reduced (claim).

STEP 2   Find the critical value. Subtract the after values (X_A) from the before values (X_B) and indicate the difference by a $+$ or $-$ sign, according to the value, as shown in the following table.

Swimmer	Before, X_B	After, X_A	Sign of difference
A	3	2	$+$
B	0	1	$-$
C	5	4	$+$
D	4	0	$+$
E	2	1	$+$
F	4	3	$+$
G	3	1	$+$
H	5	3	$+$
I	2	2	0
J	1	3	$-$

From Table J, with $n = 9$ (the total number of $+$ and $-$ signs; the 0 is not counted) and $\alpha = 0.05$ (one-tailed), at most 1 $-$ sign is needed to reject the null hypothesis because 1 is the smallest entry in the $\alpha = 0.05$ column of Table J.

STEP 3   Compute the test value. Count the number of $+$ and $-$ signs found in step 2, and use the smaller value as the test value. There are 2 $-$ signs, so the test value is 2.

STEP 4   Make the decision. There are 2 $-$ signs. The decision is do not reject the null hypothesis. The reason is that with $n = 9$ the C.V. $= 1$ and $1 < 2$.

STEP 5   Summarize the results. There is not enough evidence to support the claim that the use of earplugs reduced the number of ear infections.

When conducting a one-tailed sign test, the researcher must scrutinize the data to determine whether or not they support the null hypothesis. If the data support the null hypothesis, there is no need to conduct the test. In Example 14–3, the null hypothesis states that the number of ear infections will not be reduced. The data would support the null hypothesis if there were more $-$ signs than $+$ signs. The reason is that the before values, X_B, would be in most cases smaller than the after values, X_A, and the $X_B - X_A$ values would be negative more often than positive. This would indicate that there is not enough evidence to reject the null hypothesis. The researcher would stop here, since there is no need to continue the procedure.

On the other hand, if the number of ear infections were reduced, the X_B values, for the most part, would be larger than the X_A values, and the $X_B - X_A$ values would most often be positive, as in Example 14–3. Hence, the researcher would continue the procedure. A word of caution is in order, and a little reasoning is required.

When the sample size is 26 or more, the normal approximation can be used in the same manner as in Example 14–2. The steps for conducting the sign test for single or paired samples are given in the Procedure Table.

Procedure Table

Sign Test for Single and Paired Samples

STEP 1   State the hypotheses and identify the claim.

STEP 2   Find the critical value(s). For the single-sample test, compare each value with the conjectured median. If the value is larger than the conjectured median, replace it with a + sign. If it is smaller than the conjectured median, replace it with a − sign.

For the paired-sample sign test, subtract the after from the before values, and indicate the difference with a + or − sign, according to the value. Use Table J and n = total number of + and − signs.

Check the data to see whether or not they support the null hypothesis. If they do, do not reject the null hypothesis. If not, continue with step 3.

STEP 3   Compute the test value. Count the number of + and − signs found in step 2, and use the smaller value as the test value.

STEP 4   Make the decision. Compare the test value with the critical value in Table J. If the test value is less than or equal to the critical value, reject the null hypothesis.

STEP 5   Summarize the results.

Note: If the sample size, n, is 26 or more, use Table E and the following formula for the test value:

$$z = \frac{(X + 0.5) - (n/2)}{\sqrt{n}/2}$$

where

X = smaller number of + or − signs
n = sample size

Exercises

14–11. Why is the sign test the simplest nonparametric test to use?

14–12. What population parameter can be tested with the sign test?

14–13. In the sign test, what is used as the test value when $n < 26$?

14–14. When $n \geq 26$, what is used in place of Table J for the sign test?

For Exercises 14–15 through 14–30, perform the following steps.

a.   State the hypotheses and identify the claim.
b.   Find the critical value(s).
c.   Compute the test value.
d.   Make the decision.
e.   Summarize the results.

Use the traditional method of hypothesis testing unless otherwise specified.

14–15. An oceanographer believes that the median height of the waves at Ocean City is 2.8 feet. The wave heights are measured for a random sample of 20 days. The data are as shown here. At $\alpha = 0.05$, is there enough evidence to reject the oceanographer's claim?

3.6	2.1	2.3	2.1	2.7
3.2	3.9	3.4	3.0	2.9
2.0	1.9	3.2	3.5	2.8
1.8	2.3	3.7	3.9	4.2

14–16. A meteorologist suggests that the median temperature for the month of July in Jacksonville, Florida, is 81°F. The sample here shows the temperatures taken at noon in Jacksonville during 20 days in July. At $\alpha = 0.01$, is there enough evidence to reject the meteorologist's claim?

81	83	87	92	91
78	73	81	93	96
79	80	84	86	82
85	77	72	73	80

14–17. A real estate agent suggests that the median rent for a one-bedroom apartment in Blue View is $325 per month. A sample of 12 one-bedroom apartments shows the following monthly rent for a one-bedroom apartment. At $\alpha = 0.05$, is there enough evidence to reject the agent's claim?

$420	$460	$514	$405
320	435	531	450
560	309	312	350

14–18. A government economist estimates that the median cost per pound of beef is $5.00. A sample of 22 livestock buyers shows the following costs per pound of beef. Is there enough evidence to reject the economist's hypothesis at $\alpha = 0.10$?

$5.35	$5.16	$4.97	$4.83	$5.05	$5.19
4.78	4.93	4.86	5.00	4.63	5.06
5.19	5.00	5.05	5.10	5.16	5.25
5.16	5.42	5.13	5.27		

14–19. Fifteen out of 37 first-grade students completed a manual dexterity test in less than 7.30 minutes. At $\alpha = 0.05$ is there enough evidence to reject the claim that the median time to complete the test is 7.30 minutes?

14–20. One hundred people were placed on a special exercise program. After one month, 58 lost weight, 12 gained weight, and 30 weighed the same as before. Test the hypothesis that the exercise program is effective at $\alpha = 0.10$. (*Note:* It will be effective if fewer than 50% of the people did not lose weight.)

14–21. Of 50 students surveyed, 29 favored single-room dormitories. At $\alpha = 0.02$, test the hypothesis that more than 50% of the students favor single-room dormitories. Use the *P*-value method.

14–22. Of 75 lifeguards surveyed, 23 favored water over an electrolyte-balanced drink to maintain hydration. At $\alpha = 0.02$ test the hypothesis that less than 50% of the lifeguards favor water to maintain hydration.

14–23. One hundred students are asked if they favor increasing the school year by 20 days. The responses are 62 no, 36 yes, and 2 undecided. At $\alpha = 0.10$, test the hypothesis that 50% of the students are against extending the school year. Use the *P*-value method.

14–24. Is there enough evidence to reject the hypothesis that the median age of marathon runners is 27, if, out of 29 marathon runners, 12 are older than 27? Use $\alpha = 0.05$.

14–25. A study was conducted to see whether a certain diet medication had an effect on the weights (in pounds) of eight women. Their weights were taken before and six weeks after daily administration of the medication. The data are shown here. At $\alpha = 0.05$, can one conclude that the medication had an effect (increase or decrease) on the weights of the women?

Subject	A	B	C	D	E	F	G	H
Weight before	187	163	201	158	139	143	198	154
Weight after	178	162	188	156	133	150	175	150

14–26. Two different laboratory machines measure the sodium content (in milligrams) of the same 10 blood samples. The data are shown here. At $\alpha = 0.01$, test the claim that both machines gave the same reading.

Sample	1	2	3	4	5	6	7	8	9	10
Machine 1	138	136	142	151	154	141	140	138	132	136
Machine 2	140	136	141	150	153	144	143	136	131	138

14–27. An educator designed a reasoning skills course. Nine students were selected and given a pretest to determine their reasoning abilities. After completing the course, the same students were given an equivalent form of the test to see whether their reasoning skills had improved. The data are shown here. At $\alpha = 0.05$, did the course improve their reasoning skills?

Student	1	2	3	4	5	6	7	8	9
Pretest	80	76	74	83	92	78	91	74	88
Posttest	82	78	73	85	95	79	93	78	90

14–28. A researcher wishes to test the effects of a pill on a person's appetite. Twelve subjects are allowed to eat a meal of their choice, and their caloric intake is measured. The next day, the same subjects take the pill and eat a meal of their choice. The caloric intake of the second meal is measured. The data are shown here. At $\alpha = 0.02$, can the researcher conclude that the pill had an effect on a person's appetite?

Subject	1	2	3	4	5	6	7	8
Meal 1	856	732	900	1321	843	642	738	1005
Meal 2	843	721	872	1341	805	531	740	900

Subject	9	10	11	12
Meal 1	888	756	911	998
Meal 2	805	695	878	914

14–29. In order to test a theory that alcohol consumption can have an effect on test scores, a researcher conducts a study on 10 adults. Each is given a test. Then for one week, each subject is required to consume a certain amount of alcohol; then he or she is retested. The results are shown here. At $\alpha = 0.10$, test the claim that alcohol does not affect a person's test score.

Subject	Score before	Scored after
1	105	106
2	109	105
3	98	94
4	112	109
5	109	105
6	117	115
7	123	125
8	114	114
9	95	98
10	101	100

14–30. A manufacturer believes that if routine maintenance (cleaning and oiling of machines) is increased to once a day rather than once a week, the number of defective parts produced by the machines will decrease. Nine machines are selected, and the number of defective parts produced over a 24-hour operating period is counted. Maintenance is then increased to once a day for a week, and the number of defective parts each machine produces is again counted over a 24-hour operating period. The data are shown here. At $\alpha = 0.01$, can the manufacturer conclude that increased maintenance reduces the number of defective parts manufactured by the machines?

Machine	1	2	3	4	5	6	7	8	9
Before	6	18	5	4	16	13	20	9	3
After	5	16	7	4	18	12	14	7	1

The confidence interval for the median of a set of values less than or equal to 25 in number can be found by ordering the data from smallest to largest, finding the median, and using Table J. For example, to find the 95% confidence interval of the true median for 17, 19, 3, 8, 10, 15, 1, 23, 2, 12, order the data:

1, 2, 3, 8, 10, 12, 15, 17, 19, 23

From Table J, select $n = 10$ and $\alpha = 0.05$, and find the critical value. Use the two-tailed row. In this case, the critical value is 1. Add 1 to this value to get 2. In the ordered list, count from the left two numbers and from the right two numbers, and use these numbers to get the confidence interval, as shown:

1, 2, 3, 8, 10, 12, 15, 17, 19, 23
$$2 \leq MD \leq 19$$

Always add 1 to the number obtained from the table before counting. For example, if the critical value is 3, then count 4 values from the left and right.

For Exercises 14–31 through 14–35, find the confidence interval of the median, indicated in parentheses, for each set of data.

14–31. 3, 12, 15, 18, 16, 15, 22, 30, 25, 4, 6, 9 (95%)

14–32. 101, 115, 143, 106, 100, 142, 157, 163, 155, 141, 145, 153, 152, 147, 143, 115, 164, 160, 147, 150 (90%)

14–33. 8.2, 7.1, 6.3, 5.2, 4.8, 9.3, 7.2, 9.3, 4.5, 9.6, 7.8, 5.6, 4.7, 4.2, 9.5, 5.1 (98%)

14–34. 1, 8, 2, 6, 10, 15, 24, 33, 56, 41, 58, 54, 5, 3, 42, 31, 15, 65, 21 (99%)

14–35. 12, 15, 18, 14, 17, 19, 25, 32, 16, 47, 14, 23, 27, 42, 33, 35, 39, 41, 21, 19 (95%)

Technology Step by Step

MINITAB
Step by Step

The Sign Test

Example MT14–1

1. Type the data for Example 14–1 into a column of MINITAB. Name the column **SnowCones**.
2. Select **Stat>Nonparametrics>1 Sample Sign Test**.
3. Double click the SnowCones column in the list box.
4. Click on Test median.
5. Click in the text box for the median value and enter the hypothesized value of **40**.

Sign Test Dialog Box

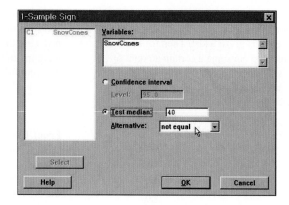

6. Click [OK].

In the session window, the *P*-value is .0075. Reject the null hypothesis.

....

Paired Sample Sign Test

Example MT14–2

1. Enter the two columns of data into a worksheet. Use Example 14-2.
2. To make a column with the differences select **Calc>Calculator.**

Calculate the Differences

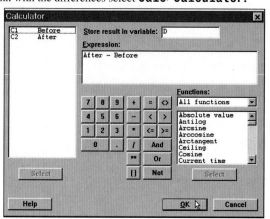

3. Type **D** in the `Store result in variable:` box.
4. Press the **TAB** key to move to the `Expression` box.
5. Click on **C2** in the list, then the minus-sign button, then **C1**.
6. Click [OK]. Minitab will calculate the differences and store them in the first available column with the name **D**.
7. Follow the instructions for the sign test above with a hypothesized value of zero.

The resulting *P*-value is .1797. Do not reject the null hypothesis.

The Wilcoxon Rank Sum Test

The sign test does not consider the magnitude of the data. For example, whether a value is 1 point or 100 points below the median, it will receive a − sign. And when one compares values in the pretest/posttest situation, the magnitude of the differences is not considered. The Wilcoxon tests consider differences in magnitudes by using ranks.

The two tests considered in this and the next section are the **Wilcoxon rank sum test,** which is used for independent samples, and the **Wilcoxon signed-rank test,** which

Objective 3. Test hypotheses using the Wilcoxon rank sum test.

is used for dependent samples. Both tests are used to compare distributions. The parametric equivalents are the z and t tests for independent samples (Sections 10–2 and 10–4) and the t test for dependent samples (Section 10–5). For the parametric tests, as stated previously, the samples must be selected from approximately normally distributed populations, but the only assumption for the Wilcoxon signed-rank tests is that the population of differences has a symmetric distribution.

In the Wilcoxon tests, the values of the data for both samples are combined and then ranked. If the null hypothesis is true—meaning that there is no difference in the population distributions—then the values in each sample should be ranked approximately the same. Therefore, when the ranks are summed for each sample, the sums should be approximately equal, and the null hypothesis will not be rejected. If there is a large difference in the sums of the ranks, then the distributions are not identical, and the null hypothesis will be rejected.

The first test to be considered is the Wilcoxon rank sum test for independent samples. For this test, both sample sizes must be greater than or equal to 10. The formulas needed for the test are given next.

Formula for the Wilcoxon Rank Sum Test when Samples Are Independent

$$z = \frac{R - \mu_R}{\sigma_R}$$

where

$$\mu_R = \frac{n_1(n_1 + n_2 + 1)}{2}$$

$$\sigma_R = \sqrt{\frac{n_1 n_2(n_1 + n_2 + 1)}{12}}$$

R = sum of the ranks for the smaller sample size (n_1)
n_1 = smaller of the sample sizes
n_2 = larger of the sample sizes
$n_1 \geq 10$ and $n_2 \geq 10$

Note that if both samples are the same size, either size can be used as n_1.

The next example illustrates the Wilcoxon rank sum test for independent samples.

Example 14–4

Two independent samples of army and marine recruits are selected, and the time in minutes it takes each recruit to complete an obstacle course is recorded as shown in the table. At $\alpha = 0.05$, is there a difference in the times it takes the recruits to complete the course?

Army	15	18	16	17	13	22	24	17	19	21	26	28	Mean = 19.67
Marines	14	9	16	19	10	12	11	8	15	18	25		Mean = 14.27

Solution

STEP 1   State the hypotheses and identify the claim.

H_0: There is no difference in the times it takes the recruits to complete the obstacle course.

H_1: There is a difference in the times it takes the recruits to complete the obstacle course (claim).

STEP 2 Find the critical value. Since $\alpha = 0.05$ and this test is a two-tailed test, use the z values of $+1.96$ and -1.96 from Table E.

STEP 3 Compute the test value.

    a. Combine the data from the two samples, arrange the combined data in order, and rank each value. Be sure to indicate the group.

Time	8	9	10	11	12	13	14	15	15	16	16	17
Group	M	M	M	M	M	A	M	A	M	A	M	A
Rank	1	2	3	4	5	6	7	8.5	8.5	10.5	10.5	12.5

Time	17	18	18	19	19	21	22	24	25	26	28
Group	A	M	A	A	M	A	A	A	M	A	A
Rank	12.5	14.5	14.5	16.5	16.5	18	19	20	21	22	23

    b. Sum the ranks of the group with the smaller sample size. (*Note:* If both groups have the same sample size, either one can be used.) In this case, the sample size for the marines is smaller.

$$R = 1 + 2 + 3 + 4 + 5 + 7 + 8.5 + 10.5 + 14.5 + 16.5 + 21$$
$$= 93$$

    c. Substitute in the formulas to find the test value.

$$\mu_R = \frac{n_1(n_1 + n_2 + 1)}{2} = \frac{(11)(11 + 12 + 1)}{2} = 132$$

$$\sigma_R = \sqrt{\frac{n_1 n_2(n_1 + n_2 + 1)}{12}} = \sqrt{\frac{(11)(12)(11 + 12 + 1)}{12}}$$
$$= \sqrt{264} = 16.2$$

$$z = \frac{R - \mu_R}{\sigma_R} = \frac{93 - 132}{16.2} = -2.41$$

STEP 4 Make the decision. The decision is to reject the null hypothesis, since $-2.41 < -1.96$.

STEP 5 Summarize the results. There is enough evidence to support the claim that there is a difference in the times it takes the recruits to complete the course.

··

The steps for the Wilcoxon rank sum test are given in the Procedure Table.

Procedure Table

Wilcoxon Rank Sum Test

STEP 1 State the hypotheses and identify the claim.

STEP 2 Find the critical value(s). Use Table E.

STEP 3 Compute the test value.

    a. Combine the data from the two samples, arrange the combined data in order, and rank each value.

continued

Procedure Table (concluded)

    b. Sum the ranks of the group with the smaller sample size. (*Note:* If both groups have the same sample size, either one can be used.)

    c. Use these formulas to find the test value.

$$\mu_R = \frac{n_1(n_1 + n_2 + 1)}{2}$$

$$\sigma_R = \sqrt{\frac{n_1 n_2 (n_1 + n_2 + 1)}{12}}$$

$$z = \frac{R - \mu_R}{\sigma_R}$$

where R is the sum of the ranks of the data in the smaller sample and n_1 and n_2 are each greater than or equal to 10.

STEP 4   Make the decision.

STEP 5   Summarize the results.

Exercises

14–36. Explain the difference between the Wilcoxon tests and the sign test.

14–37. What are the minimum samples sizes for the Wilcoxon rank sum test?

14–38. What are the parametric equivalent tests for the Wilcoxon rank sum tests?

14–39. What distribution is used for the Wilcoxon rank sum test?

For Exercises 14–40 through 14–47, use the Wilcoxon rank sum test. Assume that the samples are independent. Also perform the following steps.

    a. State the hypotheses and identify the claim.
    b. Find the critical value(s).
    c. Compute the test value.
    d. Make the decision
    e. Summarize the results.

Use the traditional method of hypothesis testing unless otherwise specified.

14–40. A random sample of men and women in prison was asked to give the length of sentence each received for a certain type of crime. At $\alpha = 0.05$, test the claim that there is no difference in the sentence received by each gender. The data (in months) are shown here.

Males	8	12	6	14	22	27	32	24	26	19	15
Females	7	5	2	3	21	26	30	9	4	17	23

Males	13
Females	12	11	16

14–41. A researcher surveyed married women and single women to ascertain whether there was a difference in the number of books each had read during the past year. The data are shown here. At $\alpha = 0.10$, test the claim that each group read the same number of books.

Married	6	8	7	4	9	12	13	7	10	18	15
Single	2	3	5	11	3	5	11	12	16	4	0	1

14–42. To test the claim that there is no difference in the lifetimes of two brands of handheld video games, a researcher selects a sample of 11 video games of each brand. The lifetimes (in months) of each brand are shown here. At $\alpha = 0.01$ can the researcher conclude that there is a difference in the distributions of lifetimes for the two brands?

Brand A	42	34	39	42	22	47	51	34	41	39	28
Brand B	29	39	38	43	45	49	53	38	44	43	32

14–43. Over the past 12 years, a statistician kept track of the total number of academic scholarships awarded to Valley View High School seniors and seniors at their rival school, Ocean View High School. The data are shown here. At $\alpha = 0.05$, is there a difference in the number of academic scholarships awarded to seniors at the schools?

Valley View	4	4	1	8	7	9	3	7	4	6	11	10
Ocean View	4	5	2	7	8	6	3	9	3	5	6	12

14–44. Two groups of employees were given a questionnaire to ascertain their degree of job satisfaction. The scale ranged from 0 to 100. The groups were divided into those who had under five years of work experience and those who had five or more years of experience. The data are shown here. At $\alpha = 0.10$, test the claim that there is no difference in the job satisfaction of the two groups, as measured by the questionnaire. Use the *P*-value method.

Under 5	78	98	83	86	75	77	72
5 and over	94	79	82	85	73	66	64

Under 5	68	56	93	97	99	93
5 and over	59	52	58	63	68	88

14–45. The results of a study of payments for flood damages awarded by insurance companies in two Texas cities are shown here. The data are given in dollars. At $\alpha = 0.05$, is there a difference in the amount of money awarded for flood damages in the two cities?

City A	$563	648	925	602	921	232
City B	$869	718	626	453	832	752

City A	$953	824	605	601	687	431
City B	$769	324	885	927	918	239

14–46. Supervisors were asked to rate the productivity of employees on their jobs. A researcher wishes to see whether married men receive higher ratings than single men. A rating scale of 1 to 50 yielded the data shown here. At $\alpha = 0.01$, is there evidence to support this claim?

Single men	48	46	42	50	38	36	40	31	28	24	49	34
Married men	44	35	41	37	42	43	29	31	37	32	36	

14–47. A study was conducted to see whether there is a difference in the time it takes employees of a factory to assemble the product. Samples of high school graduates and nongraduates were timed. At $\alpha = 0.05$, is there a difference in the distributions for the two groups in the times needed to assemble the product? The data (in minutes) are shown here.

Graduates	3.6	3.2	4.4	3.0	5.6	6.3	8.2
Nongraduates	2.7	3.8	5.3	1.6	1.9	2.4	2.9

Graduates	7.1	5.8	7.3	6.4	4.2	4.7
Nongraduates	1.7	2.6	2.0	3.1	3.4	3.9

Technology Step by Step

MINITAB
Step by Step

Wilcoxon Rank Sum (Mann Whitney) Test

The Wilcoxon rank sum test is also know as the Mann Whitney test

Example MT14–3

1. Enter the data for Example 14–4 into two columns of a worksheet. Name the columns **Army** and **Marines.**

2. Select **Stat>Nonparametrics>Mann-Whitney**

Mann-Whitney Dialog Box

3. Double click Army to select it for the First Sample.

4. Double click Marines to select it for the Second Sample.

5. Click [OK].

Mann-Whitney Test and CI: Army, Marines

```
Army      N =  12   Median =      18.500
Marines   N =  11   Median =      14.000
Point estimate for ETA1-ETA2 is       6.000
95.5 Percent CI for ETA1-ETA2 is (1.003, 9.998)
W = 183.0
Test of ETA1 = ETA2 vs ETA1 not = ETA2 is significant at 0.0178
The test is significant at 0.0177 (adjusted for ties)
```

The *P*-value for the test is 0.0177. Reject the null hypothesis. There is a difference in the times it takes the recruits to complete the course. The confidence interval estimate is between 1.003 and 9.998.

14–5

The Wilcoxon Signed-Rank Test

Objective 4. Test hypotheses using the signed-rank test.

When the samples are dependent, as they would be in a before and after test using the same subjects, the Wilcoxon signed-rank test can be used in place of the *t* test for dependent samples. Again, this test does not require the condition of normality. Table K is used to find the critical values.

The procedure for this test is shown in the next example.

Example 14–5

In a large department store, the owner wishes to see whether the number of shoplifting incidents per day will change if the number of uniformed security officers is doubled. A sample of seven days before security is increased and seven days after the increase shows the number of shoplifting incidents.

Day	Number of shoplifting incidents	
	Before	**After**
Monday	7	5
Tuesday	2	3
Wednesday	3	4
Thursday	6	3
Friday	5	1
Saturday	8	6
Sunday	12	4

Is there enough evidence to support the claim, at $\alpha = 0.05$, that there is a difference in the number of shoplifting incidents before and after the increase in security?

Solution

STEP 1   State the hypotheses and identify the claim.

H_0: There is no difference in the number of shoplifting incidents before and after the increase in security.

H_1: There is a difference in the number of shoplifting incidents before and after the increase in security (claim).

STEP 2   Find the critical value from Table K. Since $n = 7$ and $\alpha = 0.05$ for this two-tailed test, the critical value is 2. See Figure 14–2.

Figure 14–2

Finding the Critical
Value in Table K for
Example 14–5

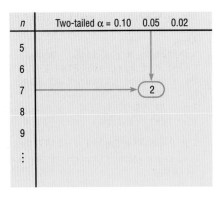

STEP 3 Find the test value.

a. Make a table as shown here.

Day	Before, X_B	After, X_A	Difference, $D = X_B - X_A$	Absolute value, $|D|$	Rank	Signed rank
Mon.	7	5				
Tues.	2	3				
Wed.	3	4				
Thurs.	6	3				
Fri.	5	1				
Sat.	8	6				
Sun.	12	4				

b. Find the differences (before − after), and place the values in the
"Difference" column.

$$7 - 5 = 2 \qquad 6 - 3 = 3 \qquad 8 - 6 = 2$$
$$2 - 3 = -1 \qquad 5 - 1 = 4 \qquad 12 - 4 = 8$$
$$3 - 4 = -1$$

c. Find the absolute value of each difference, and place the results in the
"Absolute value" column. (*Note:* The absolute value of any number
except 0 is the positive value of the number. Any differences of 0 should
be ignored.)

$$|2| = 2 \qquad |3| = 3 \qquad |2| = 2$$
$$|-1| = 1 \qquad |4| = 4 \qquad |8| = 8$$
$$|-1| = 1$$

d. Rank each absolute value from lowest to highest, and place the rankings
in the "Rank" column. In the case of a tie, assign each rank that value
plus 0.5.

Value	2	1	1	3	4	2	8
Rank	3.5	1.5	1.5	5	6	3.5	7

e. Give each rank a + or − sign, according to the sign in the "Difference"
column. The completed table is shown here.

Day	Before, X_B	After, X_A	Difference, $D = X_B - X_A$	Absolute value, $|D|$	Rank	Signed rank
Mon.	7	5	2	2	3.5	+3.5
Tues.	2	3	−1	1	1.5	−1.5
Wed.	3	4	−1	1	1.5	−1.5
Thurs.	6	3	3	3	5	+5
Fri.	5	1	4	4	6	+6
Sat.	8	6	2	2	3.5	+3.5
Sun.	12	4	8	8	7	+7

f. Find the sum of the positive ranks and the sum of the negative ranks separately.

$$\text{positive rank sum } (+3.5) + (+5) + (+6) + (+3.5) + (+7) = +25$$
$$\text{negative rank sum } (-1.5) + (-1.5) \qquad\qquad\qquad = -3$$

g. Select the smaller of the absolute values of the sums ($|-3|$), and use this absolute value as the test value, w_s. In this case, $w_s = |-3| = 3$.

STEP 4 Make the decision. Reject the null hypothesis if the test value is less than or equal to the critical value. In this case, $3 > 2$; hence, the decision is not to reject the null hypothesis.

STEP 5 Summarize the results. There is not enough evidence to support the claim that there is a difference in the number of shoplifting incidents. Hence, the security increase probably made no difference in the number of shoplifting incidents.

The rationale behind the signed-rank test can be explained by a diet example. If the diet is working, then the majority of the postweights will be smaller than the preweights. When the postweights are subtracted from the preweights, the majority of the signs will be positive, and the absolute value of the sum of the negative ranks will be small. This sum will probably be smaller than the critical value obtained from Table K, and the null hypothesis will be rejected. On the other hand, if the diet does not work, some people will gain weight, some people will lose weight, and some people will remain about the same in weight. In this case, the sum of the positive ranks and the absolute value of the sum of the negative ranks will be approximately equal and will be about half of the sum of the absolute value of all the ranks. In this case, the smaller of the absolute value of the two sums will still be larger than the critical value obtained from Table K, and the null hypothesis will not be rejected.

When $n \geq 30$, the normal distribution can be used to approximate the Wilcoxon distribution. The same critical values from Table E used for the z test for specific α values are used. The formula is

$$z = \frac{w_s - \dfrac{n(n + 1)}{4}}{\sqrt{\dfrac{n(n + 1)(2n + 1)}{24}}}$$

where

n = number of pairs where the difference is not 0
w_s = smaller sum in absolute value of the signed ranks

The steps for the Wilcoxon signed-rank test are given in the Procedure Table.

Procedure Table

Wilcoxon Signed-Rank Test

STEP 1 State the hypotheses and identify the claim.

STEP 2 Find the critical value from Table K.

STEP 3 Compute the test value.

    a. Make a table, as shown.

Before, X_B	After, X_A	Difference, $D = X_B - X_A$	Absolute value, $\lvert D \rvert$	Rank	Signed rank

    b. Find the differences (before − after), and place the values in the "Difference" column.

    c. Find the absolute value of each difference, and place the results in the "Absolute value" column.

    d. Rank each absolute value from lowest to highest, and place the rankings in the "Rank" column.

    e. Give each rank a + or − sign, according to the sign in the "Difference" column.

    f. Find the sum of the positive ranks and the sum of the negative ranks separately.

    g. Select the smaller of the absolute values of the sums, and use this absolute value as the test value, w_s.

STEP 4 Make the decision. Reject the null hypothesis if the test value is less than or equal to the critical value.

STEP 5 Summarize the results.

    Note: When $n \geq 30$, use Table E and the following test value:

$$z = \frac{w_s - \dfrac{n(n+1)}{4}}{\sqrt{\dfrac{n(n+1)(2n+1)}{24}}}$$

where

    n = number of pairs where the difference is not 0
    w_s = smaller sum in absolute value of the signed ranks

Exercises

14–48. What is the parametric equivalent test for the Wilcoxon signed-rank test?

14–49. In the Wilcoxon signed-rank test, is the smaller or larger sum, in absolute value, of the signed ranks used as the test value?

For Exercises 14–50 and 14–51, find the sum of the signed ranks. Assume that the samples are dependent. State which sum is used as the test value.

14–50.

Pretest	18	32	35	37	25	41	52	43	56	62
Posttest	20	21	26	37	29	40	31	37	51	65

Speaking of STATISTICS

The data show the number of police officers in the 10 largest cities in the United States for 1983 and 1993. The signed-rank test can be used to determine if the sizes have changed. See Exercise 14–59.

Changes in Police Force Size		
City	1983	1993
New York	23,339	29,327
Los Angeles	6,886	7,637
Chicago	12,353	12,093
Houston	3,716	4,734
Philadelphia	7,218	6,225
San Diego	1,376	1,861
Detroit	3,808	3,860
Dallas	2,084	2,807
Phoenix	1,635	1,978
San Antonio	1,159	1,662

Source: Federal Bureau of Investigation. *USA Today,* February 17, 1995.

14–51.

Pretest	108	97	115	162	156	105	153
Posttest	110	97	103	168	143	112	141

For Exercises 14–52 through 14–56, use Table K to determine whether the null hypothesis should be rejected.

14–52. $w_s = 62$, $n = 21$, $\alpha = 0.05$, two-tailed test

14–53. $w_s = 18$, $n = 15$, $\alpha = 0.02$, two-tailed test

14–54. $w_s = 53$, $n = 25$, $\alpha = 0.05$, one-tailed test

14–55. $w_s = 142$, $n = 28$, $\alpha = 0.05$, one-tailed test

14–56. $w_s = 109$, $n = 27$, $\alpha = 0.025$, one-tailed test

14–57. Eight students were given a pretest to measure their public speaking anxiety. They completed a workshop to reduce their anxiety and were then given a posttest. At $\alpha = 0.05$ can one conclude that the workshop reduced their anxiety? The pretest and posttest scores are shown here. (*Note:* A lower score indicates a lower anxiety level.)

Pretest	23	26	30	31	39	23	28	27
Posttest	22	29	27	29	33	21	25	28

14–58. In a corporation, male and female workers were matched according to years of experience working for the company. Their salaries were then compared. The data (in thousands of dollars) are shown in the table. At $\alpha = 0.10$, is there a difference in the salaries of the males and females?

Males	18	43	32	27	15	45	21	22
Females	16	38	35	29	15	46	25	28

14–59. Using the data about police forces shown in the above "Speaking of Statistics," can it be concluded that police forces have become larger in the sample of cities shown in the data? Use $\alpha = 0.05$.

1983	23,339	6,886	12,353	3,716	7,218	1,376	3,808
1993	29,327	7,637	12,093	4,734	6,225	1,861	3,860

1983	2,084	1,635	1,159
1993	2,807	1,978	1,662

Source: *USA Today,* February 17, 1995.

14–60. Eight couples are given a questionnaire designed to measure marital compatibility. After completing a workshop, they are given a second questionnaire to see whether there is a change in their attitudes toward each other. The data are as shown here. At $\alpha = 0.10$, is there any difference in the scores of the couples?

Before	43	52	37	29	51	62	57	61
After	48	59	36	29	60	68	59	72

14–61. Using the data shown in the "Speaking of Statistics," test the claim at $\alpha = 0.10$ that there is no change in the 1995 lacrosse rosters as compared to the 1994 rosters. The data are shown here.

1995	33	37	46	34	45	58	46	23	35	22	45	21	42
1994	33	29	42	32	28	55	46	26	17	30	45	21	42

Source: *Pittsburgh Tribune-Review,* May 14, 1995.

The data show the size of the rosters for schools in a Lacrosse league. The signed-rank test can be used to de- termine if the sizes have been changed. See Exer- cise 14–61.

Lacrosse by the Numbers
(Western Pennsylvania Interscholastic Lacrosse League)

School	Years	1995 roster	1994 roster	Fee
Baldwin	5	33	33	$160
Bethel Park	2	37	29	$100
Central Catholic	7	46	42	$110
Fox Chapel	8	34	32	$100
Franklin Regional	7	45	28	$175
Mt. Lebanon	9	58	55	$245
North Alleghany	6	46	46	$200
North Catholic	4	23	26	(NF)
North Hills	2	35	17	$100
Shaler Area	4	22	30	$200
Shady Side Academy	25	45	45	(NF)
Sewickley Academy	25	21	21	(NF)
Upper St. Clair	12	42	42	$200

Rosters includes JV squad, if team has one.
(NF): No registration fee charged.

Source: *Pittsburgh Tribune-Review,* May 14, 1995.

Technology Step by Step

MINITAB
Step by Step

Wilcoxon Signed-Rank Test

Example MT 14–4

1. Enter the data for Example 14–5 into two columns of a worksheet. Name the columns **Before** and **After.**

2. To calculate the differences select **Calc>Calculator.** Type **D** in the box for `Store result in variable:`. In the `Expression` box type **Before-After** and click [`OK`].

Calculate the Differences

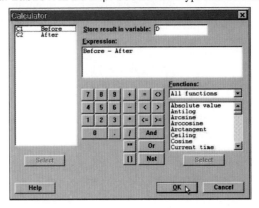

3. Select **Stat>Nonparametrics>1-Sample Wilcoxon**.

4. Double click C3 to select the differences as the Variable.

Wixcoxon Signed-Rank
Dialog Box

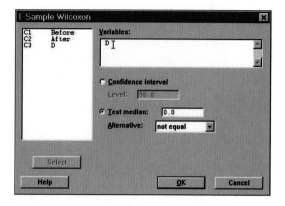

5. Click on Test median. The value should be 0.

6. Click [OK].

Wilcoxon Signed Rank Test: D

```
Test of median = 0.000000 versus median not = 0.000000

                 N for      Wilcoxon                  Estimated
          N      Test      Statistic        P          Median
D         7        7           25.0      0.076           2.250
```

The *P*-value of the test is 0.076. Do not reject the null hypothesis.

The Kruskal-Wallis Test

Objective 5. Test hypotheses using the Kruskal-Wallis test.

The analysis of variance uses the *F* test to compare the means of three or more populations. The assumptions for the ANOVA test are that the populations are normally distributed and that the population variances are equal. When these assumptions cannot be met, the nonparametric **Kruskal-Wallis test,** sometimes called the *H* test, can be used to compare three or more means.

In this test, each sample size must be five or more. In these situations, the distribution can be approximated by the chi-square distribution with $k - 1$ degrees of freedom, where $k =$ number of groups. This test also uses ranks. The formula for the test is given next.

In the Kruskal-Wallis test, one considers all the data values as a group and then ranks them. Next, the ranks are separated and the *H* formula is computed. This formula approximates the variance of the ranks. If the samples are from different populations, the sum of the ranks will be different, and the *H* value will be large; hence, the null hypothesis will be rejected if the *H* value is large enough. If the samples are from the same population, the sum of the ranks will be approximately the same, and the *H* value will be small; therefore, the null hypothesis will not be rejected. This test is always a right-tailed test. The chi-square table, Table G, with d.f. $= k - 1$, should be used for critical values.

Formula for the Kruskal-Wallis Test

$$H = \frac{12}{N(N+1)} \left(\frac{R_1^2}{n_1} + \frac{R_2^2}{n_2} + \cdots + \frac{R_k^2}{n_k} \right) - 3(N+1)$$

where

R_1 = sum of the ranks of sample 1
n_1 = size of sample 1
R_2 = sum of the ranks of sample 2
n_2 = size of sample 2
.
.
.
R_k = sum of the ranks of sample k
n_k = size of sample k
$N = n_1 + n_2 + \cdots + n_k$
k = number of samples

The next example illustrates the procedure for conducting the Kruskal-Wallis test.

Example 14–6

A researcher tests three different brands of breakfast drinks to see how many milliequivalents of potassium per quart each contains. The following data are obtained.

Brand A	Brand B	Brand C
4.7	5.3	6.3
3.2	6.4	8.2
5.1	7.3	6.2
5.2	6.8	7.1
5.0	7.2	6.6

At a = 0.05, is there enough evidence to reject the hypothesis that all brands contain the same amount of potassium?

Solution

STEP 1 State the hypotheses and identify the claim.

H_0: There is no difference in the amount of potassium contained in the brands (claim).

H_1: There is a difference in the amount of potassium contained in the brands.

STEP 2 Find the critical value. Use the chi-square table, Table G, with d.f. = $k - 1$ (k = number of groups). With $\alpha = 0.05$ and d.f. = $3 - 1 = 2$, the critical value is 5.991.

STEP 3 Compute the test value.

a. Arrange all the data from the lowest to highest, and rank each value.

Amount	Brand	Rank
3.2	A	1
4.7	A	2
5.0	A	3
5.1	A	4
5.2	A	5
5.3	B	6
6.2	C	7
6.3	C	8
6.4	B	9
6.6	C	10
6.8	B	11
7.1	C	12
7.2	B	13
7.3	B	14
8.2	C	15

b. Find the sum of the ranks of each brand.

$$\text{brand A} \quad 1 + 2 + 3 + 4 + 5 = 15$$
$$\text{brand B} \quad 6 + 9 + 11 + 13 + 14 = 53$$
$$\text{brand C} \quad 7 + 8 + 10 + 12 + 15 = 52$$

c. Substitute in the formula.

$$H = \frac{12}{N(N+1)}\left(\frac{R_1^2}{n_1} + \frac{R_2^2}{n_2} + \frac{R_3^2}{n_3}\right) - 3(N+1)$$

where

$$N = 15 \qquad R_1 = 15 \qquad R_2 = 53 \qquad R_3 = 52$$
$$n_1 = n_2 = n_3 = 5$$

Therefore,

$$H = \frac{12}{15(15+1)}\left(\frac{15^2}{5} + \frac{53^2}{5} + \frac{52^2}{5}\right) - 3(15+1) = 9.38$$

STEP 4   Make the decision. Since the test value, 9.38, is greater than the critical value, 5.991, the decision is to reject the null hypothesis.

STEP 5   Summarize the results. There is enough evidence to reject the claim that there is no difference in the amount of potassium contained in the three brands. Hence, all brands do not contain the same amount of potassium.

The steps for the Kruskal-Wallis test are given in the Procedure Table.

Procedure Table

Kruskal-Wallis Test

STEP 1 State the hypotheses and identify the claim.

STEP 2 Find the critical value. Use the chi-square table, Table G, with d.f. $= k - 1$ (k = number of groups).

STEP 3 Compute the test value.

a. Arrange the data from lowest to highest and rank each value.

b. Find the sum of the ranks of each group.

c. Substitute in the formula.

$$H = \frac{12}{N(N + 1)}\left(\frac{R_1^2}{n_1} + \frac{R_2^2}{n_2} + \cdots + \frac{R_k^2}{n_k}\right) - 3(N + 1)$$

where

$$N = n_1 + n_2 + \cdots + n_k$$
$$R_k = \text{sum of the ranks for the } k\text{th group}$$
$$k = \text{number of groups}$$

STEP 4 Make the decision.

STEP 5 Summarize the results.

Exercises

14-62. What is the parametric test that is equivalent to the Kruskal-Wallis test?

For exercises 14-63 through 14-73, perform the following steps.

a. State the hypotheses and identify the claim.
b. Find the critical value.
c. Compute the test value.
d. Make the decision.
e. Summarize the results.

Use the traditional method of hypothesis testing unless otherwise specified.

14-63. Samples of four different cereals show the following number of calories for the suggested servings of each brand. At $\alpha = 0.05$, is there a difference in the number of calories for the different brands?

Brand A	Brand B	Brand C	Brand D
112	110	109	106
120	118	116	122
135	123	125	130
125	128	130	117
108	102	128	116
121	101	132	114

14-64. A test to measure self-esteem is given to three different samples of individuals based on birth order. The scores range from 0 to 50. The data are as shown here. At $\alpha = 0.05$ is there a difference in the scores?

Oldest child	Middle child	Youngest child
48	50	47
46	49	45
42	42	46
41	43	30
37	39	32
32	28	41

14-65. A large grocery chain decides to advertise a product by three different methods (one method in each area): radio, television, and newspaper. One week's sales from randomly selected stores in each area are recorded here. At $\alpha = 0.10$, is there a difference in sales for the different types of advertising?

Radio	Television	Newspaper
$832	$1024	$329
648	996	437
562	1011	561
786	853	329
452	471	382
975		495
		262

14–66. Three brands of microwave dinners were advertised as low in sodium. Samples of the three different brands show the following milligrams of sodium. At $\alpha = 0.05$, is there a difference in the amount of sodium among the brands?

Brand A	Brand B	Brand C
810	917	893
702	912	790
853	952	603
703	958	744
892	893	623
732		743
713		609
613		

14–67. Three different types of soils are used to grow strawberries. The yields (in quarts) for plots of the same size are shown here. At $\alpha = 0.01$, is there a difference in the yields of the three plots?

Soil A	Soil B	Soil C
32	43	50
38	45	56
31	49	58
40	46	54
39	51	52

14–68. A recent study recorded the number of job offers received by newly graduated chemical engineers at three colleges. The data are shown here. At $\alpha = 0.05$, is there a difference in the average number of job offers received by the graduates at the three colleges?

College A	College B	College C
6	2	10
8	1	12
7	0	9
5	3	13
6	6	4

14–69. A meteorologist wishes to see if there is a difference in the number of deaths in the United States due to severe weather. The data from the past six years are shown here. At $\alpha = 0.10$, is there a difference in the number of deaths from the different weather conditions?

Lightning	Tornado	Flash flood	Blizzard
39	30	46	54
41	39	55	43
73	39	45	39
74	53	109	35
67	50	62	56
68	32	30	48

Source: Jack Williams, *The USA TODAY Weather Almanac, 1995* (New York: Vintage Books, 1994), p. 90.

14–70. Three brands of copy machines are used in a large office building. The monthly maintenance cost of each machine is recorded, and the results are shown here. At $\alpha = 0.05$, can one conclude that there is a difference in the monthly maintenance costs?

Brand 1	Brand 2	Brand 3
$56	$63	$82
42	72	81
48	71	79
53	74	77
51	76	55

14–71. In a large city, the number of crimes per week in five precincts is recorded for five weeks. The data are shown here. At $\alpha = 0.01$, is there a difference in the number of crimes?

Precinct 1	Precinct 2	Precinct 3	Precinct 4	Precinct 5
105	87	74	56	103
108	86	83	43	98
99	91	78	52	94
97	93	74	58	89
92	82	60	62	88

14–72. A recent study examined the number of unemployed people in five cities who are actively seeking employment. They are listed here according to the education each received. At $\alpha = 0.05$, is there a difference in the number of unemployed based on education received? Use the *P*-value method.

High school diploma	College degree	Postgraduate degree
49	23	7
43	49	38
51	54	23
108	87	52
68	28	26

14–73. Three different methods of first-aid instruction are given to students. The same final examination is given to each class. The data are shown here. At $\alpha = 0.10$, is there a difference in the final examination scores? Use the *P*-value method.

Method A	Method B	Method C
98	97	99
100	88	94
95	82	96
92	84	89
86	75	81
76	73	72
71	74	

Technology Step by Step

MINITAB
Step by Step

The Kruskal-Wallis Test

Example MT14–5

Worksheet

1. Stack the data for Example 14–6 into two columns of a worksheet.
   a. First, enter all of the potassium amounts into C1. Name it **Potassium.**

C1	C2-T
Potassium	Brand
4.7	A
3.2	A
5.1	A
5.2	A
5.0	A
5.3	B
6.4	B

   b. Enter codes **A, B,** or **C** for the Brand into C2. The first seven rows of the completed worksheet are shown.

2. Select **Stat>Nonparametrics>Kruskal-Wallis.**

Kruskal-Wallis
Dialog Box

3. Double-click on C1 Potassium to select it for Response. This variable must be quantitative so the column for brand will not be available in the list.
4. Double-click on C2 Brand to select it for Factor.
5. Click [OK].

Kruskal-Wallis Test: Potassium versus Brand

```
Kruskal-Wallis Test on Potassiu
Brand         N      Median     Ave Rank          z
A             5       5.000          3.0      -3.06
B             5       6.800         10.6       1.59
C             5       6.600         10.4       1.47
Overall      15                      8.0
H = 9.38    DF = 2    P = 0.009
```

The value $H = 9.38$ has a P-value of 0.009. Reject the null hypothesis.

14–7

The Spearman Rank Correlation Coefficient and the Runs Test

The techniques of regression and correlation were explained in Chapter 11. In order to determine whether two variables are linearly related, one uses the Pearson product moment correlation coefficient. Its values range from $+1$ to -1. One assumption for testing the hypothesis that $\rho = 0$ for the Pearson coefficient is that the populations from which the samples are obtained are normally distributed. If this requirement cannot be

Rank Correlation Coefficient

Objective 6. Compute the Spearman rank correlation coefficient.

met, the nonparametric equivalent, called the **Spearman rank correlation coefficient** (denoted by r_s), can be used when the data are ranked.

The computations for the rank correlation coefficient are simpler than those for the Pearson coefficient and involve ranking each set of data. The difference in ranks is found, and r_s is computed by using these differences. If both sets of data have the same ranks, r_s will be $+1$. If the sets of data are ranked in exactly the opposite way, r_s will be -1. If there is no relationship between the rankings, r_s will be near 0.

Formula for Computing the Spearman Rank Correlation Coefficient

$$r_s = 1 - \frac{6 \sum d^2}{n(n^2 - 1)}$$

where

$d = $ difference in the ranks

$n = $ number of data pairs

This formula is algebraically equivalent to the formula for r given in Chapter 11, except that ranks are used instead of raw data.

The computational procedure is shown in the next example. For a test of the significance of r_s, Table L is used for values of n up to 30. For larger values, the normal distribution can be used. (See Exercises 14–98 through 14–102.)

Example 14–7

Two students were asked to rate eight different textbooks for a specific course on an ascending scale from 0 to 20 points. Points were assigned for each of several categories, such as reading level, use of illustrations, and use of color. At $\alpha = 0.05$, test the hypothesis that there is a significant linear correlation between the two students' ratings. The data are shown in the table.

Textbook	Student 1's rating	Student 2's rating
A	4	4
B	10	6
C	18	20
D	20	14
E	12	16
F	2	8
G	5	11
H	9	7

Solution

STEP 1   State the hypotheses.

$$H_0\colon \rho = 0 \quad \text{and} \quad H_1\colon \rho \neq 0$$

STEP 2   Find the critical value. Use Table L to find the value for $n = 8$ and $\alpha = 0.05$. It is 0.738. See Figure 14–3.

Figure 14–3

Finding the Critical
Value in Table L for
Example 14–7

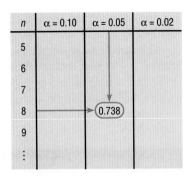

n	$\alpha = 0.10$	$\alpha = 0.05$	$\alpha = 0.02$
5			
6			
7			
8		0.738	
9			
⋮			

STEP 3   Find the test value.

a. Rank each data set, as shown in the table below.

Textbook	Student 1	Rank	Student 2	Rank
A	4	7	4	8
B	10	4	6	7
C	18	2	20	1
D	20	1	14	3
E	12	3	16	2
F	2	8	8	5
G	5	6	11	4
H	9	5	7	6

Let X_1 be the first student's rankings and X_2 be the second student's rankings.

b. Subtract the rankings $(X_1 - X_2)$.

$$7 - 8 = -1 \qquad 4 - 7 = -3 \qquad \text{etc.}$$

c. Square the differences.

$$(-1)^2 = 1 \qquad (-3)^2 = 9 \qquad \text{etc.}$$

d. Find the sum of the squares.

$$1 + 9 + 1 + 4 + 1 + 9 + 4 + 1 = 30$$

The results can be summarized in a table, as shown here.

X_1	X_2	$d = X_1 - X_2$	d^2
7	8	−1	1
4	7	−3	9
2	1	1	1
1	3	−2	4
3	2	1	1
8	5	3	9
6	4	2	4
5	6	−1	1
			$\Sigma d^2 = 30$

Use the data in the following table to rank each state according to the number of permits issued and then according to the average cost per unit. Omit the states with the code N/A. Using the Spearman rank correlation coefficient, see whether there is a significant relationship among the ranks.

Stock Permits

The Bureau of Land Management and the U.S. Forest Service together lease 270 million acres to 30,000 livestock ranchers in 16 states. Grazing units* issued and average cost per grazing unit:

State	Permits	Cost
Ariz.	1,804,369	$114
Calif.	944,597	$53
Colo.	1,597,434	$75
Idaho	2,747,787	$60
Kan.	120	N/A
Mont.	1,837,335	$76
Neb.	85,334	$140
Nev.	2,743,959	$40
N.M.	2,880,010	$103
N.D.	261,363	$56
Okla.	475	N/A
Ore.	1,442,014	$56
S.D.	95,814	N/A
Utah	2,425,300	$50
Wash.	79,315	N/A
Wyo.	2,594,592	$49

*One unit is enough forage land to feed one horse or a cow and a calf.
Source: USDA, Interior Department

Source: Based on data from the USDA, Interior Department. Copyright 1993, *USA TODAY.* Reprinted with permission.

e. Substitute in the formula to find r_s.

$$r_s = 1 - \frac{6 \Sigma d^2}{n(n^2 - 1)}$$

where n = the number of data pairs. For this problem,

$$r_s = 1 - \frac{(6)(30)}{8(8^2 - 1)} = 1 - \frac{180}{504} = 0.643$$

STEP 4   Make the decision. Do not reject the null hypothesis, since $r_s = 0.643$, which is less than the critical value of 0.738.

STEP 5   Summarize the results. There is not enough evidence to say that there is a correlation between the rankings of the two students.

The steps for finding and testing the Spearman rank correlation coefficient are given in the Procedure Table.

Procedure Table

Finding and Testing the Spearman Rank Correlation Coefficient

STEP 1   State the hypotheses.

continued

Procedure Table (concluded)

STEP 2 Rank each data set.

STEP 3 Subtract the rankings $(X_1 - X_2)$.

STEP 4 Square the differences.

STEP 5 Find the sum of the squares.

STEP 6 Substitute in the formula.

$$r_s = 1 - \frac{6 \sum d^2}{n(n^2 - 1)}$$

where

d = difference in the ranks

n = number of pairs of data

STEP 7 Find the critical value.

STEP 8 Make the decision.

STEP 9 Summarize the results.

The Runs Test

Objective 7. Test hypotheses using the runs test.

When samples are selected, one assumes that they are selected at random. How does one know if the data obtained from a sample are truly random? Before the answer to this question is given, consider the following situations for a researcher interviewing 20 people for a survey. Let their genders be denoted by M for male and F for female. Suppose the participants were chosen as follows:

situation 1   M M M M M M M M M M F F F F F F F F F F

It does not look as if the people in this sample were selected at random, since 10 men were selected first, followed by 10 females.

Consider a different selection:

situation 2   F M F M F M F M F M F M F M F M F M F M

In this case, it seems as if the researcher selected a female, then a male, etc. This selection is probably not random either.

Finally, consider the following selection:

situation 3   F F F M M F M F M M F F M M F F M M M F

This selection of data looks as if it may be random, since there is a mix of men and women and no apparent pattern to their selection.

Rather than trying to guess whether the data of a sample has been selected at random, statisticians have devised a nonparametric test to determine its randomness. This test is called the **runs test.**

A **run** is a succession of identical letters preceded or followed by a different letter or no letter at all, such as the beginning or end of the succession.

For example, the first situation presented has two runs:

run 1:   M M M M M M M M M M

run 2:   F F F F F F F F F F

The second situation has 20 runs. (Each letter constitutes one run.) The third situation has 11 runs.

run 1:	F F F	run 5:	F	run 9:	F F
run 2:	M M	run 6:	M M	run 10:	M M M
run 3:	F	run 7:	F F	run 11:	F
run 4:	M	run 8:	M M		

Example 14–8

Determine the number of runs in each sequence.

a. F F F M M F F F F M

b. H H H T T T T

c. A A B B A A B B A A B B

Solution

a. There are four runs, as shown.

$$\underbrace{F\ F\ F}_{1} \quad \underbrace{M\ M}_{2} \quad \underbrace{F\ F\ F\ F}_{3} \quad \underbrace{M}_{4}$$

b. There are two runs, as shown.

$$\underbrace{H\ H\ H}_{1} \quad \underbrace{T\ T\ T\ T}_{2}$$

c. There are six runs, as shown.

$$\underbrace{A\ A}_{1} \quad \underbrace{B\ B}_{2} \quad \underbrace{A\ A}_{3} \quad \underbrace{B\ B}_{4} \quad \underbrace{A\ A}_{5} \quad \underbrace{B\ B}_{6}$$

The test for randomness considers the number of runs rather than the frequency of the letters. For example, for data to be selected at random, there should not be too few or too many runs, as in situations 1 and 2. The runs test does not consider the questions of how many males or females were selected or how many of each are in a specific run.

To determine whether the number of runs is within the random range, use Table M in Appendix C. The values are for a two-tailed test with $\alpha = 0.05$. For a sample of 12 males and 8 females, the table values shown in Figure 14–4 mean that any number of runs from 7 to 15 would be considered random. If the number of runs is 6 or less or 16 or more, the sample is probably not random, and the null hypothesis should be rejected.

Figure 14–4

Finding the Critical Value in Table M

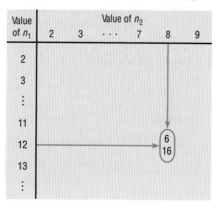

Example 14–9 shows the procedure for conducting the runs test using letters as data. Example 14–10 shows how the runs test can be used for numerical data.

Example 14–9

On a commuter train, the conductor wishes to see whether the passengers enter the train at random. He observes the first 25 people, with the following sequence of males (M) and females (F).

F F F M M F F F F M F M M M M F F F F M M F F F M M

Test for randomness at $\alpha = 0.05$.

Solution

STEP 1   State the hypotheses and identify the claim.

H_0: The passengers board the train at random, according to gender (claim).

H_1: The null hypothesis is not true.

STEP 2   Find the number of runs. Arrange the letters according to runs of males and females, as shown.

Run	Gender
1	F F F
2	M M
3	F F F F
4	M
5	F
6	M M M
7	F F F F
8	M M
9	F F F
10	M M

There are 15 females (n_1) and 10 males (n_2).

STEP 3   Find the critical value. Find the number of runs in Table M for $n_1 = 15$, $n_2 = 10$, and $\alpha = 0.05$. The values are 7 and 18. *Note:* In this situation the critical value is found after the number of runs is determined.

STEP 4   Make the decision. Compare these critical values with the number of runs. Since the number of runs is 10, and 10 is between 7 and 18, do not reject the null hypothesis.

STEP 5   Summarize the results. There is not enough evidence to reject the hypothesis that the passengers board the train at random according to gender.

Example 14–10

Twenty people enrolled in a drug abuse program. Test the claim that the ages of the people, according to the order in which they enroll, occur at random, at $\alpha = 0.05$. The data are 18, 36, 19, 22, 25, 44, 23, 27, 27, 35, 19, 43, 37, 32, 28, 43, 46, 19, 20, 22.

Solution

STEP 1   State the hypotheses and identify the claim.

H_0: The ages of the people, according to the order in which they enroll in a drug program, occur at random (claim).

H_1: The null hypothesis is not true.

STEP 2   Find the number of runs.

   a. Find the median of the data. Arrange the data in ascending order.

   18, 19, 19, 19, 20, 22, 22, 23, 25, 27, 27,

   28, 32, 35, 36, 37, 43, 43, 44, 46

   The median is 27.

   b. Replace each number in the original sequence with an A if it is above the median and with a B if it is below the median. Eliminate any numbers that are equal to the median.

   B, A, B, B, B, A, B, A, B, A, A, A, A, A, A, B, B, B

   c. Arrange the letters according to runs.

Run	Letters
1	B
2	A
3	B B B
4	A
5	B
6	A
7	B
8	A A A A A A
9	B B B

STEP 3   Find the critical value. Table M shows that with $n_1 = 9$, $n_2 = 9$, and $\alpha = 0.05$, the number of runs should be between five and 15.

STEP 4   Make the decision. Since there are nine runs, and nine falls between 5 and 15, the null hypothesis is not rejected.

STEP 5   Summarize the results. There is not enough evidence to reject the hypothesis that the ages of the people who enroll occur at random.

The steps for the runs test are given in the Procedure Table.

Procedure Table

The Runs Test

STEP 1   State the hypotheses and identify the claim.

STEP 2   Find the number of runs.
   Note: When the data are numerical, find the median. Then compare each data value with the median and classify it as above or below the median. Other methods such as odd-even can also be used. (Discard any value that is equal to the median.)

STEP 3   Find the critical value. Use Table M.

STEP 4   Make the decision. Compare the actual number of runs with the critical value.

STEP 5   Summarize the results.

Exercises

For Exercises 14–74 through 14–78, find the critical value from Table L for the rank correlation coefficient, given sample size _n_ and α. Assume that the test is two-tailed.

14–74. $n = 30$, $\alpha = 0.05$

14–75. $n = 14$, $\alpha = 0.01$

14–76. $n = 28$, $\alpha = 0.02$

14–77. $n = 10$, $\alpha = 0.05$

14–78. $n = 9$, $\alpha = 0.01$

For Exercises 14–79 through 14–88, perform the following steps.

a. Find the Spearman rank correlation coefficient.
b. State the hypotheses.
c. Find the critical value. Use $\alpha = 0.05$.
d. Make the decision.
e. Summarize the results.

Use the traditional method of hypothesis testing unless otherwise specified.

14–79. The table shows the total number of tornadoes that occurred in 10 states from 1962–91 and the record high temperatures for the same states. At $\alpha = 0.10$, is there a relationship between the number of tornadoes and the record high temperatures?

State	Tornadoes	Record high temperatures
AL	668	112
CO	781	118
FL	1590	109
IL	798	117
KS	1198	121
NY	169	108
PA	310	111
TN	360	113
VT	21	105
WI	625	114

Source: *The World Almanac and Book of Facts, 1995* (New York: Funk & Wagnalls Corporation, 1995), pp. 183, 186.

14–80. Eight dogs were treated and ranked on their ease of handling by the veterinarian and her assistant. The data are shown here. (1 is the highest ranking.) Is there a relationship between the rankings of the two people?

Dogs	1	2	3	4	5	6	7	8
Vet	4	5	6	1	7	2	3	8
Assistant	3	5	4	6	8	1	2	7

14–81. The table shows the average maximum sentence length and the actual time served in months. At $\alpha = 0.05$, is there a relationship between the two?

Crime	Sentence	Time served
Murder	227	97
Rape	120	45
Robbery	106	40
Burglary	77	26
Drug offenses	60	18
Weapons offenses	49	21

Source: *World Almanac and Book of Facts,* 1995 (New York: Funk & Wagnalls Corporation, 1995), p. 218.

14–82. Six different summer theater actors were ranked by male and female patrons on the basis of diction and appearance. The data are shown here. (One is the highest rating). At $\alpha = 0.05$ is there a relationship between the rankings?

Actors	A	B	C	D	E	F
Males	6	3	2	5	1	4
Females	4	5	1	6	3	2

14–83. Eight music videos were ranked by teenagers and their parents on style and clarity, with 1 being the highest ranking. The data are shown here. At $\alpha = 0.05$, is there a relationship between the rankings?

Music videos	1	2	3	4	5	6	7	8
Teenagers	4	6	2	8	1	7	3	5
Parents	1	7	5	4	3	8	2	6

14–84. The sociology department is selecting new textbooks for the next semester. Five instructors and three teaching assistants reviewed seven books and assigned each book rating points on a scale of 1 to 12, with 1 being the poorest and 12 being the best. Each book was rated on content, readability, etc. The data are shown here. Is there a relationship between the ratings of the instructors and assistants?

Textbooks	1	2	3	4	5	6	7
Instructors	5	8	12	3	11	9	1
Assistants	5	7	10	4	11	8	2

14–85. Six model kitchens were rated for style and convenience by independent interior designers and by potential customers. The scale ran from 1 to 100 points, with 1 being the lowest and 100 being the highest. The data are shown here. At $\alpha = 0.05$, is there a relationship between the two ratings?

Kitchens	A	B	C	D	E	F
Designers	48	76	30	88	61	93
Customers	35	44	28	50	75	85

14–86. Nine tennis players were ranked by sportswriters and by coaches. The data are shown here (1 is the highest ranking). Is there a relationship between the two rankings?

Players	A	B	C	D	E	F	G	H	I
Coaches	4	6	5	1	7	2	3	8	9
Writers	7	6	4	3	5	2	1	9	8

14–87. Twelve cars were rated for style, performance, driveability, etc., by independent automotive engineers and by potential customers. The scale ran from 1 to 100 points, with 1 being the lowest and 100 being the highest. The data are shown here. Is there a relationship between the two ratings?

Cars	A	B	C	D	E	F
Engineers	81	70	65	54	43	90
Customers	85	75	68	50	52	95

Cars	G	H	I	J	K	L
Engineers	41	88	40	85	82	35
Customers	48	100	44	90	83	20

14–88. Six watercolor paintings at an art show were ranked by professional judges and by the general public. The data are shown here (1 is the highest ranking). Is there a relationship between the rankings?

Paintings	1	2	3	4	5	6
Judges	3	5	1	2	4	6
Public	3	4	2	1	6	5

14–89. A school dentist wanted to test the claim, at $\alpha = 0.05$, that the number of cavities in fourth-grade students is random. Forty students were checked, and the number of cavities each had is shown here. Test for randomness of the values above or below the median.

0	4	6	0	6	2	5	3	1	5	1
2	2	1	3	7	3	6	0	2	6	0
2	3	1	5	2	1	3	0	2	3	7
3	1	5	1	1	2	2				

14–90. A drawing was held each day for a month. Categorize the winning numbers as odd or even. The data follow. Test for randomness, at $\alpha = 0.05$.

409	872	235	338	472	481	318	129	229
084	291	991	356	212	457	473	834	304
361	301	051	652	405	458	094	633	809
299	712	802						

14–91. The winning numbers for the Pennsylvania State Lotto drawing for April are listed here. Classify each as odd or even and test for randomness, at $\alpha = 0.05$. No drawings were held on weekends.

457, 605, 348, 927, 463, 300, 620, 261, 614, 098, 467, 961, 957, 870, 262, 571, 633, 448, 187, 462, 565, 180, 050

14–92. An irate student believes that the answers to his history professor's final true–false examination are not random. Test the claim, at $\alpha = 0.05$. The answers to the questions follow.

T T T F F T T T F F F F F F T

T T F F F T T T F T F F T T F

14–93. A machine manufactures audiocassette cases that are either defective (D) or acceptable (A). The sequence is shown here. At $\alpha = 0.05$, test for randomness.

D A A A A A A D D A D D A A A

D D A A A A A A A D D D A A A

14–94. Twenty shoppers are in a check out line at a grocery store. At $\alpha = 0.05$ test for randomness of their genders: male (M) or female (F). The data are shown here.

F M M F F M F M M F

F M M M F F F F F M

14–95. A supervisor records the number of employees absent over a 30-day period. Test for randomness, at $\alpha = 0.05$.

27	6	19	24	18	12	15	17	18	20
0	9	4	12	3	2	7	7	0	5
32	16	38	31	27	15	5	9	4	10

14–96. A ski lodge manager observes the weather for the month of February. If his customers are able to ski, he records S; if weather conditions do not permit skiing, he records N. Test for randomness at $\alpha = 0.05$.

S S S S S N N N N N N N

N S S S N N S S S S S S S

14–97. The following data are the scores on an IQ exam in the order that the students finished the test. At $\alpha = 0.05$, test for randomness.

101	98	99	110	119	121	118
106	96	88	91	97	92	106
94	93	100	89	86	95	99

When $n \geq 30$, the formula $r = \dfrac{\pm z}{\sqrt{n-1}}$ can be used to find the critical values for the rank correlation coefficient. For example, if $n = 40$ and $\alpha = 0.05$ for a two-tailed test,

$$r = \frac{\pm 1.96}{\sqrt{40-1}} = \pm 0.314$$

Hence, any r_s greater than or equal to $+0.314$ or less than or equal to -0.314 is significant.

For Exercises 14–98 through 14–102, find the critical r value for each (assume that the test is two-tailed).

14–98. $n = 50, \alpha = 0.05$

14–99. $n = 30, \alpha = 0.01$

14–100. $n = 35, \alpha = 0.02$

14–101. $n = 60, \alpha = 0.10$

14–102. $n = 40, \alpha = 0.01$

Technology Step by Step

MINITAB
Step by Step

Runs Test for Randomness

Example MT14–6

1. Sequence is important! Enter the data down C1 in the same order it was collected:

    32 18 12 54 63 17 5 63 27 9 56 58 47 9 14 16 18 50 42 37 35 29
    18 3

2. Select **Stat>Nonparametrics>Runs Test.**

Runs Test Dialog Box

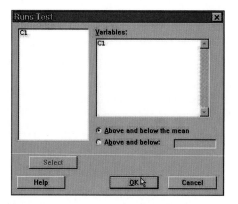

3. Double-click C1 to select the variable.
4. Click the button for `Above and below the mean`.
5. Click `[OK]`.

Runs Test: C1
```
C1
K =    30.5000
   The observed number of runs =   10
   The expected number of runs =   12.9167
   11 Observations above K    13 below
            The test is significant at 0.2201
            Cannot reject at alpha = 0.05
```

The *P*-value of the test is 0.2201. Do not reject the null hypothesis, which is that the sequence of the numbers is random.

14–8

Summary

In many research situations, the assumptions (particularly normality) for the use of parametric statistics cannot be met. Also, some statistical studies do not involve parameters such as means, variances, and proportions. For both situations, statisticians have developed nonparametric statistical methods, also called *distribution-free methods*.

STATISTICS

This study involves the relationships between pessimism and early death. Notice that other contributing factors—such as blood pressure, weight, and smoking—were controlled. How do you think the researchers controlled these factors? How do you think they measured the variable of hopelessness? What type of statistical test may have been used to obtain the results?

Pessimism Linked to Early Death

By Marilyn Elias
USA TODAY

SAN DIEGO—Middle-age men who feel hopeless about the future and their chances of attaining goals are far more likely to die early than are equally healthy but more hopeful men, suggests a large study.

Dr. Susan Everson of the Human Population Laboratory, Berkeley, Calif., reported on a six-year study of 2,428 Finnish men at the Society of Behavioral Medicine meeting here.

The men, ages 42-60, got physical and mental health exams; Everson's team controlled for risk factors such as blood pressure, weight and smoking. Six years later, hopeless men were about twice as likely to have died of any cause compared with the hopeful; the most hopeless had double the heart attack risk. Hopeless men had a higher death rate due to accidents and violence.

"Doctors should be looking at this (as a health risk), just as they do blood pressure and cholesterol," says psychologist Martin Seligman, University of Pennsylvania, Philadelphia.

There are several advantages to the use of nonparametric methods. The most important one is that no knowledge of the population distributions is required. Other advantages include ease of computation and understanding. The major disadvantage is that they are less efficient than their parametric counterparts when the assumptions for the parametric methods are met. In other words, larger sample sizes are needed to get as accurate results as given by their parametric counterparts.

The following list gives the nonparametric statistical tests presented in this chapter, along with their parametric counterparts.

Nonparametric test	Parametric test	Condition
Single-sample sign test	z or t test	One sample
Paired-sample sign test	z or t test	Two dependent samples
Wilcoxon rank sum test	z or t test	Two independent samples
Wilcoxon signed-rank test	t test	Two dependent samples
Kruskal-Wallis test	ANOVA	Three or more independent samples
Spearman rank correlation coefficient	Pearson's correlation coefficient	Relationships between variables
Runs test	None	Randomness

When the assumptions of the parametric tests can be met, the parametric tests should be used instead of their nonparametric counterparts.

Important Terms

distribution-free
statistics   584

Kruskal-Wallis test   605

nonparametric
statistics   583

parametric tests   583

ranking   585

run   614

runs test   614

sign test   586

Spearman rank
correlation
coefficient   611

Wilcoxon rank sum
test   594

Wilcoxon signed-rank
test   599

Important Formulas

Formula for the z test value in the sign test:

$$z = \frac{(X + 0.5) - (n/2)}{\sqrt{n}/2}$$

where

n = sample size (greater than or equal to 26)

X = smaller number of + or − signs

Formula for the Wilcoxon rank sum test:

$$z = \frac{R - \mu_R}{\sigma_R}$$

where

$$\mu_R = \frac{n_1(n_1 + n_2 + 1)}{2}$$

$$\sigma_R = \sqrt{\frac{n_1 n_2(n_1 + n_2 + 1)}{12}}$$

R = sum of the ranks for the smaller sample size (n_1)

n_1 = smaller of the sample sizes

n_2 = larger of the sample sizes

$n_1 \geq 10$ and $n_2 \geq 10$

Formula for the Wilcoxon signed-rank test:

$$z = \frac{w_s - \frac{n(n + 1)}{4}}{\sqrt{\frac{n(n + 1)(2n + 1)}{24}}}$$

where

n = number of pairs where the difference is not 0 and $n \geq 30$

w_s = smaller sum in absolute value of the signed ranks

Formula for the Kruskal-Wallis test:

$$H = \frac{12}{N(N + 1)}\left(\frac{R_1^2}{n_1} + \frac{R_2^2}{n_2} + \cdots + \frac{R_k^2}{n_k}\right) - 3(N + 1)$$

where

R_1 = sum of the ranks of sample 1

n_1 = size of sample 1

R_2 = sum of the ranks of sample 2

n_2 = size of sample 2

$.$

$.$

$.$

R_k = sum of the ranks of sample k

n_k = size of sample k

$N = n_1 + n_2 + \cdots + n_k$

k = number of samples

Formula for the Spearman rank correlation coefficient:

$$r_s = 1 - \frac{6 \Sigma d^2}{n(n^2 - 1)}$$

where

d = difference in the ranks

n = number of data pairs

Review   Exercises

For Exercises 14–103 through 14–115, follow this procedure:

a.   State the hypotheses and identify the claim.

b.   Find the critical value(s).

c.   Compute the test value.

d.   Make the decision.

e.   Summarize the results.

Use the traditional method of hypothesis testing unless otherwise specified.

14–103. A researcher believes that the median number of wet days per month in Yosemite National Park is five. A

sample of the number of wet days per month for last year is shown here. At $\alpha = 0.05$, is there enough evidence to reject the researcher's claim?

| 8 | 6 | 7 | 6 | 4 | 2 | 1 | 1 | 1 | 2 | 6 | 7 |

Source: Jack Williams, *The USA Today Weather Almanac, 1995* (New York: Vintage Books, 1994) p. 43.

14–104. A tire manufacturer claims that the median lifetime of a certain brand of truck tires is 40,000 miles. A sample of 30 tires shows that 12 lasted longer than 40,000 miles. Is there enough evidence to reject the claim at $\alpha = 0.05$? Use the sign test.

14–105. A special diet including hormones is fed to adult hogs to see whether they will gain weight. The before and after weights (in pounds) are given here. Use the paired-sample sign test at $\alpha = 0.05$ to determine whether there is a weight gain.

Before	320	432	456	358	371	394	362	359	319
After	333	430	459	362	381	395	367	356	315

14–106. Shown here are the record high temperatures for Dawson Creek in British Columbia, Canada, and for Whitehorse in Yukon, Canada, for 12 months. Using the Wilcoxon rank sum test at $\alpha = 0.05$, is there a difference in the record high temperatures? Use the P-value method.

Dawson Creek	52	60	57	71	86	89	94	93	88	80	66	52
Whitehorse	47	50	51	69	86	89	91	86	80	66	51	47

Source: Jack Williams, *The USA TODAY Weather Almanac, 1995* (New York: Vintage Books, 1994), p. 37.

14–107. Samples of students majoring in business and engineering are selected, and the amount (in dollars) each spent on a required textbook for the fall semester is recorded. The data are shown here. For the Wilcoxon rank sum test at $\alpha = 0.10$, is there a difference in the amount spent by each group?

Business		Engineering	
48	36	98	73
52	62	72	78
74	50	63	93
63	46	78	88
51	53	55	86
49	58	58	85
		64	

14–108. Twelve automobiles were tested to see how many miles per gallon each one obtained. Under similar driving conditions, they were tested again using a special additive. The data are shown here. At $\alpha = 0.05$, did the additive improve gas mileage? Use the Wilcoxon signed-rank test.

Before		After	
13.6	18.3	22.6	23.7
18.2	19.5	21.9	20.8
16.1	18.2	25.3	25.3
15.3	16.7	28.6	27.2
19.2	21.3	15.2	17.2
18.8	17.2	16.3	18.5

14–109. The number of sick days taken by seven assembly-line workers were recorded for one year. The owners of the company then installed brighter lighting and permitted the employees to take a 10-minute break in the morning and afternoon. The number of sick days taken were recorded for another year. The data are shown here.

From the Wilcoxon signed-rank test at $\alpha = 0.05$, can one conclude that the number of sick days taken was reduced?

Before	6	15	18	14	27	17	9
After	8	12	16	9	23	14	15

14–110. Samples of three types of ropes are tested for breaking strength. The data (in pounds) are shown here. At $\alpha = 0.05$, is there a difference in the breaking strength of the ropes? Use the Kruskal-Wallis test.

Cotton	Nylon	Hemp
230	356	506
432	303	527
505	361	581
487	405	497
451	432	459
380	378	507
462	361	562
531	399	571
366	372	499
372	363	475
453	306	505
488	304	561
462	318	532
467	322	501

14–111. Rats were fed three different diets for one month to see whether diet has any effect on learning. Each rat was then taught to traverse a simple maze. The number of trials it took each rat to learn the correct path is shown in the table. At $\alpha = 0.05$, does diet have any effect on learning? Use the Kruskal-Wallis test.

Diet 1	8	6	12	15	9	7	5	
Diet 2	2	3	6	8	7	4		
Diet 3	9	15	17	8	4	13	18	20

14–112. A statistics instructor wishes to see whether there is a relationship between the number of homework exercises a student completes and his or her exam score. The data are shown here. Using the Spearman rank correlation coefficient, test the hypothesis that there is no relationship, at $\alpha = 0.05$.

Homework problems	Exam score
63	85
55	71
58	75
87	98
89	93
52	63
46	72
75	89
105	100

14–113. Six brands of breakfast cereals were ranked according to taste by fifth grade boys and girls. The data are shown here. Using the Spearman rank correlation coefficient, test the hypothesis that there is no relationship in the rankings of boys and girls. Use $\alpha = 0.05$.

Brand	A	B	C	D	E	F
Boys	3	2	6	1	5	4
Girls	4	5	1	3	2	6

14–114. In a recent survey, 20 college students were asked, as they arrived for class, if they worked during the academic year (W) or did not work (N). At $\alpha = 0.05$ test for randomness. The data are shown here. Assume the probabilities are the same.

W	N	N	N	W	W	W	N	W	N
N	W	W	N	N	W	N	N	W	N

14–115. An instructor wishes to see whether grades of students who finish an exam occur at random. Shown here are the grades of 30 students in the order that they finished an exam. (Read from left to right across each row, and then proceed to the next row.) Test for randomness, at $\alpha = 0.05$.

87	93	82	77	64	98
100	93	88	65	72	73
56	63	85	92	95	91
88	63	72	79	55	53
65	68	54	71	73	72

Statistics Today

Too Much or Too Little? Revisited

In this case, the manufacturer would select a sequence of bottles and see how many bottles contained more than 40 ounces, denoted by $+$, and how many bottles contained less than 40 ounces, denoted by $-$. The sequence could then be analyzed according to the number of runs, as explained in Section 14–7. If the sequence were not random, then the machine would need to be checked to see if it was malfunctioning. Another method that can be used to see if machines are functioning properly is *statistical quality control*. This method is beyond the scope of this book.

Data Analysis

The Data Bank is found in Appendix D, or on the World Wide Web by following links from www.mhhe.com/math/stat/bluman

1. From the Data Bank, choose a sample and use the sign test to test one of the following hypotheses:
   a. For serum cholesterol, test H_0: median = 220 milligram percent (mg%).
   b. For systolic pressure, test H_0: median = 120 millimeters of mercury (mm Hg).
   c. For IQ, test H_0: median = 100.
   d. For sodium level, test H_0: median = 140 mEq/l.

2. From the Data Bank, select a sample of subjects. Use the Kruskal-Wallis test to see if the sodium levels of smokers and nonsmokers are equal.

3. Using the data from Data Set IX in Appendix D, test the hypothesis that the mean population of the states east of the Mississippi River is equal to the mean population of the states west of the Mississippi. You may have to consult an atlas and make some decisions about which states are in each group. Use the Wilcoxon rank sum test. Use an α value of your choice.

4. Using the runs test and the data from Data Set XIII in Appendix D, see if the numbers for the Keystone Jackpot occur at random. Use an α of your choice.

Quiz

Determine whether each statement is true or false. If the statement is false, explain why.

1. Nonparametric statistics cannot be used to test the difference between two means.

2. Nonparametric statistics are more sensitive than their parametric counterparts.

3. Nonparametric statistics can be used to test hypotheses about parameters other than means, proportions, and standard deviations.

4. Parametric tests are preferred over their nonparametric counterparts if the assumptions can be met.

Select the best answer.

5. The _____ test is used to test means when samples are dependent and the normality assumption cannot be met.
   a. Wilcoxon signed-rank
   b. Wilcoxon rank sum
   c. Sign
   d. Kruskal-Wallis

6. The Kruskal-Wallis test uses the _____ distribution.
   a. z   c. Chi-square
   b. t   d. F

7. The nonparametric counterpart of the ANOVA is the _____ test.
   a. Wilcoxon signed-rank test
   b. Sign test
   c. Runs test
   d. None of the above

8. To see if two rankings are related, one can use the _____.
   a. Runs test
   b. Spearman's correlation coefficient
   c. Sign test
   d. Kruskal-Wallis test

Complete the following statements with the best answer.

9. When the assumption of normality cannot be met, one can use _____ tests.

10. When data are _____ or _____ in nature, nonparametric methods are used.

11. To test to see whether a median is equal to a specific value, one would use the _____ test.

12. Nonparametric tests are less _____ than their parametric counterparts.

For the following exercises, use the traditional method of hypothesis testing unless otherwise specified.

13. The owner of a candy store states that she sells on average 300 candy bars per day. A random sample of 18 days shows the number of candy bars sold each day. At $\alpha = 0.10$, is the claim correct? Use the sign test.

271	297	315	282	106	297	268	215
262	305	315	256	311	375	319	297
311	299						

14. A battery manufacturer claims that the median lifetime of a certain brand of heavy-duty battery is 1200 hours; A sample of 25 batteries shows that 15 lasted longer than 1200 hours. Test the claim at $\alpha = 0.05$. Use the sign test.

15. A special diet is fed to adult turkeys to see if they will gain weight. The before and after weights (in pounds) are given here. Use the paired-sample sign test at $\alpha = 0.05$ to see if there is weight gain.

Before	28	24	29	30	32	33	25	26	28
After	30	29	31	32	32	35	29	25	31

16. Two groups of alcoholics, one group male and the other female, were asked at what age they first drank alcohol. The data are shown here. Using the Wilcoxon rank sum test at $\alpha = 0.05$, is there a difference in the ages of the sexes?

Males	6	12	14	16	17	17	13	12	10	11
Females	8	9	9	12	14	15	12	16	17	19

17. Samples of students majoring in law and nursing are selected, and the amount each spent on textbooks for the spring semester is recorded here. Using the Wilcoxon rank sum test at $\alpha = 0.10$, is there a difference in the amount spent by each group?

Law	$167	158	162	106	98	206	112	121
	133	145	151	199				
Nursing	$ 98	198	209	168	157	126	104	122
	111	138	116	201				

18. The grade point average of a group of students was recorded for one month. During the next nine-week grading period, each student attended a workshop on study skills. Their GPAs were recorded at the end of the grading period, and the data appear here. Using the Wilcoxon signed rank test at $\alpha = 0.05$, can it be concluded that the GPA increased?

Before	3.0	2.9	2.7	2.5	2.1	2.6	1.9	2.0
After	3.2	3.4	2.9	2.5	3.0	3.1	2.4	2.8

19. Samples of three different types of wrapping tapes are tested for breaking strength in pounds. The data are shown here. At $\alpha = 0.05$, is there a difference in the breaking strength of the tapes? Use the Kruskal-Wallis test.

Type A	225	332	404	387	351	280	362	431	266
Type B	256	203	261	305	232	278	261	299	272
Type C	406	427	481	397	351	409	462	471	399

Type A	353	288	362	367	272
Type B	206	206	218	222	263
Type C	405	461	432	401	375

20. Three different groups of monkeys were fed three different medications for one month to see if the medication has any effect on reaction time. Each monkey was then taught to repeat a series of steps to receive a reward. The number of trials it took each to

receive the reward is shown here. At $\alpha = 0.05$, does the medication have an effect on reaction time? Use the Kruskal-Wallis test. Use the P-value method.

Med 1	8	7	11	14	8	6	5
Med 2	3	4	6	7	9	3	4
Med 3	8	14	13	7	5	9	12

21. A chemistry instructor wishes to see if there is a relationship between the number of homework exercises a student completes and his or her unit exam score. The data are shown here. Using the Spearman rank correlation coefficient, test the hypothesis that there is no relationship at $\alpha = 0.05$.

Homework	10	12	8	6	9	14	12	15	15
Exam score	80	86	75	80	79	93	90	89	95

22. Five brands of soup were ranked according to taste by males and females. The data are shown here. Using the Spearman rank correlation coefficient, test the hypothesis that there is no relationship in the ranking of the males and females. Use $\alpha = 0.05$.

Brand	A	B	C	D	E
Males	4	2	3	1	5
Females	5	3	2	4	1

23. At the state registry of vital statistics, the birth certificates issued for males (M) and females (F) were tallied. At $\alpha = 0.05$, test for randomness. The data are shown here.

M M F F F F F F F F M M M M F F
M F M F M M M F F F

24. The output in revolutions per minute (rpm) of 10 motors was obtained. The motors were tested again under similar conditions after they had been reconditioned. The data are shown here. At $\alpha = 0.05$, did the reconditioning improve the motors' performance? Use the Wilcoxon signed-rank test.

Before	413	701	397	602	405	512	450	487	388	351
After	433	712	406	650	450	550	450	500	402	415

25. A statistician wishes to determine if a state's lottery numbers are selected at random. The winning numbers selected for the month of February are shown here. Test for randomness at $\alpha = 0.05$.

321 909 715 700 487 808 509 606 943 761
200 123 367 012 444 576 409 128 567 908
103 407 890 193 672 867 003 578

Critical Thinking Challenges

1. Two commuters ride to work together in one car. To decide who pays the toll for a bridge on the way to work, they flip a coin and the loser pays. Explain why over a period of one year, one person might have to pay the toll five days in a row. There is no toll on the return trip. (*Hint:* You may want to use random numbers.)

2. Shown on the next page are the type and number of medals each country won in the 1994 Winter Olympic Games. You are to rank the countries from highest to lowest. Gold medals are highest, followed by silver, followed by bronze. There are many different ways to rank objects and events. Here are several suggestions:

   a. Rank the countries according to the total medals won.

   b. List some advantages and disadvantages of this method.

   c. Rank each country separately for the number of gold medals won, then for the number of silver medals won, and then for the number of bronze medals won. Then rank the countries according to the sum of the *ranks* for the categories.

   d. Are the rankings of the countries the same as those in Step 1? Explain any differences.

   e. List some advantages and disadvantages of this method of ranking.

   f. A third way to rank the countries is to assign a weight to each medal. In this case, assign three points for each gold medal, two points for each silver medal, and one point for each bronze medal the country won. Multiply the number of medals by the weights for each medal and find the sum. For example, since Austria won two gold medals, three silver medals, and four bronze medals, its rank sum would be $(2 \times 3) + (3 \times 2) + (4 \times 1) = 16$. Rank the countries according to this method.

   g. Compare the ranks using this method with those using the other two methods. Are the rankings the same or different? Explain.

   h. List some advantages and disadvantages of this method.

   i. Select two of the rankings and run the Spearman rank correlation test to see if they differ significantly.

Winter Olympic Games 1994 Final Medal Standings

Country	Gold	Silver	Bronze
Austria	2	3	4
Canada	3	6	4
Germany	9	7	8
Italy	7	5	8
Norway	10	11	5
Russia	11	8	4
Switzerland	3	4	2
U.S.A.	6	3	2

Source: Reprinted with permission from the *World Almanac and Book of Facts 1995,* Copyright © 1994 Funk & Wagnalls Corporation. All rights reserved.

Data Projects

Where appropriate, use MINITAB, the TI-83, or a computer program of your choice to complete the following nonparametric exercises.

1. There are many nonparametric statistical tests. Decide on a project that will use one of the nonparametric tests and collect data from a sample.
   a. Describe the purpose of the study.
   b. Define the population.
   c. State how the sample was obtained.
   d. Select an α value
   e. State the hypotheses for the study.
   f. Decide which nonparametric test statistic will be used and compute the test value.
   g. Make the decision.
   h. Summarize the results.
   i. Conduct the corresponding parametric test and compare the results.

   j. Write a brief paragraph on which test is more appropriate and give reasons why.

2. Select a variable in which you can perform a runs test. For example, you might observe the gender of 20 or 30 individuals waiting in the cafeteria, bookstore checkout, or registration line.
   a. Conduct the runs test and decide whether or not the sequence is random.
   b. Write a brief summary of the results.

You may use the following websites to obtain raw data:

   http://www.mhhe.com/math/stat/bluman/
   http://lib.stat.cmu.edu/DASL
   http://www.oecd.org/statlist.htm
   http://www.statcan.ca/english/

Hypothesis-Testing Summary 3*

15. Test to see whether the median of a sample is a specific value when $n \geq 26$.

   Example: H_0: median = 100

Use the sign test:

$$z = \frac{(X + 0.5) - (n/2)}{\sqrt{n}/2}$$

16. Test to see whether two independent samples are obtained from populations that have identical distributions.

   Example: H_0: There is no difference in the ages of the subjects.

Use the Wilcoxon rank sum test:

$$z = \frac{R - \mu_R}{\sigma_R}$$

where

$$\mu_R = \frac{n_1(n_1 + n_2 + 1)}{2}$$

$$\sigma_R = \sqrt{\frac{n_1 n_2(n_1 + n_2 + 1)}{12}}$$

17. Test to see whether two dependent samples have identical distributions.

Example: H_0: There is no difference in the effects of a
tranquilizer on the number of hours a
person sleeps at night.

Use the Wilcoxon signed-rank test:

$$z = \frac{w_s - \dfrac{n(n + 1)}{4}}{\sqrt{\dfrac{n(n + 1)(2n + 1)}{24}}}$$

when $n \geq 30$.

18. Test to see whether three or more samples come from identical populations.

Example: H_0: There is no difference in the weights of the three groups.

Use the Kruskal-Wallis test:

$$H = \frac{12}{N(N + 1)}\left(\frac{R_1^2}{n_1} + \frac{R_2^2}{n_2} + \cdots + \frac{R_k^2}{n_k}\right) - 3(N + 1)$$

19. Rank correlation coefficient.

$$r_s = 1 - \frac{6 \Sigma d^2}{n(n^2 - 1)}$$

20. Test for randomness: Use the runs test.

*This summary is a continuation of Hypothesis-Testing Summary 2 at the end of Chapter 13.

Chapter

15 Sampling and Simulation

Objectives

After completing this chapter, you should be able to

1. Demonstrate a knowledge of the four basic sampling methods.

2. Recognize faulty questions on a survey and other factors that can bias responses.

3. Solve problems using simulation techniques.

Statistics Today

No Room in the Van?

A van contains 10 seats. The owner sells tickets in advance for seats in the van, and he realizes that 10% of the people who buy tickets do not show up each day. In order to maximize profits, the owner decides to sell 12 tickets each day. The owner wishes to find out how many times per week the van will be overbooked. Rather than waiting weeks and keeping records to find the answer, the owner can use *simulation techniques* shown in this chapter to find out the answer.

15–1

Introduction

Most people have heard of Gallup, Harris, and Nielsen. These and other pollsters gather information about the habits and opinions of the American people. Such survey firms, and the U.S. Census Bureau, gather information by selecting samples from well-defined populations. Recall from Chapter 1 that the subjects in the sample should be a subgroup of the subjects in the population. Sampling methods often use what are called *random numbers* to select samples.

Since many statistical studies use surveys and questionnaires, some information about these are presented in Section 15–3.

Random numbers are also used in *simulation techniques*. Instead of studying a real-life situation, which may be costly or dangerous, researchers create a similar situation in a laboratory or with a computer. Then, by studying the simulated situation, researchers can gain the necessary information about the real-life situation in a less expensive or safer manner. This chapter will explain some common methods used to obtain samples as well as the techniques used in simulations.

15–2

Common Sampling Techniques

Objective 1. Demonstrate a knowledge of the four basic sampling methods.

In Chapter 1, a *population* was defined as all subjects (human or otherwise) under study. Since some populations can be very large, researchers cannot use every single subject, so a sample must be selected. A *sample* is a subgroup of the population. Any subgroup of the population, technically speaking, can be called a sample. However, for researchers to make valid inferences about population characteristics, the sample must be random.

For a sample to be a **random sample,** every member of the population must have an equal chance of being selected.

When a sample is chosen at random from a population, it is said to be an **unbiased sample.** That is, the sample, for the most part, is representative of the population. Conversely, if a sample is selected incorrectly, it may be a biased sample. Samples are said to be **biased samples** when some type of systematic error has been made in the selection of the subjects.

A sample is used to get information about a population for several reasons:

1. *It saves the researcher time and money.*

2. *It enables the researcher to get information that he or she might not be able to obtain otherwise.* For example, if a person's blood is to be analyzed for cholesterol, a researcher can not analyze every single drop of blood without killing the person. Or if the breaking strength of cables is to be determined, a researcher cannot test to destruction every cable manufactured, since the company would not have any cables left to sell.

3. *It enables the researcher to get more detailed information about a particular subject.* If only a few people are surveyed, the researcher can conduct in-depth interviews by spending more time with each person, thus getting more information about the subject. This is not to say that the smaller the sample, the better; in fact, the opposite is true. In general, larger samples—if correct sampling techniques are used—give more reliable information about the population.

It would be ideal if the sample were a perfect miniature of the population in all characteristics. This ideal, however, is impossible to achieve, because there are so many human traits (height, weight, IQ, etc.). The best that can be done is to select a sample that will be representative with respect to *some* characteristics, preferably those pertaining to the study. For example, if half of the population subjects are female, then approximately half of the sample subjects should be female. Likewise, other characteristics, such as age, socioeconomic status, and IQ, should be represented proportionately. In order to obtain unbiased samples, statisticians have developed several basic sampling methods. The most common methods are *random, systematic, stratified,* and *cluster sampling.* Each method will be explained in detail in this section.

In addition to the basic methods, there are other methods used to obtain samples. Some of these methods are also explained in this section.

Random Sampling

A random sample is obtained by using methods such as random numbers, which can be generated from calculators, computers, or tables. In *random sampling,* the basic requirement is that for a sample of size *n,* all possible samples of this size must have an equal chance of being selected from the population. But before the correct method of obtaining a random sample is explained, several incorrect methods commonly used by various researchers and agencies to gain information will be discussed.

One incorrect method commonly used is to ask "the person on the street." News reporters use this technique quite often. Selecting people haphazardly on the street does not meet the requirement for simple random sampling, since all possible samples of a specific size do not have an equal chance of being selected. Many people will be at home or at work when the interview is being conducted and therefore do not have a chance of being selected.

Another incorrect technique is to ask a question either by radio or television and have the listeners or viewers call the station to give their responses or opinions. Again, this sample is not random, since only those who feel strongly for or against the issue may respond or people may not have heard or seen the program. A third erroneous method is to ask people to respond by mail. Again, only those who are concerned and who have the time are likely to respond.

These methods do not meet the requirement of random sampling, since all possible samples of a specific size do not have an equal chance of being selected. In order to meet this requirement, researchers can use one of two methods. The first method is to number each element of the population and then place the numbers on cards. Place the cards in a hat or fishbowl, mix them, and then select the sample by drawing the cards. When using this procedure, researchers must ensure that the numbers are well mixed. On occasion, when this procedure is used, the numbers are not mixed well, and the numbers chosen for the sample are those that were placed in the bowl last.

The second and preferred way of selecting a random sample is to use random numbers. Figure 15–1 shows a table of two-digit random numbers generated by a computer. A more detailed table of random numbers is found in Table D of Appendix C.

Figure 15–1

Table of Random Numbers

79	41	71	93	60	35	04	67	96	04	79	10	86
26	52	53	13	43	50	92	09	87	21	83	75	17
18	13	41	30	56	20	37	74	49	56	45	46	83
19	82	02	69	34	27	77	34	24	93	16	77	00
14	57	44	30	93	76	32	13	55	29	49	30	77
29	12	18	50	06	33	15	79	50	28	50	45	45
01	27	92	67	93	31	97	55	29	21	64	27	29
55	75	65	68	65	73	07	95	66	43	43	92	16
84	95	95	96	62	30	91	64	74	83	47	89	71
62	62	21	37	82	62	19	44	08	64	34	50	11
66	57	28	69	13	99	74	31	58	19	47	66	89
48	13	69	97	29	01	75	58	05	40	40	18	29
94	31	73	19	75	76	33	18	05	53	04	51	41
00	06	53	98	01	55	08	38	49	42	10	44	38
46	16	44	27	80	15	28	01	64	27	89	03	27
77	49	85	95	62	93	25	39	63	74	54	82	85
81	96	43	27	39	53	85	61	12	90	67	96	02
40	46	15	73	23	75	96	68	13	99	49	64	11

Many studies have been done comparing left-handed people to right-handed people. Here is one comparing injuries for each group. State some reasons for the differences. For example, why do lefties have 51% more accidents that righties when using a machine?

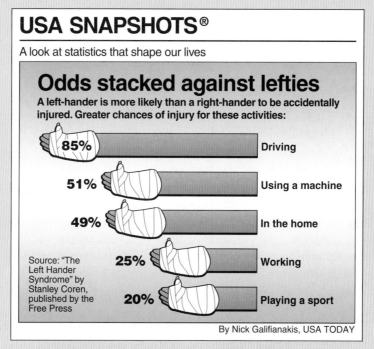

USA SNAPSHOTS®

A look at statistics that shape our lives

Odds stacked against lefties

A left-hander is more likely than a right-hander to be accidentally injured. Greater chances of injury for these activities:

85%   Driving

51%   Using a machine

49%   In the home

25%   Working

20%   Playing a sport

Source: "The Left Hander Syndrome" by Stanley Coren, published by the Free Press

By Nick Galifianakis, USA TODAY

Source: *The Left Hander Syndrome* by Stanley Coren, published by The Free Press. Copyright 1993, *USA TODAY.* Reprinted with permission.

The theory behind random numbers is that each digit, 0 through 9, has an equal probability of occurring. That is, in every sequence of 10 digits, each digit has a probability of $\frac{1}{10}$ of occurring. This does not mean that in every sequence of 10 digits, one will find each digit. Rather, it means that on the average, each digit will occur once. For example, the digit 2 may occur three times in a sequence of 10 digits, but in later sequences, it may not occur at all, thus averaging to a probability of $\frac{1}{10}$.

To obtain a sample by using random numbers, number the elements of the population sequentially and then select each person by using random numbers. This process is shown in Example 15–1.

Random samples can be selected with or without replacement. If the same member of the population cannot be used more than once in the study, then the sample is selected without replacement. That is, once a random number is selected, it cannot be used later.

Note: In the explanations and examples of the sampling procedures, a small population will be used, and small samples will be selected from this population. Small populations are used for illustrative purposes only, because the entire population could be included with little difficulty. In real life, however, researchers must usually sample from very large populations, using the procedures shown in this chapter.

Example 15–1

Suppose a researcher wants to produce a television show featuring in-depth interviews with state governors on the subject of capital punishment. Because of time constraints, the 60-minute program will have room for only 10 governors. The researcher wishes to select the governors at random. Select a random sample of 10 states from 50.

Note: This answer is not unique.

Solution

STEP 1 Number each state from 1 to 50, as shown. In this case, they are numbered alphabetically.

1. Alabama	14. Indiana	27. Nebraska	40. South Carolina
2. Alaska	15. Iowa	28. Nevada	41. South Dakota
3. Arizona	16. Kansas	29. New Hampshire	42. Tennessee
4. Arkansas	17. Kentucky	30. New Jersey	43. Texas
5. California	18. Louisiana	31. New Mexico	44. Utah
6. Colorado	19. Maine	32. New York	45. Vermont
7. Connecticut	20. Maryland	33. North Carolina	46. Virginia
8. Delaware	21. Massachusetts	34. North Dakota	47. Washington
9. Florida	22. Michigan	35. Ohio	48. West Virginia
10. Georgia	23. Minnesota	36. Oklahoma	49. Wisconsin
11. Hawaii	24. Mississippi	37. Oregon	50. Wyoming
12. Idaho	25. Missouri	38. Pennsylvania	
13. Illinois	26. Montana	39. Rhode Island	

STEP 2 Using the random numbers shown in Figure 15–1, find a starting point. To find a starting point, one generally closes one's eyes and places one's finger anywhere on the table. In this case, the first number selected was 27 in the fourth column. Going down the column and continuing on to the next column, select the first 10 numbers. They are 27, 95, 27, 73, 60, 43, 56, 34, 93, and 06. See Figure 15–2.

Figure 15–2

Selecting a Starting Point and 10 Numbers from the Random Number Table

79	41	71	93	60 ✓	35	04	67	96	04	79	10	86
26	52	53	13	43 ✓	50	92	09	87	21	83	75	17
18	13	41	30	56 ✓	20	37	74	49	56	45	46	83
19	82	02	69	34 ✓	27	77	34	24	93	16	77	00
14	57	44	30	93 ✓	76	32	13	55	29	49	30	77
29	12	18	50	06 ✓	33	15	79	50	28	50	45	45
01	27	92	67	93	31	97	55	29	21	64	27	29
55	75	65	68	65	73	07	95	66	43	43	92	16
84	95	95	96	62	30	91	64	74	83	47	89	71
62	62	21	37	82	62	19	44	08	64	34	50	11
66	57	28	69	13	99	74	31	58	19	47	66	89
48	13	69	97	29	01	75	58	05	40	40	18	29
94	31	73	19	75	76	33	18	05	53	04	51	41
00	06	53	*Start here 01	55	08	38	49	42	10	44	38	
46	16	44	(27) ✓	80	15	28	01	64	27	89	03	27
77	49	85	95 ✓	62	93	25	39	63	74	54	82	85
81	96	43	27 ✓	39	53	85	61	12	90	67	96	02
40	46	15	73 ✓	23	75	96	68	13	99	49	64	11

Now, refer to the list of states and identify the state corresponding to each number. The sample consists of the following states:

27	Nebraska	43	Texas
95		56	
27	Nebraska	34	North Dakota
73		93	
60		06	Colorado

STEP 3   Since the numbers 95, 73, 60, 56, and 93 are too large, they are disregarded. And since 27 appears twice, it is also disregarded the second time. Now, one must select six more random numbers between 1 and 50 and omit duplicates, since this sample will be selected without replacement. Make this selection by continuing down the column and moving over to the next column until a total of 10 numbers is selected. The final 10 numbers are 27, 43, 34, 06, 13, 29, 01, 39, 23, and 35. See Figure 15–3.

Figure 15–3

The Final 10 Numbers Selected

79	41	71	93	60	(35)	04	67	96	04	79	10	86
26	52	53	13	(43)	50	92	09	87	21	83	75	17
18	13	41	30	56	20	37	74	49	56	45	46	83
19	82	02	69	(34)	27	77	34	24	93	16	77	00
14	57	44	30	93	76	32	13	55	29	49	30	77
29	12	18	50	(06)	33	15	79	50	28	50	45	45
01	27	92	67	93	31	97	55	29	21	64	27	29
55	75	65	68	65	73	07	95	66	43	43	92	16
84	95	95	96	62	30	91	64	74	83	47	89	71
62	62	21	37	82	62	19	44	08	64	34	50	11
66	57	28	69	(13)	99	74	31	58	19	47	66	89
48	13	69	97	(29)	01	75	58	05	40	40	18	29
94	31	73	19	75	76	33	18	05	53	04	51	41
00	06	53	98	(01)	55	08	38	49	42	10	44	38
46	16	44	(27)	80	15	28	01	64	27	89	03	27
77	49	85	95	62	93	25	39	63	74	54	82	85
81	96	43	27	(39)	53	85	61	12	90	67	96	02
40	46	15	73	(23)	75	96	68	13	99	49	64	11

These numbers correspond to the following states:

27	Nebraska	29	New Hampshire
43	Texas	01	Alabama
34	North Dakota	39	Rhode Island
06	Colorado	23	Minnesota
13	Illinois	35	Ohio

Thus, the governors of these 10 states will constitute the sample.

Random sampling has one limitation. If the population is extremely large, it is time-consuming to number and select the sample elements. Also, notice that the random numbers in the table are two-digit numbers. If three digits are needed, then the first digit from the next column can be used, as shown in Figure 15–4. Table D in Appendix C gives five-digit random numbers.

Figure 15–4

Method for Selecting Three-Digit Numbers

79	41	71	93	60	35	04	67	96	04	79	10	86
26	52	53	13	43	50	92	09	87	21	83	75	17
18	13	41	30	56	20	37	74	49	56	45	46	83
19	82	02	69	34	27	77	34	24	93	16	77	00
14	57	44	30	93	76	32	13	55	29	49	30	77
29	12	18	50	06	33	15	79	50	28	50	45	45
01	27	92	67	93	31	97	55	29	21	64	27	29
55	75	65	68	65	73	07	95	66	43	43	92	16
84	95	95	96	62	30	91	64	74	83	47	89	71
62	62	21	37	82	62	19	44	08	64	34	50	11
66	57	28	69	13	99	74	31	58	19	47	66	89
48	13	69	97	29	01	75	58	05	40	40	18	29
94	31	73	19	75	76	33	18	05	53	04	51	41
00	06	53	98	01	55	08	38	49	42	10	44	38
46	16	44	27	80	15	28	01	64	27	89	03	27
77	49	85	95	62	93	25	39	63	74	54	82	85
81	96	43	27	39	53	85	61	12	90	67	96	02
40	46	15	73	23	75	96	68	13	99	49	64	11

Use one column and part of the next column for three digits, i.e., 404.

Systematic Sampling

A **systematic sample** is a sample obtained by numbering each element in the population and then selecting every third or fifth or tenth, etc., number from the population to be included in the sample. This is done after the first number is selected at random.

The procedure of systematic sampling is illustrated in the next example.

Example 15–2

Using the population of 50 states in Example 15–1, select a systematic sample of 10 states.

Solution

STEP 1   Number the population units as shown in Example 15–1.

STEP 2   Since there are 50 states, and 10 are to be selected, the rule is to select every fifth state. This rule was determined by dividing 50 by 10, which yields 5.

STEP 3   Using the table of random numbers, select the first digit (from 1 to 5) at random. In this case, 4 was selected.

STEP 4   Select every fifth number on the list, starting with 4. The numbers include the following:

$$1, 2, 3, ④, 5, 6, 7, 8, ⑨, 10, 11, 12, 13, ⑭ \ldots$$

The selected states are as follows:

4	Arkansas	29	New Hampshire
9	Florida	34	North Dakota
14	Indiana	39	Rhode Island
19	Maine	44	Utah
24	Mississippi	49	Wisconsin

The advantage of systematic sampling is the ease of selecting the sample elements. Also, in many cases, a numbered list of the population units may already exist. For example, the manager of a factory may have a list of employees who work for the company, or there may be an in-house telephone directory.

One disadvantage of systematic sampling that may cause a biased sample is the arrangement of the items on the list. For example, if each unit were arranged, say, as

1. Husband
2. Wife
3. Husband
4. Wife

the selection of the starting number could produce a sample of all males or all females, depending on whether the starting number is even or odd and whether the number to be added is even or odd. As another example, if the list were arranged in order of heights of individuals, one would get a different average from two samples if the first were selected by using a small starting number and the second by using a large starting number.

Stratified Sampling

A **stratified sample** is a sample obtained by dividing the population into subgroups, called *strata*, according to various homogeneous characteristics and then selecting members from each stratum for the sample.

For example, a population may consist of males and females who are smokers or nonsmokers. The researcher will want to include in the sample people from each group—that is, males who smoke, males who do not smoke, females who smoke, and females who do not smoke. To accomplish this selection, the researcher divides the population into four subgroups and then selects a random sample from each subgroup. This method ensures that the sample is representative on the basis of the characteristics of gender and smoking. Of course, it may not be representative on the basis of other characteristics.

Example 15–3

Using the population of 20 students shown in Figure 15–5, select a sample of eight students on the basis of gender (male/female) and grade level (freshman/sophomore) by stratification.

Figure 15–5

Population of Students for
Example 15–3

1. Ald, Peter	M	Fr	11. Martin, Janice	F	Fr
2. Brown, Danny	M	So	12. Meloski, Gary	M	Fr
3. Bear, Theresa	F	Fr	13. Oeler, George	M	So
4. Carson, Susan	F	Fr	14. Peters, Michele	F	So
5. Collins, Carolyn	F	Fr	15. Peterson, John	M	Fr
6. Davis, William	M	Fr	16. Smith, Nancy	F	Fr
7. Hogan, Michael	M	Fr	17. Thomas, Jeff	M	So
8. Jones, Lois	F	So	18. Toms, Debbie	F	So
9. Lutz, Harry	M	So	19. Unger, Roberta	F	So
10. Lyons, Larry	M	So	20. Zibert, Mary	F	So

Solution

STEP 1   Divide the population into two subgroups, consisting of males and females,
as shown in Figure 15–6.

Figure 15–6

Population Divided into
Subgroups by Gender

Males			Females		
1. Ald, Peter	M	Fr	1. Bear, Theresa	F	Fr
2. Brown, Danny	M	So	2. Carson, Susan	F	Fr
3. Davis, William	M	Fr	3. Collins, Carolyn	F	Fr
4. Hogan, Michael	M	Fr	4. Jones, Lois	F	So
5. Lutz, Harry	M	So	5. Martin, Janice	F	Fr
6. Lyons, Larry	M	So	6. Peters, Michele	F	So
7. Meloski, Gary	M	Fr	7. Smith, Nancy	F	Fr
8. Oeler, George	M	So	8. Toms, Debbie	F	So
9. Peterson, John	M	Fr	9. Unger, Roberta	F	So
10. Thomas, Jeff	M	So	10. Zibert, Mary	F	So

STEP 2   Divide each subgroup further into two groups of freshmen and sophomores,
as shown in Figure 15–7.

Figure 15–7

Each Subgroup Divided
into Subgroups by Grade
Level

Group 1			Group 2		
1. Ald, Peter	M	Fr	1. Bear, Theresa	F	Fr
2. Davis, William	M	Fr	2. Carson, Susan	F	Fr
3. Hogan, Michael	M	Fr	3. Collins, Carolyn	F	Fr
4. Meloski, Gary	M	Fr	4. Martin, Janice	F	Fr
5. Peterson, John	M	Fr	5. Smith, Nancy	F	Fr

Group 3			Group 4		
1. Brown, Danny	M	So	1. Jones, Lois	F	So
2. Lutz, Harry	M	So	2. Peters, Michele	F	So
3. Lyons, Larry	M	So	3. Toms, Debbie	F	So
4. Oeler, George	M	So	4. Unger, Roberta	F	So
5. Thomas, Jeff	M	So	5. Zibert, Mary	F	So

STEP 3 Determine how many students need to be selected from each subgroup to have a proportional representation of each subgroup in the sample. There are four groups, and since a total of eight students is needed for the sample, two students must be selected from each subgroup.

STEP 4 Select two students from each group by using random numbers. In this case, the random numbers are as follows:

Group 1	Students 5 and 4	Group 2	Students 5 and 2
Group 3	Students 1 and 3	Group 4	Students 3 and 4

The stratified sample then consists of the following people:

Peterson, John	M	Fr	Smith, Nancy	F	Fr
Meloski, Gary	M	Fr	Carson, Susan	F	Fr
Brown, Danny	M	So	Toms, Debbie	F	So
Lyons, Larry	M	So	Unger, Roberta	F	So

The major advantage of stratification is that it ensures representation of all population subgroups that are important to the study. There are two major drawbacks to stratification, however. First, if there are many variables of interest, dividing a large population into representative subgroups requires a great deal of effort. Second, if the variables are somewhat complex or ambiguous (such as beliefs, attitudes, or prejudices), it is difficult to separate individuals into the subgroups according to these variables.

Cluster Sampling

A **cluster sample** is a sample obtained by selecting a preexisting or natural group, called a *cluster*, and using the members in the cluster for the sample.

For example, many studies in education use already existing classes, such as the seventh grade in Wilson Junior High School. The voters of a certain electoral district might be surveyed to determine their preferences for a mayoralty candidate in the upcoming election. Or the residents of an entire city block might be polled to ascertain the percentage of households that have two or more incomes. In cluster sampling, researchers may use all units of a cluster if that is feasible, or they may select only part of a cluster to use as a sample. This selection is done by random methods.

There are three advantages to using a cluster sample instead of other types of samples: (1) A cluster sample can reduce costs; (2) it can simplify fieldwork; and (3) it is convenient. For example, in a dental study involving X-raying fourth-grade students' teeth to see how many cavities each child had, it would be a simple matter to select a single classroom and bring the X-ray equipment to the school to conduct the study. If other sampling methods were used, researchers might have to transport the machine to several different schools or transport the pupils to the dental office.

The major disadvantage of cluster sampling is that the elements in a cluster may not have the same variations in characteristics as elements selected individually from a population. The reason is that groups of people may be more homogeneous (alike) in specific clusters such as neighborhoods or clubs. For example, the people who live in a certain neighborhood tend to have similar incomes, drive similar cars, live in similar houses, and for the most part, have similar habits.

Other Types of Sampling Techniques

In addition to the four basic sampling methods, other methods are sometimes used. In **sequence sampling,** which is used in quality control, successive units taken from production lines are sampled to ensure that the products meet certain standards set by the manufacturing company.

In **double sampling,** a very large population is given a questionnaire to determine those who meet the qualifications for a study. After the questionnaires are reviewed, a second, smaller population is defined. Then a sample is selected from this group.

In **multistage sampling,** the researcher uses a combination of sampling methods. For example, suppose a research organization wants to conduct a nationwide survey for a new product being manufactured. A sample can be obtained by using the following combination of methods. First, the researchers divide the 50 states into four or five regions (or clusters). Then, several states from each region are selected at random. Next the states are divided into various areas by using large cities and small towns. Samples of these areas are then selected. Next, each city and town is divided into districts or wards. Finally, streets in these wards are selected at random, and the families living on these streets are given samples of the product to test and asked to report the results. This hypothetical example illustrates a typical multistage sampling method.

The steps for conducting a sample survey are given in the Procedure Table.

> **Interesting Fact**
>
> Folks in extra-large aerobics classes—those with 70 to 90 participants—show up more often and are more fond of their classmates than exercisers in sessions of 18 to 26 people, report researchers at the University of Arizona. Source: *Psychology Today*, July–August 1996, p. 26.

Procedure Table

Conducting a Sample Survey

STEP 1	Decide what information is needed.
STEP 2	Determine how the data will be collected (phone interview, mail survey, etc.).
STEP 3	Select the information-gathering instrument or design the questionnaire if one is not available.
STEP 4	Set up a sampling list, if possible.
STEP 5	Select the best method for obtaining the sample (random, systematic, stratified, cluster, or other).
STEP 6	Conduct the survey and collect the data.
STEP 7	Tabulate the data.
STEP 8	Conduct the statistical analysis.
STEP 9	Report the results.

Exercises

15–1. Name the four basic sampling techniques.

15–2. Why are samples used in statistics?

15–3. What is the basic requirement for a sample?

15–4. Why should random numbers be used when one is selecting a random sample?

In this study, researchers first divided the subjects into two groups on the basis of their self-esteem. They then forced both groups to do poorly on a computer word game. Finally, they tested their self-esteem again. Explain how you would feel in this situation and cite some reasons why you would feel this way. What type of statistical test might be used to compare the self-esteem of the two groups?

Help, I've Failed and I Can't Get Up

Afraid you too are suffering from the epidemic of low self-esteem? Finish this sentence: *If at first you don't succeed . . .*

If you came up with something along the lines of *you are utterly useless* there may be cause for concern. What separates the high from the low on the self-esteem meter is response to failure, says University of Washington psychologist Jonathon Brown, Ph.D.

He put 172 people—81 with high self-esteem, the rest with low—through a computer word game. Half the participants received a version too difficult to do in the time alloted, assuring their failure. Afterwards Brown asked them to evaluate their performance.

For those lacking self-esteem, failure hit like the proverbial ton of bricks. Only feelings of shame and humiliation rose from the rubble. Worse, they overgeneralized their failure, rating their intelligence and competence more negatively after a poor performance than a successful one.

People with high self-esteem did just the opposite. They rated their intelligence a bit higher after failure, compensating for their sub-par performance. This is the value of self-esteem, explains Brown: It enables us to respond to events—good or bad—in ways that bolster our sense of worth.

Because failure is so agonizing to people with low self-esteem, they are less willing to take risks and more apt to be conformists. Though some researchers feel such people go with the flow because they don't trust their judgment enough to break away from the pack, Brown says what they really fear is the dark side of nonconformity—ostracism.

Source: *Psychology Today* 28, no. 5 (September–October, 1995), p. 14. Used with permission.

15–5. List three incorrect methods that are often used to obtain a sample.

15–6. What is the principle behind random numbers?

15–7. List the advantages and disadvantages of random sampling.

15–8. List the advantages and disadvantages of systematic sampling.

15–9. List the advantages and disadvantages of stratified sampling.

15–10. List the advantages and disadvantages of cluster sampling.

Using the student survey at Utopia University, shown in Figure 15–8, as the population, complete Exercises 15–11 through 15–15.

15–11. Using the table of random numbers in Figure 15–1, select 10 students and find the sample mean (average) of the GPA, IQ, and distance traveled to school. Compare these sample means with the population means.

15–12. Select a sample of 10 students by the systematic method and compute the sample means of the GPA, IQ, and distance traveled to school of this sample. Compare these sample means with the population means.

15–13. Select a cluster of 10 students—for example, students 9 through 18—and compute the sample means of their GPA, IQ, and distance traveled to school. Compare these sample means with the population means.

15–14. Divide the 50 students into subgroups according to class rank. Then select a sample of two students from each rank and compute the means of these 10 students for the GPA, IQ, and distance traveled to school each day. Compare these sample means with the population means.

Figure 15–8

Student Survey at Utopia University (for Exercises 15–11 through 15–15)

Student number	Gen- der	Class rank	GPA	Miles traveled to school	IQ	Major field	Student number	Gen- der	Class rank	GPA	Miles traveled to school	IQ	Major field
1	M	Fr	1.4	1	104	Bio	26	M	Fr	1.1	8	100	Ed
2	M	Fr	2.3	2	95	Ed	27	F	Jr	2.1	3	101	Bus
3	M	So	2.7	6	108	Psy	28	M	Gr	3.7	5	99	Bio
4	F	So	3.2	7	119	Eng	29	M	Se	2.4	8	105	Eng
5	F	Gr	3.8	12	114	Ed	30	M	So	2.1	15	108	Bus
6	M	Jr	4.0	13	91	Psy	31	M	Gr	3.9	2	112	Ed
7	F	Jr	3.0	2	106	Eng	32	F	Jr	2.4	4	111	Psy
8	M	Jr	3.3	6	100	Bio	33	M	Se	2.7	6	107	Eng
9	F	Se	2.7	9	102	Eng	34	F	So	2.5	1	104	Bio
10	F	So	2.3	5	99	Ed	35	M	Se	3.2	3	96	Bus
11	M	Se	1.6	18	100	Bus	36	F	Fr	3.4	7	98	Bio
12	M	Gr	3.2	7	105	Psy	37	M	Gr	3.6	14	105	Ed
13	F	Gr	3.8	3	103	Bus	38	M	Jr	3.8	4	115	Psy
14	F	Se	3.1	5	97	Eng	39	F	Se	2.2	8	113	Eng
15	F	Jr	2.7	5	106	Bio	40	F	So	2.0	8	103	Psy
16	F	Fr	1.4	4	114	Bus	41	F	Fr	2.3	9	103	Eng
17	M	So	3.6	17	102	Ed	42	F	Se	2.5	10	99	Bus
18	M	Fr	2.2	1	101	Psy	43	M	Gr	3.7	13	114	Ed
19	F	Gr	4.0	7	108	Bus	44	M	Fr	3.0	11	121	Bus
20	M	Jr	2.1	4	97	Ed	45	M	Jr	2.1	10	101	Eng
21	F	Fr	2.0	3	113	Bio	46	F	Jr	3.4	2	104	Ed
22	F	So	3.6	4	104	Bio	47	M	So	3.6	9	105	Psy
23	F	Gr	3.3	16	110	Eng	48	M	Se	2.1	1	97	Psy
24	F	Se	2.5	4	99	Psy	49	F	Gr	3.3	12	111	Bio
25	M	So	3.0	5	96	Psy	50	F	Fr	2.2	11	102	Bio

15–15. In your opinion, which sampling method(s) provided the best sample to represent the population?

"O.K. You ask me, then I'll ask you."

Source: © 1988, *Scouting*. Reprinted by permission of Orlando Busino.

Figure 15–9 shows the 50 states and the number of electoral votes each state has in the presidential election. Using this listing as a population, complete Exercises 15–16 through 15–19.

15–16. Select a random sample of 10 states and find the mean number of electoral votes for this sample. Compare this mean with the population mean.

15–17. Select a systematic sample of 10 states and compute the mean number of electoral votes for the sample. Compare this mean with the population mean.

15–18. Divide the 50 states into five subgroups by geographic location, using a map of the United States. Each subgroup should include 10 states. The subgroups should be northeast, southeast, central, northwest, and southwest. Select two states from each subgroup and find the mean number of electoral votes for the sample. Compare these means with the population mean.

15–19. Select a cluster of 10 states and compute the mean number of electoral votes for the sample. Compare this mean with the population mean.

15–20. Many research studies described in newspapers and magazines do not report the sample size or the sampling method used. Try to find a research article that gives this information; state the sampling method that was used and the sample size.

"Now think carefully. The answer you give will represent the opinion of millions of Americans."

Source: Reprinted from *The Saturday Evening Post*, © 1987 BFL&MS, Inc., Indianapolis.

Figure 15–9

States and Number of Electoral Votes for Each (for Exercises 15–16 through 15–19)

1. Alabama	9	14. Indiana	12	27. Nebraska	5	40. South Carolina	8
2. Alaska	3	15. Iowa	8	28. Nevada	4	41. South Dakota	3
3. Arizona	7	16. Kansas	7	29. New Hampshire	4	42. Tennessee	11
4. Arkansas	6	17. Kentucky	9	30. New Jersey	16	43. Texas	29
5. California	47	18. Louisiana	10	31. New Mexico	5	44. Utah	5
6. Colorado	8	19. Maine	4	32. New York	36	45. Vermont	3
7. Connecticut	8	20. Maryland	10	33. North Carolina	13	46. Virginia	12
8. Delaware	3	21. Massachusetts	13	34. North Dakota	3	47. Washington	10
9. Florida	21	22. Michigan	20	35. Ohio	23	48. West Virginia	6
10. Georgia	12	23. Minnesota	10	36. Oklahoma	8	49. Wisconsin	11
11. Hawaii	4	24. Mississippi	7	37. Oregon	7	50. Wyoming	3
12. Idaho	4	25. Missouri	11	38. Pennsylvania	25		
13. Illinois	24	26. Montana	4	39. Rhode Island	4		

Technology Step by Step

MINITAB
Step by Step

Generate Random Numbers

Example MT15–1

A list of integers will be created first. Then a random sample will be selected from the list.

1. Select **Calc>Make Patterned Data>Simple Set of Numbers.**
2. Type in the name of the new column, **Integers.**
3. Press **[Tab]** to move the cursor into the next box.
4. Enter **1** for first value and **50** for last value. Leave a 1 in the remaining boxes.

Make Patterned Data
Dialog Box

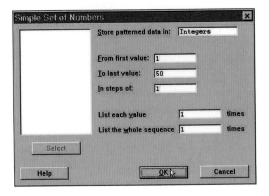

5. Click [OK].

To select a random sample from among these numbers:

6. Select **Calc>Random Data>Select From Columns.**

7. Type **10** for the number to be selected. Press **[Tab]** or click in the box for Store Samples in:.

8. Double-click C1 Integers in the list to select the from column(s). Press **[Tab]**.

Select Random Sample

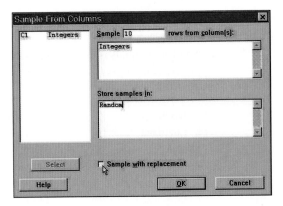

9. Type in the name of the new column, **Random.** Do not select Sample with replacement or you may get duplicate numbers.

10. Click [OK]. The list will be in the Random column of the worksheet.

Note: Steps 6 to 10 can be used to select a sample from raw data in a column of a worksheet.

TI-83
Step by Step

Generate Random Numbers

To generate random numbers from zero to one using the TI-83:

1. Press **MATH** and move the cursor to PRB and press **1** for rand, then press **ENTER**.

The calculator will generate a random decimal from 0 to 1.

2. To generate additional random numbers press **ENTER.**

To generate a list of random integers between two specific values:

1. Press **MATH** and move the cursor to PRB.

2. Press **5** for randInt(.

3. Enter the lowest value followed by a comma, then the largest value followed by a comma, then the number of random numbers desired followed by). Press **ENTER.**

Example: Generate 5 three-digit random numbers.

Enter **0, 999, 5)** at the randInt(as shown.

```
rand
            .9435974025
randInt(0,999,5)
  {908 146 514 40…
```

The calculator will generate 5 three-digit random numbers. Use the arrow keys to view the entire list.

15–3

Surveys and Questionnaire Design

Objective 2. Recognize faulty questions on a survey and other factors that can bias responses.

Many statistical studies obtain information from *surveys.* A survey is conducted when a sample of individuals is asked to respond to questions about a particular subject. There are two types of surveys, interviewer-administered and self-administered. Interviewer-administered surveys require a person to ask the questions. The interview can be conducted face to face in an office, on a street, in the mall, or via telephone.

Self-administered surveys can be done by mail or in a group setting such as a classroom.

When analyzing the results of surveys, you should be very careful about the interpretations. The way a question is phrased can influence the way people respond.

For example, when a group of people was asked if they favored a waiting period and background check before guns could be sold, 91% of the respondents were in favor of it and 7% were against it. However, when asked if there should be a national gun registration program costing about 20% of all dollars spent on crime control, only 33% of the respondents were in favor of it and 61% were against it.

As you can see, by phrasing questions in different ways, different responses can be obtained.

When writing questions for a questionnaire, it is important to avoid these common mistakes.

1. *Asking biased questions.* By asking questions in a certain way, the researcher can lead the respondents to answer in the way he or she wants them to. For example, asking a question such as "Are you going to vote for the Candidate Jones even though the latest survey indicates that he will lose the election?" instead of "Are you going to vote for Candidate Jones?" may dissuade some people from answering in the affirmative.

2. *Using confusing words.* In this case, the participant misinterprets the meaning of the words and answers the questions in a biased way. For example, the question "Do you think people would live longer if they were on a diet?" could be

misinterpreted since there are many different types of diets: weight loss diets, low salt diets, medically prescribed diets, etc.

3. *Asking double-barreled questions.* Sometimes questions contain compound sentences that require the participant to respond to two questions at the same time. For example, the question "Are you in favor of a special tax to provide national health care for the citizens of the United States?" asks two questions, "Are you in favor of a national health care program?" and "Do you favor a tax to support it?"

4. *Using double negatives in questions.* Questions with double negatives can be confusing to the respondents. For example, the question, "Do you feel that it is not appropriate to have areas where people cannot smoke?" is very confusing since not is used twice in the sentence.

5. *Ordering questions improperly.* By arranging the questions in a certain order, the researcher can lead the participant to respond in a way that he or she may otherwise not have done. For example, a question might ask the respondent "At what age should an elderly person not be permitted to drive?" A later question might ask the respondent to list some problems of elderly people. The respondent may indicate that transportation is a problem based on reading the previous question.

Other factors can also bias a survey. For example, the participant may not know anything about the subject of the question but will answer the question anyway to avoid being considered uninformed. For example, many people might respond yes or no to the following question: "Would you be in favor of giving pensions to the widows of unknown soldiers?" In this case, the question makes no sense since if the soldiers were unknown, their widows would also be unknown.

Many people will make responses on the basis of what they think the person asking the questions wants to hear. For example, if a question states, "How often do you lie?" people may *understate* the incidences of their lying.

Participants will, in some cases, respond differently to questions depending on whether or not their identity is known. This is especially true if the questions concern sensitive issues such as income, sexuality, abortion, etc. Researchers try to ensure confidentiality rather than anonymity; however, many people will be suspicious in either case.

Still other factors that could bias a survey include the time and place of the survey, and whether the questions were open-ended or closed-ended.

The time and place where a survey is conducted can influence the results. For example, if a survey on airline safety is conducted immediately after a major airline crash, the results may differ from those obtained in a year in which no major airline disasters occurred.

Finally, the type of questions asked influence the responses. In this case, the concern is whether the question is open-ended or close-ended.

An *open-ended question* would be one such as, "List three activities that you plan to spend more time on when you retire." A *closed-ended question* would be one such as, "Select three activities that you plan to spend more time on after you retire

_____ traveling

_____ eating out

_____ fishing, hunting, etc.

_____ exercising

_____ visiting relatives"

One problem with a closed-ended question is that the respondent is forced to choose the answers that the researcher gives and cannot supply his or her own. But there is also a problem with open-ended questions in that the results may be so varied that attempting to summarize them might be difficult if not impossible. Hence, you should be aware of what type of questions are being asked before drawing any conclusions from the survey.

There are several other things to consider when conducting a study that uses questionnaires. For example, a pilot study should be done to test the design and usage of the questionnaire (i.e., the *validity* of the questionnaire). The pilot study helps the researcher to pretest the questionnaire in order to determine if it meets the objectives of the study. It also helps the researcher to rewrite any questions that may be misleading, ambiguous, etc.

If the questions are being asked by an interviewer, some training should be given to that person. If the survey is being done by mail, a cover letter and clear directions should accompany the questionnaire.

Questionnaires help researchers to gather needed statistical information for their studies; however, much care must be given to proper questionnaire design and usage; otherwise, the results will be unreliable.

Exercises

Exercises 15–21 through 15–28 include questions that contain a flaw. Identify the flaw and rewrite the question following the guidelines presented in this section.

15–21. Will you continue to shop at XYZ Department Store even though it does not carry brand names?

15–22. Would you buy an ABC car even if you knew the manufacturer used imported parts?

15–23. Should banks charge their checking account customers a fee to balance their checkbooks when customers are not able to do so?

15–24. Do you feel that it is not appropriate for shopping malls to have activities for children who cannot read?

15–25. How long have you studied for this examination?

15–26. Do you think children would watch less television if they read more?

15–27. If a plane were to crash on the border of New York and New Jersey, where should the survivors be buried?

15–28. Are you in favor of imposing a tax on tobacco to pay for health care related to diseases caused by smoking?

15–29. Find a study that uses a questionnaire. Select any questions that you feel are improperly written.

15–30. Many television and radio stations have a phone vote poll. If there is one in your area, select a specific day and write a brief paragraph stating the question of the day and state if it could be misleading in any way.

15–4

Simulation Techniques

Many real-life problems can be solved by employing simulation techniques.

A **simulation technique** uses a probability experiment to mimic a real-life situation.

Instead of studying the actual situation, which might be too costly, too dangerous, or too time-consuming, scientists and researchers create a similar situation but one that is less expensive, less dangerous, or less time-consuming. For example, NASA uses space shuttle flight simulators so that its astronauts can practice flying the shuttle. Most video games use the computer to simulate real-life sports such as boxing, wrestling, baseball, and hockey.

Simulation techniques go back to ancient times when the game of chess was invented to simulate warfare. Modern techniques date to the mid-1940s when two physicists, John Von Neumann and Stanislaw Ulam, developed simulation techniques to study the behavior of neutrons in the design of atomic reactors.

Mathematical simulation techniques use probability and random numbers to create conditions similar to those of real-life problems. Computers have played an important role in simulation techniques, since they can generate random numbers, perform experiments, tally the outcomes, and compute the probabilities much faster than human beings. The basic simulation technique is called the *Monte Carlo method*. This topic is discussed in the next section.

15–5

The Monte Carlo Method

Objective 3. Solve problems using simulation techniques.

The **Monte Carlo method** is a simulation technique using random numbers. Monte Carlo simulation techniques are used in business and industry to solve problems that are extremely difficult or involve a large number of variables. The steps for simulating real-life experiments in the Monte Carlo method follow.

1. List all possible outcomes of the experiment.

2. Determine the probability of each outcome.

3. Set up a correspondence between the outcomes of the experiment and the random numbers.

4. Select random numbers from a table and conduct the experiment.

5. Repeat the experiment and tally the outcomes.

6. Compute any statistics and state the conclusions.

Before examples of the complete simulation technique are given, an illustration is needed for step 3, "Set up a correspondence between the outcomes of the experiment and the random numbers." Tossing a coin, for instance, can be simulated by using random numbers as follows: Since there are only two outcomes, heads and tails, and since each outcome has a probability of $\frac{1}{2}$, the odd digits— 1, 3, 5, 7, and 9—can be used to represent a head, and the even digits—0, 2, 4, 6, and 8—can represent a tail.

Suppose a random number, 8631, is selected. This number represents four tosses of a single coin and the results T, T, H, H. Or this number could represent one toss of four coins with the same results.

An experiment of rolling a single die can also be simulated by using random numbers. In this case, the digits 1, 2, 3, 4, 5, and 6 can represent the number of spots that appear on the face of the die. The digits 7, 8, 9, and 0 are ignored, since they cannot be rolled.

When two dice are rolled, two random digits are needed. For example, the number 26 represents a 2 on the first die and a 6 on the second die. The random number 37 represents a 3 on the first die, but the 7 cannot be used, so another digit must be selected. As another example, a three-digit daily lotto number can be simulated by using three-digit random numbers. Finally, a spinner with four numbers, as shown in Figure 15–10, can be simulated by letting the random numbers 1 and 2 represent 1 on the spinner, 3 and 4 represent 2 on the spinner, 5 and 6 represent 3 on the spinner, and 7 and 8 represent 4 on the spinner, since each number has a probability of 1 out of 4 of being selected. The random numbers 9 and 0 are ignored in this situation.

Many real-life games, such as bowling and baseball, can be simulated by using random numbers, as shown in Figure 15–11.

Figure 15–10

Spinner with Four Numbers

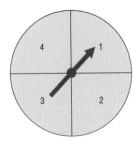

Figure 15–11

Example of Simulation of
a Game

Simulated Bowling Game

Let's use the random digit table to simulate a bowling game. Our game is much simpler than commercial simulation games.

First Ball		Second Ball			
		2-Pin Split		No split	
Digit	Results	Digit	Results	Digit	Results
1–3	Strike	1	Spare	1–3	Spare
4–5	2-pin split	2–8	Leave one pin	4–6	Leave 1 pin
6–7	9 pins down	9–0	Miss both pins	7–8	*Leave 2 pins
8	8 pins down			9	+Leave 3 pins
9	7 pins down			0	Leave all pins
0	6 pins down				

*If there are fewer than 2 pins, result is a spare.
+If there are fewer than 3 pins, those pins are left.

Here's how to score bowling:
1. There are 10 frames to a **game** or **line**.
2. You roll two balls for each frame, unless you knock all the pins down with the first ball (a **strike**).
3. Your score for a frame is the sum of the pins knocked down by the two balls, if you don't knock down all 10.
4. If you knock all 10 pins down with two balls (a **spare**, shown as ▱), your score is 10 pins plus the number knocked down with the next ball.
5. If you knock all 10 pins down with the first ball (a **strike**, shown as ⊠), your score is 10 pins plus the number knocked down by the next **two** balls.
6. A **split** (shown as 0) is when there is a big space between the remaining pins. Place in the circle the number of pins remaining after the second ball.
7. A **miss** is shown as —.

Here is how one person simulated a bowling game using the random digits 7 2 7 4 8 2 2 3 6 1 6 0 4 6 1 5 5, chosen in that order from the table.

					Frame					
	1	2	3	4	5	6	7	8	9	10
Digit(s)	7/2	7/4	8/2	2	3	6/1	6/0	4/6	1	5/5
Bowling result	9▱ 19	9— 28	8▱ 48	⊠ 77	⊠ 97	9▱ 116	9— 125	8① 134	⊠ 153	8① 162	▱ 162

Now you try several.

					Frame					
	1	2	3	4	5	6	7	8	9	10
Digit(s)										
Bowling result										

	1	2	3	4	5	6	7	8	9	10
Digit(s)										
Bowling result										

If you wish to, you can change the probabilities in the simulation to better reflect *your* actual bowling ability.

Source: Albert Shuylte, "Simulated Bowling Game," Student Math Notes, March 1986. Published by the National Council of Teachers of Mathematics. Reprinted with permission.

Example 15–4

Using random numbers, simulate the gender of children born.

Solution

There are only two possibilities, male and female. Since the probability of each outcome is 0.5, the odd digits can be used to represent male births, and the even digits to represent female births.

Example 15–5

Using random numbers, simulate the outcomes of a tennis game between two people, Bill and Mike, with the additional condition that Bill is twice as good as Mike.

Solution

Since Bill is twice as good as Mike, he will win approximately two games for every one Mike wins; hence, the probability that Bill wins will be $\frac{2}{3}$ and the probability that Mike wins will be $\frac{1}{3}$. The random digits 1 through 6 can be used to represent a game Bill wins; the random digits 7, 8, and 9 can be used to represent Mike's wins. The digit 0 is disregarded. Suppose they play five games, and the random number 86314 is selected. This number means that Bill won games 2, 3, 4, and 5, and Mike won the first game. The sequence is

8	6	3	1	4
M	B	B	B	B

More complex problems can be solved by using random numbers, as shown in the next three examples.

Example 15–6

A die is rolled until a 6 appears. Using simulation, find the average number of rolls needed. Try the experiment 20 times.

Solution

STEP 1   List all possible outcomes. They are 1, 2, 3, 4, 5, 6.

STEP 2   Assign the probabilities. Each outcome has a probability of $\frac{1}{6}$.

STEP 3   Set up a correspondence between the random numbers and the outcome. Use random numbers 1 through 6. Omit the numbers 7, 8, 9, and 0.

STEP 4   Select a block of random numbers and count each digit 1 through 6 until the first 6 is obtained. For example, the block 857236 means that it takes 4 rolls to get a 6.

8	5	7	2	3	6
	↑		↑	↑	↑
	5		2	3	6

STEP 5   Repeat the experiment 19 more times and tally the data as shown.

Trial	Random number	Number of rolls
1	8 5 7 2 3 6	4
2	2 1 0 4 8 0 1 5 1 1 0 1 5 3 6	11
3	2 3 3 6	4
4	2 4 1 3 0 4 8 3 6	7
5	4 2 1 6	4
6	3 7 5 2 0 3 9 8 7 5 8 1 8 3 7 1 6	9
7	7 7 9 2 1 0 6	3
8	9 9 5 6	2
9	9 6	1
10	8 9 5 7 9 1 4 3 4 2 6	7
11	8 5 4 7 5 3 6	5
12	2 8 9 1 8 6	3
13	6	1
14	0 9 4 2 9 9 3 9 6	4
15	1 0 3 6	3
16	0 7 1 1 9 9 7 3 3 6	5
17	5 1 0 8 5 1 2 7 6	6
18	0 2 3 6	3
19	0 1 0 1 1 5 4 0 9 2 3 3 3 6	10
20	5 2 1 6	4
		Total 96

STEP 6   Compute the results and draw a conclusion. In this case, one must find the average.

$$\bar{X} = \frac{\Sigma X}{n} = \frac{96}{20} = 4.8$$

Hence, the average is about 5 rolls.

Note: The theoretical average obtained from the expected value formula is 6. If this experiment is done many times, say 1000 times, the results should be closer to the theoretical results.

Example 15–7

A person selects a key at random from four keys to open a lock. Only one key fits. If the first key does not fit, she tries other keys until one fits. Find the average of the number of keys a person will have to try to open the lock. Try the experiment 25 times.

Solution

Assume that each key is numbered from 1 through 4 and that key 2 fits the lock. Naturally, the person doesn't know this, so she selects the keys at random. For the simulation, select a sequence of random digits using only 1 through 4 until the digit 2 is reached. The trials are shown here.

Trial	Random digit (key)	Number	Trial	Random digit (key)	Number
1	2	1	14	2	1
2	2	1	15	4 2	2
3	1 2	2	16	1 3 2	3
4	1 4 3 2	4	17	1 2	2
5	3 2	2	18	2	1
6	3 1 4 2	4	19	3 4 2	3
7	4 2	2	20	2	1
8	4 3 2	3	21	2	1
9	4 2	2	22	2	1
10	2	1	23	4 2	2
11	4 2	2	24	4 3 1 2	4
12	3 1 2	3	25	3 1 2	3
13	3 1 2	3			Total 54

Next, find the average:

$$\bar{X} = \frac{\Sigma X}{n} = \frac{1 + 1 + \cdots + 3}{25} = \frac{54}{25} = 2.16$$

The theoretical average is 2.5. Again, only 25 repetitions were used; more repetitions should give a result closer to the theoretical average.

Example 15–8

A box contains five $1 bills, three $5 bills, and two $10 bills. A person selects a bill at random. What is the expected value of the bill? Perform the experiment 25 times.

Solution

STEP 1   List all possible outcomes. They are $1, $5, and $10.

STEP 2   Assign the probabilities to each outcome:

$$P(\$1) = \tfrac{5}{10} \qquad P(\$5) = \tfrac{3}{10} \qquad P(\$10) = \tfrac{2}{10}$$

STEP 3 Set up a correspondence between the random numbers and the outcomes. Use random numbers 1 through 5 to represent a $1 bill being selected, 6 through 8 to represent a $5 bill being selected, and 9 and 0 to represent a $10 bill being selected.

STEPS 4 AND 5 Select 25 random numbers and tally the results.

Number	Results
4 5 8 2 9	$1, $1, $5, $1, $10
2 5 6 4 6	$1, $1, $5, $1, $5
9 1 8 0 3	$10, $1, $5, $10, $1
8 4 0 6 0	$5, $1, $10, $5, $10
9 6 9 4 3	$10, $5, $10, $1, $1

STEP 6 Compute the average:

$$\bar{X} = \frac{\Sigma X}{n} = \frac{\$1 + \$1 + \$5 + \cdots + \$1}{25} = \frac{\$116}{25} = \$4.64$$

Hence, the average (expected value) is $4.64.

Recall that using the expected value formula, $E(X) = \Sigma X \cdot P(X)$, gives a theoretical average of

$$E(X) = \Sigma X \cdot P(X) = (0.5)(\$1) + (0.3)(\$5) + (0.2)(\$10) = \$4.00$$

Remember that simulation techniques do not give exact results. The more times the experiment is performed, though, the closer the actual results should be to the theoretical results. (Recall the law of large numbers.)

The steps for solving problems using the Monte Carlo method are summarized in the Procedure Table.

Procedure Table

Simulating Experiments Using the Monte Carlo Method

STEP 1 List all possible outcomes of the experiment.

STEP 2 Determine the probability of each outcome.

STEP 3 Set up a correspondence between the outcomes of the experiment and the random numbers.

STEP 4 Select random numbers from a table and conduct the experiment.

STEP 5 Repeat the experiment and tally the outcomes.

STEP 6 Compute any statistics and state the conclusions.

Exercises

15–31. Define *simulation techniques.*

15–32. Give three examples of simulation techniques.

15–33. Who is responsible for the development of modern simulation techniques?

15–34. What role does the computer play in simulation?

15–35. What are the steps in the simulation of an experiment?

15–36. What purpose do random numbers play in simulation?

15–37. What happens when the number of repetitions is increased?

For Exercises 15–38 through 15–43, explain how each experiment can be simulated by using random numbers.

15–38. A spinner contains five equal areas.

15–39. A basketball player makes 80% of her shots.

15–40. A certain brand of VCRs manufactured has a 20% defective rate.

15–41. A dart thrower hits the target 70% of the time.

15–42. Two players match pennies.

15–43. Three players play odd man out. (Three coins are tossed; if all three match, the game is repeated and no one wins. If two players match, the third person wins all three coins.)

For Exercises 15–44 through 15–51, use random numbers to simulate the experiments. The number in parentheses is the number of times the experiment should be repeated.

15–44. A coin is tossed until four heads are obtained. Find the average number of tosses necessary. (50)

15–45. A die is rolled until all faces appear at least once. Find the average number of tosses. (30)

15–46. A caramel-corn company gives four different prizes, one in each box. They are placed in the boxes at random. Find the average number of boxes a person needs to buy to get all four prizes. (40)

15–47. Two teams are evenly matched. They play a tournament in which the first team to win three games wins the tournament. Find the average number of games the tournament will last. (20)

15–48. To win a certain lotto, a person must spell the word *big*. Sixty percent of the tickets contain the letter *b*, 30% contain the letter *i*, and 10% contain the letter *g*. Find the average number of tickets a person must buy to win the prize. (30)

15–49. Two shooters shoot clay pigeons. Gail has an 80% accuracy rate and Paul has a 60% accuracy rate. Paul shoots first. The first person who hits the target wins. Find the probability that each wins. (30).

15–50. In Exercise 15–49, find the average number of shots fired. (30)

15–51. A basketball player has a 60% success rate for shooting foul shots. If she gets two shots, find the probability that she will make one or both shots. (50).

15–52. Select a game such as baseball or football and write a simulation using random numbers.

15–53. Explain how cards can be used to generate random numbers.

15–54. Explain how a pair of dice can be used to generate random numbers.

15–6

Summary

To obtain information and make inferences about a large population, researchers select a sample. A sample is a subgroup of the population. Using a sample rather than a population, researchers can save time and money, get more detailed information, and get information that otherwise would be impossible to obtain.

The four most common methods researchers use to obtain samples are random, systematic, stratified, and cluster sampling methods. In random sampling, some type of random method (usually random numbers) is used to obtain the sample. In systematic sampling, the researcher selects every *k*th person or item after selecting the first one at random. In stratified sampling, the population is divided into subgroups according to various characteristics, and elements are then selected at random from the subgroups. In cluster sampling, the researcher selects an intact group to use as a sample. When the population is large, multistage sampling (a combination of methods) is used to obtain a subgroup of the population.

Researchers must use caution when conducting surveys and designing questionnaires, otherwise conclusions obtained from these will be inaccurate. Guidelines were presented in section 15–3.

Most sampling methods use random numbers, which can also be used to simulate many real-life problems or situations. The basic method of simulation is known as the Monte Carlo method. The purpose of simulation is to duplicate situations that are too dangerous, too costly, or too time-consuming to study in real life. Most simulation techniques can be done on the computer or calculator, since they can rapidly generate random numbers, count the outcomes, and perform the necessary computations.

Sampling and simulation are two techniques that enable researchers to gain information that might otherwise be unobtainable.

Important Terms

biased sample   631

cluster sample   639

double sampling   640

Monte Carlo method   648

multistage sampling   640

random sample   631

sequence sampling   640

simulation technique   647

stratified sample   637

systematic sample   636

unbiased sample   631

Review Exercises

Use Figure 15–12 for Exercises 15–55 through 15–62.

15–55. Select a random sample of 10 people and find the mean of the weights of the individuals. Compare this mean with the population mean.

15–56. Select a systematic sample of 10 people and compute the mean of their weights. Compare this mean with the population mean.

15–57. Divide the individuals into subgroups of males and females. Select five individuals from each group and find

the mean of their weights. Compare these means with the population mean.

15–58. Select a cluster of 10 people and find the mean of their weights. Compare this mean with the population mean.

15–59. Repeat Exercise 15–55 for blood pressure.

15–60. Repeat Exercise 15–56 for blood pressure.

15–61. Repeat Exercise 15–57 for blood pressure.

15–62. Repeat Exercise 15–58 for blood pressure.

Figure 15–12

Population for Exercises 15–55 through 15–62

Individual	Gender	Weight	Systolic blood pressure	Individual	Gender	Weight	Systolic blood pressure	Individual	Gender	Weight	Systolic blood pressure
1	F	122	132	18	F	118	125	35	M	172	116
2	F	128	116	19	F	107	138	36	M	175	123
3	M	183	140	20	M	214	121	37	F	101	114
4	M	165	136	21	F	114	127	38	F	123	113
5	M	192	120	22	M	119	125	39	M	186	145
6	F	116	118	23	F	125	114	40	F	100	119
7	M	206	116	24	M	182	137	41	M	202	135
8	F	131	120	25	F	127	127	42	F	117	121
9	M	155	118	26	F	132	130	43	F	120	130
10	F	106	122	27	M	198	114	44	M	193	125
11	F	103	119	28	F	135	119	45	M	200	115
12	M	169	136	29	M	183	137	46	F	118	132
13	M	173	134	30	F	140	123	47	F	121	143
14	M	195	145	31	M	189	135	48	M	189	128
15	F	107	113	32	M	165	121	49	M	114	118
16	M	201	111	33	M	211	117	50	M	174	138
17	F	114	141	34	F	111	127				

For Exercises 15–63 through 15–67, explain how to simulate each experiment by using random numbers.

15–63. A baseball player strikes out 65% of the time.

15–64. An airline overbooks 5% of its flights.

15–65. Two players roll a die, and the one who rolls the lower number wins.

15–66. One player rolls two dice, and one player rolls one die. The lowest number wins.

15–67. Two players play rock, paper, scissors. The rules are as follows: Since paper covers rock, paper wins. Since rock breaks scissors, rock wins. Since scissors cut paper, scissors win. Each person selects either rock, paper, or scissors by random numbers and then compares results.

For Exercises 15–68 through 15–72, use random numbers to simulate the experiments. The number in parentheses is the number of times the experiment should be repeated.

15–68. A football is placed on the 10-yard line and a team has four downs to score a touchdown. The team can move the ball only 0 to 5 yards per play. Find the average number of times the team will score a touchdown. (30)

15–69. In Exercise 15–68, find the average number of plays it will take to score a touchdown. Ignore the four-downs rule and keep playing until a touchdown is scored. (30)

15–70. Four dice are rolled 50 times. Find the average of the sum of the number of spots that will appear. (50)

15–71. A field-goal kicker is successful in 60% of his kicks inside the 35-yard line. Find the probability of kicking three field goals in a row. (50)

15–72. A sales representative finds that there is a 30% probability of making a sale by visiting the potential customer personally. For every 20 calls, find the probability of making three sales in a row. (50)

For Exercises 15–73 through 15–75, explain what is wrong with each question.

15–73. How often do you run red lights?

15–74. Do you think students who are not failing should not be tutored?

15–75. Do you think all automobiles should have heavy-duty bumpers, even though it will raise the price of the cars by $500?

15–76. Explain the difference between an open-ended question and a closed-ended question.

Statistics Today

No Room in the Van? Revisited

By using random number simulation techniques, the owner can generate 100 random samples of 12 digits and see how many times the digit 0 occurs. This would represent the person who does not show up. Any sample with no zeros or one zero would be an overbooked van. Thus, the probability of being overbooked can be computed by dividing the number of overbooked samples by 100.

Data Analysis

The Data Bank is found in Appendix D.

1. From the Data Bank, choose a variable. Select a random sample of 20 individuals and find the mean of the data.

2. Select a systematic sample of 20 individuals, and using the same variable as in Exercise 1, find the mean.

3. Select a cluster sample of 20 individuals, and using the same variable as in Exercise 1, find the mean.

4. Stratify the data according to marital status and gender, and sample 20 individuals. Compute the mean of the sample variable selected in Exercise 1 (use 4 groups of 5 individuals).

5. Compare all four means and decide which one is most appropriate. (*Hint:* Find the population mean.)

Quiz

Determine whether each statement is true or false. If the statement is false, explain why.

1. When researchers are sampling from large populations, such as adult citizens living in the United States, they may use a combination of sampling techniques to ensure representativeness.

2. Simulation techniques using random numbers are a substitute for performing the actual statistical experiment.

3. When researchers perform simulation experiments, they do not need to use random numbers since they can make up random numbers.

4. Random samples are said to be unbiased.

Select the best answer.

5. When all subjects under study are used, the group is called a _____.
   - *a.* Population
   - *b.* Large group
   - *c.* Sample
   - *d.* Study group

6. When a population is divided into subgroups with similar characteristics, and then a sample is obtained, this sampling method is called _____.
   - *a.* Random
   - *b.* Systematic
   - *c.* Stratified
   - *d.* Cluster

7. Interviewing selected people at a local supermarket can be considered an example of _____ sampling.
   - *a.* Random
   - *b.* Systematic
   - *c.* Convenience
   - *d.* Stratified

Complete the following statements with the best answer.

8. In general, when one conducts sampling, the _____ the sample, the more representative it will be.

9. When samples are not representative, they are said to be _____ .

10. When all residents of a street are interviewed for a survey, the sampling method used is _____ .

Use Figure 15–12 in the Review Exercises (p. 654) for Questions 11 through 14.

11. Select a random sample of 12 people and find the mean of the blood pressures of the individuals. Compare this with the population mean.

12. Select a systematic sample of 12 people and compute the mean of their blood pressures. Compare this with the population mean.

13. Divide the individuals into subgroups of six males and six females. Find the means of their blood pressures. Compare these means with the population mean.

14. Select a cluster of 12 people and find the mean of their blood pressures. Compare this with the population mean.

For Questions 15 through 19, explain how each could be simulated using random numbers.

15. A wrestler wins 35% of his matches.

16. A travel agency experiences a 3% cancellation rate.

17. Two players each draw a card, and the one who draws the higher card wins.

18. One player rolls two dice, and another player selects one card. The player who has the lower total points wins.

19. Two players each roll two dice. The one who rolls the higher number wins.

For Problems 20 through 24, use random numbers to simulate the experiments. The number in parentheses is the number of times the experiment should be done.

20. A telephone solicitor finds that there is a 15% probability of selling her product over the phone. For every 20 calls, find the probability of making two sales in a row. (100)

21. A field-goal kicker is successful in 65% of his kicks inside the 40-yard line. Find the probability of his kicking four field goals in a row. (40)

22. Two coins are tossed. Find the average number of times two tails will appear. (40)

23. A single card is drawn from a deck. Find the average number of times it takes to draw an ace. (30)

24. A bowler finds that there is a 30% probability that he will make a strike. For every 15 frames he bowls, find the probability of making two strikes. (30)

Critical Thinking Challenges

1. Explain why two different opinion polls might yield different results on a survey. Also, give an example of an opinion poll and explain how the data may have been collected.

2. Use a computer to generate random numbers to simulate the following real-life problem.

In a certain geographic region, 40% of the people have type O blood. On a certain day, the blood center needs four pints of type O blood. On average, how many donors are needed to obtain four pints?

Data Projects

Where appropriate, use MINITAB, the TI-83, or a computer program of your choice to complete the following exercises.

1. Using the rules given in Figure 15–11 on page 649 of your textbook, play the simulated bowling game at least 10 times. Each game consists of 10 frames.
   a. Analyze the results of the scores by finding the mean, median, mode, range, variance, and standard deviation.
   b. Draw a box plot and explain the nature of the distribution.
   c. Write several paragraphs explaining the results.
   d. Compare this simulation with real bowling. Do you think the game actually simulates bowling? Why or why not?

2. Select a sports game that you like to play or watch on television (for example, baseball, golf, or hockey). Write a simulated version of the game using random numbers or dice. Play the game several times and answer the following questions.
   a. Does your simulated game represent the real game accurately?
   b. Is your game purely chance or is strategy involved?
   c. What are some shortcomings of your game?
   d. What parts of the real game cannot be simulated in your game?
   e. Is there any way that you could improve your simulated game by changing some rules?

A Appendix A

Algebra Review

A–1 Factorials
A–2 Summation Notation
A–3 The Line

A–1 Factorials

Definition and Properties of Factorials

The notation called factorial notation is used in probability. *Factorial notation* uses the exclamation point and involves multiplication. For example,

$$5! = 5 \cdot 4 \cdot 3 \cdot 2 \cdot 1 = 120$$
$$4! = 4 \cdot 3 \cdot 2 \cdot 1 = 24$$
$$3! = 3 \cdot 2 \cdot 1 = 6$$
$$2! = 2 \cdot 1 = 2$$
$$1! = 1$$

In general, a factorial is evaluated as follows:

$$n! = n(n-1)(n-2) \cdots 3 \cdot 2 \cdot 1$$

Note that the factorial is the product of n factors, with the number decreased by one for each factor.

One property of factorial notation is that it can be stopped at any point by using the exclamation point. For example,

$5! = 5 \cdot 4!$	since	$4! = 4 \cdot 3 \cdot 2 \cdot 1$
$= 5 \cdot 4 \cdot 3!$	since	$3! = 3 \cdot 2 \cdot 1$
$= 5 \cdot 4 \cdot 3 \cdot 2!$	since	$2! = 2 \cdot 1$
$= 5 \cdot 4 \cdot 3 \cdot 2 \cdot 1$		

Thus, $n! = n(n-1)!$
$$= n(n-1)(n-2)!$$
$$= n(n-1)(n-2)(n-3)! \qquad \text{etc.}$$

Another property of factorials is

$$0! = 1$$

This fact is needed for formulas.

Operations with Factorials

Factorials cannot be added or subtracted directly. They must be multiplied out. Then the products can be added or subtracted.

Example A–1

Evaluate $3! + 4!$.

Solution

$$3! + 4! = (3 \cdot 2 \cdot 1) + (4 \cdot 3 \cdot 2 \cdot 1)$$
$$= 6 + 24 = 30$$

Note: $3! + 4! \neq 7!$, since $7! = 5040$.

Example A–2

Evaluate $5! - 3!$.

Solution

$$5! - 3! = (5 \cdot 4 \cdot 3 \cdot 2 \cdot 1) - (3 \cdot 2 \cdot 1)$$
$$= 120 - 6 = 114$$

Note: $5! - 3! \neq 2!$, since $2! = 2$.

Factorials cannot be multiplied directly. Again, one must multiply them out and then multiply the products.

Example A–3

Evaluate $3! \cdot 2!$.

Solution

$$3! \cdot 2! = (3 \cdot 2 \cdot 1) \cdot (2 \cdot 1) = 6 \cdot 2 = 12$$

Note: $3! \cdot 2! \neq 6!$, since $6! = 720$.

Finally, factorials cannot be divided directly unless they are equal.

Example A–4

Evaluate $6! \div 3!$.

Solution

$$\frac{6!}{3!} = \frac{6 \cdot 5 \cdot 4 \cdot 3 \cdot 2 \cdot 1}{3 \cdot 2 \cdot 1} = \frac{720}{6} = 120$$

Note: $\dfrac{6!}{3!} \neq 2!$ since $2! = 2$

But $\dfrac{3!}{3!} = \dfrac{3 \cdot 2 \cdot 1}{3 \cdot 2 \cdot 1} = \dfrac{6}{6} = 1$

In division, one can take some shortcuts, as shown:

$$\frac{6!}{3!} = \frac{6 \cdot 5 \cdot 4 \cdot 3!}{3!} \quad \text{and} \quad \frac{3!}{3!} = 1$$

$$= 6 \cdot 5 \cdot 4 = 120$$

$$\frac{8!}{6!} = \frac{8 \cdot 7 \cdot 6!}{6!} \quad \text{and} \quad \frac{6!}{6!} = 1$$

$$= 8 \cdot 7 = 56$$

Another shortcut that can be used with factorials is cancellation, after factors have been expanded. For example,

$$\frac{7!}{(4!)(3!)} = \frac{7 \cdot 6 \cdot 5 \cdot 4!}{3 \cdot 2 \cdot 1 \cdot 4!}$$

Now cancel both 4!s. Then cancel the 3 · 2 in the denominator with the 6 in the numerator.

$$\frac{7 \cdot \overset{1}{\cancel{6}} \cdot 5 \cdot \cancel{4!}}{\underset{1}{\cancel{3}} \cdot \underset{1}{\cancel{2}} \cdot 1 \cdot \underset{1}{\cancel{4!}}} = 7 \cdot 5 = 35$$

Example A–5

Evaluate $10! \div (6!)(4!)$.

Solution

$$\frac{10!}{(6!)(4!)} = \frac{10 \cdot \overset{3}{\cancel{9}} \cdot \overset{1}{\cancel{8}} \cdot 7 \cdot \cancel{6!}}{\cancel{4} \cdot \cancel{3} \cdot \cancel{2} \cdot 1 \cdot \cancel{6!}} = 10 \cdot 3 \cdot 7 \cdot = 210$$

Exercises

Evaluate each expression.

A–1. 9!

A–2. 7!

A–3. 5!

A–4. 0!

A–5. 1!

A–6. 3!

A–7. $\dfrac{12!}{9!}$

A–8. $\dfrac{10!}{2!}$

A–9. $\dfrac{5!}{3!}$

A–10. $\dfrac{11!}{7!}$

A–11. $\dfrac{9!}{(4!)(5!)}$

A–12. $\dfrac{10!}{(7!)(3!)}$

A–13. $\dfrac{8!}{(4!)(4!)}$

A–14. $\dfrac{15!}{(12!)(3!)}$

A–15. $\dfrac{10!}{(10!)(0!)}$

A–16. $\dfrac{5!}{(3!)(2!)(1!)}$

A–17. $\dfrac{8!}{(3!)(3!)(2!)}$

A–18. $\dfrac{11!}{(7!)(2!)(2!)}$

A–19. $\dfrac{10!}{(3!)(2!)(5!)}$

A–20. $\dfrac{6!}{(2!)(2!)(2!)}$

A–2 Summation Notation

In mathematics, the symbol Σ (Greek letter sigma) means to add or find the sum. For example, ΣX means to add the numbers represented by the variable X. Thus, when X represents 5, 8, 2, 4, and 6, then ΣX means $5 + 8 + 2 + 4 + 6 = 25$.

Sometimes, a subscript notation is used, such as

$$\sum_{i=1}^{5} X_i$$

This notation means to find the sum of five numbers represented by X, as shown:

$$\sum_{i=1}^{5} X_i = X_1 + X_2 + X_3 + X_4 + X_5$$

When the number of values is not known, the unknown number can be represented by n, such as

$$\sum_{i=1}^{n} X_i = X_1 + X_2 + X_3 + \cdots + X_n$$

There are several important types of summation used in statistics. The notation ΣX^2 means to square each value before summing. For example, if the values of the X's are 2, 8, 6, 1, and 4, then

$$\Sigma X^2 = 2^2 + 8^2 + 6^2 + 1^2 + 4^2$$
$$= 4 + 64 + 36 + 1 + 16 = 121$$

The notation $(\Sigma X)^2$ means to find the sum of X's and then square the answer. For instance, if the values for X are 2, 8, 6, 1, and 4, then

$$(\Sigma X)^2 = (2 + 8 + 6 + 1 + 4)^2$$
$$= (21)^2 = 441$$

Another important use of summation notation is in finding the mean (shown in Section 3–2). The mean, $\overline{X}$, is defined as

$$\overline{X} = \frac{\Sigma X}{n}$$

For example, to find the mean of 12, 8, 7, 3, and 10, use the formula and substitute the values, as shown:

$$\overline{X} = \frac{\Sigma X}{n} = \frac{12 + 8 + 7 + 3 + 10}{5} = \frac{40}{5} = 8$$

The notation $\Sigma\,(X - \bar{X})^2$ means to perform the following steps.

STEP 1 Find the mean.

STEP 2 Subtract the mean from each value.

STEP 3 Square the answers.

STEP 4 Find the sum.

Example A–6

Find the value of $\Sigma\,(X - \bar{X})^2$ for the values 12, 8, 7, 3, and 10 of X.

Solution

STEP 1 Find the mean.

$$\bar{X} = \frac{12 + 8 + 7 + 3 + 10}{5} = \frac{40}{5} = 8$$

STEP 2 Subtract the mean from each value.

$$12 - 8 = 4 \qquad 7 - 8 = -1 \qquad 10 - 8 = 2$$
$$8 - 8 = 0 \qquad 3 - 8 = -5$$

STEP 3 Square the answers.

$$4^2 = 16 \qquad 0^2 = 0 \qquad (-1)^2 = 1$$
$$(-5)^2 = 25 \qquad 2^2 = 4$$

STEP 4 Find the sum.

$$16 + 0 + 1 + 25 + 4 = 46$$

Example A–7

Find $\Sigma\,(X - \bar{X})^2$ for the following values of X: 5, 7, 2, 1, 3, 6.

Solution

Find the mean:

$$\bar{X} = \frac{5 + 7 + 2 + 1 + 3 + 6}{6} = \frac{24}{6} = 4$$

Then the steps in Example A–6 can be shortened as follows:

$$\Sigma\,(X - \bar{X})^2 = (5 - 4)^2 + (7 - 4)^2 + (2 - 4)^2$$
$$+ (1 - 4)^2 + (3 - 4)^2 + (6 - 4)^2$$
$$= 1^2 + 3^2 + (-2)^2 + (-3)^2$$
$$+ (-1)^2 + 2^2$$
$$= 1 + 9 + 4 + 9 + 1 + 4 = 28$$

Exercises

For each set of values, find $\Sigma\,X$, $\Sigma\,X^2$, $(\Sigma\,X)^2$, and $\Sigma\,(X - \bar{X})^2$.

A–21. 9, 17, 32, 16, 8, 2, 9, 7, 3, 18

A–22. 4, 12, 9, 13, 0, 6, 2, 10

A–23. 5, 12, 8, 3, 4

A–24. 6, 2, 18, 30, 31, 42, 16, 5

A–25. 80, 76, 42, 53, 77

A–26. 123, 132, 216, 98, 146, 114

A–27. 53, 72, 81, 42, 63, 71, 73, 85, 98, 55

A–28. 43, 32, 116, 98, 120

A–29. 12, 52, 36, 81, 63, 74

A–30. $-9, -12, 18, 0, -2, -15$

A–3   The Line

The following figure shows the *rectangular coordinate system* or *Cartesian plane*. This figure consists of two axes: the horizontal axis, called the x axis, and the vertical axis, called the y axis. Each axis has numerical scales. The point of intersection of the axes is called the *origin*.

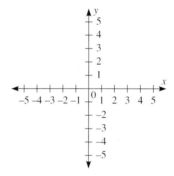

Points can be graphed by using coordinates. For example, the notation for point $P(3, 2)$ means that the x coordinate is 3 and the y coordinate is 2. Hence, P is located at the intersection of $x = 3$ and $y = 2$, as shown below.

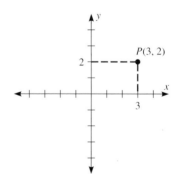

Other points, such as $Q(-5, 2)$, $R(4, 1)$, $S(-3, -4)$ can be plotted as shown in the next figure.

When a point lies on the y axis, the x coordinate is 0, as in (0, 6) (0, -3), etc. When a point lies on the x axis, the y coordinate is 0, as in (6, 0) (-8, 0), etc., as shown at the top of next page.

A

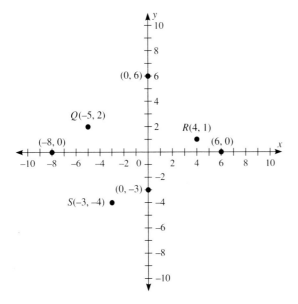

The slopes of lines can be positive, negative, or zero. A line going uphill from left to right has a positive slope. A line going downhill from left to right has a negative slope. And a line that is horizontal has a slope of zero.

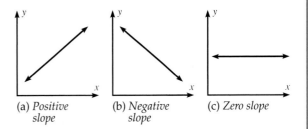

(a) *Positive slope*   (b) *Negative slope*   (c) *Zero slope*

A point b where the line crosses the x axis is called the *x intercept* and has the coordinates $(b, 0)$. A point a where the line crosses the y axis is called the *y intercept* and has the coordinates $(0, a)$.

Two points determine a line. There are two properties of a line: its slope and its equation. The *slope m* of a line is determined by the ratio of the rise (called Δy) and the run (Δx).

$$m = \frac{\text{rise}}{\text{run}} = \frac{\Delta y}{\Delta x}$$

For example, the slope of the line shown below is $\frac{3}{2}$, or 1.5, since the height Δy is 3 units and the run Δx is 2 units.

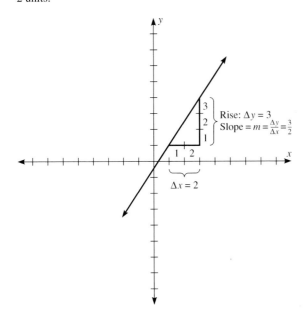

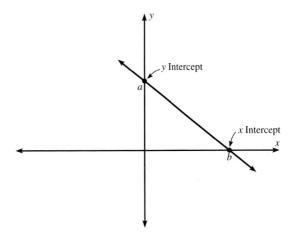

Every line has a unique equation of the form $y = a + bx$. For example, the equations

$$y = 5 + 3x$$
$$y = 8.6 + 3.2x$$
$$y = 5.2 - 6.1x$$

all represent different, unique lines. The number represented by a is the y intercept point; the number represented by b is the slope. The line whose equation is $y = 3 + 2x$ has a y intercept at 3 and a slope of 2, or 2/1. This line can be shown as in the following graph.

A

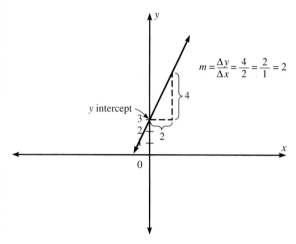

If two points are known, then the graph of the line can be plotted. For example, to find the graph of a line passing through the points $P(2, 1)$ and $Q(3, 5)$, plot the points and connect them as shown below.

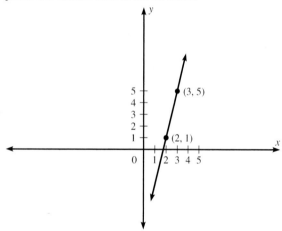

Given the equation of a line, one can graph the line by finding two points and then plotting them.

Example A–8

Plot the graph of the line whose equation is $y = 3 + 2x$.

Solution

Select any number as an x value and substitute it in the equation to get the corresponding y value. Let $x = 0$.

Then,

$$y = 3 + 2x = 3 + 2(0) = 3$$

Hence, when $x = 0$, then $y = 3$, and the line passes through the point $(0, 3)$.

Now select any other value of x, say $x = 2$.

$$y = 3 + 2x = 3 + 2(2) = 7$$

Hence, a second point is $(2, 7)$. Then plot the points and graph the line.

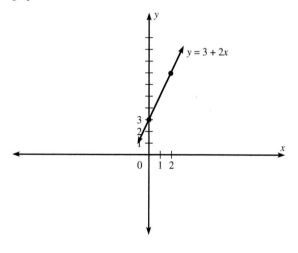

Exercises

Plot the line passing through each set of points.

A–31. $P(3, 2), Q(1, 6)$ A–34. $P(-1, -2), Q(-7, 8)$

A–32. $P(0, 5), Q(8, 0)$ A–35. $P(6, 3), Q(10, 3)$

A–33. $P(-2, 4), Q(3, 6)$

Find at least two points on each line, and then graph the line containing these points.

A–36. $y = 5 + 2x$ A–39. $y = -2 - 2x$

A–37. $y = -1 + x$ A–40. $y = 4 - 3x$

A–38. $y = 3 + 4x$

Appendix B–1

Writing the Research Report

After conducting a statistical study, a researcher must write a final report explaining how the study was conducted and giving the results. The formats of research reports, theses, and dissertations vary from school to school; however, they tend to follow the general format explained here.

Front materials

The front materials typically include the following items:

Title page
Copyright page
Acknowledgments
Table of contents
Table of appendixes
List of tables
List of figures

Chapter 1: Nature and background of the study

This chapter should introduce the reader to the nature of the study and present some discussion on the background. It should contain the following information:

Introduction
Statement of the problem
Background of the problem
Rationale for the study
Research questions and/or hypotheses
Assumptions, limitations, and deliminations
Definitions of terms

Chapter 2: Review of literature

This chapter should explain what has been done in previous research related to the study. It should contain the following information:

Prior research
Related literature

Chapter 3: Methodology

This chapter should explain how the study was conducted. It should contain the following information:

Development of questionnaires, tests, survey instruments, etc.
Definition of the population
Sampling methods used
How the data were collected
Research design used
Statistical tests that will be used to analyze the data

Chapter 4: Analysis of data

This chapter should explain the results of the statistical analysis of the data. It should state whether or not the null hypothesis should be rejected. Any statistical tables used to analyze the data should be included here.

Chapter 5: Summary, conclusions, and recommendations

This chapter summarizes the results of the study and explains any conclusions that have resulted from the statistical analysis of the data. The researchers should cite and explain any shortcomings of the study. Recommendations obtained from the study should be included here, and further studies should be suggested.

Appendix B–2

Bayes's Theorem

Objective B–1 Find the probability of an event using Bayes's theorem.

Given two dependent events, A and B, the previous formulas for conditional probability allow one to find $P(A$ and $B)$, or $P(B|A)$. Related to these formulas is a rule developed by the English Presbyterian minister Thomas Bayes (1702–61). The rule is known as **Bayes's theorem.**

It is possible, given the outcome of the second event in a sequence of two events, to determine the probability of various possibilities for the first event. In Example 5–31 of Section 5–4, in Chapter 5, there were two boxes, each containing red balls and blue balls. A box was selected and a ball was drawn. The example asked for the probability that the ball selected was red. Now, a different question can be asked: "If the ball is red, what is the probability it came from box 1?" In this case, the outcome is known, a red ball was selected, and one is asked to find the probability that it is a result of a previous event, that it came from box 1.

Bayes's theorem can enable one to compute this probability and can be explained by using tree diagrams.

The tree diagram for the solution of Example 5–31 is shown in Figure B–1, along with the appropriate notation and the corresponding probabilities. In this case, A_1 is the event of selecting box 1, A_2 is the event of selecting box 2, R is the event of selecting a red ball, and B is the event of selecting a blue ball.

To answer the question, "If the ball selected is red, what is the probability that it came from box 1?" the two previous formulas,

$$P(B|A) = \frac{P(A \text{ and } B)}{P(A)} \tag{1}$$

$$P(A \text{ and } B) = P(A) \cdot P(B|A) \tag{2}$$

can be used. The notation that will be used is that of Example 5–31, shown in Figure B–1. Finding the probability that box 1 was selected given that the ball selected was red can be written symbolically as $P(A_1|R)$. By Formula 1,

$$P(A_1|R) = \frac{P(R \text{ and } A_1)}{P(R_1)}$$

Note: $P(R \text{ and } A_1) = P(A_1 \text{ and } R)$.

Figure B–1

Tree Diagram for
Example 5–31

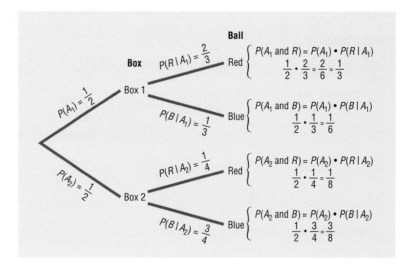

By Formula 2,

$$P(A_1 \text{ and } R) = P(A_1) \cdot P(R|A_1)$$

and

$$P(R) = P(A_1 \text{ and } R) + P(A_2 \text{ and } R)$$

as shown in Figure B–1; $P(R)$ was found by adding the products of the probabilities of the branches in which a red ball was selected. Now,

$$P(A_1 \text{ and } R) = P(A_1) \cdot P(R|A_1)$$
$$P(A_2 \text{ and } R) = P(A_2) \cdot P(R|A_2)$$

Substituting these values in the original formula for $P(A_1|R)$, one gets

$$P(A_1|R) = \frac{P(A_1) \cdot P(R|A_1)}{P(A_1) \cdot P(R|A_1) + P(A_2) \cdot P(R|A_2)}$$

Refer to Figure B–1. The numerator of the fraction is the product of the top branch of the tree diagram, which consists of selecting a red ball and selecting box 1. And the denominator is the sum of the products of the two branches of the tree where the red ball was selected.

Using this formula and the probability values shown in Figure B–1, one can find the probability that box 1 was selected given that the ball was red, as shown.

$$P(A_1|R) = \frac{P(A_1) \cdot P(R|A_1)}{P(A_1) \cdot P(R|A_1) + P(A_2) \cdot P(R|A_2)}$$

$$= \frac{\frac{1}{2} \cdot \frac{2}{3}}{\frac{1}{2} \cdot \frac{2}{3} + \frac{1}{2} \cdot \frac{1}{4}} = \frac{\frac{1}{3}}{\frac{1}{3} + \frac{1}{8}} = \frac{\frac{1}{3}}{\frac{8}{24} + \frac{3}{24}} = \frac{\frac{1}{3}}{\frac{11}{24}}$$

$$= \frac{1}{3} \div \frac{11}{24} = \frac{1}{\underset{1}{\cancel{3}}} \cdot \frac{\overset{8}{\cancel{24}}}{11} = \frac{8}{11}$$

This formula is a simplified version of Bayes's theorem.

Before Bayes's theorem is stated, another example is shown.

Example B–1

A shipment of two boxes, each containing six telephones, is received by a store. Box 1 contains one defective phone and box 2 contains two defective phones. After the boxes are unpacked, a phone is selected and found to be defective. Find the probability that it came from box 2.

Solution

STEP 1   Select the proper notation. Let A_1 represent box 1 and A_2 represent box 2. Let D represent a defective phone and ND represent a phone that is not defective.

STEP 2   Draw a tree diagram and find the corresponding probabilities for each branch. The probability of selecting box 1 is $\frac{1}{2}$, and the probability of selecting box 2 is $\frac{1}{2}$. Since there is one defective phone in box 1, the probability of selecting it is $\frac{1}{6}$. The probability of selecting a nondefective phone from box 1 is $\frac{5}{6}$.

Since there are two defective phones in box 2, the probability of selecting a defective phone from box 2 is $\frac{2}{6}$, or $\frac{1}{3}$; and the probability of selecting a nondefective phone is $\frac{4}{6}$, or $\frac{2}{3}$. The tree diagram is shown in Figure B–2.

STEP 3   Write the corresponding formula. Since the example is asking for the probability that, given a defective phone, it came from box 2, the corresponding formula is as shown.

Figure B–2

Tree Diagram for Example B–1

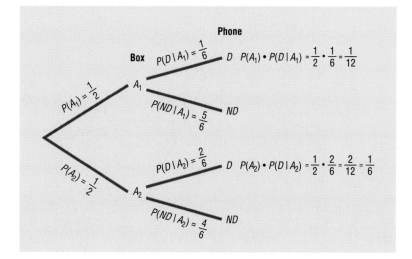

$$P(A_2|D) = \frac{P(A_2) \cdot P(D|A_2)}{P(A_1) \cdot P(D|A_1) + P(A_2) \cdot P(D|A_2)}$$

$$= \frac{\frac{1}{2} \cdot \frac{2}{6}}{\frac{1}{2} \cdot \frac{1}{6} + \frac{1}{2} \cdot \frac{2}{6}} = \frac{\frac{1}{6}}{\frac{1}{12} + \frac{2}{12}} = \frac{\frac{1}{6}}{\frac{3}{12}}$$

$$= \frac{1}{6} \div \frac{3}{12} = \frac{1}{\cancel{6}} \cdot \frac{\overset{2}{\cancel{12}}}{3} = \frac{2}{3}$$

Bayes's theorem can be generalized to events with three or more outcomes and formally stated as in the next box.

Bayes's theorem For two events, A and B, where event B follows event A, event A can occur in $A_1, A_2, \ldots, A_n$ mutually exclusive ways, and event B can occur in $B_1, B_2, \ldots, B_m$ mutually exclusive ways,

$$P(A_1|B_1) = \frac{P(A_1) \cdot P(B_1|A_1)}{\begin{array}{l} P(A_1) \cdot P(B_1|A_1) + P(A_2) \cdot P(B_1|A_2) \\ + \cdots + P(A_n) \cdot P(B_1|A_n) \end{array}}$$

for any specific events A_1 and B_1.

The numerator is the product of the probabilities on the branch of the tree that consists of outcomes A_1 and B_1. The denominator is the sum of the products of the probabilities of the branches containing B_1 and A_1, B_1 and A_2, . . . ,B_1 and A_n.

Example B–2

On a game show, a contestant can select one of four boxes. Box 1 contains one $100 bill and nine $1 bills. Box 2 contains two $100 bills and eight $1 bills. Box 3 contains three $100 bills and seven $1 bills. Box 4 contains five $100 bills and five $1 bills. The contestant selects a box at random and selects a bill from the box at random. If a $100 bill is selected, find the probability that it came from box 4.

Solution

STEP 1 Select the proper notation. Let B_1, B_2, B_3, and B_4 represent the boxes and 100 and 1 represent the values of the bills in the boxes.

STEP 2 Draw a tree diagram and find the corresponding probabilities. The probability of selecting each box is $\frac{1}{4}$, or 0.25. The probabilities of selecting the $100 bill from each box, respectively, are $\frac{1}{10} = 0.1$, $\frac{2}{10} = 0.2$, $\frac{3}{10} = 0.3$, and $\frac{5}{10} = 0.5$. The tree diagram is shown in Figure B–3.

STEP 3 Using Bayes's theorem, write the corresponding formula. Since the example asks for the probability that box 4 was selected, given that

Figure B–3

Tree Diagram for Example B–2

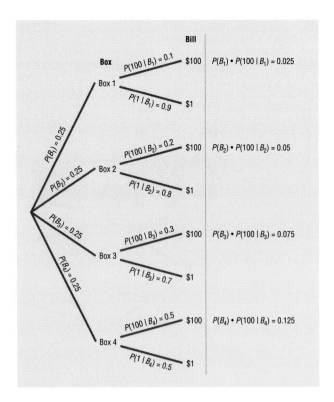

$100 was obtained, the corresponding formula is as follows:

$$P(B_4|100) = \frac{P(B_4) \cdot P(100|B_4)}{P(B_1) \cdot P(100|B_1) + P(B_2) \cdot P(100|B_2) + P(B_3) \cdot P(100|B_3) + P(B_4) \cdot P(100|B_4)}$$

$$= \frac{0.125}{0.025 + 0.05 + 0.075 + 0.125}$$

$$= \frac{0.125}{0.275} = 0.455$$

In Example B–2, the original probability of selecting box 4 was 0.25. However, once additional information was obtained—and the condition was considered, that a $100 bill was selected—the revised probability of selecting box 4 became 0.455.

Bayes's theorem can be used to revise probabilities of events once additional information becomes known. Bayes's theorem is used as the basis for a branch of statistics called *Bayesian decision making,* which includes the use of subjective probabilities in making statistical inferences.

A summary of the probability rules is shown in Table B–1.

Exercises

B–1. An appliance store purchases electric ranges from two companies. From company A, 500 ranges are purchased, and 2% are defective. From company B, 850 ranges are purchased, and 2% are defective. Given that a range is defective, find the probability that it came from company B.

B–2. Two manufacturers supply blankets to emergency relief organizations. Manufacturer A supplies 3,000 blankets, and 4% are irregular in workmanship. Manufacturer B supplies 2,400 blankets, and 7% are found to be irregular. Given that a blanket is irregular, find the probability that it came from manufacturer B.

B–3. A test for a certain disease is found to be 95% accurate, meaning that it will correctly diagnose the disease in 95 out of 100 people who have the ailment. For a certain segment of the population, the incidence of the disease is 9%. If a person tests positive, find the probability that the person actually has the disease. The test is also 95% accurate for a negative result.

B–4. Using the test in Exercise B–3, if a person tests negative for the disease, find the probability that the person actually has the disease. Remember, 9% of the population have the disease.

B–5. A corporation has three methods of training employees. Because of time, space, and location, it sends 20% of its employees to location A, 35% to location B, and 45% to location C. Location A has an 80% success rate. That is, 80% of the employees who complete the course will pass the licensing exam. Location B has a 75% success rate, and location C has a 60% success rate. If a person has passed the exam, find the probability that the person went to location B.

B–6. Exercise B–5, if a person failed the exam, find the probability that the person went to location C.

B–7. A store purchases baseball hats from three different manufacturers. In manufacturer A's box, there are 12 blue hats, 6 red hats, and 6 green hats. In manufacturer B's box, there are 10 blue hats, 10 red hats, and 4 green hats. In manufacturer C's box, there are 8 blue hats, 8 red hats, and 8 green hats. A box is selected at random and a hat is selected at random from that box. If the hat is red, find the probability that it came from manufacturer A's box.

B–8. In Exercise B–7, if the hat selected is green, find the probability that it came from manufacturer B's box.

B–9. A driver has three ways to get from one city to another. There is an 80% probability of encountering a traffic jam on route 1, a 60% probability on route 2, and a 30% probability on route 3. Because of other factors, such as distance and speed limits, the driver uses route 1 50% of the time and routes 2 and 3 each 25% of the time. If the driver calls the dispatcher to inform him that she is in a traffic jam, find the probability that she has selected route 1.

B–10. In Exercise B–9, if the driver did not encounter a traffic jam, find the probability that she selected route 3.

B–11. A store owner purchases telephones from two companies. From company A, 350 telephones are purchased, and 2% are defective. From company B, 1050 telephones are purchased, and 4% are defective. Given that a phone is defective, find the probability that it came from company B.

B–12. Two manufacturers supply food to a large cafeteria. Manufacturer A supplies 2400 cans of soup, and 3% are found to be dented. Manufacturer B supplies 3600 cans, and 1% are found to be dented. Given that a can of soup is dented, find the probability that it came from manufacturer B.

C Appendix C

Tables

Table A	Factorials
n	*n*!
0	1
1	1
2	2
3	6
4	24
5	120
6	720
7	5,040
8	40,320
9	362,880
10	3,628,800
11	39,916,800
12	479,001,600
13	6,227,020,800
14	87,178,291,200
15	1,307,674,368,000
16	20,922,789,888,000
17	355,687,428,096,000
18	6,402,373,705,728,000
19	121,645,100,408,832,000
20	2,432,902,008,176,640,000

Table B The Binomial Distribution

n	x	0.05	0.1	0.2	0.3	0.4	0.5	0.6	0.7	0.8	0.9	0.95
2	0	0.902	0.810	0.640	0.490	0.360	0.250	0.160	0.090	0.040	0.010	0.002
	1	0.095	0.180	0.320	0.420	0.480	0.500	0.480	0.420	0.320	0.180	0.095
	2	0.002	0.010	0.040	0.090	0.160	0.250	0.360	0.490	0.640	0.810	0.902
3	0	0.857	0.729	0.512	0.343	0.216	0.125	0.064	0.027	0.008	0.001	
	1	0.135	0.243	0.384	0.441	0.432	0.375	0.288	0.189	0.096	0.027	0.007
	2	0.007	0.027	0.096	0.189	0.288	0.375	0.432	0.441	0.384	0.243	0.135
	3		0.001	0.008	0.027	0.064	0.125	0.216	0.343	0.512	0.729	0.857
4	0	0.815	0.656	0.410	0.240	0.130	0.062	0.026	0.008	0.002		
	1	0.171	0.292	0.410	0.412	0.346	0.250	0.154	0.076	0.026	0.004	
	2	0.014	0.049	0.154	0.265	0.346	0.375	0.346	0.265	0.154	0.049	0.014
	3		0.004	0.026	0.076	0.154	0.250	0.346	0.412	0.410	0.292	0.171
	4			0.002	0.008	0.026	0.062	0.130	0.240	0.410	0.656	0.815
5	0	0.774	0.590	0.328	0.168	0.078	0.031	0.010	0.002			
	1	0.204	0.328	0.410	0.360	0.259	0.156	0.077	0.028	0.006		
	2	0.021	0.073	0.205	0.309	0.346	0.312	0.230	0.132	0.051	0.008	0.001
	3	0.001	0.008	0.051	0.132	0.230	0.312	0.346	0.309	0.205	0.073	0.021
	4			0.006	0.028	0.077	0.156	0.259	0.360	0.410	0.328	0.204
	5				0.002	0.010	0.031	0.078	0.168	0.328	0.590	0.774
6	0	0.735	0.531	0.262	0.118	0.047	0.016	0.004	0.001			
	1	0.232	0.354	0.393	0.303	0.187	0.094	0.037	0.010	0.002		
	2	0.031	0.098	0.246	0.324	0.311	0.234	0.138	0.060	0.015	0.001	
	3	0.002	0.015	0.082	0.185	0.276	0.312	0.276	0.185	0.082	0.015	0.002
	4		0.001	0.015	0.060	0.138	0.234	0.311	0.324	0.246	0.098	0.031
	5			0.002	0.010	0.037	0.094	0.187	0.303	0.393	0.354	0.232
	6				0.001	0.004	0.016	0.047	0.118	0.262	0.531	0.735
7	0	0.698	0.478	0.210	0.082	0.028	0.008	0.002				
	1	0.257	0.372	0.367	0.247	0.131	0.055	0.017	0.004			
	2	0.041	0.124	0.275	0.318	0.261	0.164	0.077	0.025	0.004		
	3	0.004	0.023	0.115	0.227	0.290	0.273	0.194	0.097	0.029	0.003	
	4		0.003	0.029	0.097	0.194	0.273	0.290	0.227	0.115	0.023	0.004
	5			0.004	0.025	0.077	0.164	0.261	0.318	0.275	0.124	0.041
	6				0.004	0.017	0.055	0.131	0.247	0.367	0.372	0.257
	7					0.002	0.008	0.028	0.082	0.210	0.478	0.698
8	0	0.663	0.430	0.168	0.058	0.017	0.004	0.001				
	1	0.279	0.383	0.336	0.198	0.090	0.031	0.008	0.001			
	2	0.051	0.149	0.294	0.296	0.209	0.109	0.041	0.010	0.001		
	3	0.005	0.033	0.147	0.254	0.279	0.219	0.124	0.047	0.009		
	4		0.005	0.046	0.136	0.232	0.273	0.232	0.136	0.046	0.005	
	5			0.009	0.047	0.124	0.219	0.279	0.254	0.147	0.033	0.005
	6			0.001	0.010	0.041	0.109	0.209	0.296	0.294	0.149	0.051
	7				0.001	0.008	0.031	0.090	0.198	0.336	0.383	0.279
	8					0.001	0.004	0.017	0.058	0.168	0.430	0.663

C

C

Table B (continued)

n	x	0.05	0.1	0.2	0.3	0.4	0.5	0.6	0.7	0.8	0.9	0.95
9	0	0.630	0.387	0.134	0.040	0.010	0.002					
	1	0.299	0.387	0.302	0.156	0.060	0.018	0.004				
	2	0.063	0.172	0.302	0.267	0.161	0.070	0.021	0.004			
	3	0.008	0.045	0.176	0.267	0.251	0.164	0.074	0.021	0.003		
	4	0.001	0.007	0.066	0.172	0.251	0.246	0.167	0.074	0.017	0.001	
	5		0.001	0.017	0.074	0.167	0.246	0.251	0.172	0.066	0.007	0.001
	6			0.003	0.021	0.074	0.164	0.251	0.267	0.176	0.045	0.008
	7				0.004	0.021	0.070	0.161	0.267	0.302	0.172	0.063
	8					0.004	0.018	0.060	0.156	0.302	0.387	0.299
	9						0.002	0.010	0.040	0.134	0.387	0.630
10	0	0.599	0.349	0.107	0.028	0.006	0.001					
	1	0.315	0.387	0.268	0.121	0.040	0.010	0.002				
	2	0.075	0.194	0.302	0.233	0.121	0.044	0.011	0.001			
	3	0.010	0.057	0.201	0.267	0.215	0.117	0.042	0.009	0.001		
	4	0.001	0.011	0.088	0.200	0.251	0.205	0.111	0.037	0.006		
	5		0.001	0.026	0.103	0.201	0.246	0.201	0.103	0.026	0.001	
	6			0.006	0.037	0.111	0.205	0.251	0.200	0.088	0.011	0.001
	7			0.001	0.009	0.042	0.117	0.215	0.267	0.201	0.057	0.010
	8				0.001	0.011	0.044	0.121	0.233	0.302	0.194	0.075
	9					0.002	0.010	0.040	0.121	0.268	0.387	0.315
	10						0.001	0.006	0.028	0.107	0.349	0.599
11	0	0.569	0.314	0.086	0.020	0.004						
	1	0.329	0.384	0.236	0.093	0.027	0.005	0.001				
	2	0.087	0.213	0.295	0.200	0.089	0.027	0.005	0.001			
	3	0.014	0.071	0.221	0.257	0.177	0.081	0.023	0.004			
	4	0.001	0.016	0.111	0.220	0.236	0.161	0.070	0.017	0.002		
	5		0.002	0.039	0.132	0.221	0.226	0.147	0.057	0.010		
	6			0.010	0.057	0.147	0.226	0.221	0.132	0.039	0.002	
	7			0.002	0.017	0.070	0.161	0.236	0.220	0.111	0.016	0.001
	8				0.004	0.023	0.081	0.177	0.257	0.221	0.071	0.014
	9				0.001	0.005	0.027	0.089	0.200	0.295	0.213	0.087
	10					0.001	0.005	0.027	0.093	0.236	0.384	0.329
	11							0.004	0.020	0.086	0.314	0.569
12	0	0.540	0.282	0.069	0.014	0.002						
	1	0.341	0.377	0.206	0.071	0.017	0.003					
	2	0.099	0.230	0.283	0.168	0.064	0.016	0.002				
	3	0.017	0.085	0.236	0.240	0.142	0.054	0.012	0.001			
	4	0.002	0.021	0.133	0.231	0.213	0.121	0.042	0.008	0.001		
	5		0.004	0.053	0.158	0.227	0.193	0.101	0.029	0.003		
	6			0.016	0.079	0.177	0.226	0.177	0.079	0.016		
	7			0.003	0.029	0.101	0.193	0.227	0.158	0.053	0.004	
	8			0.001	0.008	0.042	0.121	0.213	0.231	0.133	0.021	0.002
	9				0.001	0.012	0.054	0.142	0.240	0.236	0.085	0.017
	10					0.002	0.016	0.064	0.168	0.283	0.230	0.099
	11						0.003	0.017	0.071	0.206	0.377	0.341
	12							0.002	0.014	0.069	0.282	0.540

Table B (continued)

n	x	p										
		0.05	0.1	0.2	0.3	0.4	0.5	0.6	0.7	0.8	0.9	0.95
13	0	0.513	0.254	0.055	0.010	0.001						
	1	0.351	0.367	0.179	0.054	0.011	0.002					
	2	0.111	0.245	0.268	0.139	0.045	0.010	0.001				
	3	0.021	0.100	0.246	0.218	0.111	0.035	0.006	0.001			
	4	0.003	0.028	0.154	0.234	0.184	0.087	0.024	0.003			
	5		0.006	0.069	0.180	0.221	0.157	0.066	0.014	0.001		
	6		0.001	0.023	0.103	0.197	0.209	0.131	0.044	0.006		
	7			0.006	0.044	0.131	0.209	0.197	0.103	0.023	0.001	
	8			0.001	0.014	0.066	0.157	0.221	0.180	0.069	0.006	
	9				0.003	0.024	0.087	0.184	0.234	0.154	0.028	0.003
	10				0.001	0.006	0.035	0.111	0.218	0.246	0.100	0.021
	11					0.001	0.010	0.045	0.139	0.268	0.245	0.111
	12						0.002	0.011	0.054	0.179	0.367	0.351
	13							0.001	0.010	0.055	0.254	0.513
14	0	0.488	0.229	0.044	0.007	0.001						
	1	0.359	0.356	0.154	0.041	0.007	0.001					
	2	0.123	0.257	0.250	0.113	0.032	0.006	0.001				
	3	0.026	0.114	0.250	0.194	0.085	0.022	0.003				
	4	0.004	0.035	0.172	0.229	0.155	0.061	0.014	0.001			
	5		0.008	0.086	0.196	0.207	0.122	0.041	0.007			
	6		0.001	0.032	0.126	0.207	0.183	0.092	0.023	0.002		
	7			0.009	0.062	0.157	0.209	0.157	0.062	0.009		
	8			0.002	0.023	0.092	0.183	0.207	0.126	0.032	0.001	
	9				0.007	0.041	0.122	0.207	0.196	0.086	0.008	
	10				0.001	0.014	0.061	0.155	0.229	0.172	0.035	0.004
	11					0.003	0.022	0.085	0.194	0.250	0.114	0.026
	12					0.001	0.006	0.032	0.113	0.250	0.257	0.123
	13						0.001	0.007	0.041	0.154	0.356	0.359
	14							0.001	0.007	0.044	0.229	0.488
15	0	0.463	0.206	0.035	0.005							
	1	0.366	0.343	0.132	0.031	0.005						
	2	0.135	0.267	0.231	0.092	0.022	0.003					
	3	0.031	0.129	0.250	0.170	0.063	0.014	0.002				
	4	0.005	0.043	0.188	0.219	0.127	0.042	0.007	0.001			
	5	0.001	0.010	0.103	0.206	0.186	0.092	0.024	0.003			
	6		0.002	0.043	0.147	0.207	0.153	0.061	0.012	0.001		
	7			0.014	0.081	0.177	0.196	0.118	0.035	0.003		
	8			0.003	0.035	0.118	0.196	0.177	0.081	0.014		
	9			0.001	0.012	0.061	0.153	0.207	0.147	0.043	0.002	
	10				0.003	0.024	0.092	0.186	0.206	0.103	0.010	0.001
	11				0.001	0.007	0.042	0.127	0.219	0.188	0.043	0.005
	12					0.002	0.014	0.063	0.170	0.250	0.129	0.031
	13						0.003	0.022	0.092	0.231	0.267	0.135
	14							0.005	0.031	0.132	0.343	0.366
	15								0.005	0.035	0.206	0.463

Table B (continued)

n	x	p 0.05	0.1	0.2	0.3	0.4	0.5	0.6	0.7	0.8	0.9	0.95
16	0	0.440	0.185	0.028	0.003							
	1	0.371	0.329	0.113	0.023	0.003						
	2	0.146	0.275	0.211	0.073	0.015	0.002					
	3	0.036	0.142	0.246	0.146	0.047	0.009	0.001				
	4	0.006	0.051	0.200	0.204	0.101	0.028	0.004				
	5	0.001	0.014	0.120	0.210	0.162	0.067	0.014	0.001			
	6		0.003	0.055	0.165	0.198	0.122	0.039	0.006			
	7			0.020	0.101	0.189	0.175	0.084	0.019	0.001		
	8			0.006	0.049	0.142	0.196	0.142	0.049	0.006		
	9			0.001	0.019	0.084	0.175	0.189	0.101	0.020		
	10				0.006	0.039	0.122	0.198	0.165	0.055	0.003	
	11				0.001	0.014	0.067	0.162	0.210	0.120	0.014	0.001
	12					0.004	0.028	0.101	0.204	0.200	0.051	0.006
	13					0.001	0.009	0.047	0.146	0.246	0.142	0.036
	14						0.002	0.015	0.073	0.211	0.275	0.146
	15							0.003	0.023	0.113	0.329	0.371
	16								0.003	0.028	0.185	0.440
17	0	0.418	0.167	0.023	0.002							
	1	0.374	0.315	0.096	0.017	0.002						
	2	0.158	0.280	0.191	0.058	0.010	0.001					
	3	0.041	0.156	0.239	0.125	0.034	0.005					
	4	0.008	0.060	0.209	0.187	0.080	0.018	0.002				
	5	0.001	0.017	0.136	0.208	0.138	0.047	0.008	0.001			
	6		0.004	0.068	0.178	0.184	0.094	0.024	0.003			
	7		0.001	0.027	0.120	0.193	0.148	0.057	0.009			
	8			0.008	0.064	0.161	0.185	0.107	0.028	0.002		
	9			0.002	0.028	0.107	0.185	0.161	0.064	0.008		
	10				0.009	0.057	0.148	0.193	0.120	0.027	0.001	
	11				0.003	0.024	0.094	0.184	0.178	0.068	0.004	
	12				0.001	0.008	0.047	0.138	0.208	0.136	0.017	0.001
	13					0.002	0.018	0.080	0.187	0.209	0.060	0.008
	14						0.005	0.034	0.125	0.239	0.156	0.041
	15						0.001	0.010	0.058	0.191	0.280	0.158
	16							0.002	0.017	0.096	0.315	0.374
	17								0.002	0.023	0.167	0.418

Table B   (continued)

n	x	0.05	0.1	0.2	0.3	0.4	0.5	0.6	0.7	0.8	0.9	0.95
18	0	0.397	0.150	0.018	0.002							
	1	0.376	0.300	0.081	0.013	0.001						
	2	0.168	0.284	0.172	0.046	0.007	0.001					
	3	0.047	0.168	0.230	0.105	0.025	0.003					
	4	0.009	0.070	0.215	0.168	0.061	0.012	0.001				
	5	0.001	0.022	0.151	0.202	0.115	0.033	0.004				
	6		0.005	0.082	0.187	0.166	0.071	0.015	0.001			
	7		0.001	0.035	0.138	0.189	0.121	0.037	0.005			
	8			0.012	0.081	0.173	0.167	0.077	0.015	0.001		
	9			0.003	0.039	0.128	0.185	0.128	0.039	0.003		
	10			0.001	0.015	0.077	0.167	0.173	0.081	0.012		
	11				0.005	0.037	0.121	0.189	0.138	0.035	0.001	
	12				0.001	0.015	0.071	0.166	0.187	0.082	0.005	
	13					0.004	0.033	0.115	0.202	0.151	0.022	0.001
	14					0.001	0.012	0.061	0.168	0.215	0.070	0.009
	15						0.003	0.025	0.105	0.230	0.168	0.047
	16						0.001	0.007	0.046	0.172	0.284	0.168
	17							0.001	0.013	0.081	0.300	0.376
	18								0.002	0.018	0.150	0.397
19	0	0.377	0.135	0.014	0.001							
	1	0.377	0.285	0.068	0.009	0.001						
	2	0.179	0.285	0.154	0.036	0.005						
	3	0.053	0.180	0.218	0.087	0.017	0.002					
	4	0.011	0.080	0.218	0.149	0.047	0.007	0.001				
	5	0.002	0.027	0.164	0.192	0.093	0.022	0.002				
	6		0.007	0.095	0.192	0.145	0.052	0.008	0.001			
	7		0.001	0.044	0.153	0.180	0.096	0.024	0.002			
	8			0.017	0.098	0.180	0.144	0.053	0.008			
	9			0.005	0.051	0.146	0.176	0.098	0.022	0.001		
	10			0.001	0.022	0.098	0.176	0.146	0.051	0.005		
	11				0.008	0.053	0.144	0.180	0.098	0.071		
	12				0.002	0.024	0.096	0.180	0.153	0.044	0.001	
	13				0.001	0.008	0.052	0.145	0.192	0.095	0.007	
	14					0.002	0.022	0.093	0.192	0.164	0.027	0.002
	15					0.001	0.007	0.047	0.149	0.218	0.080	0.011
	16						0.002	0.017	0.087	0.218	0.180	0.053
	17							0.005	0.036	0.154	0.285	0.179
	18							0.001	0.009	0.068	0.285	0.377
	19								0.001	0.014	0.135	0.377

Table B (concluded)

n	x	0.05	0.1	0.2	0.3	0.4	0.5	0.6	0.7	0.8	0.9	0.95
20	0	0.358	0.122	0.012	0.001							
	1	0.377	0.270	0.058	0.007							
	2	0.189	0.285	0.137	0.028	0.003						
	3	0.060	0.190	0.205	0.072	0.012	0.001					
	4	0.013	0.090	0.218	0.130	0.035	0.005					
	5	0.002	0.032	0.175	0.179	0.075	0.015	0.001				
	6		0.009	0.109	0.192	0.124	0.037	0.005				
	7		0.002	0.055	0.164	0.166	0.074	0.015	0.001			
	8			0.022	0.114	0.180	0.120	0.035	0.004			
	9			0.007	0.065	0.160	0.160	0.071	0.012			
	10			0.002	0.031	0.117	0.176	0.117	0.031	0.002		
	11				0.012	0.071	0.160	0.160	0.065	0.007		
	12				0.004	0.035	0.120	0.180	0.114	0.022		
	13				0.001	0.015	0.074	0.166	0.164	0.055	0.002	
	14					0.005	0.037	0.124	0.192	0.109	0.009	
	15					0.001	0.015	0.075	0.179	0.175	0.032	0.002
	16						0.005	0.035	0.130	0.218	0.090	0.013
	17						0.001	0.012	0.072	0.205	0.190	0.060
	18							0.003	0.028	0.137	0.285	0.189
	19								0.007	0.058	0.270	0.377
	20								0.001	0.012	0.122	0.358

Note: All values of 0.0005 or less are omitted.

Source: John E. Freund, *Modern Elementary Statistics,* 8th ed., © 1992. Reprinted by permission of Prentice Hall, Inc., Upper Saddle River, New Jersey.

Table C   The Poisson Distribution

	λ									
x	0.1	0.2	0.3	0.4	0.5	0.6	0.7	0.8	0.9	1.0
0	.9048	.8187	.7408	.6703	.6065	.5488	.4966	.4493	.4066	.3679
1	.0905	.1637	.2222	.2681	.3033	.3293	.3476	.3595	.3659	.3679
2	.0045	.0164	.0333	.0536	.0758	.0988	.1217	.1438	.1647	.1839
3	.0002	.0011	.0033	.0072	.0126	.0198	.0284	.0383	.0494	.0613
4	.0000	.0001	.0003	.0007	.0016	.0030	.0050	.0077	.0111	.0153
5	.0000	.0000	.0000	.0001	.0002	.0004	.0007	.0012	.0020	.0031
6	.0000	.0000	.0000	.0000	.0000	.0000	.0001	.0002	.0003	.0005
7	.0000	.0000	.0000	.0000	.0000	.0000	.0000	.0000	.0000	.0001

	λ									
x	1.1	1.2	1.3	1.4	1.5	1.6	1.7	1.8	1.9	2.0
0	.3329	.3012	.2725	.2466	.2231	.2019	.1827	.1653	.1496	.1353
1	.3662	.3614	.3543	.3452	.3347	.3230	.3106	.2975	.2842	.2707
2	.2014	.2169	.2303	.2417	.2510	.2584	.2640	.2678	.2700	.2707
3	.0738	.0867	.0998	.1128	.1255	.1378	.1496	.1607	.1710	.1804
4	.0203	.0260	.0324	.0395	.0471	.0551	.0636	.0723	.0812	.0902
5	.0045	.0062	.0084	.0111	.0141	.0176	.0216	.0260	.0309	.0361
6	.0008	.0012	.0018	.0026	.0035	.0047	.0061	.0078	.0098	.0120
7	.0001	.0002	.0003	.0005	.0008	.0011	.0015	.0020	.0027	.0034
8	.0000	.0000	.0001	.0001	.0001	.0002	.0003	.0005	.0006	.0009
9	.0000	.0000	.0000	.0000	.0000	.0000	.0001	.0001	.0001	.0002

	λ									
x	2.1	2.2	2.3	2.4	2.5	2.6	2.7	2.8	2.9	3.0
0	.1225	.1108	.1003	.0907	.0821	.0743	.0672	.0608	.0550	.0498
1	.2572	.2438	.2306	.2177	.2052	.1931	.1815	.1703	.1596	.1494
2	.2700	.2681	.2652	.2613	.2565	.2510	.2450	.2384	.2314	.2240
3	.1890	.1966	.2033	.2090	.2138	.2176	.2205	.2225	.2237	.2240
4	.0992	.1082	.1169	.1254	.1336	.1414	.1488	.1557	.1622	.1680
5	.0417	.0476	.0538	.0602	.0668	.0735	.0804	.0872	.0940	.1008
6	.0146	.0174	.0206	.0241	.0278	.0319	.0362	.0407	.0455	.0504
7	.0044	.0055	.0068	.0083	.0099	.0118	.0139	.0163	.0188	.0216
8	.0011	.0015	.0019	.0025	.0031	.0038	.0047	.0057	.0068	.0081
9	.0003	.0004	.0005	.0007	.0009	.0011	.0014	.0018	.0022	.0027
10	.0001	.0001	.0001	.0002	.0002	.0003	.0004	.0005	.0006	.0008
11	.0000	.0000	.0000	.0000	.0000	.0001	.0001	.0001	.0002	.0002
12	.0000	.0000	.0000	.0000	.0000	.0000	.0000	.0000	.0000	.0001

	λ									
x	3.1	3.2	3.3	3.4	3.5	3.6	3.7	3.8	3.9	4.0
0	.0450	.0408	.0369	.0334	.0302	.0273	.0247	.0224	.0202	.0183
1	.1397	.1304	.1217	.1135	.1057	.0984	.0915	.0850	.0789	.0733
2	.2165	.2087	.2008	.1929	.1850	.1771	.1692	.1615	.1539	.1465
3	.2237	.2226	.2209	.2186	.2158	.2125	.2087	.2046	.2001	.1954
4	.1734	.1781	.1823	.1858	.1888	.1912	.1931	.1944	.1951	.1954

Table C *(continued)*

					λ					
x	**3.1**	**3.2**	**3.3**	**3.4**	**3.5**	**3.6**	**3.7**	**3.8**	**3.9**	**4.0**
5	.1075	.1140	.1203	.1264	.1322	.1377	.1429	.1477	.1522	.1563
6	.0555	.0608	.0662	.0716	.0771	.0826	.0881	.0936	.0989	.1042
7	.0246	.0278	.0312	.0348	.0385	.0425	.0466	.0508	.0551	.0595
8	.0095	.0111	.0129	.0148	.0169	.0191	.0215	.0241	.0269	.0298
9	.0033	.0040	.0047	.0056	.0066	.0076	.0089	.0102	.0116	.0132
10	.0010	.0013	.0016	.0019	.0023	.0028	.0033	.0039	.0045	.0053
11	.0003	.0004	.0005	.0006	.0007	.0009	.0011	.0013	.0016	.0019
12	.0001	.0001	.0001	.0002	.0002	.0003	.0003	.0004	.0005	.0006
13	.0000	.0000	.0000	.0000	.0001	.0001	.0001	.0001	.0002	.0002
14	.0000	.0000	.0000	.0000	.0000	.0000	.0000	.0000	.0000	.0001

					λ					
x	**4.1**	**4.2**	**4.3**	**4.4**	**4.5**	**4.6**	**4.7**	**4.8**	**4.9**	**5.0**
0	.0166	.0150	.0136	.0123	.0111	.0101	.0091	.0082	.0074	.0067
1	.0679	.0630	.0583	.0540	.0500	.0462	.0427	.0395	.0365	.0337
2	.1393	.1323	.1254	.1188	.1125	.1063	.1005	.0948	.0894	.0842
3	.1904	.1852	.1798	.1743	.1687	.1631	.1574	.1517	.1460	.1404
4	.1951	.1944	.1933	.1917	.1898	.1875	.1849	.1820	.1789	.1755
5	.1600	.1633	.1662	.1687	.1708	.1725	.1738	.1747	.1753	.1755
6	.1093	.1143	.1191	.1237	.1281	.1323	.1362	.1398	.1432	.1462
7	.0640	.0686	.0732	.0778	.0824	.0869	.0914	.0959	.1002	.1044
8	.0328	.0360	.0393	.0428	.0463	.0500	.0537	.0575	.0614	.0653
9	.0150	.0168	.0188	.0209	.0232	.0255	.0280	.0307	.0334	.0363
10	.0061	.0071	.0081	.0092	.0104	.0118	.0132	.0147	.0164	.0181
11	.0023	.0027	.0032	.0037	.0043	.0049	.0056	.0064	.0073	.0082
12	.0008	.0009	.0011	.0014	.0016	.0019	.0022	.0026	.0030	.0034
13	.0002	.0003	.0004	.0005	.0006	.0007	.0008	.0009	.0011	.0013
14	.0001	.0001	.0001	.0001	.0002	.0002	.0003	.0003	.0004	.0005
15	.0000	.0000	.0000	.0000	.0001	.0001	.0001	.0001	.0001	.0002

					λ					
x	**5.1**	**5.2**	**5.3**	**5.4**	**5.5**	**5.6**	**5.7**	**5.8**	**5.9**	**6.0**
0	.0061	.0055	.0050	.0045	.0041	.0037	.0033	.0030	.0027	.0025
1	.0311	.0287	.0265	.0244	.0225	.0207	.0191	.0176	.0162	.0149
2	.0793	.0746	.0701	.0659	.0618	.0580	.0544	.0509	.0477	.0446
3	.1348	.1293	.1239	.1185	.1133	.1082	.1033	.0985	.0938	.0892
4	.1719	.1681	.1641	.1600	.1558	.1515	.1472	.1428	.1383	.1339

Table C   (continued)

	λ									
x	**5.1**	**5.2**	**5.3**	**5.4**	**5.5**	**5.6**	**5.7**	**5.8**	**5.9**	**6.0**
5	.1753	.1748	.1740	.1728	.1714	.1697	.1678	.1656	.1632	.1606
6	.1490	.1515	.1537	.1555	.1571	.1584	.1594	.1601	.1605	.1606
7	.1086	.1125	.1163	.1200	.1234	.1267	.1298	.1326	.1353	.1377
8	.0692	.0731	.0771	.0810	.0849	.0887	.0925	.0962	.0998	.1033
9	.0392	.0423	.0454	.0486	.0519	.0552	.0586	.0620	.0654	.0688
10	.0200	.0220	.0241	.0262	.0285	.0309	.0334	.0359	.0386	.0413
11	.0093	.0104	.0116	.0129	.0143	.0157	.0173	.0190	.0207	.0225
12	.0039	.0045	.0051	.0058	.0065	.0073	.0082	.0092	.0102	.0113
13	.0015	.0018	.0021	.0024	.0028	.0032	.0036	.0041	.0046	.0052
14	.0006	.0007	.0008	.0009	.0011	.0013	.0015	.0017	.0019	.0022
15	.0002	.0002	.0003	.0003	.0004	.0005	.0006	.0007	.0008	.0009
16	.0001	.0001	.0001	.0001	.0001	.0002	.0002	.0002	.0003	.0003
17	.0000	.0000	.0000	.0000	.0000	.0000	.0001	.0001	.0001	.0001

	λ									
x	**6.1**	**6.2**	**6.3**	**6.4**	**6.5**	**6.6**	**6.7**	**6.8**	**6.9**	**7.0**
0	.0022	.0020	.0018	.0017	.0015	.0014	.0012	.0011	.0010	.0009
1	.0137	.0126	.0116	.0106	.0098	.0090	.0082	.0076	.0070	.0064
2	.0417	.0390	.0364	.0340	.0318	.0296	.0276	.0258	.0240	.0223
3	.0848	.0806	.0765	.0726	.0688	.0652	.0617	.0584	.0552	.0521
4	.1294	.1249	.1205	.1162	.1118	.1076	.1034	.0992	.0952	.0912
5	.1579	.1549	.1519	.1487	.1454	.1420	.1385	.1349	.1314	.1277
6	.1605	.1601	.1595	.1586	.1575	.1562	.1546	.1529	.1511	.1490
7	.1399	.1418	.1435	.1450	.1462	.1472	.1480	.1486	.1489	.1490
8	.1066	.1099	.1130	.1160	.1188	.1215	.1240	.1263	.1284	.1304
9	.0723	.0757	.0791	.0825	.0858	.0891	.0923	.0954	.0985	.1014
10	.0441	.0469	.0498	.0528	.0558	.0588	.0618	.0649	.0679	.0710
11	.0245	.0265	.0285	.0307	.0330	.0353	.0377	.0401	.0426	.0452
12	.0124	.0137	.0150	.0164	.0179	.0194	.0210	.0227	.0245	.0264
13	.0058	.0065	.0073	.0081	.0089	.0098	.0108	.0119	.0130	.0142
14	.0025	.0029	.0033	.0037	.0041	.0046	.0052	.0058	.0064	.0071
15	.0010	.0012	.0014	.0016	.0018	.0020	.0023	.0026	.0029	.0033
16	.0004	.0005	.0005	.0006	.0007	.0008	.0010	.0011	.0013	.0014
17	.0001	.0002	.0002	.0002	.0003	.0003	.0004	.0004	.0005	.0006
18	.0000	.0001	.0001	.0001	.0001	.0001	.0001	.0002	.0002	.0002
19	.0000	.0000	.0000	.0000	.0000	.0000	.0000	.0001	.0001	.0001

Table C *(continued)*

						λ					
x	*7.1*	*7.2*	*7.3*	*7.4*	*7.5*	*7.6*	*7.7*	*7.8*	*7.9*	*8.0*	
0	.0008	.0007	.0007	.0006	.0006	.0005	.0005	.0004	.0004	.0003	
1	.0059	.0054	.0049	.0045	.0041	.0038	.0035	.0032	.0029	.0027	
2	.0208	.0194	.0180	.0167	.0156	.0145	.0134	.0125	.0116	.0107	
3	.0492	.0464	.0438	.0413	.0389	.0366	.0345	.0324	.0305	.0286	
4	.0874	.0836	.0799	.0764	.0729	.0696	.0663	.0632	.0602	.0573	
5	.1241	.1204	.1167	.1130	.1094	.1057	.1021	.0986	.0951	.0916	
6	.1468	.1445	.1420	.1394	.1367	.1339	.1311	.1282	.1252	.1221	
7	.1489	.1486	.1481	.1474	.1465	.1454	.1442	.1428	.1413	.1396	
8	.1321	.1337	.1351	.1363	.1373	.1382	.1388	.1392	.1395	.1396	
9	.1042	.1070	.1096	.1121	.1144	.1167	.1187	.1207	.1224	.1241	
10	.0740	.0770	.0800	.0829	.0858	.0887	.0914	.0941	.0967	.0993	
11	.0478	.0504	.0531	.0558	.0585	.0613	.0640	.0667	.0695	.0722	
12	.0283	.0303	.0323	.0344	.0366	.0388	.0411	.0434	.0457	.0481	
13	.0154	.0168	.0181	.0196	.0211	.0227	.0243	.0260	.0278	.0296	
14	.0078	.0086	.0095	.0104	.0113	.0123	.0134	.0145	.0157	.0169	
15	.0037	.0041	.0046	.0051	.0057	.0062	.0069	.0075	.0083	.0090	
16	.0016	.0019	.0021	.0024	.0026	.0030	.0033	.0037	.0041	.0045	
17	.0007	.0008	.0009	.0010	.0012	.0013	.0015	.0017	.0019	.0021	
18	.0003	.0003	.0004	.0004	.0005	.0006	.0006	.0007	.0008	.0009	
19	.0001	.0001	.0001	.0002	.0002	.0002	.0003	.0003	.0003	.0004	
20	.0000	.0000	.0001	.0001	.0001	.0001	.0001	.0001	.0001	.0002	
21	.0000	.0000	.0000	.0000	.0000	.0000	.0000	.0000	.0001	.0001	

						λ					
x	*8.1*	*8.2*	*8.3*	*8.4*	*8.5*	*8.6*	*8.7*	*8.8*	*8.9*	*9.0*	
0	.0003	.0003	.0002	.0002	.0002	.0002	.0002	.0002	.0001	.0001	
1	.0025	.0023	.0021	.0019	.0017	.0016	.0014	.0013	.0012	.0011	
2	.0100	.0092	.0086	.0079	.0074	.0068	.0063	.0058	.0054	.0050	
3	.0269	.0252	.0237	.0222	.0208	.0195	.0183	.0171	.0160	.0150	
4	.0544	.0517	.0491	.0466	.0443	.0420	.0398	.0377	.0357	.0337	
5	.0882	.0849	.0816	.0784	.0752	.0722	.0692	.0663	.0635	.0607	
6	.1191	.1160	.1128	.1097	.1066	.1034	.1003	.0972	.0941	.0911	
7	.1378	.1358	.1338	.1317	.1294	.1271	.1247	.1222	.1197	.1171	
8	.1395	.1392	.1388	.1382	.1375	.1366	.1356	.1344	.1332	.1318	
9	.1256	.1269	.1280	.1290	.1299	.1306	.1311	.1315	.1317	.1318	

Table C   (continued)

x					λ					
	8.1	**8.2**	**8.3**	**8.4**	**8.5**	**8.6**	**8.7**	**8.8**	**8.9**	**9.0**
10	.1017	.1040	.1063	.1084	.1104	.1123	.1140	.1157	.1172	.1186
11	.0749	.0776	.0802	.0828	.0853	.0878	.0902	.0925	.0948	.0970
12	.0505	.0530	.0555	.0579	.0604	.0629	.0654	.0679	.0703	.0728
13	.0315	.0334	.0354	.0374	.0395	.0416	.0438	.0459	.0481	.0504
14	.0182	.0196	.0210	.0225	.0240	.0256	.0272	.0289	.0306	.0324
15	.0098	.0107	.0116	.0126	.0136	.0147	.0158	.0169	.0182	.0194
16	.0050	.0055	.0060	.0066	.0072	.0079	.0086	.0093	.0101	.0109
17	.0024	.0026	.0029	.0033	.0036	.0040	.0044	.0048	.0053	.0058
18	.0011	.0012	.0014	.0015	.0017	.0019	.0021	.0024	.0026	.0029
19	.0005	.0005	.0006	.0007	.0008	.0009	.0010	.0011	.0012	.0014
20	.0002	.0002	.0002	.0003	.0003	.0004	.0004	.0005	.0005	.0006
21	.0001	.0001	.0001	.0001	.0001	.0002	.0002	.0002	.0002	.0003
22	.0000	.0000	.0000	.0000	.0001	.0001	.0001	.0001	.0001	.0001

x					λ					
	9.1	**9.2**	**9.3**	**9.4**	**9.5**	**9.6**	**9.7**	**9.8**	**9.9**	**10.0**
0	.0001	.0001	.0001	.0001	.0001	.0001	.0001	.0001	.0001	.0000
1	.0010	.0009	.0009	.0008	.0007	.0007	.0006	.0005	.0005	.0005
2	.0046	.0043	.0040	.0037	.0034	.0031	.0029	.0027	.0025	.0023
3	.0140	.0131	.0123	.0115	.0107	.0100	.0093	.0087	.0081	.0076
4	.0319	.0302	.0285	.0269	.0254	.0240	.0226	.0213	.0201	.0189
5	.0581	.0555	.0530	.0506	.0483	.0460	.0439	.0418	.0398	.0378
6	.0881	.0851	.0822	.0793	.0764	.0736	.0709	.0682	.0656	.0631
7	.1145	.1118	.1091	.1064	.1037	.1010	.0982	.0955	.0928	.0901
8	.1302	.1286	.1269	.1251	.1232	.1212	.1191	.1170	.1148	.1126
9	.1317	.1315	.1311	.1306	.1300	.1293	.1284	.1274	.1263	.1251
10	.1198	.1210	.1219	.1228	.1235	.1241	.1245	.1249	.1250	.1251
11	.0991	.1012	.1031	.1049	.1067	.1083	.1098	.1112	.1125	.1137
12	.0752	.0776	.0799	.0822	.0844	.0866	.0888	.0908	.0928	.0948
13	.0526	.0549	.0572	.0594	.0617	.0640	.0662	.0685	.0707	.0729
14	.0342	.0361	.0380	.0399	.0419	.0439	.0459	.0479	.0500	.0521
15	.0208	.0221	.0235	.0250	.0265	.0281	.0297	.0313	.0330	.0347
16	.0118	.0127	.0137	.0147	.0157	.0168	.0180	.0192	.0204	.0217
17	.0063	.0069	.0075	.0081	.0088	.0095	.0103	.0111	.0119	.0128
18	.0032	.0035	.0039	.0042	.0046	.0051	.0055	.0060	.0065	.0071
19	.0015	.0017	.0019	.0021	.0023	.0026	.0028	.0031	.0034	.0037

C

C

Table C (continued)

					λ					
x	9.1	9.2	9.3	9.4	9.5	9.6	9.7	9.8	9.9	10.0
20	.0007	.0008	.0009	.0010	.0011	.0012	.0014	.0015	.0017	.0019
21	.0003	.0003	.0004	.0004	.0005	.0006	.0006	.0007	.0008	.0009
22	.0001	.0001	.0002	.0002	.0002	.0002	.0003	.0003	.0004	.0004
23	.0000	.0001	.0001	.0001	.0001	.0001	.0001	.0001	.0002	.0002
24	.0000	.0000	.0000	.0000	.0000	.0000	.0000	.0001	.0001	.0001

					λ					
x	11	12	13	14	15	16	17	18	19	20
0	.0000	.0000	.0000	.0000	.0000	.0000	.0000	.0000	.0000	.0000
1	.0002	.0001	.0000	.0000	.0000	.0000	.0000	.0000	.0000	.0000
2	.0010	.0004	.0002	.0001	.0000	.0000	.0000	.0000	.0000	.0000
3	.0037	.0018	.0008	.0004	.0002	.0001	.0000	.0000	.0000	.0000
4	.0102	.0053	.0027	.0013	.0006	.0003	.0001	.0001	.0000	.0000
5	.0224	.0127	.0070	.0037	.0019	.0010	.0005	.0002	.0001	.0001
6	.0411	.0255	.0152	.0087	.0048	.0026	.0014	.0007	.0004	.0002
7	.0646	.0437	.0281	.0174	.0104	.0060	.0034	.0018	.0010	.0005
8	.0888	.0655	.0457	.0304	.0194	.0120	.0072	.0042	.0024	.0013
9	.1085	.0874	.0661	.0473	.0324	.0213	.0135	.0083	.0050	.0029
10	.1194	.1048	.0859	.0663	.0486	.0341	.0230	.0150	.0095	.0058
11	.1194	.1144	.1015	.0844	.0663	.0496	.0355	.0245	.0164	.0106
12	.1094	.1144	.1099	.0984	.0829	.0661	.0504	.0368	.0259	.0176
13	.0926	.1056	.1099	.1060	.0956	.0814	.0658	.0509	.0378	.0271
14	.0728	.0905	.1021	.1060	.1024	.0930	.0800	.0655	.0514	.0387
15	.0534	.0724	.0885	.0989	.1024	.0992	.0906	.0786	.0650	.0516
16	.0367	.0543	.0719	.0866	.0960	.0992	.0963	.0884	.0772	.0646
17	.0237	.0383	.0550	.0713	.0847	.0934	.0963	.0936	.0863	.0760
18	.0145	.0256	.0397	.0554	.0706	.0830	.0909	.0936	.0911	.0844
19	.0084	.0161	.0272	.0409	.0557	.0699	.0814	.0887	.0911	.0888
20	.0046	.0097	.0177	.0286	.0418	.0559	.0692	.0798	.0866	.0888
21	.0024	.0055	.0109	.0191	.0299	.0426	.0560	.0684	.0783	.0846
22	.0012	.0030	.0065	.0121	.0204	.0310	.0433	.0560	.0676	.0769
23	.0006	.0016	.0037	.0074	.0133	.0216	.0320	.0438	.0559	.0669
24	.0003	.0008	.0020	.0043	.0083	.0144	.0226	.0328	.0442	.0557
25	.0001	.0004	.0010	.0024	.0050	.0092	.0154	.0237	.0336	.0446
26	.0000	.0002	.0005	.0013	.0029	.0057	.0101	.0164	.0246	.0343
27	.0000	.0001	.0002	.0007	.0016	.0034	.0063	.0109	.0173	.0254
28	.0000	.0000	.0001	.0003	.0009	.0019	.0038	.0070	.0117	.0181
29	.0000	.0000	.0001	.0002	.0004	.0011	.0023	.0044	.0077	.0125

Table C *(concluded)*

					λ					
x	*11*	*12*	*13*	*14*	*15*	*16*	*17*	*18*	*19*	*20*
30	.0000	.0000	.0000	.0001	.0002	.0006	.0013	.0026	.0049	.0083
31	.0000	.0000	.0000	.0000	.0001	.0003	.0007	.0015	.0030	.0054
32	.0000	.0000	.0000	.0000	.0001	.0001	.0004	.0009	.0018	.0034
33	.0000	.0000	.0000	.0000	.0000	.0001	.0002	.0005	.0010	.0020
34	.0000	.0000	.0000	.0000	.0000	.0000	.0001	.0002	.0006	.0012
35	.0000	.0000	.0000	.0000	.0000	.0000	.0000	.0001	.0003	.0007
36	.0000	.0000	.0000	.0000	.0000	.0000	.0000	.0001	.0002	.0004
37	.0000	.0000	.0000	.0000	.0000	.0000	.0000	.0000	.0001	.0002
38	.0000	.0000	.0000	.0000	.0000	.0000	.0000	.0000	.0000	.0001
39	.0000	.0000	.0000	.0000	.0000	.0000	.0000	.0000	.0000	.0001

Table D Random Numbers

10480	15011	01536	02011	81647	91646	67179	14194	62590	36207	20969	99570	91291	90700
22368	46573	25595	85393	30995	89198	27982	53402	93965	34095	52666	19174	39615	99505
24130	48360	22527	97265	76393	64809	15179	24830	49340	32081	30680	19655	63348	58629
42167	93093	06243	61680	07856	16376	39440	53537	71341	57004	00849	74917	97758	16379
37570	39975	81837	16656	06121	91782	60468	81305	49684	60672	14110	06927	01263	54613
77921	06907	11008	42751	27756	53498	18602	70659	90655	15053	21916	81825	44394	42880
99562	72905	56420	69994	98872	31016	71194	18738	44013	48840	63213	21069	10634	12952
96301	91977	05463	07972	18876	20922	94595	56869	69014	60045	18425	84903	42508	32307
89579	14342	63661	10281	17453	18103	57740	84378	25331	12566	58678	44947	05584	56941
85475	36857	43342	53988	53060	59533	38867	62300	08158	17983	16439	11458	18593	64952
28918	69578	88231	33276	70997	79936	56865	05859	90106	31595	01547	85590	91610	78188
63553	40961	48235	03427	49626	69445	18663	72695	52180	20847	12234	90511	33703	90322
09429	93969	52636	92737	88974	33488	36320	17617	30015	08272	84115	27156	30613	74952
10365	61129	87529	85689	48237	52267	67689	93394	01511	26358	85104	20285	29975	89868
07119	97336	71048	08178	77233	13916	47564	81056	97735	85977	29372	74461	28551	90707
51085	12765	51821	51259	77452	16308	60756	92144	49442	53900	70960	63990	75601	40719
02368	21382	52404	60268	89368	19885	55322	44819	01188	65255	64835	44919	05944	55157
01011	54092	33362	94904	31273	04146	18594	29852	71585	85030	51132	01915	92747	64951
52162	53916	46369	58586	23216	14513	83149	98736	23495	64350	94738	17752	35156	35749
07056	97628	33787	09998	42698	06691	76988	13602	51851	46104	88916	19509	25625	58104
48663	91245	85828	14346	09172	30168	90229	04734	59193	22178	30421	61666	99904	32812
54164	58492	22421	74103	47070	25306	76468	26384	58151	06646	21524	15227	96909	44592
32639	32363	05597	24200	13363	38005	94342	28728	35806	06912	17012	64161	18296	22851
29334	27001	87637	87308	58731	00256	45834	15398	46557	41135	10367	07684	36188	18510
02488	33062	28834	07351	19731	92420	60952	61280	50001	67658	32586	86679	50720	94953
81525	72295	04839	96423	24878	82651	66566	14778	76797	14780	13300	87074	79666	95725
29676	20591	68086	26432	46901	20849	89768	81536	86645	12659	92259	57102	80428	25280
00742	57392	39064	66432	84673	40027	32832	61362	98947	96067	64760	64584	96096	98253
05366	04213	25669	26422	44407	44048	37937	63904	45766	66134	75470	66520	34693	90449
91921	26418	64117	94305	26766	25940	39972	22209	71500	64568	91402	42416	07844	69618
00582	04711	87917	77341	42206	35126	74087	99547	81817	42607	43808	76655	62028	76630
00725	69884	62797	56170	86324	88072	76222	36086	84637	93161	76038	65855	77919	88006
69011	65797	95876	55293	18988	27354	26575	08625	40801	59920	29841	80150	12777	48501
25976	57948	29888	88604	67917	48708	18912	82271	65424	69774	33611	54262	85963	03547
09763	83473	73577	12908	30883	18317	28290	35797	05998	41688	34952	37888	38917	88050
91567	42595	27958	30134	04024	86385	29880	99730	55536	84855	29080	09250	79656	73211
17955	56349	90999	49127	20044	59931	06115	20542	18059	02008	73708	83517	36103	42791
46503	18584	18845	49618	02304	51038	20655	58727	28168	15475	56942	53389	20562	87338
92157	89634	94824	78171	84610	82834	09922	25417	44137	48413	25555	21246	35509	20468
14577	62765	35605	81263	39667	47358	56873	56307	61607	49518	89656	20103	77490	18062
98427	07523	33362	64270	01638	92477	66969	98420	04880	45585	46565	04102	46880	45709
34914	63976	88720	82765	34476	17032	87589	40836	32427	70002	70663	88863	77775	69348
70060	28277	39475	46473	23219	53416	94970	25832	69975	94884	19661	72828	00102	66794
53976	54914	06990	67245	68350	82948	11398	42878	80287	88267	47363	46634	06541	97809
76072	29515	40980	07391	58745	25774	22987	80059	39911	96189	41151	14222	60697	59583
90725	52210	83974	29992	65831	38857	50490	83765	55657	14361	31720	57375	56228	41546
64364	67412	33339	31926	14883	24413	59744	92351	97473	89286	35931	04110	23726	51900
08962	00358	31662	25388	61642	34072	81249	35648	56891	69352	48373	45578	78547	81788
95012	68379	93526	70765	10593	04542	76463	54328	02349	17247	28865	14777	62730	92277
15664	10493	20492	38391	91132	21999	59516	81652	27195	48223	46751	22923	32261	85653

Reprinted with permission from W. H. Beyer, *Handbook of Tables for Probability and Statistics,* 2nd ed. Copyright CRC Press, Boca Raton, Florida, 1986.

Table E The Standard Normal Distribution

z	.00	.01	.02	.03	.04	.05	.06	.07	.08	.09
0.0	.0000	.0040	.0080	.0120	.0160	.0199	.0239	.0279	.0319	.0359
0.1	.0398	.0438	.0478	.0517	.0557	.0596	.0636	.0675	.0714	.0753
0.2	.0793	.0832	.0871	.0910	.0948	.0987	.1026	.1064	.1103	.1141
0.3	.1179	.1217	.1255	.1293	.1331	.1368	.1406	.1443	.1480	.1517
0.4	.1554	.1591	.1628	.1664	.1700	.1736	.1772	.1808	.1844	.1879
0.5	.1915	.1950	.1985	.2019	.2054	.2088	.2123	.2157	.2190	.2224
0.6	.2257	.2291	.2324	.2357	.2389	.2422	.2454	.2486	.2517	.2549
0.7	.2580	.2611	.2642	.2673	.2704	.2734	.2764	.2794	.2823	.2852
0.8	.2881	.2910	.2939	.2967	.2995	.3023	.3051	.3078	.3106	.3133
0.9	.3159	.3186	.3212	.3238	.3264	.3289	.3315	.3340	.3365	.3389
1.0	.3413	.3438	.3461	.3485	.3508	.3531	.3554	.3577	.3599	.3621
1.1	.3643	.3665	.3686	.3708	.3729	.3749	.3770	.3790	.3810	.3830
1.2	.3849	.3869	.3888	.3907	.3925	.3944	.3962	.3980	.3997	.4015
1.3	.4032	.4049	.4066	.4082	.4099	.4115	.4131	.4147	.4162	.4177
1.4	.4192	.4207	.4222	.4236	.4251	.4265	.4279	.4292	.4306	.4319
1.5	.4332	.4345	.4357	.4370	.4382	.4394	.4406	.4418	.4429	.4441
1.6	.4452	.4463	.4474	.4484	.4495	.4505	.4515	.4525	.4535	.4545
1.7	.4554	.4564	.4573	.4582	.4591	.4599	.4608	.4616	.4625	.4633
1.8	.4641	.4649	.4656	.4664	.4671	.4678	.4686	.4693	.4699	.4706
1.9	.4713	.4719	.4726	.4732	.4738	.4744	.4750	.4756	.4761	.4767
2.0	.4772	.4778	.4783	.4788	.4793	.4798	.4803	.4808	.4812	.4817
2.1	.4821	.4826	.4830	.4834	.4838	.4842	.4846	.4850	.4854	.4857
2.2	.4861	.4864	.4868	.4871	.4875	.4878	.4881	.4884	.4887	.4890
2.3	.4893	.4896	.4898	.4901	.4904	.4906	.4909	.4911	.4913	.4916
2.4	.4918	.4920	.4922	.4925	.4927	.4929	.4931	.4932	.4934	.4936
2.5	.4938	.4940	.4941	.4943	.4945	.4946	.4948	.4949	.4951	.4952
2.6	.4953	.4955	.4956	.4957	.4959	.4960	.4961	.4962	.4963	.4964
2.7	.4965	.4966	.4967	.4968	.4969	.4970	.4971	.4972	.4973	.4974
2.8	.4974	.4975	.4976	.4977	.4977	.4978	.4979	.4979	.4980	.4981
2.9	.4981	.4982	.4982	.4983	.4984	.4984	.4985	.4985	.4986	.4986
3.0	.4987	.4987	.4987	.4988	.4988	.4989	.4989	.4989	.4990	.4990

Note: Use 0.4999 for z values above 3.09.

Source: Frederick Mosteller and Robert E. K. Rourke, *Sturdy Statistics,* Table A–1 (Reading, Mass.: Addison-Wesley, 1973). Reprinted with permission of the copyright owners.

Area given in table

C

Table F The *t* Distribution

d.f.	Confidence intervals	50%	80%	90%	95%	98%	99%
	One tail, α	0.25	0.10	0.05	0.025	0.01	0.005
	Two tails, α	0.50	0.20	0.10	0.05	0.02	0.01
1		1.000	3.078	6.314	12.706	31.821	63.657
2		.816	1.886	2.920	4.303	6.965	9.925
3		.765	1.638	2.353	3.182	4.541	5.841
4		.741	1.533	2.132	2.776	3.747	4.604
5		.727	1.476	2.015	2.571	3.365	4.032
6		.718	1.440	1.943	2.447	3.143	3.707
7		.711	1.415	1.895	2.365	2.998	3.499
8		.706	1.397	1.860	2.306	2.896	3.355
9		.703	1.383	1.833	2.262	2.821	3.250
10		.700	1.372	1.812	2.228	2.764	3.169
11		.697	1.363	1.796	2.201	2.718	3.106
12		.695	1.356	1.782	2.179	2.681	3.055
13		.694	1.350	1.771	2.160	2.650	3.012
14		.692	1.345	1.761	2.145	2.624	2.977
15		.691	1.341	1.753	2.131	2.602	2.947
16		.690	1.337	1.746	2.120	2.583	2.921
17		.689	1.333	1.740	2.110	2.567	2.898
18		.688	1.330	1.734	2.101	2.552	2.878
19		.688	1.328	1.729	2.093	2.539	2.861
20		.687	1.325	1.725	2.086	2.528	2.845
21		.686	1.323	1.721	2.080	2.518	2.831
22		.686	1.321	1.717	2.074	2.508	2.819
23		.685	1.319	1.714	2.069	2.500	2.807
24		.685	1.318	1.711	2.064	2.492	2.797
25		.684	1.316	1.708	2.060	2.485	2.787
26		.684	1.315	1.706	2.056	2.479	2.779
27		.684	1.314	1.703	2.052	2.473	2.771
28		.683	1.313	1.701	2.048	2.467	2.763
(z) ∞		.674	1.282[a]	1.645[b]	1.960	2.326[c]	2.576[d]

[a]This value has been rounded to 1.28 in the textbook.
[b]This value has been rounded to 1.65 in the textbook.
[c]This value has been rounded to 2.33 in the textbook.
[d]This value has been rounded to 2.58 in the textbook.

Source: Adapted from W. H. Beyer, *Handbook of Tables for Probability and Statistics,* 2nd ed., CRC Press, Boca Raton, Florida, 1986. Reprinted with permission.

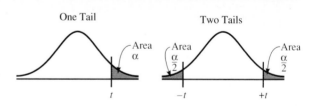

T~~ails~~ 1

USe this side

C

Table G The Chi-Square Distribution

Degrees of freedom	α									
	0.995	0.99	0.975	0.95	0.90	0.10	0.05	0.025	0.01	0.005
1	—	—	0.001	0.004	0.016	2.706	3.841	5.024	6.635	7.879
2	0.010	0.020	0.051	0.103	0.211	4.605	5.991	7.378	9.210	10.597
3	0.072	0.115	0.216	0.352	0.584	6.251	7.815	9.348	11.345	12.838
4	0.207	0.297	0.484	0.711	1.064	7.779	9.488	11.143	13.277	14.860
5	0.412	0.554	0.831	1.145	1.610	9.236	11.071	12.833	15.086	16.750
6	0.676	0.872	1.237	1.635	2.204	10.645	12.592	14.449	16.812	18.548
7	0.989	1.239	1.690	2.167	2.833	12.017	14.067	16.013	18.475	20.278
8	1.344	1.646	2.180	2.733	3.490	13.362	15.507	17.535	20.090	21.955
9	1.735	2.088	2.700	3.325	4.168	14.684	16.919	19.023	21.666	23.589
10	2.156	2.558	3.247	3.940	4.865	15.987	18.307	20.483	23.209	25.188
11	2.603	3.053	3.816	4.575	5.578	17.275	19.675	21.920	24.725	26.757
12	3.074	3.571	4.404	5.226	6.304	18.549	21.026	23.337	26.217	28.299
13	3.565	4.107	5.009	5.892	7.042	19.812	22.362	24.736	27.688	29.819
14	4.075	4.660	5.629	6.571	7.790	21.064	23.685	26.119	29.141	31.319
15	4.601	5.229	6.262	7.261	8.547	22.307	24.996	27.488	30.578	32.801
16	5.142	5.812	6.908	7.962	9.312	23.542	26.296	28.845	32.000	34.267
17	5.697	6.408	7.564	8.672	10.085	24.769	27.587	30.191	33.409	35.718
18	6.265	7.015	8.231	9.390	10.865	25.989	28.869	31.526	34.805	37.156
19	6.844	7.633	8.907	10.117	11.651	27.204	30-144	32.852	36.191	38.582
20	7.434	8.260	9.591	10.851	12.443	28.412	31.410	34.170	37.566	39.997
21	8.034	8.897	10.283	11.591	13.240	29.615	32.671	35.479	38.932	41.401
22	8.643	9.542	10.982	12.338	14.042	30.813	33.924	36.781	40.289	42.796
23	9.262	10.196	11.689	13.091	14.848	32.007	35.172	38.076	41.638	44.181
24	9.886	10.856	12.401	13.848	15.659	33.196	36.415	39.364	42.980	45.559
25	10.520	11.524	13.120	14.611	16.473	34.382	37.652	40.646	44.314	46.928
26	11.160	12.198	13.844	15.379	17.292	35.563	38.885	41.923	45.642	48.290
27	11.808	12.879	14.573	16.151	18.114	36.741	40.113	43.194	46.963	49.645
28	12.461	13.565	15.308	16.928	18.939	37.916	41.337	44.461	48.278	50.993
29	13.121	14.257	16.047	17.708	19.768	39.087	42.557	45.722	49.588	52.336
30	13.787	14.954	16.791	18.493	20.599	40.256	43.773	46.979	50.892	53.672
40	20.707	22.164	24.433	26.509	29.051	51.805	55.758	59.342	63.691	66.766
50	27.991	29.707	32.357	34.764	37.689	63.167	67-505	71.420	76.154	79.490
60	35.534	37.485	40.482	43.188	46.459	74.397	79.082	83.298	88.379	91.952
70	43.275	45.442	48.758	51.739	55.329	85.527	90.531	95.023	100.425	104.215
80	51.172	53.540	57.153	60.391	64.278	96.578	101.879	106.629	112.329	116.321
90	59.196	61.754	65.647	69.126	73.291	107.565	113.145	118.136	124.116	128.299
100	67.328	70.065	74.222	77.929	82.358	118.498	124.342	129.561	135.807	140.169

Source: Donald B. Owen. *Handbook of Statistics Tables,* © 1962, by Addison-Wesley Publishing Co., Inc.. Reading, Massachusetts. Table A–5. Reprinted with permission of Addison-Wesley Longman.

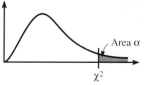

Table H   The F Distribution

$\alpha = 0.005$

d.f.D.: degrees of freedom, denominator	d.f.N.: Degrees of freedom, numerator																		
	1	2	3	4	5	6	7	8	9	10	12	15	20	24	30	40	60	120	∞
1	16211	20000	21615	22500	23056	23437	23715	23925	24091	24224	24426	24630	24836	24940	25044	25148	25253	25359	25465
2	198.5	199.0	199.2	199.2	199.3	199.3	199.4	199.4	199.4	199.4	199.4	199.4	199.4	199.5	199.5	199.5	199.5	199.5	199.5
3	55.55	49.80	47.47	46.19	45.39	44.84	44.43	44.13	43.88	43.69	43.39	43.08	42.78	42.62	42.47	42.31	42.15	41.99	41.83
4	31.33	26.28	24.26	23.15	22.46	21.97	21.62	21.35	21.14	20.97	20.70	20.44	20.17	20.03	19.89	19.75	19.61	19.47	19.32
5	22.78	18.31	16.53	15.56	14.94	14.51	14.20	13.96	13.77	13.62	13.38	13.15	12.90	12.78	12.66	12.53	12.40	12.27	12.14
6	18.63	14.54	12.92	12.03	11.46	11.07	10.79	10.57	10.39	10.25	10.03	9.81	9.59	9.47	9.36	9.24	9.12	9.00	8.88
7	16.24	12.40	10.88	10.05	9.52	9.16	8.89	8.68	8.51	8.38	8.18	7.97	7.75	7.65	7.53	7.42	7.31	7.19	7.08
8	14.69	11.04	9.60	8.81	8.30	7.95	7.69	7.50	7.34	7.21	7.01	6.81	6.61	6.50	6.40	6.29	6.18	6.06	5.95
9	13.61	10.11	8.72	7.96	7.47	7.13	6.88	6.69	6.54	6.42	6.23	6.03	5.83	5.73	5.62	5.52	5.41	5.30	5.19
10	12.83	9.43	8.08	7.34	6.87	6.54	6.30	6.12	5.97	5.85	5.66	5.47	5.27	5.17	5.07	4.97	4.86	4.75	4.64
11	12.23	8.91	7.60	6.88	6.42	6.10	5.86	5.68	5.54	5.42	5.24	5.05	4.86	4.76	4.65	4.55	4.44	4.34	4.23
12	11.75	8.51	7.23	6.52	6.07	5.76	5.52	5.35	5.20	5.09	4.91	4.72	4.53	4.43	4.33	4.23	4.12	4.01	3.90
13	11.37	8.19	6.93	6.23	5.79	5.48	5.25	5.08	4.94	4.82	4.64	4.46	4.27	4.17	4.07	3.97	3.87	3.76	3.65
14	11.06	7.92	6.68	6.00	5.56	5.26	5.03	4.86	4.72	4.60	4.43	4.25	4.06	3.96	3.86	3.76	3.66	3.55	3.44
15	10.80	7.70	6.48	5.80	5.37	5.07	4.85	4.67	4.54	4.42	4.25	4.07	3.88	3.79	3.69	3.58	3.48	3.37	3.26
16	10.58	7.51	6.30	5.64	5.21	4.91	4.69	4.52	4.38	4.27	4.10	3.92	3.73	3.64	3.54	3.44	3.33	3.22	3.11
17	10.38	7.35	6.16	5.50	5.07	4.78	4.56	4.39	4.25	4.14	3.97	3.79	3.61	3.51	3.41	3.31	3.21	3.10	2.98
18	10.22	7.21	6.03	5.37	4.96	4.66	4.44	4.28	4.14	4.03	3.86	3.68	3.50	3.40	3.30	3.20	3.10	2.99	2.87
19	10.07	7.09	5.92	5.27	4.85	4.56	4.34	4.18	4.04	3.93	3.76	3.59	3.40	3.31	3.21	3.11	3.00	2.89	2.78
20	9.94	6.99	5.82	5.17	4.76	4.47	4.26	4.09	3.96	3.85	3.68	3.50	3.32	3.22	3.12	3.02	2.92	2.81	2.69
21	9.83	6.89	5.73	5.09	4.68	4.39	4.18	4.01	3.88	3.77	3.60	3.43	3.24	3.15	3.05	2.95	2.84	2.73	2.61
22	9.73	6.81	5.65	5.02	4.61	4.32	4.11	3.94	3.81	3.70	3.54	3.36	3.18	3.08	2.98	2.88	2.77	2.66	2.55
23	9.63	6.73	5.58	4.95	4.54	4.26	4.05	3.88	3.75	3.64	3.47	3.30	3.12	3.02	2.92	2.82	2.71	2.60	2.48
24	9.55	6.66	5.52	4.89	4.49	4.20	3.99	3.83	3.69	3.59	3.42	3.25	3.06	2.97	2.87	2.77	2.66	2.55	2.43
25	9.48	6.60	5.46	4.84	4.43	4.15	3.94	3.78	3.64	3.54	3.37	3.20	3.01	2.92	2.82	2.72	2.61	2.50	2.38
26	9.41	6.54	5.41	4.79	4.38	4.10	3.89	3.73	3.60	3.49	3.33	3.15	2.97	2.87	2.77	2.67	2.56	2.45	2.33
27	9.34	6.49	5.36	4.74	4.34	4.06	3.85	3.69	3.56	3.45	3.28	3.11	2.93	2.83	2.73	2.63	2.52	2.41	2.29
28	9.28	6.44	5.32	4.70	4.30	4.02	3.81	3.65	3.52	3.41	3.25	3.07	2.89	2.79	2.69	2.59	2.48	2.37	2.25
29	9.23	6.40	5.28	4.66	4.26	3.98	3.77	3.61	3.48	3.38	3.21	3.04	2.86	2.76	2.66	2.56	2.45	2.33	2.24
30	9.18	6.35	5.24	4.62	4.23	3.95	3.74	3.58	3.45	3.34	3.18	3.01	2.82	2.73	2.63	2.52	2.42	2.30	2.18
40	8.83	6.07	4.98	4.37	3.99	3.71	3.51	3.35	3.22	3.12	2.95	2.78	2.60	2.50	2.40	2.30	2.18	2.06	1.93
60	8.49	5.79	4.73	4.14	3.76	3.49	3.29	3.13	3.01	2.90	2.74	2.57	2.39	2.29	2.19	2.08	1.96	1.83	1.69
120	8.18	5.54	4.50	3.92	3.55	3.28	3.09	2.93	2.81	2.71	2.54	2.37	2.19	2.09	1.98	1.87	1.75	1.61	1.43
∞	7.88	5.30	4.28	3.72	3.35	3.09	2.90	2.74	2.62	2.52	2.36	2.19	2.00	1.90	1.79	1.67	1.53	1.36	1.00

Table H *(continued)*

α = 0.01

d.f.D.: degrees of freedom, denominator vs **d.f.N.: Degrees of freedom, numerator**

d.f.D.	1	2	3	4	5	6	7	8	9	10	12	15	20	24	30	40	60	120	∞
1	4052	4999.5	5403	5625	5764	5859	5928	5982	6022	6056	6106	6157	6209	6235	6261	6287	6313	6339	6366
2	98.50	99.00	99.17	99.25	99.30	99.33	99.36	99.37	99.39	99.40	99.42	99.43	99.45	99.46	99.47	99.47	99.48	99.49	99.50
3	34.12	30.82	29.46	28.71	28.24	27.91	27.67	27.49	27.35	27.23	27.05	26.87	26.69	26.60	26.50	26.41	26.32	26.22	26.13
4	21.20	18.00	16.69	15.98	15.52	15.21	14.98	14.80	14.66	14.55	14.37	14.20	14.02	13.93	13.84	13.75	13.65	13.56	13.46
5	16.26	13.27	12.06	11.39	10.97	10.67	10.46	10.29	10.16	10.05	9.89	9.72	9.55	9.47	9.38	9.29	9.20	9.11	9.02
6	13.75	10.92	9.78	9.15	8.75	8.47	8.26	8.10	7.98	7.87	7.72	7.56	7.40	7.31	7.23	7.14	7.06	6.97	6.88
7	12.25	9.55	8.45	7.85	7.46	7.19	6.99	6.84	6.72	6.62	6.47	6.31	6.16	6.07	5.99	5.91	5.82	5.74	5.65
8	11.26	8.65	7.59	7.01	6.63	6.37	6.18	6.03	5.91	5.81	5.67	5.52	5.36	5.28	5.20	5.12	5.03	4.95	4.86
9	10.56	8.02	6.99	6.42	6.06	5.80	5.61	5.47	5.35	5.26	5.11	4.96	4.81	4.73	4.65	4.57	4.48	4.40	4.31
10	10.04	7.56	6.55	5.99	5.64	5.39	5.20	5.06	4.94	4.85	4.71	4.56	4.41	4.33	4.25	4.17	4.08	4.00	3.91
11	9.65	7.21	6.22	5.67	5.32	5.07	4.89	4.74	4.63	4.54	4.40	4.25	4.10	4.02	3.94	3.86	3.78	3.69	3.60
12	9.33	6.93	5.95	5.41	5.06	4.82	4.64	4.50	4.39	4.30	4.16	4.01	3.86	3.78	3.70	3.62	3.54	3.45	3.36
13	9.07	6.70	5.74	5.21	4.86	4.62	4.44	4.30	4.19	4.10	3.96	3.82	3.66	3.59	3.51	3.43	3.34	3.25	3.17
14	8.86	6.51	5.56	5.04	4.69	4.46	4.28	4.14	4.03	3.94	3.80	3.66	3.51	3.43	3.35	3.27	3.18	3.09	3.00
15	8.68	6.36	5.42	4.89	4.56	4.32	4.14	4.00	3.89	3.80	3.67	3.52	3.37	3.29	3.21	3.13	3.05	2.96	2.87
16	8.53	6.23	5.29	4.77	4.44	4.20	4.03	3.89	3.78	3.69	3.55	3.41	3.26	3.18	3.10	3.02	2.93	2.84	2.75
17	8.40	6.11	5.18	4.67	4.34	4.10	3.93	3.79	3.68	3.59	3.46	3.31	3.16	3.08	3.00	2.92	2.83	2.75	2.65
18	8.29	6.01	5.09	4.58	4.25	4.01	3.84	3.71	3.60	3.51	3.37	3.23	3.08	3.00	2.92	2.84	2.75	2.66	2.57
19	8.18	5.93	5.01	4.50	4.17	3.94	3.77	3.63	3.52	3.43	3.30	3.15	3.00	2.92	2.84	2.76	2.67	2.58	2.49
20	8.10	5.85	4.94	4.43	4.10	3.87	3.70	3.56	3.46	3.37	3.23	3.09	2.94	2.86	2.78	2.69	2.61	2.52	2.42
21	8.02	5.78	4.87	4.37	4.04	3.81	3.64	3.51	3.40	3.31	3.17	3.03	2.88	2.80	2.72	2.64	2.55	2.46	2.36
22	7.95	5.72	4.82	4.31	3.99	3.76	3.59	3.45	3.35	3.26	3.12	2.98	2.83	2.75	2.67	2.58	2.50	2.40	2.31
23	7.88	5.66	4.76	4.26	3.94	3.71	3.54	3.41	3.30	3.21	3.07	2.93	2.78	2.70	2.62	2.54	2.45	2.35	2.26
24	7.82	5.61	4.72	4.22	3.90	3.67	3.50	3.36	3.26	3.17	3.03	2.89	2.74	2.66	2.58	2.49	2.40	2.31	2.21
25	7.77	5.57	4.68	4.18	3.85	3.63	3.46	3.32	3.22	3.13	2.99	2.85	2.70	2.62	2.54	2.45	2.36	2.27	2.17
26	7.72	5.53	4.64	4.14	3.82	3.59	3.42	3.29	3.18	3.09	2.96	2.81	2.66	2.58	2.50	2.42	2.33	2.23	2.13
27	7.68	5.49	4.60	4.11	3.78	3.56	3.39	3.26	3.15	3.06	2.93	2.78	2.63	2.55	2.47	2.38	2.29	2.20	2.10
28	7.64	5.45	4.57	4.07	3.75	3.53	3.36	3.23	3.12	3.03	2.90	2.75	2.60	2.52	2.44	2.35	2.26	2.17	2.06
29	7.60	5.42	4.54	4.04	3.73	3.50	3.33	3.20	3.09	3.00	2.87	2.73	2.57	2.49	2.41	2.33	2.23	2.14	2.03
30	7.56	5.39	4.51	4.02	3.70	3.47	3.30	3.17	3.07	2.98	2.84	2.70	2.55	2.47	2.39	2.30	2.21	2.11	2.01
40	7.31	5.18	4.31	3.83	3.51	3.29	3.12	2.99	2.89	2.80	2.66	2.52	2.37	2.29	2.20	2.11	2.02	1.92	1.80
60	7.08	4.98	4.13	3.65	3.34	3.12	2.95	2.82	2.72	2.63	2.50	2.35	2.20	2.12	2.03	1.94	1.84	1.73	1.60
120	6.85	4.79	3.95	3.48	3.17	2.96	2.79	2.66	2.56	2.47	2.34	2.19	2.03	1.95	1.86	1.76	1.66	1.53	1.38
∞	6.63	4.61	3.78	3.32	3.02	2.80	2.64	2.51	2.41	2.32	2.18	2.04	1.88	1.79	1.70	1.59	1.47	1.32	1.00

C

Table H *(continued)*

$\alpha = 0.025$

d.f.D.: degrees of freedom, denominator

d.f.N.: Degrees of freedom, numerator

d.f.D.	1	2	3	4	5	6	7	8	9	10	12	15	20	24	30	40	60	120	∞
1	647.8	799.5	864.2	899.6	921.8	937.1	948.2	956.7	963.3	968.6	976.7	984.9	993.1	997.2	1001	1006	1010	1014	1018
2	38.51	39.00	39.17	39.25	39.30	39.33	39.36	39.37	39.39	39.40	39.41	39.43	39.45	39.46	39.46	39.47	39.48	39.49	39.50
3	17.44	16.04	15.44	15.10	14.88	14.73	14.62	14.54	14.47	14.42	14.34	14.25	14.17	14.12	14.08	14.04	13.99	13.95	13.90
4	12.22	10.65	9.98	9.60	9.36	9.20	9.07	8.98	8.90	8.84	8.75	8.66	8.56	8.51	8.46	8.41	8.36	8.31	8.26
5	10.01	8.43	7.76	7.39	7.15	6.98	6.85	6.76	6.68	6.62	6.52	6.43	6.33	6.28	6.23	6.18	6.12	6.07	6.02
6	8.81	7.26	6.60	6.23	5.99	5.82	5.70	5.60	5.52	5.46	5.37	5.27	5.17	5.12	5.07	5.01	4.96	4.90	4.85
7	8.07	6.54	5.89	5.52	5.29	5.12	4.99	4.90	4.82	4.76	4.67	4.57	4.47	4.42	4.36	4.31	4.25	4.20	4.14
8	7.57	6.06	5.42	5.05	4.82	4.65	4.53	4.43	4.36	4.30	4.20	4.10	4.00	3.95	3.89	3.84	3.78	3.73	3.67
9	7.21	5.71	5.08	4.72	4.48	4.32	4.20	4.10	4.03	3.96	3.87	3.77	3.67	3.61	3.56	3.51	3.45	3.39	3.33
10	6.94	5.46	4.83	4.47	4.24	4.07	3.95	3.85	3.78	3.72	3.62	3.52	3.42	3.37	3.31	3.26	3.20	3.14	3.08
11	6.72	5.26	4.63	4.28	4.04	3.88	3.76	3.66	3.59	3.53	3.43	3.33	3.23	3.17	3.12	3.06	3.00	2.94	2.88
12	6.55	5.10	4.47	4.12	3.89	3.73	3.61	3.51	3.44	3.37	3.28	3.18	3.07	3.02	2.96	2.91	2.85	2.79	2.72
13	6.41	4.97	4.35	4.00	3.77	3.60	3.48	3.39	3.31	3.25	3.15	3.05	2.95	2.89	2.84	2.78	2.72	2.66	2.60
14	6.30	4.86	4.24	3.89	3.66	3.50	3.38	3.29	3.21	3.15	3.05	2.95	2.84	2.79	2.73	2.67	2.61	2.55	2.49
15	6.20	4.77	4.15	3.80	3.58	3.41	3.29	3.20	3.12	3.06	2.96	2.86	2.76	2.70	2.64	2.59	2.52	2.46	2.40
16	6.12	4.69	4.08	3.73	3.50	3.34	3.22	3.12	3.05	2.99	2.89	2.79	2.68	2.63	2.57	2.51	2.45	2.38	2.32
17	6.04	4.62	4.01	3.66	3.44	3.28	3.16	3.06	2.98	2.92	2.82	2.72	2.62	2.56	2.50	2.44	2.38	2.32	2.25
18	5.98	4.56	3.95	3.61	3.38	3.22	3.10	3.01	2.93	2.87	2.77	2.67	2.56	2.50	2.44	2.38	2.32	2.26	2.19
19	5.92	4.51	3.90	3.56	3.33	3.17	3.05	2.96	2.88	2.82	2.72	2.62	2.51	2.45	2.39	2.33	2.27	2.20	2.13
20	5.87	4.46	3.86	3.51	3.29	3.13	3.01	2.91	2.84	2.77	2.68	2.57	2.46	2.41	2.35	2.29	2.22	2.16	2.09
21	5.83	4.42	3.82	3.48	3.25	3.09	2.97	2.87	2.80	2.73	2.64	2.53	2.42	2.37	2.31	2.25	2.18	2.11	2.04
22	5.79	4.38	3.78	3.44	3.22	3.05	2.93	2.84	2.76	2.70	2.60	2.50	2.39	2.33	2.27	2.21	2.14	2.08	2.00
23	5.75	4.35	3.75	3.41	3.18	3.02	2.90	2.81	2.73	2.67	2.57	2.47	2.36	2.30	2.24	2.18	2.11	2.04	1.97
24	5.72	4.32	3.72	3.38	3.15	2.99	2.87	2.78	2.70	2.64	2.54	2.44	2.33	2.27	2.21	2.15	2.08	2.01	1.94
25	5.69	4.29	3.69	3.35	3.13	2.97	2.85	2.75	2.68	2.61	2.51	2.41	2.30	2.24	2.18	2.12	2.05	1.98	1.91
26	5.66	4.27	3.67	3.33	3.10	2.94	2.82	2.73	2.65	2.59	2.49	2.39	2.28	2.22	2.16	2.09	2.03	1.95	1.88
27	5.63	4.24	3.65	3.31	3.08	2.92	2.80	2.71	2.63	2.57	2.47	2.36	2.25	2.19	2.13	2.07	2.00	1.93	1.85
28	5.61	4.22	3.63	3.29	3.06	2.90	2.78	2.69	2.61	2.55	2.45	2.34	2.23	2.17	2.11	2.05	1.98	1.91	1.83
29	5.59	4.20	3.61	3.27	3.04	2.88	2.76	2.67	2.59	2.53	2.43	2.32	2.21	2.15	2.09	2.03	1.96	1.89	1.81
30	5.57	4.18	3.59	3.25	3.03	2.87	2.75	2.65	2.57	2.51	2.41	2.31	2.20	2.14	2.07	2.01	1.94	1.87	1.79
40	5.42	4.05	3.46	3.13	2.90	2.74	2.62	2.53	2.45	2.39	2.29	2.18	2.07	2.01	1.94	1.88	1.80	1.72	1.64
60	5.29	3.93	3.34	3.01	2.79	2.63	2.51	2.41	2.33	2.27	2.17	2.06	1.94	1.88	1.82	1.74	1.67	1.58	1.48
120	5.15	3.80	3.23	2.89	2.67	2.52	2.39	2.30	2.22	2.16	2.05	1.94	1.82	1.76	1.69	1.61	1.53	1.43	1.31
∞	5.02	3.69	3.12	2.79	2.57	2.41	2.29	2.19	2.11	2.05	1.94	1.83	1.71	1.64	1.57	1.48	1.39	1.27	1.00

Table H   (continued)

α = 0.05

d.f.D.: degrees of freedom, denominator	d.f.N.: Degrees of freedom, numerator																		
	1	2	3	4	5	6	7	8	9	10	12	15	20	24	30	40	60	120	∞
1	161.4	199.5	215.7	224.6	230.2	234.0	236.8	238.9	240.5	241.9	243.9	245.9	248.0	249.1	250.1	251.1	252.2	253.3	254.3
2	18.51	19.00	19.16	19.25	19.30	19.33	19.35	19.37	19.38	19.40	19.41	19.43	19.45	19.45	19.46	19.47	19.48	19.49	19.50
3	10.13	9.55	9.28	9.12	9.01	8.94	8.89	8.85	8.81	8.79	8.74	8.70	8.66	8.64	8.62	8.59	8.57	8.55	8.53
4	7.71	6.94	6.59	6.39	6.26	6.16	6.09	6.04	6.00	5.96	5.91	5.86	5.80	5.77	5.75	5.72	5.69	5.66	5.63
5	6.61	5.79	5.41	5.19	5.05	4.95	4.88	4.82	4.77	4.74	4.68	4.62	4.56	4.53	4.50	4.46	4.43	4.40	4.36
6	5.99	5.14	4.76	4.53	4.39	4.28	4.21	4.15	4.10	4.06	4.00	3.94	3.87	3.84	3.81	3.77	3.74	3.70	3.67
7	5.59	4.74	4.35	4.12	3.97	3.87	3.79	3.73	3.68	3.64	3.57	3.51	3.44	3.41	3.38	3.34	3.30	3.27	3.23
8	5.32	4.46	4.07	3.84	3.69	3.58	3.50	3.44	3.39	3.35	3.28	3.22	3.15	3.12	3.08	3.04	3.01	2.97	2.93
9	5.12	4.26	3.86	3.63	3.48	3.37	3.29	3.23	3.18	3.14	3.07	3.01	2.94	2.90	2.86	2.83	2.79	2.75	2.71
10	4.96	4.10	3.71	3.48	3.33	3.22	3.14	3.07	3.02	2.98	2.91	2.85	2.77	2.74	2.70	2.66	2.62	2.58	2.54
11	4.84	3.98	3.59	3.36	3.20	3.09	3.01	2.95	2.90	2.85	2.79	2.72	2.65	2.61	2.57	2.53	2.49	2.45	2.40
12	4.75	3.89	3.49	3.26	3.11	3.00	2.91	2.85	2.80	2.75	2.69	2.62	2.54	2.51	2.47	2.43	2.38	2.34	2.30
13	4.67	3.81	3.41	3.18	3.03	2.92	2.83	2.77	2.71	2.67	2.60	2.53	2.46	2.42	2.38	2.34	2.30	2.25	2.21
14	4.60	3.74	3.34	3.11	2.96	2.85	2.76	2.70	2.65	2.60	2.53	2.46	2.39	2.35	2.31	2.27	2.22	2.18	2.13
15	4.54	3.68	3.29	3.06	2.90	2.79	2.71	2.64	2.59	2.54	2.48	2.40	2.33	2.29	2.25	2.20	2.16	2.11	2.07
16	4.49	3.63	3.24	3.01	2.85	2.74	2.66	2.59	2.54	2.49	2.42	2.35	2.28	2.24	2.19	2.15	2.11	2.06	2.01
17	4.45	3.59	3.20	2.96	2.81	2.70	2.61	2.55	2.49	2.45	2.38	2.31	2.23	2.19	2.15	2.10	2.06	2.01	1.96
18	4.41	3.55	3.16	2.93	2.77	2.66	2.58	2.51	2.46	2.41	2.34	2.27	2.19	2.15	2.11	2.06	2.02	1.97	1.92
19	4.38	3.52	3.13	2.90	2.74	2.63	2.54	2.48	2.42	2.38	2.31	2.23	2.16	2.11	2.07	2.03	1.98	1.93	1.88
20	4.35	3.49	3.10	2.87	2.71	2.60	2.51	2.45	2.39	2.35	2.28	2.20	2.12	2.08	2.04	1.99	1.95	1.90	1.84
21	4.32	3.47	3.07	2.84	2.68	2.57	2.49	2.42	2.37	2.32	2.25	2.18	2.10	2.05	2.01	1.96	1.92	1.87	1.81
22	4.30	3.44	3.05	2.82	2.66	2.55	2.46	2.40	2.34	2.30	2.23	2.15	2.07	2.03	1.98	1.94	1.89	1.84	1.78
23	4.28	3.42	3.03	2.80	2.64	2.53	2.44	2.37	2.32	2.27	2.20	2.13	2.05	2.01	1.96	1.91	1.86	1.81	1.76
24	4.26	3.40	3.01	2.78	2.62	2.51	2.42	2.36	2.30	2.25	2.18	2.11	2.03	1.98	1.94	1.89	1.84	1.79	1.73
25	4.24	3.39	2.99	2.76	2.60	2.49	2.40	2.34	2.28	2.24	2.16	2.09	2.01	1.96	1.92	1.87	1.82	1.77	1.71
26	4.23	3.37	2.98	2.74	2.59	2.47	2.39	2.32	2.27	2.22	2.15	2.07	1.99	1.95	1.90	1.85	1.80	1.75	1.69
27	4.21	3.35	2.96	2.73	2.57	2.46	2.37	2.31	2.25	2.20	2.13	2.06	1.97	1.93	1.88	1.84	1.79	1.73	1.67
28	4.20	3.34	2.95	2.71	2.56	2.45	2.36	2.29	2.24	2.19	2.12	2.04	1.96	1.91	1.87	1.82	1.77	1.71	1.65
29	4.18	3.33	2.93	2.70	2.55	2.43	2.35	2.28	2.22	2.18	2.10	2.03	1.94	1.90	1.85	1.81	1.75	1.70	1.64
30	4.17	3.32	2.92	2.69	2.53	2.42	2.33	2.27	2.21	2.16	2.09	2.01	1.93	1.89	1.84	1.79	1.74	1.68	1.62
40	4.08	3.23	2.84	2.61	2.45	2.34	2.25	2.18	2.12	2.08	2.00	1.92	1.84	1.79	1.74	1.69	1.64	1.58	1.51
60	4.00	3.15	2.76	2.53	2.37	2.25	2.17	2.10	2.04	1.99	1.92	1.84	1.75	1.70	1.65	1.59	1.53	1.47	1.39
120	3.92	3.07	2.68	2.45	2.29	2.17	2.09	2.02	1.96	1.91	1.83	1.75	1.66	1.61	1.55	1.50	1.43	1.35	1.25
∞	3.84	3.00	2.60	2.37	2.21	2.10	2.01	1.94	1.88	1.83	1.75	1.67	1.57	1.52	1.46	1.39	1.32	1.22	1.00

C

Table H (concluded)

$\alpha = 0.10$

d.f.N.: Degrees of freedom, numerator

d.f.D.: degrees of freedom, denominator	1	2	3	4	5	6	7	8	9	10	12	15	20	24	30	40	60	120	∞
1	39.86	49.50	53.59	55.83	57.24	58.20	58.91	59.44	59.86	60.19	60.71	61.22	61.74	62.00	62.26	62.53	62.79	63.06	63.33
2	8.53	9.00	9.16	9.24	9.29	9.33	9.35	9.37	9.38	9.39	9.41	9.42	9.44	9.45	9.46	9.47	9.47	9.48	9.49
3	5.54	5.46	5.39	5.34	5.31	5.28	5.27	5.25	5.24	5.23	5.22	5.20	5.18	5.18	5.17	5.16	5.15	5.14	5.13
4	4.54	4.32	4.19	4.11	4.05	4.01	3.98	3.95	3.94	3.92	3.90	3.87	3.84	3.83	3.82	3.80	3.79	3.78	3.76
5	4.06	3.78	3.62	3.52	3.45	3.40	3.37	3.34	3.32	3.30	3.27	3.24	3.21	3.19	3.17	3.16	3.14	3.12	3.10
6	3.78	3.46	3.29	3.18	3.11	3.05	3.01	2.98	2.96	2.94	2.90	2.87	2.84	2.82	2.80	2.78	2.76	2.74	2.72
7	3.59	3.26	3.07	2.96	2.88	2.83	2.78	2.75	2.72	2.70	2.67	2.63	2.59	2.58	2.56	2.54	2.51	2.49	2.47
8	3.46	3.11	2.92	2.81	2.73	2.67	2.62	2.59	2.56	2.54	2.50	2.46	2.42	2.40	2.38	2.36	2.34	2.32	2.29
9	3.36	3.01	2.81	2.69	2.61	2.55	2.51	2.47	2.44	2.42	2.38	2.34	2.30	2.28	2.25	2.23	2.21	2.18	2.16
10	3.29	2.92	2.73	2.61	2.52	2.46	2.41	2.38	2.35	2.32	2.28	2.24	2.20	2.18	2.16	2.13	2.11	2.08	2.06
11	3.23	2.86	2.66	2.54	2.45	2.39	2.34	2.30	2.27	2.25	2.21	2.17	2.12	2.10	2.08	2.05	2.03	2.00	1.97
12	3.18	2.81	2.61	2.48	2.39	2.33	2.28	2.24	2.21	2.19	2.15	2.10	2.06	2.04	2.01	1.99	1.96	1.93	1.90
13	3.14	2.76	2.56	2.43	2.35	2.28	2.23	2.20	2.16	2.14	2.10	2.05	2.01	1.98	1.96	1.93	1.90	1.88	1.85
14	3.10	2.73	2.52	2.39	2.31	2.24	2.19	2.15	2.12	2.10	2.05	2.01	1.96	1.94	1.91	1.89	1.86	1.83	1.80
15	3.07	2.70	2.49	2.36	2.27	2.21	2.16	2.12	2.09	2.06	2.02	1.97	1.92	1.90	1.87	1.85	1.82	1.79	1.76
16	3.05	2.67	2.46	2.33	2.24	2.18	2.13	2.09	2.06	2.03	1.99	1.94	1.89	1.87	1.84	1.81	1.78	1.75	1.72
17	3.03	2.64	2.44	2.31	2.22	2.15	2.10	2.06	2.03	2.00	1.96	1.91	1.86	1.84	1.81	1.78	1.75	1.72	1.69
18	3.01	2.62	2.42	2.29	2.20	2.13	2.08	2.04	2.00	1.98	1.93	1.89	1.84	1.81	1.78	1.75	1.72	1.69	1.66
19	2.99	2.61	2.40	2.27	2.18	2.11	2.06	2.02	1.98	1.96	1.91	1.86	1.81	1.79	1.76	1.73	1.70	1.67	1.63
20	2.97	2.59	2.38	2.25	2.16	2.09	2.04	2.00	1.96	1.94	1.89	1.84	1.79	1.77	1.74	1.71	1.68	1.64	1.61
21	2.96	2.57	2.36	2.23	2.14	2.08	2.02	1.98	1.95	1.92	1.87	1.83	1.78	1.75	1.72	1.69	1.66	1.62	1.59
22	2.95	2.56	2.35	2.22	2.13	2.06	2.01	1.97	1.93	1.90	1.86	1.81	1.76	1.73	1.70	1.67	1.64	1.60	1.57
23	2.94	2.55	2.34	2.21	2.11	2.05	1.99	1.95	1.92	1.89	1.84	1.80	1.74	1.72	1.69	1.66	1.62	1.59	1.55
24	2.93	2.54	2.33	2.19	2.10	2.04	1.98	1.94	1.91	1.88	1.83	1.78	1.73	1.70	1.67	1.64	1.61	1.57	1.53
25	2.92	2.53	2.32	2.18	2.09	2.02	1.97	1.93	1.89	1.87	1.82	1.77	1.72	1.69	1.66	1.63	1.59	1.56	1.52
26	2.91	2.52	2.31	2.17	2.08	2.01	1.96	1.92	1.88	1.86	1.81	1.76	1.71	1.68	1.65	1.61	1.58	1.54	1.50
27	2.90	2.51	2.30	2.17	2.07	2.00	1.95	1.91	1.87	1.85	1.80	1.75	1.70	1.67	1.64	1.60	1.57	1.53	1.49
28	2.89	2.50	2.29	2.16	2.06	2.00	1.94	1.90	1.87	1.84	1.79	1.74	1.69	1.66	1.63	1.59	1.56	1.52	1.48
29	2.89	2.50	2.28	2.15	2.06	1.99	1.93	1.89	1.86	1.83	1.78	1.73	1.68	1.65	1.62	1.58	1.55	1.51	1.47
30	2.88	2.49	2.28	2.14	2.05	1.98	1.93	1.88	1.85	1.82	1.77	1.72	1.67	1.64	1.61	1.57	1.54	1.50	1.46
40	2.84	2.44	2.23	2.09	2.00	1.93	1.87	1.83	1.79	1.76	1.71	1.66	1.61	1.57	1.54	1.51	1.47	1.42	1.38
60	2.79	2.39	2.18	2.04	1.95	1.87	1.82	1.77	1.74	1.71	1.66	1.60	1.54	1.51	1.48	1.44	1.40	1.35	1.29
120	2.75	2.35	2.13	1.99	1.90	1.82	1.77	1.72	1.68	1.65	1.60	1.55	1.48	1.45	1.41	1.37	1.32	1.26	1.19
∞	2.71	2.30	2.08	1.94	1.85	1.77	1.72	1.67	1.63	1.60	1.55	1.49	1.42	1.38	1.34	1.30	1.24	1.17	1.00

From M. Merrington and C. M. Thompson (1943). Table of Percentage Points of the Inverted Beta (*F*) Distribution. *Biometrika 33*, pp. 74–87. Reprinted with permission from Biometrika.

Table I   Critical Values for PPMC

Reject H_0: $\rho = 0$ if the absolute value of r is greater than the value given in the table. The values are for a two-tailed test; d.f. $= n - 2$.

d.f	$\alpha = 0.05$	$\alpha = 0.01$
1	0.999	0.999
2	0.950	0.999
3	0.878	0.959
4	0.811	0.917
5	0.754	0.875
6	0.707	0.834
7	0.666	0.798
8	0.632	0.765
9	0.602	0.735
10	0.576	0.708
11	0.553	0.684
12	0.532	0.661
13	0.514	0.641
14	0.497	0.623
15	0.482	0.606
16	0.468	0.590
17	0.456	0.575
18	0.444	0.561
19	0.433	0.549
20	0.423	0.537
25	0.381	0.487
30	0.349	0.449
35	0.325	0.418
40	0.304	0.393
45	0.288	0.372
50	0.273	0.354
60	0.250	0.325
70	0.232	0.302
80	0.217	0.283
90	0.205	0.267
100	0.195	0.254

Source: From *Biometrika Tables for Statisticians* Vol. 1 (1962) p. 138. Reprinted with permission.

Table J   Critical Values for the Sign Test

Reject the null hypothesis if the smaller number of $+$ or $-$ signs is less than or equal to the value in the table.

	One-tailed, $\alpha = 0.005$	$\alpha = 0.01$	$\alpha = 0.025$	$\alpha = 0.05$
n	Two-tailed, $\alpha = 0.01$	$\alpha = 0.02$	$\alpha = 0.05$	$\alpha = 0.10$
8	0	0	0	1
9	0	0	1	1
10	0	0	1	1
11	0	1	1	2
12	1	1	2	2
13	1	1	2	3
14	1	2	3	3
15	2	2	3	3
16	2	2	3	4
17	2	3	4	4
18	3	3	4	5
19	3	4	4	5
20	3	4	5	5
21	4	4	5	6
22	4	5	5	6
23	4	5	6	7
24	5	5	6	7
25	5	6	6	7

Note: Table J is for one-tailed or two-tailed tests. The term n represents the total number of $+$ and $-$ signs. The test value is the number of less frequent signs.

Source: From *Journal of American Statistical Association* Vol. 41 (1946) pp. 557–66. W. J. Dixon and A. M. Mood.

C

Table K Critical Values for the Wilcoxon Signed-Rank Test

Reject the null hypothesis if the test value is less than or equal to the value given in the table.

n	One-tailed, $\alpha = 0.05$ Two-tailed, $\alpha = 0.10$	$\alpha = 0.025$ $\alpha = 0.05$	$\alpha = 0.01$ $\alpha = 0.02$	$\alpha = 0.005$ $\alpha = 0.01$
5	1			
6	2	1		
7	4	2	0	
8	6	4	2	0
9	8	6	3	2
10	11	8	5	3
11	14	11	7	5
12	17	14	10	7
13	21	17	13	10
14	26	21	16	13
15	30	25	20	16
16	36	30	24	19
17	41	35	28	23
18	47	40	33	28
19	54	46	38	32
20	60	52	43	37
21	68	59	49	43
22	75	66	56	49
23	83	73	62	55
24	92	81	69	61
25	101	90	77	68
26	110	98	85	76
27	120	107	93	84
28	130	117	102	92
29	141	127	111	100
30	152	137	120	109

Source: From *Some Rapid Approximate Statistical Procedures,* Copyright 1949, 1964 Lerderle Laboratories, American Cyanamid Co., Wayne, N.J. Reprinted with permission.

Table L Critical Values for the Rank Correlation Coefficient

Reject H_0: $\rho = 0$ if the absolute value of r_s is greater than the value given in the table.

n	$\alpha = 0.10$	$\alpha = 0.05$	$\alpha = 0.02$	$\alpha = 0.01$
5	0.900	—	—	—
6	0.829	0.886	0.943	—
7	0.714	0.786	0.893	0.929
8	0.643	0.738	0.833	0.881
9	0.600	0.700	0.783	0.833
10	0.564	0.648	0.745	0.794
11	0.536	0.618	0.709	0.818
12	0.497	0.591	0.703	0.780
13	0.475	0.566	0.673	0.745
14	0.457	0.545	0.646	0.716
15	0.441	0.525	0.623	0.689
16	0.425	0.507	0.601	0.666
17	0.412	0.490	0.582	0.645
18	0.399	0.476	0.564	0.625
19	0.388	0.462	0.549	0.608
20	0.377	0.450	0.534	0.591
21	0.368	0.438	0.521	0.576
22	0.359	0.428	0.508	0.562
23	0.351	0.418	0.496	0.549
24	0.343	0.409	0.485	0.537
25	0.336	0.400	0.475	0.526
26	0.329	0.392	0.465	0.515
27	0.323	0.385	0.456	0.505
28	0.317	0.377	0.488	0.496
29	0.311	0.370	0.440	0.487
30	0.305	0.364	0.432	0.478

Source: From N. L. Johnson and F. C. Leone, *Statistical and Experimental Design,* Vol. I (1964), p. 412. Reprinted with permission from the Institute of Mathematical Statistics.

C

Table M   Critical Values for the Number of Runs

This table gives the critical values at $\alpha = 0.05$ for a two-tailed test. Reject the null hypothesis if the number of runs is less than or equal to the smaller value or greater than or equal to the larger value.

Value of n_1	2	3	4	5	6	7	8	9	10	11	12	13	14	15	16	17	18	19	20
2	1	1	1	1	1	1	1	1	1	1	2	2	2	2	2	2	2	2	2
	6	6	6	6	6	6	6	6	6	6	6	6	6	6	6	6	6	6	6
3	1	1	1	1	2	2	2	2	2	2	2	2	2	3	3	3	3	3	3
	6	8	8	8	8	8	8	8	8	8	8	8	8	8	8	8	8	8	8
4	1	1	1	2	2	2	3	3	3	3	3	3	3	3	4	4	4	4	4
	6	8	9	9	9	10	10	10	10	10	10	10	10	10	10	10	10	10	10
5	1	1	2	2	3	3	3	3	3	4	4	4	4	4	4	4	5	5	5
	6	8	9	10	10	11	11	12	12	12	12	12	12	12	12	12	12	12	12
6	1	2	2	3	3	3	3	4	4	4	4	5	5	5	5	5	5	6	6
	6	8	9	10	11	12	12	13	13	13	13	14	14	14	14	14	14	14	14
7	1	2	2	3	3	3	4	4	5	5	5	5	5	6	6	6	6	6	6
	6	8	10	11	12	13	13	14	14	14	14	15	15	15	16	16	16	16	16
8	1	2	3	3	3	4	4	5	5	5	6	6	6	6	6	7	7	7	7
	6	8	10	11	12	13	14	14	15	15	16	16	16	16	17	17	17	17	17
9	1	2	3	3	4	4	5	5	5	6	6	6	7	7	7	7	8	8	8
	6	8	10	12	13	14	14	15	16	16	16	17	17	18	18	18	18	18	18
10	1	2	3	3	4	5	5	5	6	6	7	7	7	7	8	8	8	8	9
	6	8	10	12	13	14	15	16	16	17	17	18	18	18	19	19	19	20	20
11	1	2	3	4	4	5	5	6	6	7	7	7	8	8	8	9	9	9	9
	6	8	10	12	13	14	15	16	17	17	18	19	19	19	20	20	20	21	21
12	2	2	3	4	4	5	6	6	7	7	7	8	8	8	9	9	9	10	10
	6	8	10	12	13	14	16	16	17	18	19	19	20	20	21	21	21	22	22
13	2	2	3	4	5	5	6	6	7	7	8	8	9	9	9	10	10	10	10
	6	8	10	12	14	15	16	17	18	19	19	20	20	21	21	22	22	23	23
14	2	2	3	4	5	5	6	7	7	8	8	9	9	9	10	10	10	11	11
	6	8	10	12	14	15	16	17	18	19	20	20	21	22	22	23	23	23	24
15	2	3	3	4	5	6	6	7	7	8	8	9	9	10	10	11	11	11	12
	6	8	10	12	14	15	16	18	18	19	20	21	22	22	23	23	24	24	25
16	2	3	4	4	5	6	6	7	8	8	9	9	10	10	11	11	11	12	12
	6	8	10	12	14	16	17	18	19	20	21	21	22	23	23	24	25	25	25
17	2	3	4	4	5	6	7	7	8	9	9	10	10	11	11	11	12	12	13
	6	8	10	12	14	16	17	18	19	20	21	22	23	23	24	25	25	26	26
18	2	3	4	5	5	6	7	8	8	9	9	10	10	11	11	12	12	13	13
	6	8	10	12	14	16	17	18	19	20	21	22	23	24	25	25	26	26	27
19	2	3	4	5	6	6	7	8	8	9	10	10	11	11	12	12	13	13	13
	6	8	10	12	14	16	17	18	20	21	22	23	23	24	25	26	26	27	27
20	2	3	4	5	6	6	7	8	9	9	10	10	11	12	12	13	13	13	14
	6	8	10	12	14	16	17	18	20	21	22	23	24	25	25	26	27	27	28

Value of n_2

Source: Adapted from C. Eisenhardt and F. Swed, "Tables for Testing Randomness of Grouping in a Sequence of Alternatives," *The Annals of Statistics* 14 (1943), pp. 83–86. Reprinted with permission of the Institute of Mathematical Statistics and of the Benjamin/Cummings Publishing Company, in whose publication, *Elementary Statistics,* 3rd ed. (1989), by Mario F. Triola, this table appears.

C

Table N Critical Values for the Tukey Test

$\alpha = 0.01$

k \ v	2	3	4	5	6	7	8	9	10	11	12	13	14	15	16	17	18	19	20
1	90.03	135.0	164.3	185.6	202.2	215.8	227.2	237.0	245.6	253.2	260.0	266.2	271.8	277.0	281.8	286.3	290.4	294.3	298.0
2	14.04	19.02	22.29	24.72	26.63	28.20	29.53	30.68	31.69	32.59	33.40	34.13	34.81	35.43	36.00	36.53	37.03	37.50	37.95
3	8.26	10.62	12.17	13.33	14.24	15.00	15.64	16.20	16.69	17.13	17.53	17.89	18.22	18.52	18.81	19.07	19.32	19.55	19.77
4	6.51	8.12	9.17	9.96	10.58	11.10	11.55	11.93	12.27	12.57	12.84	13.09	13.32	13.53	13.73	13.91	14.08	14.24	14.40
5	5.70	6.98	7.80	8.42	8.91	9.32	9.67	9.97	10.24	10.48	10.70	10.89	11.08	11.24	11.40	11.55	11.68	11.81	11.93
6	5.24	6.33	7.03	7.56	7.97	8.32	8.61	8.87	9.10	9.30	9.48	9.65	9.81	9.95	10.08	10.21	10.32	10.43	10.54
7	4.95	5.92	6.54	7.01	7.37	7.68	7.94	8.17	8.37	8.55	8.71	8.86	9.00	9.12	9.24	9.35	9.46	9.55	9.65
8	4.75	5.64	6.20	6.62	6.96	7.24	7.47	7.68	7.86	8.03	8.18	8.31	8.44	8.55	8.66	8.76	8.85	8.94	9.03
9	4.60	5.43	5.96	6.35	6.66	6.91	7.13	7.33	7.49	7.65	7.78	7.91	8.03	8.13	8.23	8.33	8.41	8.49	8.57
10	4.48	5.27	5.77	6.14	6.43	6.67	6.87	7.05	7.21	7.36	7.49	7.60	7.71	7.81	7.91	7.99	8.08	8.15	8.23
11	4.39	5.15	5.62	5.97	6.25	6.48	6.67	6.84	6.99	7.13	7.25	7.36	7.46	7.56	7.65	7.73	7.81	7.88	7.95
12	4.32	5.05	5.50	5.84	6.10	6.32	6.51	6.67	6.81	6.94	7.06	7.17	7.26	7.36	7.44	7.52	7.59	7.66	7.73
13	4.26	4.96	5.40	5.73	5.98	6.19	6.37	6.53	6.67	6.79	6.90	7.01	7.10	7.19	7.27	7.35	7.42	7.48	7.55
14	4.21	4.89	5.32	5.63	5.88	6.08	6.26	6.41	6.54	6.66	6.77	6.87	6.96	7.05	7.13	7.20	7.27	7.33	7.39
15	4.17	4.84	5.25	5.56	5.80	5.99	6.16	6.31	6.44	6.55	6.66	6.76	6.84	6.93	7.00	7.07	7.14	7.20	7.26
16	4.13	4.79	5.19	5.49	5.72	5.92	6.08	6.22	6.35	6.46	6.56	6.66	6.74	6.82	6.90	6.97	7.03	7.09	7.15
17	4.10	4.74	5.14	5.43	5.66	5.85	6.01	6.15	6.27	6.38	6.48	6.57	6.66	6.73	6.81	6.87	6.94	7.00	7.05
18	4.07	4.70	5.09	5.38	5.60	5.79	5.94	6.08	6.20	6.31	6.41	6.50	6.58	6.65	6.73	6.79	6.85	6.91	6.97
19	4.05	4.67	5.05	5.33	5.55	5.73	5.89	6.02	6.14	6.25	6.34	6.43	6.51	6.58	6.65	6.72	6.78	6.84	6.89
20	4.02	4.64	5.02	5.29	5.51	5.69	5.84	5.97	6.09	6.19	6.28	6.37	6.45	6.52	6.59	6.65	6.71	6.77	6.82
24	3.96	4.55	4.91	5.17	5.37	5.54	5.69	5.81	5.92	6.02	6.11	6.19	6.26	6.33	6.39	6.45	6.51	6.56	6.61
30	3.89	4.45	4.80	5.05	5.24	5.40	5.54	5.65	5.76	5.85	5.93	6.01	6.08	6.14	6.20	6.26	6.31	6.36	6.41
40	3.82	4.37	4.70	4.93	5.11	5.26	5.39	5.50	5.60	5.69	5.76	5.83	5.90	5.96	6.02	6.07	6.12	6.16	6.21
60	3.76	4.28	4.59	4.82	4.99	5.13	5.25	5.36	5.45	5.53	5.60	5.67	5.73	5.78	5.84	5.89	5.93	5.97	6.01
120	3.70	4.20	4.50	4.71	4.87	5.01	5.12	5.21	5.30	5.37	5.44	5.50	5.56	5.61	5.66	5.71	5.75	5.79	5.83
∞	3.64	4.12	4.40	4.60	4.76	4.88	4.99	5.08	5.16	5.23	5.29	5.35	5.40	5.45	5.49	5.54	5.57	5.61	5.65

Table N (continued)

$\alpha = 0.05$

v \ k	2	3	4	5	6	7	8	9	10	11	12	13	14	15	16	17	18	19	20
1	17.97	26.98	32.82	37.08	40.41	43.12	45.40	47.36	49.07	50.59	51.96	53.20	54.33	55.36	56.32	57.22	58.04	58.83	59.56
2	6.08	8.33	9.80	10.88	11.74	12.44	13.03	13.54	13.99	14.39	14.75	15.08	15.38	15.65	15.91	16.14	16.37	16.57	16.77
3	4.50	5.91	6.82	7.50	8.04	8.48	8.85	9.18	9.46	9.72	9.95	10.15	10.35	10.53	10.69	10.84	10.98	11.11	11.24
4	3.93	5.04	5.76	6.29	6.71	7.05	7.35	7.60	7.83	8.03	8.21	8.37	8.52	8.66	8.79	8.91	9.03	9.13	9.23
5	3.64	4.60	5.22	5.67	6.03	6.33	6.58	6.80	6.99	7.17	7.32	7.47	7.60	7.72	7.83	7.93	8.03	8.12	8.21
6	3.46	4.34	4.90	5.30	5.63	5.90	6.12	6.32	6.49	6.65	6.79	6.92	7.03	7.14	7.24	7.34	7.43	7.51	7.59
7	3.34	4.16	4.68	5.06	5.36	5.61	5.82	6.00	6.16	6.30	6.43	6.55	6.66	6.76	6.85	6.94	7.02	7.10	7.17
8	3.26	4.04	4.53	4.89	5.17	5.40	5.60	5.77	5.92	6.05	6.18	6.29	6.39	6.48	6.57	6.65	6.73	6.80	6.87
9	3.20	3.95	4.41	4.76	5.02	5.24	5.43	5.59	5.74	5.87	5.98	6.09	6.19	6.28	6.36	6.44	6.51	6.58	6.64
10	3.15	3.88	4.33	4.65	4.91	5.12	5.30	5.46	5.60	5.72	5.83	5.93	6.03	6.11	6.19	6.27	6.34	6.40	6.47
11	3.11	3.82	4.26	4.57	4.82	5.03	5.20	5.35	5.49	5.61	5.71	5.81	5.90	5.98	6.06	6.13	6.20	6.27	6.33
12	3.08	3.77	4.20	4.51	4.75	4.95	5.12	5.27	5.39	5.51	5.61	5.71	5.80	5.88	5.95	6.02	6.09	6.15	6.21
13	3.06	3.73	4.15	4.45	4.69	4.88	5.05	5.19	5.32	5.43	5.53	5.63	5.71	5.79	5.86	5.93	5.99	6.05	6.11
14	3.03	3.70	4.11	4.41	4.64	4.83	4.99	5.13	5.25	5.36	5.46	5.55	5.64	5.71	5.79	5.85	5.91	5.97	6.03
15	3.01	3.67	4.08	4.37	4.59	4.78	4.94	5.08	5.20	5.31	5.40	5.49	5.57	5.65	5.72	5.78	5.85	5.90	5.96
16	3.00	3.65	4.05	4.33	4.56	4.74	4.90	5.03	5.15	5.26	5.35	5.44	5.52	5.59	5.66	5.73	5.79	5.84	5.90
17	2.98	3.63	4.02	4.30	4.52	4.70	4.86	4.99	5.11	5.21	5.31	5.39	5.47	5.54	5.61	5.67	5.73	5.79	5.84
18	2.97	3.61	4.00	4.28	4.49	4.67	4.82	4.96	5.07	5.17	5.27	5.35	5.43	5.50	5.57	5.63	5.69	5.74	5.79
19	2.96	3.59	3.98	4.25	4.47	4.65	4.79	4.92	5.04	5.14	5.23	5.31	5.39	5.46	5.53	5.59	5.65	5.70	5.75
20	2.95	3.58	3.96	4.23	4.45	4.62	4.77	4.90	5.01	5.11	5.20	5.28	5.36	5.43	5.49	5.55	5.61	5.66	5.71
24	2.92	3.53	3.90	4.17	4.37	4.54	4.68	4.81	4.92	5.01	5.10	5.18	5.25	5.32	5.38	5.44	5.49	5.55	5.59
30	2.89	3.49	3.85	4.10	4.30	4.46	4.60	4.72	4.82	4.92	5.00	5.08	5.15	5.21	5.27	5.33	5.38	5.43	5.47
40	2.86	3.44	3.79	4.04	4.23	4.39	4.52	4.63	4.73	4.82	4.90	4.98	5.04	5.11	5.16	5.22	5.27	5.31	5.36
60	2.83	3.40	3.74	3.98	4.16	4.31	4.44	4.55	4.65	4.73	4.81	4.88	4.94	5.00	5.06	5.11	5.15	5.20	5.24
120	2.80	3.36	3.68	3.92	4.10	4.24	4.36	4.47	4.56	4.64	4.71	4.78	4.84	4.90	4.95	5.00	5.04	5.09	5.13
∞	2.77	3.31	3.63	3.86	4.03	4.17	4.29	4.39	4.47	4.55	4.62	4.68	4.74	4.80	4.85	4.89	4.93	4.97	5.01

C

Table N *(concluded)*

α = 0.10

k \ v	2	3	4	5	6	7	8	9	10	11	12	13	14	15	16	17	18	19	20
1	8.93	13.44	16.36	18.49	20.15	21.51	22.64	23.62	24.48	25.24	25.92	26.54	27.10	27.62	28.10	28.54	28.96	29.35	29.71
2	4.13	5.73	6.77	7.54	8.14	8.63	9.05	9.41	9.72	10.01	10.26	10.49	10.70	10.89	11.07	11.24	11.39	11.54	11.68
3	3.33	4.47	5.20	5.74	6.16	6.51	6.81	7.06	7.29	7.49	7.67	7.83	7.98	8.12	8.25	8.37	8.48	8.58	8.68
4	3.01	3.98	4.59	5.03	5.39	5.68	5.93	6.14	6.33	6.49	6.65	6.78	6.91	7.02	7.13	7.23	7.33	7.41	7.50
5	2.85	3.72	4.26	4.66	4.98	5.24	5.46	5.65	5.82	5.97	6.10	6.22	6.34	6.44	6.54	6.63	6.71	6.79	6.86
6	2.75	3.56	4.07	4.44	4.73	4.97	5.17	5.34	5.50	5.64	5.76	5.87	5.98	6.07	6.16	6.25	6.32	6.40	6.47
7	2.68	3.45	3.93	4.28	4.55	4.78	4.97	5.14	5.28	5.41	5.53	5.64	5.74	5.83	5.91	5.99	6.06	6.13	6.19
8	2.63	3.37	3.83	4.17	4.43	4.65	4.83	4.99	5.13	5.25	5.36	5.46	5.56	5.64	5.72	5.80	5.87	5.93	6.00
9	2.59	3.32	3.76	4.08	4.34	4.54	4.72	4.87	5.01	5.13	5.23	5.33	5.42	5.51	5.58	5.66	5.72	5.79	5.85
10	2.56	3.27	3.70	4.02	4.26	4.47	4.64	4.78	4.91	5.03	5.13	5.23	5.32	5.40	5.47	5.54	5.61	5.67	5.73
11	2.54	3.23	3.66	3.96	4.20	4.40	4.57	4.71	4.84	4.95	5.05	5.15	5.23	5.31	5.38	5.45	5.51	5.57	5.63
12	2.52	3.20	3.62	3.92	4.16	4.35	4.51	4.65	4.78	4.89	4.99	5.08	5.16	5.24	5.31	5.37	5.44	5.49	5.55
13	2.50	3.18	3.59	3.88	4.12	4.30	4.46	4.60	4.72	4.83	4.93	5.02	5.10	5.18	5.25	5.31	5.37	5.43	5.48
14	2.49	3.16	3.56	3.85	4.08	4.27	4.42	4.56	4.68	4.79	4.88	4.97	5.05	5.12	5.19	5.26	5.32	5.37	5.43
15	2.48	3.14	3.54	3.83	4.05	4.23	4.39	4.52	4.64	4.75	4.84	4.93	5.01	5.08	5.15	5.21	5.27	5.32	5.38
16	2.47	3.12	3.52	3.80	4.03	4.21	4.36	4.49	4.61	4.71	4.81	4.89	4.97	5.04	5.11	5.17	5.23	5.28	5.33
17	2.46	3.11	3.50	3.78	4.00	4.18	4.33	4.46	4.58	4.68	4.77	4.86	4.93	5.01	5.07	5.13	5.19	5.24	5.30
18	2.45	3.10	3.49	3.77	3.98	4.16	4.31	4.44	4.55	4.65	4.75	4.83	4.90	4.98	5.04	5.10	5.16	5.21	5.26
19	2.45	3.09	3.47	3.75	3.97	4.14	4.29	4.42	4.53	4.63	4.72	4.80	4.88	4.95	5.01	5.07	5.13	5.18	5.23
20	2.44	3.08	3.46	3.74	3.95	4.12	4.27	4.40	4.51	4.61	4.70	4.78	4.85	4.92	4.99	5.05	5.10	5.16	5.20
24	2.42	3.05	3.42	3.69	3.90	4.07	4.21	4.34	4.44	4.54	4.63	4.71	4.78	4.85	4.91	4.97	5.02	5.07	5.12
30	2.40	3.02	3.39	3.65	3.85	4.02	4.16	4.28	4.38	4.47	4.56	4.64	4.71	4.77	4.83	4.89	4.94	4.99	5.03
40	2.38	2.99	3.35	3.60	3.80	3.96	4.10	4.21	4.32	4.41	4.49	4.56	4.63	4.69	4.75	4.81	4.86	4.90	4.95
60	2.36	2.96	3.31	3.56	3.75	3.91	4.04	4.16	4.25	4.34	4.42	4.49	4.56	4.62	4.67	4.73	4.78	4.82	4.86
120	2.34	2.93	3.28	3.52	3.71	3.86	3.99	4.10	4.19	4.28	4.35	4.42	4.48	4.54	4.60	4.65	4.69	4.74	4.78
∞	2.33	2.90	3.24	3.48	3.66	3.81	3.93	4.04	4.13	4.21	4.28	4.35	4.41	4.47	4.52	4.57	4.61	4.65	4.69

Source: "Tables of Range and Studentized Range," *Annals of Mathematical Statistics*, 31, no. 4. Reprinted with permission of the Institute of Mathematical Sciences.

Appendix D

Data Bank

Data Bank Values

The following list explains the values given for the categories in the Data Bank.

1. "Age" is given in years.
2. "Educational level" values are defined as follows:

   0 = no high school degree   2 = college graduate

   1 = high school graduate   3 = graduate degree
3. "Smoking status" values are defined as follows:

   0 = does not smoke

   1 = smokes less than one pack per day

   2 = smokes one or more than one pack per day
4. "Exercise" values are defined as follows:

   0 = none   2 = moderate

   1 = light   3 = heavy

5. "Weight" is given in pounds.
6. "Serum cholesterol" is given in milligram percent (mg%).
7. "Systolic pressure" is given in millimeters of mercury (mm Hg).
8. "IQ" is given in standard IQ test score values.
9. "Sodium" is given in milliequivalents per liter (mEq/1).
10. "Gender" is listed as male (M) or female (F).
11. "Marital status" values are defined as follows:

    M = married   S = single

    W = widowed   D = divorced

Data Bank

ID number	Age	Educational level	Smoking status	Exercise	Weight	Serum cholesterol	Systolic pressure	IQ	Sodium	Gender	Marital status
01	27	2	1	1	120	193	126	118	136	F	M
02	18	1	0	1	145	210	120	105	137	M	S
03	32	2	0	0	118	196	128	115	135	F	M
04	24	2	0	1	162	208	129	108	142	M	M
05	19	1	2	0	106	188	119	106	133	F	S
06	56	1	0	0	143	206	136	111	138	F	W
07	65	1	2	0	160	240	131	99	140	M	W
08	36	2	1	0	215	215	163	106	151	M	D
09	43	1	0	1	127	201	132	111	134	F	M
10	47	1	1	1	132	215	138	109	135	F	D

Data Bank *(concluded)*

ID number	Age	Educational level	Smoking status	Exercise	Weight	Serum cholesterol	Systolic pressure	IQ	Sodium	Gender	Marital status
11	48	3	1	2	196	199	148	115	146	M	D
12	25	2	2	3	109	210	115	114	141	F	S
13	63	0	1	0	170	242	149	101	152	F	D
14	37	2	0	3	187	193	142	109	144	M	M
15	40	0	1	1	234	208	156	98	147	M	M
16	25	1	2	1	199	253	135	103	148	M	S
17	72	0	0	0	143	288	156	103	145	F	M
18	56	1	1	0	156	164	153	99	144	F	D
19	37	2	0	2	142	214	122	110	135	M	M
20	41	1	1	1	123	220	142	108	134	F	M
21	33	2	1	1	165	194	122	112	137	M	S
22	52	1	0	1	157	205	119	106	134	M	D
23	44	2	0	1	121	223	135	116	133	F	M
24	53	1	0	0	131	199	133	121	136	F	M
25	19	1	0	3	128	206	118	122	132	M	S
26	25	1	0	0	143	200	118	103	135	M	M
27	31	2	1	1	152	204	120	119	136	M	M
28	28	2	0	0	119	203	118	116	138	F	M
29	23	1	0	0	111	240	120	105	135	F	S
30	47	2	1	0	149	199	132	123	136	F	M
31	47	2	1	0	179	235	131	113	139	M	M
32	59	2	2	0	206	260	151	99	143	M	W
33	36	2	1	0	191	201	148	118	145	M	D
34	59	0	1	1	156	235	142	100	132	F	W
35	35	1	0	0	122	232	131	106	135	F	M
36	29	2	0	2	175	195	129	121	148	M	M
37	43	3	0	3	194	211	138	129	146	M	M
38	44	1	2	0	132	240	130	109	132	F	S
39	63	2	2	1	188	255	156	121	145	M	M
40	36	2	1	1	125	220	126	117	140	F	S
41	21	1	0	1	109	206	114	102	136	F	M
42	31	2	0	2	112	201	116	123	133	F	M
43	57	1	1	1	167	213	141	103	143	M	W
44	20	1	2	3	101	194	110	111	125	F	S
45	24	2	1	3	106	188	113	114	127	F	D
46	42	1	0	1	148	206	136	107	140	M	S
47	55	1	0	0	170	257	152	106	130	F	M
48	23	0	0	1	152	204	116	95	142	M	M
49	32	2	0	0	191	210	132	115	147	M	M
50	28	1	0	1	148	222	135	100	135	M	M
51	67	0	0	0	160	250	141	116	146	F	W
52	22	1	1	1	109	220	121	103	144	F	M
53	19	1	1	1	131	231	117	112	133	M	S
54	25	2	0	2	153	212	121	119	149	M	D
55	41	3	2	2	165	236	130	131	152	M	M

Data Bank   *(concluded)*

ID number	Age	Educational level	Smoking status	Exercise	Weight	Serum cholesterol	Systolic pressure	IQ	Sodium	Gender	Marital status
56	24	2	0	3	112	205	118	100	132	F	S
57	32	2	0	1	115	187	115	109	136	F	S
58	50	3	0	1	173	203	136	126	146	M	M
59	32	2	1	0	186	248	119	122	149	M	M
60	26	2	0	1	181	207	123	121	142	M	S
61	36	1	1	0	112	188	117	98	135	F	D
62	40	1	1	0	130	201	121	105	136	F	D
63	19	1	1	1	132	237	115	111	137	M	S
64	37	2	0	2	179	228	141	127	141	F	M
65	65	3	2	1	212	220	158	129	148	M	M
66	21	1	2	2	99	191	117	103	131	F	S
67	25	2	2	1	128	195	120	121	131	F	S
68	68	0	0	0	167	210	142	98	140	M	W
69	18	1	1	2	121	198	123	113	136	F	S
70	26	0	1	1	163	235	128	99	140	M	M
71	45	1	1	1	185	229	125	101	143	M	M
72	44	3	0	0	130	215	128	128	137	F	M
73	50	1	0	0	142	232	135	104	138	F	M
74	63	0	0	0	166	271	143	103	147	F	W
75	48	1	0	3	163	203	131	103	144	M	M
76	27	2	0	3	147	186	118	114	134	M	M
77	31	3	1	1	152	228	116	126	138	M	D
78	28	2	0	2	112	197	120	123	133	F	M
79	36	2	1	2	190	226	123	121	147	M	M
80	43	3	2	0	179	252	127	131	145	M	D
81	21	1	0	1	117	185	116	105	137	F	S
82	32	2	1	0	125	193	123	119	135	F	M
83	29	2	1	0	123	192	131	116	131	F	D
84	49	2	2	1	185	190	129	127	144	M	M
85	24	1	1	1	133	237	121	114	129	M	M
86	36	2	0	2	163	195	115	119	139	M	M
87	34	1	2	0	135	199	133	117	135	F	M
88	36	0	0	1	142	216	138	88	137	F	M
89	29	1	1	1	155	214	120	98	135	M	S
90	42	0	0	2	169	201	123	96	137	M	D
91	41	1	1	1	136	214	133	102	141	F	D
92	29	1	1	0	112	205	120	102	130	F	M
93	43	1	1	0	185	208	127	100	143	M	M
94	61	1	2	0	173	248	142	101	141	M	M
95	21	1	1	3	106	210	111	105	131	F	S
96	56	0	0	0	149	232	142	103	141	F	M
97	63	0	1	0	192	193	163	95	147	M	M
98	74	1	0	0	162	247	151	99	151	F	W
99	35	2	0	1	151	251	147	113	145	F	M
100	28	2	0	3	161	199	129	116	138	M	M

D

Data Set I Record Temperatures

Record high temperatures by state in degrees Fahrenheit

112	100	128	120	134
118	105	114	106	109
112	100	118	117	116
118	121	114	114	105
109	107	112	114	115
118	117	118	125	106
110	122	108	110	121
113	120	119	111	104
111	120	113	120	117
105	110	118	112	114

Record low temperatures by state in degrees Fahrenheit

−27	−80	−40	−29	−45
−61	−32	−17	−66	−2
−17	12	−60	−35	−36
−47	−40	−34	−16	−48
−40	−35	−51	−59	−19
−40	−70	−47	−50	−46
−34	−50	−52	−34	−60
−39	−27	−54	−42	−23
−19	−58	−32	−23	−69
−50	−30	−48	−37	−54

Source: Reprinted with permission from the *World Almanac and Book of Facts 1996*. Copyright © 1995 K-III Reference Corporation. All rights reserved.

Data Set II Crime by Neighborhood

The data show the number of crimes per 1000 people in 88 neighborhoods located in the Pittsburgh, Pennsylvania, area. The crimes include murder, manslaughter, rape, robbery, aggravated assault, burglary, larceny-theft, vehicle theft, and arson. The rates have been rounded to the nearest whole number.

120	44	86	42	29	64
159	108	77	118	219	18
50	93	76	30	60	31
52	29	20	14	82	51
38	1900	57	48	49	138
17	19	49	13	56	169
34	89	43	57	105	26
24	436	69	467	22	22
48	37	137	65	57	
65	124	171	20	36	
87	42	48	20	26	
46	60	53	40	43	
32	83	76	58	12	
20	42	115	38	1758	
97	83	83	87	19	
30	80	54	85	18	

Source: Copyright *The Pittsburgh Post-Gazette*, September 29, 1996, all rights reserved. Reprinted with permission.

Data Set III Length of Major North American Rivers, in Miles

729	610	325	392	524
1459	450	465	605	330
923	906	329	290	1000
600	1450	862	1243	890
407	525	720	390	850
649	730	532	306	420
710	340	352	259	250
470	724	693	332	2340
560	1025	625	1171	3710
2315	2540	774	410	460
431	800	618	981	1310
500	790	605	1210	411
926	435	531	310	569
383	375	1290	545	445
1900	380	300	380	377
425	800	420	350	360
538	800	865	314	610
540	1038	424	313	447
301	659	652	886	525
360	512	500	722	309
336	430	692	1979	

Source: Reprinted with permission from the *World Almanac and Book of Facts 1996*. Copyright © 1995 K-III Reference Corporation. All rights reserved.

Data Set IV Heights of 130 Tallest Buildings in New York City, in Feet

1368	1362	1250	1414	1046
950	927	915	850	814
813	808	729	778	757
752	750	750	745	743
739	730	725	724	716
707	705	700	697	687
687	685	682	680	679
674	673	673	670	664
653	650	650	648	648
645	640	640	634	630
630	630	630	628	628
628	625	625	620	620
618	615	615	610	610
609	603	600	592	590
588	587	587	580	580
577	576	575	575	575
575	574	574	572	570
570	565	565	563	562
561	560	557	555	555
552	552	550	550	550
550	547	546	545	540
540	540	540	540	533
530	525	525	525	525
525	522	522	520	520
520	518	515	515	513
513	512	512	512	512

Source: Reprinted with permission from the *World Almanac and Book of Facts 1996*. Copyright © 1995 K-III Reference Corporation. All rights reserved.

Data Set V Life Expectancy by Region, 1995

African countries

48	47	59	55	44
70	47	46	47	51
42	44	56	46	56
47	48	52	54	52
66	62	63	63	52
68	61	61	59	63
46	48	56	45	44
52	55	46	48	47
53	49	43	55	41

Latin American countries

76	68	57	74	75	73
71	76	66	65	66	71
66	72	69	67	67	61
65	72	70	70	67	

North American countries

77	76

Asian countries

71	78	79	71	71
71	64	51	63	51
71	58	65	74	69
64	43	53	48	60
67	54	59	72	66
77	68	75	69	70
69	67	67	71	53

European countries

72	73	70	72	70	72
76	76	71	76	75	76
71	73	77	78	76	
73	78	77	75	78	
77	78	76	77	76	

Oceanic countries

77	76	59	56

Former Soviet Union countries

71	70	72	72	69
68	68	70	70	65
71				

Source: *The Universal Almanac 1995*, pp. 356–57.

Data Set VI Acreage of U.S. National Parks, in Thousands of Acres

41	66	233	775	169
36	338	223	46	64
183	4724	61	1449	7075
1013	3225	1181	308	77
520	77	27	217	5
539	3575	650	462	1670
2574	106	52	52	236
505	913	94	75	265
402	196	70	13	132
28	7656	2220	760	143

Source: *The Universal Almanac 1995*, p. 45.

Data Set VII Number of Franchises for the Top 50 Franchisers, 1993

10,604	9,770	8,013	5,903	5,891
5,304	4,948	4,131	3,793	3,557
3,523	3,389	3,339	3,162	3,019
2,880	2,821	2,735	2,713	2,571
2,515	2,335	2,208	2,106	2,013
1,737	1,575	1,448	1,314	1,296
1,129	1,105	1,058	1,051	1,017
989	960	954	930	925
898	895	884	843	837
833	816	816	801	768

Source: *The Universal Almanac 1995*, p. 277.

Data Set VIII Oceans of the World

Ocean	Area (thousands of square miles)	Maximum depth (feet)
Arctic	5,400	17,881
Caribbean Sea	1,063	25,197
Mediterranean Sea	967	16,470
Norwegian Sea	597	13,189
Gulf of Mexico	596	14,370
Hudson Bay	475	850
Greenland Sea	465	15,899
North Sea	222	2,170
Black Sea	178	7,360
Baltic Sea	163	1,440
Atlantic Ocean	31,830	30,246
South China Sea	1,331	18,241
Sea of Okhotsk	610	11,063
Bering Sea	876	13,750
Sea of Japan	389	12,280
East China Sea	290	9,126
Yellow Sea	161	300
Pacific Ocean	63,800	36,200
Arabian Sea	1,492	19,029
Bay of Bengal	839	17,251
Red Sea	169	7,370
Indian Ocean	28,360	24,442

Source: *The Universal Almanac 1995*, p. 330.

Data Set IX Population of the States for 1995, in Millions

AL	4.25	FL	14.17	LA	3.43	NE	1.64
AK	0.60	GA	7.20	ME	1.24	NV	1.53
AZ	4.22	HI	1.19	MD	5.04	NH	1.15
AR	2.48	ID	1.16	MA	6.07	NJ	18.14
CA	31.59	IL	11.83	MI	9.55	NM	1.69
CO	3.75	IN	5.80	MN	4.61	NY	18.14
CT	3.28	IA	2.84	MS	2.70	NC	7.20
DE	0.72	KS	2.57	MO	5.32	ND	0.64
DC	0.55	KY	3.86	MT	0.87	OH	11.15
OK	3.28	SC	3.67	UT	1.95	WV	1.83

OR	3.14	SD	0.73	VT	0.59	WI	5.12
PA	12.07	TN	5.26	VA	6.62	WY	0.48
RI	0.99	TX	18.72	WA	5.43		

Source: *USA Today,* October 23, 1996.

Data Set X Child Support Collections for 1996, in Millions of Dollars

AL	$158	IL	244	MT	29	RI	35
AK	59	IN	193	NE	100	SC	118
AZ	112	IA	150	NV	56	SD	27
AR	79	KS	103	NH	48	TN	166
CA	903	KY	143	NJ	501	TX	546
CO	105	LA	144	NM	27	UT	75
CT	124	ME	63	NY	691	VT	25
DE	36	MD	289	NC	260	VA	258
DC	27	MA	249	ND	28	WA	410
FL	407	MI	967	OH	972	WV	84
GA	260	MN	320	OK	74	WI	444
HI	54	MS	84	OR	179	WY	24
ID	44	MO	264	PA	972		

Source: *USA Today,* October 23, 1996. Copyright 1996. Reprinted with permission.

Data Set XI Commuter and Rapid Rail Systems in the United States

System	Stations	Miles	Vehicles operated
Long Island RR	134	638.2	947
N.Y. Metro North	108	535.9	702
New Jersey Transit	158	926.0	582
Chicago RTA	117	417.0	358
Chicago & NW Transit	62	309.4	277
SEPTA Philadelphia	181	442.8	263
Boston Amtrak/MBTA	101	529.8	291
Chicago, Burlington, Northern	27	75.0	139
NW Indiana CTD	18	134.8	39
New York City TA	469	492.9	4923
Washington Metro Area TA	70	162.1	534
Metro Boston TA	53	76.7	368
Chicago TA	137	191.0	924
Philadelphia SEPTA	76	75.8	300
San Francisco BART	34	142.0	415
Metro Atlantic RTA	29	67.0	136
New York PATH	13	28.6	282
Miami/Dade Co TA	21	42.2	82
Baltimore MTA	12	26.6	48
Philadelphia PATCO	13	31.5	102
Cleveland RTA	18	38.2	30
New York, Staten Island RT	22	28.6	36

Source: *The Universal Almanac 1995,* p. 287.

Data Set XII Deaths Related to Weather

Year	1993	1992	1991	1990	1989	1988
Lightning	39	41	73	74	67	68
Tornadoes	30	39	39	53	50	32
Winds	19	13	32	39	30	18
Extreme heat	16	14	13	13	121	17
Extreme cold	11	8	36	32	6	41
Flash floods	46	55	45	109	62	30
River floods	36	7	16	33	23	1
Hurricanes	1	27	13	0	38	9
Snow/blizzards	54	43	37	35	56	48
Ice storms	7	16	8	13	7	7
High winds	38	15	32	32	12	22

Source: *The USA Weather Almanac 1995,* p. 90. Copyright 1995, reprinted with permission.

Data Set XIII Keystone Jackpot Analysis*

Ball	Times drawn	Ball	Times drawn	Ball	Times drawn
1	11	12	10	23	7
2	5	13	11	24	8
3	10	14	5	25	13
4	11	15	8	24	11
5	7	16	14	27	7
6	13	17	8	28	10
7	8	18	11	29	11
8	10	19	10	30	5
9	16	20	7	31	7
10	12	21	11	32	8
11	10	22	6	33	11

*Times each number has been selected in the regular drawings of the Pennsylvania Lottery since November 4, 1995.

Source: Copyright *Pittsburgh Post-Gazette,* October 31, 1996, all rights reserved. Reprinted with permission.

Data Set XIV Weights of Professional Football Players, in Pounds

Pittsburgh Steelers

215	202	212	216	201
195	196	193	193	200
180	185	233	243	210
240	196	203	213	246
246	235	226	224	250
287	288	297	295	330
305	290	335	302	282
300	187	193	202	208
245	248	258	190	192
311	293	245	300	240
305	298	264		

St. Louis Rams

210	220	220	210	225
185	230	199	206	185
185	212	225	215	250
209	184	189	229	200

235	230	225	242	226
242	239	295	300	345
315	311	300	305	315
306	280	304	305	186
260	196	255	255	195
195	270	280	295	290
285				

Source: *The Daily News,* November 2, 1996. Courtesy of *The Daily News,* McKeesport, PA.

Data Set XV   Earthquakes Measured on the Richter Scale for a Selected Sample during 1996

Western Hemisphere

3.4	4.5	5.1	5.8	4.3
5.1	5.8	5.6	2.5	6.3
5.7	4.8	4.9	5.3	5.3
3.8	5.1	6.1	2.8	5.3
5.0	2.0	6.4	2.9	5.5

Eastern Hemisphere

4.2	3.3	5.3	4.1	5.5	4.2	5.4
3.5	4.7	5.7	4.0	4.8	6.0	5.8
5.2	3.7	5.7	4.1	6.4	4.0	4.1
4.0	5.8	4.2	4.7	5.5	6.6	5.5
5.0	3.2	6.4	4.6	4.9	4.1	4.5
5.0	4.6	5.7	4.3	3.2	6.2	5.7
5.2	4.8	5.9	5.7	6.6	5.9	5.1
3.8	4.8	4.8	3.8	4.3	6.8	5.3
5.7	6.3	4.0	6.2	5.6	5.4	4.5
4.4	3.7	5.0	5.8	4.8	4.3	

Source: *Pittsburgh Tribune-Review,* selected weeks during 1996. Used with permission.

Data Set XVI   Hospital Data*

Number	Number of beds	Admissions	Payroll ($000)	Personnel
1	235	6,559	18,190	722
2	205	6,237	17,603	692
3	371	8,915	27,278	1,187
4	342	8,659	26,722	1,156
5	61	1,779	5,187	237
6	55	2,261	7,519	247
7	109	2,102	5,817	245
8	74	2,065	5,418	223
9	74	3,204	7,614	326
10	137	2,638	7,862	362
11	428	18,168	70,518	2,461
12	260	12,821	40,780	1,422
13	159	4,176	11,376	465
14	142	3,952	11,057	450
15	45	1,179	3,370	145
16	42	1,402	4,119	211
17	92	1,539	3,520	158

Data Set XVI   Hospital Data*   *(continued)*

Number	Number of beds	Admissions	Payroll ($000)	Personnel
18	28	503	1,172	72
19	56	1,780	4,892	195
20	68	2,072	6,161	243
21	206	9,868	30,995	1,142
22	93	3,642	7,912	305
23	68	1,558	3,929	180
24	330	7,611	33,377	1,116
25	127	4,716	13,966	498
26	87	2,432	6,322	240
27	577	19,973	60,934	1,822
28	310	11,055	31,362	981
29	49	1,775	3,987	180
30	449	17,929	53,240	1,899
31	530	15,423	50,127	1,669
32	498	15,176	49,375	1,549
33	60	565	5,527	251
34	350	11,793	34,133	1,207
35	381	13,133	49,641	1,731
36	585	22,762	71,232	2,608
37	286	8,749	28,645	1,194
38	151	2,607	12,737	377
39	98	2,518	10,731	352
40	53	1,848	4,791	185
41	142	3,658	11,051	421
42	73	3,393	9,712	385
43	624	20,410	72,630	2,326
44	78	1,107	4,946	139
45	85	2,114	4,522	221
46	120	3,435	11,479	417
47	84	1,768	4,360	184
48	667	22,375	74,810	2,461
49	36	1,008	2,311	131
50	598	21,259	113,972	4,010
51	1,021	40,879	165,917	6,264
52	233	4,467	22,572	558
53	205	4,162	21,766	527
54	80	469	8,254	280
55	350	7,676	58,341	1,525
56	290	7,499	57,298	1,502
57	890	31,812	134,752	3,933
58	880	31,703	133,836	3,914
59	67	2,020	8,533	280
60	317	14,595	68,264	2,772
61	123	4,225	12,161	504
62	285	7,562	25,930	952
63	51	1,932	6,412	472
64	34	1,591	4,393	205
65	194	5,111	19,367	753
66	191	6,729	21,889	946
67	227	5,862	18,285	731
68	172	5,509	17,222	680

D

Data Set XVI   Hospital Data*   *(continued)*

Number	Number of beds	Admissions	Payroll ($000)	Personnel
69	285	9,855	27,848	1,180
70	230	7,619	29,147	1,216
71	206	7,368	28,592	1,185
72	102	3,255	9,214	359
73	76	1,409	3,302	198
74	540	396	22,327	788
75	110	3,170	9,756	409
76	142	4,984	13,550	552
77	380	335	11,675	543
78	256	8,749	23,132	907
79	235	8,676	22,849	883
80	580	1,967	33,004	1,059
81	86	2,477	7,507	309
82	102	2,200	6,894	225
83	190	6,375	17,283	618
84	85	3,506	8,854	380
85	42	1,516	3,525	166
86	60	1,573	15,608	236
87	485	16,676	51,348	1,559
88	455	16,285	50,786	1,537
89	266	9,134	26,145	939
90	107	3,497	10,255	431
91	122	5,013	17,092	589
92	36	519	1,526	80
93	34	615	1,342	74
94	37	1,123	2,712	123
95	100	2,478	6,448	265
96	65	2,252	5,955	237
97	58	1,649	4,144	203
98	55	2,049	3,515	152
99	109	1,816	4,163	194
100	64	1,719	3,696	167
101	73	1,682	5,581	240

Data Set XVI   Hospital Data*   *(concluded)*

Number	Number of beds	Admissions	Payroll ($000)	Personnel
102	52	1,644	5,291	222
103	326	10,207	29,031	1,074
104	268	10,182	28,108	1,030
105	49	1,365	4,461	215
106	52	763	2,615	125
107	106	4,629	10,549	456
108	73	2,579	6,533	240
109	163	201	5,015	260
110	32	34	2,880	124
111	385	14,553	52,572	1,724
112	95	3,267	9,928	366
113	339	12,021	54,163	1,607
114	50	1,548	3,278	156
115	55	1,274	2,822	162
116	278	6,323	15,697	722
117	298	11,736	40,610	1,606
118	136	2,099	7,136	255
119	97	1,831	6,448	222
120	369	12,378	35,879	1,312
121	288	10,807	29,972	1,263
122	262	10,394	29,408	1,237
123	94	2,143	7,593	323
124	98	3,465	9,376	371
125	136	2,768	7,412	390
126	70	824	4,741	208
127	35	883	2,505	142
128	52	1,279	3,212	158

*This information was obtained from a sample of hospitals in a selected state. The hospitals are identified by number instead of name.

Source: *American Hospital Association Guide to Health Care Field,* 1993 edition, copyright by the American Hospital Association.

Appendix E

Glossary

adjusted R^2 used in multiple regression when n and k are approximately equal to provide a more realistic value of R^2

alpha the probability of a type I error, represented by the Greek letter α

alternative hypothesis a statistical hypothesis that states a difference between a parameter and a specific value or states that there is a difference between two parameters

analysis of variance (ANOVA) a statistical technique used to test a hypothesis concerning the means of three or more populations

ANOVA summary table the table used to summarize the results of an ANOVA test

Bayes's theorem a theorem that allows one to compute the revised probability of an event that occurred before another event when the events are dependent

beta the probability of a type II error, represented by the Greek letter β

between-group variance a variance estimate using the means of the groups or between the groups in an F test

biased sample a sample for which some type of systematic error has been made in the selection of subjects for the sample

binomial distribution the outcomes of a binomial experiment and the corresponding probabilities of these outcomes

binomial experiment a probability experiment in which each trial has only two outcomes, there is a fixed number of trials, the outcomes of the trials are independent, and the probability of success remains the same for each trial

boxplot a graph used to represent a data set when the data set contains a small number of values

categorical frequency distribution a frequency distribution used when the data are categorical (nominal)

central limit theorem a theorem that states that as the sample size increases, the shape of the distribution of the sample means taken from the population with mean μ and standard deviation σ will approach a normal distribution; the distribution will have a mean μ and a standard deviation σ

Chebyshev's theorem a theorem that states that the proportion of values from a data set that fall within k standard deviations of the mean will be at least $1 - 1/k^2$, where k is a number greater than 1

chi-square distribution a probability distribution obtained from the values of $(n - 1)s^2/\sigma^2$ when random samples are selected from a normally distributed population whose variance is σ^2

class boundaries the upper and lower values of a class for a grouped frequency distribution whose values have one additional decimal place more than the data and end in the digit 5

class midpoint a value for a class in a frequency distribution obtained by adding the lower and upper class boundaries (or the lower and upper class limits) and dividing by 2

class width the difference between the upper class boundary and the lower class boundary for a class in a frequency distribution

classical probability the type of probability that uses sample spaces to determine the numerical probability that an event will happen

cluster sample a sample obtained by selecting a preexisting or natural group, called a cluster, and using the members in the cluster for the sample

coefficient of determination a measure of the variation of the dependent variable that is explained by the regression line and the independent variable; the ratio of the explained variation to the total variation

coefficient of variation the standard deviation divided by the mean; the result is expressed as a percentage

combination a selection of objects without regard to order

complement of an event the set of outcomes in the sample space that are not in the outcomes of the event itself

compound event an event that consists of two or more outcomes or simple events

conditional probability the probability that an event B occurs after an event A has already occurred

confidence interval   a specific interval estimate of a parameter determined by using data obtained from a sample and the specific confidence level of the estimate

confidence level   the probability that a parameter will fall within the specified interval estimate of the parameter

confounding variable   a variable that influences the outcome variable but cannot be separated from the other variables that influence the outcome variable

consistent estimator   an estimator whose value approaches the value of the parameter estimated as the sample size increases

contingency table   data arranged in table form for the chi-square independence test, with R rows and C columns

continuous variable   a variable that can assume all values between any two specific values; a variable obtained by measuring

control group   a group in an experimental study that is not given any special treatment

convenience sample   sample of subjects used because they are convenient and available

correction for continuity   a correction employed when a continuous distribution is used to approximate a discrete distribution

correlation   a statistical method used to determine whether a relationship exists between variables

correlation coefficient   a statistic or parameter that measures the strength and direction of a relationship between two variables

critical or **rejection region**   the range of values of the test value that indicates that there is a significant difference and the null hypothesis should be rejected in a hypothesis test

critical value (C.V.)   a value that separates the critical region from the noncritical region in a hypothesis test

cumulative frequency   the sum of the frequencies accumulated up to the upper boundary of a class in a frequency distribution

data   measurements or observations for a variable

data array   a data set that has been ordered

data set   a collection of data values

data value or **datum**   a value in a data set

decile   a location measure of a data value; it divides the distribution into 10 groups

degrees of freedom   the number of values that are free to vary after a sample statistic has been computed; used when a distribution (such as the t distribution) consists of a family of curves

dependent events   events for which the outcome or occurrence of the first event affects the outcome or occurrence of the second event in such a way that the probability is changed

dependent samples   samples in which the subjects are paired or matched in some way; i.e., the samples are related

dependent variable   a variable in correlation and regression analysis that cannot be controlled or manipulated

descriptive statistics   a branch of statistics that consists of the collection, organization, summarization, and presentation of data

discrete variable   a variable that assumes values that can be counted

disordinal interaction   an interaction between variables in ANOVA, indicated when the graphs of the lines connecting the mean intersect

distribution-free statistics   *see* nonparametric statistics

double sampling   a sampling method in which a very large population is given a questionnaire to determine those who meet the qualifications for a study; the questionnaire is reviewed, a second smaller population is defined, and a sample is selected from this group

empirical probability   the type of probability that uses frequency distributions based on observations to determine numerical probabilities of events

empirical rule   a rule that states that when a distribution is bell-shaped (normal), approximately 68% of the data values will fall within one standard deviation of the mean; approximately 95% of the data values will fall within two standard deviations of the mean; and approximately 99.7% of the data values will fall within three standard deviations of the mean

equally likely events   the events in the sample space that have the same probability of occurring

estimation   the process of estimating the value of a parameter from information obtained from a sample

estimator   a statistic used to estimate a parameter

event   outcomes of a probability experiment

expected frequency   the frequency obtained by calculation (as if there were no preference) and used in the chi-square test

expected value   the theoretical average of a variable that has a probability distribution

experimental study   a study in which the researcher manipulates one of the variables and tries to determine how the manipulation influences other variables

explanatory variable   a variable that is being manipulated by the researcher in order to see if it affects the outcome variable

exploratory data analysis   the act of analyzing data to determine what information can be obtained using stem and leaf plots, medians, interquartile ranges, and box plots

F distribution   the sampling distribution of the variances when two independent samples are selected from two normally distributed populations in which the variances are equal and the variances s_1^2 and s_2^2 are compared as $s_1^2 \div s_2^2$

F test   a statistical test used to compare two variances or three or more means

factors   the independent variables in ANOVA tests

finite population correction factor   a correction factor used to correct the standard error of the mean when the sample size is greater than 5% of the population size

five-number summary   five specific values for a data set that consist of the lowest and highest values, Q_1 and Q_3, and the median

frequency   the number of values in a specific class of a frequency distribution

frequency distribution   an organization of raw data in table form, using classes and frequencies

frequency polygon   a graph that displays the data by using lines that connect points plotted for the frequencies at the midpoints of the classes

goodness-of-fit test   a chi-square test used to see whether a frequency distribution fits a specific pattern

grouped frequency distribution   a distribution used when the range is large and classes of several units in width are needed

Hawthorne effect   an effect on an outcome variable caused by the fact that subjects of the study know that they are participating in the study

histogram   a graph that displays the data by using vertical bars of various heights to represent the frequencies of a distribution

homogeneity of proportions test   a test used to determine the equality of three or more proportions

hypergeometric distribution   the distribution of a variable that has two outcomes when sampling is done without replacement

hypothesis testing   a decision-making process for evaluating claims about a population

independence test   a chi-square test used to test the independence of two variables when data are tabulated in table form in terms of frequencies

independent events   events for which the probability of the first occurring does not affect the probability of the second occurring

independent samples   samples that are not related

independent variable   a variable in correlation and regression analysis that can be controlled or manipulated

inferential statistics   a branch of statistics that consists of generalizing from samples to populations, performing

hypothesis testing, determining relationships among variables, and making predictions

interaction effect   the effect of two or more variables on each other in a two-way ANOVA study

interquartile range $Q_3 - Q_1$

interval estimate   a range of values used to estimate a parameter

interval level of measurement   a measurement level that ranks data and in which precise differences between units of measure exist. *See also* nominal, ordinal, and ratio levels of measurement

Kruskal-Wallis test   a nonparametric test used to compare three or more means

law of large numbers   when a probability experiment is repeated a large number of times, the relative frequency probability of an outcome will approach its theoretical probability

left-tailed test   a test used on a hypothesis when the critical region is on the left side of the distribution

level   a treatment in ANOVA for a variable

level of significance   the maximum probability of committing a type I error in hypothesis testing

lower class limit   the lower value of a class in a frequency distribution that has the same decimal place value as the data

main effect   the effect of the factors or independent variables when there is a nonsignificant interaction effect in a two-way ANOVA study

maximum error of estimate   the maximum likely difference between the point estimate of a parameter and the actual value of the parameter

mean   the sum of the values divided by the total number of values

mean square   the variance found by dividing the sum of the squares of a variable by the corresponding degrees of freedom; used in ANOVA

measurement scales   a type of classification that tells how variables are categorized, counted, or measured; the four types of scales are nominal, ordinal, interval, and ratio

median   the midpoint of a data array

midrange   the sum of the lowest and highest data values divided by 2

modal class   the class with the largest frequency

mode   the value that occurs most often in a data set

Monte Carlo method   a simulation technique using random numbers

multinomial distribution   a probability distribution for an experiment in which each trial has more than two outcomes

E

multiple correlation coefficient a measure of the strength of the relationship between the independent variables and the dependent variable in a multiple regression study

multiple regression a study that seeks to determine if several independent variables are related to a dependent variable

multiple relationship a relationship in which many variables are under study

multistage sampling a sampling technique that uses a combination of sampling methods

mutually exclusive events probability events that cannot occur at the same time

negative relationship a relationship between variables such that as one variable increases, the other variable decreases, and vice versa

negatively skewed or **left skewed distribution** a distribution in which the majority of the data values fall to the right of the mean

nominal level of measurement a measurement level that classifies data into mutually exclusive (nonoverlapping) exhaustive categories in which no order or ranking can be imposed on them. *See also* interval, ordinal, and ratio levels of measurement

noncritical or **nonrejection region** the range of values of the test value that indicates that the difference was probably due to chance and the null hypothesis should not be rejected

nonparametric statistics a branch of statistics for use when the population from which the samples are selected is not normally distributed and for use in testing hypotheses that do not involve specific population parameters

nonrejection region *see* noncritical region

normal distribution a continuous, symmetric, bell-shaped distribution of a variable

null hypothesis a statistical hypothesis that states that there is no difference between a parameter and a specific value or that there is no difference between two parameters

observational study a study in which the researcher merely observes what is happening or what has happened in the past and draws conclusions based on these observations

observed frequency the actual frequency value obtained from a sample and used in the chi-square test

ogive a graph that represents the cumulative frequencies for the classes in a frequency distribution

one-tailed test a test that indicates that the null hypothesis should be rejected when the test statistic value is in the critical region on one side of the mean

one-way ANOVA a study used to test for differences among means for a single independent variable when there are three or more groups

open-ended distribution a frequency distribution that has no specific beginning value or no specific ending value

ordinal interaction an interaction between variables in ANOVA, indicated when the graphs of the lines connecting the means do not intersect

ordinal level of measurement a measurement level that classifies data into categories that can be ranked; however, precise differences between the ranks do not exist. *See also* interval, nominal, and ratio levels of measurement

outcome the result of a single trial of a probability experiment

outcome variable a variable which is studied in order to see if it has changed significantly due to the manipulation of the explanatory variable

outlier an extreme value in a data set; it is omitted from a boxplot

parameter a characteristic or measure obtained by using all the data values for a specific population

parametric tests statistical tests for population parameters such as means, variances, and proportions that involve assumptions about the populations from which the samples were selected

Pareto chart chart that uses vertical bars to represent frequencies for a categorical variable

Pearson product moment correlation coefficient (PPMCC) a statistic used to determine the strength of a relationship when the variables are normally distributed

percentile a location measure of a data value; it divides the distribution into 100 groups

permutation an arrangement of n objects in a specific order

pie graph a circle that is divided into sections or wedges according to the percentage of frequencies in each category of the distribution

point-biserial correlation coefficient a statistic used to determine the strength of a relationship when one variable is a true dichotomous variable and the other variable is continuous and normally distributed

point estimate a specific numerical value estimate of a parameter

Poisson distribution a probability distribution used when n is large and p is small and when the independent variables occur over a period of time

pooled estimate of the variance a weighted average of the variance using the two sample variances and their respective degrees of freedom as the weights

population the totality of all subjects possessing certain common characteristics that are being studied

population correlation coefficient the value of the correlation coefficient computed by using all possible pairs of data values (x, y) taken from a population

positive relationship   a relationship between two variables such that as one variable increases, the other variable increases or as one variable decreases, the other decreases

positively skewed or **right skewed distribution**   a distribution in which the majority of the data values fall to the left of the mean

power of a test   the probability of rejecting the null hypothesis when it is false

prediction interval   a confidence interval for a predicted value y

probability   the chance of an event occurring

probability distribution   the values a random variable can assume and the corresponding probabilities of the values

probability experiment   a chance process that leads to well-defined results called outcomes

proportion   a part of a whole, represented by a fraction, a decimal, or a percentage

***P*-value**   the actual probability of getting the sample mean value if the null hypothesis is true

qualitative variable   a variable that can be placed into distinct categories, according to some characteristic or attribute

quantitative variable   a variable that is numerical in nature and that can be ordered or ranked

quartile   a location measure of a data value; it divides the distribution into four groups

quasi-experimental study   a study that uses intact groups rather than random assignment of subjects to groups

random sample   a sample obtained by using random or chance methods; a sample for which every member of the population has an equal chance of being selected

random variable   a variable whose values are determined by chance

random variation   *see* chance variation

range   the highest data value minus the lowest data value

ranking   the positioning of a data value in a data array according to some rating scale

ratio level of measurement   a measurement level that possesses all the characteristics of interval measurement and a true zero; it also has true ratios between different units of measure. *See also* interval, nominal, and ordinal levels of measurement

raw data   data collected in original form

regression   a statistical method used to describe the nature of the relationship between variables—i.e., a positive or negative, linear or nonlinear relationship

regression line   the line of best fit of the data

rejection region   *see* critical region

relative frequency graph   a graph using proportions instead of raw data as frequencies

relatively efficient estimator   an estimator that has the smallest variance from among all the statistics that can be used to estimate a parameter

resistant statistic   a statistic that is not affected by an extremely skewed distribution

right-tailed test   a test used on a hypothesis when the critical region is on the right side of the distribution

run   a succession of identical letters preceded by or followed by a different letter or no letter at all, such as the beginning or end of the succession

runs test   a nonparametric test used to determine whether data are random

sample   a group of subjects selected from the population

sample space   the set of all possible outcomes of a probability experiment

sampling distribution of sample means   a distribution obtained by using the means computed from random samples taken from a population

sampling error   the difference between the sample measure and the corresponding population measure due to the fact that the sample is not a perfect representation of the population

scatter plot   a graph of the independent and dependent variables in regression and correlation analysis

Scheffé test   a test used after ANOVA, if the null hypothesis is rejected, to locate significant differences in the means

sequence sampling   a sampling technique used in quality control in which successive units are taken from production lines and tested to see whether they meet the standards set by the manufacturing company

sign test   a nonparametric test used to test the value of the median for a specific sample or to test sample means in a comparison of two dependent samples

simple event   an outcome that results from a single trial of a probability experiment

simple relationship   a relationship in which only two variables are under study

simulation techniques   techniques that use probability experiments to mimic real-life situations

Spearman rank correlation coefficient   the nonparametric equivalent to the correlation coefficient, used when the data are ranked

standard deviation   the square root of the variance

standard error of estimate   the standard deviation of the observed y values about the predicted y' values in regression and correlation analysis

standard error of the mean   the standard deviation of the sample means for samples taken from the same population

standard normal distribution   a normal distribution for which the mean is equal to 0 and the standard deviation is equal to 1

standard score the difference between a data value and the mean divided by the standard deviation

statistic a characteristic or measure obtained by using the data values from a sample

statistical hypothesis a conjecture about a population parameter, which may or may not be true

statistical test a test that uses data obtained from a sample to make a decision about whether or not the null hypothesis should be rejected

statistics the science of conducting studies to collect, organize, summarize, analyze, and draw conclusions from data

stem and leaf plot a data plot that uses part of a data value as the stem and part of the data value as the leaf to form groups or classes

stratified sample a sample obtained by dividing the population into subgroups, called strata, according to various homogeneous characteristics and then selecting members from each stratum

subjective probability the type of probability that uses a probability value based on an educated guess or estimate, employing opinions and inexact information

sum of squares between groups a statistic computed in the numerator of the fraction used to find the between-group variance in ANOVA

sum of squares within groups a statistic computed in the numerator of the fraction used to find the within-group variance in ANOVA

symmetrical distribution a distribution in which the data values are uniformly distributed about the mean

systematic sample a sample obtained by numbering each element in the population and then selecting every kth number from the population to be included in the sample

t **distribution** a family of bell-shaped curves based on degrees of freedom, similar to the standard normal distribution with the exception that the variance is greater than 1; used when one is testing small samples and when the population standard deviation is unknown

t **test** a statistical test for the mean of a population, used when the population is normally distributed, the population standard deviation is unknown, and the sample size is less than 30

test value the numerical value obtained from a statistical test, computed from (observed value − expected value) ÷ standard error

time series graph a graph that represents data that occur over a specific period of time

treatment group a group in an experimental study that has received some type of treatment

treatment groups the groups used in an ANOVA study

tree diagram a device used to list all possibilities of a sequence of events in a systematic way

Tukey test a test used to make pairwise comparisons of means in an ANOVA study when samples are the same size

two-tailed test a test that indicates that the null hypothesis should be rejected when the test value is in either of the two critical regions

two-way ANOVA a study used to test the effects of two or more independent variables and the possible interaction between them

type I error the error that occurs if one rejects the null hypothesis when it is true

type II error the error that occurs if one does not reject the null hypothesis when it is false

unbiased estimator an estimator whose value approximates the expected value of a population parameter, used for the variance or standard deviation when the sample size is less than 30; an estimator whose expected value or mean must be equal to the mean of the parameter being estimated

unbiased sample a sample chosen at random from the population that is, for the most part, representative of the population

ungrouped frequency distribution a distribution that uses individual data and has a small range of data

upper class limit the upper value of a class in a frequency distribution that has the same decimal place value as the data

variable a characteristic or attribute that can assume different values

variance the average of the squares of the distance each value is from the mean

Venn diagram a diagram used as a pictorial representative for a probability concept or rule

weighted mean the mean found by multiplying each value by its corresponding weight and dividing by the sum of the weights

Wilcoxon rank sum test a nonparametric test used to test independent samples and compare distributions

Wilcoxon signed-rank test a nonparametric test used to test dependent samples and compare distributions

within-group variance a variance estimate using all the sample data for an F test; it is not affected by differences in the means

z **distribution** *see* standard normal distribution

z **score** *see* standard score

z **test** a statistical test for means and proportions of a population, used when the population is normally distributed and the population standard deviation is known or the sample size is 30 or more

z **value** same as z score

Appendix F

Bibliography

Aczel, Amir D. *Complete Business Statistics,* 3rd ed. Chicago: Irwin, 1996.

Beyer, William H. *CRC Handbook of Tables for Probability and Statistics,* 2nd ed. Boca Raton, Fla.: CRC Press, 1986.

Brase, Charles, and Corrinne P. Brase. *Understanding Statistics,* 5th ed. Lexington, Mass.: D.C. Heath, 1995.

Chao, Lincoln L. *Introduction to Statistics.* Monterey, Calif.: Brooks/Cole, 1980.

Daniel, Wayne W., and James C. Terrell. *Business Statistics,* 4th ed. Boston: Houghton Mifflin, 1986.

Edwards, Allan L. *An Introduction to Linear Regression and Correlation,* 2nd ed. New York: Freeman, 1984.

Eves, Howard. *An Introduction to the History of Mathematics,* 3rd ed. New York: Holt, Rinehart and Winston, 1969.

Famighetti, Robert, ed. *The World Almanac and Book of Facts 1996.* New York: Pharos Books, 1995.

Freund, John E., and Gary Simon. *Statistics—A First Course,* 6th ed. Englewood Cliffs, N.J.: Prentice Hall, 1995.

Gibson, Henry R. *Elementary Statistics.* Dubuque, Iowa: Wm. C. Brown Publishers, 1994.

Glass, Gene V., and Kenneth D. Hopkins. *Statistical Methods in Education and Psychology,* 2nd ed. Englewood Cliffs, N.J.: Prentice Hall, 1984.

Guilford, J. P. *Fundamental Statistics in Psychology and Education,* 4th ed. New York: McGraw-Hill, 1965.

Haack, Dennis G. *Statistical Literacy: A Guide to Interpretation.* Boston: Duxbury Press, 1979.

Hartwig, Frederick, with Brian Dearing. *Exploratory Data Analysis.* Newbury Park, Calif.: Sage Publications, 1979.

Henry, Gary T. *Graphing Data: Techniques for Display and Analysis.* Thousand Oaks, Calif.: Sage Publications, 1995.

Isaac, Stephen, and William B. Michael. *Handbook in Research and Evaluation,* 2nd ed. San Diego: EdITS, 1990.

Johnson, Robert. *Elementary Statistics,* 6th ed. Boston: PWS–Kent, 1992.

Kachigan, Sam Kash. *Statistical Analysis.* New York: Radius Press, 1986.

Khazanie, Ramakant. *Elementary Statistics in a World of Applications,* 3rd ed. Glenview, Ill.: Scott, Foresman, 1990.

Kuzma, Jan W. *Basic Statistics for the Health Sciences.* Mountain View, Calif.: Mayfield, 1984.

Lapham, Lewis H., Michael Pollan, and Eric Ethridge. *The Harper's Index Book.* New York: Henry Holt, 1987.

Lipschultz, Seymour. *Schaum's Outline of Theory and Problems of Probability.* New York: McGraw-Hill, 1968.

Marascuilo, Leonard A., and Maryellen McSweeney. *Nonparametric and Distribution-Free Methods for the Social Sciences.* Monterey, Calif.: Brooks/Cole, 1977.

Marzillier, Leon F. *Elementary Statistics.* Dubuque, Iowa: Wm. C. Brown Publishers, 1990.

Mason, Robert D., Douglas A. Lind, and William G. Marchal. *Statistics: An Introduction.* New York: Harcourt Brace Jovanovich, 1988.

MINITAB. *MINITAB Reference Manual.* State College, Pa.: MINITAB, Inc., 1994.

Minium, Edward W. *Statistical Reasoning in Psychology and Education.* New York: Wiley, 1970.

Moore, David S. *The Basic Practice of Statistics.* New York: W. H. Freeman and Co., 1995.

Moore, Davis S., and George P. McCabe. *Introduction to the Practice of Statistics,* 3rd ed. New York: W.H. Freeman, 1999.

Newmark, Joseph. *Statistics and Probability in Modern Life.* New York: Saunders, 1988.

Pagano, Robert R. *Understanding Statistics,* 3rd ed. New York: West, 1990.

F

Phillips, John L., Jr. *How to Think about Statistics.* New York: Freeman, 1988.

Reinhardt, Howard E., and Don O. Loftsgaarden. *Elementary Probability and Statistical Reasoning.* Lexington, Mass.: Heath, 1977.

Roscoe, John T. *Fundamental Research Statistics for the Behavioral Sciences,* 2nd ed. New York: Holt, Rinehart and Winston, 1975.

Rossman, Allan J. *Workshop Statistics, Discovery with Data.* New York: Springer, 1996.

Runyon, Richard P., and Audrey Haber. *Fundamentals of Behavioral Statistics,* 6th ed. New York: Random House, 1988.

Shulte, Albert P., 1981 yearbook editor, and James R. Smart, general yearbook editor. *Teaching Statistics and Probability, 1981 Yearbook.* Reston, Va.: National Council of Teachers of Mathematics, 1981.

Smith, Gary. *Statistical Reasoning.* Boston: Allyn and Bacon, 1985.

Spiegel, Murray R. *Schaum's Outline of Theory and Problems of Statistics.* New York: McGraw-Hill, 1961.

Texas Instruments. *TI-83 Graphing Calculator Guidebook.* Temple, Tex.: Texas Instruments, 1996.

Triola, Mario G. *Elementary Statistics,* 7th ed. Reading, Mass.: Addison-Wesley, 1998.

Wardrop, Robert L. *Statistics: Learning in the Presence of Variation.* Dubuque, Iowa: Wm. C. Brown Publishers, 1995.

Warwick, Donald P., and Charles A. Lininger. *The Sample Survey: Theory and Practice.* New York: McGraw-Hill, 1975.

Weiss, Daniel Evan. *100% American.* New York: Poseidon Press, 1988.

Williams, Jack. *The USA Today Weather Almanac 1995.* New York: Vintage Books, 1994.

Wright, John W., ed. *The Universal Almanac 1995.* Kansas City, Mo.: Andrews & McMeel, 1994.

Appendix G

Photo Credits

Chapter 1
Page 2: © Zane Williams/Tony Stone Images
Page 11: © Bruce Ayres/Tony Stone Images

Chapter 2
Page 31: © Peter Gridley/FPG International
Page 47: © VCG/FPG International

Chapter 3
Page 81(top): © Barbara Filet/Tony Stone Images
Page 81(bottom): © Steve Elmore/Tony Stone Images
Page 84: © Robert Daemmrich/Tony Stone Images

Chapter 4
Page 150: © PhotoDisk/Business Today
Page 158: © Michael Newman/PhotoEdit

Chapter 5
Page 167: © Peter Saloutos/Tony Stone Images
Page 199: © Betts Anderson/Unicorn Stock Photos

Chapter 6
Page 210: © FPG International
Page 224: PhotoDisk/Sports & Recreation

Chapter 7
Page 248: © Jeff Kaufman/FPG International
Page 266: © David-Young Wolff/PhotoEdit

Chapter 8
Page 297: © Tom McCarthy/PhotoEdit
Page 325: © C. Orrico/Superstock

Chapter 9
Page 336: © Robert F. Houser
Page 363: © Willinger/FPG International
Page 387: UPI/Corbis-Bettmann

Chapter 10
Page 399: © Michelle Bridwell/PhotoEdit

Chapter 11
Page 463: © Tony Freeman/PhotoEdit
Page 474: © Leland Bobbe/Tony Stone Images
Page 497: © M. Kagan/Monkmeyer

Chapter 12
Page 511: © W.A. Banszewski/Visuals Unlimited

Chapter 13
Page 544: © Jim Pickerell/Tony Stone Images
Page 550: © Joe Sohm/Unicorn Stock Photos
Page 561: © Martha McBride/Unicorn Stock Photos

Chapter 14
Page 583: © Tony Freeman/PhotoEdit

Chapter 15
Page 630: © Michael Newman/PhotoEdit

Selected Answers*

Chapter 1

1–1. Descriptive statistics describe the data set. Inferential statistics use the data to draw conclusions about the population.

1–3. Answers will vary.

1–5. Samples are used to save time and money when the population is large and when the units must be destroyed to gain information.

1–6. *a.* Inferential   *e.* Inferential
    b. Descriptive   *f.* Inferential
    c. Descriptive   *g.* Descriptive
    d. Descriptive   *h.* Inferential

1–7. *a.* Ratio   *d.* Ratio   *g.* Ratio   *i.* Ordinal
    b. Ordinal   *e.* Ratio   *h.* Ratio   *j.* Ratio
    c. Interval   *f.* Nominal

1–8. *a.* Qualitative   *e.* Quantitative
    b. Quantitative   *f.* Quantitative
    c. Qualitative   *g.* Quantitative
    d. Quantitative

1–9. *a.* Discrete   *d.* Continuous   *f.* Continuous
    b. Continuous   *e.* Continuous   *g.* Discrete
    c. Discrete

1–11. Random, systematic, stratified, cluster

1–12. *a.* Cluster   *c.* Random   *e.* Stratified
     b. Systematic   *d.* Systematic

1–13. Answers will vary.

1–15. Answers will vary.

1–17. *a.* Experimental   *c.* Observational
     b. Observational   *d.* Experimental

1–19. Possible answers:
    a. Workplace of subjects, smoking habits, etc.
    b. Gender, age, etc.
    c. Diet, type of job, etc.
    d. Exercise, heredity, age, etc.

1–21. Answers will vary.

1–23. Answers will vary.

Note: These answers to odd-numbered and selected even-numbered exercises include *all* quiz answers.

*Answers may vary due to rounding.

Quiz—Chapter 1

1. True
2. False
3. False
4. False
5. False
6. True
7. False
8. *c.*
9. *b.*
10. *d.*
11. *a.*
12. *c.*
13. *a.*
14. Descriptive, inferential
15. Gambling, insurance
16. Population
17. Sample
18. *a.* Saves time
     b. Saves money
     c. Use when population is infinite
19. *a.* Random
     b. Systematic
     c. Cluster
     d. Stratified
20. quasi-experimental
21. random
22. *a.* Inferential
     b. Descriptive
     c. Inferential
     d. Descriptive
     e. Inferential
23. *a.* Ratio
     b. Ordinal
     c. Interval
     d. Ratio
     e. Nominal

24. *a.* Continuous
  b. Discrete
  c. Discrete
  d. Continuous
  e. Continuous
  f. Discrete

25. *a.* 3.15–3.25
  b. 17.5–18.5
  c. 8.5–9.5
  d. 0.265–0.275
  e. 35.5–36.5

Chapter 2

2–1. To organize data in a meaningful way, to determine the shape of the distribution, to facilitate computational procedures for statistics, to make it easier to draw charts and graphs, to make comparisons among different sets of data.

2–3. *a.* 10.5–15.5; 13; 5
  b. 16.5–39.5; 28; 23
  c. 292.5–353.5; 323; 61
  d. 11.75–14.75; 13.25; 3
  e. 3.125–3.935; 3.53; 0.81

2–5. *a.* Class width is not uniform.
  b. Class limits overlap, and class width is not uniform.
  c. A class has been omitted.
  d. Class width is not uniform.

2–7.

Class	f	cf
15130	5	5
15131	3	8
15132	3	11
15133	7	18
15134	2	20
	20	

2–9.

	Class	f	cf
0	−0.5–0.5	5	5
1	0.5–1.5	8	13
2	1.5–2.5	10	23
3	2.5–3.5	2	25
4	3.5–4.5	3	28
5	4.5–5.5	2	30
		30	

2–11.

Limits	Boundaries	f	cf
21–53	20.5–53.5	2	2
54–86	53.5–86.5	0	2
87–119	86.5–119.5	1	3
120–152	119.5–152.5	1	4
153–185	152.5–185.5	2	6
186–218	185.5–218.5	10	16
219–251	218.5–251.5	9	25
252–284	251.5–284.5	9	34
285–317	284.5–317.5	5	39
318–350	317.5–350.5	4	43
351–383	350.5–383.5	5	48
384–416	383.5–416.5	1	49
		49	

2–13.

Limits	Boundaries	f	cf
27–33	26.5–33.5	7	7
34–40	33.5–40.5	14	21
41–47	40.5–47.5	15	36
48–54	47.5–54.5	11	47
55–61	54.5–61.5	3	50
62–68	61.5–68.5	3	53
69–75	68.5–75.5	2	55
		55	

2–15.

Limits	Boundaries	f	cf
0–19	−0.5–19.5	13	13
20–39	19.5–39.5	18	31
40–59	39.5–59.5	10	41
60–79	59.5–79.5	5	46
80–99	79.5–99.5	3	49
100–119	99.5–119.5	1	50
		50	

2–17.

Limits	Boundaries	f	cf
150–1,276	149.5–1,276.5	2	2
1,277–2,403	1,276.5–2,403.5	2	4
2,404–3,530	2,403.5–3,530.5	5	9
3,531–4,657	3,530.5–4,657.5	8	17
4,658–5,784	4,657.5–5,784.5	7	24
5,785–6,911	5,784.5–6,911.5	3	27
6,912–8,038	6,911.5–8,038.5	7	34
8,039–9,165	8,038.5–9,165.5	3	37
9,166–10,292	9,165.5–10,292.5	3	40
10,293–11,419	10,292.5–11,419.5	2	42
		42	

2–19.

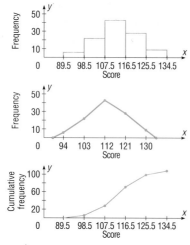

Eighty applicants do not need to enroll in the developmental programs.

2–21.

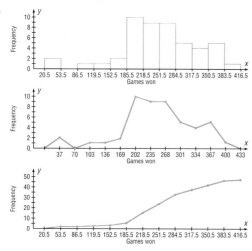

The majority of the data values fall above 185.5. There is a gap in the histogram between 53.5 and 86.5.

2–23.

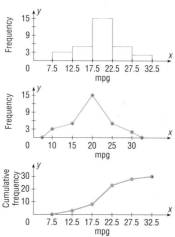

2–25.

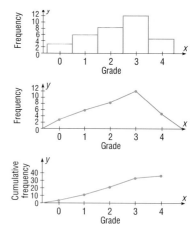

Yes, 26 out of 35 students can enroll in the course.

2–27.

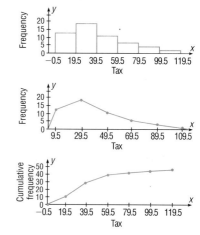

The data is clustered to the left of the distribution. The majority of the states have a tax of less than \$0.40 per pack. The peak is on the left of the distribution.

2–29.

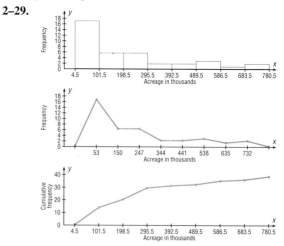

The peak is in the first class, and then the histogram is rather uniform after the first class. Most of the parks have less than 101.5 thousand acres as compared with any other class of values.

2–31.

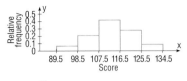

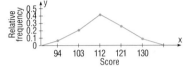

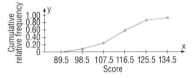

The proportion of applicants who need to enroll in the developmental program is about 0.26.

2–33.

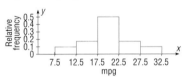

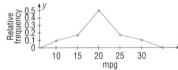

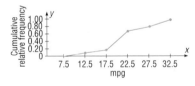

The proportion of automobiles that have a fuel efficiency of 17.5 miles per gallon or higher is 0.73.

2–35.

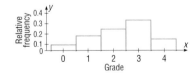

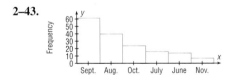

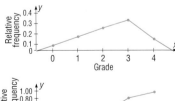

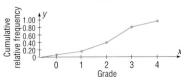

The proportion of students who cannot meet the requirement for the next course is about 0.26.

2–37. *a.*

Limits	Boundaries	f	Midpoints	cf
22–24	21.5–24.5	1	23	1
25–27	24.5–27.5	3	26	4
28–30	27.5–30.5	0	29	4
31–33	30.5–33.5	6	32	10
34–36	33.5–36.5	5	35	15
37–39	36.5–39.5	3	38	18
40–42	39.5–42.5	2	41	20

b.

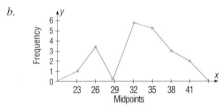

c.

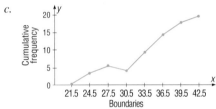

2–39.

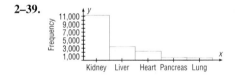

2–41.

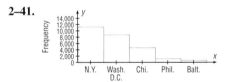

2–43.

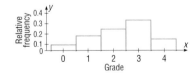

It might be advisable to avoid August and September if possible.

2–45.

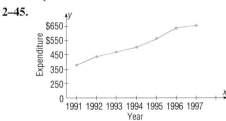

2–47.

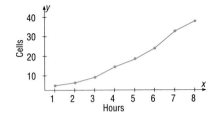

2–49.

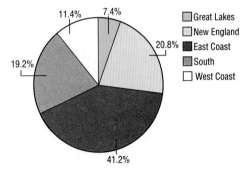

About 41.2% of the people plan to visit the East Coast.

2–51.

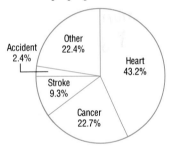

Yes, if one wanted to make comparisons using heights of bars.

2–53.

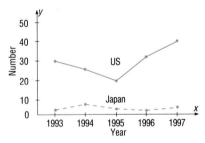

The United States has many more launches than Japan. The number of space launches by Japan is relatively stable for the period. The number of launches for the United States dropped in 1995 and then increased after that.

2–55.

Class	f
Newspaper	7
Television	5
Radio	7
Magazine	6
	25

2–57.

Class	f
Baseballs	4
Golf balls	5
Tennis balls	6
Soccer balls	5
Footballs	5
	25

2–59.

Class	f	cf
11	1	1
12	2	3
13	2	5
14	2	7
15	1	8
16	2	10
17	4	14
18	2	16
19	2	18
20	1	19
21	0	19
22	1	20
	20	

2–61.

Class limits	Class boundaries	f	cf
1910–1919	1909.5–1919.5	1	1
1920–1929	1919.5–1929.5	2	3
1930–1939	1929.5–1939.5	15	18
1940–1949	1939.5–1949.5	12	30
1950–1959	1949.5–1959.5	20	50
1960–1969	1959.5–1969.5	18	68
1970–1979	1969.5–1979.5	18	86
1980–1989	1979.5–1989.5	6	92
1990–1999	1989.5–1999.5	8	100
		100	

2–63.

Class limits	Class boundaries	f	cf
170–188	169.5–188.5	11	11
189–207	188.5–207.5	9	20
208–226	207.5–226.5	4	24
227–245	226.5–245.5	5	29
246–264	245.5–264.5	0	29
265–283	264.5–283.5	0	29
284–302	283.5–302.5	0	29
303–321	302.5–321.5	1	30
		30	

2–65.

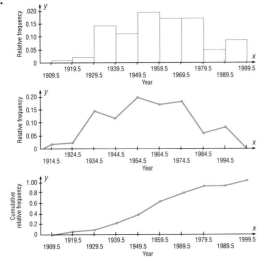

2–67.

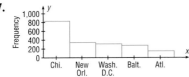

2–69.

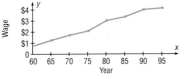

The minimum wage has increased over the years with the largest increase occurring between 1975 and 1980.

2–71.

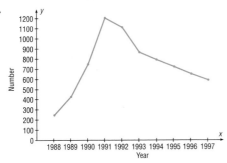

The sale of sports cards increased rapidly from 1988 to 1991, and then began to fall after that.

2–73.

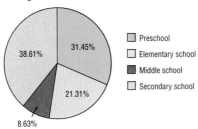

The smallest number of students were enrolled in middle school field, and more students were enrolled in the secondary school field than any other field.

Quiz—Chapter 2

1. False
2. False
3. False
4. True
5. True
6. False
7. False
8. *c.*
9. *c.*
10. *b.*
11. *b.*
12. Categorical, ungrouped, grouped

13. 5, 20
14. Categorical
15. Time series
16. Raw
17. Vertical or *y*
18.

	f	cf
H	6	6
A	5	11
M	6	17
C	$\frac{8}{25}$	25

19.

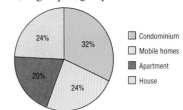

20.

Class boundaries	f	cf
0.5–1.5	1	1
1.5–2.5	5	6
2.5–3.5	3	9
3.5–4.5	4	13
4.5–5.5	2	15
5.5–6.5	6	21
6.5–7.5	2	23
7.5–8.5	3	26
8.5–9.5	$\frac{4}{30}$	30

21.

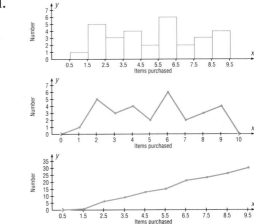

H

22.

Class limits	Class boundaries	Midpoints	f	cf
102–116	101.5–116.5	109	4	4
117–131	116.5–131.5	124	3	7
132–146	131.5–146.5	139	1	8
147–161	146.5–161.5	154	4	12
162–176	161.5–176.5	169	11	23
177–191	176.5–191.5	184	7	30
			30	

23.

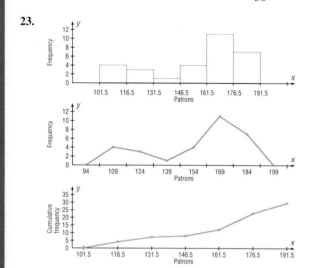

The distribution is somewhat U-shaped with a peak for the class 161.5–176.5.

24.

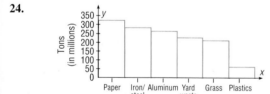

25.

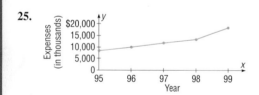

The amount of money spent has increased every year and is more than twice what it was in 1995.

Chapter 3

3–1. *a.* 8.8   *b.* 8   *c.* 8   *d.* 10.5
3–3. *a.* 15.1   *b.* 7   *c.* 3   *d.* 31

The median is probably the best measure of average because 61 is an extremely large data value and makes the mean artificially high.
3–5. *a.* 105   *b.* 110   *c.* 110   *d.* 97.5
3–7. *a.* 6.63   *b.* 6.45   *c.* none   *d.* 6.7
3–9. *a.* 5678.9   *b.* 5342   *c.* 4450   *d.* 5781.5

The distribution is skewed to the right.
3–11. 1995: *a.* 922.6   *b.* 527   *c.* none   *d.* 2130.5
1996: *a.* 911.7   *b.* 485   *c.* 1430   *d.* 2055

The mean, median, and midrange data for 1996 are somewhat less than those values for the data for 1995, indicating the number of fatalities has decreased.
3–13. *a.* 19.8   *b.* 15.5–18.5
3–15. *a.* 7.1   *b.* 5.5–8.5
3–17. *a.* 85.1   *b.* 74.5–85.5
3–19. *a.* 33.8   *b.* 26.5–33.5
3–21. *a.* 23.7   *b.* 21.5–24.5
3–23. *a.* 44.8   *b.* 40.5–47.5
3–25. *a.* 37.5   *b.* 19.5–39.5
3–27. 2.896
3–29. $545,666.67
3–31. 82.7
3–33. *a.* Median   *c.* Mode   *e.* Mode
b. Mean   *d.* Mode   *f.* Mean
3–35. Greek letters, μ
3–37. 6
3–39. *a.* 36 mph   *b.* 30.77 mph   *c.* $16.67
3–41. 5.48
3–43. The square root of the variance is the standard deviation.
3–45. σ^2; σ
3–47. When the sample size is less than 30, the formula for the variance of the sample will underestimate the population variance.
3–49. 9; 6.2; 2.5
3–51. 10; 11.3; 3.4
3–53. 30; 77.1; 8.8
3–55. 5; 3.19; 1.79
3–57. For 1995   For 1996
$R = 4123$ $R = 3970$
$s^2 = 1,030,817.6$ $s^2 = 1,019,853.8$
$s = 1015.3$ $s = 1009.9$
The data for 1995 are more variable.
3–59. 11,263; 7,436, 475.0; 2727.0
3–61. 133.6; 11.6
3–63. 9.6; 3.1
3–65. 211.2; 14.5
3–67. 80.3; 9.0

3–69. 11.7; 3.4

3–71. 10%; 10%; they are equal

3–73. 23.1%; 12.9%; age is more variable

3–75. *a.* 96%   *b.* 93.75%

3–77. $4.84–$5.20

3–79. 89–101

3–81. 86%

3–83. All the data values fall within two standard deviations of the mean.

3–85. 56%; 75%; 84%; 88.89%; 92%

3–87. 4.36

3–89. A z score tells how many standard deviations the data value is above or below the mean.

3–91. A percentile is a relative measurement of position; a percentage is an absolute measure of the part to the total.

3–93. $Q_1 = P_{25}$; $Q_2 = P_{50}$; $Q_3 = P_{75}$

3–95. $D_1 = P_{10}$; $D_2 = P_{20}$; $D_3 = P_{30}$; etc.

3–97. *a.* 1.5   *b.* 2.4   *c.* −0.7
d. 0   *e.* −1.5

3–99. *a.* 0.75   *b.* −1.25   *c.* 2.25
d. −2   *e.* −0.5

3–101. *a.* 1   *b.* 0.6; grade in part *a* is higher

3–103. *a.* −0.93   *b.* −0.85
c. −1.4; score in part *b* is highest

3–105. *a.* 21st   *b.* 58th   *c.* 77th   *d.* 29th

3–106. *a.* 7   *b.* 25   *c.* 64   *d.* 76   *e.* 93

3–107. *a.* 235   *b.* 255   *c.* 261   *d.* 275   *e.* 283

3–108. *a.* 376   *b.* 389   *c.* 432   *d.* 473   *e.* 498

3–109. *a.* 17th   *b.* 39th   *c.* 53rd   *d.* 79th   *e.* 91st

3–111. 82

3–113. 47

3–115. 12

3–117. *a.* 12; 20.5; 32; 22; 20   *b.* 62; 94; 99; 80.5; 37

3–119. Stem and leaf plot, median, interquartile range and boxplot

3–121. Resistant statistics are relatively more unaffected by outliers than are nonresistant statistics.

3–123.
```
3 | 8
4 | 1
5 | 0 0 2 3 3 6 8 9
6 | 6 8 9 9
7 | 0 0 3 4 5 8
8 | 0 1 3 3 4 4 4 5 7 9 9 9
9 | 0 2 4
```
The distribution has two peaks and the data are grouped somewhat toward the numerically higher end of the distribution.

3–125.

Females		Males
5	0	3
	1	5 9
	2	2
7 4 3 2 0	3	1 1
6	4	1 4 6 6
9 6 3 0	5	2 6 6 6 9
8 5	6	0 0 6 6
7 2 0	7	7
8 7 6 6 0 0	8	7 8
4 2	9	6 8

The variation for the distribution for males is larger than the variation for the distribution for females. Overall, the percentage of unemployed females is larger than the percentage of unemployed males worldwide.

3–127.

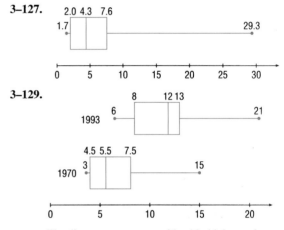

3–129.

The divorce rates are considerably higher and more variable in 1993 than those in 1970.

3–131. *a.* 1203.9   *d.* 1400.5   *g.* 689.6
b. 1164   *e.* 1885
c. No mode   *f.* 475,610.1

3–133. *a.* 7.3   *b.* 6.5–9.5   *c.* 10.0   *d.* 3.2

3–135. *a.* 55.5   *c.* 566.1
b. 57.5–72.5   *d.* 23.8

3–137. 1.1

3–139. 6

3–141. Magazine variance: 0.214; year variance: 0.417; years are more variable

3–143. *a.*

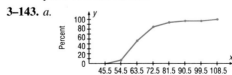

b. 60; 67; 75
c. 4th; 9th; 42nd

H

3–145. 0.26–0.38

3–147. 56%

3–149. 88.89%

3–151.

```
2 | 9  9
3 | 2  4  5  6  8  8
4 | 1  2  3  7  7
5 | 1  3  5  8
6 | 2  2  2  3  7
7 | 2  3
```

3–153.

```
20 | 0  4  9
21 | 0  1  2  7  8  8
22 | 2  7  7  7  8
23 | 0  1  3  7  8
24 | 1  2  2  3  7
25 | 1  1  3  4  6
26 | 0
```

Quiz—Chapter 3

1. True

2. True

3. False

4. False

5. False

6. False

7. False

8. False

9. False

10. *c.*

11. *c.*

12. *a.* and *b.*

13. *b.*

14. *d.*

15. *b.*

16. Statistic

17. Parameters, statistics

18. Standard deviation

19. σ

20. Midrange

21. Positively

22. Outlier

23. *a.* 84.1   *c.* none   *e.* 12   *g.* 4.1
      b. 85   *d.* 84   *f.* 17.1

24. *a.* 6.4   *b.* 5.5–8.5   *c.* 11.6   *d.* 3.4

25. *a.* 51.4   *b.* 35.5–50.5   *c.* 451.5   *d.* 21.2

26. *a.* 8.2   *b.* 6.5–9.5   *c.* 21.6   *d.* 4.6

27. 1.6

28. 4.5

29. 0.33; 0.162; newspapers

30. 0.3125; 0.229; brands

31. −0.75; −1.67; science

32. *a.* 0.5   *b.* 1.6   *c.* 15, *c* is higher

33. *a.* 6; 19; 31; 44; 56; 69; 81; 94
     b. 27
     c.

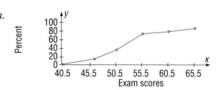

34. *a.*

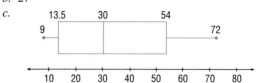

     b. 47; 53; 65
     c. 60th, 6th, 98th percentile

35.

```
1 | 5  9
2 | 6  8
3 | 1  5  8  8  9
4 | 1  7  8
5 | 3  3  4
6 | 2  3  7  8
7 | 6  9
8 | 6  8  9
9 | 8
```

36.

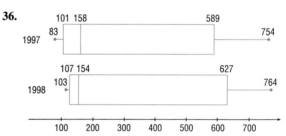

The distributions are very similar. The medians are
almost equal. The range is slightly larger for 1997.

Chapter 4

4–1.

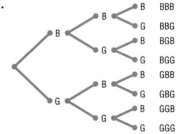

4–3.

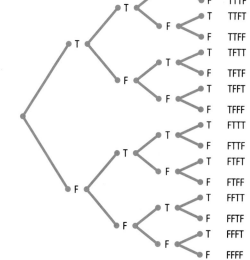
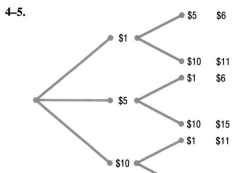

4–5.

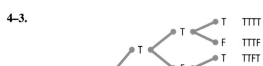

4–7.

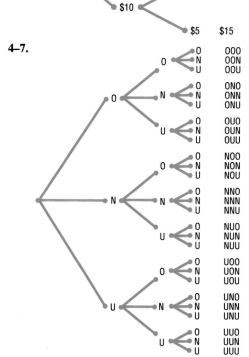

4–9.

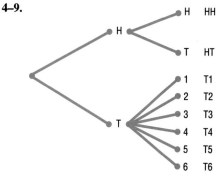

4–11.

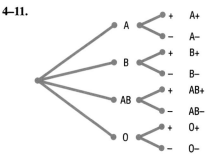

4–13. 100,000; 30,240

4–15. 5,040

4–17. 40,320

4–19. 120

4–21. 1000; 72

4–23. 10

4–25. 64

4–27. 15

4–29. 7 chickens, 8 cows

4–31. *a.* 40,320   *c.* 24   *e.* 1
    b. 3,628,800   *d.* 1

4–32. *a.* 56   *c.* 11,880   *e.* 1   *g.* 1   *i.* 990
    b. 2,520   *d.* 60   *f.* 720   *h.* 40,320   *j.* 30

4–33. 840

4–35. 151,200

4–37. 24

4–39. 120

4–41. 120

4–43. 1,860,480

4–45. 2,520

4–46. *a.* 10   *e.* 15   *i.* 66
    b. 56   *f.* 1   *j.* 4
    c. 35   *g.* 1
    d. 15   *h.* 36

4–47. 22,100

4–49. 41,580

4–51. 120

4–53. 462

H

4–55. 166,320

4–57. 14,400

4–59. 194,040

4–61. 53,130

4–63. 126

4–65. 45

4–67. 24,310

4–69. *a.* 4   *b.* 40   *c.* 624   *d.* 3,744

4–71. 120

4–73. 78 (*Note:* Only two cards are drawn.)

4–75. 792

4–77.

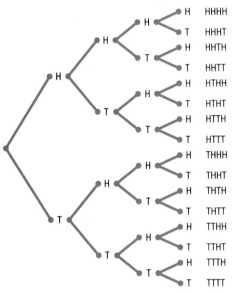

4–79. 40,320

4–81. 800

4–83.

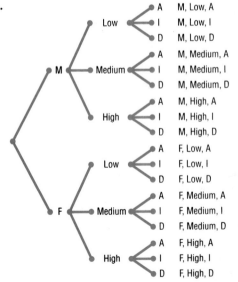

4–85. 720

4–87. 24

4–89. 25; 20

4–91. 4

4–93. 30

4–95. 495

4–97. 15,504

Quiz—Chapter 4

1. False

2. False

3. True

4. True

5. False

6. *b.*

7. *d.*

8. *d.*

9. *b.*

10. *b.*

11. Tree diagram

12. 12

13. n^k

14. $n!$

15. Permutations

16. 1,188,137,600; 710,424,000

17. 720

18. 33,554,432

19. 35

20. 8

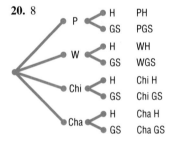

21. 2,646

22. 40,320

23.

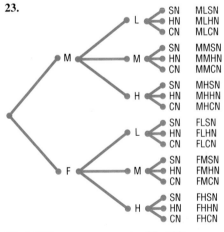

24. 1,365

25. 64; 24

26. 4, or 6 if he wears a sweater over the shirt

27. 120

28. 5005

29. 60

30. 15

Chapter 5

5–1. A probability experiment is a chance process that leads to well-defined outcomes.

5–3. An outcome is the result of a single trial of a probability experiment, but an event can consist of more than one outcome.

5–5. The range of values is 0 to 1 inclusive.

5–7. 0

5–9. 0.15

5–11. *a.* Empirical   *d.* Classical   *f.* Empirical
   b. Classical   *e.* Empirical   *g.* Subjective
   c. Empirical

5–12. *a.* $\frac{1}{6}$ *c.* $\frac{1}{3}$ *e.* 1   *f.* $\frac{5}{6}$ *g.* $\frac{1}{6}$
   b. $\frac{1}{2}$ *d.* 1

5–13. *a.* $\frac{5}{36}$ *b.* $\frac{1}{6}$ *c.* $\frac{2}{9}$ *d.* $\frac{1}{6}$ *e.* $\frac{1}{6}$

5–14. *a.* $\frac{1}{13}$ *c.* $\frac{1}{52}$ *e.* $\frac{4}{13}$ *g.* $\frac{1}{2}$ *i.* $\frac{7}{13}$
   b. $\frac{1}{4}$ *d.* $\frac{2}{13}$ *f.* $\frac{4}{13}$ *h.* $\frac{1}{26}$ *j.* $\frac{1}{26}$

5–15. *a.* $\frac{1}{2}$ *b.* $\frac{3}{10}$ *c.* $\frac{7}{10}$ *d.* $\frac{7}{10}$ *e.* $\frac{1}{2}$

5–17. $\frac{7}{50}$

5–19. 47%

5–21. *a.* $\frac{1}{8}$ *b.* $\frac{1}{4}$ *c.* $\frac{3}{4}$ *d.* $\frac{3}{4}$

5–23. $\frac{1}{9}$

5–25. *a.* $\frac{9}{19}$ *b.* $\frac{11}{38}$ *c.* $\frac{7}{19}$

5–27. 18

5–29. $\frac{1}{36}$

5–31. *a.* 0.08   *b.* 0.01   *c.* 0.35   *d.* 0.36

5–33. The statement is probably not based on empirical probability, and is probably not true.

5–35. Actual outcomes will differ; however, each number should occur approximately $\frac{1}{6}$ of the time.

5–37. *a.* 1:5, 5:1   *d.* 1:1, 1:1   *g.* 1:1, 1:1
   b. 1:1, 1:1   *e.* 1:12, 12:1
   c. 1:3, 3:1   *f.* 1:3, 3:1

5–39. Answers will vary.

5–41. $\frac{1}{6}$

5–43. $\frac{11}{19}$

5–45. *a.* $\frac{8}{17}$ *b.* $\frac{6}{17}$ *c.* $\frac{9}{17}$ *d.* $\frac{12}{17}$

5–47. 0.93

5–49. *a.* $\frac{6}{7}$ *b.* $\frac{4}{7}$ *c.* 1

5–51. *a.* $\frac{67}{118}$ *b.* $\frac{81}{118}$ *c.* $\frac{44}{59}$

5–53. *a.* $\frac{38}{45}$ *b.* $\frac{22}{45}$ *c.* $\frac{2}{3}$

5–55. *a.* $\frac{14}{31}$ *b.* $\frac{23}{31}$ *c.* $\frac{19}{31}$

5–57. *a.* $\frac{1}{15}$ *b.* $\frac{1}{3}$ *c.* $\frac{5}{6}$ *d.* $\frac{5}{6}$ *e.* $\frac{1}{3}$

5–59. *a.* $\frac{5}{12}$ *b.* $\frac{1}{8}$ *c.* $\frac{2}{3}$ *d.* $\frac{23}{24}$

5–61. *a.* $\frac{3}{13}$ *b.* $\frac{3}{4}$ *c.* $\frac{19}{52}$ *d.* $\frac{7}{13}$ *e.* $\frac{15}{26}$

5–63. $\frac{7}{10}$

5–65. 0.06

5–67. 0.30

5–69. *a.* Independent   *e.* Independent
   b. Dependent   *f.* Dependent
   c. Dependent   *g.* Dependent
   d. Dependent   *h.* Independent

5–71. 0.462

5–73. 0.003

5–75. $\frac{1}{12}$

5–77. $\frac{1}{1728}$

5–79. $\frac{1}{133,225}$

5–81. $\frac{4}{15}$

5–83. $\frac{243}{1024}$

5–85. $\frac{5}{28}$

5–87. $\frac{51}{145}$

5–89. 0.116

5–91. 0.03

5–93. $\frac{49}{72}$

5–95. 0.6

5–97. 89%

5–99. 70%

5–101. 82%

5–103. *a.* $\frac{18}{47}$ *b.* $\frac{14}{23}$

5–105. $\frac{55}{56}$

5–107. $\frac{31}{32}$

5–109. $\frac{31}{32}$

5–111. $\frac{14,498}{20,825}$

5–113. 26.6%

5–115. $\frac{63}{64}$

5–117. $\frac{4651}{7776}$

5–119. $\frac{7}{8}$

5–121. $\frac{11}{221}$

5–123. *a.* $\frac{4}{35}$ *b.* $\frac{1}{35}$ *c.* $\frac{12}{35}$ *d.* $\frac{18}{35}$

5–125. *a.* $\frac{11}{102}$ *b.* $\frac{7}{612}$ *c.* $\frac{77}{204}$ *d.* $\frac{77}{612}$ *e.* $\frac{77}{204}$

5–127. $\frac{1}{1225}$

5–129. *a.* $\frac{10}{143}$ *b.* $\frac{60}{143}$ *c.* $\frac{15}{1001}$ *d.* $\frac{160}{1001}$ *e.* $\frac{48}{143}$

5–131. $\frac{1}{100}$

5–133. $\frac{5}{72}$

5–135. $\frac{1}{60}$

5–137. *a.* $\frac{1}{6}$ *b.* $\frac{1}{6}$ *c.* $\frac{2}{3}$

5–139. $\frac{16}{45}$

5–141. $\frac{17}{30}$

5–143. *a.* $\frac{1}{10}$ *b.* $\frac{11}{30}$ *c.* $\frac{13}{15}$ *d.* $\frac{13}{15}$

5–145. 0.98

5–147. 28.9%

5–149. *a.* $\frac{2}{17}$ *b.* $\frac{11}{850}$ *c.* $\frac{1}{5525}$

5–151. $\frac{5}{13}$

5–153. 0.4

5–155. 0.51

H

5–157. 57.3%

5–159. *a.* $\frac{6}{67}$ *b.* $\frac{7}{50}$

5–161. 99.7%

5–163. 55.6%

5–165. $\frac{1}{8}$

Quiz—Chapter 5

1. False

2. False

3. True

4. False

5. False

6. False

7. *b.*

8. *b.* and *d.*

9. *d.*

10. *b.*

11. *c.*

12. Sample space

13. Zero, one

14. Zero

15. One

16. Mutually exclusive

17. *a.* $\frac{1}{13}$ *b.* $\frac{1}{13}$ *c.* $\frac{4}{13}$

18. *a.* $\frac{1}{4}$ *b.* $\frac{4}{13}$ *c.* $\frac{1}{52}$ *d.* $\frac{1}{13}$ *e.* $\frac{1}{2}$

19. *a.* $\frac{12}{31}$ *b.* $\frac{12}{31}$ *c.* $\frac{27}{31}$ *d.* $\frac{24}{31}$

20. *a.* $\frac{11}{36}$ *b.* $\frac{5}{18}$ *c.* $\frac{11}{36}$ *d.* $\frac{1}{3}$ *e.* 0   *f.* $\frac{11}{12}$

21. 0.84

22. 0.002

23. *a.* $\frac{253}{9996}$ *b.* $\frac{33}{66,640}$ *c.* 0

24. 0.54

25. 0.53

26. 0.81

27. 0.056

28. *a.* $\frac{1}{2}$ *b.* $\frac{3}{7}$

29. 0.99

30. 0.518

31. 0.9999886

32. $\frac{1}{4}$

33. $\frac{3}{14}$

34. $\frac{12}{55}$

Chapter 6

6–1. A random variable is a variable whose values are determined by chance. Examples will vary.

6–3. The number of commercials a radio station plays during each hour. The number of times a student uses his or her calculator during a mathematics exam. The number of leaves on a specific type of tree.

6–5. A probability distribution is a distribution that consists of the values a random variable can assume along with the corresponding probabilities of these values.

6–7. Yes

6–9. Yes

6–11. No, the probability values cannot be greater than one.

6–13. Discrete

6–15. Continuous

6–17. Discrete

6–19.

X	0	1	2	3
P(X)	$\frac{6}{15}$	$\frac{5}{15}$	$\frac{3}{15}$	$\frac{1}{15}$

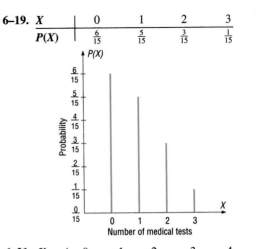

6–21.

X	0	1	2	3	4	5
P(X)	0.75	0.17	0.04	0.025	0.01	0.005

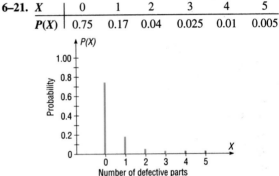

6–23.

X	1	2	3	4	5	6
P(X)	$\frac{1}{2}$	$\frac{1}{6}$	$\frac{1}{12}$	$\frac{1}{12}$	$\frac{1}{12}$	$\frac{1}{12}$

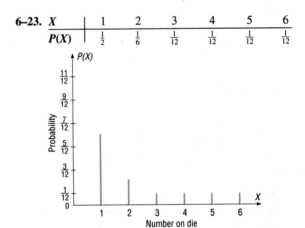

6–25.

X	1	2	3	4	5
P(X)	0.1	0.25	0.25	0.2	0.2

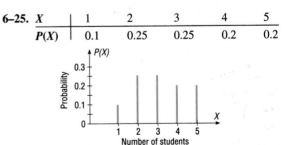

H

6–27.

X	$1	$5	$10	$20
P(X)	$\frac{3}{7}$	$\frac{2}{7}$	$\frac{1}{7}$	$\frac{1}{7}$

6–29.

X	1	2	3	4
P(X)	$\frac{1}{4}$	$\frac{1}{4}$	$\frac{3}{8}$	$\frac{1}{8}$

6–31.

X	1	2	3
P(X)	$\frac{1}{6}$	$\frac{1}{3}$	$\frac{1}{2}$

Yes

6–33.

X	3	4	7
P(X)	$\frac{3}{6}$	$\frac{4}{6}$	$\frac{7}{6}$

No

6–35.

X	1	2	4
P(X)	$\frac{1}{7}$	$\frac{2}{7}$	$\frac{4}{7}$

Yes

6–37. 0.2; 0.3; 0.6; 2

6–39. 1.3, 0.9, 1 No; on average, each person has about one credit card.

6–41. 2.0; 1.6; 1.3; 200

6–43. 6.6; 1.3; 1.1

6–45. 13.9; 1.3; 1.1

6–47. −$3.00; yes, they will make $7500.

6–49. $0.83

6–51. −$1.00

6–53. −$0.50, −$0.52

6–55. All answers are −$0.05

6–57. 10.5

6–59. Answers will vary.

6–61. Answers will vary.

6–63. *a.* Yes   *c.* Yes   *e.* No   *g.* Yes   *i.* No
b. Yes   *d.* No   *f.* Yes   *h.* Yes   *j.* Yes

6–64. *a.* 0.420   *c.* 0.590   *e.* 0.000   *g.* 0.418   *i.* 0.246
b. 0.346   *d.* 0.251   *f.* 0.250   *h.* 0.176

6–65. *a.* 0.0005   *c.* 0.342   *e.* 0.173
b. 0.131   *d.* 0.007

6–67. 0.377; no. Your score will be about 40%.

6–69. 0.267

6–71. 0.071

6–73. *a.* 0.346   *b.* 0.913   *c.* 0.663   *d.* 0.683

6–75. *a.* 0.878   *b.* 0.201   *c.* 0.033

6–76. *a.* 75; 18.8; 4.3   *e.* 100; 90; 9.5
b. 90; 63; 7.9   *f.* 125; 93.8; 9.7
c. 10; 5; 2.2   *g.* 20; 12; 3.5
d. 8; 1.6; 1.3   *h.* 6; 5; 2.2

6–77. 8; 7.9; 2.8

6–79. 10; 9.8; 3.1

6–81. 210; 165.9; 12.9

6–83. 0.199

6–85. 0.559

6–87. 0.018

6–89. 0.770; yes. The probability is high, 77%.

6–91.

X	0	1	2	3	4	5
P(X)	0.328	0.410	0.205	0.051	0.006	0.00

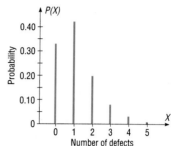

6–93. *a.* 0.135   *c.* 0.0096   *e.* 0.0112
b. 0.0324   *d.* 0.18

6–95. 0.06

6–97. $\frac{1}{108}$

6–99. *a.* 0.1563   *c.* 0.0504   *e.* 0.1241
b. 0.1465   *d.* 0.071

6–101. *a.* 0.0183   *b.* 0.0733   *c.* 0.1465   *d.* 0.7619

6–103. 0.3554

6–105. 0.9502

6–107. 0.1563

6–109. 0.38

6–111. 0.13

6–113. 0.597

6–115. No; the sum of the probabilities is greater than one.

6–117. No; the sum of the probabilities is greater than one.

6–119.

X	0	1	2	3	4
P(X)	0.05	0.30	0.45	0.12	0.08

6–121.

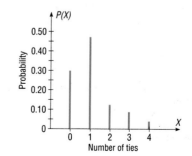

H

6–123. 9.9; 1.5; 1.2

6–125. 24.2; 1.5; 1.2

6–127. $7.23; $7.23

6–129. *a.* 0.122   *b.* 0.989   *c.* 0.043

6–131. 135; 33.8; 5.8

6–133. 0.886

6–135. 0.190

6–137. 0.008

6–139. 0.050

6–141. *a.* 0.5543   *b.* 0.8488   *c.* 0.4457

6–143. 0.27

6–145. *a.* $\frac{21}{44}$ *b.* $\frac{1}{22}$ *c.* $\frac{7}{22}$

Quiz—Chapter 6

1. True   **8.** *c.*

2. False   **9.** *c.*

3. False   **10.** *d.*

4. True   **11.** No, since $\Sigma P(X) > 1$

5. Chance   **12.** Yes

6. $n \cdot p$ **13.** Yes

7. One   **14.** Yes

15.

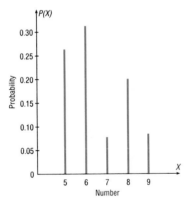

16.

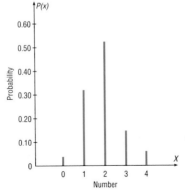

X	0	1	2	3	4
P(X)	0.02	0.3	0.48	0.13	0.07

17. 2.0; 1.3; 1.1

18. 32.2; 1.1; 1.0

19. 5.2

20. $9.65

21. 0.124

22. *a.* 0.075

     b. 0.872

     c. 0.125

23. 240; 48; 6.9

24. 9; 7.9; 2.8

25. 0.008

26. 0.0003

27. 0.061

28. 0.122

29. *a.* 0.5471

     b. 0.9863

     c. 0.4529

30. 0.128

31. *a.* 0.160

     b. 0.42

     c. 0.07

Chapter 7

7–1. The characteristics of the normal distribution are as follows:

   a. It is bell-shaped.

   b. It is symmetric about the mean.

   c. Its mean, median, and mode are equal.

   d. It is continuous.

   e. It never touches the *x* axis.

   f. The area under the curve is equal to 1.

   g. It is unimodal.

7–3. 1, or 100%

7–5. 68%; 95%; 99.7%

7–7. 0.2123

7–9. 0.4808

7–11. 0.4090

7–13. 0.0764

7–15. 0.1145

7–17. 0.0258

7–19. 0.8417

7–21. 0.9826

7–23. 0.5714

7–25. 0.3574

7–27. 0.2486

7–29. 0.4418

7–31. 0.0023

7–33. 0.0655

7–35. 0.9522

7–37. 0.0706

7–39. 0.9222

7–41. −1.94

7–43. −2.13

7–45. −1.26

7–47. *a.* −2.28   *b.* −0.92   *c.* −0.27

7–49. *a.* $z = +1.96$ and $z = -1.96$

     b. $z = +1.65$ and $z = -1.65$, approximately

     c. $z = +2.58$ and $z = -2.58$, approximately

7–51.

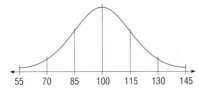

7–53.

X	−2	−1.5	−1	−0.5	0	0.5	1	1.5	2
y	0.05	0.13	0.24	0.35	0.4	0.35	0.24	0.13	0.05

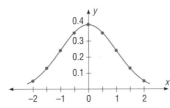

7–55. *a.* 0.3859   *b.* 0.0838
7–57. *a.* 0.1056   *b.* 0.0183
7–59. *a.* 0.0049   *b.* 0.1624   *c.* 0.7486
7–61. *a.* 0.1525   *b.* 0.7745   *c.* 0.1865
7–63. *a.* 0.9772   *b.* 0.6915
7–65. *a.* 0.5691   *b.* 0.6412
7–67. *a.* 0.9938   *b.* 0.1894   *c.* 0.9637
7–69. *a.* 0.5987   *b.* 0.8413   *c.* 0.2432
7–71. 89.95
7–73. The maximum size is 1927.76 square feet; the minimum size is 1692.24 square feet.
7–75. 54.988 minutes
7–77. The maximum price is $9,222 and the minimum price is $7,290.
7–79. 76.18
7–81. $18,840.48
7–83. 18.6 months
7–85. *a.* $\mu = 120, \sigma = 20$ *b.* $\mu = 15, \sigma = 2.5$
c. $\mu = 30, \sigma = 5$
7–87. There are several mathematics tests that can be used.
7–89. 2.59
7–91. $\mu = 45, \sigma = 1.34$
7–93. The distribution is called the sampling distribution of sample means.
7–95. The mean of the sample means is equal to the population mean.
7–97. The distribution will be approximately normal when the sample size is large.
7–99. $z = \dfrac{\bar{X} - \mu}{\sigma/\sqrt{n}}$
7–101. 0.2486
7–103. 0.2327

7–105. 0.0571
7–107. 0.8239
7–109. 0.2033
7–111. 0.1254
7–113. *a.* 0.3446
        b. 0.0023
        c. Yes, since it is within one standard deviation of the mean.
        d. Very unlikely
7–115. *a.* 0.3707   *b.* 0.0475
7–117. *a.* 0.1815   *c.* Means are less variable
        b. 0.3854   than individual data.
7–119. 0.0143
7–121. $\sigma_{\bar{X}} = 1.5, n = 25$
7–123. When *p* is approximately 0.5, as *n* increases, the shape of the binomial distribution becomes similar to the normal distribution. The conditions are $n \cdot p$ and $n \cdot q$ are both ≥ 5. The correction is necessary because the normal distribution is continuous and the binomial distribution is discrete.
7–124. *a.* 0.0811   *c.* 0.1052   *e.* 0.2327
        b. 0.0516   *d.* 0.1711   *f.* 0.9988
7–125. *a.* Yes   *c.* No   *e.* Yes
        b. No   *d.* Yes   *f.* No
7–127. 0.166
7–129. *a.* 0.9793   *b.* 0.0094
7–131. 0.1034
7–133. 0.6844
7–135. 0.9875
7–137. *a.* 0.4744   *e.* 0.2139   *i.* 0.0183
        b. 0.1443   *f.* 0.8284   *j.* 0.9535
        c. 0.0590   *g.* 0.0233
        d. 0.8329   *h.* 0.9131
7–139. *a.* 0.2039   *b.* 0.0099   *c.* 0.0918
7–141. *a.* 0.4649   *b.* 0.0228   *c.* 0.1894
7–143. *a.* 0.5745   *b.* 0.2090   *c.* 0.4483
7–145. 92.2—107.8
7–147. 0.0351
7–149. 0.2033
7–151. 0.0465

Quiz—Chapter 7

1. False   10. *b.*
2. True   11. *c.*
3. True   12. 0.5
4. True   13. Sampling error
5. False   14. The population mean
6. False   15. Standard error of the mean
7. *a.*   16. 5
8. *a.*   17. 5%
9. *b.*

H

18. *a.* 0.4332   *d.* 0.1029   *g.* 0.0401   *j.* 0.9131
b. 0.3944   *e.* 0.2912   *h.* 0.8997
c. 0.0344   *f.* 0.8284   *i.* 0.017

19. *a.* 0.4846   *d.* 0.0188   *g.* 0.0089   *j.* 0.8461
b. 0.4693   *e.* 0.7461   *h.* 0.9582
c. 0.9334   *f.* 0.0384   *i.* 0.9788

20. *a.* 0.0531   *b.* 0.1056   *c.* 0.1056   *d.* 0.0994

21. *a.* 0.0668   *b.* 0.0228   *c.* 0.4649   *d.* 0.0934

22. *a.* 0.4525   *b.* 0.3707   *c.* 0.3707   *d.* 0.019

23. *a.* 0.0013   *b.* 0.5   *c.* 0.0081   *d.* 0.5511

24. *a.* 0.0037   *b.* 0.0228   *c.* 0.5   *d.* 0.3232

25. 8.804 centimeters

26. 121.24 is the lowest acceptable score.

27. 0.015   **30.** 0.0630

28. 0.9738   **31.** 0.0336

29. 0.2296   **32.** 0.0838

Chapter 8

8–1. A point estimate of a parameter specifies a specific value, such as $\mu = 87$; an interval estimate specifies a range of values for the parameter, such as $84 < \mu < 90$. The advantage of an interval estimate is that a specific confidence level (say 95%) can be selected, and one can be 95% confident that the interval contains the parameter that is being estimated.

8–3. The maximum error of estimate is the likely range of values to the right or left of the statistic which may contain the parameter.

8–5. A good estimator should be unbiased, consistent, and relatively efficient.

8–7. For one to be able to determine sample size, the maximum error of estimate and the degree of confidence must be specified and the population standard deviation must be known.

8–9. *a.* 2.58   *b.* 2.33   *c.* 1.96
d. 1.65   *e.* 1.88

8–11. *a.* $77 < \mu < 87$ *b.* $75 < \mu < 89$
c. The 99% confidence interval is larger because the confidence level is larger.

8–13. *a.* $11.9 < \mu < 13.3$
b. It would be highly unlikely, since this is far larger than 13.3.

8–15. $18.13 < \mu < \$18.87$

8–17. $\bar{X} = 3222.4$; $s = 3480.1$; $2314.5 < \mu < 4130.3$

8–19. $59.5 < \mu < 62.9$

8–21. 45

8–23. 25

8–25. 5

8–27. W. S. Gossett

8–29. The t distribution should be used when σ is unknown and $n < 30$.

8–30. *a.* 2.898   *c.* 2.624   *e.* 2.093
b. 2.074   *d.* 1.833

8–31. $15 < \mu < 17$

8–33. $\bar{X} = 33.4$; $s = 28.7$; $21.2 < \mu < 45.6$; The point estimate is 33.4, and it is close to 32. Also, the interval does indeed contain $\mu = 32$. The data value 132 is unusually large (an outlier). The mean may not be the best estimate in this case.

8–35. $11,990 < \mu < 12,410$

8–37. $8.7 < \mu < 9.9$

8–39. $17.29 < \mu < \$19.77$

8–41. $109 < \mu < 121$

8–43. $7.9 < \mu < 165.9$

8–45. $58,197.00 < \mu < \$58,241.00$

8–47. *a.* 0.5, 0.5   *c.* 0.46, 0.54   *e.* 0.45, 0.55
b. 0.45, 0.55   *d.* 0.58, 0.42

8–48. *a.* 0.12, 0.88   *c.* 0.65, 0.35   *e.* 0.67, 0.33
b. 0.29, 0.71   *d.* 0.53, 0.47

8–49. $0.365 < p < 0.415$

8–51. $0.557 < p < 0.743$

8–53. $0.797 < p < 0.883$

8–55. $0.153 < p < 0.307$

8–57. $0.337 < p < 0.543$

8–59. $0.149 < p < 0.351$

8–61. *a.* 3121   *b.* 4161

8–63. *a.* 99   *b.* 273

8–65. 95%

8–67. chi-square

8–69. *a.* 6.262; 27.488   *d.* 15.308; 44.461
b. 0.711; 9.488   *e.* 5.892; 22.362
c. 8.643; 42.796

8–71. $30.9 < \sigma^2 < 78.2$
$5.6 < \sigma < 8.8$

8–73. $0.40 < \sigma^2 < 2.25$
$0.63 < \sigma < 1.50$

8–75. $1.2 < \sigma^2 < 7.7$
$1.1 < \sigma < 2.8$

8–77. $4.1 < \sigma < 7.1$

8–79. $16.2 < \sigma < 19.8$

8–81. $2.5 < \mu < 2.7$

8–83. $7.46 < \mu < 7.54$

8–85. $25 < \mu < 31$

8–87. 28

8–89. $0.343 < p < 0.457$

8–91. 1418

8–93. $0.25 < \sigma < 0.51$

8–95. $5.1 < \sigma^2 < 18.3$

Quiz—Chapter 8

1. True
2. True
3. False
4. True
5. *b.*
6. *a.*
7. *b.*
8. Unbiased, consistent, relatively efficient
9. Maximum error of estimate
10. Point
11. 90; 95; 99
12. $22.79 < \mu < 24.11$
13. $\$43.15 < \mu < \$46.45$
14. $3954 < \mu < 4346$
15. $45.7 < \mu < 51.5$
16. $418 < \mu < 458$
17. $26 < \mu < 36$
18. 180
19. 25
20. $0.604 < p < 0.810$
21. $0.295 < p < 0.425$
22. $0.342 < p < 0.547$
23. 545
24. $7 < \sigma < 13$
25. $30.9 < \sigma^2 < 78.2$
    $5.6 < \sigma < 8.8$
26. $1.8 < \sigma < 3.2$

Chapter 9

Note: For Chapters 9–14, specific *P*-values are given in parentheses after the *P*-value intervals. When the specific *P*-value is extremely small, it is not given.

9–1. The null hypothesis states that there is no difference between a parameter and a specific value or that there is no difference between two parameters. The alternative hypothesis states that there is a specific difference between a parameter and a specific value or that there is a difference between two parameters. Examples will vary.

9–3. A statistical test uses the data obtained from a sample to make a decision about whether or not the null hypothesis should be rejected.

9–5. The critical region is the range of values of the test statistic that indicates that there is a significant difference and the null hypothesis should be

rejected. The noncritical region is the range of values of the test statistic that indicates that the difference was probably due to chance and the null hypothesis should not be rejected.

9–7. α, β

9–9. A one-tailed test should be used when a specific direction, such as greater than or less than, is being hypothesized; when no direction is specified, a two-tailed test should be used.

9–11. Hypotheses can be proved true only when the entire population is used to compute the test statistic. In most cases, this is impossible.

9–12. *a.* ± 2.58 *d.* -1.28 *g.* -2.33 *i.* $+2.05$
       b. $+1.65$ *e.* ± 1.96 *h.* ± 1.65 *j.* ± 2.33
       c. -2.58 *f.* $+1.75$

9–13. *a.* H_0: $\mu = 36.3$ (claim), H_1: $\mu \neq 36.3$
       b. H_0: $\mu = \$36,250$ (claim), H_1: $\mu \neq \$36,250$
       c. H_0: $\mu \leq 27.6$, H_1: $\mu > 27.6$ (claim)
       d. H_0: $\mu \geq 72$, H_1: $\mu < 72$ (claim)
       e. H_0: $\mu \geq 100$, H_1: $\mu < 100$ (claim)
       f. H_0: $\mu = \$297.75$ (claim), H_1: $\mu \neq \$297.75$
       g. H_0: $\mu \leq \$52.98$, H_1: $\mu > \$52.98$ (claim)
       h. H_0: $\mu \leq 300$ (claim), H_1: $\mu > 300$
       i. H_0: $\mu \geq 3.6$ (claim), H_1: $\mu < 3.6$

9–15. H_0: $\mu = \$69.21$ (claim) and H_1: $\mu \neq \$69.21$; C.V. $= \pm 1.96$; $z = -1.15$; do not reject. There is not enough evidence to reject the claim that the average cost of a hotel stay in Atlanta is $69.21.

9–17. H_0: $\mu \leq \$24$ billion and H_1: $> \$24$ billion (claim); C.V. $= 1.65$; $z = 1.85$; reject. There is enough evidence to support the claim that the average revenue is greater than $24 billion.

9–19. H_0: $\mu \geq 14$ and H_1: $\mu < 14$ (claim); C.V. $= -2.33$; $z = -4.89$; reject. There is enough evidence to support the claim that the average age of the planes in the executive's airline is less than the national average.

9–21. H_0: $\mu = 70$ (claim) and H_1: $\mu \neq 70$; C.V. $= \pm 1.96$; $z = -2.59$; reject. There is enough evidence to reject the claim that the average age of the purchasers is 70 years old.

9–23. H_0: $\mu = 36$ (claim) and H_1: $\mu \neq 36$; C.V. $= \pm 2.58$; $z = -3.54$; reject. Yes; there is enough evidence to reject the claim that the average lifetime of the lightbulbs is 36 months.

9–25. H_0: $\mu \geq 240$ and H_1: $\mu < 240$ (claim); C.V. $= -2.33$; $z = -3.87$; reject. Yes; there is enough evidence to support the claim that the medication lowers the cholesterol level.

9–27. H_0: $\mu = \$24.44$ and H_1: $\mu \neq \$24.44$ (claim); C.V. $= \pm 2.33$; $z = -2.28$; do not reject. There is not enough evidence to support the claim that the

H

amount spent at a local mall is not equal to the national average of $24.44.

9–29. *a.* Do not reject. *d.* Reject.
b. Do not reject. *e.* Reject.
c. Do not reject.

9–31. H_0: $\mu \geq 264$ and H_1: $\mu < 264$ (claim); $z = -2.53$; P-value = 0.0057; reject. There is enough evidence to support the claim that the average stopping distance is less than 264 ft.

9–33. H_0: $\mu \leq 84$ and H_1: $\mu > 84$ (claim); $z = 1.1$; P-value = 0.1357; do not reject. There is not enough evidence to support the claim that the average lifetime is more than 84 months.

9–35. H_0: $\mu = 6.32$ (claim) and H_1: $\mu \neq 6.32$; $z = 2.49$; P-value = 0.0128; reject. There is enough evidence to reject the claim that the average wage is $6.32.

9–37. H_0: $\mu = 30{,}000$ (claim) and H_1: $\mu \neq 30{,}000$; $z = 1.71$; P-value = 0.0872; reject. There is enough evidence to reject the claim that the customers are adhering to the recommendation. Yes, the 0.10 level is appropriate.

9–39. H_0: $\mu \geq 10$ and H_1: $\mu < 10$ (claim); $z = -8.67$, P-value < 0.0001; since P-value < 0.05, reject. Yes; there is enough evidence to support the claim that the average number of days missed per year is less than 10.

9–41. H_0: $\mu = 8.65$ (claim) and H_1: $\mu \neq 8.65$; C.V. = ± 1.96; $z = -1.35$; do not reject. Yes; there is not enough evidence to reject the claim that the average hourly wage of the employees is $8.65.

9–43. The degrees of freedom are the number of values that are free to vary after a sample statistic has been computed. They tell the researcher which specific curve to use when a distribution consists of a family of curves.

9–45. *a.* +1.833 *c.* −3.365 *e.* ±2.145 *g.* ±2.771
b. ±1.740 *d.* +2.306 *f.* −2.819 *h.* ±2.583

9–46. Specific P-values are in parentheses.
a. 0.01 < P-value < 0.025 (0.018)
b. 0.05 < P-value < 0.10 (0.062)
c. 0.10 < P-value < 0.25 (0.123)
d. 0.10 < P-value < 0.20 (0.138)
e. P-value < 0.005 (0.003)
f. 0.10 < P-value < 0.25 (0.158)
g. P-value = 0.05 (0.05)
h. P-value > 0.25 (0.261)

9–47. H_0: $\mu \geq 11.52$ and H_1: $\mu < 11.52$ (claim); C.V. = -1.833; d.f. = 9; $t = -9.97$; reject. There is enough evidence to support the claim that the amount of rainfall is below average.

9–49. H_0: $\mu = 800$ (claim) and H_1: $\mu \neq 800$; C.V. = ± 2.262; d.f. = 9; $t = 9.96$; reject. Yes;

there is enough evidence to reject the claim that the average rent that small-business establishments pay in Eagle City is $800.

9–51. H_0: $\mu \geq 700$ (claim) and H_1: $\mu < 700$; C.V. = -2.262; d.f. = 9; $t = -2.71$; reject. There is enough evidence to reject the claim that the average height of the buildings is at least 700 feet.

9–53. H_0: $\mu = \$750$ (claim) and H_1: $\mu \neq \$750$; C.V. = ± 3.106; $t = -3.67$; reject. Yes; there is enough evidence to reject the claim that the average rent is $750.

9–55. H_0: $\mu \leq 350$ and H_1: $\mu > 350$ (claim); C.V. = 1.796; d.f. = 11; $t = 1.732$; do not reject. No; there is not enough evidence to support the claim that the average fine is higher than $350.

9–57. H_0: $\mu \geq 37$ (claim) and H_1: $\mu < 37$; d.f. = 28; $t = -1.88$; 0.025 < P-value < 0.05 (0.035); since P-value < 0.05, reject. There is enough evidence to reject the claim that the average household receives at least 37 phone calls per month.

9–59. H_0: $\mu = 75$ (claim) and H_1: $\mu \neq 75$; d.f. = 19; $t = -2.83$; 0.01 < P-value < 0.02 (0.011); since P-value > 0.01, do not reject. There is not enough evidence to reject the claim that the average score on the real estate exam is 75.

9–61. Answers will vary.

9–63. $np \geq 5$ and $nq \geq 5$

9–65. H_0: $p \geq 0.23$ (claim) and H_1: $p < 0.23$; C.V. = -1.65; $z = -0.83$; do not reject. There is not enough evidence to reject the claim that at least 23% of the 14-year-old residents own a skateboard.

9–67. H_0: $p \geq 0.40$ (claim) and H_1: $p < 0.40$; C.V. = -1.28; $z = -0.457$; do not reject. There is not enough evidence to reject the claim that at least 40% of the arsonists are under 21 years old.

9–69. H_0: $p = 0.63$ (claim) and H_1: $p \neq 0.63$; C.V. = ± 1.96; $z = -0.88$; do not reject. There is not enough evidence to reject the claim that the percentage is the same.

9–71. H_0: $p \geq 0.15$ (claim) and H_1: $p < 0.15$; C.V. = -1.65; $z = -0.94$; do not reject. There is not enough evidence to reject the claim that at least 15% of all eighth-grade students are overweight.

9–73. H_0: $p \leq 0.30$ and H_1: $p > 0.30$ (claim); $z = +1.85$; P-value = 0.0322; since P-value < 0.05, reject. Yes; there is enough evidence to support the claim that more than 30% of the customers have at least two telephones.

9–75. H_0: $p = 0.18$ (claim) and H_1: $p \neq 0.18$; $z = -0.60$; P-value = 0.5486; since P-value > 0.05, do not reject. There is not enough evidence to reject the claim that 18% of all high school students smoke at least a pack of cigarettes a day.

9–77. No

9–79. *a.* H_0: $\sigma^2 \leq 225$ and H_1: $\sigma^2 > 225$; C.V. = 27.587; d.f. = 17

b. H_0: $\sigma^2 \geq 225$ and H_1: $\sigma^2 < 225$; C.V. = 14.042; d.f. = 22

c. H_0: $\sigma^2 = 225$ and H_1: $\sigma^2 \neq 225$; C.V. = 5.629; 26.119; d.f. = 14

d. H_0: $\sigma^2 = 225$ and H_1: $\sigma^2 \neq 225$; C.V. = 2.167; 14.067; d.f. = 7

e. H_0: $\sigma^2 \leq 225$ and H_1: $\sigma^2 > 225$; C.V. = 32.000; d.f. = 16

f. H_0: $\sigma^2 \geq 225$ and H_1: $\sigma^2 < 225$; C.V. = 8.907; d.f. = 19

g. H_0: $\sigma^2 = 225$ and H_1: $\sigma^2 \neq 225$; C.V. = 3.074; 28.299; d.f. = 12

h. H_0: $\sigma^2 \geq 225$ and H_1: $\sigma^2 < 225$; C.V. = 15.308; d.f. = 28

9–80. *a.* $0.01 < P\text{-value} < 0.025$ (0.015)

b. $0.005 < P\text{-value} < 0.01$ (0.006)

c. $0.01 < P\text{-value} < 0.025$ (0.012)

d. $P\text{-value} < 0.005$ (0.003)

e. $0.025 < P\text{-value} < 0.05$ (0.037)

f. $0.10 < P\text{-value} < 0.20$ (0.088)

g. $0.05 < P\text{-value} < 0.10$ (0.066)

h. $P\text{-value} < 0.01$ (0.007)

9–81. H_0: $\sigma = 60$ (claim) and H_1: $\sigma \neq 60$; C.V. = 8.672, 27.587; d.f. = 17; $\chi^2 = 19.707$; do not reject. There is not enough evidence to reject the claim that the standard deviation is 60.

9–83. H_0: $\sigma^2 \leq 25$ (claim) and H_1: $\sigma^2 > 25$; C.V. = 27.204; $\alpha = 0.10$; d.f. = 19; $\chi^2 = 27.36$; reject. There is enough evidence to reject the claim that the variance is less than or equal to 25.

9–85. H_0: $\sigma \leq 1.2$ (claim) and H_1: $\sigma > 1.2$; $\alpha = 0.01$; d.f. = 14; $\chi^2 = 31.5$; $P\text{-value} < 0.005$ (0.0047); since $P\text{-value} < 0.01$, reject. There is enough evidence to reject the claim that the standard deviation is less than or equal to 1.2 minutes.

9–87. H_0: $\sigma \leq 2$ and H_1: $\sigma > 2$ (claim); C.V. = 24.725; $\alpha = 0.01$; d.f. = 11; $\chi^2 = 22.02$; do not reject. The lot is acceptable, since there is not enough evidence to support the claim that the standard deviation is greater than 2 pounds.

9–89. H_0: $\mu = 1800$ (claim) and H_1: $\mu \neq 1800$; C.V. = ± 1.96; $z = 0.47$; $1706.04 < \mu < 1953.96$; do not reject. There is not enough evidence to reject the claim that the average of the sales is $1800.

9–91. H_0: $\mu = 86$ (claim) and H_1: $\mu \neq 86$; C.V. = ± 2.58; $z = -1.29$; $80.00 < \mu < 88.00$; do not reject. There is not enough evidence to reject the claim that the average monthly maintenance is $86.

9–93. H_0: $\mu = 22$ and H_1: $\mu \neq 22$ (claim); C.V. = ± 2.58; $z = -2.32$; $19.47 < \mu < 22.13$; do not reject. There is not enough evidence to support the claim that the average has changed.

9–95. The power of a statistical test is the probability of rejecting the null hypothesis when it is false.

9–97. The power of a test can be increased by increasing α or selecting a larger sample size.

9–99. H_0: $\mu = 98°$ (claim) and H_1: $\mu \neq 98°$; C.V. = ± 1.96; $z = -2.02$; reject. There is enough evidence to reject the claim that the average high temperature in the United States is 98°.

9–101. H_0: $\mu = 16.3$ (claim) and H_1: $\mu \neq 16.3$; $z = 11.3$; $P\text{-value} < 0.01$; reject. There is enough evidence to reject the claim that the average age is 16.3.

9–103. H_0: $\mu \leq 67$ and H_1: $\mu > 67$ (claim); C.V. = 1.383; d.f. = 9; $t = 7.47$; reject. There is enough evidence to support the claim that 1995 was warmer than average.

9–105. H_0: $\mu = 6$ and H_1: $\mu \neq 6$ (claim); $t = 1.835$; d.f. = 9; C.V. = ± 2.821; do not reject. There is not enough evidence to support the claim that the attendance has changed.

9–107. H_0: $p \geq 0.60$ (claim) and H_1: $p < 0.60$; C.V. = -1.28; $z = -1.22$; do not reject. No. There is not enough evidence to reject the claim that at least 60% of adults eat eggs for breakfast at least four times a week.

9–109. H_0: $p = 0.65$ (claim) and H_1: $p \neq 0.65$; $z = 1.17$; $P\text{-value} = 0.242$; since $P\text{-value} > 0.05$, do not reject. There is not enough evidence to reject the claim that 65% of teenagers own their own radios.

9–111. H_0: $\mu \geq 10$ and H_1: $\mu < 10$ (claim); $z = -2.22$; $P\text{-value} = 0.0132$; reject. There is enough evidence to support the claim that the average time is less than 10 minutes.

9–113. H_0: $\sigma \geq 4.3$ (claim) and H_1: $\sigma < 4.3$; d.f. = 19; $\chi^2 = 6.95$; $0.005 < P\text{-value} < 0.01$ (0.006); since $P\text{-value} < 0.05$, reject. Yes, there is enough evidence to reject the claim that the standard deviation is greater than or equal to 4.3 miles per gallon.

9–115. H_0: $\sigma = 18$ (claim) and H_1: $\sigma \neq 18$; C.V. = 11.143 and 0.484; d.f. = 4; $\chi^2 = 5.44$; do not reject. There is not enough evidence to reject the claim that the standard deviation is 18 minutes.

9–117. H_0: $\mu = 4$ and H_1: $\mu \neq 4$ (claim); C.V. = ± 2.58; $z = 1.49$; $3.85 < \mu < 4.55$; do not reject. There is not enough evidence to support the claim that the growth has changed.

H

Quiz—Chapter 9

1. True

2. True

3. False

4. True

5. False

6. *b.*

7. *d.*

8. *c.*

9. *b.*

10. Type I

11. β

12. Statistical hypothesis

13. Right

14. $n - 1$

15. H_0: $\mu = 28.6$ (claim) and H_1: $\mu \neq 28.6$; $z = 2.14$; C.V. $= \pm 1.96$; reject. There is enough evidence to reject the claim that the average age of the mothers is 28.6 years.

16. H_0: $\mu = \$6500$ (claim) and H_1: $\mu \neq \$6500$; $z = 5.27$; C.V. $= \pm 1.96$; reject. There is enough evidence to reject the agent's claim.

17. H_0: $\mu \leq 8$ and H_1: $\mu > 8$ (claim); $z = 6$; C.V. $= 1.65$; reject. There is enough evidence to support the claim that the average is greater than 8.

18. H_0: $\mu = 21$ (claim) and H_1: $\mu \neq 21$; d.f. $= 16$; $t = -2.06$; C.V. $= \pm 2.921$; do not reject. There is not enough evidence to reject the claim that the average age of the dropouts is 21.

19. H_0: $\mu \geq 67$ and H_1: $\mu < 67$ (claim); $t = -3.1568$; P-value < 0.005 (0.003); since P-value < 0.05, reject. There is enough evidence to support the claim that the average height is less than 67 inches.

20. H_0: $\mu \geq 12.4$ and H_1: $\mu < 12.4$ (claim); $t = -2.324$; C.V. $= -1.345$; reject. There is enough evidence to support the claim that the average is less than the company claimed.

21. H_0: $\mu \leq 63.5$ and H_1: $\mu > 63.5$ (claim); $t = 0.47075$; P-value > 0.25 (0.322); since P-value > 0.05, do not reject. There is not enough evidence to support the claim that the average is greater than 63.5.

22. H_0: $\mu = 26$ (claim) and H_1: $\mu \neq 26$; $t = -1.5$; C.V. $= \pm 2.492$; do not reject. There is not enough evidence to reject the claim that the average is 26.

23. H_0: $p \geq 0.25$ (claim) and H_1: $p < 0.25$; $z = -0.6928$; C.V. $= -1.65$; do not reject. There is not enough evidence to reject the claim that the proportion is at least 0.25.

24. H_0: $p \geq 0.55$ (claim) and H_1: $p < 0.55$; $z = -0.8989$; C.V. $= -1.28$; do not reject. There is not enough evidence to reject the survey's claim.

25. H_0: $p = 0.7$ (claim) and H_1: $p \neq 0.7$; $z = 0.7968$; C.V. $= \pm 2.33$; do not reject. There is not enough evidence to reject the claim that the proportion is 0.7.

26. H_0: $p = 0.75$ (claim) and H_1: $p \neq 0.75$; $z = 2.6833$; C.V. $= \pm 2.58$; reject. There is enough evidence to reject the claim.

27. P-value $= 0.246$

28. P-value $= 0.0002$

29. H_0: $\sigma \leq 6$ and H_1: $\sigma > 6$ (claim); $\chi^2 = 54$; C.V. $= 36.415$; reject. There is enough evidence to support the claim.

30. H_0: $\sigma = 8$ (claim) and H_1: $\sigma \neq 8$; $\chi^2 = 33.2$; C.V. $= 27.991, 79.490$; do not reject. There is not enough evidence to reject the claim that $\sigma = 8$.

31. H_0: $\sigma \geq 2.3$ and H_1: $\sigma < 2.3$ (claim); $\chi^2 = 13$; C.V. $= 10.117$; do not reject. There is not enough evidence to support the claim that the standard deviation is less than 2.3.

32. H_0: $\sigma = 9$ (claim) and H_1: $\sigma \neq 9$; $\chi^2 = 13.4$; P-value > 0.20 (0.291); since P-value > 0.05, do not reject. There is not enough evidence to reject the claim that $\sigma = 9$.

33. $28.3 < \mu < 30.1$

34. $\$6562.81 < \mu < \$6637.19$

Chapter 10

10–1. Testing a single mean involves comparing a sample mean to a specific value such as $\mu = 100$; testing the difference between two means involves comparing the means of two samples, such as $\mu_1 = \mu_2$.

10–3. The populations must be independent of each other, and they must be normally distributed; s_1 and s_2 can be used in place of σ_1 and σ_2 when σ_1 and σ_2 are unknown and both samples are each greater than or equal to 30.

10–5. H_0: $\mu_1 = \mu_2$ (claim) and H_1: $\mu_1 \neq \mu_2$; C.V. $= \pm 2.58$; $z = -0.856$; do not reject. There is not enough evidence to reject the claim that the average lengths of the major rivers are the same.

10–7. H_0: $\mu_1 \leq \mu_2$ and H_1: $\mu_1 > \mu_2$ (claim); C.V. $= +1.65$; $z = 2.56$; reject. Yes, there is enough evidence to support the claim that pulse rates of smokers are higher than pulse rates of nonsmokers.

10–9. H_0: $\mu_1 \leq \mu_2$ and H_1: $\mu_1 > \mu_2$ (claim); C.V. $= +2.05$; $z = 1.12$; do not reject. There is not enough evidence to support the claim that the noise levels in the corridors are higher than the noise levels in the clinics.

10–11. H_0: $\mu_1 \geq \mu_2$ and H_1: $\mu_1 < \mu_2$ (claim); C.V. $= -1.65$; $z = -2.01$; reject. There is enough evidence to support the claim that the stayers had a higher grade point average.

10–13. H_0: $\mu_1 \leq \mu_2$ and H_1: $\mu_1 > \mu_2$ (claim); C.V. $= +2.33$; $z = +1.09$; do not reject. There is not enough evidence to support the claim that colleges spend more money on male sports than they spend on female sports.

10–15. H_0: $\mu_1 = \mu_2$ and H_1: $\mu_1 \neq \mu_2$ (claim); $z = 1.01$; P-value $= 0.3124$; do not reject. There is not enough evidence to support the claim that there is a difference in self-esteem scores.

10–17. $2.8 < \mu_1 - \mu_2 < 6.0$

10–19. $-7.3 < \mu_1 - \mu_2 < -1.3$

10–21. H_0: $\mu_1 - \mu_2 \leq 8$ (claim) and H_1: $\mu_1 - \mu_2 > 8$; C.V. $= +1.65$; $z = -0.73$; do not reject. There is not enough evidence to reject the claim that private-school students have exam scores that are at most 8 points higher than that of students in public schools.

10–23. The larger variance is placed in the numerator of the formula.

10–25. One degree of freedom is used for the variance associated with the numerator, and one is used for the variance associated with the denominator.

10–27. *a.* d.f.N. $= 15$, d.f.D. $= 22$; C.V. $= 3.36$
b. d.f.N. $= 24$, d.f.D. $= 13$; C.V. $= 3.59$
c. d.f.N. $= 45$, d.f.D. $= 29$; C.V. $= 2.03$
d. d.f.N. $= 20$, d.f.D. $= 16$; C.V. $= 2.28$
e. d.f.N. $= 10$, d.f.D. $= 10$; C.V. $= 2.98$

10–28. Specific P-values are in parentheses.
a. $0.025 < P$-value < 0.05 (0.033)
b. $0.05 < P$-value < 0.10 (0.072)
c. P-value $= 0.05$
d. $0.005 < P$-value < 0.01 (0.006)
e. P-value $= 0.05$
f. P-value > 0.10 (0.112)
g. $0.05 < P$-value < 0.10 (0.068)
h. $0.01 < P$-value < 0.02 (0.015)

10–29. H_0: $\sigma_1^2 \leq \sigma_2^2$; H_1: $\sigma_1^2 > \sigma_2^2$ (claim); $\alpha = 0.05$; C.V. $= 2.23$; d.f.N. $= 19$; d.f.D. $= 19$; $F = 1.41$; do not reject. There is not enough evidence to support the claim that variance of the exam scores of the students who had the word processing will be larger than the variance of the exam scores of the students who did not have word processing in conjunction with a composition course.

10–31. H_0: $\sigma_1^2 = \sigma_2^2$ and H_1: $\sigma_1^2 \neq \sigma_2^2$ (claim); C.V. $= 2.86$; d.f.N. $= 15$; d.f.D. $= 15$; $F = 7.85$; reject. There is enough evidence to support the claim that the variances are different.

10–33. H_0: $\sigma_1^2 = \sigma_2^2$; H_1: $\sigma_1^2 \neq \sigma_2^2$ (claim); $\alpha = 0.10$; C.V. $= 2.53$; d.f.N. $= 14$; d.f.D. $= 14$; $F = 2.09$; do not reject. There is not enough evidence to support the claim that variations of the lengths of newborn males differ from the variations of the lengths of newborn females.

10–35. H_0: $\sigma_1^2 \leq \sigma_2^2$; H_1: $\sigma_1^2 > \sigma_2^2$ (claim); C.V. $= 2.66$; $\alpha = 0.01$; d.f.N. $= 27$; d.f.D. $= 24$; $F = 5.27$; reject. There is enough evidence to support the claim that the variation of blood pressure of overweight individuals is greater than the variation of blood pressure of normal-weight individuals.

10–37. H_0: $\sigma_1^2 \leq \sigma_2^2$; H_1: $\sigma_1^2 > \sigma_2^2$ (claim); $\alpha = 0.05$; C.V. $= 2.90$; d.f.N. $= 9$; d.f.D. $= 11$; $F = 1.72$; do not reject. There is not enough evidence to support the claim that the variation in pounds lost following diet A is greater than the variation in pounds lost from diet B.

10–39. H_0: $\sigma_1^2 = \sigma_2^2$ (claim) and H_1: $\sigma_1^2 \neq \sigma_2^2$; C.V. $= 3.87$; d.f.N. $= 6$; d.f.D. $= 7$; $F = 3.18$; do not reject. There is not enough evidence to reject the claim that the variances of the heights are equal.

10–41. H_0: $\sigma_1^2 = \sigma_2^2$ (claim) and H_1: $\sigma_1^2 \neq \sigma_2^2$; $F = 5.32$; d.f.N. $= 14$; d.f.D. $= 14$; P-value < 0.01 (0.004); reject. There is enough evidence to reject the claim that the variances of the weights are equal.

10–43. If H_0: $\sigma_1^2 = \sigma_2^2$; C.V. $= 4.03$; $F = 1.93$; do not reject. If H_0: $\mu_1 = \mu_2$ and H_1: $\mu_1 \neq \mu_2$ (claim); C.V. $= \pm 2.101$; d.f. $= 18$; $t = -4.02$; reject. There is enough evidence to support the claim that there is a significant difference in the values of the homes based on the appraisers' values. $-\$7762 < \mu_1 - \mu_2 < -\$2434$

10–45. If H_0: $\sigma_1^2 = \sigma_2^2$; C.V. $= 2.72$; $F = 5.06$; reject. If H_0: $\mu_1 \geq \mu_2$ and H_1: $\mu_1 < \mu_2$ (claim); C.V. $= -1.363$; d.f. $= 11$; $t = -1.24$; do not reject. There is not enough evidence to support the claim that math majors can write programs faster than business majors.

10–47. H_0: $\sigma_1^2 = \sigma_2^2$; C.V. $= 14.94$; $F = 1.41$; do not reject. H_0: $\mu_1 < \mu_2$ and H_1: $\mu_1 > \mu_2$ (claim); C.V. $= 2.764$; d.f. $= 10$; $t = 1.45$; do not reject. There is not enough evidence to support the claim that the average number of family day care homes is greater than the average number of day care centers.

10–49. H_0: $\sigma_1^2 = \sigma_2^2$; C.V. $= 4.19$; $F = 1.70$; do not reject. H_0: $\mu_1 = \mu_2$ and H_1: $\mu_1 \neq \mu_2$ (claim); C.V. $= \pm 2.508$; d.f. $= 22$; $t = -2.97$; reject. There is enough evidence to support the claim that there is a difference in the average times of the two groups. $-11.1 < \mu_1 - \mu_2 < -0.93$

10–51. H_0: $\sigma_1^2 = \sigma_2^2$; $F = 3.67$. The P-value for the F test is $0.02 < P$-value < 0.05 (0.028). Reject since P-value < 0.05. H_0: $\mu_1 \geq \mu_2$ and H_1: $\mu_1 < \mu_2$ (claim); d.f. $= 11$; $t = -4.98$. The P-value for the t test is P-value < 0.005. Reject since P-value < 0.05. There is enough evidence to support the claim that the nurses pay more for insurance than the administrators.

10–53. H_0: $\sigma_1^2 = \sigma_2^2$; C.V. = 7.15; $F = 1.23$; do not reject. H_0: $\mu_1 = \mu_2$ and H_1: $\mu_1 \neq \mu_2$ (claim); C.V. = ±2.228; d.f. = 10; $t = 0.119$; do not reject. There is not enough evidence to support the claim that the color of the mice made a difference. $-5.9 < \mu_1 - \mu_2 < 6.5$

10–55. a. Dependent   d. Dependent
b. Dependent   e. Independent
c. Independent

10–57. H_0: $\mu_D \geq 0$ and H_1: $\mu_D < 0$ (claim); C.V. = −1.397; d.f. = 8; $t = -2.8$; reject. There is enough evidence to support the claim that the seminar increased the number of hours students studied.

10–59. H_0: $\mu_D \leq 0$ and H_1: $\mu_D > 0$ (claim); C.V. = +1.833; d.f. = 9; $t = 5.435$; reject. There is enough evidence to support the claim that the diet reduced the sodium level of the patients.

10–61. H_0: $\mu_D \leq 0$ and H_1: $\mu_D > 0$ (claim); C.V. = 2.571; d.f. = 5; $t = 2.24$; do not reject. There is not enough evidence to support the claim that the errors have been reduced.

10–63. H_0: $\mu_D = 0$ and H_1: $\mu_D \neq 0$ (claim); d.f. = 7; $t = 0.978$; $0.20 < P\text{-value} < 0.50$ (0.361). Do not reject since $P\text{-value} > 0.01$. There is not enough evidence to support the claim that there is a difference in the pulse rates. $-3.23 < \mu_D < 5.73$

10–65. Using the previous problem $\bar{D} = -1.5625$, whereas the mean of the before values is 95.375 and the mean of the after values is 96.9375; hence, $\bar{D} = 95.375 - 96.9375 = -1.5625$.

10–67. a. 16   b. 4   c. 4.8   d. 104   e. 30

10–69. $\hat{p}_1 = 0.533$; $\hat{p}_2 = 0.3$; $\bar{p} = 0.44$; $\bar{q} = 0.56$; H_0: $p_1 = p_2$ and H_1: $p_1 \neq p_2$ (claim); C.V. = ±1.96; $z = 3.64$; reject. There is enough evidence to support the claim that there is a significant difference in the proportions.

10–71. $\hat{p}_1 = 0.43$; $\hat{p}_2 = 0.58$; $\bar{p} = 0.505$; $\bar{q} = 0.495$; H_0: $p_1 = p_2$ and H_1: $p_1 \neq p_2$ (claim); C.V. = ±1.96; $z = -2.12$; reject. There is enough evidence to support the claim that the proportions are different.

10–73. $\hat{p}_1 = 0.83$; $\hat{p}_2 = 0.75$; $\bar{p} = 0.79$; $\bar{q} = 0.21$; H_0: $p_1 = p_2$ (claim) and H_1: $p_1 \neq p_2$; C.V. = ±1.96; $z = 1.39$; do not reject. There is not enough evidence to reject the claim that the proportions are equal. $-0.032 < p_1 - p_2 < 0.192$

10–75. $\hat{p}_1 = 0.55$; $\hat{p}_2 = 0.45$; $\bar{p} = 0.497$; $\bar{q} = 0.503$; H_0: $p_1 = p_2$ and H_1: $p_1 \neq p_2$ (claim); C.V. = ±2.58; $z = 1.302$; do not reject. There is not enough

evidence to support the claim that the proportions are different. $-0.097 < p_1 - p_2 < 0.297$

10–77. $\hat{p}_1 = 0.5625$; $\hat{p}_2 = 0.525$; $\bar{p} = 0.54$; $\bar{q} = 0.46$; H_0: $p_1 = p_2$ and H_1: $p_1 \neq p_2$ (claim); C.V. = ±1.96; $z = 0.521$; do not reject. There is not enough evidence to support the claim that there is a difference in the proportions. $-1.03 < p_1 - p_2 < 0.178$

10–79. $\hat{p}_1 = 0.25$; $\hat{p}_2 = 0.31$; $\bar{p} = 0.286$; $\bar{q} = 0.714$; H_0: $p_1 = p_2$ and H_1: $p_1 \neq p_2$ (claim); C.V. = ±2.58; $z = -1.45$; do not reject. There is not enough evidence to support the claim that the proportions are different. $-0.165 < p_1 - p_2 < 0.045$

10–81. $0.077 < p_1 - p_2 < 0.323$

10–83. No, p_1 could equal p_2.

10–85. H_0: $\mu_1 \leq \mu_2$ and H_1: $\mu_1 > \mu_2$ (claim); C.V. = 2.33; $z = 0.59$; do not reject. There is not enough evidence to support the claim that single drivers do more pleasure driving than married drivers.

10–87. H_0: $\sigma_1 = \sigma_2$ and H_1: $\sigma_1 \neq \sigma_2$ (claim); C.V. = 2.77; $\alpha = 0.10$; d.f.N. = 23; d.f.D. = 10; $F = 10.365$; reject. There is enough evidence to support the claim that there is a difference in the standard deviations.

10–89. H_0: $\sigma_1^2 \leq \sigma_2^2$ and H_1: $\sigma_1^2 > \sigma_2^2$ (claim); $\alpha = 0.05$; d.f.N. = 9; d.f.D. = 9; $F = 5.06$. The P-value for the F test is $0.01 < P\text{-value} < 0.025$ (0.012); reject since $P\text{-value} < 0.05$. There is enough evidence to support the claim that the variance of the number of speeding tickets issued on Route 19 is greater than the variance of the number of speeding tickets issued on Route 22.

10–91. H_0: $\sigma_1^2 \leq \sigma_2^2$ and H_1: $\sigma_1^2 > \sigma_2^2$ (claim); C.V. = 1.47; $\alpha = 0.10$; d.f.N. = 64; d.f.D. = 41; $F = 2.32$; reject. There is enough evidence to support the claim that the variation in the number of days factory workers miss per year due to illness is greater than the variation in the number of days hospital workers miss per year.

10–93. H_0: $\sigma_1^2 = \sigma_2^2$; C.V. = 1.98; $F = 1.11$; do not reject. H_0: $\mu_1 \leq \mu_2$ and H_1: $\mu_1 > \mu_2$ (claim); C.V. = 1.28; $t = 1.31$; reject. There is enough evidence to support the claim that it is warmer in Birmingham.

10–95. H_0: $\sigma_1^2 = \sigma_2^2$; $F = 39.7$; reject since $P\text{-value} < 0.05$. H_0: $\mu_1 \leq \mu_2$ and H_1: $\mu_1 > \mu_2$ (claim); d.f. = 11; $t = 3.74$; $P\text{-value} < 0.005$; reject since $P\text{-value} < 0.05$. There is enough evidence to support the claim that incomes of city residents are greater than incomes of suburban residents.

10–97. H_0: $\mu_D \geq 0$ and H_1: $\mu_D < 0$ (claim);
C.V. $= -1.895$; d.f. $= 7$; $t = -2.73$; reject.
There is enough evidence to support the claim
that the music has increased production.

10–99. $\hat{p}_1 = 0.64$; $\hat{p}_2 = 0.40$; $\bar{p} = 0.509$; $\bar{q} = 0.491$;
H_0: $p_1 = p_2$ (claim) and H_1: $p_1 \neq p_2$; C.V. $= \pm 1.96$;
$z = +2.51$; reject. There is enough evidence to
reject the claim that the proportions are equal.
$-0.058 < p_1 - p_2 < 0.422$

Quiz—Chapter 10

1. False
2. False
3. True
4. False
5. *d.*
6. *a.*
7. *c.*

8. *b.*
9. $\mu_1 = \mu_2$
10. Pooled
11. Normal
12. Negative
13. $\dfrac{s_1^2}{s_2^2}$

14. H_0: $\mu_1 = \mu_2$ and H_1: $\mu \neq \mu_2$ (claim); $z = -3.69274$;
C.V. $= \pm 2.58$; reject. There is enough evidence to
support the claim that there is a difference in the
cholesterol levels of the two groups. $-10.2 < \mu_1 - \mu_2$
< -1.8.

15. H_0: $\mu_1 \leq \mu_2$ and H_1: $\mu > \mu_2$ (claim); C.V. $= 1.28$;
$z = 1.60$; reject. There is enough evidence to support
the claim that the average rental fees for the apartments
in the East is greater than the average rental fees for the
apartments in the West.

16. H_0: $\sigma_1^2 = \sigma_2^2$ and H_1: $\sigma_1^2 \neq \sigma_2^2$ (claim); $F = 1.637$;
d.f.N. $= 17$; d.f.D. $= 14$; P-value > 0.20 (0.357). Do
not reject since P-value > 0.05. There is not enough
evidence to support the claim that the variances
are different.

17. H_0: $\sigma_1^2 = \sigma_2^2$ and H_1: $\sigma_1^2 \neq \sigma_2^2$ (claim); $F = 1.296$;
C.V. $= 1.90$; do not reject. There is not enough
evidence to support the claim that the variances are
different.

18. H_0: $\sigma_1 = \sigma_2$ (claim) and H_1: $\sigma_1 \neq \sigma_2$; d.f.N. $= 11$;
d.f.D. $= 11$; C.V. $= 3.53$; $F = 1.13$; do not reject.
There is not enough evidence to reject the claim that
the standard deviations of the number of hours of
television viewing is the same.

19. H_0: $\sigma_1^2 = \sigma_2^2$ and H_1: $\sigma_1^2 \neq \sigma_2^2$ (claim); $F = 1.94$;
C.V. $= 3.01$; do not reject. There is not enough
evidence to support the claim that the variances are not
equal.

20. H_0: $\sigma_1^2 \leq \sigma_2^2$ and H_1: $\sigma_1^2 > \sigma_2^2$ (claim); $F = 1.474$;
C.V. $= 1.44$; reject. There is enough evidence to
support the claim that the variance of the days missed

by teachers is greater than the variance of the days
missed by nurses.

21. H_0: $\sigma_1 = \sigma_2$ and H_1: $\sigma_1 \neq \sigma_2$ (claim); $F = 1.65$;
C.V. $= 2.46$; do not reject. There is not enough
evidence to support the claim that the standard
deviations are different.

22. H_0: $\sigma_1^2 = \sigma_2^2$; C.V. $= 5.05$; $F = 1.23$; do not reject.
H_0: $\mu_1 = \mu_2$ and H_1: $\mu_1 \neq \mu_2$ (claim); $t = 10.922$;
C.V. $= \pm 2.779$; reject. There is enough evidence to
support the claim that the average prices are different.
$0.298 < \mu_1 - \mu_2 < 0.502$

23. H_0: $\sigma_1^2 = \sigma_2^2$; C.V. $= 9.60$; $F = 5.71$; do not reject.
H_0: $\mu_1 \geq \mu_2$ and H_1: $\mu_1 < \mu_2$ (claim); C.V. $= -1.860$;
d.f. $= 8$; $t = -4.05$; reject. There is enough evidence
to support the claim that accidents have increased.

24. H_0: $\sigma_1^2 = \sigma_2^2$; C.V. $= 4.02$; $F = 6.155$; reject. H_0:
$\mu_1 = \mu_2$ and H_1: $\mu_1 \neq \mu_2$ (claim); $t = 9.807$;
C.V. $= \pm 2.718$; reject. There is enough evidence to
support the claim that the salaries are different. $6,653
< \mu_1 - \mu_2 < 11,757$

25. H_0: $\sigma_1^2 = \sigma_2^2$; $F = 23.08$; reject since P-value < 0.05.
H_0: $\mu_1 \leq \mu_2$ and H_1: $\mu_1 > \mu_2$ (claim); d.f. $= 10$;
$t = 0.874$; $0.10 < P$-value < 0.25 (0.198); do not
reject since P-value > 0.05. There is not enough
evidence to support the claim that the incomes of city
residents are greater than the incomes of the rural
residents.

26. H_0: $\mu_1 \geq \mu_2$ and H_1: $\mu_1 < \mu_2$ (claim); $t = -4.172$;
C.V. $= -2.821$; reject. There is enough evidence to
support the claim that the sessions improved math
skills.

27. H_0: $\mu_1 \geq \mu_2$ and H_1: $\mu_1 < \mu_2$ (claim); $t = -1.714$;
C.V. $= -1.833$; do not reject. There is not enough
evidence to support the claim that egg production was
increased.

28. H_0: $p_1 = p_2$ and H_1: $p_1 \neq p_2$ (claim); $z = -0.69$;
C.V. $= \pm 1.65$; do not reject. There is not enough
evidence to support the claim that the proportions are
different. $-0.101 < p_1 - p_2 < 0.041$

29. H_0: $p_1 = p_2$ (claim) and H_1: $p_1 \neq p_2$; $z = 2.58$;
C.V. $= \pm 1.96$; reject. There is enough evidence
to reject the claim that the proportions are equal.
$0.067 < p_1 - p_2 < 0.445$

Chapter 11

11–1. Two variables are related when a discernible
pattern exists between them.

11–3. r, ρ (rho)

11–5. A positive relationship means that as x increases,
y increases. A negative relationship means that as
x increases, y decreases.

H

11–7. Answers will vary.

11–9. Pearson product moment correlation coefficient

11–11. There are many other possibilities, such as chance, or relationship to a third variable.

11–13. $H_0: \rho = 0$ and $H_1: \rho \neq 0$; $r = -0.832$; C.V. $= \pm 0.811$; reject. There is a significant relationship between a person's age and the number of hours he or she exercises.

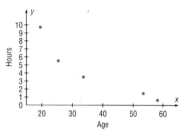

11–15. $H_0: \rho = 0$ and $H_1: \rho \neq 0$; $r = -0.883$; C.V. $= \pm 0.811$; d.f. $= 4$; reject. There is a significant relationship between a person's age and his or her contribution.

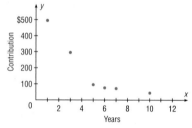

11–17. $H_0: \rho = 0$ and $H_1: \rho \neq 0$; $r = 0.716$; C.V. $= \pm 0.707$; d.f. $= 6$; reject. There is a significant relationship between test scores and GPA.

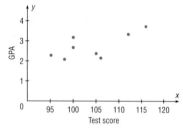

11–19. $H_0: \rho = 0$ and $H_1: \rho \neq 0$; $r = 0.814$; C.V. $= \pm 0.878$; d.f. $= 3$; do not reject. There is not a significant relationship between the variables.

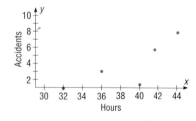

11–21. $H_0: \rho = 0$ and $H_1: \rho \neq 0$; $r = 0.143$; C.V. $= \pm 0.754$; do not reject. There is no significant relationship between the calories of a sandwich and its sodium content.

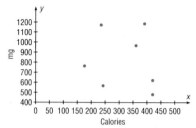

11–23. $H_0: \rho = 0$ and $H_1: \rho \neq 0$; $r = 0.913$; C.V. $= \pm 0.754$; reject. There is a significant relationship between the EER and the cost of the air conditioner.

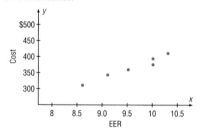

11–25. $H_0: \rho = 0$ and $H_1: \rho \neq 0$; $r = +0.873$; C.V. $= \pm 0.811$; d.f. $= 4$; reject. There is a significant relationship between the IQs of the girls and the boys.

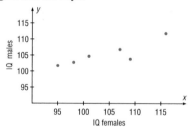

11–27. $H_0: \rho = 0$ and $H_1: \rho \neq 0$; $r = -0.909$; C.V. $= \pm 0.811$; d.f. $= 4$; reject. There is a significant relationship between the years of service and the number of resignations.

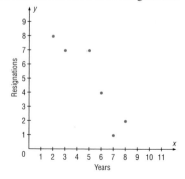

H

11–29. $r = 1.00$: All values fall in a straight line. $r = 1.00$: The relationship between x and y is the same when the values are interchanged.

11–31. A scatter plot should be drawn and the value of the correlation coefficient should be tested to see whether it is significant.

11–33. $y' = a + bx$

11–35. It is the line that is drawn through the points on the scatter plot such that the sum of the squares of the vertical distances from each point to the line is at a minimum.

11–37. When r is positive, b will be positive. When r is negative, b will be negative.

11–39. The closer r is to $+1$ or -1, the more accurate the predicted value will be.

11–41. $y' = 10.499 - 0.18x$; 4.2

11–43. $y' = 453.176 - 50.439x$; 251.42

11–45. $y' = -3.759 + 0.063x$; 2.793

11–47. Since r is not significant, no regression analysis should be done.

11–49. No regression should be done.

11–51. $y' = -167.07 + 55.99x$; 336.84

11–53. $y' = 63.193 + 0.405x$; 105.313

11–55. $y' = 10.770 - 1.149x$; 6.174

11–57. $r = +0.956$; $H_0: \rho = 0$ and $H_1: \rho \neq 0$; C.V. $= \pm0.754$; d.f. $= 5$; $y' = -10.944 + 1.969x$; when $x = 30$, $y' = 48.126$; reject. There is a significant relationship between the amount of lung damage and the number of years a person has been smoking.

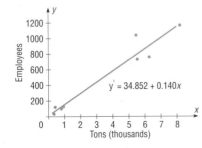

11–59. $H_0: \rho = 0$; $H_1: \rho \neq 0$; $r = 0.9$; C.V. $= \pm0.707$; d.f. $= 6$; $y' = 34.852 + 0.140x$; reject; when $x = 500$, $y' = 104.8$. There is a significant relationship between the number of tons of coal produced and the number of employees.

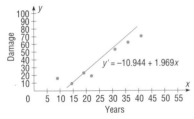

11–61. $H_0: \rho = 0$ and $H_1: \rho \neq 0$; $r = -0.981$; C.V. $= \pm0.811$; d.f. $= 4$; reject. There is a significant relationship between the number of absences and the final grade; $y' = 96.784 - 2.668x$.

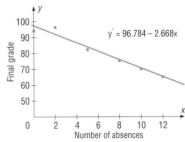

11–63. $H_0: \rho = 0$; $H_1: \rho \neq 0$; d.f. $= 3$; $r = 0.491$; $t = 0.976$; $0.20 < P\text{-value} < 0.50$ (0.401); do not reject. There is no significant relationship between the years of experience and the number of cars sold per month. No regression should be done.

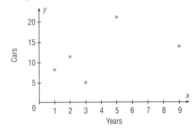

11–65. 453.173, 21.1, -3.7895

11–67. Explained variation is the variation due to the relationship. It is computed by $\Sigma (y' - \bar{y})^2$.

11–69. Total variation is the sum of the squares of the vertical distances of the points from the mean. It is computed by $\Sigma (y - \bar{y})^2$.

11–71. The coefficient of determination is found by squaring the value of the correlation coefficient.

11–73. The coefficient of nondetermination is found by subtracting r^2 from 1.

11–75. $r^2 = 0.49$; 49% of the variation of y is due to the variation of x; 51% of the variation of y is due to chance.

11–77. $r^2 = 0.1369$; 13.69% of the variation of y is due to the variation of x; 86.31% of the variation of y is due to chance.

11–79. $r^2 = 0.0025$; 0.25% of the variation of y is due to the variation of x; 99.75% of the variation of y is due to chance.

11–81. 2.092*

11–83. 94.22*

11–85. $1.59 < y < 12.21$*

11–87. $\$30.46 < y < \$472.38$*

H

11–89. Simple regression has one dependent variable and one independent variable. Multiple regression has one dependent variable and two or more independent variables.

11–91. The relationship would include all variables in one equation.

11–93. They will all be smaller.

11–95. 3.48 or 3

11–97. $39,390

11–99. R is the strength of the relationship between the dependent variable and all the independent variables.

11–101. R^2 is the coefficient of multiple determination. R^2_{adj} is adjusted for sample size.

11–103. The F test

11–105. $r = -0.388$; H_0: $\rho = 0$ and H_1: $\rho \neq 0$; C.V. $= \pm 0.875$; d.f. $= 5$; do not reject. There is not a significant relationship between the number of hours a person watches television and the person's GPA. No regression should be done.

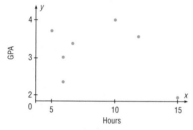

11–107. $r = 0.873$; H_0: $\rho = 0$ and H_1: $\rho \neq 0$; C.V. $= \pm 0.834$; d.f. $= 6$; reject. There is a significant relationship between the mother's age and the number of children she has; $y' = -2.457 + 0.187x$; $y' = 3.9$.

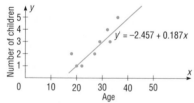

11–109. $r = -0.974$; H_0: $\rho = 0$ and H_1: $\rho \neq 0$; C.V. $= \pm 0.708$; d.f. $= 10$; reject. There is a significant relationship between speed and time; $y' = 14.086 - 0.137x$; $y' = 4.2222$.

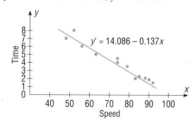

11–111. $r = -0.397$; H_0: $\rho = 0$ and H_1: $\rho \neq 0$; C.V. $= \pm 0.798$; d.f. $= 7$; do not reject. Since r is not significant, no regression analysis should be done.

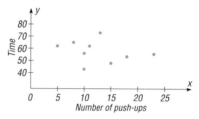

11–113. 0.468*

11–115. $3.34 < y < 5.10$*

11–117. 22.01*

11–119. $R^2_{adj} = 0.643$*

Quiz—Chapter 11

1. False
2. True
3. True
4. False
5. False
6. False
7. *a.*
8. *a.*
9. *d.*
10. *c.*
11. *b.*
12. Scattered diagram
13. Independent
14. -1, $+1$
15. *b.*
16. Line of best fit
17. $+1$, -1

18. $r = 0.857$; H_0: $\rho = 0$ and H_1: $\rho \neq 0$; C.V. $= \pm 0.707$; reject. $y' = -2.818 + 0.19x$; 3.84 or 4.

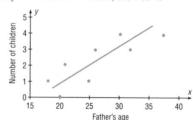

19. $r = -0.078$; H_0: $\rho = 0$ and H_1: $\rho \neq 0$; C.V. $= \pm 0.754$; do not reject. No regression should be done.

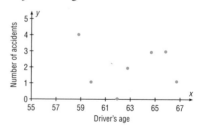

20. $r = 0.842$; H_0: $\rho = 0$ and H_1: $\rho \neq 0$; C.V. = ± 0.811; reject. $y' = -1.918 + 0.551x$; 4.14 or 4.

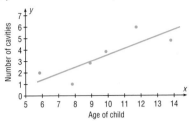

21. $r = 0.602$; H_0: $\rho = 0$ and H_1: $\rho \neq 0$; C.V. = ± 0.707; do not reject. No regression should be done.

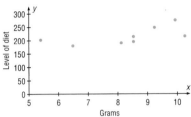

22. 1.129*

23. 29.5*

24. $0 < y < 5$*

25. 217.5 (average of y' values is used since there is no significant relationship)

26. 119.9*

27. $R = 0.729$*

28. $R^2_{adj} = 0.439$*

*These answers may vary due to method of calculation or rounding.

Chapter 12

12–1. The variance test compares a sample variance with a hypothesized population variance; the goodness-of-fit test compares a distribution obtained from a sample with a hypothesized distribution.

12–3. The expected values are computed on the basis of what the null hypothesis states about the distribution.

12–5. H_0: The number of accidents is equally distributed throughout the week (claim). H_1: The number of accidents is not equally distributed throughout the week. C.V. = 12.592; d.f. = 6; $\chi^2 = 28.887$; reject. There is enough evidence to reject the claim that the number of accidents is equally distributed during the week.

12–7. H_0: There is no preference for sherbet flavors (claim). H_1: There is a preference for sherbet flavors. C.V. = 11.345; d.f. = 3; $\chi^2 = 8.625$; do not reject. There is not enough evidence to reject the claim that there is no preference for sherbet flavors.

12–9. H_0: The blood types of the individuals are 42% type 0, 44% type A, 10% type B, and 4% type AB

(claim). H_1: The null hypothesis is not true. C.V. = 6.251; d.f. = 3; $\chi^2 = 100.275$; reject. There is enough evidence to reject the claim that the distribution of blood types is the same as the national distribution.

12–11. H_0: The types of loans are distributed as follows: 21% for home mortgages, 39% for automobile purchases, 20% for credit card, 12% for real estate, and 8% for miscellaneous (claim). H_1: The distribution is different from that stated in the null hypothesis. C.V. = 9.488; d.f. = 4; $\chi^2 = 4.936$; do not reject. There is not enough evidence to reject the claim that the distribution is the same as reported in the newspaper.

12–13. H_0: The types of payments of adult shoppers for purchases are distributed as follows: 53% pay cash, 30% use checks, 16% use credit cards, and 1% have no preference (claim). H_1: The distribution is not the same as stated in the null hypothesis. C.V. = 11.345; d.f. = 3; $\chi^2 = 36.8897$; reject. There is enough evidence to reject the claim that the distribution at the large store is the same as in the survey.

12–15. H_0: The distribution of college majors is as follows: 40% business, 25% computer science, 15% science, 10% social science, 5% liberal arts, and 5% general studies (claim). H_1: The null hypothesis is not true. C.V. = 9.236; d.f. = 5; $\chi^2 = 5.613$; do not reject. There is not enough evidence to reject the dean's hypothesis.

12–17. H_0: 50% of customers purchase word-processing programs, 25% purchase spreadsheet programs, and 25% purchase data base programs (claim). H_1: The null hypothesis is not true. $\chi^2 = 0.6$; d.f. = 2; $\alpha = 0.05$; P-value > 0.10 (0.741); do not reject since P-value > 0.05. There is not enough evidence to reject the store owner's assumption.

12–19. Answers will vary.

12–21. d.f. = (rows − 1) (columns − 1)

12–23. H_0: The variables are independent (or not related). H_1: The variables are dependent (or related).

12–25. The expected values are computed as (row total × column total) ÷ grand total.

12–27. H_0: $p_1 = p_2 = p_3 = p_4 = \cdots = p_n$. H_1: At least one proportion is different from the others.

12–29. H_0: Coffee consumption is independent of age. H_1: Coffee consumption is dependent on age (claim). C.V. = 16.812; d.f. = 6; $\chi^2 = 15.824$; do not reject. There is not enough evidence to support the claim that coffee consumption is dependent on age.

12–31. H_0: The price of the automobile is independent of the age of the purchaser. H_1: The price of the automobile is dependent upon the age of the purchaser (claim). C.V. = 12.592; d.f. = 6; $\chi^2 = 24.004$;

H

reject. There is enough evidence to support the claim that the price of the automobile is dependent upon the age of the purchaser.

12–33. H_0: The number of ads people think that they've seen or heard in the media is independent of the gender of the individual. H_1: The number of ads people think that they've seen or heard in the media is dependent upon the gender of the individual (claim). C.V. = 13.277; d.f. = 4; χ^2 = 9.562; Do not reject. There is not enough evidence to support the claim that the number of ads people think they've seen or heard is related to the gender of the individual.

12–35. H_0: The student's rating of the instructor is independent of the type of degree the instructor has. H_1: The rating is dependent on the type of degree the instructor has (claim). C.V. = 7.779; d.f. = 4; χ^2 = 19.507; reject. Yes, there is enough evidence to support the claim that the instructors' degrees are related to students' opinions about their effectiveness.

12–37. H_0: The type of video rented is independent of the person's age. H_1: The type of video rented is dependent on the person's age (claim). C.V. = 13.362; d.f. = 8; χ^2 = 46.733; reject. Yes, there is enough evidence to support the claim that the type of movie selected is related to the age of the customer.

12–39. H_0: The type of snack purchased is independent of the gender of the consumer (claim). H_1: The type of snack purchased is dependent upon the gender of the consumer. C.V. = 4.605; d.f. = 2; χ^2 = 6.342; reject. There is enough evidence to reject the claim that the type of snack is independent of the gender of the consumer.

12–41. H_0: The type of book purchased by an individual is independent of the gender of the individual (claim). H_1: The type of book purchased by an individual is dependent upon the gender of the individual. d.f. = 2; α = 0.05; χ^2 = 19.43; P-value < 0.005; reject since P-value < 0.05. There is enough evidence to reject the claim that the type of book purchased by an individual is independent of the gender of the individual.

12–43. H_0: $p_1 = p_2 = p_3 = p_4$ (claim). H_1: At least one proportion is different. C.V. = 7.851; d.f. = 3; χ^2 = 5.317; do not reject. There is not enough evidence to reject the claim that the proportions are equal.

12–45. H_0: $p_1 = p_2 = p_3 = p_4$ (claim). H_1: At least one proportion is different. C.V. = 7.815; d.f. = 3; χ^2 = 3.40; do not reject. There is not enough evidence to reject the claim that the proportions are equal.

12–47. H_0: $p_1 = p_2 = p_3 = p_4$ (claim). H_1: At least one proportion is different. C.V. = 6.251; d.f. = 3;

χ^2 = 12.755; reject. There is enough evidence to reject the claim that the proportions are equal.

12–49. H_0: $p_1 = p_2 = p_3 = p_4$ (claim). H_1: At least one proportion is different. d.f. = 3; χ^2 = 1.734; α = 0.05; P-value > 0.10 (0.629); do not reject since P-value > 0.05. There is not enough evidence to reject the claim that the proportions are equal.

12–51. H_0: $p_1 = p_2 = p_3$ (claim). H_1: At least one proportion is different. C.V. = 4.605; d.f. = 2; χ^2 = 2.401; do not reject. There is not enough evidence to reject the claim that the proportions are equal.

12–53. χ^2 = 1.075

12–55. H_0: The number of sales is equally distributed over five regions (claim). H_1: The null hypothesis is not true. C.V. = 9.488; d.f. = 4; χ^2 = 87.14; reject. No, there is enough evidence to reject the claim that the number of items sold in each region is the same.

12–57. H_0: Opinion is independent of gender. H_1: Opinion is dependent on gender (claim). C.V. = 4.605; d.f. = 2; χ^2 = 6.163; reject. There is enough evidence to support the claim that opinion is dependent on gender.

12–59. H_0: The type of investment is independent of the age of the investor. H_1: The type of investment is dependent on the age of the investor (claim). C.V. = 9.488; d.f. = 4; χ^2 = 25.6; reject. There is enough evidence to support the claim that the type of investment is dependent on the age of the investor.

12–61. H_0: $p_1 = p_2 = p_3$ (claim). H_1: At least one proportion is different. χ^2 = 4.912; d.f. = 2; α = 0.01; 0.05 < P-value < 0.10 (0.086); do not reject since P-value > 0.01. There is not enough evidence to reject the claim that the proportions are equal.

12–63. H_0: $p_1 = p_2 = p_3 = p_4$ (claim). H_1: At least one proportion is different. C.V. = 6.251; d.f. = 3; χ^2 = 6.70; reject. There is enough evidence to reject the claim that the proportions are equal.

Quiz—Chapter 12

1. False

2. True

3. False

4. *c.*

5. *b.*

6. *d.*

7. 6

8. Independent

9. Right

10. At least five

11. H_0: The number of ads is equally distributed over five geographic regions (claim). H_1: The number of ads is not equally distributed over five geographic regions. χ^2 = 45.4; C.V. = 9.488; reject. There is enough evidence to reject the claim that the number of ads is equally distributed over five geographic regions.

12. H_0: The ads produced the same number of responses (claim). H_1: The ads produced different numbers of responses. $\chi^2 = 12.6$; C.V. $= 13.277$; do not reject. There is not enough evidence to reject the claim that the ads produced the same number of responses.

13. H_0: College students show the same preference for shopping channels as those surveyed. H_1: College students show a different preference for shopping channels (claim). C.V. $= 7.815$; d.f. $= 3$; $\alpha = 0.05$; $\chi^2 = 21.789$; reject. There is enough evidence to support the claim that college students show a different preference for shopping channels.

14. H_0: All gifts were purchased with the same frequency (claim). H_1: The gifts were not purchased with the same frequency. $\chi^2 = 73.1$; C.V. $= 9.21$; reject. There is enough evidence to reject the claim that all gifts were purchased with the same frequency.

15. H_0: The type of novel purchased is independent of the gender of the purchaser (claim). H_1: The type of novel purchased is dependent on the gender of the purchaser. $\chi^2 = 132.9$; C.V. $= 5.991$; reject. There is enough evidence to reject the claim that the type of novel purchased is independent of the gender of the purchaser.

16. H_0: The type of pizza ordered is independent of the age of the individual who purchases it. H_1: The type of pizza ordered is dependent on the age of the individual who purchases it (claim). $\chi^2 = 107.3$; d.f. $= 9$; $\alpha = 0.10$; P-value < 0.005; reject since P-value < 0.10. There is enough evidence to support the claim that the pizza purchased is related to the age of the purchaser.

17. H_0: The color of the pennant purchased is independent of the gender of the purchaser (claim). H_1: The color of the pennant purchased is dependent on the gender of the purchaser. $\chi^2 = 5.6$; C.V. $= 4.605$; reject. There is enough evidence to reject the claim that the color of the pennant purchased is independent of the gender of the purchaser.

18. H_0: The opinion of the children on the use of the tax credit is independent of the gender of the children. H_1: The opinion of the children on the use of the tax credit is dependent upon the gender of the children (claim). C.V. $= 4.605$; d.f. $= 2$; $\chi^2 = 1.534$; do not reject. There is not enough evidence to support the claim that the opinion of the children on the use of the tax is dependent on their gender.

Chapter 13

13–1. The analysis of variance using the F test can be used to compare three or more means.

13–3. The populations from which the samples were obtained must be normally distributed. The samples must be independent of each other. The variances of the populations must be equal.

13–5. $F = \dfrac{s_B^2}{s_W^2}$

13–7. The Scheffé test and the Tukey test.

13–9. H_0: $\mu_1 = \mu_2 = \mu_3$. H_1: At least one mean is different from the others (claim). C.V. $= 3.35$; $\alpha = 0.05$; d.f.N. $= 2$; d.f.D $= 27$; $F = 1.17$; do not reject. There is not enough evidence to support the claim that at least one mean is different from the others.

13–11. H_0: $\mu_1 = \mu_2 = \mu_3$. H_1: At least one mean is different from the others (claim). C.V. $= 3.55$; $\alpha = 0.05$; d.f.N. $= 2$; d.f.D. $= 18$; $F = 19.05$; reject. Tukey test: $\bar{X}_1 = 5.57$, $\bar{X}_2 = 13.57$, $\bar{X}_3 = 17$; C.V. $= 3.61$; $\bar{X}_1$ vs. $\bar{X}_2$: $q = -5.97$; $\bar{X}_1$ vs. $\bar{X}_3$: $q = -8.53$; $\bar{X}_2$ vs. $\bar{X}_3$: $q = -2.56$. There is a significant difference between $\bar{X}_1$ and $\bar{X}_2$ and between $\bar{X}_1$ and $\bar{X}_3$.

13–13. H_0: $\mu_1 = \mu_2 = \mu_3$. H_1: At least one mean is different from the others (claim). C.V. $= 2.64$; $\alpha = 0.10$; d.f.N. $= 2$; d.f.D. $= 17$; $F = 1.28$; do not reject. There is not enough evidence to support the claim that at least one mean is different from the rest.

13–15. H_0: $\mu_1 = \mu_2 = \mu_3 = \mu_4$. H_1: At least one mean is different from the others. C.V. $= 2.28$; $\alpha = 0.10$; d.f.N. $= 3$; d.f.D. $= 31$; $F = 234.5$; reject. Scheffé test: C.V. $= 6.84$; $\bar{X}_1 = 101.5$; $\bar{X}_2 = 69.125$; $\bar{X}_3 = 47.9$; $\bar{X}_4 = 22.91$; $\bar{X}_1$ vs. $\bar{X}_2$: $F_S = 95.85$; $\bar{X}_1$ vs. $\bar{X}_3$: $F_S = 287.36$; $\bar{X}_1$ vs. $\bar{X}_4$: $F_S = 639.57$; $\bar{X}_2$ vs $\bar{X}_3$: $F_S = 53.41$; $\bar{X}_2$ vs. $\bar{X}_4$: $F_S = 263.85$; $\bar{X}_3$ vs. $\bar{X}_4$: $F_S = 87.25$. There is a significant difference between the pairs of means.

13–17. H_0: $\mu_1 = \mu_2 = \mu_3$. H_1: At least one mean is different from the others (claim). $F = 10.118$; P-value $= 0.00102$; reject. Scheffé test: C.V. $= 5.22$; $\bar{X}_1$ vs. $\bar{X}_2$: $F = 2.91$; $\bar{X}_1$ vs. $\bar{X}_3$: $F = 19.3$; $\bar{X}_2$ vs. $\bar{X}_3$: $F = 8.40$. There is a significant difference between $\bar{X}_1$ and $\bar{X}_3$ and $\bar{X}_2$ and $\bar{X}_3$;

13–19. H_0: $\mu_1 = \mu_2 = \mu_3 = \mu_4$. H_1: At least one mean is different from the others (claim). C.V. $= 4.13$; d.f.N. $= 3$; d.f.D. $= 64$; $\alpha = 0.01$; $F = 1.11$; do not reject. There is not enough evidence to support the claim that at least one mean is different from the others.

13–21. The main effects are the effects of the independent variables taken separately. The interaction effect occurs when one independent variable affects the dependent variable differently at different levels of the other independent variable.

13–23. The mean square values are computed by dividing the sum of squares by the corresponding degrees of freedom.

13–25. *a.* For factor A d.f.$_A = 2$ *c.* d.f.$_{A \times B} = 2$
    b. For factor B d.f.$_B = 1$ *d.* d.f.$_{within} = 24$

13–27. The two types of interactions that can occur are ordinal and disordinal.

13–29. *a.* The lines will be parallel or approximately parallel. They may also coincide.

b. The lines will not intersect and they will not be parallel.

c. The lines will intersect.

13–31. H_0: There is no interaction effect between the time of day and the type of diet on a person's sodium level. H_1: There is an interaction effect between the time of day and the type of diet on a person's sodium level. H_0: There is no difference between the means for the sodium level for the times of day. H_1: There is a difference between the means for the sodium level for the times of day. H_0: There is no difference between the means for the sodium level for the type of diet. H_1: There is a difference between the means for the sodium level for the type of diet.

ANOVA Summary Table

Source	SS	d.f.	MS	F
Time	1800	1	1800	25.806
Diet	242	1	242	3.470
Interaction	264.5	1	264.5	3.792
Within	279	4	69.750	
Total	2585.5	7		

The critical value at $\alpha = 0.05$ with d.f.N. = 1 and d.f.D. = 4 is 7.71 for F_A, F_B, and $F_{A \times B}$. Since the only F test value that exceeds 7.71 is the one for the time, 25.806, it can be concluded that there is a difference in the means for the sodium level taken at two different times.

13–33. H_0: There is no interaction effect between the gender of the individual and the duration of the training on the test scores. H_1: There is an interaction effect between the gender of the individual and the duration of the training on the test scores. H_0: There is no difference between the means of the test scores for the males and females. H_1: There is a difference between the means of the test scores for the males and females. H_0: There is no difference between the means of the test scores for the two different durations. H_1: There is a difference between the means of the test scores for the two different durations.

ANOVA Summary Table

Source	SS	d.f.	MS	F
Gender	57.042	1	57.042	0.835
Duration	7.042	1	7.042	0.103
Interaction	3978.375	1	3978.375	58.270
Within	1365.5	20	68.275	
Total	5407.959	23		

The critical value at $\alpha = 0.10$ with d.f.N. = 1 and d.f.D. = 20 is 2.97. Since the F test value for the interaction is greater than the critical value, it can be concluded that gender affects test scores differently for the duration levels.

13–35. H_0: There is no interaction effect between the ages of the salespeople and the products they sell on the monthly sales. H_1: There is an interaction effect between the ages of the salespeople and the products they sell on the monthly sales. H_0: There is no difference in the means of the monthly sales of the two age groups. H_1: There is a difference in the means of the monthly sales of the two age groups. H_0: There is no difference among the means of the sales for the different products. H_1: There is a difference among the means of the sales for the different products.

ANOVA Summary Table

Source	SS	d.f.	MS	F
Age	168.033	1	168.033	1.567
Product	1,762.067	2	881.034	8.215
Interaction	7,955.267	2	3,977.634	37.087
Error	2,574	24	107.250	
Total	12,459.367	29		

At $\alpha = 0.05$, the critical values are: for age, d.f.N. = 1; d.f.D. = 24; C.V. = 4.26; for product and interaction, d.f.N. = 2 and d.f.D. = 24; C.V. = 3.40. The null hypotheses for the interaction effect and for the type of product sold are rejected since the F test values exceed the critical value, 3.40. The cell means are:

Age \ Product	Pools	Spas	Saunas
Over 30	38.8	28.6	55.4
30 and under	21.2	68.6	18.8

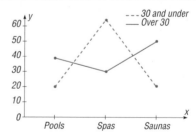

Since the lines cross, there is a disordinal interaction; hence, there is an interaction effect between the ages of salespeople and the type of products sold.

13–37. H_0: $\mu_1 = \mu_2 = \mu_3$. H_1: At least one mean is different from the others (claim). C.V. = 3.81;

$\alpha = 0.05$; d.f.N. $= 2$; d.f.D. $= 13$; $F = 0.533$; do not reject. There is not enough evidence to support the claim that at least one mean is different from the others.

13–39. H_0: $\mu_1 = \mu_2 = \mu_3$. H_1: At least one mean is different from the others. C.V. $= 3.89$; $\alpha = 0.05$; d.f.N. $= 2$; d.f.D. $= 12$; $F = 6.141$; reject. Scheffé test: $F_s = 7.78$; $\bar{X}_1$ vs. $\bar{X}_2$: $F_s = 12.167$; $\bar{X}_1$ vs. $\bar{X}_3$: $F_s = 2.849$; $\bar{X}_2$ vs. $\bar{X}_3$: $F = 2.509$. There is a significant difference between $\bar{X}_1$ and $\bar{X}_2$.

13–41. H_0: $\mu_1 = \mu_2 = \mu_3$. H_1: At least one mean is different (claim). C.V. $= 3.89$; $\alpha = 0.05$; d.f.N. $= 2$; d.f.D. $= 12$; $F = 3.673$; do not reject. There is not enough evidence to conclude that there is a difference in the means.

13–43. H_0: $\mu_1 = \mu_2 = \mu_3$. H_1: At least one mean is different from the others (claim). C.V. $= 6.36$; $\alpha = 0.01$; d.f.N. $= 2$; d.f.D. $= 15$; $F = 2.704$; do not reject.

13–45. H_0: There is no interaction effect between the type of exercise program and the type of diet on a person's glucose level. H_1: There is an interaction effect between type of exercise program and the type of diet on a person's glucose level. H_0: There is no difference in the means for the glucose levels of the people in the two exercise programs. H_1: There is a difference in the means for the glucose levels of the people in the two exercise programs. H_0: There is no difference in the means for the glucose levels of the people in the two diet programs. H_1: There is a difference in the means for the glucose levels of the people in the two diet programs.

ANOVA Summary Table

Source	SS	d.f.	MS	F
Exercise	816.75	1	816.75	60.5
Diet	102.083	1	102.083	7.56
Interaction	444.083	1	444.083	32.89
Within	108	8	13.5	
Total	1470.916	11		

At $\alpha = 0.05$, d.f.N. $= 1$, and d.f.D. $= 8$, the critical value is 5.32 for each F_A, F_B, and $F_{A \times B}$. Hence, all three null hypotheses are rejected. The cell means should be calculated.

Exercise / Diet	A	B
I	64	57.667
II	68.333	86.333

Since the means for exercise program I are both smaller than those for exercise program II and the vertical differences are not the same, the interaction

is ordinal. Hence one can say that there is a difference for exercise and diet, and that an interaction effect is present.

Quiz—Chapter 13

1. False
2. False
3. False
4. True
5. *d.*
6. *a.*
7. *a.*
8. *c.*
9. ANOVA
10. Tukey
11. Two

12. H_0: $\mu_1 = \mu_2 = \mu_3$. H_1: At least one mean is different from the others. $F = 0.119$; C.V. $= 3.55$; do not reject.

13. H_0: $\mu_1 = \mu_2 = \mu_3 = \mu_4$. H_1: At least one mean is different from the others. $F = 4.19$; C.V. $= 3.10$; reject. Tukey test: C.V. $= 3.96$; 1 vs. 2 $q = 0.41$; 2 vs. 3 $q = 1.15$; 1 vs. 3 $q = 0.75$; 2 vs. 4 $q = 4.48^*$; 1 vs. 4 $q = 4.08^*$; 3 vs. 4 $q = 3.33$. (*Significant.)

14. H_0: $\mu_1 = \mu_2 = \mu_3$. H_1: At least one mean is different from the others. $F = 0.708$; C.V. $= 6.36$; do not reject.

15. H_0: $\mu_1 = \mu_2 = \mu_3$. H_1: At least one mean is different from the others. $F = 0.049$; C.V. $= 3.63$; do not reject.

16. H_0: $\mu_1 = \mu_2 = \mu_3$. H_1: At least one mean is different from the others (claim). $F = 49.689$; C.V. $= 3.89$; reject. Tukey test: C.V. $= 3.77$; 1 vs. 2 $q = 11.8^*$; 1 vs. 3 $q = 5.59^*$; 2 vs. 3 $q = 17.4^*$ (*Significant.)

17. *a.* Two-way ANOVA
   b. Diet and exercise program
   c. 2
   d. H_0: There is no interaction effect between the type of exercise program and the type of diet on a person's weight loss. H_1: There is an interaction effect between the type of exercise program and the type of diet on a person's weight loss. H_0: There is no difference in the means of the weight losses of people in the exercise programs. H_1: There is a difference in the means of the weight losses of people in the exercise programs. H_0: There is no difference in the means of the weight losses of people in the diet programs. H_1: There is a difference in the means of the weight losses of people in the diet programs.
   e. Diet: $F = 21$, significant; exercise program: $F = 0.429$, not significant; interaction: $F = 0.429$, not significant
   f. Reject the null hypothesis for the diets.

Chapter 14

14–1. *Nonparametric* means hypotheses other than those using population parameters can be tested; *distribution-free* means no assumptions about the population distributions have to be satisfied.

H

14–3. Nonparametric methods have the following advantages:
  a. They can be used to test population parameters when the variable is not normally distributed.
  b. They can be used when data are nominal or ordinal.
  c. They can be used to test hypotheses other than those involving population parameters.
  d. The computations are easier in some cases than the computations of the parametric counterparts.
  e. They are easier to understand.

  The disadvantages are as follows:
  a. They are less sensitive than their parametric counterparts.
  b. They tend to use less information than their parametric counterparts.
  c. They are less efficient than their parametric counterparts.

14–5.

Data	21	31	34	41	41	61	65	72
Rank	1	2	3	4.5	4.5	6	7	8

14–7.

Data	3	5	5	6	7	8	8	9	12	14	15	17
Rank	1	2.5	2.5	4	5	6.5	6.5	8	9	10	11	12

14–9.

Data	187	190	190	236	321	532	673
Rank	1	2.5	2.5	4	5	6	7

14–11. The sign test uses only $+$ or $-$ signs.

14–13. The smaller number of $+$ or $-$ signs.

14–15. H_0: median $= 2.8$ (claim) and H_1: median $\neq 2.8$; test value $= 8$; C.V. $= 4$; do not reject. There is not enough evidence to reject the claim that the median is 2.8.

14–17. H_0: median $= \$325$ (claim) and H_1: median $\neq \$325$; C.V. $= 2$; test value $= 3$; do not reject. There is not enough evidence to reject the claim that the median rent is $325.

14–19. H_0: median $= 7.30$ (claim) and H_1: median $\neq 7.30$; C.V. $= \pm 1.96$; $z = -0.99$; do not reject. There is not enough evidence to reject the claim that the median is 7.30.

14–21. H_0: $p \leq 50\%$ and H_1: $p > 50\%$ (claim); $z = -0.99$; P-value $= 0.1611$; do not reject. There is not enough evidence to support the claim that more than 50% of the students favor single-room dormitories.

14–23. H_0: median $= 50$ (claim) and H_1: median $\neq 50$; $z = -2.3$; P-value $= 0.0057$; reject. There is enough evidence to reject the claim that 50% of the students are against extending the school year.

14–25. H_0: The medication has no effect on weight loss. H_1: The medication affects weight loss (claim). C.V. $= 0$; test value $= 1$; do not reject. There is not enough evidence to support the claim that the medication affects weight loss.

14–27. H_0: Reasoning ability will not be affected by the course. H_1: Reasoning ability increased after the course (claim). C.V. $= 1$; test value $= 1$; reject. There is enough evidence to support the claim that reasoning ability has increased after the course.

14–29. H_0: Alcohol has no effect on a person's IQ test score. H_1: Alcohol does affect a person's IQ test score (claim). C.V. $= 1$; test value $= 3$; do not reject. There is not enough evidence to reject the claim that alcohol has no effect on a person's IQ score.

14–31. $6 \leq$ median ≤ 22

14–33. $4.7 \leq$ median ≤ 9.3

14–35. $17 \leq$ median ≤ 33

14–37. n_1 and n_2 are each greater than or equal to 10.10.

14–39. The standard normal distribution

14–41. H_0: There is no difference in the number of books each group read (claim). H_1: There is a difference in the number of books each group read. C.V. $= \pm 1.65$; $z = +1.97$; reject. There is enough evidence to reject the claim that there is no difference in the number of books read by each group.

14–43. H_0: There is no difference in the number of academic scholarships awarded to the seniors at the two schools and H_1: There is a difference in the number of scholarships awarded to the seniors at the two schools (claim). C.V. $= \pm 1.96$; $z = 0.38$; do not reject. There is not enough evidence to support the claim that there is a difference in the number of scholarships awarded at the two schools.

14–45. H_0: There is no difference in the amount of money awarded to the cities. H_1: There is a difference in the amount of money awarded to the cities (claim). C.V. $= \pm 1.96$; $z = -0.520$; do not reject. There is not enough evidence to support the claim that there is a difference in the amount of money awarded to the cities.

14–47. H_0: There is no difference in the times needed to assemble the product. H_1: There is a difference in the times needed to assemble the product (claim). C.V. $= \pm 1.96$; $z = +3.56$; reject. There is enough evidence to support the claim that there is a difference in the productivity of the two groups.

14–49. The smaller value in absolute terms is used.

14–51. Sum of $-$ ranks is -6; sum of $+$ ranks is $+15$.

14–53. C.V. $= 20$; reject

14–55. C.V. $= 130$; do not reject

14–57. H_0: The workshop did not reduce the public speaking anxiety of the subjects and H_1: The workshop did reduce the public speaking anxiety of the subjects (claim). C.V. $= 6$; $w_s = 7.5$; do not reject. There is not enough evidence to support the claim that the workshop reduced the anxiety of the subjects.

14–59. H_0: The sizes of the police forces have decreased or remained the same. H_1: The sizes of the police forces have increased (claim). C.V. = 11; w_s = 10; reject. There is enough evidence to support the claim that the police forces have increased.

14–61. H_0: There is no change in the size of the rosters (claim). H_1: The size of the rosters has changed. C.V. = 6; w_s = 6.5; do not reject. There is not enough evidence to reject the claim that there is no change in the rosters.

14–63. H_0: There is no difference in the number of calories. H_1: There is a difference in the number of calories (claim). C.V. = 7.815; H = 2.842; do not reject. There is not enough evidence to support the claim that there is a difference in the number of calories.

14–65. H_0: There is no difference in the sales of the stores. H_1: There is a difference in the sales of the stores (claim). C.V. = 4.605; H = 10.8; reject. There is enough evidence to support the claim that there is a difference in sales.

14–67. H_0: There is no difference in the yields of the three plots. H_1: There is a difference in the yields of the three plots (claim). C.V. = 9.210; H = 12.020; reject. There is enough evidence to support the claim that the yields of the three plots are different.

14–69. H_0: There is no difference in the number of deaths due to severe weather. H_1: There is a difference in the number of deaths due to severe weather (claim). C.V. = 6.251; H = 5.537; do not reject. There is not enough evidence to support the claim that there is a difference in the number of deaths due to severe weather.

14–71. H_0: There is no difference in the number of crimes in the five precincts. H_1: There is a difference in the number of crimes in the five precincts (claim). C.V. = 13.277; H = 20.753; reject. There is enough evidence to support the claim that there is a difference in the number of crimes in the five precincts.

14–73. H_0: There is no difference in the final exam scores of the three groups. H_1: There is a difference in the final exam scores of the three groups (claim). H = 1.710; P-value > 0.10 (0.425); do not reject. There is not enough evidence to support the claim that there is a difference in the final exam scores of the three groups.

14–75. 0.716

14–77. 0.648

14–79. r_S = 0.612. H_0: ρ = 0 and H_1: $\rho \neq 0$. C.V. = ±0.564; reject. There is a significant relationship between the number of temperatures and the record high temperatures.

14–81. r_S = 0.943. H_0: ρ = 0 and H_1: $\rho \neq 0$. C.V. = 0.886; reject. There is a significant relationship between the sentence and the time period.

14–83. r_S = 0.5. H_0: ρ = 0 and H_1: $\rho \neq 0$. C.V. = ±0.738; do not reject. There is not enough evidence to say that a significant correlation exists.

14–85. r_S = 0.829. H_0: ρ = 0 and H_1: $\rho \neq 0$. C.V. = ±0.886; do not reject. There is not enough evidence to say that a significant correlation exists.

14–87. r_S = 0.979. H_0: ρ = 0 and H_1: $\rho \neq 0$. C.V. = ±0.591; reject. There is a significant relationship between the rankings.

14–89. H_0: The number of cavities in a person occurs at random. H_1: The null hypothesis is not true. There are 21 runs; the expected number of runs is between 10 and 22. Therefore, do not reject the null hypothesis; the number of cavities in a person occurs at random.

14–91. H_0: The Lotto numbers occur at random. H_1: The null hypothesis is not true. There are 14 runs, and this value is between 7 and 18. Hence, do not reject the null hypothesis; the Lotto occurs at random.

14–93. H_0: The number of defective cassette cases manufactured by the machine occurs at random. H_1: The null hypothesis is not true. There are 10 runs; and since this value is between 9 and 20, do not reject the null hypothesis; the number of defective cassette cases manufactured by a machine occurs at random.

14–95. H_0: The number of absences of employees occurs at random over a 30-day period. H_1: The null hypothesis is not true. There are only 6 runs, and this value does not fall within the 9-to-21 range. Hence, the null hypothesis is rejected; the absences do not occur at random.

14–97. H_0: The IQs of students who complete the tests are random. H_1: The null hypothesis is not true. Do not reject, since there are 7 runs, and this value is within the 6-to-16 range; the IQs occur at random.

14–99. ±0.479

14–101. ±0.215

14–103. H_0: median = 5 (claim) and H_1: median $\neq$ 5; test value = 6; C.V. = 2. Do not reject; there is not enough evidence to reject the claim that the median is 5.

14–105. H_0: The special diet has no effect on weight. H_1: The diet increases weight (claim). C.V. = 1; test value = 3. Do not reject; there is not enough

H

evidence to support the claim that there was an increase in weight.

14–107. H_0: There is no difference in the amount of money the groups spent for the textbook. H_1: There is a difference in the amount of money the groups spent for the textbook (claim). C.V. = ±1.65; z = −3.59. Reject; there is enough evidence to support the claim that there is a difference in the amount of money the groups spent on the textbooks.

14–109. H_0: The number of sick days workers used was not reduced. H_1: The number of sick days workers used was reduced (claim). C.V. = 4; w_s = 8.5. Do not reject; there is not enough evidence to support the claim that the number of sick days was reduced.

14–111. H_0: The diet has no effect on learning. H_1: The diet affects learning (claim). C.V. = 5.991; H = 8.5. Reject; there is enough evidence to support the claim that the diets do affect learning.

14–113. H_0: $\rho = 0$ and H_1: $\rho \neq 0$; C.V. = ±0.886; r_S = −0.486; do not reject. There is not enough evidence to say that there is a correlation between the rankings of the boys and girls.

14–115. H_0: The grades of students who finish the exam occur at random. H_1: The null hypothesis is not true. Since there are 8 runs, and this value does not fall between the 9-to-21 interval, the null hypothesis is rejected. The grades do not occur at random.

Quiz—Chapter 14

1. False
2. False
3. True
4. True
5. *a.*
6. *c.*
7. *d.*
8. *b.*
9. Nonparametric
10. Nominal, ordinal
11. Sign
12. Sensitive

13. H_0: median = 300 (claim) and H_1: median ≠ 300. There are seven plus signs. Do not reject since 7 is greater than the critical value, 5. There is not enough evidence to reject the claim that the median is 300.

14. H_0: median = 1200 (claim) and H_1: median ≠ 1200. There are 10 − signs. Do not reject since the 10 is greater than the critical value 6. There is not enough evidence to reject the claim that the median is 1200.

15. H_0: There will be no change in the weight of the turkeys after the special diet. H_1: The turkeys will weigh more after the special diet (claim). There is one plus sign; hence, the null hypothesis is rejected. There is enough evidence to support the claim that the turkeys gained weight on the special diet.

16. H_0: The distributions are the same. H_1: The distributions are different (claim). z = −0.56; C.V. = ±1.96; do not

reject the null hypothesis. There is not enough evidence to reject the claim that the distributions are the same.

17. H_0: The distributions are the same. H_1: The distributions are different (claim). z = −0.14434; C.V. = ±1.65; do not reject the null hypothesis. There is not enough evidence to support the claim that the distributions are the different.

18. H_0: There is no difference in the GPA of the students before and after the workshop. H_1: There is a difference in the GPA of the students before and after the workshop (claim). Test statistic = 0; C.V. = 2; reject the null hypothesis. There is enough evidence to support the claim that there is a difference in the GPAs of the students.

19. H_0: There is no difference in the breaking strengths of the tapes. H_1: There is a difference in the breaking strengths of the tapes (claim). H = 91.04; χ^2 = 5.991; reject the null hypothesis. There is enough evidence to support the claim that there is a difference in the breaking strengths of the tapes.

20. H_0: There is no difference in the reaction times of the monkeys. H_1: There is a difference in the reaction times of the monkeys (claim). H = 6.9; 0.025 < P-value < 0.05 (0.032); reject the null hypothesis. There is enough evidence to support the claim that there is a difference in the reaction times of the monkeys.

21. H_0: $\rho = 0$ and H_1: $\rho \neq 0$. r_s = 0.846; C.V. = 0.700; reject the null hypothesis. There is a significant relationship between the number of homework exercises and the exam scores.

22. H_0: $\rho = 0$ and H_1: $\rho \neq 0$. r_s = −0.25; C.V. = 0.900; do not reject the null hypothesis. There is not enough evidence to say that a correlation exists between the rankings of the brands by males and females.

23. H_0: The births of babies occur at random according to gender. H_1: The null hypothesis is not true. There are 10 runs, and since this is between 8 and 19, the null hypothesis is not rejected. There is not enough evidence to reject the null hypothesis that the gender occurs at random.

24. H_0: There is no difference in the rpm's of the motors before and after the reconditioning. H_1: There is a difference in the rpm's of the motors before and after the reconditioning (claim). Test statistics = 0; C.V. = 6; do not reject the null hypothesis. There is not enough evidence to support the claim that there is a difference in the rpm's of the motors before and after reconditioning.

25. H_0: The numbers occur at random. H_1: The null hypothesis is not true. There are 16 runs, and since this is between 9 and 21, the null hypothesis is not rejected. There is not enough evidence to reject the null hypothesis that the numbers occur at random.

Chapter 15

15–1. Random, systematic, stratified, cluster

15–3. A sample must be randomly selected.

15–5. Talking to people on the street, calling people on the phone, and asking one's friends are three incorrect ways of obtaining a sample.

15–7. Random sampling has the advantage that each unit of the population has an equal chance of being selected. One disadvantage is that the units of the population must be numbered; if the population is large, this could be somewhat time-consuming.

15–9. An advantage of stratified sampling is that it ensures representation for the groups used in stratification; however, it is virtually impossible to stratify the population so that all groups are represented.

15–11. Answers will vary.

15–13. Answers will vary.

15–15. Answers will vary.

15–17. Answers will vary.

15–19. Answers will vary.

15–21. Flaw—asking a biased question. Do you think XYZ Department Store should carry brand name merchandise?

15–23. Flaw—asking a biased question. Should banks charge a fee to balance their customers' checkbooks?

15–25. Flaw—confusing words. How many hours did you study for this exam?

15–27. Flaw—confusing words. If a plane were to crash on the border of New York and New Jersey, where should the victims be buried?

15–29. Answers will vary.

15–31. Simulation involves setting up probability experiments that mimic the behavior of real-life events.

15–33. John Von Neumann and Stanislaw Ulam

15–35. The steps are as follows:
a. List all possible outcomes.
b. Determine the probability of each outcome.
c. Set up a correspondence between the outcomes and the random numbers.
d. Conduct the experiment by using random numbers.
e. Repeat the experiment and tally the outcomes.
f. Compute any statistics and state the conclusions.

15–37. When the repetitions increase, there is a higher probability that the simulation will yield more precise answers.

15–39. Use random numbers 1 through 8 to make a shot and 9 and 0 to represent a missed shot.

15–41. Use random numbers 1 through 7 to represent a hit and 8, 9, and 0 to represent a miss.

15–43. Let an odd number represent heads and an even number represent tails. Then each person selects a digit at random.

15–45 through 15–61. Answers will vary.

15–63. Use two-digit random numbers 01 through 65 to represent a strikeout and 66 through 99 and 00 to represent something other than a strikeout.

15–65. Select two digits between 1 and 6 to represent the dice.

15–67. Let the digits 1–3 represent rock, let 4–6 represent paper, let 7–9 represent scissors, and omit 0.

15–69. Answers will vary.

15–71. Answers will vary.

15–73. Flaw—asking a biased question. Have you ever driven through a red light?

15–75. Flaw—asking a double-barreled question. Do you think all automobiles should have heavy duty bumpers?

Quiz—Chapter 15

1. True	**8.** Larger
2. True	**9.** Biased
3. False	**10.** Cluster
4. True	**11.** Answers will vary.
5. *a.*	**12.** Answers will vary.
6. *c.*	**13.** Answers will vary.
7. *c.*	**14.** Answers will vary.

15. Use two-digit random numbers; 01–35 constitute a win, and 36 and 00 constitute a loss.

16. Use two-digit random numbers; 01–03 constitute a cancellation.

17. Use two-digit random numbers 01–13 for cards.

18. Use random numbers 1–6 to simulate the roll of a die and random numbers 01–13 to simulate the cards.

19. Use random numbers 1–6 to simulate a roll of a die.

20. Answers will vary.

21. Answers will vary.

22. Answers will vary.

23. Answers will vary.

24. Answers will vary.

Appendix A

A–1.	362,880	**A–11.**	126
A–3.	120	**A–13.**	70
A–5.	1	**A–15.**	1
A–7.	1320	**A–17.**	560
A–9.	20	**A–19.**	2520

H

A–21. 121; 2181; 14,641; 716.9

A–23. 32; 258; 1024; 53.2

A–25. 328; 22,678; 107,584; 1161.2

A–27. 693; 50,511; 480,249; 2486.1

A–29. 318; 20,150; 101,124; 3296

A–31.

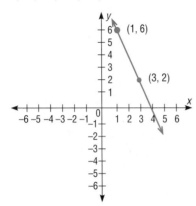

A–33.

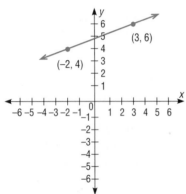

A–35.

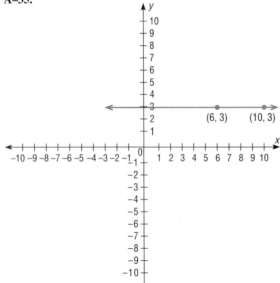

A–37.

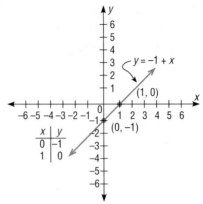

A–39.

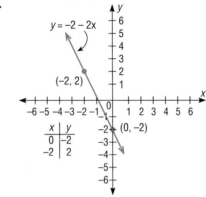

Appendix B–2

B–1. 0.65

B–3. 0.653

B–5. 0.379

B–7. $\frac{1}{4}$

B–9. 0.64

B–11. 0.857

Index

Glossary of Symbols

a	y intercept of a line
α	Probability of a type I error
b	Slope of a line
β	Probability of a type II error
C	Column frequency
cf	Cumulative frequency
$_nC_r$	Number of combinations of n objects taking r at a time
C.V.	Critical value
CVar	Coefficient of variation
D	Difference; decile
$\bar{D}$	Mean of the differences
d.f.	Degrees of freedom
d.f.N.	Degrees of freedom, numerator
d.f.D.	Degrees of freedom, denominator
E	Event; expected frequency; maximum error of estimate
$\bar{E}$	Complement of an event
e	Euler's constant ≈ 2.7183
$E(X)$	Expected value
f	Frequency
F	F test value; failure
F'	Critical value for the Scheffé test
MD	Median

MR	Midrange
MS_B	Mean square between groups
MS_W	Mean square within groups (error)
n	Sample size
N	Population size
$n(E)$	Number of ways E can occur
$n(S)$	Number of outcomes in the sample space
O	Observed frequency
P	Percentile; Probability
p	Probability; population proportion
$\hat{p}$	Sample proportion
$\bar{p}$	Weighted estimate of p
$P(B\vert A)$	Conditional probability
$P(E)$	Probability of an event E
$P(\bar{E})$	Probability of the complement of E
$_nP_r$	Number of permutations of n objects taking r at a time
π	Pi ≈ 3.14
Q	Quartile
q	$1 - p$; test value for Tukey test
$\hat{q}$	$1 - \hat{p}$
$\bar{q}$	$1 - \bar{p}$
R	Range; rank sum
F_S	Scheffé test value
GM	Geometric mean